TOOLS FOR LEARNING

TOOLS FOR LEARNING is an on-line review area that provides a variety of activities designed to help you study for your class. You will find chapter outlines, review questions, animated flashcards, figure labeling exercises, helpful hints for answering math problems, and research tips for some of the critical thinking questions.

[Web Navigator]
Look for the OceanLink icon that appears next to topics in the book. It indicates that you can find links to other Web sites offering valuable extra information.

[The Ocean Sciences]
Icons in *The Ocean Sciences* boxes in the book identify topics that are matched to Web sites picked by the author for their additional coverage of the concept and their examples of science in use.

[Oceanography in the News]
We've linked to useful sites that provide daily or weekly reports on new developments in the study of oceanography.

[Interactive Glossary]
This is a searchable version of the text's Glossary.

[Lecture Outlines]
We created a point-by-point summary based on each chapter's main section titles and subheads that you can use to study the chapter's material.

[Practice Quizzes]
The review questions test your knowledge of the important concepts and applications in each chapter and provide feedback.

[Math Tutor]
Math Tutor gently guides you through the problem solving process for selected end-of-chapter math exercises with clues, but no answers.

[Critical Thinking Essays]
We've provided "Web Research Tips," links to other Web sites, and suggestions to help you answer selected Critical Thinking Essays from the end of each chapter in the book.

[Figure Labeling]
You can practice identifying and labeling physical and geological features on the same art that appears in the text.

[Animated Flashcards]
Studying oceanography requires learning new terms. Virtual flashcards help master the new vocabulary from each chapter.

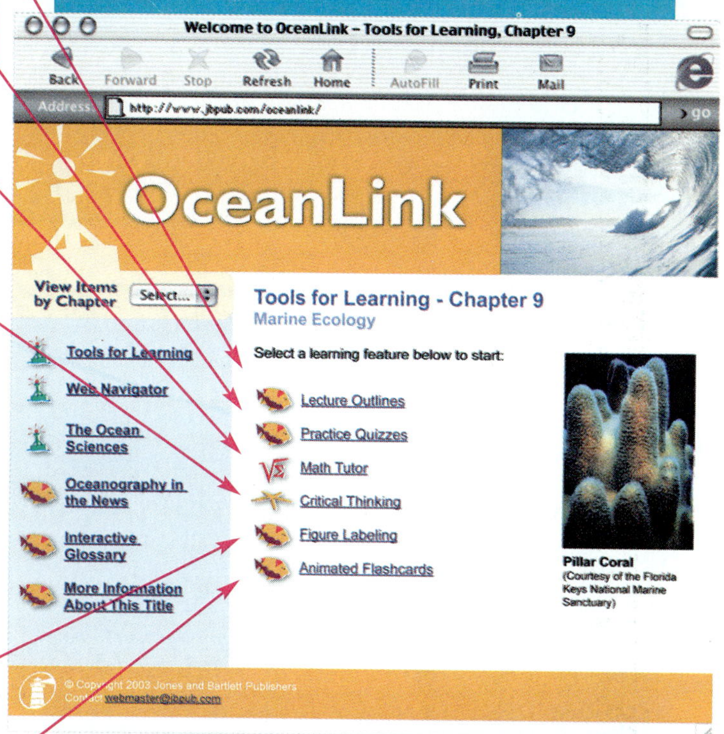

To find out more about OceanLink, please e-mail *oceanlink@jbpub.com*, or call your Jones and Bartlett sales representative at

800 832-0034.

Invitation to Oceanography

THIRD EDITION

Paul R. Pinet
COLGATE UNIVERSITY

JONES AND BARTLETT PUBLISHERS
Sudbury, Massachusetts
BOSTON TORONTO LONDON SINGAPORE

DEDICATION

Marita E. Hyman

A wise, passionate, caring partner who shares her life with me living on a special parcel of land on a large Earth in a vast universe

World Headquarters
Jones and Bartlett Publishers
40 Tall Pine Drive
Sudbury, MA 01776
978-443-5000
info@jbpub.com
www.jbpub.com

Jones and Bartlett Publishers Canada
2406 Nikanna Road
Mississauga, ON L5C 2W6
CANADA

Jones and Bartlett Publishers International
Barb House, Barb Mews
London W6 7PA
UK

PRODUCTION CREDITS

Chief Executive Officer: Clayton Jones
Chief Operating Officer: Don W. Jones, Jr.
Executive V.P. & Publisher: Robert W. Holland, Jr.
V.P., Design and Production: Anne Spencer
V.P., Sales and Marketing: William Kane
V.P., Manufacturing and Inventory Control: Therese Bräuer
Executive Editor, Science: Stephen L. Weaver
Managing Editor, Science: Dean W. DeChambeau
Senior Marketing Manager: Nathan Schultz
Senior Production Editor: Louis C. Bruno, Jr.
Composition: Philip Regan
Printing and Binding: Courier Kendallville
Cover Printing: Lehigh Press

Library of Congress Cataloging-in-Publication Data
Pinet, Paul R.
Invitation to oceanography / Paul R. Pinet.—3rd ed.
p. cm.
Includes bibliographical references and index.
ISBN 0-7637-2136-0
1. Oceanography. I. Title.
GC11.2 .P55 2003 2002033996

Printed in the United States of America
07 06 05 04 03 10 9 8 7 6 5 4 3 2 1

ILLUSTRATION CREDITS

Artwork: Randy Miyake, Miyake Illustration; Patti Isaacs, Parrot Graphics; Precision Graphics; Magellan Geographix; Seventeenth Street Studios

PHOTO CREDITS

[Text photo credits are located on page 556, which constitutes a continuation of this copyright page.]

Contents in Brief

CHAPTER 1

The Growth of Oceanography 3

CHAPTER 2

The Planet Oceanus 31

CHAPTER 3

The Origin of Ocean Basins 61

CHAPTER 4

Marine Sedimentation 91

CHAPTER 5

The Properties of Seawater 133

CHAPTER 6

Wind and Ocean Circulation 185

CHAPTER 7

Waves in the Ocean 227

CHAPTER 8

Tides 257

CHAPTER 9

Marine Ecology 285

CHAPTER 10

Biological Productivity in the Ocean 323

CHAPTER 11

The Dynamic Shoreline 365

CHAPTER 12

Coastal Habitats 399

CHAPTER 13

Ocean Habitats and Their Biota 435

CHAPTER 14

The Ocean's Resources 455

CHAPTER 15

The Human Presence in the Ocean 475

Appendixes 517

Contents

CHAPTER 1

The Growth of Oceanography 3

Preview 3

1.1 Oceanography: What Is It? 4

1.2 Historical Review of Oceanography 4

Ocean Exploration 4

Early Scientific Investigations 17

Modern Oceanography 20

1.3 Current and Future Oceanographic Research 24

Study Guide 27

FEATURES

THE PROCESS OF SCIENCE
THE SCIENTIFIC PROCESS 14

SCIENCE BY NUMBERS
GRAPHS 17

THE OCEAN SCIENCES: PHYSICS
MARINE ARCHEOLOGY 23

SCIENCE BY NUMBERS
CONVERSIONS 26

CHAPTER 2

The Planet Oceanus 31

Preview 31

2.1 The Earth's Structure 32

The Earth's Interior Spheres 32

The Earth's Exterior Envelopes 34

2.2 The Physiography of the Ocean Floor 34

Bathymetric Provinces 35

Continental Margins 36

Continental Shelf 36

Continental Slope 36

Continental Rise 36

Deep-Ocean Basins 38

Midocean Ridges 38

2.3 Geologic Differences between Continents and Ocean Basins 39

The Earth's Topography and Bathymetry 39

Mass Balance and Isostasy 41

2.4 The Physiography of the Western North Atlantic Ocean 43

A Trek to the Crest of the Mid-Atlantic Ridge 47

Study Guide 57

FEATURES

SCIENCE BY NUMBERS
POWERS OF 10 41

THE OCEAN SCIENCES: GEOLOGY
PROBING THE SEA FLOOR 44

CHAPTER 3

The Origin of Ocean Basins 61

Preview 61

3.1 Continental Drift 62

3.2 Sea-Floor Spreading 64
The Geomagnetic Field 66
Spreading Ocean Ridges 68

3.3 Global Plate Tectonics 70
Subduction Zones 70
Plate-Tectonic Model 72
The Opening and Closing of Ocean Basins 75
A Summary of Global Plate Tectonics 80

3.4 Future Discoveries 81
Study Guide 86

FEATURES

THE OCEAN SCIENCES: GEOLOGY
THE SAN ANDREAS FAULT 77

THE PROCESS OF SCIENCE
THE SCIENTIFIC PROCESS: SEA-FLOOR SPREADING 80

THE OCEAN SCIENCES: GEOLOGY
THE RED SEA 82

SCIENCE BY NUMBERS
SEA-FLOOR SPREADING RATES 85

CHAPTER 4

Marine Sedimentation 91

Preview 91

4.1 Sediment in the Sea 92
Classification of Marine Sediment 92
Factors that Control Sedimentation 92

4.2 Sedimentation in the Ocean 97
Shelf Sedimentation 97
Worldwide Distribution of Shelf Sediments 99
Geologic Controls of Continental Shelf Sedimentation 100
Carbonate Shelves 107
Deep-Sea Sedimentation 110
Sources of Sediment to the Deep Sea 110
Sedimentation Processes in the Deep Sea 110
Bulk Emplacement 110
Pelagic Sediment 112
Authigenic Deposits 116
Global Distribution of Deep-Sea Sediments 119
Surface Deposits 119
Deep-Sea Stratigraphy 120

4.3 Future Discoveries 122
Study Guide 128

FEATURES

THE OCEAN SCIENCES: GEOLOGY
PROBING THE SEA FLOOR 94

THE OCEAN SCIENCES: GEOLOGY
DUST STORMS 102

THE PROCESS OF SCIENCE
CLIMATE VARIABILITY AND CHANGE 106

THE OCEAN SCIENCES: GEOLOGY
THE DRYING UP OF THE MEDITERRANEAN SEA 124

SCIENCE BY NUMBERS
SEDIMENTATION RATES 127

CHAPTER 5

The Properties of Seawater 133

Preview 133

5.1 Basic Chemical Notions 134

5.2 Basic Physical Notions 135

5.3 The Water Molecule 136

The Solutes of Seawater 140

Major Constituents 140

Nutrients 142

Gases 142

Trace Elements 142

Organic Compounds 142

5.4 Salinity 143

Principle of Constant Proportion 143

Factors that Regulate the Salinity of Seawater 146

Effects of Salinity on the Properties of Water 150

Freezing Point 150

Density 151

Vapor Pressure 151

5.5 Chemical and Physical Structure of the Oceans 151

Temperature 151

Salinity 153

Density 157

Surface Layer 157

Pycnocline Layer 157

Deep Layer 157

5.6 Gases in Seawater 158

Oxygen 161

Carbon Dioxide 162

5.7 The Ocean as a Physical Chemical System 167

Reservoirs of Water 172

The Global Water Cycle 172

The Ocean as a Biogeochemical System 174

Study Guide 179

FEATURES

SCIENCE BY NUMBERS
PARTS PER THOUSAND 141

THE OCEAN SCIENCES: CHEMISTRY
CHEMICAL TECHNIQUES 144

THE OCEAN SCIENCES: CHEMISTRY
DESALINATION 164

THE OCEAN SCIENCES: PHYSICS
OTHER PHYSICAL PROPERTIES OF WATER 168

THE OCEAN SCIENCES: CHEMISTRY
THE SEA-SURFACE MICROLAYER 176

SCIENCE BY NUMBERS
ORDER OF MAGNITUDE 178

CHAPTER 6

Wind and Ocean Circulation 185

Preview 185

6.1 Atmospheric Processes 186

Air Pressure 186

Coriolis Deflection 186

General Wind Circulation 189

6.2 Surface Ocean Currents 192

The Wind-Driven Currents of the Sea Surface 192

Wind Drag 192

Pressure Gradients 192

Coriolis Deflection 197

Types of Surface Flows 199

A Model of Geostrophic Flow 201

Refinement of the Geostrophic-Flow Model 203

6.3 Deep-Ocean Circulation 208

Water Masses and Density-Driven Water Flow 208

A General Model of Thermohaline Circulation 210

Atlantic Ocean 212

Pacific Ocean 212

Indian Ocean 213

6.4 Water Flow in Semienclosed Seaways 215

6.5 Future Discoveries 222

Study Guide 222

FEATURES

THE OCEAN SCIENCES: PHYSICS
CURRENT-MEASURING TECHNIQUES 194

THE OCEAN SCIENCES: PHYSICS
HURRICANES AND TYPHOONS 206

SCIENCE BY NUMBERS
VOLUME TRANSPORT 210

THE OCEAN SCIENCES: PHYSICS
UNDERWATER WEATHER AND WATERFALLS 218

CHAPTER 7

Waves in the Ocean 227

Preview 227

7.1 Properties of Ocean Waves 228
Wind Generation of Waves 228

7.2 Wave Motions 232
The Motion of Water Particles beneath Waves 232
Motion of the Wave Form 235

7.3 The Life History of Ocean Waves 236
Growth of Waves in the Fetch Area 236
Storm Waves outside the Generating Area 237
Waves in Shallow Water 237
Shallow-Water Wave Transformations 239
Wave Refraction 240
Shore Breakers 240
Shoreline Under Storm Conditions 243

7.4 Standing Waves 246

7.5 Other Types of Progressive Waves 249
Internal Waves 249
Tsunamis 250
Study Guide 252

FEATURES

THE OCEAN SCIENCES: PHYSICS
WAVE-MEASURING TECHNIQUES 230

SCIENCE BY NUMBERS
WAVE CELERITY 234

THE OCEAN SCIENCES: PHYSICS
TINY WAVES AND GIANT WAVES 244

CHAPTER 8

Tides 257

Preview 257

8.1 Tidal Characteristics 258

8.2 Origin of the Tides 260
Equilibrium Model of Tides 263
Dynamic Model of the Tides 265

8.3 Tides in Small and Elongated Basins 272

8.4 Tidal Currents 275

8.5 Power from the Tides 279
Study Guide 281

FEATURES

THE OCEAN SCIENCES: PHYSICS
CURRENTS THROUGH TIDAL INLETS 268

SCIENCE BY NUMBERS
SPEED OF THE TIDE 273

THE OCEAN SCIENCES: BIOLOGY
TIDAL RHYTHMS IN ORGANISMS 276

SCIENCE BY NUMBERS
DEPTH OF THE BAY OF FUNDY 278

CHAPTER 9

Marine Ecology 285

Preview 285

9.1 Ocean Habitats 286

9.2 Classification of Organisms 288
Kingdom Monera 288
Kingdom Protista 288
Kingdom Fungi 291
Kingdom Metaphytae 291
Kingdom Metazoa 291

9.3 Classification by Lifestyle 291

9.4 Basic Ecology 294
Temperature 295
Salinity 299
Hydrostatic Pressure 301

9.5 Selected Adaptive Strategies 302
Life Cycles of Plankton 303
Functional Morphology of Fishes (Nekton) 307
Benthic Communities 312

9.6 Future Discoveries 317
Study Guide 318

FEATURES

OCEAN SCIENCES: BIOLOGY **SAMPLING THE BIOTA** 296

THE OCEAN SCIENCES: BIOLOGY **KILLER WHALES** 311

THE OCEAN SCIENCES: BIOLOGY **ECOLOGY OF THE GIANT KELP COMMUNITY** 314

CHAPTER 10

Biological Productivity in the Ocean 323

Preview 323

10.1 Food Webs and Trophic Dynamics 324
Energy and Trophic Dynamics 325
Plants as Primary Producers 328
Animals as Consumers 328
Bacteria as Decomposers 331
Food Chains and Energy Transfer 333

10.2 General Marine Productivity 338
Primary Production in the Ocean 338
Solar Radiation 338
Nutrient Concentrations 339
Upwelling and Turbulence 340
Grazing 341
Water Turbidity 341
Variations in Productivity 341

10.3 Global Patterns of Productivity 344
Spatial Variations 345
Estimates of Plant and Fish Production 348

10.4 Biological Productivity of Upwelling Water 354

10.5 Future Discoveries 359
Study Guide 360

FEATURES

SCIENCE BY NUMBERS **DOUBLING RATES** 326

THE OCEAN SCIENCES: BIOLOGY **LARGE SHARKS** 330

THE OCEAN SCIENCES: BIOLOGY **VOLCANIC VENT COMMUNITIES** 334

THE OCEAN SCIENCES: BIOLOGY **SATELLITE OCEANOGRAPHY** 342

THE OCEAN SCIENCES: BIOLOGY **MIGRANTS** 350

THE OCEAN SCIENCES: PHYSICS **WHAT CAUSES EL NIÑO?** 356

CHAPTER 11

The Dynamic Shoreline 365

Preview 365

11.1 Coastal Water Movement 366

Shoaling Waves and Refraction 366

Circulation in the Surf Zone 366

11.2 Beaches 367

Beach Profiles 367

Sand Budgets 370

11.3 Coastal Dunes 371

11.4 Barrier Islands 377

Barrier-Island Landscape 378

Origin of Barrier Islands 379

Storm Effects 379

11.5 Cliffed Coasts 381

11.6 Deltas 386

11.7 Impact of People on the Coastline 388

11.8 Future Discoveries 393

Study Guide 394

FEATURES

THE OCEAN SCIENCES: GEOLOGY
CAPE COD, MASSACHUSETTS 372

THE OCEAN SCIENCES: GEOLOGY
THE OREGON COASTLINE 374

THE OCEAN SCIENCES: GEOLOGY
PADRE ISLAND, TEXAS 382

THE OCEAN SCIENCES: GEOLOGY
SAN DIEGO COUNTY, CALIFORNIA 385

THE OCEAN SCIENCES: GEOLOGY
OCEAN CITY AND ASSATEAGUE ISLAND, MARYLAND 390

SCIENCE BY NUMBERS
SAND-BUDGET STUDY 392

CHAPTER 12

Coastal Habitats 399

Preview 399

12.1 Estuaries 400

Origin of Estuaries 400

Circulation and Sedimentation in Estuaries 402

Salt-Wedge Estuaries 402

Partially Mixed Estuaries 402

Well-Mixed Estuaries 405

The Biology of Estuaries 405

12.2 Lagoons 409

12.3 Salt Marshes 411

12.4 Mangrove Swamps 416

12.5 Coral Reefs 421

Biology of Corals 421

Ecology of Coral Reefs 424

Geology of Coral Reefs 428

Study Guide 430

FEATURES

THE OCEAN SCIENCES: BIOLOGY **CHESAPEAKE BAY** 406

THE OCEAN SCIENCES: GEOLOGY **SALT-MARSH EVOLUTION** 414

THE OCEAN SCIENCES: BIOLOGY **SAN FRANCISCO BAY** 418

THE OCEAN SCIENCES: BIOLOGY **GLOBAL DECLINE IN CORAL REEFS** 422

SCIENCE BY NUMBERS **RESIDENCE TIME** 430

CHAPTER 13

Ocean Habitats and Their Biota 435

Preview 435

13.1 Biology of the Continental Shelf 436

Neritic Zone 436

Sublittoral Zone 437

13.2 Biology of the Open Ocean and the Deep Sea 440

The Oceanic Realm 441

The Deep-sea Bottom 445

13.3 Future Discoveries 450

Study Guide 451

FEATURES

THE OCEAN SCIENCES: BIOLOGY **PENGUINS** 438

THE OCEAN SCIENCES: BIOLOGY **BOTTOM FEEDING BY WHALES** 442

THE OCEAN SCIENCES: BIOLOGY **SARGASSUM GULFWEED** 446

THE OCEAN SCIENCES: BIOLOGY **SQUIDS** 448

CHAPTER 14

The Ocean's Resources 455

Preview 455

14.1 Law of the Sea 456
Law of the Sea Treaty 456
Exclusive Economic Zones 457

14.2 Mineral Resources 457
Oil and Natural Gas 457
Gas Hydrates 459
Sand and Gravel 459
Manganese Nodules 461
Cobalt-Rich Oceanic Crusts 462
Phosphate Deposits 463

14.3 Living Resources 463
Fisheries 463
Mariculture 467
Study Guide 471

FEATURES

THE OCEAN SCIENCES: GEOLOGY
OFFSHORE OIL AND GAS IN THE GULF OF MEXICO 460

THE PROCESS OF SCIENCE
FISH FARMING AND "SUPER" FISH 464

THE OCEAN SCIENCES: BIOLOGY
ANTARCTIC KRILL 468

CHAPTER 15

The Human Presence in the Ocean 475

Preview 475

15.1 Pollution: What Is It? 476

15.2 Hydrocarbons in the Sea 477

15.3 Municipal and Industrial Effluent 484
Sewage 486
Metals 489
Artificial Biocides 491

15.4 Ocean Dredging and Mining 492
Dredging 493
Ocean Mining 494

15.5 Overfishing 496

15.6 Climate Change 500

15.7 The Ocean's Future 506

15.8 Future Discoveries 512
Study Guide 512

FEATURES

THE OCEAN SCIENCES: BIOLOGY
THE *TORREY CANYON* DISASTER 478

THE OCEAN SCIENCES: BIOLOGY
BIOREMEDIATION 484

THE OCEAN SCIENCES: BIOLOGY
THE *EXXON VALDEZ* OIL SPILL 486

THE OCEAN SCIENCES: BIOLOGY
COLLAPSE OF THE NEW ENGLAND FISHERIES 498

THE PROCESS OF SCIENCE
THE GLOBAL RISE OF SEA LEVEL AND ITS EFFECTS 504

THE OCEAN SCIENCES: BIOLOGY
RED TIDES 508

SCIENCE BY NUMBERS
STEADY STATE 510

APPENDIXES

I -Properties of the Earth 517

II-Conversion Factors 518

III-Geologic Time 519

IV-Maps and Topographic Profiles 520
Contour Maps 521
Topographic Profiles 522
Oceanographic Profiles 523

Glossary 525

Index 541

Preface

This book deals with the workings of the ocean, the dynamic processes that affect its water, sea floor, and abundant life forms. The approach used is a broad one, relying on basic concepts to explain the ocean's many mysteries. Anybody—whether sailor, surfer, beachcomber, or student—can learn about the processes and creatures of the oceans by reading this visually exciting book. No background in science is required to grasp the many important ideas that are relevant to the working of the oceans. Wherever appropriate, the underlying science is first explained clearly, and only then is it used to account for ocean processes. These overarching scientific concepts are summarized conveniently as "Key Concepts" at the end of every chapter. In order to help those unfamiliar with the practice of science, a series of "process of science" boxes provides an explanation of how scientists reason and draw conclusions about the natural world. In the glossary, important words are clearly defined and are accompanied by page numbers that refer you to the critical section of the book where the term in question was first introduced.

The figures and their accompanying captions do not merely illustrate, but also supplement the written text. Many new photographs have been added to this edition. All the drawings have been beautifully and accurately rendered by a team of talented artists and illustrators in order to present in visual form ideas that are at times necessarily abstract. They should be studied carefully before advancing to the next section of the chapter, because they help provide concreteness to the ideas discussed. It has been the author's experience that those students who truly understand the "ins and outs" of the illustrations tend to have a solid grasp of the chapters' main concepts. This will take a bit of time, but it is time well invested.

ORGANIZATION

The third edition of *Invitation to Oceanography* incorporates new and updated material, based on the many valuable suggestions made by faculty and students who have worked with the previous editions of the book. This means that the organization of the material, the development of the ideas, and the quality of the prose and illustrations are better than ever. We are always working to improve each succeeding version of the book, and so we welcome all comments and criticisms from our readers. Both faculty and students agree that the development of key oceanographic concepts flows logically and systematically from chapter to chapter, as well as from section to section. The first two chapters review the long history of ocean exploration and research and the fundamental structure of the Earth's interior and its exterior ocean basins. An imaginary trek across the sea bottom of the North Atlantic Ocean, beginning in New Jersey and ending at the very top of the Mid-Atlantic Ridge, is the highlight of Chapter 2 for many readers including the author. Chapters 3 through 10 examine the geology, chemistry, physics, and biology of the sea, highlighting the key scientific concepts and latest discoveries in these subdisciplines of oceanography. In some sense, the material and concepts in these seven chapters represent the core ideas of the ocean sciences, and when comprehended and synthesized, they provide the framework for understanding ocean habitats as whole, functional systems, the chapter topics of the remainder of the book. For example, Chapters 11 and 12 examine the intriguing intricacies of dynamic coastal environments, including beaches, dunes, barrier islands, estuaries, deltas, salt marshes, mangrove swamps, lagoons, and coral reefs. Two chapters are devoted to coastal ecosystems, because we are most familiar and come in regular contact with the shoreline rather than the open ocean. It is likely that many of us as voting citizens will be in a position to influence regulatory legislation and management practices of these fragile habitats. Chapter 13 provides an overview of the many fascinating and exotic ecosystems that are found far offshore, either in open water or on the deep-sea floor. Chapter 14 surveys the ocean's abundant resources, both living (fish) and nonliving (petroleum, metals, phosphate), that are vital for the modern human world. The final Chapter 15 provides a balanced appraisal of the environmental stresses brought about by human activity, showing the nature and alarming extent of this impact and providing examples of groups of concerned citizens who are striving hard and successfully to reverse environmental despoilment. Throughout the book, local and regional examples are drawn from all parts of the U.S. coastline, including the Pacific coast as far north as Alaska, the Atlantic seaboard as well as maritime Canada, and the Gulf of Mexico. Examples from foreign seas are used where appropriate.

WEB ENHANCEMENT

Invitation to Oceanography's extensive web page, *OceanLink*, has also been upgraded. Students can reach the new *OceanLink* home page by entering the URL http://www.jbpub.com/oceanlink into a World Wide Web browser and click on the book cover of this edition.

OceanLink also includes "Tools for Learning," a free, on-line student review area that provides a variety of activities designed to help students study for their class. Students will find new lecture outlines, review questions, key term reviews, animated vocabulary flashcards, and figure labeling exercises. The math tutorials and critical thinking exercises from the original *OceanLink* site are here as well.

As you look through the pages of *Invitation to Oceanography*, Third Edition, you'll see three distinctive icons that call out the book's connection to the *OceanLink* web page:

Web Navigator

Critical Thinking on the Web

Math Tutor on the Web

You'll find the Web Navigator buoy icon next to section titles in the chapters and at the end of "The Ocean Sciences" boxes. The buoy identifies topics that are matched to relevant independent web sites that students and instructors can visit through *OceanLink*. These web sites, run by research institutes, scientists, educational programs, and government agencies reinforce and enhance the topics in the book.

The starfish icon can be found next to many critical thinking questions at the end of each chapter. The icon indicates that students can find—within Tools for Learning—tips for researching the question and links to additional resources

The same is true for the "Discovering with Numbers" questions at the end of each chapter. The Math Tutor icon means that Tools for Learning has an area in which the author, Paul Pinet, patiently guides the student through the mathematical solving process without giving away the answer. He believes the effort in seeking the solution must still come from the student's own work.

OceanLink offers various web links and activities to make the study of oceanography more current and more enjoyable. In addition to the areas described above, *OceanLink* has an on-line glossary and links to sites offering the latest oceanography news. For links to other sites, we provide a brief description to place the link in context *before* the student connects to the site. Jones and Bartlett Publishers constantly monitors the links to ensure there will always be a working and appropriate site on line. An illustrated description of *OceanLink* is provided just inside the front cover of this book.

FEATURED BOXES

Featured boxes, **The Ocean Sciences,** abound in all of the chapters. They consist of four types, based on the principal subfields of oceanography: biology, chemistry, geology, and physics. Each is identified as such by a colorful and distinctive logo placed near the title of the box.

The boxes serve several purposes. Some review common research techniques employed by oceanographers to investigate the seas. Some flesh out a concept merely outlined in the text. Others spotlight case histories in which the oceanography of a specific place is presented in concrete terms from the standpoint of an idea introduced in the text. A few featured boxes review a concept that is simply interesting, and that otherwise could not be integrated easily into the main text of the chapter. They are like eating dessert after finishing the main course of a meal. Enjoy them! Five new boxes have been added to this edition. Here is a list of each box in the book and where each appears in the respective chapters.

CHAPTER 1

THE PROCESS OF SCIENCE: **THE SCIENTIFIC PROCESS** 14
SCIENCE BY NUMBERS: **GRAPHS** 17
PHYSICS: **MARINE ARCHEOLOGY** 23
SCIENCE BY NUMBERS: **CONVERSIONS** 26

CHAPTER 2

SCIENCE BY NUMBERS: **POWERS OF 10** 41
GEOLOGY: **PROBING THE SEA FLOOR** 44
A TREK TO THE CREST OF THE MID-ATLANTIC RIDGE 47

CHAPTER 3

GEOLOGY: **THE SAN ANDREAS FAULT** 77
THE PROCESS OF SCIENCE: **SEA-FLOOR SPREADING** 80
GEOLOGY: **THE RED SEA** 82
SCIENCE BY NUMBERS: **SEA-FLOOR SPREADING RATES** 85

CHAPTER 4

GEOLOGY: **PROBING THE SEA FLOOR** 94
GEOLOGY: **DUST STORMS** 102
THE PROCESS OF SCIENCE: **CLIMATE VARIABILITY AND CHANGE** 106
GEOLOGY: **THE DRYING UP OF THE MEDITERRANEAN SEA** 124
SCIENCE BY NUMBERS: **SEDIMENTATION RATES** 127

CHAPTER 5

SCIENCE BY NUMBERS: **PARTS PER THOUSAND** 141
CHEMISTRY: **CHEMICAL TECHNIQUES** 144
CHEMISTRY: **DESALINATION** 164
PHYSICS: **OTHER PHYSICAL PROPERTIES OF WATER** 168
CHEMISTRY: **THE SEA-SURFACE MICROLAYER** 176
SCIENCE BY NUMBERS: **ORDER OF MAGNITUDE** 178

CHAPTER 6

PHYSICS: **CURRENT-MEASURING TECHNIQUES** 194
PHYSICS: **HURRICANES AND TYPHOONS** 206
SCIENCE BY NUMBERS: **VOLUME TRANSPORT** 210
PHYSICS: **UNDERWATER WEATHER AND WATERFALLS** 218

CHAPTER 7

PHYSICS: **WAVE-MEASURING TECHNIQUES** 230
SCIENCE BY NUMBERS: **WAVE CELERITY** 234
PHYSICS: **TINY WAVES AND GIANT WAVES** 244

CHAPTER 8

PHYSICS: **CURRENTS THROUGH TIDAL INLETS** 268
SCIENCE BY NUMBERS: **SPEED OF THE TIDE** 273
BIOLOGY: **TIDAL RHYTHMS IN ORGANISMS** 276
SCIENCE BY NUMBERS: **DEPTH OF THE BAY OF FUNDY** 278

CHAPTER 9

OCEAN SCIENCES: BIOLOGY: **SAMPLING THE BIOTA** 296
BIOLOGY: **KILLER WHALES** 311
BIOLOGY: **ECOLOGY OF THE GIANT KELP COMMUNITY** 314

CHAPTER 10

SCIENCE BY NUMBERS: **DOUBLING RATES** 326
BIOLOGY: **LARGE SHARKS** 330
BIOLOGY: **VOLCANIC VENT COMMUNITIES** 334
BIOLOGY: **SATELLITE OCEANOGRAPHY** 342
BIOLOGY: **MIGRANTS** 350
PHYSICS: **WHAT CAUSES EL NIÑO?** 356

CHAPTER 11

GEOLOGY: **CAPE COD, MASSACHUSETTS** 372
GEOLOGY: **THE OREGON COASTLINE** 374
GEOLOGY: **PADRE ISLAND, TEXAS** 382
GEOLOGY: **SAN DIEGO COUNTY, CALIFORNIA** 385
GEOLOGY: **OCEAN CITY AND ASSATEAGUE ISLAND, MARYLAND** 390
SCIENCE BY NUMBERS: **SAND-BUDGET STUDY** 392

CHAPTER 12

BIOLOGY: **CHESAPEAKE BAY** 406
GEOLOGY: **SALT-MARSH EVOLUTION** 414
BIOLOGY: **SAN FRANCISCO BAY** 418

BIOLOGY: **GLOBAL DECLINE OF CORAL REEFS** 422
SCIENCE BY NUMBERS: **RESIDENCE TIME** 430

CHAPTER 13

BIOLOGY: **PENGUINS** 438
BIOLOGY: **BOTTOM FEEDING BY WHALES** 442
BIOLOGY: **SARGASSUM GULFWEED** 446
BIOLOGY: **SQUIDS** 448

CHAPTER 14

GEOLOGY: **OFFSHORE OIL AND GAS IN THE GULF OF MEXICO** 460
THE PROCESS OF SCIENCE: **FISH FARMING AND "SUPER" FISH** 464
BIOLOGY: **ANTARCTIC KRILL** 468

CHAPTER 15

BIOLOGY: **THE TORREY CANYON DISASTER** 478
BIOLOGY: **BIOREMEDIATION** 484
BIOLOGY: **THE EXXON VALDEZ OIL SPILL** 486
BIOLOGY: **COLLAPSE OF THE NEW ENGLAND FISHERIES** 498
THE PROCESS OF SCIENCE: **THE GLOBAL RISE OF SEA LEVEL AND ITS EFFECTS** 504
BIOLOGY: **RED TIDES** 508
SCIENCE BY NUMBERS: **STEADY STATE** 510

END-OF-CHAPTER FEATURES

Most chapters conclude with a series of questions arranged into three groupings. The first set, the **Review Questions,** is just that. The questions address the main notions developed in the chapter. The second set, the **Critical Thinking Essays,** requires more thought because you must synthesize ideas, sometimes drawing from concepts developed in previous chapters. In other words, verbatim answers might not be found anywhere in the book. However, you can develop an answer by thinking deeply about the question posed and applying common sense and logic to the information provided in the book. The third set of questions, **Discovering with Numbers,** deals with making straightforward calculations about ocean processes. The questions rely on basic mathematics, the kind that any high-school graduate has mastered. In order to assist you, math boxes, **Science by Numbers,** teach the art of computation and are included in each chapter. Each box deals with a basic mathematical concept and

SCIENCE BY NUMBERS

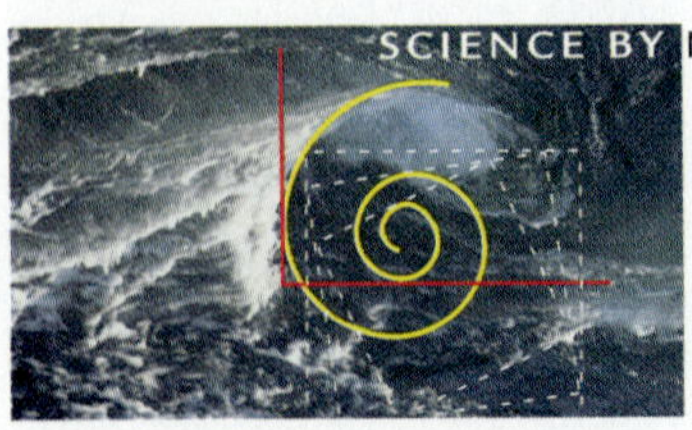

provides a step-by-step solution to a specific problem. The trick to answering math questions is to understand conceptually what it is you are trying to solve. These math boxes will help you upgrade your math skills and develop self-assurance about reasoning with numbers. With the proper learning attitude, the math problems actually become fun to solve and provide the insights into ocean processes that only numerical calculations reveal.

A **reading list** is provided at the end of each chapter and includes both classical, but still relevant, references and more recent writings on the ocean's dynamic processes and diverse habitats. Some are books; most are articles. They should prove valuable for delving deeper into an area of oceanography that intrigues you and for writing term papers. Also, the **appendixes** at the end of the book provide important ancillary material, including conversion factors, a geologic time chart, and map-reading techniques.

ANCILLARY MATERIALS

To assist you in teaching this course and supplying your students with the best in teaching aids, Jones and Bartlett Publishing has prepared a complete supplemental package available to all adopters. Additional information and review copies of any of the following items are available through your Jones and Bartlett Sales Representative.

INSTRUCTOR'S TOOLKIT CD-ROM

The *Instructor's ToolKit* CD-ROM provides adopters with the following traditional ancillaries. All the materials are cross-platform for Windows and Macintosh systems.

The *Instructor's Manual,* provided as a text file, includes chapter outlines, teaching tips, sample syllabi, learning objectives, and additional concept and essay questions.

The *Test Bank* is available as straight text files and as part of the Diploma™ Test Generator software included on the CD. The test bank contains approximately 1,500 multiple-choice, fill-in-the-blank, essay, and research questions. The Test Generator software helps you choose an appropriate variety of questions, create multiple versions of tests, even administer and grade tests on-line.

The *PowerPoint™ Lecture Outline Slides* presentation package provides lecture notes, graphs, and images for each chapter of *Invitation to*

Oceanography. Instructors with the Microsoft PowerPoint™ software can customize the outlines, art, and order of presentation.

The *Kaleidoscope Media Viewer* provides a library of all the art, photos, and tables in the text to which Jones and Bartlett Publishers has the rights to include. The Kaleidoscope Media Viewer uses your browser—Internet Explorer or Netscape Navigator—to enable you to project images from the text in the classroom, insert images into PowerPoint presentations, or print your own acetates.

The CD also includes the answers for the Laboratory Exercises manual.

TRANSPARENCY ACETATES

A full-color transparency acetate set provide clear, effective illustrations of 100 of the most important pieces of artwork from the text.

FOR THE STUDENT

STUDY GUIDE TO ACCOMPANY INVITATION TO OCEANOGRAPHY, THIRD EDITION.

The study guide contains chapter overviews, complete chapter outlines, learning objectives, diagram labeling exercises, graph and table interpretation exercises, and practice exams organized by chapter sections.

LABORATORY EXERCISES TO ACCOMPANY INVITATION TO OCEANOGRAPHY, THIRD EDITION.

The exercises in this laboratory manual are designed to make use of safe, readily available, inexpensive, and reusable materials. Many of the labs are group-based activities that demonstrate principles typically discussed in lecture. The exercises require just minimal knowledge of science and math.

Jones and Bartlett makes these ancillaries available with the text, or they can be purchased separately.

ACKNOWLEDGMENTS

Jones and Bartlett Publishers have worked hard and indefatigably to produce the best possible introductory textbook in oceanography, and they've succeeded beyond my most fanciful expectations. The current edition was overseen by Louis Bruno, the Senior Production Editor, who did a superb job at moving us forward expeditiously, but gracefully, to the publication of this current edition. The collaborative work, under Lou's guidance and vision, went flawlessly despite tight deadlines. This continues the tradition prescribed by Brian McKean, the Executive Editor who oversaw the publication of the previous edition of the book. I extend special thanks to Dean DeChambeau, the Managing Editor with whom I have worked most closely and consistently since my association with Jones and Bartlett Publishers. From my perspective, Dean is the lynchpin in this entire enterprise, providing careful, critical editorial commentary, and helping unselfishly to answer my constant questions and covering my numerous shortfalls as an author. His work with the book's illustrations is unsurpassed. I thank Philip Regan, who expertly typeset the new material, composed the pages, and set the new photos for best appearance. Also, I am grateful to many others who have helped extensively in previous editions. These include Lianne Ames, Dean Weatherbee, Michael DeFronzo, and the creative staff at Seventeenth Street Studios. I am privileged to be an author working with Jones and Bartlett Publishers.

Many colleagues at numerous institutions reviewed and constructively criticized drafts of the various editions, vastly improving their quality. Those who were particularly helpful and generous with their time and expertise over the years include:

Marsha Bollinger
  Winthrop University
Chuck Breitsprecher
  American River College
Kathleen M. Browne
  Rider University
William H. Busch
  University of New Orleans
Steve Calvert
  University of British Columbia
Barry Cameron
  Acadia University
William Chaisson
  University of Massachusetts–Amherst
Karl M. Chauffe
  St. Louis University

G. Kent Colbath
Cerritos Community College
Ronadh Cox
Williams College
Charles M. Drake
Dartmouth College
Walter C. Dudley
University of Hawaii at Hilo
Brent Dugolinsky
State University of New York College at Oneonta
John Ehleiter
West Chester University
William Frazier
Columbus College
Grant Gardner
Memorial University of Newfoundland
Tom Garrison
Orange Coast College
Jack C. Hall
University of North Carolina–Wilmington
Robert W. Hinds
Slippery Rock University
Ralph Hitz
Tacoma Community College
William Hoyt
University of Northern Colorado
R. Grant Ingram
University of British Columbia
Dennis Kelly
Orange Coast College
John A. Klasik
California State Polytechnic University–Pomona
Charles E. Knowles
North Carolina State University
William Kohland
Middle Tennessee State University
Stephen Lebsack
Linn-Benton Community College
Phillip Levin
University of California, Santa Cruz
Jonathan B. Martin
University of Florida
Heyward Mathews
St. Petersburg Junior College
James H. Meyer
Winona State University
A. Lee Meyerson
Kean College of New Jersey
John E. Mylroie
Mississippi State University
Joseph Nadeau
Rider University
Marlon Nance
California State University–Sacramento
Peter Nielsen
Keene State College
James Ogg
Purdue University
Harold Pestana
Colby College
Curt D. Peterson
Portland State University
William Prothero
University of California at Santa Barbara
C. Nicholas Raphael
Eastern Michigan University
Howard Reith
North Shore Community College
Joseph P. Richardson
Savannah State University
Richard Robinson
Santa Monica College
Peter A. Rona
Rutgers–The State University of New Jersey
Jill Singer
State University College at Buffalo
Charles R. Singler
Youngstown State University
Jon Sloan
California State University–Northridge
Alan Trujillo
Palomar Community College
Terri Woods
East Carolina University
Dawn Wright
Oregon State University
Richard Yuretich
University of Massachusetts–Amherst

Invitation to Oceanography

1

The Growth of Oceanography

What is the sense of owning a good boat if you hang around in home waters?
—*George Miller,* Oyster River

web navigator

critical thinking on the web

math tutor on the web

www.jbpub.com/oceanlink

PREVIEW

A COMPLETE HISTORICAL account of oceanographic exploration and research would be a massive undertaking. The record stretches back over several millennia to the time when ancient mariners built boats and ventured boldly onto the sea to explore the unknown. However, a brief sketch of maritime history is needed in a book that deals with the physical, chemical, geological, and biological processes of the ocean in a scientifically rigorous manner. First and foremost, this reminds us that for eons there have been people in the field of "oceanography"—people with an insatiable desire to make the unknown familiar. Knowledge that is commonplace today required painstaking investigations by numerous seafarers throughout centuries of exploration. Many intended to become rich by exploiting resources and by controlling sea routes for commerce. All were driven by a yearning to understand the mysteries of the Earth and its seas.

Today's oceanographers (modern sea explorers) carry forward this quest to satisfy humankind's curiosity. They owe a huge debt to the courage and vision of earlier mariners, who by slow increments replaced ignorance and myth with knowledge.

1-1

Oceanography: What Is It?

Before delving into the science of **oceanography**, we should understand exactly what the word means. The first part of the term is coined from the Greek word *okeanos*, or Oceanus, the name of the Titan son of the gods Uranus and Gaea, who was father of the ocean nymphs (the Oceanids). Eventually *oceanus* was applied to the sea beyond the Pillars of Hercules, the North Atlantic Ocean. The second part of the term comes from the Greek word *graphia*, which refers to the act of recording and describing. In fact, the word *oceanography* is inadequate to describe the science of the seas, because scientists do much more than merely record and describe the ocean's physical, chemical, geological, and biological characteristics. Oceanographers investigate, interpret, and model all aspects of ocean processes, using the most modern and sophisticated techniques of scientific and mathematical enquiry. The term **oceanology** (the suffix *ology* meaning "the science of") is etymologically more accurate. The distinction between oceanography and oceanology is similar to that made between geography (the physical description of the world and its biota) and geology (the scientific study of the Earth and its processes). The word *oceanology* has not, however, displaced *oceanography*, because the latter term is solidly entrenched in the minds of the laypeople as well as the Western practitioners of the science. Hence, this book will follow convention, using the more familiar term to denote the scientific study of the oceans.

A common misconception is that oceanography is a pure science in its own right, practiced by women and men who are specifically and narrowly instructed in its investigative methods. Most oceanographers are, in fact, trained in one of the traditional sciences (physics, chemistry, biology, and geology) or a related field (engineering, meteorology, mathematics, statistics, or computer science) and choose to apply their research expertise to the study of the oceans. After obtaining undergraduate training in a traditional science, they gain experience conducting oceanographic research in graduate school or at a marine institute. Recently, new career opportunities in oceanography have developed in marine policy and management, marine law, resource and environmental assessment, and other related fields. Marine studies commonly rely on collaboration among many types of scientists, mathematicians, engineers, technicians, and policymakers.

It is customary to subdivide oceanography into the four fields of physical, geological, chemical, and biological oceanography (**Figure 1–1**). These fields are in turn linked to one another by the cross-disciplines of geochemistry, biochemistry, geophysics and biophysics.

1-2

Historical Review of Oceanography

Our perception and understanding of the oceans have changed markedly over time. Although this book stresses the most current ideas championed by marine scientists, these attitudes and impressions did not suddenly appear out of an intellectual vacuum. They grew out of—and evolved from—the ideas and deductions of prior generations of ocean explorers and scientists. Marine scientists are well aware of the fact that all of their work rests on the contributions of the innumerable investigators that came before them. But, obviously, this does not mean that all the conclusions of those early investigators have been validated. Rather, as the science of oceanography has matured and as research vessels, sampling devices, and electronic instrumentation have become increasingly sophisticated and more widely applied to probe the ocean's secrets, many beliefs of the past have been disproved. The lesson from history is clear-cut. Our ideas of the oceans today, which seem so appealingly final and are written about and taught with so much fervor and certainty, will be refined by the findings and thoughts of future generations of marine scientists.

The history of oceanography has not always been the gradual and systematic development of a body of thought. Rather, bold concepts and opin-

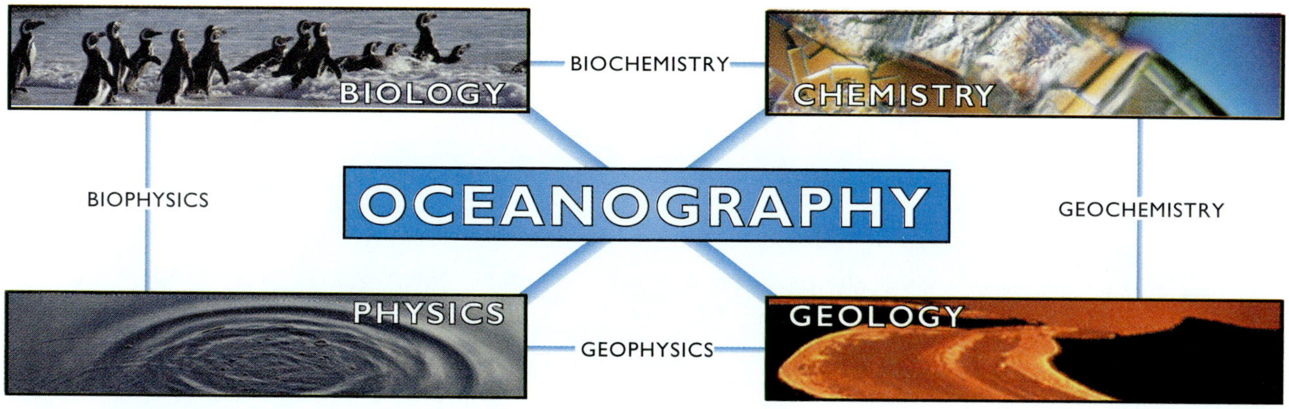

FIGURE 1-1

The field of oceanography. This diagram organizes oceanography into four principal categories—biological, geological, physical, and chemical oceanography—that are linked to one another by cross-disciplines.

ions have often burst onto the scene, necessitating critical reexamination of the wisdom of the past, and stimulating fresh insights into the workings of the oceans.

A practical means of organizing the historical record of oceanography is to arrange the events into three broad stages. The first includes the early efforts of individual mariners as they attempted to describe the geography of the Earth's oceans and landmasses. During this time of ocean exploration, the very limits of the world were sought. The second includes the early systematic attempts to use a truly scientific approach to investigate the oceans. The third covers the growth of modern oceanography that has resulted from the widespread application of state-of-the-art technology and the international collaboration of scientists. We will conclude this historical review with an assessment of future prospects and try to predict the nature of oceanographic investigations in the early part of the twenty-first century.

A limited number of the innumerable events that contribute to the rich history of oceanography can be highlighted in a single chapter. Although the details of only a few of the many important research cruises and studies are elaborated here, synopses of many others are cataloged chronologically in **Table 1–1**. Also, books that discuss the historical context of ocean exploration and the science of oceanography are listed at the end of the chapter.

OCEAN EXPLORATION

In many respects the Polynesian migration to the many large and small islands of the Pacific Ocean (**Figure 1–2**), completed well before the birth of Christ, ranks as one of the most spectacular exploration feats ever. Their canoes, which they sailed and paddled, were made by hollowing out logs or by lashing planking together with braided ropes. These seaworthy vessels were built with simple tools made of rock, bone, and coral. In order to travel safely from one island to the next, these Pacific seafarers relied on sound seamanship, extensive navigational skills, and detailed local knowledge, all of which—in the absence of a written language—was passed on to others orally in the recitation of epic poems. Polynesian seafarers could depend on accurate, detailed lore of local wind, wave, current, and weather patterns as well as on the position of key navigational stars in making a planned landfall after a deep-sea crossing of hundreds, even thousands, of kilometers.

The ability to explore and chart the seas safely depends on navigation. Records of sailing vessels indicate maritime activity in Egypt as far back as 4000 B.C. It is likely that the extent of these voyages was restricted, with mariners remaining well in sight of land, probably in the immediate vicinity of the Nile River and the shores of the eastern

(a) POLYNESIA

(b) POLYNESIAN CANOE

FIGURE 1-2

Polynesia. (a) Polynesians settled these Pacific Islands, navigating across an ocean area the size of a continent. (b) Polynesians used canoes made of hollowed-out logs or planks.

Mediterranean Sea. By the sixth century B.C., however, Phoenicians had established sea routes for trading throughout the entire Mediterranean region and had even ventured westward into the Atlantic Ocean, sailing as far north as the coast of Cornwall in England. Historians suspect that Phoenicians, around 600 B.C., were probably the first to circumnavigate the continent of Africa. True ocean navigation was difficult at the time. Navigators charted the courses of their vessels according to the stars. Undoubtedly, sailors steered their craft in sight of the coastline whenever possible, relying on distinctive landmarks to find their way and establish their position. This process is called **piloting**.

By the third century B.C., the flourishing Greek civilization, , plying the Mediterranean for trade as it established its influence and control over the entire region, was highly dependent on its maritime prowess. A notable sea adventurer of the time was **Pytheas**, the first Greek to circumnavigate England and gauge the length of its shoreline. Although his travels are not documented by first-hand accounts, some historians believe that Pytheas may have voyaged as far north as Norway and as far west as Iceland. If he did, this stands as an incredible navigational accomplishment. Historians have established that Greek mariners estimated latitude (Appendix IV) by the length of the day, correcting for the time of the year. However, without mechanical timepieces (accurate chronometers) it was impossible for them to determine longitude. Pytheas's discovery that the tides of the Atlantic Ocean vary regularly with the phases of the moon underscores his deep understanding of ocean processes.

A map compiled by **Herodotus** in 450 B.C. shows the extent of the Greeks' understanding of world geography (**Figure 1–3**). The Mediterranean Sea prominently occupies the center of the map and is

FIGURE **1-3**

The Greek world. Herodotus compiled a map of the known world around 450 B.C., showing the Mediterranean Sea surrounded by the land masses known as Europa, Asia, and Libya. Large tracts of ocean in turn surrounded this land, extending to the very edges of the world. (Reproduced from an 1895 *Challenger Report*, published in Great Britain.)

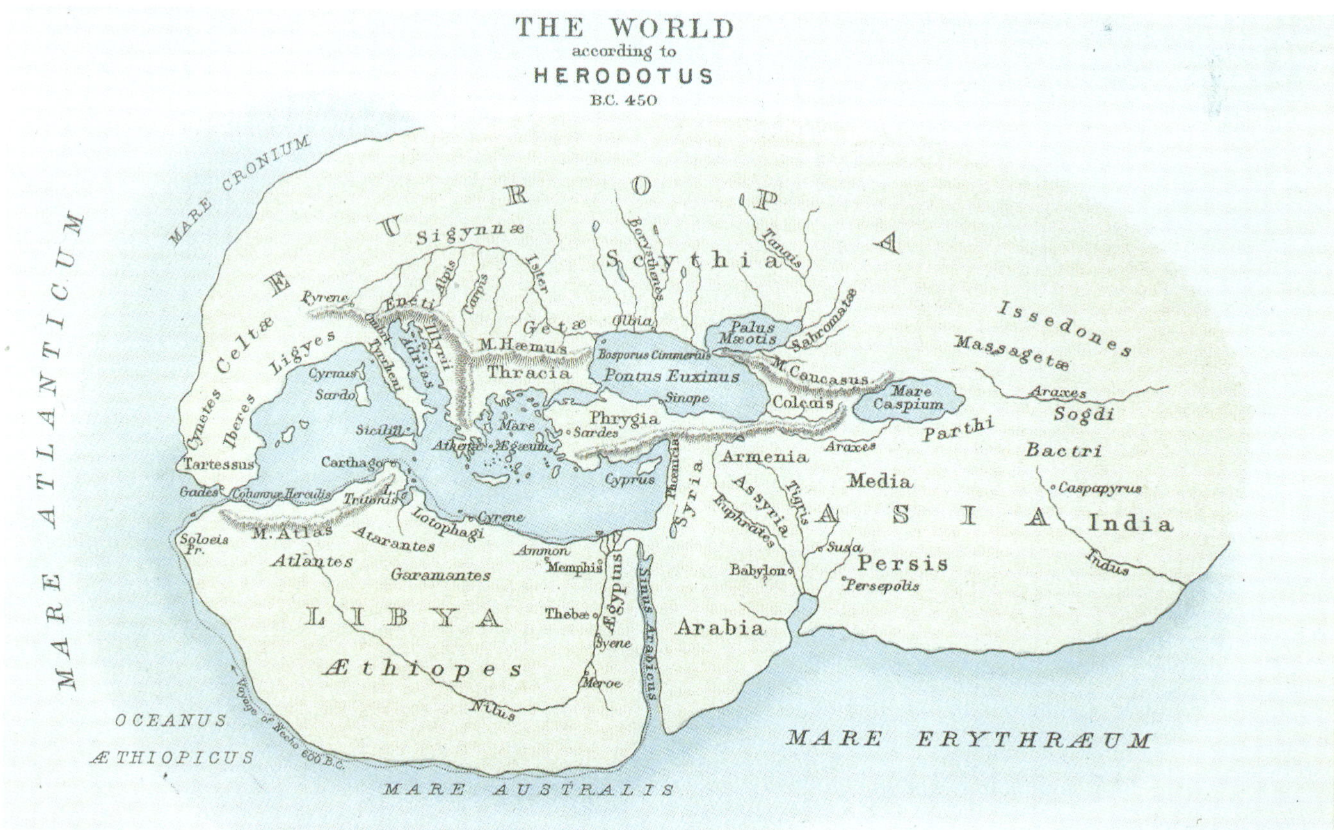

TABLE 1-1

A Chronology of Ocean Exploration

4000 | 3000 | 2000 | 1000 | B.C. A.D. | 1000 | 1100 | 1200 | 1300

ca. 4000 B.C.
Egyptians developed the arts of shipbuilding and coastal piloting.

ca. 2000–500 B.C.
Most islands of the Pacific Ocean settled by Polynesians.

ca. 1000–600 B.C.
Phoenicians explored the entire Mediterranean Sea, sailed into the Atlantic to Cornwall, England, and probably circumnavigated Africa. They navigated by familiar coastal landmarks and by the stars.

450 B.C.
The Greek Herodotus compiled a map of the known world that centered on the Mediterranean region (see Figure 1-3).

325 B.C.
The Greek Pytheas explored the coasts of England, Norway, and perhaps Iceland. He developed a means of determining latitude from the angular distance of the North Star and proposed a connection between the phases of the Moon and the tides. Aristotle published *Meteorologica*, in which he described the geography and physical structure of the Greek world, and *Historia Animalium*, the first known treatise on marine biology.

276–192 B.C.
The Greek Eratosthenes, a scholar at Alexandria, determined the circumference of the Earth with remarkable accuracy using trigonometry and noting the specific angle of sunlight that occurred at Alexandria and at Aswan (then known as Syene) in Egypt.

54 B.C.–A.D. 30
The Roman Seneca devised the hydrologic cycle to show that, despite the inflow of river water, the level of the ocean remained stable because of evaporation.

ca. A.D. 150
The Greek Ptolemy compiled a map of the entire Roman World that showed latitudes and longitudes.

A.D. 673–735
The English monk Bede published *De Temporum Ratione,* in which he discussed the lunar control of the tides and recognized monthly tidal variations and the effect of wind drag on tidal height.

A.D. 982
The Norseman Eric the Red completed the first transatlantic crossing and discovered Baffin Island in the Arctic region of Canada.

A.D. 995
Leif Ericson, son of Eric the Red, established the North American settlement of Vinland in what is now Newfoundland.

Landmark Events in Early Ocean Exploration

1400 ▼ 1425 ▼ 1450 ▼ 1475 ▼ 1500 ▼ 1525 ▼ 1550 ▼ 1575 ▼ 1600 ▼ 1625 ▼ 1650 ▼ 1675 ▼ 1700 ▼

1452–1519
Leonardo da Vinci observed, recorded, and interpreted details about currents and waves and noted that fossils in the mountains of Italy implied that the level of the sea had been higher in the ancient past.

1492
Christopher Columbus rediscovered North America by sailing to the islands of the West Indies.

1500
Pedro Alvares Cabral discovered and explored Brazil.

1513
Juan Ponce de Leon described the swift and powerful Florida current.

1513–1518
Vasco Núñez de Balboa crossed the Isthmus of Panama and sailed in the Pacific Ocean.

1515
Peter Martyr proposed an origin for the Gulf Stream.

1519–1522
Ferdinand Magellan embarked on a circumnavigation of the globe; Sebastian del Cano completed the voyage.

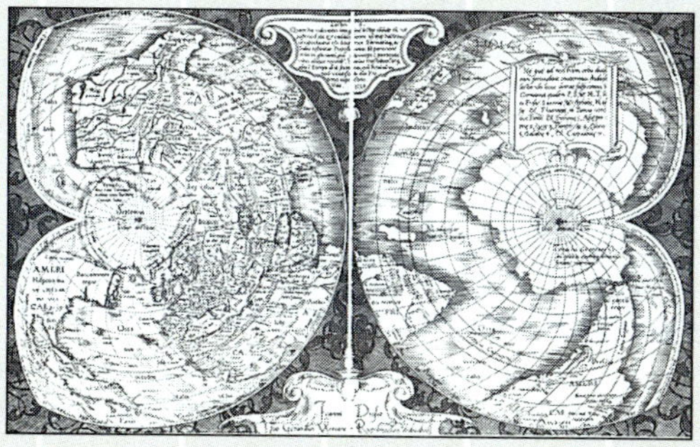

1569
Geradus Mercator constructed a map projection of the world that was adapted to navigational charts.

1674
Robert Boyle investigated the relation among temperature, salinity, and pressure with depth and reported his findings in "Observations and Experiments on the Saltiness of the Sea."

Adapted from D. E. Ingmarson and W. J. Wallace, *Oceanography: An Introduction,* Table 1.2 (Belmont, Calif.: Wadsworth, 1979); B. H. McConnaughey and R. Zottoli, *Introduction to Marine Biology*, Chapter 24 (St. Louis, Miss.: Mosby, 1983); J. C. McCormick and J. V. Thiruvathukal, *Elements of Oceanography*, Tables 1.1 and 1.2 (New York: Saunders College Publishing, 1981); H. S. Parker, *Exploring the Oceans*, Chapter 1 (Englewood Cliffs, N.J.: Prentice-Hall, 1985); and H. V. Thurman, *Introduction to Oceanography*, Chapter 1 (Columbus, Ohio: Merrill, 1988).

A Chronology of Ocean Exploration

1700 | 1725 | 1750 | 1775 | 1800 | 1825 | 1850

1725
Luigi Marsigli compiled *Histoire Physique de la Mer,* the first book to deal entirely with the science of the sea.

1740
Leonhard Euler calculated the magnitude of the forces that generate ocean tides and related them to the attractive force of the Moon.

1769–1770
Benjamin Franklin published the first ocean chart of the Gulf Stream, which shippers consulted extensively as they crossed the North Atlantic Ocean.

1768–1771, 1772–1775, 1778–1779
Captain James Cook commanded three major ocean voyages, gathering extensive data on the geography, geology, biota, currents, tides, and water temperatures of all of the principal oceans.

1802
Nathaniel Bowditch published the *New American Practical Navigator,* a superb navigational resource that continues to be revised and published to this day.

1807
President Thomas Jefferson mandated coastal charting of the entire United States and established the U.S. Coast and Geodetic Survey.

1817–1818
Sir John Ross ventured into the Arctic Ocean to explore Baffin Island, where he sounded the bottom successfully and recovered starfish and mud worms from a depth of 1.8 kilometer.

1820
Alexander Marcet, a London physician, noted that the proportion of the chemical ingredients in seawater is unvarying in all oceans.

1831–1836
The epic journey of Charles Darwin aboard the HMS *Beagle* led to a theory of atoll formation and later to the theory of organic evolution by natural selection.

1839–1843
Sir James Ross, nephew of the Arctic explorer Sir John Ross, led a scientific expedition to Antarctica, recovering samples of deep-sea bottom life down to a maximum depth of 7 kilometers.

1841, 1854
Sir Edward Forbes published *The History of British Star-Fishes* (1841) and then his influential book, *Distribution of Marine Life* (1854), in which he argued that sea life cannot exist below about 600 meters (the so-called azoic zone).

Milestones in Early Oceanography

1855 ▼ 1860 ▼ 1865 ▼ 1870 ▼ 1875 ▼ 1880 ▼ 1885 ▼ 1890 ▼ 1895 ▼ 1900 ▼

1855
Matthew Fontaine Maury compiled and standardized the wind and current data recorded in U.S. Navy ship logs and summarized his findings in *The Physical Geography of the Sea.*

1868–1870
Charles Wyville Thomson, aboard the HMS *Lightning* and HMS *Porcupine,* made the first series of deep-sea temperature measurements and collected ample life from great depths, disproving Forbes's azoic zone.

1871
The U.S. Fish Commission was established with a modern laboratory at Woods Hole, Massachusetts.

1872–1876
Under the leadership of Charles Wyville Thomson, the HMS *Challenger* conducted worldwide scientific expeditions, collecting data and specimens that were later analyzed in over fifty large volumes of the *Challenger Reports.*

1873
Charles Wyville Thomson published a general and popular book on oceanography called *The Depths of the Sea.*

1877–1880
Alexander Agassiz, an American naturalist, extensively sampled life in the deep sea while aboard the U.S. Coast and Geodetic Survey ship *Blake.* He also founded the Museum of Comparative Zoology at Harvard University and the first U.S. marine station, the Anderson School of Natural History, on Penikese Island, Buzzards Bay, Massachusetts.

1884–1901
The USS *Albatross* was designed and constructed specifically to conduct scientific research at sea and undertook numerous oceanographic cruises.

1888
The Marine Biological Laboratory was established at Woods Hole, Massachusetts, and Dr. Charles Otis Whitman served as its first director.

1893
The Norwegian Fridtjof Nansen had the *Fram* constructed with a reinforced hull for use in sea ice; he confirmed the general circulation pattern of the Arctic Ocean and the absence of a northern continent.

A Chronology of Ocean Exploration

1900 1905 1910 1915 1920 1925 1930 1935 1940 1945 1950

1902
Danish scientists with government backing established the International Council for the Exploration of the Sea (ICES) to investigate oceanographic conditions that influence North Atlantic fisheries. Council representatives were from Great Britain, Germany, Sweden, Norway, Denmark, Holland, and the Soviet Union.

1903
The Friday Harbor Oceanographic Laboratory was established by the University of Washington, Seattle.

1903
The Scripps Institution of Biological Research, which later became the Scripps Institution of Oceanography, was founded at La Jolla, California.

1912
The German meteorologist Alfred Wegener proposed his theory of continental drift.

1925–1927
A German expedition aboard the research vessel *Meteor* studied the physical oceanography of the Atlantic Ocean as never before, heralding the modern age of oceanographic investigation. Scientists used an echo sounder extensively for the first time.

1930
The Woods Hole Oceanographic Institution was established on the southwestern shore of Cape Cod, Massachusetts.

1932
The International Whaling Commission was organized to collect data on whale species and to enforce voluntary regulations on the whaling industry.

1942
Harald Sverdrup, Richard Fleming, and Martin Johnson published the scientific classic *The Oceans,* which is still consulted today.

1949
The Lamont (later changed to Lamont Doherty) Geological Observatory at Columbia University in New York was established at Torrey Cliffs Palisades on the bedrock cliffs of the Hudson River.

The Era of Modern Oceanography

1957–1958
The International Geophysical Year (IGY) was organized—an ambitious international effort to coordinate the geophysical investigation of the Earth, including its oceans.

1958
The nuclear submarine USS *Nautilus,* commanded by C.D.R. Andersen, reached the North Pole under the ice.

1959–1965
The International Indian Ocean Expedition was established under the auspices of the United Nations to make a systematic investigation of the oceanography of the Indian Ocean.

1966
The U.S. Congress adopted the Sea Grant College and Programs Act to provide nonmilitary funding for education and research in the marine sciences.

1968, 1975
The U.S. National Science Foundation organized the Deep Sea Drilling Project (DSDP) to core through the sediments and rocks of the oceans. This effort was reorganized in 1975 as the International Program of Ocean Drilling, which continues to be active in all of the world's oceans today.

1970
The U.S. government created the department of the National Oceanic and Atmospheric Administration (NOAA) to oversee and coordinate government activities that have a bearing on oceanography and meteorology.

1970s
The United Nations initiated the International Decade of Ocean Exploration (IDOE) to improve our scientific knowledge of all aspects of the oceans.

1972
The Geochemical Ocean Sections Study (GEOSECS) was organized to obtain accurate measurements of seawater chemistry in an effort to explain the nature of ocean circulation and mixing and the biogeochemical recycling of chemical substances.

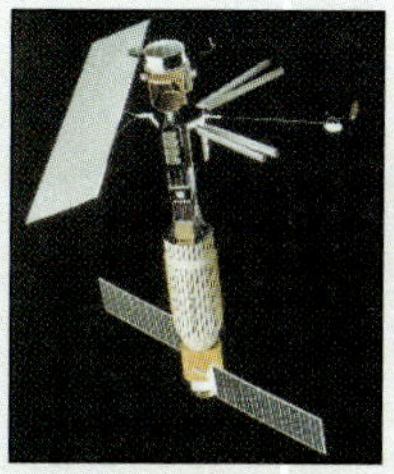

1978
Seasat-A, the first oceanographic satellite, was launched, demonstrating the utility of remote sensing in the study of the oceans.

1980s–1990s
The Coordinated Ocean Research and Exploration Section program (CORES) was organized to continue the scientific work of the IDOE into the 1980s. The Ocean Drilling Program (ODP) continues the geological exploration of the oceans.

1992
NASA launched the TOPEX/Poseidon satellite to monitor sea level and to keep track of changes in current patterns as climate fluctuates.

1998
International Year of the Ocean is organized to educate the public about the value and importance of the ocean's resources.

2001
Joint launching of Jason-1 satellite by NASA and the French Space Agency to improve forecasting of currents and climate. Implementation of GLOBEC (GLOBal Ocean ECosystem Dynamics), an international research program designed by oceanographers, marine ecologists, and fishery scientists.

THE PROCESS OF SCIENCE

The Scientific Process

As this chapter on the history of oceanography indicates, scientists make statements about the natural world; they assume that natural processes are orderly and therefore knowable by a rational mind. Statements made by scientists are not merely random opinions about the workings of the world. Rather they are logical explanations, termed **hypotheses,** that are grounded solidly on a set of observations and tested rigorously in order to evaluate their credibility.

Scientific investigations are begun typically by people who develop an interest in answering a question about the natural world. Examples of such questions in oceanography might be:

- What is the geologic origin of a particular estuary?
- How does the chemistry of the seawater in this estuary vary over time?
- What is the water current pattern in this estuary and what controls it?
- What effect does lead dissolved in the water have on a species of clam in this estuary?

The questions can be general or specific, theoretical or applied, abstract or concrete.

Scientists interested in a question then conduct laboratory, field, or modeling (mathematical) experiments in order to generate accurate facts (observations) that bear on an answer to the question being investigated. A legitimate answer (the hypothesis) to a scientific question is one that can be tested. A hypothesis is always considered to be a tentative explanation. Scientist first and foremost are skeptics, trying to disprove hypotheses in order to eliminate falsehoods from the scientific understanding of the natural world.

Depending on the results of tests, hypotheses may be verified, rejected, or modified. When a hypothesis is tested repeatedly in different ways and not disproved, scientists then assume that it is "correct" and the hypothesis becomes a theory, as new facts continue to support it. For example, Charles Darwin proposed his hypothesis of biological evolution by natural selection during the middle part of the nineteenth century. Today, after repeated tests and countless facts that support the idea, his hypothesis of biological evolution by natural selection has been elevated to the status of a theory.

In summary, scientists are not, as many believe, primarily concerned about discovering and gathering facts. Rather, researchers ask crucial questions about the natural world and then try to answer them by proposing hypotheses—creative insights about what the truthful responses to those questions might be. What really separates the scientific method from other ways of knowing is its reliance on the

surrounded by three landmasses of enormous proportions—Libya (northern Africa), Europe, and Asia. The polar limits and coastline configurations of the latter two continents were unexplored at the time and are not marked on the map. All of the familiar land is surrounded by enormous expanses of ocean that the Greeks believed extended to the very ends of the world.

Throughout the Middle Ages (between 500 and 1450 A.D.), there was little ocean exploration by Europeans, with the notable exception of the Viking seafarers. Between the ninth and the twelfth century, Scandinavians extended their influence over Europe and across the Atlantic Ocean by acquiring new lands. The Norse ventured boldly to Iceland, Greenland, and the Baffin Islands, for example, and established a North American settlement known as Vinland in the area that we now call Newfoundland. These Viking outposts eventually were abandoned because of the harsh climates. Also, the onset of the "Little Ice Age" (A.D. 1430 to 1850) caused the extensive buildup of sea ice that cut off the northern sea routes from Scandinavia.

The Norsemen—the most adept and experienced navigators in the Western world at that time—sailed westward by maintaining a course on a predetermined line of latitude. They accomplished this navigational feat by sailing to a coastal point along Norway and measuring the angular height of the North Star. They then kept it at the same angle on the starboard beam of the vessel throughout the night. Their daytime navigation relied on the careful calculation of the sun's position for the time of year. A map dated at about 1570 rigorous testing of each hypothesis by experimentation or by the gathering of additional observations; the explicit intent of the test is to determine whether the hypothesis is false or true. If the test results disagree with the prediction, then the hypothesis being evaluated is disproved, meaning that it cannot be a legitimate account of reality. Then it is either modified into a new hypothesis that is compatible with the test findings or discarded altogether and replaced by other, still-to-be-tested hypotheses. Keep in mind, however, that agreement between expected and experimental test results is not proof that the hypothesis is true. Rather, it means only that the hypothesis continues to be a valid version of reality for the time being. It may not survive the next test. If a hypothesis repeatedly avoids falsification, then scientists regard it as a close approximation of the truth. A flow diagram of this version of the scientific method is presented as **Figure B1-1**.

In this book, we describe the results of a long-standing interest among scientists in answering questions about the workings of the oceans. It is a current update of the facts, hypotheses, and theories of ocean processes. Undoubtedly, as oceanographers continue to conduct scientific work in the world's oceans, some of these ideas will be disproved and replaced by other hypotheses. This is the way it must be; this is the scientific process.

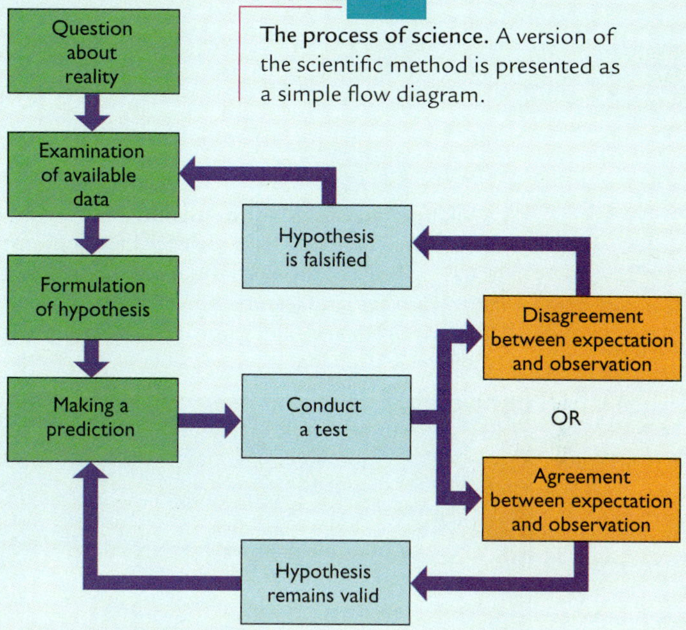

FIGURE **B1-1**

The process of science. A version of the scientific method is presented as a simple flow diagram.

Visit www.jbpub.com/oceanlink for more information.

shows the remarkable state of the Viking's geographic knowledge of the North Atlantic Ocean (**Figure 1–4**).

Economic, political, and religious motives encouraged western Europeans to undertake long sea explorations in the fifteenth and sixteenth centuries; they crossed the Atlantic and ventured into the Pacific Ocean. Portuguese sailors were particularly successful explorers during this time. In 1487 and 1488 **Bartholomew Diaz** rounded the Cape of Good Hope at the southern tip of Africa. After sailing around the Cape of Good Hope in 1498, **Vasco da Gama** continued as far eastward as India.

Perhaps the crowning achievement of this age is the circumnavigation of the globe by **Ferdinand Magellan**. Departing from Spain in late September of 1519, Magellan proceeded southwestward with his flotilla of five age-worn ships to the northeastern coast of Brazil (**Figure 1–5**). There he began to search for a seaway to the Pacific and, in the process, lost two of his vessels, one by desertion. Almost one year after his departure from Spain, Magellan located the 500-kilometer-wide (~310 miles) passage that now bears his name and sailed around South America and into the Pacific Ocean. The following three months were desperate for Magellan's crew, who endured starvation, disease, and doubtless suffered much from fear of the unknown. They eventually reached Guam on 6 March 1521. After proceeding to the Philippines later that month, Magellan was killed on 27 April on the small island of Mactan while participating in a dispute among local tribes. **Sebastian del Cano** eventually completed the circumnavigation under

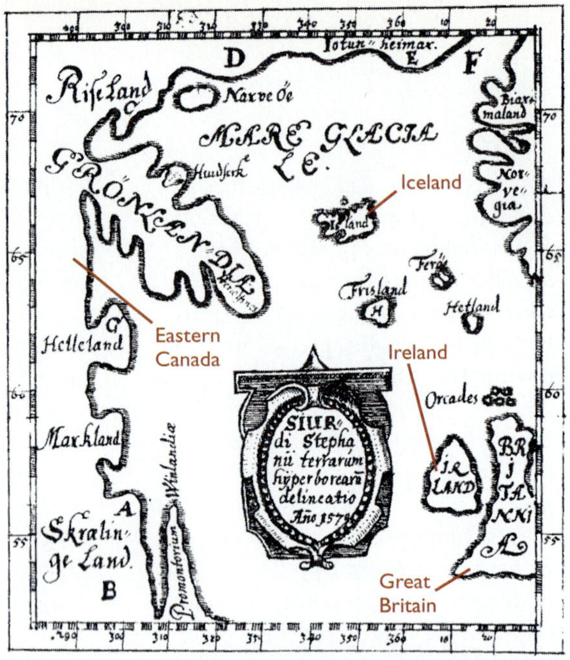

FIGURE 1-4

A Viking chart of the North Atlantic Ocean. This Viking map, dated at about 1570, demonstrates how extensive the knowledge of the North Atlantic ocean was at that time. Familiar land features include Great Britain, Ireland, Iceland, Greenland, and a portion of the northeastern shoreline of Canada.

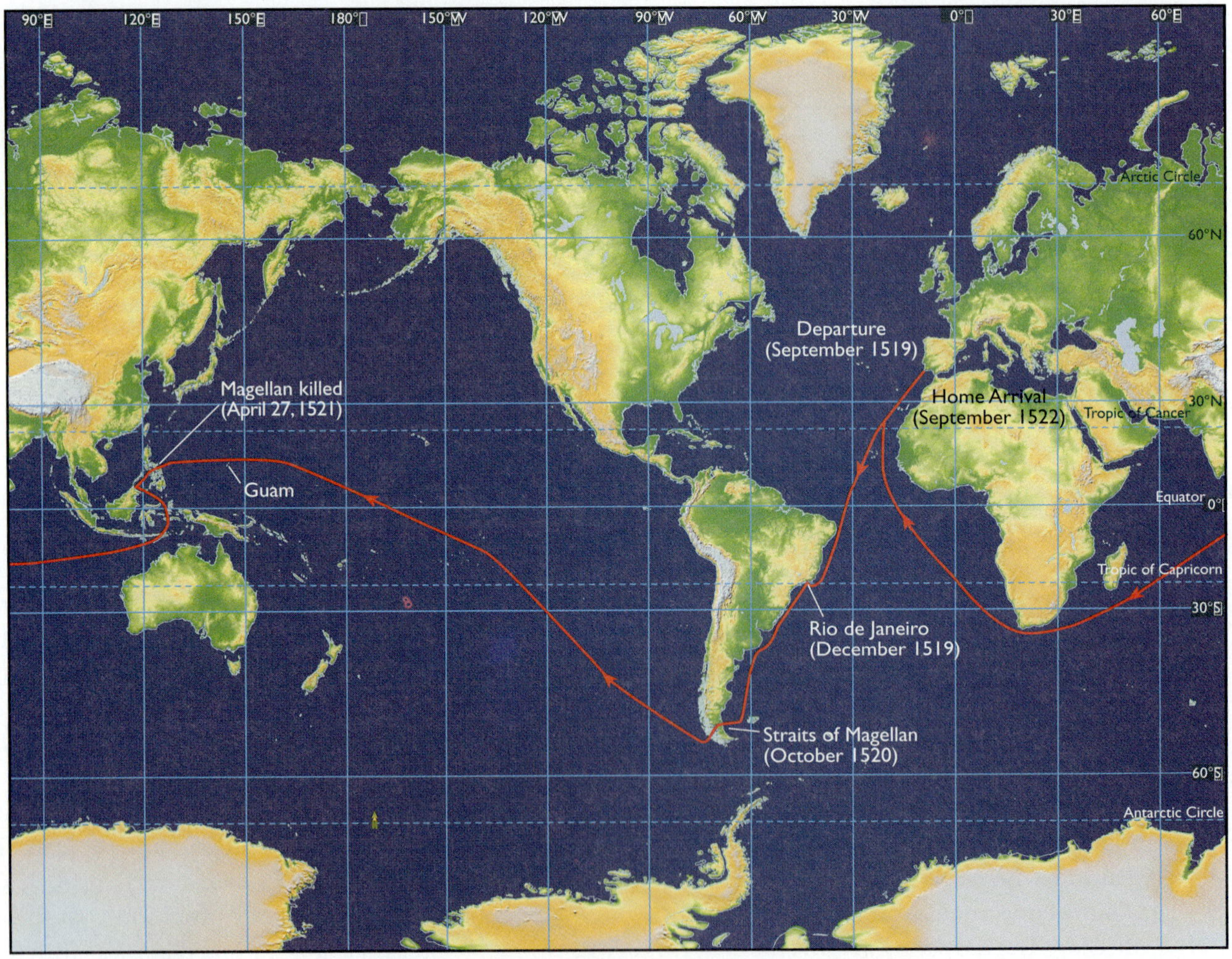

FIGURE 1-5

The circumglobal voyage of Magellan. Ferdinand Magellan embarked on a three-year-long voyage in 1519, intent on discovering a seaway to the East Indies. In 1520 he rounded the Straits of Magellan and continued to the Philippines, where he was killed during a skirmish with natives. Sebastion del Cano completed the journey as leader of the expedition.

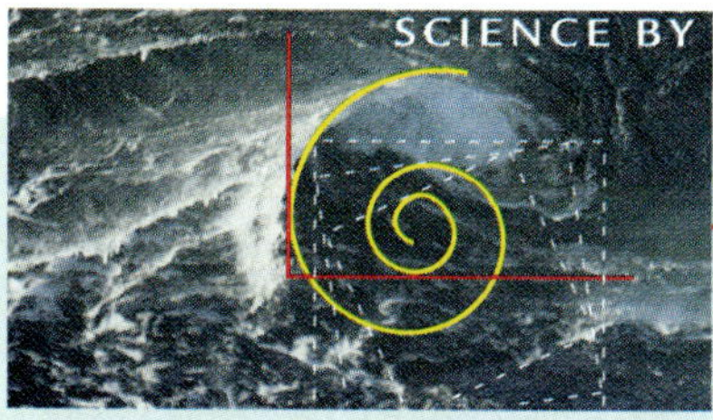

SCIENCE BY NUMBERS

Graphs

As you know, scientists strive to display information clearly and accurately. A useful and revealing way to display data is to plot the information on a graph. A data plot summarizes the information quickly and reveals trends and relationships among variables. Sometimes the relationship between variables is direct, meaning that an increase in one leads to an increase in the value of the other, and a decrease leads to a corresponding decrease. Other times, the relationship is inverse, meaning that an increase in one leads to *decrease* in the other, and vice versa.

In many cases relationships between variables are not linear, but are curved in some complicated way. Also, scientists may choose not to use linear scales for one or both variables plotted on a graph. Oceanographers, for example, commonly graph a variable as a power of 10. Let's look at **Figure B1–2**. At first glance, it appears as if the concentration of tiny zooplankton suspended in the water above the Kurile-Kamchatka Trench of the northwestern Pacific Ocean decreases regularly with water depth, except near the sea surface, where the zooplankton are a bit more abundant than they are lower down in the water column (the vertical extent of water from sea surface to sea bottom). The relationship between water depth and zooplankton abundance is inverse. To the unsuspecting, the plot suggests that zooplankton concentrations decrease gradually downward and that significant numbers of zooplankton exist even at depths below 5 kilometers. This impression is not, however, correct, because the zooplankton abundances are plotted on a scale that varies as the power of 10. Thus, the minute quantities of zooplankton found in water deeper than 1 kilometer are amplified by the scale chosen to plot them.

A conventional plot of the data using a linear scale for zooplankton concentrations shows a very different graph (**Figure B1–3**). This reveals that zooplankton are essentially confined to water that is no deeper than 1 kilometer below the sea surface. Below 2 kilometers, there are virtually no zooplankton. Also, the sharp downward curve of the graph between 1 and 2 kilometers is masked by plotting the data along a scale that varies as a power of 10 (compare Figures B1-2 and B1-3). The lesson to be learned from this is that one must always note the scale intervals that are used for all the variables plotted on a graph.

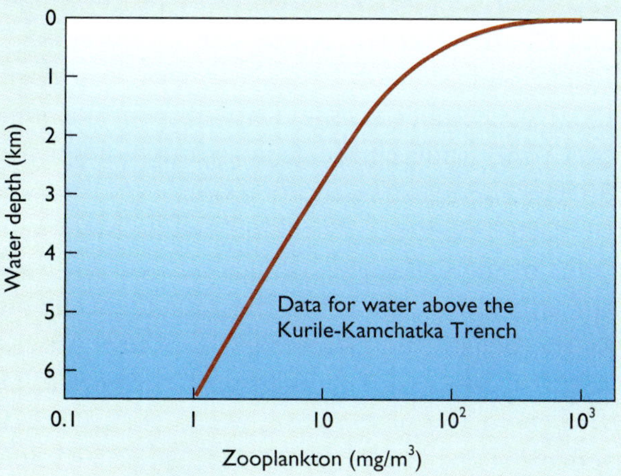

FIGURE **B1-2**

Nonlinear plot.

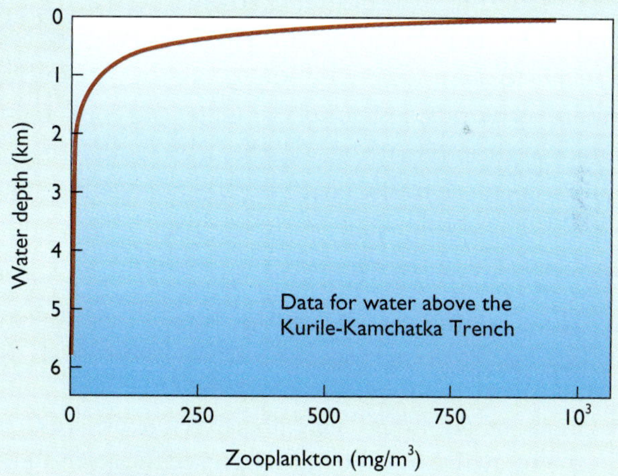

FIGURE **B1-3**

Linear plot.

tremendous hardship, reaching Spain on 8 September 1522, in the last remaining vessel of the expedition, the *Victoria*. Of the original 230 seamen, only 18 reached Seville and completed their three-year-long circumnavigation of the globe.

EARLY SCIENTIFIC INVESTIGATIONS

A number of remarkably sophisticated scientific probings of the ocean's secrets were made in the eighteenth and nineteenth centuries. The British were preeminent during this stage of ocean investigation. Through government sponsorship, and often under the auspices of major scientific societies such as the Royal Society of London, they expanded their geographic and scientific knowledge about the world's seas, which was vital if they were to uphold their maritime and economic superiority.

Captain **James Cook** best represents the British seafaring adventurer of that day. Cook constructed accurate charts of coastlines and made important observations about the geology and biology of unexplored regions, as well as of the customs of native populations. In 1768, on his first major voyage commanding the HMS *Endeavour,* Cook sighted the coast of New Zealand and charted much of its shoreline. He demonstrated convincingly that it was not part of Terra Australis (a large continent then believed to extend into the polar latitudes, conjectured on the conviction there was an equal proportion of land and ocean on the Earth). He then proceeded westward to Australia and explored and mapped its eastern coast, almost foundering on the Great Barrier Reef.

During his second major voyage between 1772 and 1775, commanding the HMS *Adventure* and the HMS *Resolution*, Cook used the prevailing westerly winds to round the Cape of Good Hope and circumnavigate the globe. He maintained a course as close to the latitude 60°S as possible, continually avoiding icebergs. In the final report of his findings, Cook wrote:

> Thus, I flatter myself that the intention of the voyage has in every respect been fully answered, the Southern Hemisphere sufficiently explored and a final end put to the searching after a Southern Continent, which has at times engrossed the attention of some of the maritime powers for near two centuries past, and the geographers of all ages. That there may be a continent or large tract of land near the pole, I will not deny. On the contrary I am of the opinion there is. (John R. Hale, *Age of Exploration* [New York: Time, Inc., 1966,] 192)

Cook's final voyage (1778–79) led him to the Pacific Ocean once again, where he discovered numerous islands, including the Hawaiian Islands. Becoming the first mariner to sail the polar seas of both hemispheres, Cook also ventured northward into the Bering Sea until stopped by pack ice at a north latitude of 70°44′. After returning to Hawaii, Cook was killed while attempting to recover a large boat stolen by a group of natives.

Important work in marine science during the mid-nineteenth century was conducted by **Matthew Fontaine Maury**, director of the U.S. Naval Depot of Charts and Instruments. While compiling *Wind and Current Charts,* a task that began in 1842, Maury realized the need for international cooperation in making ocean measurements: "[A]s these American materials are not sufficient to enable us to construct wind and current charts of all parts of the ocean, it has been judged advisable to enlist the cooperation of the other maritime powers in the same work." In 1855 Maury published an important and successful book, *The Physical Geography of the Sea,* to familiarize the general public with the most recent scientific findings about the ocean. His book went through eight editions in the United States and nineteen editions in England and was translated into several languages. This first book dedicated entirely to the science of oceanography earned him the title, "father of physical oceanography."

One of the best known ocean expeditions of the nineteenth century was the cruise of the HMS *Beagle,* with Captain **Robert Fitzroy** as commander and **Charles Darwin** as the ship's naturalist. The *Beagle* embarked on a five-year voyage, beginning in late December of 1831; Darwin spent the bulk of that time studying the geology and biology of the South American coastline (**Figure 1–6**). He was particularly impressed by the unique animal populations of the Galápagos Islands off Ecuador and by the latitudinal changes in the makeup of the biota of the coastal environments of South America. After the successful completion of the voyage, Darwin spent the next twenty years examining and reflecting on his copious data. He eventually developed a most elegant theory of organic evolution, suggesting that the appearance and evolution of new species result by natural selection,which operates slowly over very long periods of deep geologic time. His arguments, observations, and conclusions led to the publication of his seminal work, *On the Origin of Species,* in 1859. In addition, Darwin's numerous observations on the morphology of coral reefs in the Pacific Ocean resulted in an insightful theory of their geological development that remains the accepted explanation today.

One of the more successful and significant scientific voyages of the nineteenth century was

FIGURE **1-6**

The voyage of the HMS *Beagle*. (a) Drawing depicting the HMS *Beagle*, which was commanded by Captain Robert Fitzroy. (b) Charles Darwin, who occupied the post of naturalist aboard the *Beagle,* made astute observations and ample collections of the biota and rocks that he encountered everywhere during the five-year voyage. Reflection on these data later led him to postulate the evolution of all organisms by natural selection. (c) Route of the HMS *Beagle.*

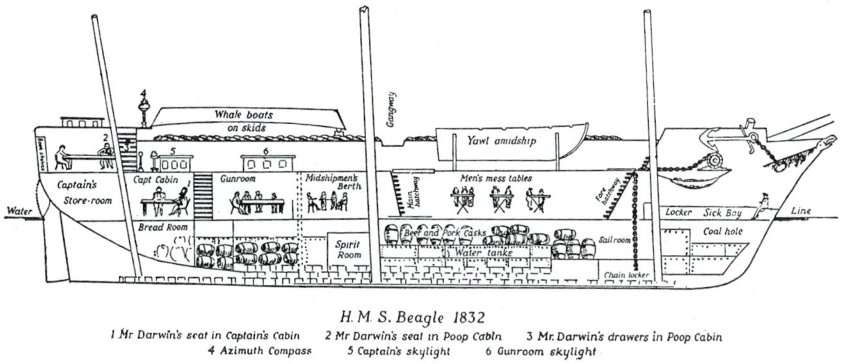

(a) HMS *BEAGLE*

(b) CHARLES DARWIN

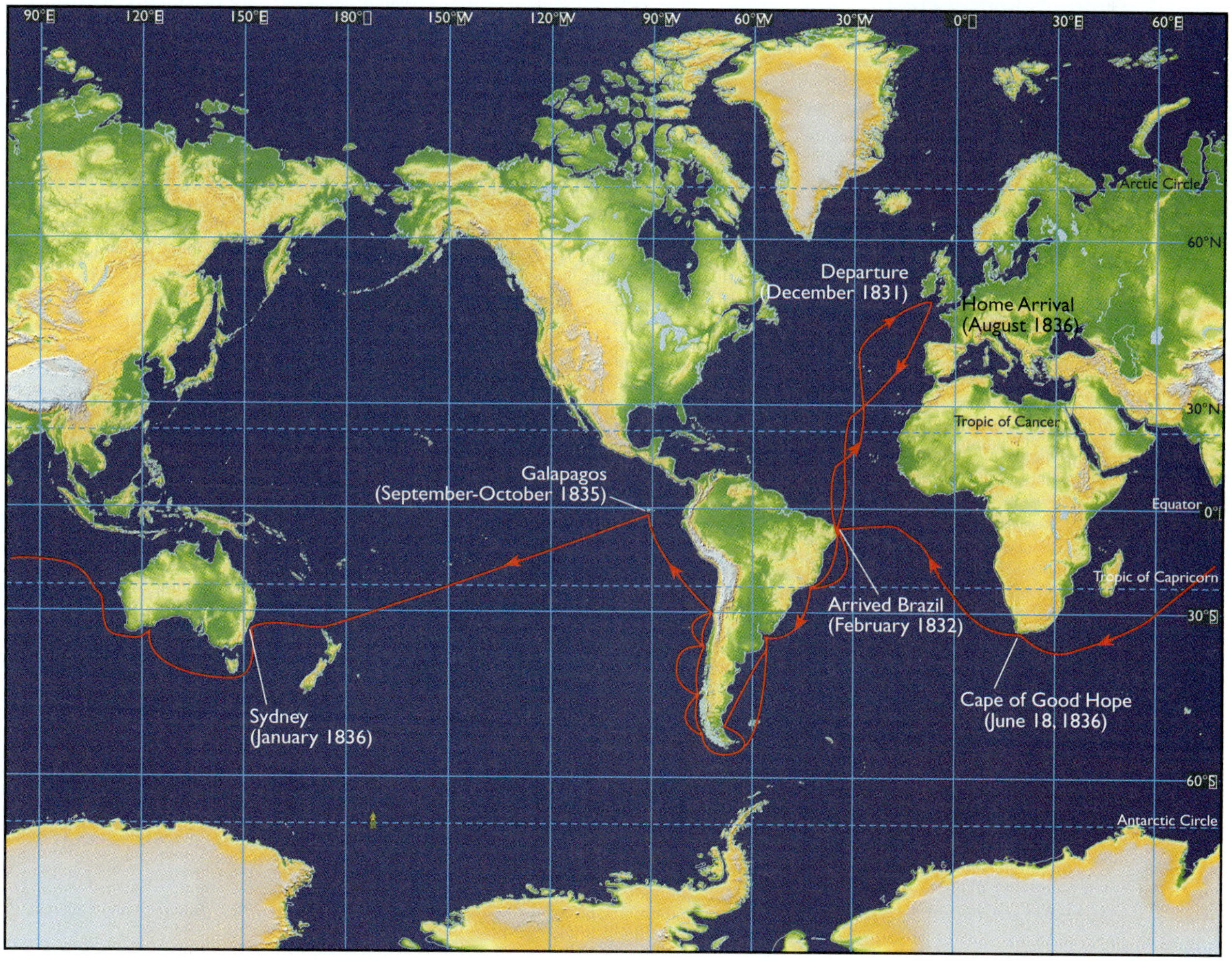

(c) ROUTE OF HMS *BEAGLE*

directed by **C. Wyville Thomson** aboard the 2,360-ton corvette, the HMS *Challenger* (**Figure 1–7a**). Between 1872 and 1876, the *Challenger* completed a globe-encircling voyage, covering almost 125,000 kilometers (~77,500 miles) (**Figure 1–7b**). A primary goal of the cruise was to resolve the controversy about whether or not life existed in the abyss of the oceans. **Edward Forbes** (1815–1854), an influential English naturalist, maintained that the ocean depths below 550 meters (~1,750 feet) were azoic (lifeless). A staff of six scientists tirelessly followed the dictates of the Royal Society of London, determining the chemical composition of seawater and the distribution of life forms at all depths, conducting observations of coastal and ocean currents, and describing the nature of the sedimentary deposits that blanket the sea floor. This global approach to ocean studies represented a fundamental step in the evolution of marine science and heralded a new era in ocean exploration.

Researchers were jubilant with the scientific success of the *Challenger* Expedition. The crew completed more than 360 deep-sea soundings and raised an equal number of dredged samples off the bottom (**Figure 1–8a**). They obtained no fewer than 7,000 sea-life specimens, some from as great a depth as 9 kilometers (~5.6 miles). Each specimen was described, catalogued carefully, and preserved for later laboratory analysis. The findings of the *Challenger* crew left no doubt that organisms lived at all depths in the ocean, finally demolishing the age-old belief championed by Forbes that the cold temperature, darkness, and high water pressure of the deep sea precluded life there. Almost 5,000 new species of marine organisms were identified and described. For the first time, preliminary charts that delineated sea-bottom topography and the distribution of deep-sea sedimentary deposits for much of the ocean were sketched out. More than twenty-three years were required to analyze all of the data and specimens collected by the *Challenger* Expedition. These findings, including observations of indigenous populations, were published in fifty large volumes (**Figure 1–8b**) that marine researchers still consult today.

Near the end of the nineteenth century, the Norwegian explorer, **Fridtjof Nansen**, embarked on a remarkable journey in an effort to study the circulation of the Arctic Ocean and to be the first man to reach the North Pole. Nansen's scheme was to construct a robust, hardy research vessel that could be frozen into the sea ice and drift safely in this icy grip for three years or more. Despite considerable opposition from scientists and mariners alike, Nansen obtained funding and built the *Fram*, a 38-meter (~125 feet), three-masted schooner with the unheard-of hull thickness of 1.2 meters (~4 feet) (**Figure 1–9a**). In late September of 1893, the *Fram*, with its crew of thirteen men and provisions for five years, was successfully locked into the sea ice north of Siberia. There it remained trapped in the ice for three years, slowly drifting at an average rate of 2 kilometers (~1.2 miles) per day, and got as close as 400 kilometers (~248 miles) to the North Pole (**Figure 1–9b**). When the ice-locked *Fram* drifted to a north latitude of 84°, Nansen and **Frederick Johansen** left the vessel in a courageous attempt to reach the North Pole by dogsled. After much hardship, they abandoned their quest after fourteen months, getting only as far as 86°14′. Fortunately, they were sighted and picked up by a British expedition on Franz Josef Land. The crew members aboard the *Fram* made many oceanographic and atmospheric observations during their sojourn, establishing the absence of a polar continent, the water depths along the drift path, and the water-mass structure of the Arctic Ocean. Today the *Fram* can be seen on display in Oslo, Norway.

MODERN OCEANOGRAPHY

Modern oceanographic research is rather arbitrarily taken to begin sometime in the twentieth century, with the design of elaborate experiments involving a truly interdisciplinary approach and a reliance on highly complex instruments and sampling devices. A case in point is the expedition in 1925–27 of the *Meteor* to the South Atlantic Ocean. For twenty-five months, the German scientists used highly developed oceanographic equipment to complete an unprecedented survey of an ocean. They delineated, as never before, the rugged bottom topography of the deep sea and gathered vertical profiles of salinity, water temperature, and dissolved oxygen at numerous hydrographic stations. No data of such quality or density had ever before been gathered from the ocean. From that day onward, many large ocean surveys patterned themselves after the cruise of the *Meteor*.

(a) HMS *CHALLENGER*

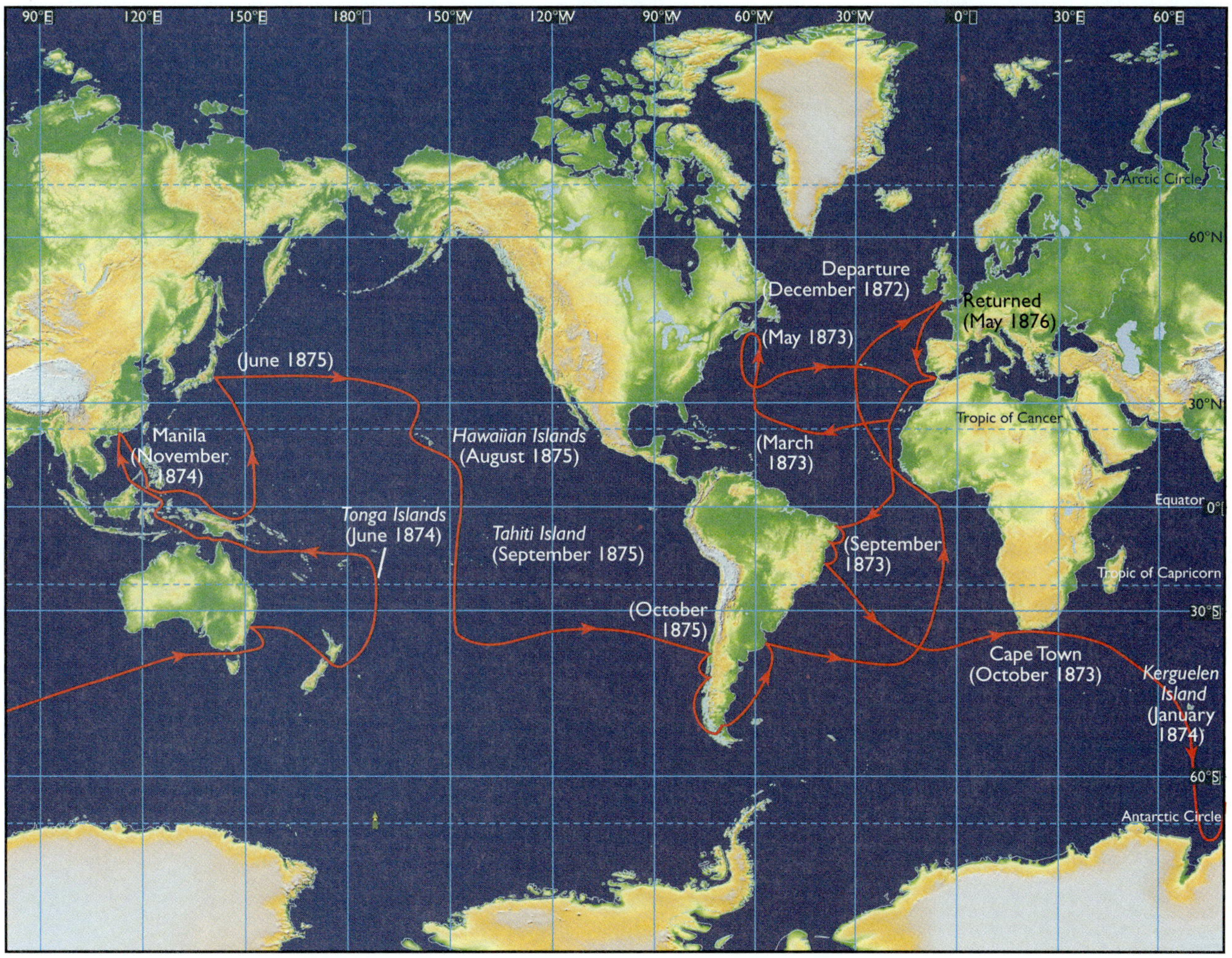

(b) ROUTE OF THE *CHALLENGER* EXPEDITION

FIGURE **1-7**

The *Challenger* expedition. (a) This painting shows the HMS *Challenger* plying the seas between 1872 and 1876. (b) The track of the *Challenger* expedition shows that measurements were made in all parts of the world's oceans, except for the northern Indian Ocean and the Arctic Ocean.

The world wars had important effects on the development of marine research. The advent of modern warfare, with its reliance on sophisticated vessels, weaponry, and electronic instruments, made the U.S. Navy aware of an urgent need to understand the nature of ocean processes. Civilian scientists were recruited, and the navy enacted a program to finance basic oceanographic research, with an emphasis on physical rather than biological problems. This financial support by a government

FIGURE 1-8

The *Challenger* expedition. (a) A small collection of the 360 dredge samples taken during the *Challenger*'s circumnavigation of the world, 1872–76. (b) Aru Island villagers photographed during the expedition.

(a) COLLECTION JARS

(b) ARU ISLAND VILLAGERS

FIGURE 1-9

The Arctic voyage of the *Fram*. (a) The Norwegian vessel, *Fram*, locked in the ice of the Arctic Ocean. (b) The *Fram*, gripped solidly in sea ice, drifted for almost three years across the Arctic Ocean.

(a) THE *FRAM*

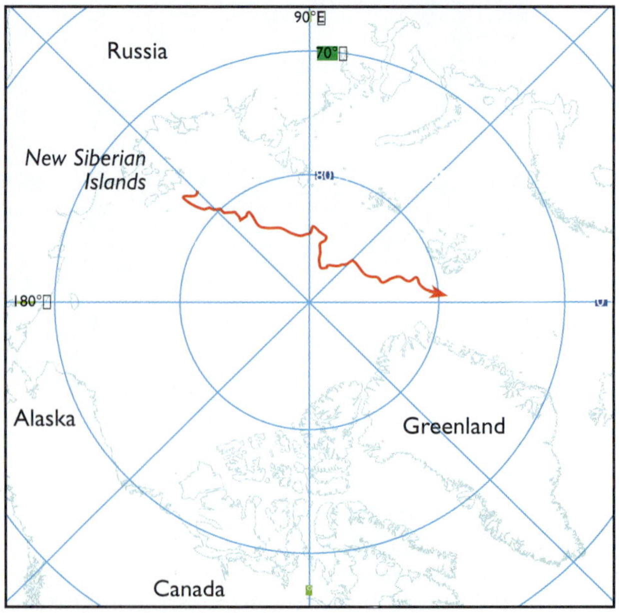

(b) DRIFT ROUTE OF THE *FRAM*

Marine Archeology

Searching Davy Jones' locker for ancient treasures is not an easy matter. The depths of the sea are mysterious and dangerous, and yet hauntingly beautiful and captivating. Exploration of this dark underwater world requires all manner of sophisticated equipment and techniques, lots of specialized multidisciplinary expertise, collaborative efforts among government, industrial, and academic experts, a deep funding pocket, and considerable luck. Typically, a project begins on land with years of extended study, searching historical documents and ancient maps for specific clues about the location of a shipwreck. After consulting current, tide, and bathymetric charts, the appropriate vessel, geophysical gear, and technical and support personnel are deployed in the project area when the weather conditions are best for maximizing the chances of success. The indirect search of the sea bottom typically relies on sophisticated geophysical equipment, such as echo sounders, magnetometers, high-resolution side-scan sonar and sub-bottom profilers, manned submersibles, and remotely operated vehicles (ROVs). Once a wreck is located, the site and its artifacts are explored systematically and mapped accurately, relying heavily on navigational and global positioning system (GPS) technology. If artifacts are retrieved, they must be treated immediately to protect wood and metal from deteriorating rapidly once exposed to air.

There is nothing more exhilarating than successfully finding and then studying an ancient shipwreck on the sea floor. To experience vicariously the frustrations and elations of underwater explorers, you are encouraged to search the web and investigate these ambitious projects.

THE JEREMY PROJECT:

This ambitious venture, a collaborative effort by the National Park Service, NASA, the U.S. Navy and Coast Guard, the Alaska State and Historic Preservation Office, and the Minerals Management Service, is the search in the cold, dark waters of the Chukchi Sea of Alaska for the remains of a nineteenth century New Bedford, Massachusetts whaling fleet that in 1871 got trapped by sea ice; all 32 vessels were crushed and sank. Several of these whaling ships have been located, using a telepresence remotely operated vehicle (TROV), a stereoscopic video device that generates a three-dimensional image of objects on the sea bottom.

USS MONITOR

In 1973, Duke University marine scientists located the USS Monitor, a Union Civil War vessel, which was the first steam-powered ironclad warship constructed without masts and sails (**Figure B1-4**). She was found in the "Graveyard of the Atlantic," some 25 km (15.6 miles) off Cape Hatteras, North Carolina in about 70 m (~230 ft) of water where she sank during an 1862 gale while being towed to port. During the summer of 2001, U.S. Navy divers in collaboration with NASA personnel recovered the Monitor's unique steam engine and donated it to The Mariners Museum in Newport News, Virginia, where it was electrochemically treated for preservation and public display. Now, NOAA divers are attempting the difficult task of salvaging the Monitor's silt-filled gun turret.

THE SCAPA FLOW MARINE ARCHEOLOGY PROJECT (SCAPAMAP)

Scapa Flow, located in the isolated Orkney Islands off the northeast shores of Scotland, is a site where 52 WWI German warships, part of the German High Seas Fleet of the time, were scuttled and sank to the rough sea bottom of Scapa Flow where they had been anchored during the early summer of 1919. The warships included the cruisers Brummer, Koln, and Dresden and the battleships Wilhelm, Konig, Markgraff, and Kronprinz. Since their sinking, there have been salvaging efforts by numerous groups that, in some cases, have damaged and weakened some of the vessels. Also, the site has always been a mecca for sports divers, attracting thousand each year. In an effort to protect the vessels for posterity, the ScapaMap Acoustic Consortium (SAC) was founded. Using a state-of-the-art multibeam echo sounder, SAC is surveying the wrecks and has obtained a remarkable set of images of many of these warships (**Figure B1-5**), some lying on their sides, others overturned with keels pointing upward.

Visit www.jbpub.com/oceanlink for more information.

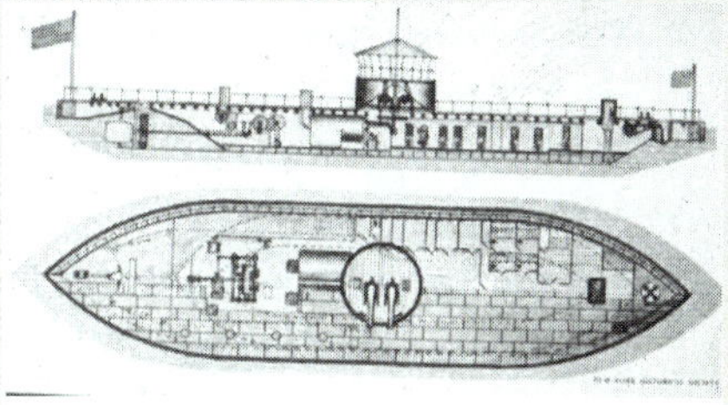

FIGURE **B1-4**

USS *Monitor*. This image of the *Monitor* shows how the steam-powered ironclad was constructed. At the moment, NOAA salvagers are trying to raise the main gun turret from the seabed off Cape Hatteras where the vessel sank while being towed during an 1862 gale.

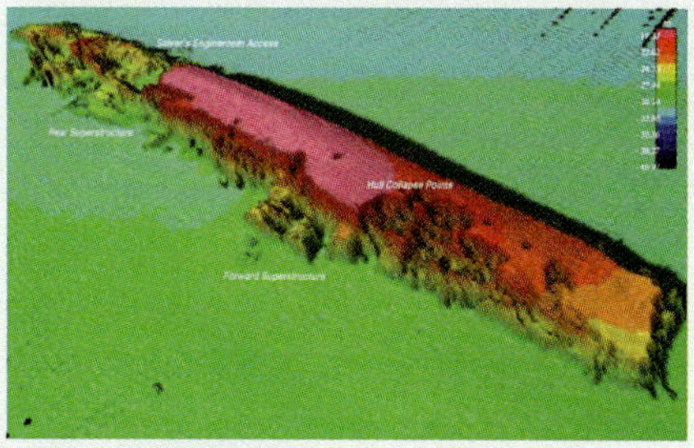

FIGURE **B1-5**

Sunken WWI German vessels. This multibeam, echo-sounding image displays in remarkable three-dimensional detail one of the many German warships that were scuttled in 1919 offshore Scotland at Scapa Flow.

agency stimulated large-scale research enterprises, and restricted the activities of many oceanographers to problems that were of interest mainly to the military. Postwar government-sponsored support led not only to great and rapid advances in instrumentation, but also eventually to the establishment of sea-grant colleges, patterned after already existing land-grant colleges that conducted important agricultural research.

A new development for promoting and facilitating oceanographic research was the establishment of marine institutions. In North America such research centers encouraged and supported both small- and large-scale, local and foreign research by providing funds, laboratory, and library facilities, equipment, research vessels, and scientific expertise. Furthermore, marine institutions gave young people the opportunity to obtain graduate training and valuable experience in conducting science at sea. In the United States, the first such center—**The Scripps Institution of Biological Research**, which later became **The Scripps Institution of Oceanography**—was founded at La Jolla by the University of California in 1903. Two oceanographic centers were later established on the East Coast: the **Woods Hole Oceanographic Institution** in 1930 on the south shore of Cape Cod, Massachusetts, and the **Lamont Doherty Geological Observatory** in 1949 (now known as **Lamont Earth Observatory**) above the massive basalt cliffs of the Hudson River in New York. Today several universities have major oceanographic programs and large, sophisticated seagoing research vessels.

The trend recently has been to organize major collaborations among marine scientists from many disciplines and nations. Three noteworthy programs of this type were the 1957–58 **International Geophysical Year** (IGY), the 1959–65 **International Indian Ocean Expedition** under the auspices of the United Nations, and the **International Decade of Ocean Exploration** (IDOE) of the 1970s, which was supported jointly by the United Nations and National Science Foundation of the United States. Research became less descriptive and more quantitative, and instruments, sampling techniques, and data storage and analysis became increasingly more complex. In fact, many of the concepts we will examine in the remainder of this book are the direct result of such cooperative efforts by teams of scientists.

Beginning in the 1960s, the National Science Foundation organized and generously funded the 1968–75 **Deep-Sea Drilling Project** (DSDP). The goals of this ambitious program included drilling into the sediments and rocks of the deep sea to confirm sea-floor spreading and global-plate tectonics, which were at the time recent theories about the mobility of the oceanic crust. Furthermore, scientists were to assess the oceans resources for the benefit of humankind. The *Glomar Challenger*—a 10,500-ton-displacement vessel (**Figure 1–10a**) designed and built to serve as a drilling platform—employed the latest electronic equipment for dynamic positioning over a borehole. Samples of sediment and rock obtained by drilling below the sea bed helped geologists reconstruct the history of the Earth and its oceans. The success of the DSDP venture, from both an engineering and a scientific perspective, exceeded the expectations of even its most optimistic supporters. In 1975 the program was reconstituted as the **International Program of Ocean Drilling** with the support and active participation of France, the United Kingdom, the Soviet Union, Japan, and the Federal Republic of Germany, as well as the United States. The *Glomar Challenger* was retired in 1983 and another drilling vessel, the *Joides Resolution* (**Figure 1–10b**), continues the geologic exploration of the oceans.

1-3

Current and Future Oceanographic Research

The methods of oceanographic investigation are changing drastically. Without doubt, this trend will continue (probably at an even more accelerated pace) as technology is applied to the study of the sea in many new and ingenious ways.

The future directions that marine research will take are manifold. A greater reliance on international efforts involving many scientists and flotillas of research vessels is an inevitable result of increases in the magnitude and complexity of scientific problems and the accompanying price tag for such ambitious undertakings at sea. The successes of such large-scale endeavors as the IDOE

(a) THE *GLOMAR CHALLENGER*

(b) THE *JOIDES RESOLUTION*

FIGURE **1-10**

Ships designed for deep-sea drilling. (a) The *Glomar Challenger,* a unique drilling vessel 122 meters long, could manage about 7.6 kilometers of drill pipe. (b) The *Joides Resolution,* about 300 meters long, can handle over 10.5 kilometers of drill pipe and operate safely in heavier seas and winds than the *Glomar Challenger* could.

and DSDP assure that they will continue in the future. To cite examples, JOIDES is developing multidisciplinary research strategies and identifying specific drilling sites to investigate climate variability over short- and long-term time scales, the dyanimcs of the Earth's crust and interior, the evolution and paleobiology of the marine biosphere, the nature of catastrophic processes such as earthquakes, volcanic eruptions, and meteorite impacts, and past variations in the sea-ice cover of the Arctic Ocean. Also, the use of submersibles (**Figure 1–11a**), both manned and unmanned, for probing the depths of the sea, will undoubtedly increase as the technology and the design of such crafts continue to improve.

Perhaps the newest research development is a much greater dependence on remote-sensing techniques. Many marine scientists in the future will never go to sea; they will remain in laboratories (some located far inland away from the coastline), and satellites will continually transmit data to them from oceanographic buoys and unmanned platforms at sea at an unprecedented rate. Some of these research techniques are already in use. Sophisticated electronic instruments have been installed in satellites that can accurately detect sea-surface temperatures and can estimate concentrations of microscopic plants and the topography of the sea surface. For example, the TOPEX/Poseidon satellite (**Figure 1–11b**), launched by NASA in 1992, can determine the level of the sea surface to within an accuracy of 13cm (~5.1 inches). These remote techniques enable scientists to survey large tracks of ocean quickly, efficiently, and at reasonable cost. Large computers are also playing an increasingly more important role in ocean research, not only as a tool for storing, sorting, and analyzing the large quantities of information being generated, but also for modeling the ocean's processes and conducting experiments to trace changes over time, ranging from time scales of a few years (El Niño cycles) to millions of years (the opening of ocean basins). The possibilities remain limitless and exciting. The findings of these future research programs will enhance our understanding of the workings of the planet and perhaps even contribute to the very survival of humankind!

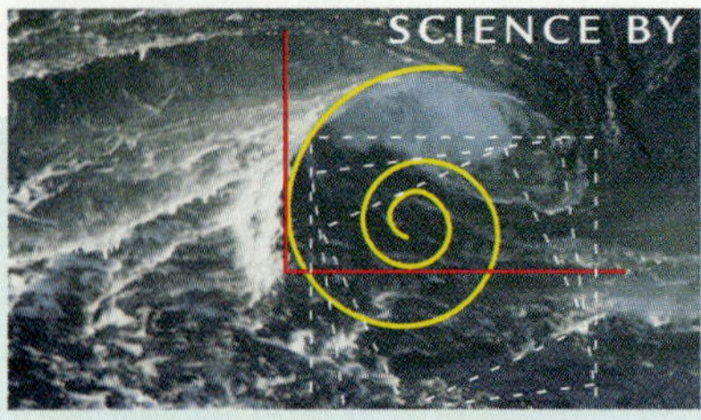

SCIENCE BY NUMBERS

Conversions

Units are defined for the measurement of length, mass, and time. Most Americans use the English system of units whereby length is expressed as inches, feet, and miles; mass as ounces, pounds, and tons; and time as seconds, minutes, hours and years. Much of the rest of the world, including scientists, uses the metric system. In this scheme, length is measured in centimeters, meters, and kilometers, and mass is expressed in grams, kilograms, and metric tons. The units of time in the metric system are identical to those of the English system. Because this is a book of science, it uses the metric system of measurement throughout, but English equivalents are included in parentheses. Appendix II lists the conversion factors that link the two systems of measurement.

A very useful technique is the conversion of a unit from one system of measurement into another system. For example, you may want to express a water depth of 1,200 meters in feet or miles. How does one do this? It is a simple matter. The key is knowing the conversion factors (Appendix II) and *keeping track of the units.* Let's try a few problems.

Convert 1,233 meters into kilometers. According to Appendix II, 1 kilometer = 1,000 meters. Dividing both sides by 1 kilometer yields

$$\frac{1 \text{ km}}{1 \text{ km}} = \frac{1{,}000 \text{ m}}{1 \text{ km}} \quad \text{or} \quad 1 = \frac{1{,}000 \text{ m}}{1 \text{ km}}$$

Dividing both sides by 1,000 m yields

$$\frac{1 \text{ km}}{1{,}000\text{m}} = \frac{1{,}000 \text{ m}}{1{,}000 \text{ m}} \quad \text{or} \quad \frac{1 \text{ km}}{1{,}000 \text{ m}} = 1$$

FIGURE 1-11

New technology for probing the sea. (a) Submersibles, such as *Alvin*, are useful for the close examination and sampling of the fauna, sediment, and rock of the deep sea. (b) The TOPEX/Poseidon satellite launched by NASA in 1992 has provided detailed, accurate data on the level of the sea surface, crucial for predicting changes in current and climate patterns.

(a) *ALVIN*

(b) TOPEX/POSEIDON SATELLITE

This is not surprising, because if the two units are equal, dividing one by the other must equal 1. This means that multiplying a value by either ratio does not change the value, because you are multiplying it by 1, and any value times 1 is that value. So in order to convert 1,200 meters into kilometers, we multiply 1,200 meters by the proper conversion ratio that eliminates the meter units. Let's do this:

$$(1{,}200\,\cancel{\text{m}})\,(1\text{ km}/1{,}000\,\cancel{\text{m}}) = 1{,}200\text{ km}/1{,}000 = 1.2\text{ km}.$$

Notice that if we use the other conversion ratio, the meter units will not cancel out:

$$(1{,}200\text{ m})(1{,}000\text{ m}/1\text{ km}) = (1{,}200)\,(1{,}000)\text{ m}^2/\text{km}.$$

Now, let's convert 1,200 meters into miles. We know from the above conversion that 1,200 meters = 1.2 kilometers. According to Appendix II, 1 kilometer = 0.621 miles. This means that

$$\frac{1\text{ km}}{1\text{ km}} = \frac{0.621\text{ miles}}{1\text{ km}} = 1 \quad\text{or}\quad \frac{1\text{ km}}{0.621\text{ miles}} = \frac{0.621\text{ miles}}{0.621\text{ miles}} = 1$$

Therefore,

$$(1.2\,\cancel{\text{km}})\,(0.621\text{ miles}/1\,\cancel{\text{km}}) = (1.2)\,(0.621)\text{ miles} = 0.745\text{ miles}.$$

It bears repeating that the key to accurate conversion is *keeping careful track of the units.*

STUDY GUIDE

KEY CONCEPTS

1. Oceanographers are well-trained scientists who investigate the ocean, its organisms, and its processes. Oceanographic work is often multidisciplinary in character, involving the collaboration of many types of scientists, mathematicians, engineers, technicians, and policymakers.

2. Early efforts to learn about the oceans involved exploration by ship. The geography of the world, both its landmasses and oceans, was mapped in increasingly more accurate detail as techniques in piloting, navigation, and surveying were developed and refined. The preeminent sea voyagers were the Egyptians, Phoenicians, Greeks, and Norsemen (Table 1–1), their pioneering explorations culminating with Magellan's epic circumnavigation of the globe between 1519 and 1521.

3. During the eighteenth and nineteenth centuries, long, large-scale expeditions were organized to sample sea life, chart the sea bottom, measure currents, and determine the chemical makeup of seawater in all parts of the world. Notable scientific achievements were made by Cook (1768–1779), Darwin (1831–1836), Thomson (1872–1876), and Nansen (1893–1895), among others (see Table 1–1).

4. Modern oceanography (Table 1–1), which began in earnest with the cruise of the German ship, the

Meteor, to the Atlantic Ocean in 1927, relies on sophisticated instruments to make accurate and efficient measurements of the ocean's properties. Also, marine institutions have been established specifically to promote research in the sea. The newest development is the organization of major collaborative programs involving marine scientists from many nations. The International Decade of Ocean Exploration (1970s), the Deep-Sea Drilling Project (1968–75), and the International Program of Ocean Drilling (1975–present) are among them.

5. Future scientific studies of the ocean will rely more and more on large international programs and remote-sensing techniques, including measurements from satellites, ocean buoys, and unmanned platforms at sea. Computers for handling and processing the enormous quantities of data and for modeling ocean processes are playing an ever-increasing role in ocean research.

QUESTIONS

1. What exactly is oceanography, and how does it differ from other fields of science?
2. Briefly describe the successes of the Egyptians and Phoenicians in ocean exploration.
3. What distinguishes modern oceanography from earlier scientific investigations of the oceans?
4. Briefly discuss the scientific achievements (consult Table 1–1) of the following:
   a. Pytheas
   b. Geradus Mercator
   c. Seneca
   d. Sir John Ross
   e. Nathaniel Bowditch
   f. Matthew Maury
   g. C. Wyville Thomson
5. In what ways are future oceanographic research techniques likely to differ from present ones?
6. What exactly is the scientific method? Can scientists "prove" their hypotheses?

SELECTED READINGS

Ashley, S. 2001. Warp drive underwater, *Scientific American*, 284 (5): 70–79.

Bailey, H. S., Jr., 1953. The voyage of the *Challenger*. *Scientific American* 188 (5): 88–94.

Bailey, H. S., Jr. 1972. The background of the *Challenger* expedition. *American Scientist* 60: 550–60.

Ballard, R. D. with Hively, W. 2000. *The Eternal Darkness.* Princeton, N.J.: Princeton University Press.

Bellwood, P. S. 1980. The peopling of the Pacific. *Scientific American* 243 (5): 174–85.

Boorstin, D. J. 1985. *The Discoverers.* New York: Random House.

Briggs, P. 1968. *Men in the Sea.* New York: Simon and Schuster.

Burgess, R. F. 1975. *Ships Beneath the Sea: A History of Subs and Submersibles.* New York: McGraw-Hill.

Carey, S. S. 1994. *A Beginner's Guide to the Scientific Method.* Belmont, Calif.: Wadsworth.

Cicin-Sain, B., and Knecht, R. W. 2000. *The Future of U.S. Ocean Policy: Choices for the New Century.* Washington, D.C.: Island Press.

Deacon, M. 1962. *Seas, Maps, and Men: An Atlas-History of Man's Exploration of the Oceans.* Garden City, N.Y.: Doubleday.

Deacon, M. 1971. *Scientists and the Sea 1650–1900: A Study of Marine Science.* New York: Academic Press.

Dietz, R. S., and Emery, K. O. 1976. Early days of oceanography. *Oceanus* 19 (4): 19–22.

Ewing, G. C. 1981. From antiquity onward. *Oceanus* 24 (3): 6–8.

Gordon, B. L., ed. 1970. *Man and the Sea: Classical Accounts of Marine Exploration.* Garden City, N.Y.: Natural History Press.

Hammond, A. L. 1970. Deep sea drilling: A giant step in geological research. *Science* 170 (3957): 520–21.

Herbert, S. 1986. Darwin as a geologist. *Scientific American* 254 (5): 116–23.

Idyll, C. P., ed. 1969. *Exploring the Ocean World: A History of Oceanography.* New York: Crowell.

Kane, H. 1991. *Voyagers.* Bellevue, Wash.: Whalesong.

Linklater, E. 1972. *The Voyage of the Challenger.* Garden City, N.Y.: Doubleday.

MacLeish, W. H., ed. 1980, A decade of big ocean science. *Oceanus* 23 (1).

Morrison, G. K. 1984. The development of the oceanographic industry in New England. *Oceanus* 27 (1): 32–34.

Oceanus. 1985. The oceans and national security. Special issue 28 (2).

Oceanus. 1988–89. DSV Alvin: 25 years of discovery. Special issue 31 (4).

Oceanus. 1993–94. 25 years of ocean drilling. Special issue 36 (4).

Oceanus. 1997. Access to the sea. Special issue 40 (1).

Oceanus. 2000. Ocean observatories. Special Issue 42 (1).

Ryan, P. R., ed. 1982. Research vessels. *Oceanus* 25 (1).

Schlee, S. 1973. *The Edge of the Unfamiliar World: A History of Oceanography.* New York: Dutton.

Schlee, S. 1975. *History of Oceanography.* London: Robert Hale.

Schuessler, R. 1984. Ferdinand Magellan: The greatest voyager of them all. *Sea Frontiers* 30 (5): 299–307.

Sears, M., and Merriman, D. 1980. *Oceanography: The Past.* New York: Springer-Verlag.

Thorne-Miller, B. 1998. *The Living Ocean: Understanding and Protecting Marine Biodiversity.* Washington, D.C.: Island Press.

Van Andel, T, 1981. *Science at Sea: Tales of an Old Ocean.* San Francisco: W. H. Freeman.

TOOLS FOR LEARNING

Tools for Learning is an on-line review area located at this book's web site OceanLink (**www.jbpub.com/oceanlink**). The review area provides a variety of activities designed to help you study for your class. You will find chapter outlines, review questions, hints for some of the book's math questions (identified by the math icon), web research tips for selected Critical Thinking Essay questions, key term reviews, and figure labeling exercises.

2

The Planet Oceanus

Why upon your first voyage as a passenger, did you yourself feel such a mystical vibration, when first told that you and your ship were now out of sight of land? Why did the old Persians hold the sea holy? Why did the Greeks give it a separate deity, and make him the own brother of Jove? Surely all this is not without meaning.

—Herman Melville,
Moby Dick

web navigator

critical thinking on the web

math tutor on the web

www.jbpub.com/oceanlink

PREVIEW

THE GLOBAL SEAS, which seem "as if they belonged to another planet" to us land dwellers, represent the very fiber and essence of the Earth. If life had evolved on some other planet, scientists peering through telescopes there undoubtedly would have named our planet "Oceanus" (the water planet). Seen from space, the Earth appears as a globe draped in blue hues, disrupted here and there by swirls of white and gray clouds and irregular patches of brown and green landmasses. What clearly sets Earth apart from the other eight planets in our solar system is its vast, commanding expanse of oceans dominating its outer surface and obscuring most of its crustal landscape.

The intent of this chapter is to acquaint you with some of these unique qualities, including the distribution of water, the physical nature of the Earth's vast submarine topography, and its deep subcrustal structure. Then we can begin to investigate the dynamic physical, chemical, geological, and biological forces that interact in countless ways in the oceans—the very fabric of the science of oceanography. It is difficult to imagine the Earth without oceans. Although we call our planet "Earth," it truly is a water planet, with oceans covering almost 71 percent of its surface. The ocean contains about 1,360,000 cubic kilometers (324,126 cubic miles) of liquid water, an enormous quantity. Water—the vital substance that sets the Earth apart from the remaining eight planets in the solar system—is an absolutely necessary ingredient for the development and sustenance of living organisms, including *Homo sapiens.* Because the oceans cover so much of the Earth, they have a great deal to tell us about the development of the planet and the origin and evolution of life forms on it. As human beings (land-bound, air-breathing animals preoccupied with understanding our own terrestrial domain), we forget how unrepresentative our perspective of "land" really is. Ambrose Bierce, it seems, did not: "Ocean, a body of water occupying about two-thirds of the world made for man—who has no gills." Marine habitats, marine topography, marine organisms, and marine processes dominate the planet Earth. Let's begin our overview of the oceans by examining the planet's subcrustal structure.

2-1

The Earth's Structure

If we could slice the Earth in half to expose its innards, we would see a regular, concentric arrangement of interior shells, not unlike the layering in an onion. These interior spheres each consist of materials with distinct chemical compositions and physical properties. Girdling the solid mass of the Earth are two exterior spheres: a concentric envelope of water (the oceans), surrounded by a thicker envelope of gases (the atmosphere).

THE EARTH'S INTERIOR SPHERES

Geologic and astronomical evidence indicates that the Earth, at one time in its distant past, was entirely molten, so elements and compounds simply segregated according to their densities. **Density** is a fundamental property of matter; it is defined as the amount of mass contained in a volume, expressed as grams (mass) per cubic centimeter (volume); that is, g/cm^3. The greater the quantity of mass in a cubic centimeter, the higher the density. Back in the Earth's history, iron and nickel, very dense materials, sank to the center of the planet because of the influence of gravity and produced the core. The lighter substances, such as aluminum and silicon, rose to the surface where they solidified and formed a thin crust over the denser material of the mantle. As the Earth evolved, gases of all types, being light and buoyant, were vented to the surface. There they slowly accumulated to produce an envelope of air or condensed to form an envelope of surface water; and, like the material composing the Earth's interior spheres, they separated from one another according to their density.

But this evolutionary model for the Earth is a very general one. What exactly constitutes the planet's interior? To satisfy our curiosity, it would be best to examine the Earth's internal organization directly with our own eyes. This is not possible, because even our deepest mines extend downward no more than 4 or so kilometers (~2.5 miles) and the deepest drill holes are a bit more than 12 kilometers (~7.5 miles). These holes are mere pinpricks, given that the Earth's very center lies 6,370 kilometers (~3,860 miles) below its surface. What to do? The answer, obviously, must result from indirect observation. This is like tapping a wall to find where studs are located, or thumping a metal container to see how much liquid it contains. Geologists rely on assorted geophysical measurements to deduce the structure of the planet's interior. These techniques include tracing the path and speed of earthquake waves traveling through the Earth, noting variations in the Earth's gravitational and magnetic fields, and evaluating the amount of heat escaping from the planet's interior. By integrating all this information into a general picture, we can "see" the Earth's internal structure to be made up of three distinct layers: the crust, the mantle, and the core (**Figure 2–1a**).

The outermost shell, the **crust**, is very thin in comparison with the deeper layers and consists of the rocks we build our homes on and mine ores from. These low-density rocks are composed of minerals rich in aluminum, silicon, and oxygen, and we know a great deal about them because they are exposed as outcrops. Below the Earth's crust and extending downward for about 2,900 kilometers (~1,750 miles), nearly halfway to the planet's center, is the **mantle**. Its relatively dense, hot rocks are made of magnesium, iron, silicon, and oxygen. The innermost zone of the Earth, the **core**, begins at the base of the mantle and extends almost 3,500 kilometers (~2,100 miles) to the very center of the planet. The detailed analysis of earthquake waves passing through this region indicates that its structure is fairly complex. It consists of a molten **outer core** that surrounds a solid **inner core**; both are composed of very dense alloys of iron and some nickel .

The picture that emerges from geophysical studies is that the Earth is made up of three concentric layers, each with a distinct chemical composition, and each becoming increasingly more dense with depth beneath the surface. Composition is not, however, the only factor that controls the physical character of the rocks in the Earth's interior. Both temperature and pressure increase with depth, and these affect the properties of rocks as well. Pressure causes the melting temperature of substances to increase, meaning that they remain solid at a temperature at which they would normally melt. So, as you increase the pressure on a material, it must be

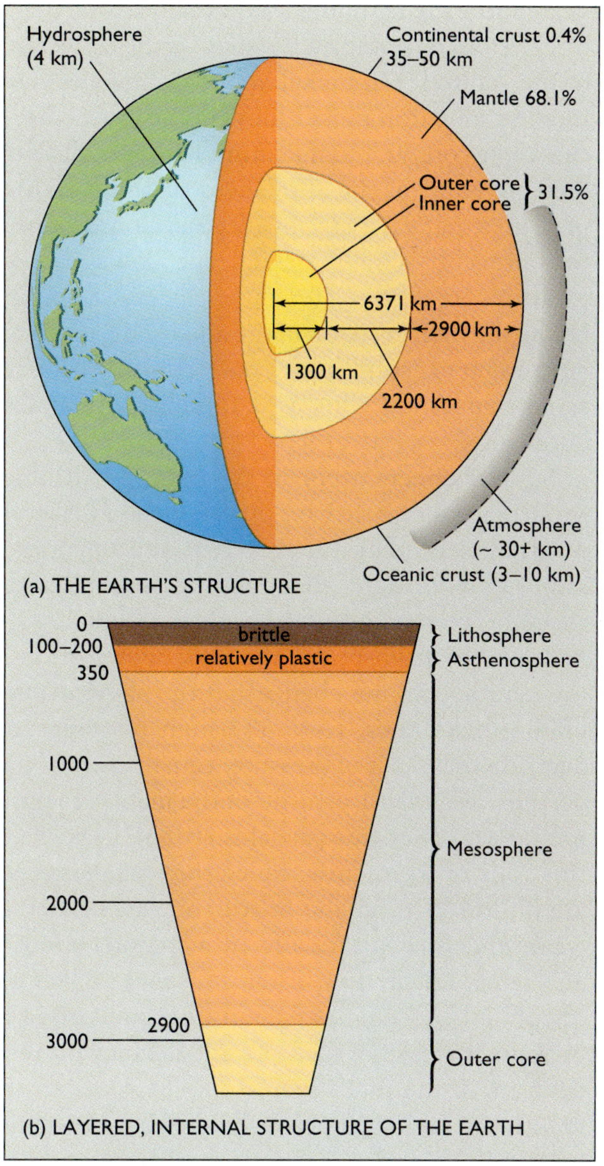

FIGURE 2-1

The Earth's structure. (a) The earth's internal mass is arranged into concentric shells. They consist of an outer crust, a middle mantle, and a center core (composed of a solid inner core surrounded by a molten outer core). The percentage values are based on volume. The solid earth is enveloped by a band of water (the *hydrosphere*) and a band of air (the *atmosphere*). (b) Based on physical characteristics, the earth's interior can be divided into a rigid, brittle *lithosphere;* a weaker, underlying *asthenosphere,* which is relatively plastic; and a more rigid mantle termed the *mesosphere*. The lithosphere includes the outer crust and the upper portion of the mantle down to a depth of 100 to 200 kilometers. The asthenosphere includes the partially melted upper portion of the mantle down to a depth of about 350 kilometers below the Earth's surface; below this is the more rigid mantle termed the mesosphere.

heated to a higher temperature than normal before it melts. An example of this effect is the conversion of the molten iron-nickel alloys of the outer core into the solid iron-nickel alloys of the inner core. In the outer core, the temperature is higher than the melting point of the alloy, even at the very high pressures that exist at those depths. Consequently, the outer core is molten. But deeper down, because pressures are even greater, the temperature remains below the melting point of the alloy. This results in an unmelted (solid) inner core composed of the same iron-nickel alloy as is found in the molten outer core.

Similar effects occur in the mantle. Although the crust and mantle have different chemical compositions, the rocks of the crust and uppermost mantle are all hard, rigid, and brittle and, taken together, comprise the **lithosphere** (**Figure 2–1b**). Beneath the lithosphere, the rock composition of the mantle does not change, but the physical state of the rocks does. Here, the high temperatures override the effect of pressure, causing a very tiny portion (less than 1 percent) of the mantle rocks to melt. Because of this partial melting, the rocks in this zone of the mantle, called the **asthenosphere**, are no longer rigid and brittle, but are relatively weak and flow plastically in a manner crudely analogous to the flow of tar or asphalt. Then the lower asthenosphere gradually becomes rigid as pressure increases with depth and the melting temperature of the rocks increases. The rocks in the mantle are no longer plastic below about 350 kilometers (~217 miles) beneath the Earth's surface, which is taken to be the bottom of the asthenosphere. Deeper down is a more rigid zone of the mantle called the **mesosphere** (see Figure 2–1b), which reaches down to the top of the outer core. Study Figure 2–1 carefully to be certain you understand the two distinct schemes that geologists use to subdivide the Earth's interior.

THE EARTH'S EXTERIOR ENVELOPES

Like the solid Earth, the two distinct bands of matter that surround the planet's surface are separated on the basis of density. The more dense of the two, the **hydrosphere** (see Figure 2–1a), is composed of water, and includes the oceans, lakes, rivers, **groundwater** (water beneath the ground surface), ice and snow, and the relatively small amount of water vapor that exists in the air. About 97 percent of the water at the Earth's surface is in the oceans. However, this enormous volume represents a mere 10 percent of the entire reservoir of water contained in the Earth; the bulk of it (90 percent) is chemically bonded to the solid minerals that comprise the rocks of the crust and mantle and is not in liquid form; hence, it cannot move. The other external envelope, the **atmosphere** (see Figure 2–1a), is a mixture of gases, primarily nitrogen and oxygen, with minor amounts of argon, carbon dioxide, methane, and variable quantities of dust, ozone, and water vapor. Water currents and air currents are constantly mixing the chemical compounds in the fluids of the hydrosphere and atmosphere, respectively.

Many biologists designate a third external envelope of matter known as the **biosphere**, which includes all organic matter—living and nonliving, large and small, simple and complex. It is an extremely thin but dynamic band of matter. The composition of the biosphere is unique, consisting largely of carbon, hydrogen, and oxygen compounds. These chemicals are extracted by organisms temporarily from the Earth's spheres, and, upon death, are returned to and then recycled through the lithosphere, the hydrosphere, and the atmosphere. All the elements in your body once belonged to rocks.

2-2

The Physiography of the Ocean Floor

Now that we are acquainted with the Earth's structure, let's examine the irregularities at the top of the crust, its topography. Being a land-dwelling organism, you are already quite aware of various terrestrial landforms. Less familiar perhaps is the nature of underwater landscapes. Exploration of the oceans' **bathymetry** (submarine topography) did not begin in earnest until after World War II, when researchers began systematically to survey and chart the ocean bottom using new echo-sounding technology (see boxed feature, "Probing the Sea Floor"). The discoveries were startling. The deep-sea bottom is not, as was widely believed, monotonously flat and featureless everywhere. Rather, parts of it are as uneven and as rugged as the familiar mountain topography of the land. Let's begin the analysis by classifying the ocean's bathymetry.

BATHYMETRIC PROVINCES

The many fine charts of the oceans (**Figure 2–2**) reveal that their bathymetry can be broadly subdivided into three large physiographic provinces—continental margins, deep-ocean basins, and midocean ridges—each having unique landform characteristics. The system of broad **midocean ridges**, a continuous submarine mountain range, winds its way through all the oceans. Midocean ridges are separated from the drowned edges of the continents, the **continental margins**, by large, intervening tracts of **deep-ocean basins**, the ocean floor that lies deeper than 2,000 meters (~6,600 feet) below sea level. Take a moment to study **Table 2–1** and the bathymetric chart in Figure 2–2 and trace out as best you can the general boundaries between these provinces in each of the oceans. Remember that as we leave the continent, we cross the continental margin, then the deep-ocean basin, and far at sea, encounter the midocean ridge. You might notice that some midocean ridges are not positioned in the center of the ocean basin. Can you find an example of such a ridge on the charts provided?

What follows is a description of the common underwater landforms in each of these provinces. Table 2–1 summarizes some of their important physical dimensions, so that you get a sense of their morphologic shape and relative size.

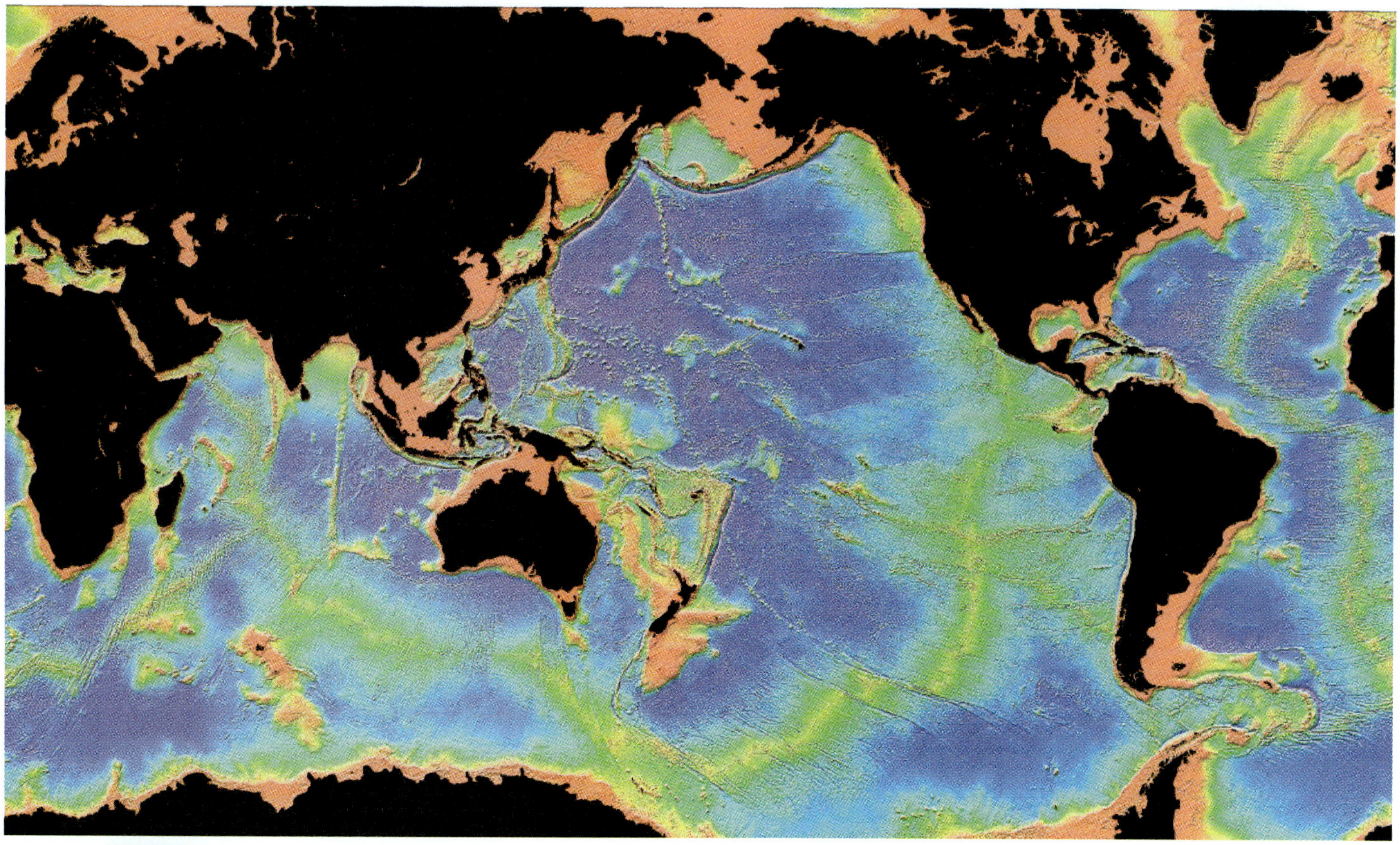

FIGURE 2-2

Ocean bathymetry. This map shows the major bathymetric features of the ocean basins of the Pacific Ocean, the Atlantic Ocean, and the Indian Ocean.

TABLE 2-1

Dimensions of bathymetric features.

Feature	Width	Relief	Water Depth	Bottom Gradient
	TYPICAL DIMENSIONS			
Continental shelf	<300 km	<20 m	<150 m	<1:1,000 (~0.5°)*
Continental slope	<150 km	locally >2 km	drops from 100+–2000+ m	~1:40 (3–6°)
Continental rise	<300 km	<40 m	1.5–5 km	1:1,000–1:700 (0.5°–1°)
Submarine canyon	1–15 km	20–2,000 m	20–2,000 m	<1:40 (3–6°)
Deep-sea trench	30–100 km	>2 km	5,000–12,000 m	___
Abyssal hills	100–100,000 m (100 km)	1–1,000 m	variable	___
Seamounts	2–100 km	>1,000 m	variable	___
Abyssal plains	1–1,000 km	0	>3 km	1:1,000–1:10,000 (<0.5°)
Midocean ridge flank	500–1,500 km	<1 km	>3 km	___
Midocean ridge crest	500–1,000 km	<2 km	2–4 km	___

*A bottom gradient of 1:1,000 means that the slope rises 1 m vertically across a horizontal distance of 1,000m.
Source: Adapted from B. C. Heezen and L. Wilson, Submarine geomorphology in *Encyclopedia of Geomorphology*, R. W. Fairbridge, ed. (New York: Reinhold, 1968); and C. D. Ollier, *Tectonics and Landforms* (New York: Longman, 1981).

CONTINENTAL MARGINS

The amount of seawater currently on Earth exceeds the capacity of the ocean basins; they are overfilled, and seawater spills out of them, flooding the edges of the continents. At those points of flooding, sand and mud, eroded from the continents and transported to the shores by rivers and glaciers, accumulate and become shaped by ocean processes into a vast and thick sedimentary wedge. These immense deposits of sediment at continental edges comprise the continental margins (**Figure 2–3a**). This ocean province is divisible into three parts: the continental shelf, slope, and rise.

■ CONTINENTAL SHELF The nearly flat plains, or terraces, at the top of the sedimentary wedge beneath the drowned edges of the continents are **continental shelves**. They average about 60 kilometers (~37 miles) in width, although local and regional variations are common, ranging from more than 1,000 kilometers (~620 miles) in the Arctic Ocean to a few kilometers (~1 mile) along the Pacific coasts of North and South America. The shelf bottom slopes gently seaward at an angle of about 0.5 degrees. This is very slight and impossible to detect with the naked eye. Continental shelves end on their ocean side at the **shelf break** where the sea bottom steepens appreciably. Shelf breaks occur, on average, at a water depth of about 130 meters (~430 feet), but can be found deeper than 200 meters (~660 feet). The bottom slopes of the shelf break are in the order of 1 to 4 degrees, which doesn't seem like much, but is two to eight times steeper than the near-horizontal surface of the continental shelf.

■ CONTINENTAL SLOPE Seaward of the shelf break is a steeper **continental slope**, which is inclined at an average angle of 4 degrees; its base lies at water depths of 2 to 3 kilometers (~1.2 to 1.9 miles). As with the continental shelf, an enormous pile of mud and sand eroded from the continent lies underneath the continental slope. Huge **submarine canyons** are cut into the sedimentary deposits of many continental slopes. These canyons have steep sides, V-shaped profiles, and topographic relief of up to 2 kilometers (~1.2 miles), making them one of the most deeply incised landforms anywhere on the Earth. For example, the Hudson Submarine Canyon off New York state is wider, deeper, and longer than the Grand Canyon! Submarine canyons serve as chutes (passageways) for the transfer of sediment from the continental margins to the deep-ocean basins by a variety of transport processes discussed in Chapter 4.

■ CONTINENTAL RISE At the base of many continental slopes, the ocean bottom flattens out to a mere gradient of ~1 degree. This broad underwater plain of sediment is called the **continental rise**. Some of

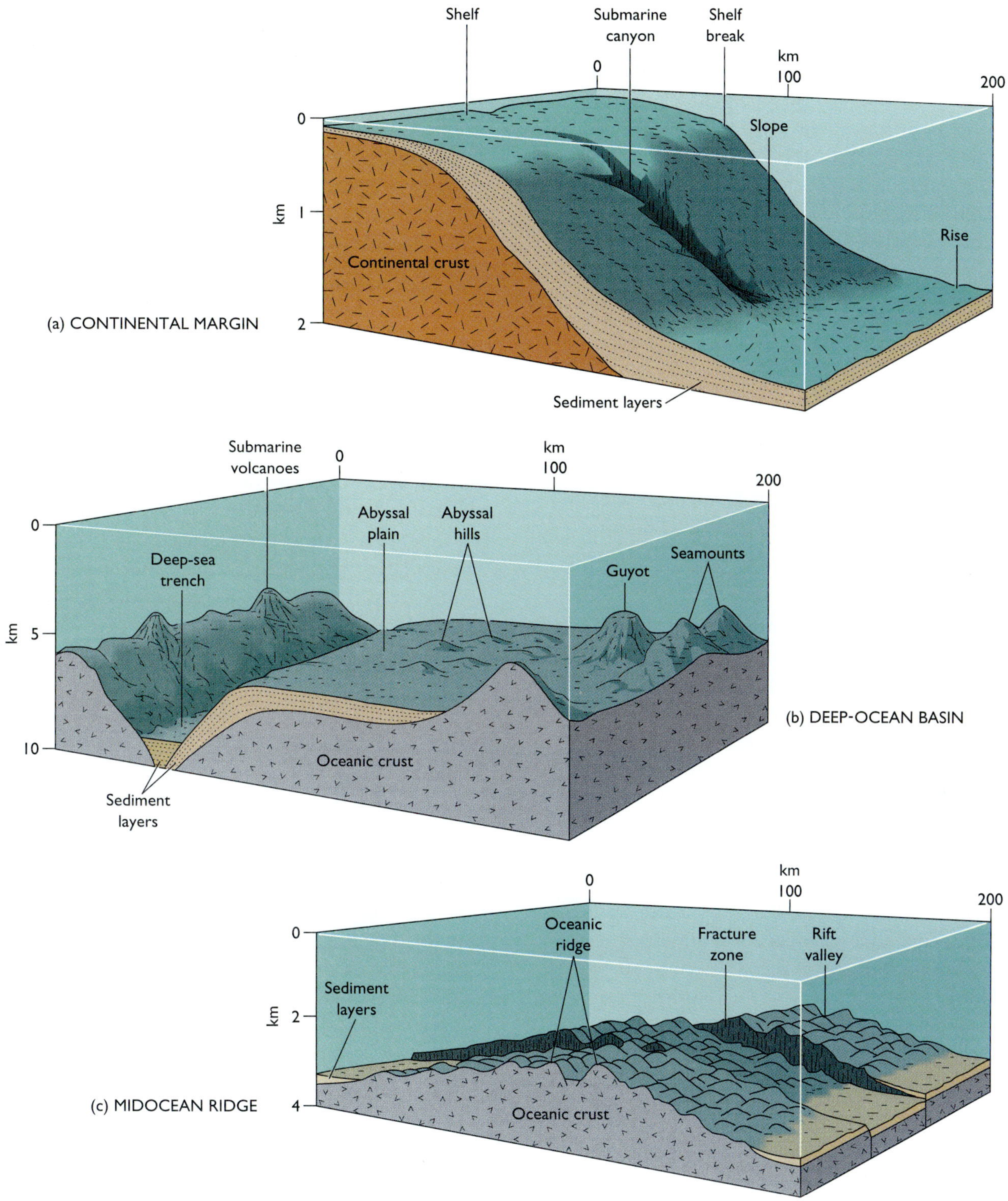

FIGURE 2-3

Landscapes of the ocean floor. (a) The continental margin consists of the continental shelf, the shelf break, the continental slope, and the continental rise. Submarine canyons are cut into continental margin. (b) The floor of the deep-ocean basin is composed of abyssal plains, abyssal hills, seamounts, guyots, and deep-sea trenches. (c) The flanks and crests of the midocean ridges are offset by a system of fractures. All of these diagrams are distorted by vertical exaggeration (Appendix IV); Table 2-1 lists the actual dimensions of each feature.

the larger continental rises extend seaward more than 500 kilometers (~310 miles) from the base of the continental slope to water depths approaching 4 kilometers (~2.4 miles). The topographic relief on the continental rise is not great, except where submarine canyons extend their channels onto the rise. Continental rises, like the shelf and slope, are underlain by deposits of sediment that are thousands of meters thick; this debris has been derived from the erosion of rock and sediment from the nearby landmasses and the adjoining continental shelf and slope.

DEEP-OCEAN BASINS

Beyond the continental margins lie the deep-ocean basins (**Figure 2–3b**). These display varied topography, ranging from flat plains to towering, steep-sided mountain peaks (see Table 2–1). The floor of each ocean basin has some or all of the following bathymetric features:

1. **Abyssal plains** are the flattest areas found anywhere on the Earth, having a regional slope of 0.5 degrees—less than one meter per kilometer. This means that across a horizontal distance of 1 kilometer (~0.6 miles) the floor of the abyssal plain drops vertically by less than 1 meter (~3.3 feet). These abyssal plains are broad aprons of mainly land-derived sediments that have been deposited and have buried the irregular volcanic topography of the solid crust of the ocean. They lie in water depths of 3 to 5 kilometers (~1.9 to 3.1 miles) and consist of unconsolidated layers of sediment ranging in thickness from 100 meters (~330 feet) to more than 1,000 meters (~3,300 feet). Most of this sediment is debris that has moved through the submarine canyons of the continental slope and spilled out to cover the crust of the deep sea.
2. **Abyssal hills** are domes, or elongated hills, that are no higher than 1,000 meters (~3,300 feet) and range in width between 100 meters (~330 feet) and 100 kilometers (~62 miles). Most are composed of volcanic rocks and may be covered by a thin layer of fine-grained sediment that has settled down through the water from above.
3. **Seamounts** are any type of abyssal mountain. Many are extinct or active volcanoes with conical tops and steep sides that rise more than 1,000 meters (~3,300 feet) above the adjoining ocean floor but do not reach or extend above the ocean's water surface. Seamounts may occur as isolated peaks or be clustered into groups. Flat-topped seamounts, called **guyots**, were once active volcanoes, but their tops were leveled and flattened by wave erosion. The edges of many of these volcanoes were colonized by coral reefs before the islands sank completely under water.
4. **Deep-sea trenches** are relatively steep-sided, long, and narrow depressions, or basins, some 3 to 5 kilometers (~1.9 to 3 miles) deeper than the surrounding ocean floor. Trenches are the deepest regions on Earth. Surprisingly, these deeps are not found anywhere near the middle of the ocean basins; rather, deep-sea trenches are close to land, nestled against continental margins or chains of volcanic islands. They are associated with active volcanoes and strong earthquakes, and invariably are partially filled with sediment that has been eroded from the nearby landmasses that tower above them.

MIDOCEAN RIDGES

Submarine mountain ranges, the *midocean ridges* mentioned earlier, are the most striking physiographic feature of the ocean floor (**Figure 2–3c**). They are all connected, and they represent the longest, most continuous mountain belt on the globe, extending for over 60,000 kilometers (~36,365 miles). This colossal range occupies almost one-third of the ocean floor, and its cragginess is comparable to the major mountain belts on land. Although termed "midocean" ridges, many are not centered in the basin. The summit of each ocean ridge is either broadly convex (curved upward) or occupied by a **rift valley**. This is a feature formed by geologic forces that pulled apart and broke the solid rock, a process called **faulting**, so that part of the ridge crest dropped down to create a valley flanked by steeply rising, fault-bounded shoulders. In fact, the midocean ridges are geologically active, characterized by frequent, shallow earthquakes (<35 kilometers, or ~22 miles, below the Earth's surface), many faults, and widespread volcanism. Furthermore, the axis of the midocean ridges is not continuous but is segmented by a series of geologically active **transform faults**, where fractured rocks slide past one another. This causes the ridge axis to be offset, creating a zigzag pattern. **Fracture zones** represent the inactive arms of transform faults that extend far into the deep-ocean basin. These gashes, anywhere from 10 to 1,000 kilometers or more (~6 to 620 miles) in length, consist of linear valleys and elongated, faulted hills and volcanoes that are oriented at 90 degrees relative to the axis of the midocean ridges (see Figure 2-2).

2-3

Geologic Differences between Continents and Ocean Basins

Now that our general survey of the ocean's bathymetry is complete, we can begin to answer a fundamental question about the Earth's crust: Except for the obvious elevation contrasts, is the geology of ocean basins fundamentally different from the geology of continental masses? Stated differently, why do land and oceans exist? Providing a satisfactory answer to this question is not an easy matter. It requires detailed measurements of land elevations and water depths, as well as knowledge of the rock composition of the Earth's crust on land and beneath the oceans. Let's take both of these steps and see what we can learn.

THE EARTH'S TOPOGRAPHY AND BATHYMETRY

A useful means of summarizing large quantities of data is a *frequency distribution*, a plot of the number of times (**frequency**) a given value occurs for a variable. For example, you can easily create a frequency distribution of the height of the people in your class. All you need to do is measure the height of each of your classmates and count the number of measurements that fall into height classes, such as 5'1'', 5'2'', . . . 6', 6'1'', etc. There might be one individual who is 5'2'' tall, one that is 5'3'', three that are 5'4'', seven that are 5'5'', continuing on to the tallest member of the class. A plot of this information would give you the frequency distribution of the people's heights in your class (**Figure 2–4a**).

Earth scientists have done the same type of analysis on the height of land above sea level (topography) and on the depth of the ocean floor (bathymetry) below sea level. The frequency plot in Figure 2–4b shows that only about 29 percent of the Earth's crust projects above sea level. This means that we humans inhabit and know most about the area of the Earth that is unrepresentative of its surface, the dry land! These data also reveal that mountains and deep-sea trenches are uncommon landforms despite their impressive size. Furthermore, the frequency data show a bimodal distribution. A *mode* is the value of a variable which is at a maximum in the frequency distribution; in other words, it is the most frequently occurring value of the variable. A *bimodal distribution* has two distinct maxima or modes. If you examine **Figure 2–4b**, you will see a mode associated with land (0 to 1 kilometers above sea level) and a mode associated with water (4 to 5 kilometers below sea level). Now, referring again to the frequency plot of the height of your classmates, note that it is more than likely the distribution would not be unimodal (one-mode) as shown in Figure 2–4a, but would be bimodal. Why? Well, there would be a separate mode for the females and males, because the height distribution of humans is gender-dependent. This same kind of reasoning applied to our elevation/depth data implies that the crust of the continents and ocean basins must be different in some fundamental way. Otherwise, we would expect a simple unimodal (one-mode) distribution. File this away in your mind, because we will come back to the significance of this very point shortly.

Extensive sampling by geologists indicate that two basic rock types characterize the Earth's crust: **granite** on land and **basalt** in the ocean (**Table 2–2**). Basalt is a dark-colored, fine-grained volcanic rock made up largely of minerals rich in oxygen (O), silicon (Si), and magnesium (Mg). Granite, by contrast, is a light-colored, coarse-grained **igneous rock** (formed from a melt, i.e., a molten rock) composed of minerals containing abundant oxygen, silicon, and aluminum (Al). To geologists, they are as different from one another as apples are from oranges. Geophysical measurements indicate that the basalt of the ocean crust is significantly more dense than the granite of the continental crust (2.9 versus 2.7 to

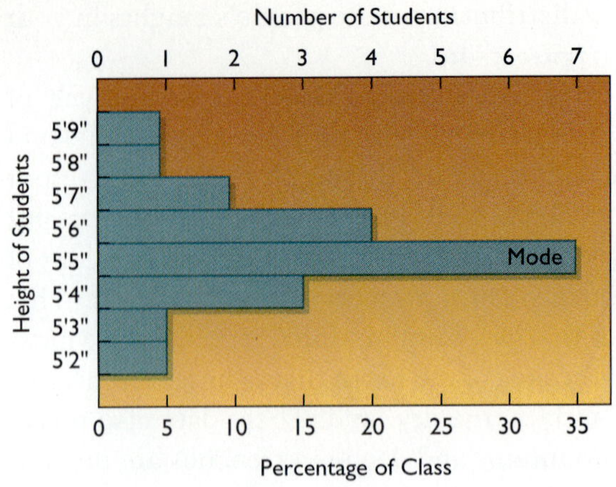

(a) FREQUENCY OF THE HEIGHT OF CLASSMATES

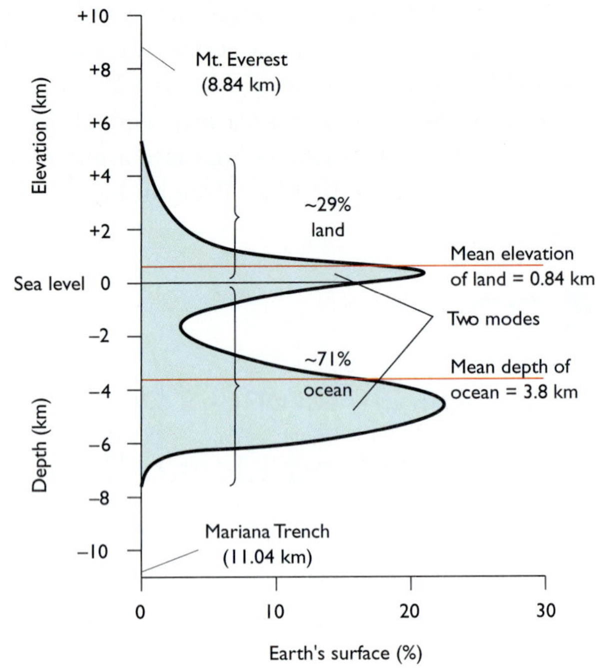

(b) FREQUENCY PLOT OF TOPOGRAPHY AND BATHYMETRY

FIGURE 2-4

Frequency plots. (a) A frequency plot of the height distribution of a class of 20 students might look like this. The most common height, the mode, is 5'5''. (b) Here is a frequency plot showing the relative distribution of the Earth's land elevations and ocean depths. The bimodal distribution of the Earth's topography and bathymetry indicates that the continental crust is fundamentally different from the oceanic crust.

TABLE 2-2

Rock properties

Rock type	Texture	Color	Minerals	Density (g/cm3)	Association
Granite	Coarse	Light	Quartz Potassium feldspar	2.7–2.8	Continental crust
Andesite	Fine	Intermediate	Sodium feldspar Amphibole	2.8+	Volcanic arcs; Andean-type mountains
Basalt	Fine/coarse	Dark	Calcium feldspar Pyroxene Olivine	2.9	Ocean crust

2.8 g/cm^3), and that both overlie mantle rocks that are even denser (3.3 g/cm^3). The boundary between the crust and the upper mantle (Figure 2–1a) is a sharp one; it is called the **Moho**, which, fortunately, is a shortened version of its discoverer's name, Andrija Mohorovicić, a Yugoslavian geologist. The Moho lies deeper below the ground beneath the continents than it does beneath the ocean basins, because the crustal thickness averages 30 to 40 kilometers (~19 to 25 miles) for land, but only 4 to 10 kilometers (~2.4 to 6.2 miles) for ocean.

Let's quickly summarize what we have learned so far. The continental crust is composed of thick, relatively low-density granite, and the ocean crust is made up of thin, relatively high-density basalt. Both crusts are in contact with denser mantle rocks at the Moho (Figure 2–5c). Now that we have these ideas in mind, let's consider a simple theory and conduct an elementary experiment that will get us back to our original question: Why do landmasses and ocean basins exist?

SCIENCE BY NUMBERS

Powers of 10

An easy means of expressing large numbers is to use powers of 10. In this scheme, 100 is expressed as 10^2 (10×10); 1,000 as 10^3 $(10 \times 10 \times 10)$; 10,000 as 10^4 $(10 \times 10 \times 10 \times 10)$, etc. Let's construct a table of such numbers.

1	$= 10^0$	1,000,000	$= 10^6$ (1 million)
10	$= 10^1$	10,000,000	$= 10^7$
100	$= 10^2$	100,000,000	$= 10^8$
1,000	$= 10^3$ (1 thousand)	1,000,000,000	$= 10^9$ (1 billion)
10,000	$= 10^4$	etc.	
100,000	$= 10^5$		

The number 1,220,000, then, equals 1.22×10^6, or 1.22 multiplied by 10 six times.

Numbers less than 1 can be expressed with this notation as well. A table of such numbers looks like this:

1	$= 10^0$	0.000001	$= 10^{-6}$ (1 millionth)
0.1	$= 10^{-1}$	0.0000001	$= 10^{-7}$
0.01	$= 10^{-2}$	0.00000001	$= 10^{-8}$
0.001	$= 10^{-3}$ (1 thousandth)	0.000000001	$= 10^{-9}$ (1 billionth)
0.0001	$= 10^{-4}$	etc.	
0.00001	$= 10^{-5}$		

The number 0.0000122 can be expressed as 1.22×10^{-5}, or 1.22 multiplied by 0.1 (i.e., 10^{-1}) five times.

Before you can add or subtract numbers expressed in exponential notation, you must convert them to equal powers of 10, as the following examples demonstrate:

Addition: $(3.5 \times 10^6) + (1.5 \times 10^5)$

$$\begin{array}{r} 3.50 \times 10^6 \\ +\ 0.15 \times 10^6 \\ \hline 3.65 \times 10^6 \end{array}$$

Subtraction: $(3.5 \times 10^6) - (1.5 \times 10^5)$

$$\begin{array}{r} 3.50 \times 10^6 \\ -\ 0.15 \times 10^6 \\ \hline 3.35 \times 10^6 \end{array}$$

To multiply or divide numbers expressed as powers of 10, you simply add or subtract the exponents, respectively, as shown below:

Multiplication (add exponents):

$$(3.5 \times 10^6)\,(1.5 \times 10^5) = (3.5)\,(1.5)\,(10^{6+5}) = 5.25 \times 10^{11}$$

Division (subtract exponents):

$$3.5 \times 10^6/1.5 \times 10^5 = (3.5/1.5)\,(10^{6-5}) = 2.33 \times 10^1 = 23.3$$

MASS BALANCE AND ISOSTASY

Imagine a bucket filled with water. The pressure, P, at the bottom of the bucket is proportional to the height (h) and density (ρ) of the water and to the strength of gravity (g). This relationship is expressed simply as

$$P = \rho g h$$

Intuitively, this formula makes a great deal of sense. It means that unless water is disturbed, it lies still in the bucket and exerts equal pressure everywhere on the bottom. This seems reasonable, because if there were unequal pressures in the bucket, water would flow from zones of high pressure to zones of low pressure in response to the imbalance. In other words, the water would redistribute its mass in order to reestablish an equilibrium. If we remove water, h (the height of water) decreases and so must the pressure at the bottom of the bucket. On the other hand, if we keep h constant and increase the density of the water by cooling it (the control that temperature has on density is explained in Chapter 5), the pressure exerted by the water at the bottom of the bucket will rise by a corresponding amount. Thus, increasing or decreasing one of the variables on the right side of the formula causes P, representing pressure, to rise or drop accordingly on the left side of the equation. The pressure at the bottom of our pail can be increased by either adding more water (increasing h) or cooling the water (increasing ρ).

Now, let's conduct an experiment (**Figure 2–5a**). Placing a block of wood in the bucket of water disrupts the mass balance, because the addition of a mass of floating wood causes a localized increase in pressure at the bottom of the bucket beneath the block. Being a fluid, the water instantly flows outward from beneath the block in response to the high pressure. Eventually, the block of wood stops bobbing up and down, water stops flowing, and everything becomes still. This, of course, means that the pressure at the bottom of the bucket everywhere, including beneath the block, is equal, *despite the fact that the wooden block's upper surface rises conspicuously above the water line.* If we go back to our formula, then a constant P everywhere at the bottom of the bucket implies that where h is greater, as it is where the block of wood is floating, density must be less. (The variable h in this case represents the combined height of the wood and water beneath it; that is $h = h_{wood} + h_{water}$.)

This is the only way we can balance the equation, because gravity has not changed. And this is exactly what has happened; the average density of the column where the block is present is less (because of the presence of the low-density wood) than everywhere else in the bucket. The apparent excess mass of wood rising above the water's surface is compensated by the submerged part of the block of wood, which is less dense than the water. The thicker the block of wood, then, the higher its top must stand above the water line and the deeper its base must penetrate into the water to make the total mass and, hence, pressure above the bottom of the bucket be the same everywhere. Do you agree? Think of the formula and run through the logic of this statement. This principle of mass balance is termed **isostasy**.

Icebergs floating in the polar seas (**Figure 2–5b**) provide an excellent example of a system in isostatic balance. Most icebergs are small in size and float low in the water. The larger icebergs are more than 100 kilometers (~62 miles) in breadth and tower 80 meters (~264 feet) or more above sea level. The higher the iceberg rises above the ocean surface, the deeper its base projects into the water. The exact depth of immersion depends on the difference in density between the ice and seawater, but it will be three to nine times the iceberg's height above the water line.

Although we seem to have strayed far from our original question, the principle of isostasy has a direct bearing on the crustal nature of continents and ocean basins. We can explain the striking bimodal distribution of the earth's topography (see Figure 2–4b) in terms of our bucket experiment. If we apply the concept of isostasy, we can imagine the continental and oceanic crusts as two rock masses floating on a denser, plastic mantle. Recall that these two distinctive types of crustal rocks—granite on the continents and basalt in the ocean basins—have their own characteristic elevation modes: granite at about 840 meters (~2,812 feet or ~0.5 miles) above sea level, and basalt at 3,800 meters (12,540 feet or ~2.4 miles) below sea level (see Figure 2–4b). Using our bucket experiment as an analogy, granite and basalt would be equivalent to two blocks of different types of wood floating in water (equivalent to the mantle) at levels appropriate to their densities. The continents rise high above the ocean because of their light, thick granite crust, and the ocean basins are low because of their dense, thin basalt crust (**Figure 2–5c**). (Imagine two blocks of different types of wood—a thin, dense block of teak and a thick, less dense block of pine. How would they float in the water? Draw a sketch and make the comparison.) Furthermore, if our analogy holds up, the granite crust must penetrate deeper into the mantle than the basalt crust

FIGURE 2-5

Isotasy. (a) Blocks of wood with different lengths are at rest in a bucket of water. This means that the pressure everywhere at the bottom of the bucket is equal. This is proven by the sample calculations to the right of the sketch. The apparent excess mass of wood above the water line is compensated for by the lower density associated with the submerged part of the block of wood. The longer the block, the higher its top rises above the water surface and the deeper its base extends into the water. (b) The higher an iceberg rises above the sea surface, the deeper the ice must penetrate into the water in order to maintain isostatic balance. (c) The oceanic crust is denser and thinner than the continental crust. The high mountains of continents are balanced isostatically by having deep crustal roots.

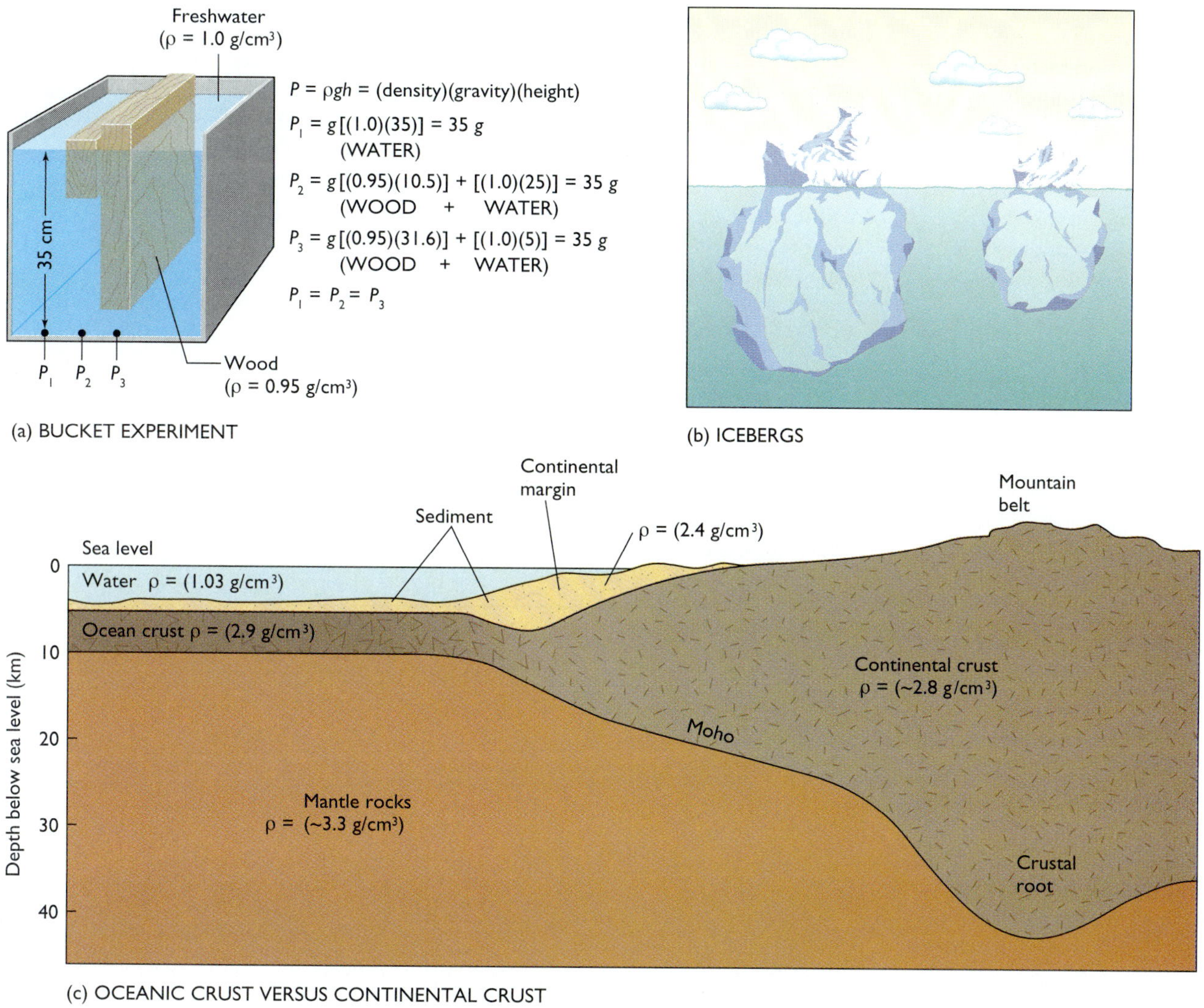

in order to get the proper mass balance. And it does, because we know that the Moho, the boundary separating the base of the crust from the top of the mantle, is three to four times deeper beneath the continents than beneath the ocean basins.

Now that we have a technical understanding of the crust and the physiography of the sea floor, let's deepen our familiarity with the ocean's bathymetry. We'll do this by ending the chapter on a whimsical note, by rereading a diary of an imaginary journey across the sea bottom.

2-4

The Physiography of the Western North Atlantic Ocean

The charts and tables in this chapter will, we hope, have helped you become familiar with the varied ocean landscape. Oceanographers

Probing the Sea Floor

You should be familiar with important geophysical techniques that geological oceanographers use to probe the ocean's bottom. Many of these techniques rely on sound waves; by noting changes in the speed that sound travels, it is possible to determine some of the physical properties of the material (such as density) through which the sound is passing. A brief description of several of these important geophysical techniques follows.

ECHO SOUNDING

Before the twentieth century, measurements of water depth were made with a "lead and line." A lead weight attached to a rope or, more recently, to piano wire was lowered to the ocean floor, and the total length of line used was an estimate of water depth. The technique was slow, because it took many hours to get one sounding of the bottom in deep water. Furthermore, sounding by lead and line gave only a rough measure of water depth, because the cable was often deflected and bent by currents or by the drifting vessel itself.

Electronic technology developed by private organizations after World War II revolutionized the sounding of the sea bottom, a process termed **echo sounding.** A sound transmitter, mounted on the bottom of a vessel, sends a sound pulse out into the water. This sound wave is reflected off the ocean floor back to the sea surface, where it is recorded by a listening device called a **hydrophone** (**Figure B2–1**). The water depth is equal to one-half the time it takes the sound pulse to leave the vessel, reflect off the sea bottom, and return to the hydrophone multiplied by the speed of sound in water, which averages about 1,460 meters per second (~3275 miles per hour) in seawater of normal salinity. (At first glance this seems complicated; it isn't. Think about it this way. If you traveled at 50 miles per hour in your car, and traveled from home to a store and back home in one hour without stopping, how far away is the store from your house?) Today, echo sounding is done continuously and recorded on graph paper to yield a topographic profile (see Appendix IV) of the sea floor.

SIDESCAN SONAR

Engineers reasoned that sound could also be directed sideways as well as downward, such that echoes would reveal the bathymetry to either side of a surveying ship as well as the water depth below its bottom. This technique, called **sidescanning,** produces a map of the portion of the sea bottom traversed by the ship (**Figure B2–2**), rather than merely a profile of the bottom. Considerable experience is needed to interpret these maps properly and accurately. Sidescan sonar has become valuable for locating sunken ships because so much of the sea floor that lies to either side of the surveying vessel is revealed in the record.

SEISMIC REFLECTION

Seismic reflection enables marine geologists to examine not only the bottom of the sea, but also within the bottom, revealing the shape and thickness of sediment and rocks beneath the sea floor—the subbottom. Seismic reflection uses stronger sound pulses and lower frequencies (number of wave cycles per second) than echo

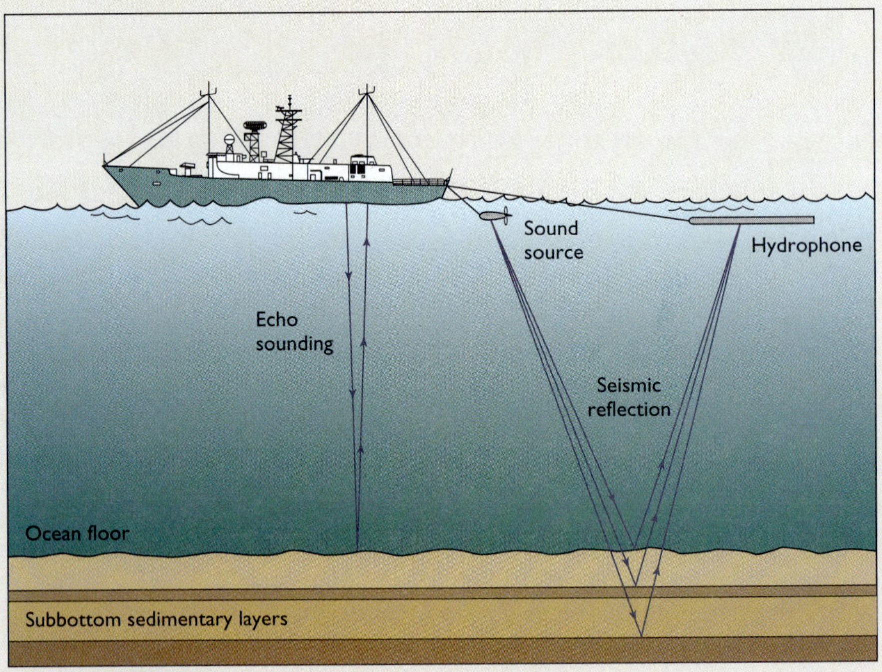

FIGURE B2-1

Echo sounding and seismic reflection. Reflecting sound pulses off the ocean floor reveal the depth of water beneath a vessel. Some sound can penetrate the ocean floor and reflect off rock layers, providing a siesmic-reflection profile of the subbottom.

sounding does. Some fraction of the strong sound pulse penetrates the ocean floor and is reflected back upward by layers of sediment and rock. These echoes are detected by a hydrophone and are displayed graphically to create a seismic-reflection profile—a display of the subbottom geologic structure along the ship's course (**Figure B2–3**). If you're having trouble imagining what this looks like, think about this analogy. The subbottom profile is equivalent to cutting a cake and removing a slice so that you can see the interior layers: what you are looking at is essentially a "subbottom" profile of the inside of the cake.

SEISMIC REFRACTION

When sound waves enter the sea bottom, they are affected in two principal ways. First, they are reflected by sediment and rock layers, and these reflections, when collected by hydrophones, yield the seismic-reflection profile just discussed. Second, some of the sound waves are *refracted* (bent, so that they curve rather than follow a straight line), as they travel through material having different physical properties. Refracted waves follow more complex pathways through sediment and rock (**Figure B2–4**) than do reflected waves. This **seismic refraction** enables marine geologists to calculate the densities, depths, and thicknesses of rock layers by recording the time it takes the refracted sound wave to travel from the sound source to the hydrophone. Because seismic refraction relies on higher energies and lower frequencies (pulses per second) than those used in seismic reflection, the wave penetrates much deeper, and can reveal the geologic structure of not only the crust, but also the mantle.

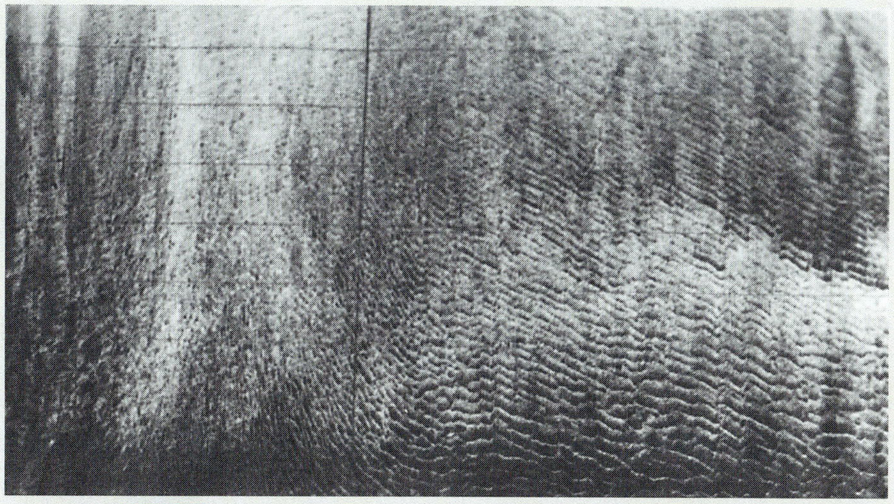

FIGURE **B2-2**

Sidescan sonar. Sound can be reflected not only from the bottom beneath a vessel, but also from both sides of the ship. The resulting map reveals bottom irregularities. In this photograph, dark areas are bottom slopes that face the vessel and reflect the sound pulse. The light areas are flat bottoms or slopes that face away from the vessel and do not reflect sound energy.

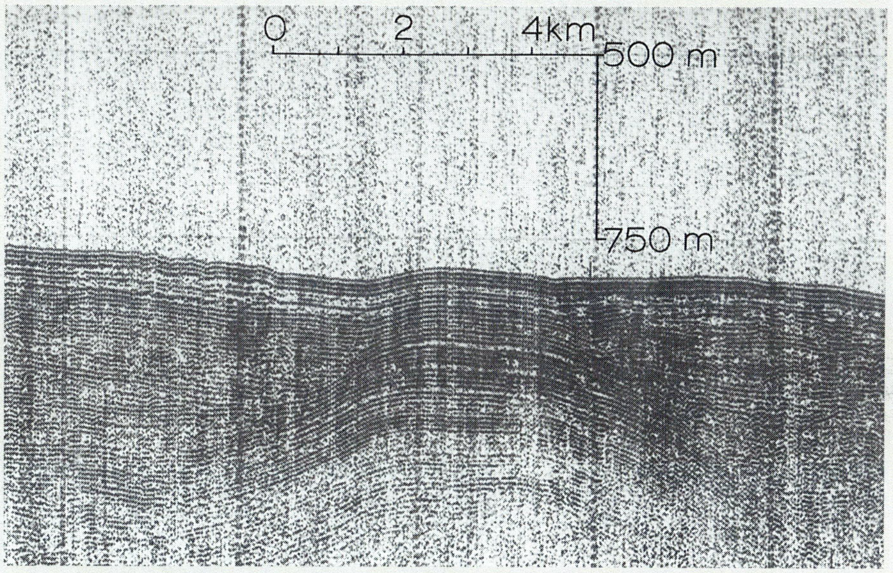

FIGURE **B2-3**

Seismic reflection profile. The continuous reflection of sound from subbottom layers (see Figure B2-1) produces a geologic cross section. In this profile, layers of sediment have been deformed into a broad fold.

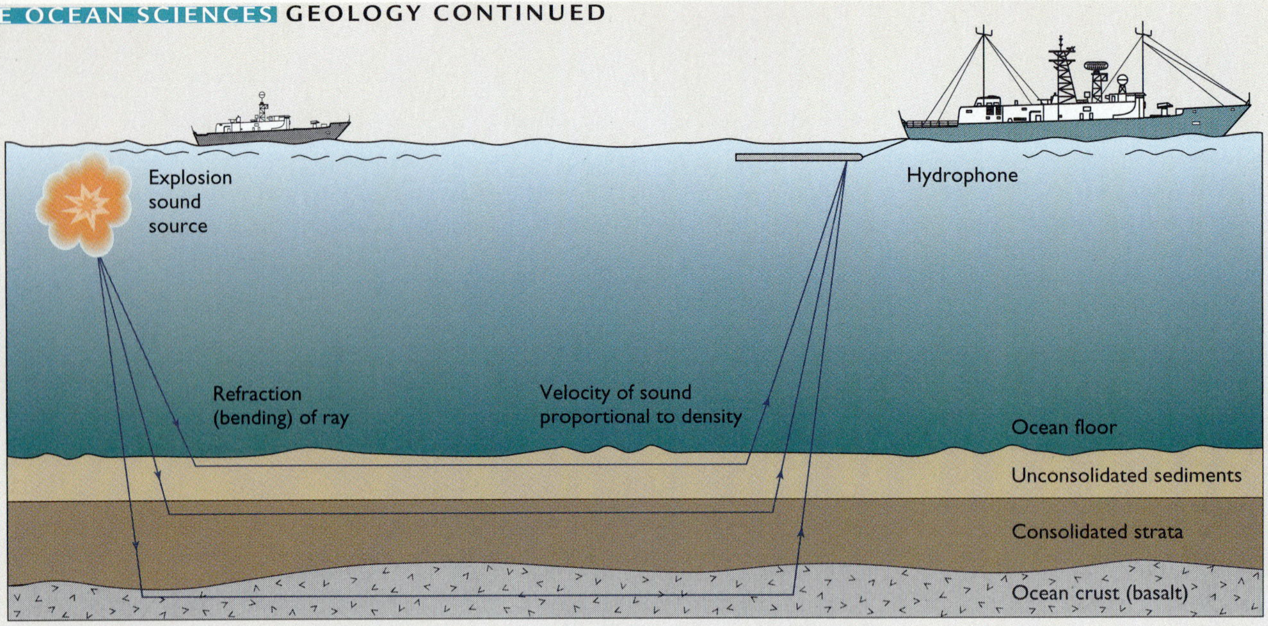

FIGURE B2-4

Seismic refraction. The refraction (bending) of sound as it travels through different rock types reveals the shape and density structure of the underlying rock masses.

ALTIMETRY

A new technique called **altimetry** has been used to survey the world's oceans. Orbiting satellites, such as SEASAT and TOPEX/Poseidon, using radar altimeters, can measure rapidly and accurately the height of the sea surface to within 3 to 5 cm (~1.2 to 2.0 inches). What surprises many people is that variations in the height of the ocean's surface can reflect the bathymetry below. Rocks are denser than water is, so any undersea mountain will "draw" water to it from all sides because of gravitational attraction. This creates a detectable mound of water over the submarine mountain. This information can be used to map the ocean's large-scale bathymetric features—seamounts, fracture zones, submarine ridges—that had not previously been detected, particularly in remote areas of the ocean where few shipborne echo-sounding surveys have been made (**Figure B2-5**).

Visit www.jbpub.com/oceanlink for more information.

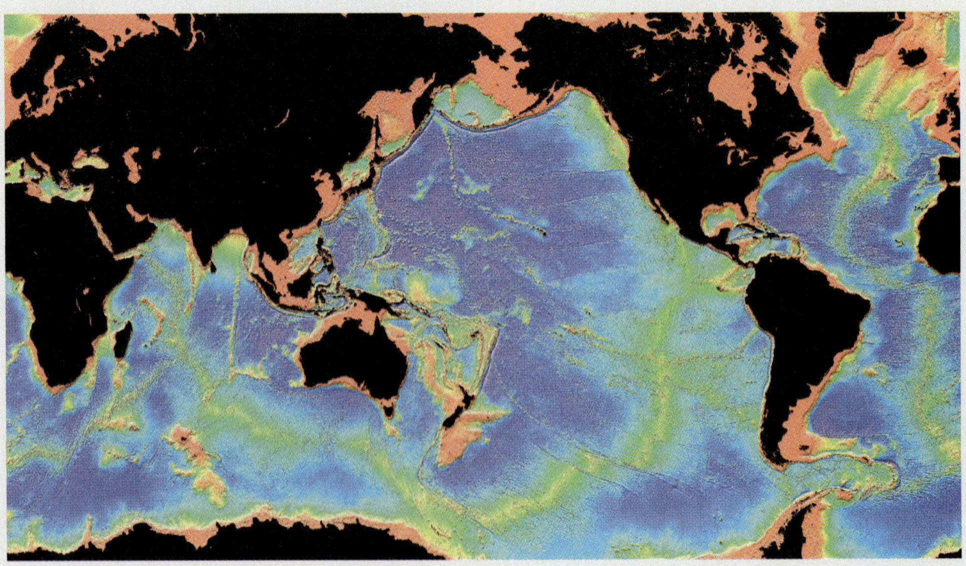

FIGURE B2-5

Altimetry. Variations in the height of the sea surface reflect the presence of large bathymetric features that, because of gravity, affect the elevation of the sea surface.

have delineated all of this bathymetry by sounding the ocean floor and reconstructing its features from those depth readings as a map. Unfortunately, we are unable to view the magnificence of the ocean bottom at first hand because it is obscured by water—a very opaque medium that keeps the depths of the sea in perpetual darkness. We can, however, overcome this difficulty by imagining that the oceans have been drained entirely of water and that their landforms are dry, exposed to the sunshine and, most fortuitously, exposed to our feet and eyes. (Actually, this daydream is not as preposterous as it seems: large basins, such as the Mediterranean Sea, may have dried up in the geologic past, as described in the box, "The Drying Up of the Mediterranean Sea" in Chapter 4.) This imaginary journey provides a personal understanding of the magnitude and diversity of this imposing ocean topography in a more effective way than that offered by studies of charts and reviews of tables of basic facts.

What follows are excerpts from the journal of the scientist who led an ambitious expedition that took a doughty party from the beaches of New Jersey eastward to the spiny summit of the Mid-Atlantic Ridge, a long round-trip journey of over 3,000 kilometers (~1,830 miles). The effort was strenuous at times. They trekked across vast plains, scrambled over knobby topography, and even climbed precipitous slopes to reach the high mountain summits.

A TREK TO THE CREST OF THE MID-ATLANTIC RIDGE

DAY 1

Noon Break—We left New Jersey very early this morning before the sun had risen. Using headlamps and guided by our compass, we started out on an easterly course designed to take us eventually out to the "deep sea" (Figure 2—6). We spent the first few hours of our trek navigating our way carefully across the dark, flat floor of the near-shore zone. Sunrise slowly brought into our view a panorama that is sensational in its flatness. I must report that the gently rolling terrain is shockingly devoid of a ground cover of plants and extends eastward as far as the eye can see. Looking back to the west, we gain some measure of comfort from a glimpse of the familiar landscape and the early morning twinkling city lights of New Jersey.

Evening Camp—Throughout the day, we've walked mainly on unconsolidated sand, which occasionally was molded into sand ripples and even into a few fields of long, sinuous sand dunes that were raised and shaped by waves and currents when this ground was covered by water. The topography reminded some of us of the smooth and sandy terrain of the Coastal Plain of the eastern and southern United States. This impression is quite valid, given that the Coastal Plain deposits are "stranded" shelf sediments—remnants of a higher and more extensive sea in the geologic past.

Today the walking was quite easy and we maintained a good, steady pace of about 4.5 kilometers an hour (2.8 mph), interspersed with many rest stops. It's now nightfall. The group is tired, but we're pleased with our progress, having covered almost 35 kilometers (21.7 miles).

DAY 4

Evening Camp—The past three days have been rather uneventful. We lost sight of New Jersey sometime during our second day out. We seem to be floating on a flat and limitless sea of sediment made up mainly of sand and some gravel, with surprisingly little mud. The monotony and boundless expanse of this "sandscape" has given us a very unsettling feeling, distorting time and space for most of us. We've been making numerous measurements of bottom gradients and have found that the shelf is sloping <0.5 degrees on average. We are dropping in elevation ever so slowly, by our estimates slightly more than one vertical meter for every horizontal kilometer of distance. We cannot see this slight bottom inclination with the naked eye; we must continually keep track of the sun or consult the compass to stay on our "seaward" course.

DAY 6

10:00 a.m.—We've reached the outer limit of the shelf. Our party has traveled almost 160 kilometers (~99 miles) and dropped 130 meters (~426 feet). From our vantage point high above the ocean basin, the vista to the east is truly spectacular; on the distant horizon, we can just make out the mountains fronted by the flat, expansive deep-ocean floor of the Atlantic basin (Figure 2—7). The ground directly to the east of our overlook drops, not steeply but rather

Figure 2-6 An ocean trek. (a) We begin our journey in New Jersey and proceed across the continental margin to the edge of the Hudson Canyon and onto the abyssal plain of the deep-ocean floor of the Atlantic. (b) The second part of our trek takes us across the broad expanse of the Atlantic basin and up the flank of the Mid-Atlantic Ridge to its very summit. [Adapted from B. C. Heezen, M. Tharp, and M. Ewing, The Floor of the Ocean, vol. 1: The North Atlantic, Special Paper 65 (New York: Geological Society of America, 1959).]

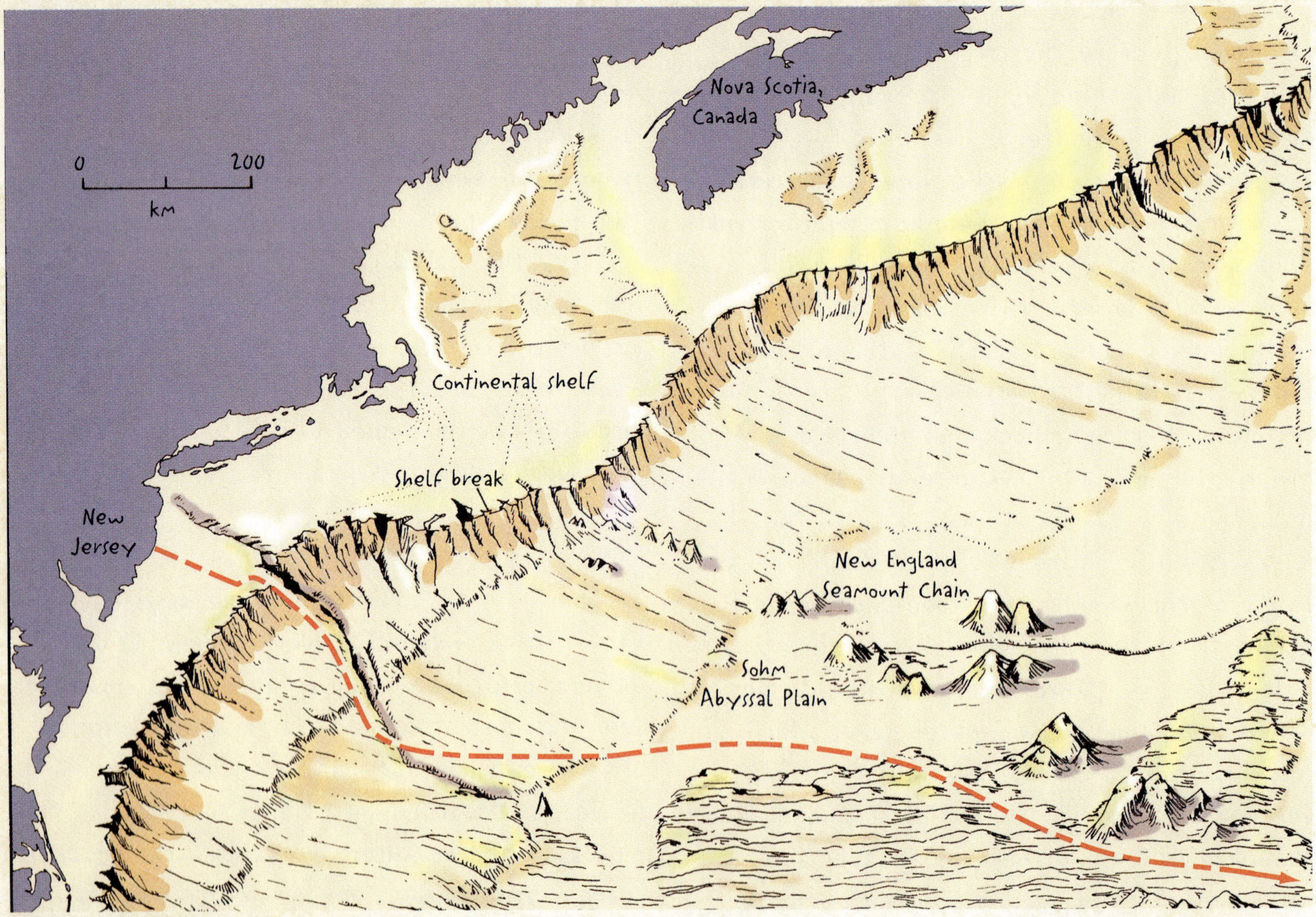

(a) FIRST PART OF TREK

gradually. We had expected a break to separate the steep cliffs of the continental slope from the flat plain of the continental shelf. The change in the bottom gradient is barely noticeable; we find it impossible to establish with any certainty the exact location of the shelf break, which appears as a knife edge on our map (see Figure 2—6a).

The answer, of course, has to do with vertical exaggeration—a technique used by cartographers to enhance subtle but important changes in bottom slopes by distorting the vertical dimensions (see Appendix IV). A vertical exaggeration of 1:10 in a physiographic map or a topographic profile, for example, indicates that elevation has been magnified by a factor of 10 relative to horizontal distance; therefore, a gradient of 4 degrees, which is used worldwide to represent continental slopes, appears on a chart to be a cliff precipice.

Evening Camp—Although everyone in the group measured a clear change in bottom slope as we proceeded across the shelf break, we were all disappointed because the bottom gradient did not vary dramatically at the break. Despite its unimposing bottom gradient, however, the continental slope remains one of the most continuous topographic boundaries of the entire earth.

DAY 7

Evening Camp—After resting for part of this morning, we resumed our trek, proceeding down the continental slope. The footing was difficult at times today. The 4-degree slope seemed much more significant than it did yesterday when we were merely gazing over the shelf "edge." In fact, a simple calculation shows that we are dropping 40 or 50 meters (~131 to 164 feet) for every kilometer we travel eastward—a slope

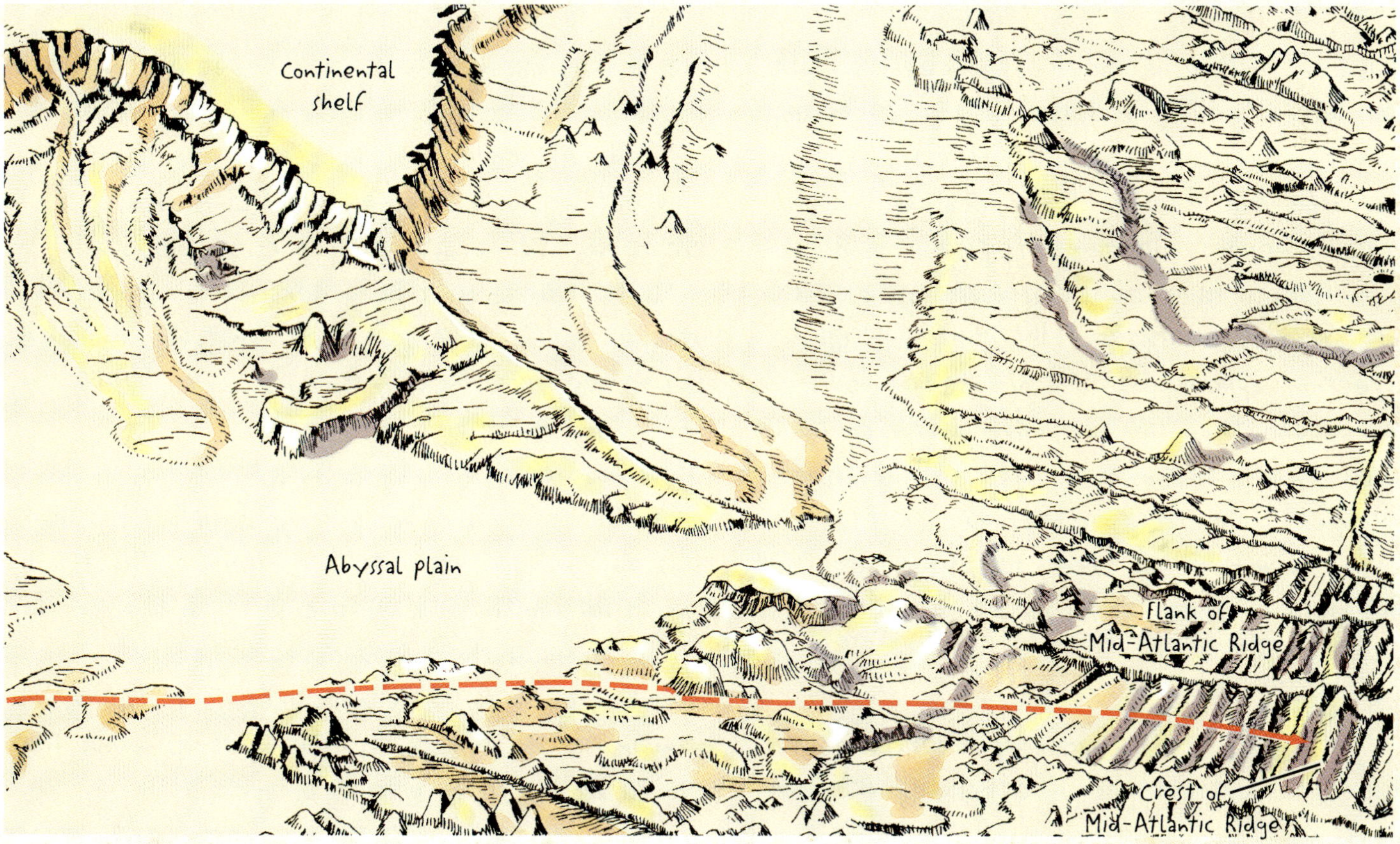

(b) SECOND PART OF TREK

Figure 2-7 The western North Atlantic Ocean. This imaginative sketch portrays the view of the deep-Atlantic landscape that would be visible from the seaward edge of the continental shelf. The distant mountains form the Mid-Atlantic Ridge.

equivalent to the gradient of a steep mountain road. We're finding that the surface of the continental slope, like that of the shelf, is quite smooth for the most part. An occasional short section steepens noticeably to perhaps 8 degrees or so. About midday, we clambered over some small, irregularly shaped hills that, on closer inspection, appeared to be composed of sedimentary debris that has slumped or slid from above. We've set up camp near some rock outcroppings. Judging by our maps, our best guess is that so far we have dropped over 1,100 meters (~3,608 feet) into the deep abyss.

DAY 8

Evening Camp—This morning, several of us woke up early to inspect the rock ledge before breakfast. The outcroppings consist of beds of semiconsolidated sandstones (compacted sand) and minor mudstones (compacted mud) that are bare of sediment cover because of the exposure. By projecting these outcrop layers landward, as we learned to do in our geology class, we found that the shelf must be underlain by a similar sequence of sedimentary beds that have become compacted into rock by the pressure of the overlying deposits.

We were underway by 7 A.M., but the hiking has been slow and unexpectedly difficult today. Instead of going eastward, we decided to strike out a bit northward, more or less parallel to the slope, to find the Hudson Canyon (see Figure 2—6a). Our new route has taken us across gullied and furrowed topography. Many times we've had to go around steep pitches of rock.

DAY 9

Evening Camp—We eventually reached the Hudson Canyon at about midafternoon. Several of us cautiously approached the lip of the canyon and peered down its steep, jagged wall. The view is definitely not for the faint-hearted! The precipitous canyon wall must be 300 or 400 meters (~984 to 1,312 feet) high. It is not, however, a sheer wall, but consists of a series of rock ledges, sloping terraces with rubble, and ridges and irregular pinnacles of rock. The opposite wall, some 10 kilometers away, rises steeply from the canyon floor (Figure 2—8). Deep below, freshwater streaming from the Hudson River rushes through the V-shaped valley. Our map indicates that the Hudson Canyon rises landward onto the shelf as a branching valley system, with a main channel directly off the Hudson River. In the absence of ocean water, the Hudson River occupies the canyon.

A few of the more adventuresome members of our group dropped down to a rock terrace some 35 meters (~115 feet) below the lip of the canyon. Returning safely, they reported that the footing was treacherous at times, because the wall was made up of mudstones and sandstones that could crumble easily under a person's weight. They also noticed that many beds of soft shale were pockmarked with holes of various shapes and sizes, obviously dug out by burrowing animals such as crabs, lobsters, fish, and clams. These holes must be undermining the stability of some of the cliffs, for we have noted numerous landslides that have swept wall material down to the canyon floor.

DAY 14

Evening Camp—We've needed more than five days of strenuous hiking to reach the mouth of the canyon at the toe of the continental shelf. The toe of the slope lies about 3,000 meters (~1.9 miles) deeper than the former position of sea level. This afternoon we skirted the southwestern shoulder of the canyon, occasionally climbing down into its more accessible parts, but no more than a few hundred meters or so, to examine the rocks of its stark walls more closely. The route was more difficult than dangerous.

Near the foot of the continental slope, the bottom gradients becomes noticeably gentler. The canyon opens up onto a broad submarine fan composed of a large accumulation of sediment that has been carried down the canyon by currents and dumped here as bottom gradients have flattened. In plan view, the fan is roughly cone shaped, with its apex jutting into the mouth of the canyon proper. Superimposed on the fan is a branching network of channels with embankments (levees) winding their way toward the deep-ocean floor.

Looking upslope, we could see a most impressive sight; a profile view of the Hudson Canyon (Figure 2—9), including its precipitous walls, towering relief, and narrow floor. We marveled at the power of ocean currents that are capable of carving such a monumental bathymetric feature out of lithified sedimentary beds.

Beyond the seaward end of the submarine fan lies the continental rise, a gently sloping pile of sediment that leads to the deep ocean floor. The topography of the rise is rather unspectacular compared with the mountainous relief of the Hudson Canyon. What impresses us most, however, is its vastness; this plain is very broad and seems to go on forever. This evening, as I look out across the ocean floor from our camp, my sense of infinity reminds me of Walt Whitman's comment about the vast, unfolding plains of the western United States: "Even their simplest statistics are sublime."

Figure 2-8 Hudson Canyon. This drawing depicts the view that we would see if we could peer into the Hudson Canyon from a perch high on its southern shoulder. The opposite canyon wall is about 10 km away.

DAY 23

Evening Camp—It has taken us nine days of constant plodding to reach the seaward edge of the continental rise. We estimate that, off the coast of the northeastern United States, the rise is wider than 200 kilometers (~124 miles). Despite its enormous breadth, the vertical relief is rather modest (5 to 10 meters [16.4 to 32.8 feet] being common; 20 to 30 meters [65.6 to 98.4 feet], the exception). Like the shelf, the pitch of the rise is almost imperceptible, with a drop of slightly more than 1 meter for every kilometer traveled seaward. We reached the true floor of the deep Atlantic when we trudged onto an arm of the Sohm Abyssal Plain this afternoon. The plain is an exceedingly broad and flat apron of sediment (see Figure 2—6a.). Stark and boundless, the abyssal plain is eerie. . None of us looks forward to the formidable task of walking across this plain; it has little topographic variety, only a monotonous and seemingly infinite levelness. But the prospect of seeing the volcanic mountains and the Mid-Atlantic Ridge beyond the abyssal plain entices us to attempt this crossing. We begin tomorrow at dawn.

DAY 24

Evening Camp—Today, the first day of our journey across the abyssal plain, we found few topographic features. We spent many hours slogging along, but it's hard to overcome the feeling that we're not covering any distance when we look eastward toward the center of the Atlantic basin. But we are clearly making progress toward the mountains, because we can no longer see the continental slope and the

Figure 2-9 The floor of the Hudson Canyon. This profile view portrays the majestic character of the Hudson Canyon, including its precipitous side slopes, soaring relief, and narrow valley floor.

continental rise behind us. This flat horizon that encircles us completely is so wearisome that we tend to stare at the ground rather than concentrate on the skyline (sealine?). Periodically during the day, we took short cores of the sediment to give ourselves something to do. We discovered that that plain is composed of a mixture of sand- and silt-sized grains of rock and mineral fragments and of the tiny shells of invertebrate organisms that once floated in seawater above the ocean bottom. The particles of rock and minerals must represent sedimentary debris that has eroded from adjacent landmasses and been swept here by strong bottom currents.

DAY 28

Evening Camp—It's been three days of plodding across the flattest type of plain found anywhere on earth. But, at last, today we encountered scattered abyssal hills, which become more abundant as we advance toward the basin. These rounded, knobby knolls, varying from hundreds of meters to several kilometers across, also vary in height: some are no more than a kilometer above the adjacent ocean floor, others are tens of meters high. Most of the larger hills are composed of lava rock and appear to be extinct volcanoes; the tops of many of the abyssal hills are blanketed by a layer of ocean sediment. A few have sides that drop steeply. We've decided that these are geologic faults, actual breaks in the lava rocks along which there has been crustal slipping.

In the distance toward the northern horizon, we sighted a series of singular peaks with broad bases rising several kilometers above the ocean floor. According to our charts, these are seamounts that belong to the New England Seamount Chain, a line of volcanoes that have long ceased to be active (Figure 2—10). Several people wanted to change course and explore the Chain. We took a vote, and the majority opted to continue our quest for the distant midocean mountains. Had we not been able to make out the mountain crests on the far horizon, if barely, I suspect that the vote might have gone with the more visible goal.

This evening, to the south we can see with the help of binoculars the broad, dark volcanic base of the Bermuda Rise; its shoulders soar skyward and merge into white peaks of coral—the former Bermuda Islands. I remember sailing out of Bermuda one summer, wondering about the landscapes that lay beneath the deep blue waters of the Atlantic. Now, I am standing on the ocean floor looking upward at the towering conical peaks of the Bermuda Rise. We've taken twenty-eight days to reach our present position.

DAY 38

Evening Camp—A week and a half ago we encountered the first abyssal hills. Since then, the morphology of the ocean floor has changed gradually as we left the abyssal plain. A week ago we started ascending the foothills of the Mid-Atlantic Ridge (Figure 2—6b). The relief is moderate to rugged and ranges locally from between zero and 100 meters (~33 to 328 feet) to more than 1,000 meters (~3,280 feet) for a few spectacular peaks. The topography is very uneven, with clusters of volcanic peaks, many of which are faulted, alternating with irregularly shaped, flat valley floors and terraces that prove to be ideal spots for our campsites. The valleys are formed of ponded sediment composed mainly of the minute shells of organisms; no mineral and rock fragments derived from land are evident.

DAY 62

Evening Camp—We've been climbing up and down the flanking ridges of the Mid-Atlantic Ridge Crest for the past three and a half weeks.. The days have blurred into a seemingly endless time span during this interminable journey. It's been difficult climbing this immense chain of mountains that runs down the center of the Atlantic Ocean. We are constantly forced to retreat from steep, dangerous pitches or inaccessible walls and to reconnoiter new, safer routes to the top of a ridge—a most difficult and time-consuming activity, given the lack of accurate topographic detail on our maps. The slow going is beginning to take its toll on the group. I worry that their fatigue will overcome their desire to reach the crest, but I am sure we are getting closer. As we have ascended each new higher ridge, we can see the sediment blanketing the dark volcanic rocks of the ridges' flank becoming thinner and more discontinuous. Today, the rocks have been starkly barren of sediment.

DAY 63

Noon—Success at last! Just a short while ago, we rounded the base of a rugged rock pinnacle

Figure 2-10 The New England Seamount Chain. The towering rock peaks in the distance represent a series of volcanic mountains that have long ceased to be active. The distribution of these volcanic peaks, which belong to the New England Seamount Chain, is depicted in Figure 2-6a.

and unexpectedly emerged on the lip of a ridge so high that no others rose above us. The whole party was surprised—and elated to realize that we have reached the very top of the Mid-Atlantic Ridge.

As I write this, we stand here gazing at a scene of unexpected beauty. A broad, rugged chasm, about 1 or 2 kilometers (0.6 to 1.2 miles) deep and 10 to 20 kilometers (~6.2 to 12.4 miles) wide appears chiseled into the rock. The panorama of this rift valley is primordial: it consists of starkly naked and dark volcanic rocks in countless shapes and forms, some rounded and some jagged; all kinds of volcanic features, both small and large; and all manner of topographic slopes, ranging from precipitous cliffs and spurs to gently rolling or even flat ground. (Figure 2—11). This must resemble the face of the earth billions of years ago when no life existed and its primitive crust had just begun to form as the molten, flowing, turbulent mass lost heat to the primeval atmosphere and solidified into rock. It is difficult to imagine a more splendid and awe-inspiring sight! My fellow trekkers remain speechless, each of us lost in the raw primitiveness of the landscape. On this ridge, we remain alert but quiet and contemplative. Few humans have ever witnessed such splendor.

We'll make camp here high on the shoulder of the rift valley and rest for a few days. During this time, members of the party can make short excursions along the ridge line in an attempt to find a route that leads down to the valley floor.

DAY 66

Valley Floor Base Camp—This morning, we chose the most likely route and, fortunately,

Figure 2-11 The crest of the Mid-Atlantic Ridge. This sketch is a panoramic rendition of the axial rift valley of the Mid-Atlantic Ridge, including fault-bounded cliffs and steep to gently rolling, even flat, volcanic terrain.

once we started, the descent was rapid and uneventful. After rappelling (using ropes to descend) a few steep rock pitches, we reached the valley bottom by early afternoon. I estimate we dropped more than 1,300 meters (~0.8 miles) in elevation (see Figure 2—6b).

Now, in a very real sense, we have entered a young geologic terrain, a laboratory where the floor and crust of the Atlantic Ocean are in the process of forming. Signs of its juvenile state abound: unweathered volcanic flows, rocks that are very hot to the touch, deep fissures cleaving the rock, and exposures of recently formed faults that have fragmented and twisted the volcanic deposits

DAY 10

Valley Floor Base Camp—Late last night, we were startled awake by a mild, short-lived earthquake and a few aftershocks that shook our camp. Luckily, no one was hurt. This morning, while exploring the center of the rift valley, we discovered active fissures issuing incandescent lava streams that accumulated into pools—another reminder that we were venturing in a dangerous, unstable land. As the lava cools on reaching the earth's exterior, its surface solidifies, forming a thin crustal rind of rock.

DAY 72

Valley Floor Base Camp—Today we attempted to ascend a small, magnificently shaped conical peak. We had to abandon the effort after being driven back by noxious sulfuric gases escaping from narrow clefts in the volcano's side.

Having satisfied our curiosity to explore this terrain, we've decided to head home. Tomorrow, we'll begin the climb up a rock buttress that leads to the top of the steep shoulder of the rift valley. After this two-and-a-half-month-long excursion, each of us now feels more familiar with the morphology that most typifies the globe—its ocean depths. For us, this knowledge symbolizes the beginning of a new understanding of the complex workings of the planet Oceanus.

STUDY GUIDE

KEY CONCEPTS

1. The earth is a water planet; 71 percent of its surface is covered by vast tracts of *oceans.* Internally, the earth consists of a dense iron-nickel *core,* a *mantle* rich in iron, magnesium, and silicon, and a thin outer *crust* of aluminum, silicon, and other elements. In terms of its mechanical properties, the layered earth is divided into a brittle *lithosphere,* which includes the crust and uppermost mantle; an *athenosphere,* composed of partially melted mantle rocks that deform and become relatively plastic; a deeper, rigid mantle zone called the *mesosphere;* a large molten *outer core;* and a rigid *inner core.*

2. The solid earth is surrounded by two exterior envelopes of fluids: the *hydrosphere* (an ocean of liquid water) and the *atmosphere* (an ocean of air, primarily composed of nitrogen and oxygen). The *biosphere*—all organic matter, living and nonliving—is a third, exceedingly thin, envelope of matter.

3. The morphology of the ocean floor can be divided into three physiographic provinces: the *continental margins,* the *deep-sea basins,* and the *mid-ocean ridges.*

4. The continental margins consist of a *continental shelf* (a flat sediment terrace that borders on the continents), a *continental slope* (a sloping ocean floor that plunges seaward to depths of between 2 and 3 kilometers (~1.2 to 1.8 miles) at an average gradient of about 4 degrees, and a *continental rise* (a broad, gently sloping plain of sediment that begins at the base of the continental slope and ends on the deep-ocean floor).

5. The ocean-basin floor includes *abyssal plains* (flat aprons of sediment that extend from the mouths of submarine canyons), *deep-sea trenches* (deep and long, steep-sided narrow basins that occur in association with volcanoes and earthquakes), *abyssal hills* (relatively small volcanic hills no higher than 1,000 meters [~3,283 feet]), and *seamounts* (large volcanic mountains rising more than 1,000 meters [~3,283 feet] above the ocean floor).

6. The midocean ridges form an enormous undersea mountain belt that winds through all of the ocean basins. The summits of these ridges are broadly convex or capped by a fault-bounded *rift valley*. These mountains are composed entirely of volcanic rocks. The axis of each ocean ridge is offset repeatedly by a series of *fracture zones* oriented perpendicular to the ridge axis.

7. The *basalt crust* of the ocean is denser and thinner than is the *granite* crust of the continents. According to the principle of *isostasy*, each crustal type seeks a characteristic level, so that rock masses are balanced in much the same way as an iceberg is balanced while floating in water.

KEY WORDS

abyssal hill
abyssal plain
asthenosphere
atmosphere
bathymetry
biosphere
continental margin
continental shelf
continental slope
continental rise
core
crust
deep-ocean basin
deep-sea trench
fault
fracture zone
guyot
hydrosphere
igneous rock
isostasy
lithosphere
mantle
mesosphere
midocean ridge
rift valley
seamount
shelf break
submarine canyon

QUESTIONS

1. How does the Earth's mantle differ from its core and crust? What is the relationship of the mantle to the lithosphere, asthenosphere, and mesosphere?

2. Examine the floor of the Indian Ocean (Figure 2–2), and identify the following physiographic elements: continental shelf, ocean ridge, seamount, continental slope, fracture zone, and continental rise.

3. Sketch a generalized, topographic profile of the trek across the western North Atlantic ocean floor that is described in this chapter. Use vertical exaggeration (see Appendix IV).

4. Assume that a continental slope with an average bottom gradient of 4 degrees begins at a 100-meter (~328-foot) water depth and ends at a 200-meter (~656-foot) water depth. Sketch three topographic profiles: one with no vertical exaggeration, one with a vertical exaggeration of 1:10, and one with a vertical exaggeration of 1:100 (see Appendix IV). What happens to the appearance of the slope on each profile?

5. Using the principle of isostasy, describe and account for the crustal structure that underlies a high mountain belt.

6. Describe the principal surveying techniques used for determining bathymetry.

SELECTED READINGS

Allégre, C. J., and S. H. Schneider. 1995. The evolution of the Earth. *Scientific American* 271 (4): 66–75.

Ambroggi, R. P. 1977. Underground reservoirs to control the water cycle. *Scientific American* 235 (5): 21–27.

Atlas of the Oceans. 1977. New York: Rand McNally.

Ballard, R. D. 1975. Dive into the great rift. *National Geographic* 147 (5): 604–15.

Ballard, R. D. 1984. *Exploring Our Living Planet.* Washington, D.C.: National Geographic Society.

Dietz, R. S. 1952. The Pacific floor. *Scientific American* 192 (4): 19–23.

Fryer, P. 1992. Mud volcanoes of the Marianas. *Scientific American* 266 (2): 46–52.

Heezen, B. C. 1956. The origin of submarine canyons. *Scientific American* 195 (2): 36–41.

Heezen, B. C., and C. D. Hollister. 1971. *The Face of the Deep.* New York: Oxford University Press.

Heezen, B. C., M. Tharp, and M. Ewing. 1959. *The Floor of the Oceans,* vol. 1: *The North Atlantic.* Special paper 65. New York: Geological Society of America.

Jeanloz, R., and T. Lay. 1993. The core-mantle boundary. *Scientific American* 268 (5): 48–55.

Lonsdale, P., and C. Small. 1991/92. Ridges and rises: A global view. *Oceanus* 34 (4): 26–35.

Macdonal, K. C., and B. P. Luyendyk. 1981. The crest of the East Pacific Rise. *Scientific American* 244 (5): 100–16.

Menard, H. W. 1969. The deep-ocean floor. *Scientific American* 221 (3): 126–45.

Pinter, N., and M. T. Brandon. 1997. How erosion builds mountains. *Scientific American* 276 (4): 74–79.

Powell, C. S. 1991. Peering inward. *Scientific American* 264 (6): 100–11.

Pratson, L. F., and W. F. Haxby. 1997. Panoramas of the seafloor. *Scientific American* 276 (6): 82–87.

Taylor, S. R., and S. M. McLennan. 1996. The evolution of continental crust. *Scientific American* 274 (1): 76–81.

Watson, L. 1988. *The Water Planet.* New York: Crown.

White, R. S., and D. P. McKenzie. 1989. Volcanism at rifts. *Scientific American* 261 (1): 62–71.

TOOLS FOR LEARNING

Tools for Learning is an on-line review area located at this book's web site OceanLink (**www.jbpub.com/oceanlink**). The review area provides a variety of activities designed to help you study for your class. You will find chapter outlines, review questions, hints for some of the book's math questions (identified by the math icon), web research tips for selected Critical Thinking Essay questions, key term reviews, and figure labeling exercises.

3

The Origin of Ocean Basins

There rolls the deep where grew the tree
O earth, what changes hast thou seen!
There where the long street roars hath been
The stillness of the central sea.
The hills are shadows, and they flow
From form to form, and nothing stands;
They melt like mist, the solid lands,
Like clouds they shape themselves and go.
—*Alfred Lord Tennyson,*
In Memoriam, *1850*

web navigator

critical thinking on the web

math tutor on the web

www.jbpub.com/oceanlink

PREVIEW

ONCE YOU HAVE COMPLETED this chapter, you will understand Tennyson's meaning: that the Earth's surface—its rocks and topography—is not dormant, but is actively changing and evolving. The Earth is a pulsating planetary body, engaging in grand geologic cycles as its immense ocean basins expand and contract, and its towering mountains are raised upward and beveled downward. At first, it's quite difficult to understand that mountains and ocean basins grow; they seem so solid and inert over our lifetimes. But change they do, as we will discover. According to current theories, the oceans and continents are continually created and recreated. "All that is solid melts into air" when viewed across dumfounding intervals of geologic time. This colossal geologic drama is the central topic of this chapter.

The chapter begins with an examination of the drift patterns of continents (continental drift) and ocean basins (seafloor spreading). Then, we'll inspect the details of a remarkable scientific idea—global plate tectonics—that accounts for the geologic and geophysical properties of the ocean floor and the land's mountain ranges.

One of the main concerns of marine geologists is tracing the development of the ocean basins since their formation on the Earth. Geologists use the term **ocean basin** to refer to the large portion of the oceans' floor that lies deeper than 2,000 meters (~6,600 feet) below sea level. In other words, ocean basins are huge topographic depressions, literally gigantic holes in the Earth's surface. Despite the fact that these basins are composed of rigid rock, their shape and size change slowly but surely with time. How is this possible? Aren't rocks as solid and unchangeable a natural material as exist anywhere? Indeed they are—from year to year, and even from generation to generation. But when viewed across enormous time spans, over millions of years, they no longer appear rigid, but flow ever so slowly (somewhat like a syrupy fluid) in reaction to the high temperatures and pressures that exist in the Earth's interior. Even the continents are not anchored lastingly to one spot on the Earth's surface as we portray them on maps, but are wandering across the globe with the passage of geologic time. As you read this, the building where you are is drifting slowly, but relentlessly, with the continent on which it is located. Our maps represent merely still photographs of continents in motion over geologic time; the geography we see as so permanent is not.

3-1

Continental Drift

A look at any world map shows the striking jigsaw-puzzle fit of the coastlines on either side of the Atlantic Ocean. Prominent bulges of land are matched across the sea by equally imposing embayments with similar geometries. This parallelism is most noticeable in the opposing edges of Africa and South America (**Figure 3–1**). In fact, careful matching of the edges of all the continents shows that they can be reassembled into a single large landmass of immense size. This implies that the continents have moved vast distances relative to one another, making the present geography of the Earth quite different from what it was in the geologic past.

In 1915 Alfred Wegener (1880–1930), a German meteorologist, published a book in which he proposed a bold new hypothesis for understanding the Earth's history—**continental drift**. According to Wegener, geologic and **paleontological** (fossil) information indicated that the present continents had been parts of a much larger landmass more than 200 million years ago. He surmised that some 100 million to 150 million years before the present, this large landmass, which he called **Pangaea** (Figure 3–1), splintered apart, and the fragments (the present-day continents) slowly drifted away from one another, opening new ocean basins between them.

Relying on detailed geographic and geologic reconstructions and abundant fossil and paleoclimatological (ancient climates) evidence, Wegener proposed that Pangaea broke apart along a global system of fractures, or geologic faults, that shattered the granitic crust of the land. The granite of the resulting continental fragments, Wegener surmised, "plowed" through the basalt crust of the oceans. Concurrently, fresh basalt was injected into the widening gap between the Americas and Africa-Europe, creating a juvenile Atlantic Ocean. During this drifting episode, the leading, western edges of North and South America were buckled into the Rocky and Andes mountains when the dense basalt crust of the Pacific Ocean resisted the drift. Just what caused the continents to drift was not apparent to Wegener.

The notion that these huge continental masses of granite were adrift struck many scientists of the time as being a bit far-fetched, more like science fiction than natural science. They particularly objected to Wegener's assertion that the light granitic crust of the continents could "plow" its way through the denser, and therefore stronger, basalt crust of the oceans. Besides, the geophysicists showed that the driving mechanism for continental drift proposed by Wegener was not possible according to their calculations. Though some reactions among geologists to such a new, bold concept bordered more on fascination than outrage, geologic orthodoxy prevailed, and Wegener's theory of continental drift was ignored for more than half a century. Later work by geological oceanographers confirmed the mobility of the continents, elevating the continental-drift idea from the realm of the impossible to a status as certain as any theory in science can hold.

(a) FITTING THE CONTINENTS TOGETHER

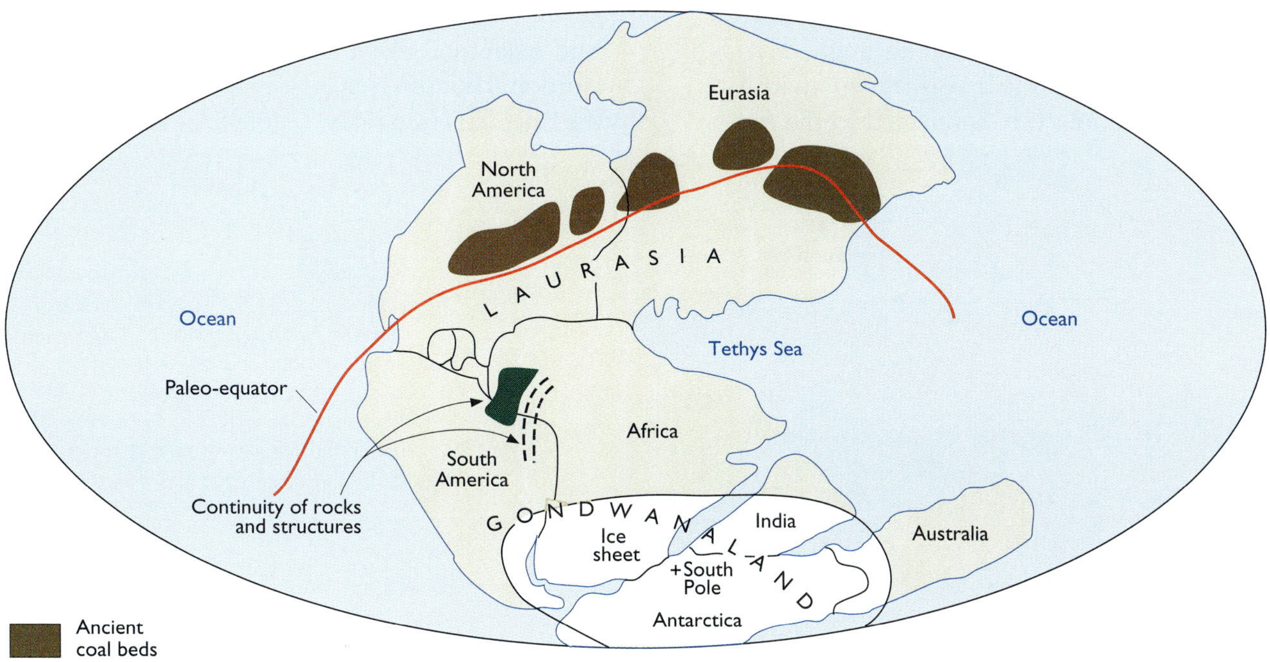

(b) PANGAEA, 200 TO 300 MYBP (MILLIONS OF YEARS BEFORE THE PRESENT)

FIGURE **3-1**

Pangaea. (a) This map, published by Antonio Snider in 1858, shows the rearrangement of the continents into a large landmass. (b) If the continents are arranged into the megacontinent Pangaea (Laurasia to the north and Gondwanaland to the south), coal beds and glacial deposits that are 200 to 300 million years old fall into latitudinal belts instead of being scattered about everywhere.

The key to unraveling this mystery turned out to be the way that ocean basins are created, a process called **sea-floor spreading**. Let's review the scientific evidence that forced geologists and geophysicists to reconsider their ideas about the mobility of continents and the ocean floor.

3-2

Sea-Floor Spreading

Systematic soundings of the deep sea led to a dramatic discovery—the presence of a submerged, continuous mountain range that rises as much as 3 to 4 kilometers (~1.9 to 2.5 miles) above the surrounding ocean floor and girdles the entire Earth. Its total length of over 60,000 kilometers (~36,365 miles) dwarfs the more familiar mountain belts on land; yet, humans were unaware of this majestic, world-encircling mountain range until the mid-twentieth century! What to make of it? How and when was it formed? Was this submarine mountain belt in some way connected to the formation of mountains on land, of which a great deal was known? Or did it originate in an entirely new and independent way? Much needed to be learned.

It was apparent that the midocean ridge in the Atlantic Ocean, aptly called the Mid-Atlantic Ridge because of its position along the center line of the Atlantic basin, mimics clearly the shape of the continental edges of the bordering continents (see Figure 2–2). Detailed geophysical surveys of this midocean ridge revealed a number of remarkable features. For example, the enormous flanks of the Mid-Atlantic Ridge rise to a sharp crest. A prominent valley that is 50 kilometers (31 miles) wide and 1 kilometer (0.6 miles) deep runs along the crest line of the ridge. The floor of this valley is composed of freshly crystallized young basalt and bounded by prominent **normal faults**—topographic **scarps** (steep rock faces) where crustal rocks have broken and dropped past one another creating the valley (**Figure 3–2a**). The amount of heat escaping from the Earth's interior at the ridge's crest is as much as ten times greater than it is for most of the Earth. Also, the Mid-Atlantic Ridge is cut by long, linear transform faults and fracture zones (see Figure 2–2) that divide the ridge axis into many segments, each offset from

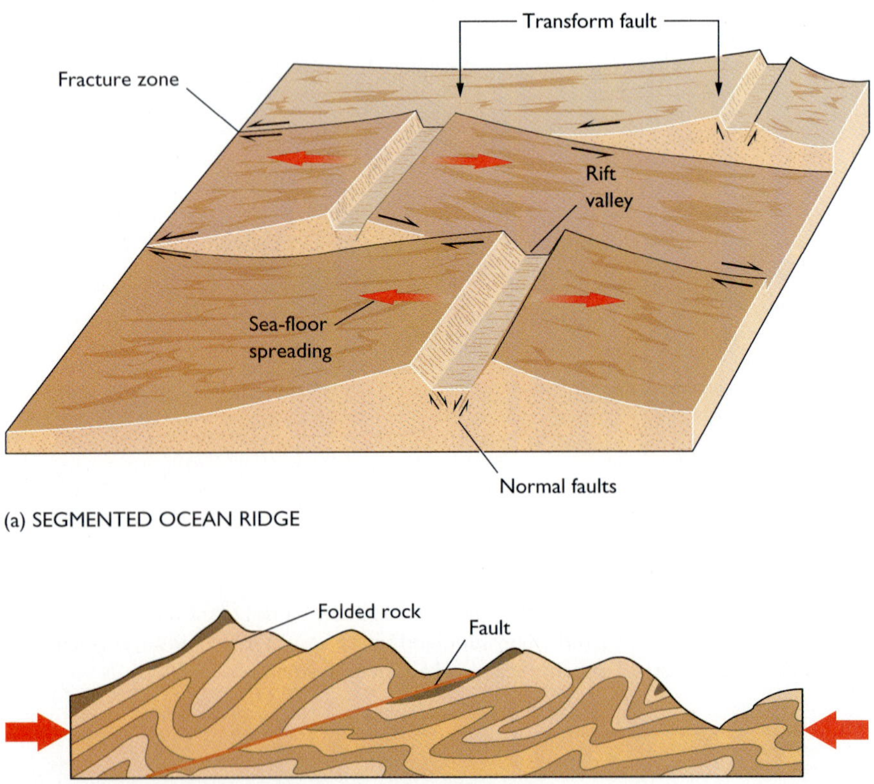

FIGURE 3-2

Crustal motions. (a) The rift valley at the crest of midocean ridges is formed by tension. Faulting at the ridge axis consists of normal faults of the rift valley and transform faults that offset the ridge crest. (b) Compression has crushed flat-lying sedimentary rocks, creating faulted and folded mountains on land.

the other. Furthermore, earthquakes are common and originate at shallow depths of less than 35 kilometers (~22 miles) along both the ridge crests and the transform faults. Clearly, the Mid-Atlantic Ridge, as well as other midocean ridges, is geologically active, being the site of widespread volcanism, faulting, and earthquakes.

On land, large mountain belts such as the Appalachians, the Rockies, the Alps, and the Himalayas are the result of tremendous pressures that squeeze rocks together, causing them to be folded and faulted. Folds result from **compressional** forces (**Figure 3–2b**). This is analogous to laying the palms of your hands flat on a tablecloth and bringing them together. This action results in the cloth being folded by compression. The rocks of the midocean ridges are not folded. Rather, all the geologic evidence indicates that these mountains have been stretched and pulled apart, creating the normal faults at the shoulders of the crestal rift valley and allowing molten (melted) basalt to rise along the cracks and spill out on the surface as lava. This "pull-apart" force, called **tension**, is equivalent to grabbing a tablecloth with both hands and pulling it apart, ripping the fabric. Unlike the normal faults, where rocks have moved vertically, a careful examination of the transform faults indicates that the sense of relative motion of the broken rocks is lateral, or as geologists say, **strike-slip**.

There are two distinct types of fault systems associated with midocean ridges (see Figure 3–2a), and the two should not be confused. The normal faults that occur along the edges of the rift valleys are zones where the crustal rocks are displaced vertically in a relative sense. The transform faults offset the ridge axis and represent fractures in the crust where the movement of rocks is essentially horizontal.

The oddest and most problematic feature of the midocean ridges pertains to their magnetic properties. To investigate these properties, **magnetometers**, instruments that detect and measure the intensity of magnetism, were towed by ships back and forth across the crests of the midocean ridges. Such a survey was conducted across the axis of the Reykjanes Ridge , part of the Mid-Atlantic Ridge just southwest of Iceland (**Figure 3-3**). Each magnetic profile across this ridge showed alternating high magnetic readings (positive peaks) and low magnetic readings (negative peaks). These high and low variations are termed **magnetic anomalies**, because each is higher or lower than the predicted value for the Earth's magnetic field at that locality and, hence, are considered to be anomalous (not normal). This means simply that positive magnetic anomalies are stronger than expected, negative readings weaker than expected.

Even more baffling to geologists investigating this issue was the fact that the magnetic highs and magnetic lows could be traced from profile to profile, so that a pattern of magnetic "stripes" of alternating positive and negative intensity became evident. This is the case for the Reykjanes Ridge (see Figure 3–3). Here distinct magnetic anomaly stripes are evident, running parallel to the ridge line. More surprisingly, these magnetic bands are

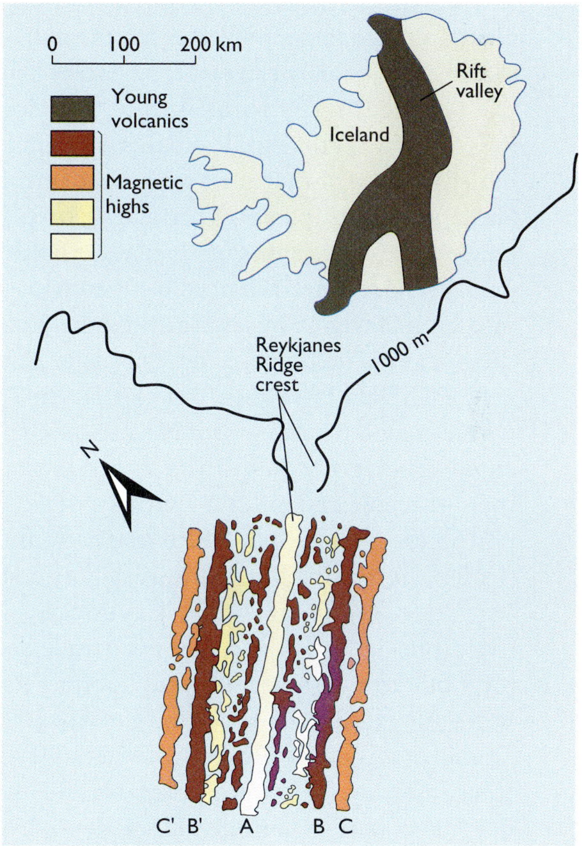

FIGURE 3-3

Magnetic anomaly stripes. Magnetic anomaly stripes run parallel and are symmetrically arranged on both sides of the midocean ridge axis. For example, anomalies B and C on the eastern ridge flank have counterparts B' and C' on the western flank that are the same distance from the crestal anomaly A. [Adapted from J. R. Heirtzler, X. LePichon, and J. C. Bason, *Deep Sea Research* 13 (1966): 427–33.]

symmetrically distributed about the ridge axis. That is, each magnetic anomaly on one flank of the ridge has a counterpart on the opposite flank at the same distance from the ridge crest. In Figure 3–3 note that the central anomaly A, located over the ridge crest, is flanked by anomalies B and B', C and C', each pair being equidistant from the ridge line. This regular, symmetric pattern of magnetic anomaly stripes has been found along most of the midocean-ridge system, and is seen as one of its fundamental properties. What created such regularity, and what could possibly be its significance? Answers to these questions awaited the study of the magnetic properties of rocks on land.

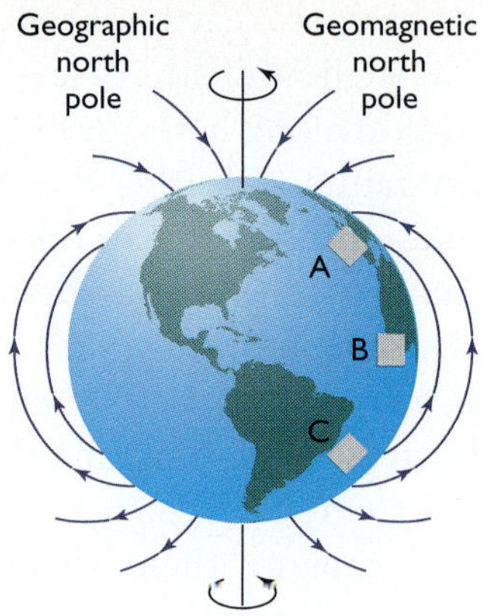

THE GEOMAGNETIC FIELD

The Earth's magnetic (geomagnetic) field can be compared to a large, powerful bar magnet, a dipole (having two poles) placed at the center of the Earth and oriented nearly parallel to its rotational axis (**Figure 3–4a**). This dipole creates a magnetic field, invisible lines of magnetic force that surround the Earth and are capable of attracting or deflecting magnetized material. This magnetic field is what "pulls" the magnetized needle of a compass, allowing you to navigate by determining the direction of the north magnetic pole. The intensity of the geomagnetic field is strongest at the magnetic poles, where magnetic lines of force are vertically oriented. They point out of the Earth's surface near the south pole and into the Earth's surface near the north pole (**Figure 3–4b**). The magnetic field lines (which, remember, are invisible) intersect the Earth's surface at angles that vary directly with latitude (see Figure 3–4b). At 45° N latitude, for example, your compass needle would point toward the north magnetic pole and would be tilted into the ground at an angle of about 45 degrees. At 45° S latitude, the needle would point toward magnetic north and be tilted into the air at about 45 degrees (see top lava layer in column C of Figure 3–4a). Near the equator, the magnetic lines of force lie parallel to the Earth's surface; so the needle has no tilt and merely points to the north magnetic pole.

When lava extrudes and cools, minerals crystallize systematically out of solution as the lava solidifies into a rock, in the same way that solid ice

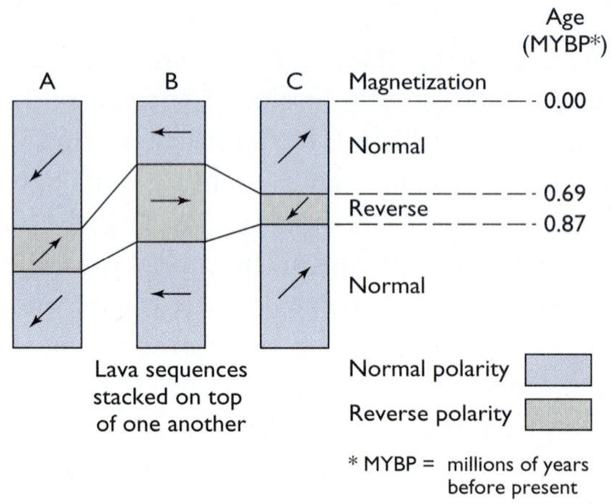

(a) MAGNETIZATION OF LAVAS

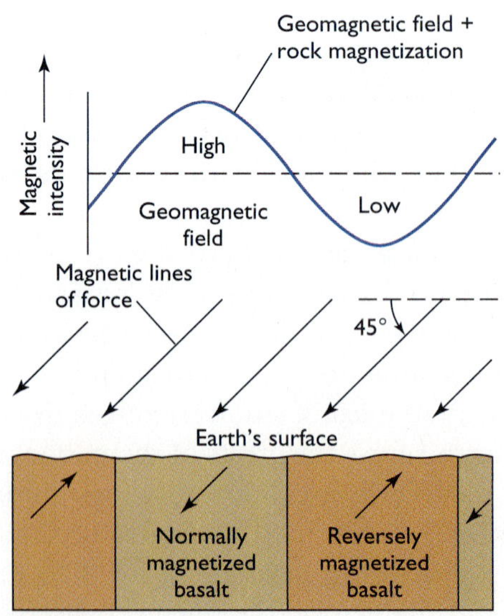

(d) MAGNETIC ANOMALIES

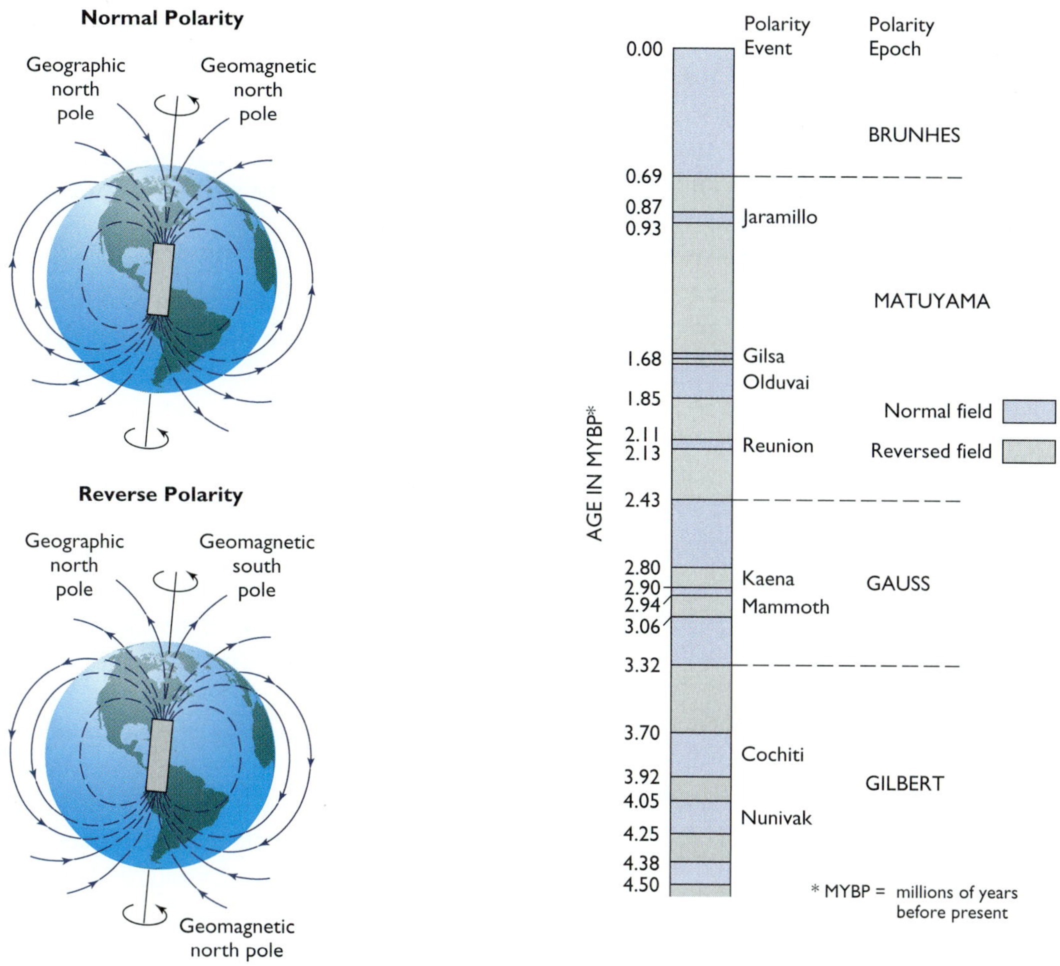

(b) GEOMAGNETIC POLARITY REVERSALS

(c) GEOMAGNETIC POLARITY TIME SCALE

FIGURE 3-4

Paleomagnetism. (a) The lava sequences in areas A, B, and C have alternating directions of rock magnetization. The lavas at each site are stacked on top of one another, with the oldest lava on the bottom and the youngest on top of the sequence. Note that the volcanic rocks of the same age in all areas are either all normally or all reversely magnetized. This effect results from the geomagnetic dipole flipping its polarity back and forth over geologic time. (b) During normal polarity, the geomagnetic north pole lies near the geographic north pole. During a reversal, the geomagnetic north pole flips with the geomagnetic south pole so that it lies adjacent to the geographic south pole. (c) Based on numerous analyses of lava flows on land, the pattern of polarity reversals of the Earth's magnetic field has been accurately established. [Adapted from A. Cox, *Science* 163 (1969): 237–45.] (d) A magnetometer measures simultaneously both the Earth's magnetic field and the fossil magnetization in the rocks. A magnetic high results from the rock magnetization reinforcing the Earth's magnetic field, so that the reading is higher than expected. A magnetic low results from rock magnetization opposing the earth's magnetic field, so that the reading is lower than expected.

crystals form in liquid water that is cooled below its freezing temperature. A few of these early forming minerals are magnetic and tend to align themselves with the geomagnetic field in a way that is similar to the pull on your compass needle. As other minerals crystallize, they lock in and trap, so to speak, the alignment of these magnetic grains. This means that the rocks record the strength and direction of the geomagnetic field at the time they crystallized into a solid. This "fossil" magnetization, called **paleomagnetization**, is quite stable over geologic time, particularly in basalts, and provides a clear magnetic record unless the rock has been reheated or deformed.

As paleomagnetic measurements on sequences of basalt flows were collected from a particular area, something quite odd appeared in the data. Many of the samples from the same area, but of different ages, seemed to be magnetized in the "wrong" way; that is, they showed a reverse magnetization, a magnetic orientation exactly opposite to that of the present-day field. For example, in an area located at 50°N latitude, the paleomagnetic orientation of the basalts either pointed into the ground at an angle of 50 degrees (normal magnetization, i.e., in proper orientation with the present-day magnetic lines of force) or out of the ground at an angle of 50 degrees (reverse magnetization, i.e., in direct opposition to the present-day magnetic force field). What could this possibly mean? Furthermore, paleomagnetic studies in many volcanic terrains showed that basalts of a particular age are either all normally or all reversely magnetized, regardless of where they erupted on the Earth's surface (see Figure 3–4a). All these data taken together could mean only one thing—the geomagnetic field had flipped repeatedly back and forth over time (see Figure 3–4b), sometimes oriented as it is at present (**normal polarity**) and sometimes oriented in the opposite direction (**reverse polarity**). This information led to the development of a paleomagnetic polarity time scale that simply depicts the times when the geomagnetic field had a normal or reverse polarity (**Figure 3–4c**). During any reverse polarity, the north-seeking compass needle would point to the south rather than to the north magnetic pole. As we shall see, this knowledge of magnetic field reversals finally helped geologists understand the significance of magnetic anomaly stripes on the sea floor.

SPREADING OCEAN RIDGES

Harry Hess, a geophysicist at Princeton University, proposed in 1960 that the ocean floor was broken into large blocks that were moving relative to one another. Few geologists accepted this radical hypothesis. By the mid-1960s, however, geologists and geophysicists studying all the data pertaining to the midocean ridges proposed a bold new hypothesis: new ocean floor and crust are created continuously by the intrusion and extrusion of basalt at the crest of all midocean ridges. According to the British geophysicists Fred Vine and Drummond Matthews, as the basalt extrusions cool and solidify, they become magnetized and record the Earth's magnetic-field orientation. Gradually, the newly formed crust moves laterally down the flanks of the midocean ridges, making space at the crest for the formation of more basalt crust. This process, called *sea-floor spreading,* leads to ocean basins that widen with time. Vine and Matthews hypothesized that, as the sea floor spreads, the basalts become normally or reversely magnetized depending on the orientation of the geomagnetic field at their time of cooling (see Figure 3–4b). All basalts that are normally magnetized have a fossil magnetization that is aligned with the Earth's present-day magnetic field. This magnetic parallelism causes a reinforcement, so that a magnetometer measures a higher than normal reading, resulting in a strong or positive magnetic anomaly (**Figure 3–4d**). In contrast, all rocks that formed during a period of reverse polarity have a fossil magnetic orientation opposite to that of the present-day geomagnetic field. This contrary alignment reduces the magnetic reading over reversely magnetized basalt, producing a low or negative magnetic anomaly. Because sea-floor spreading is symmetrical, the basalt crust is split at the crest into halves, creating pairs of magnetic anomaly stripes of about equal width that are symmetrically disposed about the ridge axis (**Figure 3-5a**). In other words, the magnetic profiles on either flank of the midocean ridges are mirror images of each other. In effect, the pattern of magnetic anomaly stripes is equivalent to a tape recording of the Earth's magnetic-field reversals of the past.

In the vicinity of the Mid-Atlantic Ridge, sea-floor spreading is causing symmetrical expansion of the Atlantic Ocean basin; the ocean is growing in size as fresh basalt is extruded at its center. In the process, continents on either side of the basin are traveling along with the moving ocean floor as part of the plate—which explains continental drift as Wegener described it, but does away with the idea of light granitic crust plowing through denser oceanic crust. Rather, the granite of the continents is simply being carried along by the expanding sea floor. Moreover, if we reverse the process of sea-floor spreading and go back into the distant geologic past, the Atlantic basin shrinks in size until it disappears, and all the continents rejoin one another into a much larger continental landmass, Pangaea. When in the distant past did this colossal megacontinent break apart? At what rate does the sea floor spread?

Actually, it turns out to be easy to measure rates of spreading using magnetic profiles measured perpendicular to the crests of midocean ridges (**Figure 3–5b**). The timing of geomagnetic polarity reversals is easily established by dating basalt flows. This, then, makes it possible to infer the age of the magnetic anomaly stripes of the sea floor, because each of these formed during unique magnetic reversals. For example, if a particular anomaly from a strip of ocean floor formed during a reversal that occurred 10 million years ago, then that piece of sea floor, now located 100 kilometers (~62 miles) from the ridge crest, must have moved 100 kilometers flankward since it was extruded and solidified at the ridge axis. Dividing that distance of transport (100 kilometers) by the age of the sea floor (10 million years) yields an average spreading rate of 10 kilometers per million years or 1 centimeter per year for each flank of the ridge. Many such determinations indicate that spreading rates for the midocean ridges range between 1 and 10 centimeters per year (~0.5 and 4 inches per year) for each flank, a speed comparable to the growth rate of a human fingernail! Every 70 years, roughly a human life span, ocean basins widen by 3 to 14 meters (~10 to 45 feet).

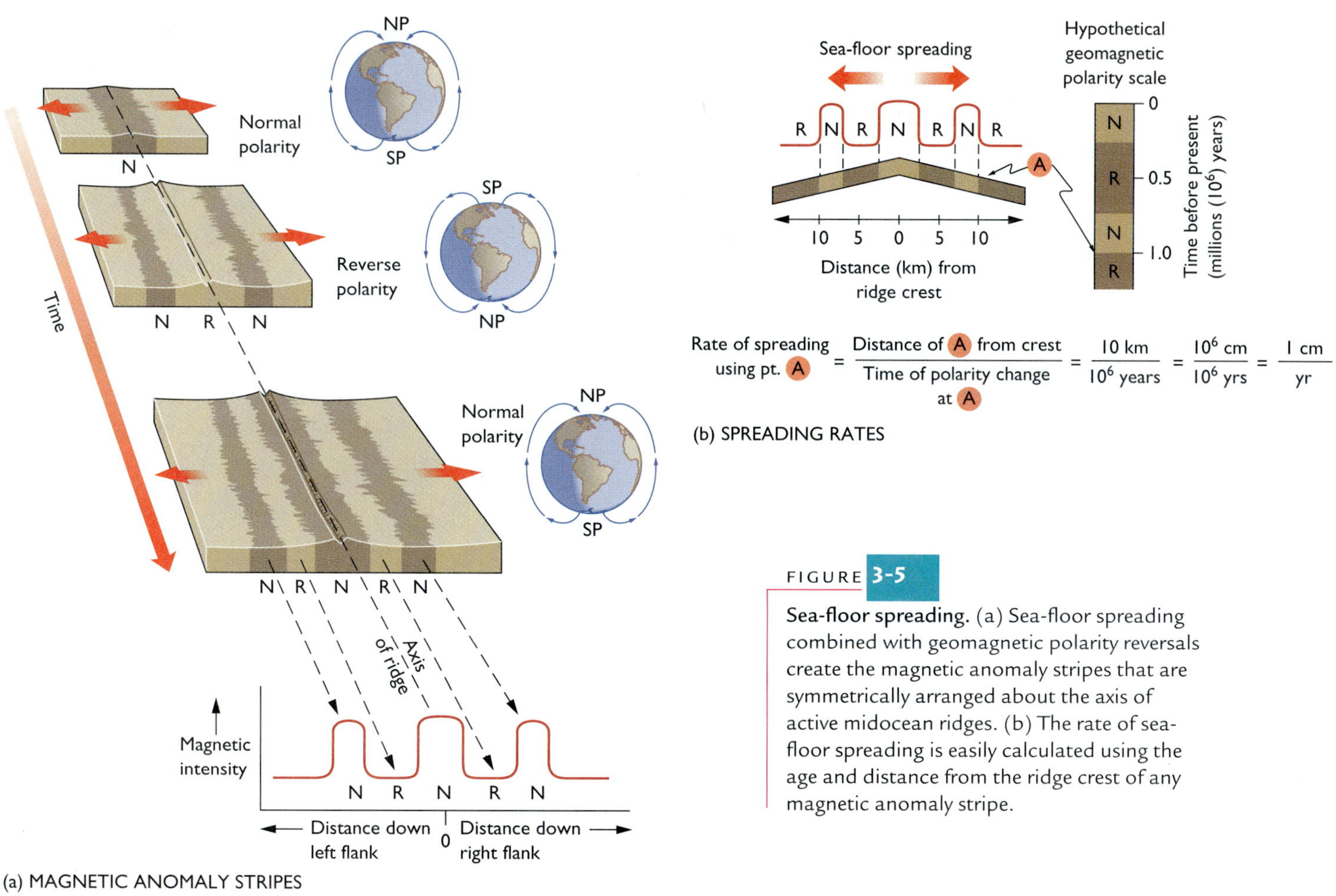

FIGURE 3-5

Sea-floor spreading. (a) Sea-floor spreading combined with geomagnetic polarity reversals create the magnetic anomaly stripes that are symmetrically arranged about the axis of active midocean ridges. (b) The rate of sea-floor spreading is easily calculated using the age and distance from the ridge crest of any magnetic anomaly stripe.

Several important findings are explained by the sea-floor spreading model. For instance, sea-floor spreading means that the oceanic crust that lies to either side of the ridge moves apart. This separation produces tensional forces that create normal faults and rift valleys at ridge crests. Also, spreading of the ocean floor implies that the basaltic crust becomes increasingly older with distance from the ridge line. Eventually, as the basaltic crust is transported down the ridge flank, it becomes covered with sediment. This sedimentary cover gets steadily thicker with distance from the ridge axis.

3-3

Global Plate Tectonics

The axis of midocean ridges is a narrow, linear zone where basaltic crust forms and then moves away from the ridge crest at a rate of several centimeters per year. Since the breakup of Pangaea during Jurassic time (some 150 million years ago), the Atlantic Ocean has grown from a young, narrow basin to its current size. This has been true for the other ocean basins as well. The presence of magnetic anomaly stripes parallel to the crest and flanks of midocean ridges in the Indian, Arctic, and Pacific Oceans indicates that their floors are spreading and that these basins must be widening as well. If all these oceans are expanding, then the Earth's diameter must be increasing proportionately in order to accommodate all this new surface area. Think about an orange. If you wanted to add more peel, you would need to make the orange bigger, that is, increase its diameter. However, geologists have determined that the diameter of the planet has not changed appreciably for hundreds of millions, if not a billion, years. What to do? There is one way around this problem. If the Earth's size is fixed, then the addition of new ocean floor by sea-floor spreading must somehow be offset by the destruction of an equivalent area of ocean floor somewhere else. If that were the case, then sea-floor spreading could occur on the Earth with a fixed diameter. Well, let's assume that this is the case. Where, then, are the areas in which oceanic crust is being destroyed?

SUBDUCTION ZONES

It stands to reason that if crustal rocks are being crushed and destroyed at a rate comparable to their formation at midocean ridges, this activity should generate earthquakes, a shaking of the ground caused by the deformation of rocks. An examination of world **seismicity**—the frequency (number), magnitude (strength), and distribution of earthquakes—reveals two distinct groupings of earthquakes on the Earth (**Figure 3–6a**). One grouping consists of a narrow clustering of shallow, relatively weak disturbances that closely follow the crest line of the spreading midocean ridges and their transform faults. The earthquakes along the ridge axis result from volcanism and normal faulting along the crestal rift valley; those along the transform faults are generated by the strike-slip motion of the crust to either side of the fault.

But, what is the significance of the second grouping of seismic events, which appears as a broad band of strong, shallow-to-deep earthquakes that follow the western edges of North and South America, arc around the western and northwestern Pacific, and extend across the southern Asian mainland through the Himalayas and across the European Alps (see Figure 3–6a)? The frequency and magnitude of the earthquakes in this seismic belt signify intense **tectonism**, a term that denotes deformation (buckling, folding, faulting, crushing) of the crust. Let's examine this second band of seismicity more closely, and see whether or not it represents zones where significant amounts of crustal rocks are being destroyed.

Around the Pacific Ocean, earthquakes are associated with deep-sea trenches and volcanic arcs. The distinctive volcanic landmasses have been built up by the abundant extrusion of **andesite**, a lava with a chemical composition intermediate between granite and basalt (see Table 2–2). Furthermore, sedimentary and volcanic deposits, which lie between the active volcanoes and the deep-sea trench, are highly deformed, buckled, and fractured, implying that the oceanic crust here is being shortened by powerful compressional forces, crudely analogous to the effect of pushing together the two sides of a tablecloth and creating folds in the fabric.

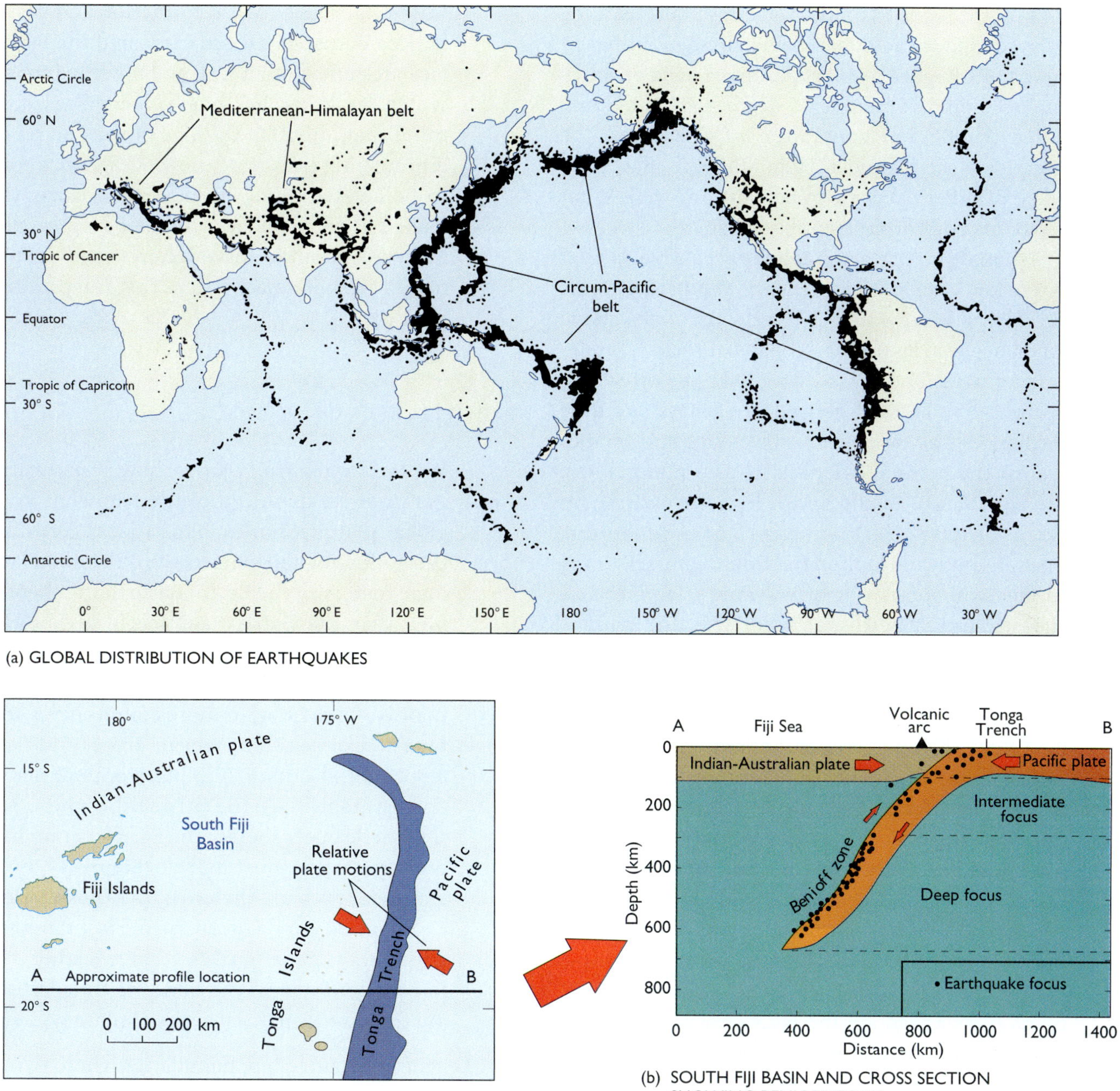

(a) GLOBAL DISTRIBUTION OF EARTHQUAKES

(b) SOUTH FIJI BASIN AND CROSS SECTION SHOWING BENIOFF ZONE

The distribution of earthquakes provides another crucial piece of evidence to indicate that the Earth's crust is being compressed and shortened at these tectonic zones. The accurate recording of seismicity at volcanic arc-trench systems discloses that earthquakes are not randomly distributed here, but are arranged in quite an orderly pattern. Although shallow earthquakes are scattered throughout the belt, intermediate and deep earthquakes originate within an inclined zone that is tilted away from the deep-sea trench toward the volcanic arc and that extends downward to depths as great as 700 kilometers

FIGURE 3-6

Global seismicity (1961–1967). (a) Most earthquakes (represented by black dots) clearly coincide with mid-ocean ridges, transform faults, and deep-sea trenches. [Adapted from B. Isacks, J. Oliver, and L. R. Sykes, *Journal of Geophysical Research* 73 (1968): 5855–900.] (b) Sketch map of the south Fiji Basin and cross section showing a plot of earthquakes that clearly defines a Benioff zone. This indicates that the Pacific plate is being subducted beneath the south Fiji Basin. [Adapted from P. J. Wyllie, *The Way the Earth Works* (New York: John Wiley, 1976).]

(~434 miles). For example, seismicity plotted on a cross section of the Tonga Trench in the southwestern Pacific (**Figure 3–6b**) appears as a band of earthquakes dipping at about 45 degrees. This feature, called the **Benioff zone** after its discoverer, the American seismologist Hugo Benioff, has been found at other deep-sea trench sites as well. What could this odd pattern possibly mean?

An analysis of earthquake waves reveals that the rocks immediately beneath the Benioff zone are sliding downward relative to the rocks above them, suggesting that large slabs of rocks are converging, with one mass riding over the other (see Figure 3–6b). The slab going down generates strong earthquakes as its upper surface slips and scrapes against the rocks above it. Also, as it plunges into the hot interior of the Earth, it melts partially at depths of 100 to 200 kilometers (~62 to 124 miles). The hot, buoyant molten fraction then rises to the surface and spews out of volcanic island arcs as andesite lava. These sites where basaltic crust is being destroyed are called **subduction zones**.

This is exactly what we hoped to find. The volcanic arc-trench systems are consuming the ocean floor that is being created at the spreading midocean ridges. Interestingly, few subduction zones are evident in the Atlantic, Indian, and Arctic oceans; here the spreading midocean ridges are the principal tectonic features. Only along the edges of the Pacific Ocean do we find nearly continuous subduction zones, where rates of sea-floor consumption are estimated to be about 15 to 45 centimeters per year (~6 to 18 inches per year). These rates of subduction are much higher than are the rates of ocean-floor production at midocean ridges, because there are more spreading sites than subduction zones on the Earth's surface. Calculations indicate that the rates of production and destruction of ocean floor for the entire planet are about equal, indicating that the Earth's diameter and surface area have been constant over geologic time.

One additional conclusion should be obvious to you as well. The absence of subduction zones in the Atlantic, Indian, and Arctic oceans indicates that these basins are expanding with time, occupying more and more of the Earth's surface as they grow progressively larger. This, in turn, implies that the Pacific Ocean is shrinking rapidly in a geologic sense at a pace equal to the combined production rates of the entire midocean ridge system. Otherwise a balance between formation and destruction of ocean crust could not exist, and the Earth's surface area and diameter would have had to increase, which we know has not happened. Also, bear in mind that although the size of the Pacific basin is diminishing over time, new ocean floor continues to be created along its midocean ridge, the East Pacific Rise, as clearly indicated by earthquakes and fresh basalt at the ridge crest and the presence of magnetic anomaly stripes that run parallel to the ridge axis.

PLATE-TECTONIC MODEL

We can combine all that we have learned about the sea floor so far into a unified concept known as **global plate tectonics**. This theory formulated in the 1960s revolutionized thinking about the geologic history of the Earth. Basic to the theory is the idea that the Earth's surface is divisible into a series of **plates** with edges that are defined by seismicity (**Figure 3–7a**). These plates may consist mainly of sea floor, or more commonly some combination of sea floor (ocean crust) and landmass (continental crust). The plate boundaries extend downward through the entire lithosphere, which is the brittle outer shell of the Earth that includes the crust and uppermost mantle (**Figure 3–7b**). Geologists refer to them, therefore, as **lithospheric plates**.

Seismicity and volcanism are not randomly distributed over the Earth's surface; rather, they are largely confined to the edges of the plates. There are three fundamental types of plate boundaries:

1. *Midocean ridges* are boundaries where two plates under tension move apart from one another. Each plate grows by the process of sea-floor spreading, which adds new lithosphere (crust plus upper mantle) to the trailing edges of the two diverging plates.

2. *Subduction zones* are plate boundaries where compression is dominant, as two plates converge, one overriding and destroying the other. The ocean floor can be thrust downward beneath another ocean plate, common in the western Pacific, or beneath a continent, as along western South America, where the andesite volcanoes, rather than being submarine landforms, appear as volcanic peaks in the high Andes. Subduction

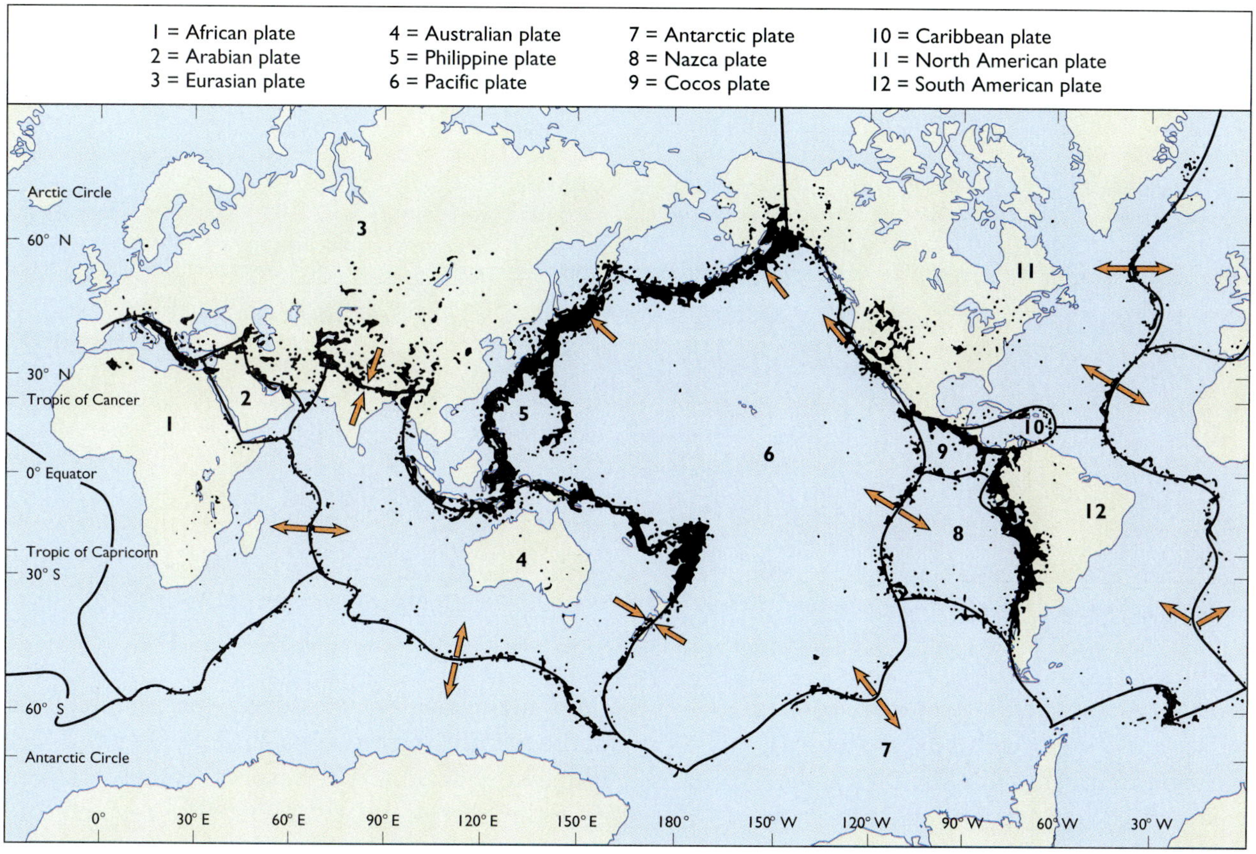

(a) LITHOSPHERIC PLATES

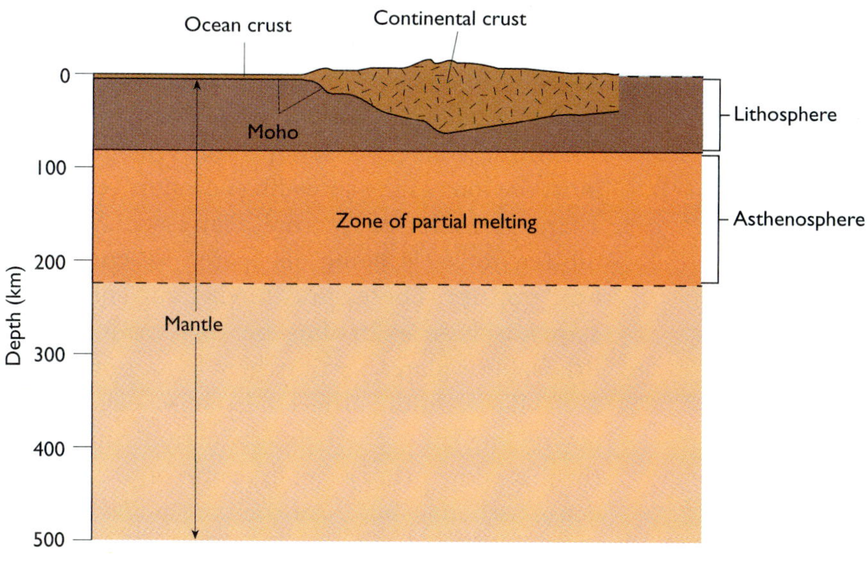

(b) THE LITHOSPHERE AND THE ASTHENOSPHERE

FIGURE 3-7

Lithospheric plates. (a) The edges of large lithospheric plates are indicated by bands of seismicity. The arrows indicate relative plate motions. [Adapted from K. S. Stowe, *Ocean Science* (New York: John Wiley, 1983).] (b) A brittle lithospheric plate, which includes the crust and the upper mantle, overlies and moves relative to the plasticlike asthenoshpere. [Adapted from J. F. Dewey, *Scientific American* 266 (1972): 56–58.]

FIGURE 3-8

Plate mechanics. Convection in the asthenosphere drags lithospheric plates away from the crests of ocean ridges. Also, the leading edge of the plate in subduction zones is cold and dense and may be "pulling" the plate down into the asthenoshpere.

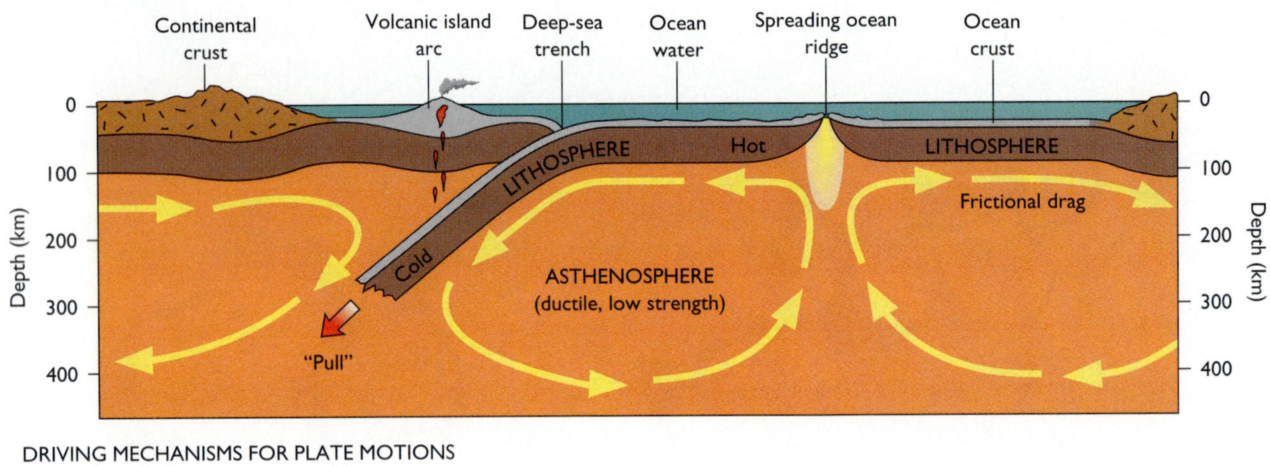

zones are also present where two or more continental masses are actively colliding, as along the Himalayan mountain range of Asia. On a globe with a fixed surface area over geologic time, subduction (plate destruction) at convergent boundaries is balanced by sea-floor spreading (plate growth) at divergent edges.

3. *Transform faults* are plate boundaries where ocean floor is neither created nor destroyed. Rather, the lithosphere along transform faults is conserved as plates slide laterally (strike-slip motion) past one another. Although there are various transform boundaries, the most common type is the one that connects two midocean ridge segments such that they are offset from each other. An example of such a transform boundary slices across southern California (see boxed feature, "The San Andreas Fault").

Thus, the distribution of seismicity indicates clearly that the Earth's outer shell is fragmented into a mosaic of plates (see Figure 3-7a) that diverge (spreading ridges), converge (subduction zones), or slide past one another (transform faults). The plate edges are, however, not merely surface ruptures. They extend downward through the entire lithosphere, which includes the crust and the uppermost mantle. The lithosphere, which is 100 to 150 kilometers (~62 to 93 miles) thick, has appreciable strength and rigidity and it overlies a hotter, partially melted, and hence ductile (plasticlike) layer of the mantle, the asthenosphere (see Figure 3–7b). If drawn to scale, the lithospheric plates are pancake thin, because they are between ten and fifty times wider than they are thick.

Although the actual driving mechanism that causes the plates to move is still being investigated, there is little doubt that **thermal convection** (heat transfer by fluid motion) of the plasticlike rocks of the asthenosphere plays an important role. As heat builds up in the Earth's interior, the rocks in the asthenosphere become less dense than those above and convect (rise) upward. This process is most obvious at the spreading ocean ridges, where a large quantity of internal heat associated with molten rocks is being dissipated by convective heat transfer. These slow-moving currents in the asthenosphere exert drag on the bottom of the lithospheric plates, setting them in motion (**Figure 3–8**). Furthermore, the cold, dense edge of the downgoing lithosphere at subduction zones pulls the plate downward as it sinks into hotter and less dense asthenosphere. An accurate understanding of these driving mechanisms must await additional theoretical and experimental studies.

Most of the volcanic outpourings on the Earth's surface occur at the plate boundaries. Basalt lava spews out of the spreading ocean ridges, and andesite lava is produced at subduction zones. Less common, but impressive, outpourings of lava also occur in the interior of plates, thousands of kilometers away from the plate edges. A case in point is a west-to-northwest-trending linear chain of volcanoes located in the center of the Pacific plate, the Hawaiian Islands (**Figure 3–9a**). These volcanic mountains are created as the Pacific plate drifts slowly over a deep-seated "hot spot" called a **mantle plume** (**Figure 3–9b**). Mantle plumes are places where molten rock originates deep below the asthenosphere. This molten rock rises and melts its way through the lithosphere, spilling out as lava on the top of the plate (**Figure 3–9c**). With time, large quantities of lava are added to the pile, creating a volcanic cone that towers above the ocean floor. Many such cones rise out of the water as islands. Eventually, the motion of the plate, as a result of sea-floor spreading, transports the newly formed island beyond the mantle plume cutting off its supply of lava. At this stage, the island stops growing, and erosion begins to wear down its rocks. Concurrently, a new volcanic island forms updrift over the plume and grows in size until drift takes it, too, beyond its source of lava. The growth of volcanic islands by mantle-plume injection results in the formation of a linear chain, such as the Hawaiian Islands. The islands become older, more eroded, and lower in elevation downdrift. Chains of volcanic islands that trace the path of a plate over a hot spot have been discovered all over the world.

THE OPENING AND CLOSING OF OCEAN BASINS

During the course of several hundred million years, ocean basins have evolved through distinct stages of development that are directly linked to global plate tectonics (**Figure 3–10**). The initial stage in the *Wilson cycle* (named after its originator) of basin evolution begins with splintering of the

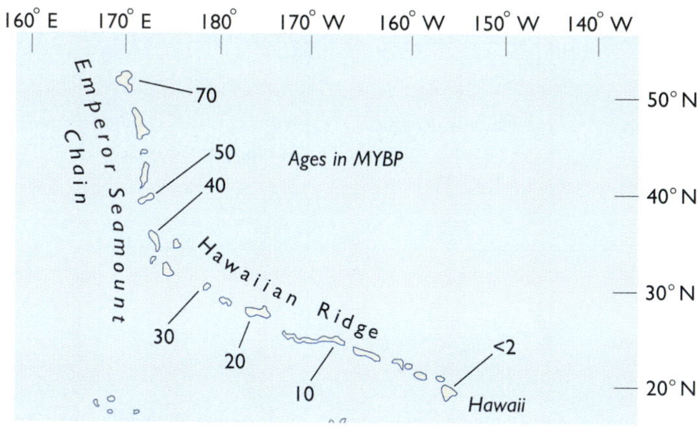

(a) VOLCANIC CHAINS

(c) KILAUEA, HAWAII

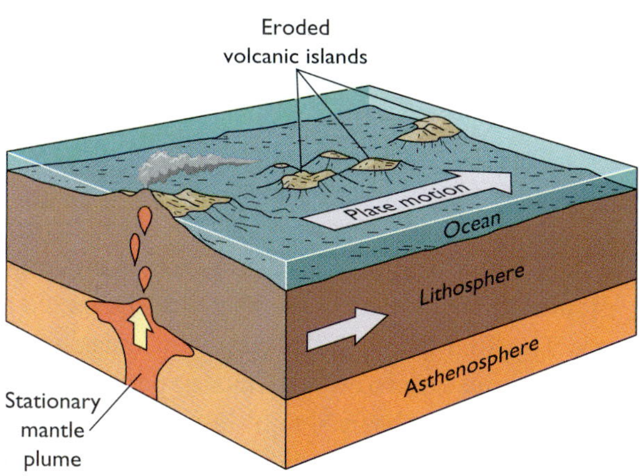

(b) MANTLE PLUME

FIGURE 3-9

Volcanic chains and mantle plumes. (a) Some volcanoes are not associated with the edges of plates. They form long, linear chains with the age of the volcanoes increasing systematically from one end to the other, as illustrated here for the Hawaiian Islands and their submarine continuation, the Emperor Seamount Chain. [Adapted from G. B. Dalrymple, E. A. Silver, and E. D. Jack, *American Scientist* 6 (1973): 294–308; D. A. Claque and R. D. Jarrard, *Geological Society of America Bulletin* 84 (1975): 1135–54.] (b) The linear volcanic chains trace the drift path of a plate over a mantle plume. The volcanoes on each island become extinct and erode once they are transported away from the mantle plume. (c) Kilauea, an active volcano spewing out lava and ash, is located at the southeastern end of the island of Hawaii. The very hot lava is igniting vegetation in its path, but some trees can withstand the flow briefly.

STAGE	MOTION	PHYSIOGRAPHY	EXAMPLE
EMBRYONIC 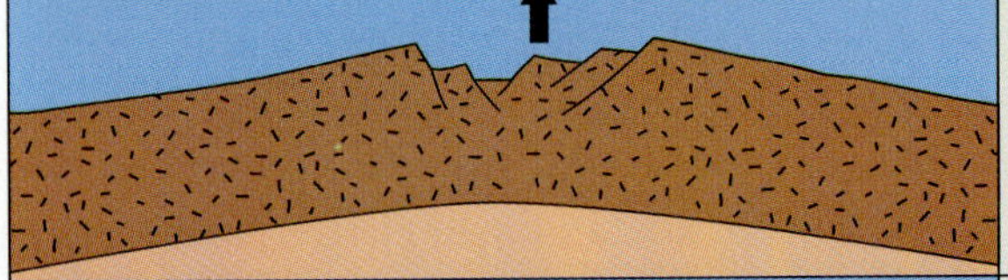	Uplift	Complex system of linear rift valleys on continent	East African rift valleys
JUVENILE 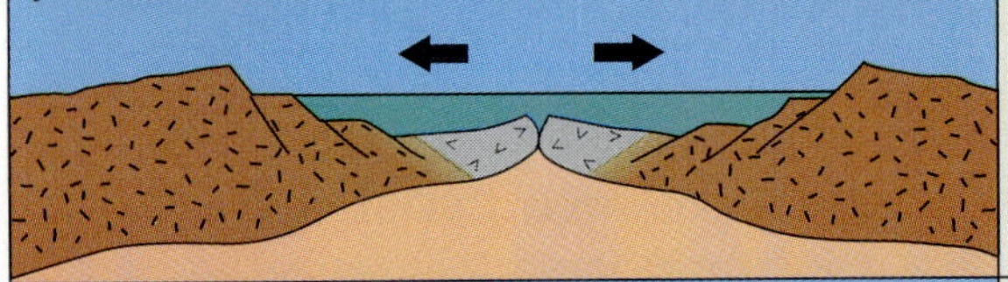	Divergence (spreading)	Narrow seas with matching coasts	Red Sea
MATURE 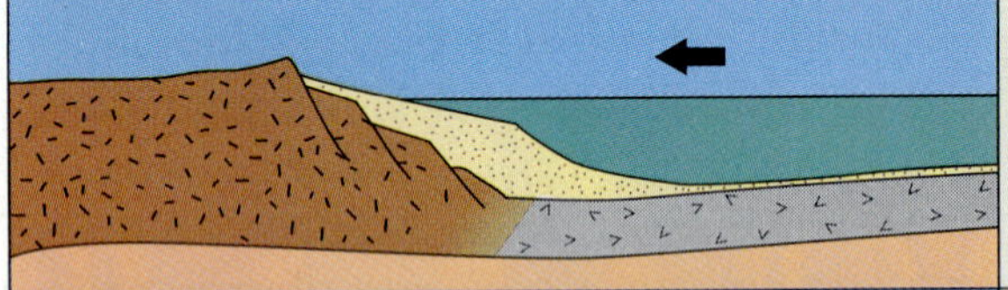	Divergence (spreading)	Ocean basin with continental margins	Atlantic, Indian, and Arctic oceans
DECLINING 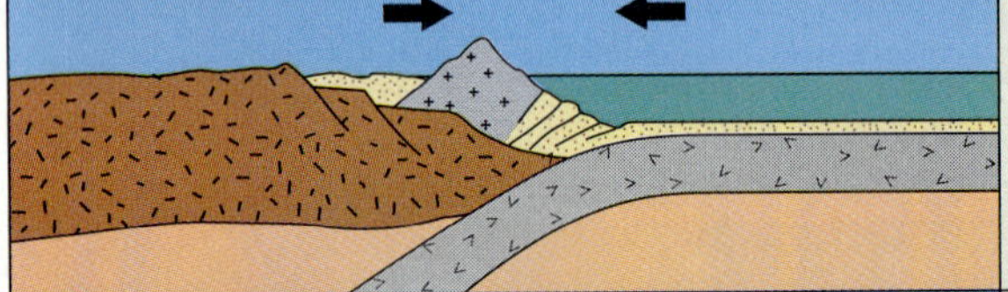	Convergence (subduction)	Island arcs and trenches around basin edge	Pacific Ocean
TERMINAL 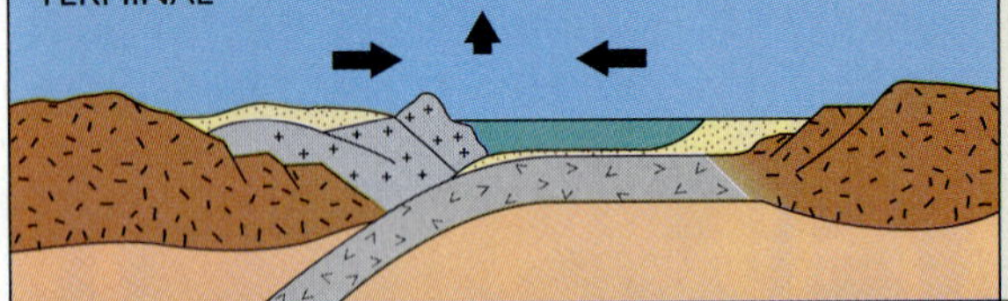	Convergence (collision) and uplift	Narrow, irregular seas with young mountains	Mediterranean Sea
SUTURING 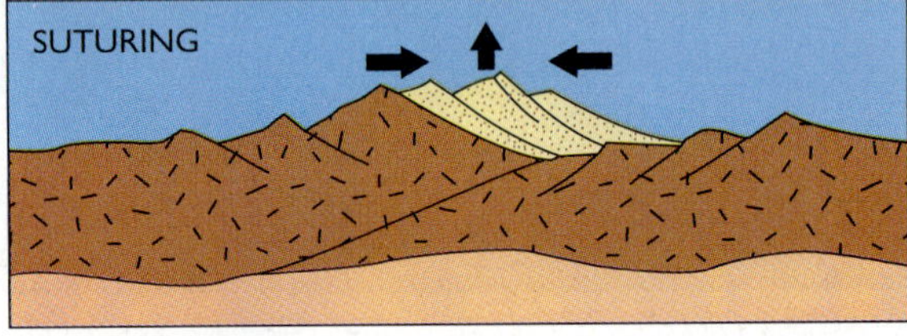	Convergence and uplift	Young to mature mountain belts	Himalayas

FIGURE 3-10

Ocean-basin evolution. The Wilson cycle depicts ocean-basin development as proceeding through a sequence of distinct stages. [Adapted from J. T. Wilson, *American Philosophical Society Proceedings* 112 (1968): 309–20; J. A. Jacobs, R. D. Russell, and J. T. Wilson, *Physics and Geology* (New York: McGraw-Hill, 1974).]

granitic crusts of continents. This results in the formation of long, linear rift valleys, a process that is now occurring in East Africa. The basin at this time is *embryonic*. The continent is fractured by a system of normal faults, and basalt escapes to the surface and spills out onto the floor of the rift valley. The next stage, a *juvenile* ocean basin, occurs once the continents are separated into two independent masses. Basaltic crust forms between them along a young, spreading ocean ridge. This stage is exemplified at present by the Red Sea (see boxed feature, "The Red Sea"). With continued sea-floor spreading, these narrow seaways expand into *mature*, broad ocean basins, such as the present Atlantic Ocean. Continued widening of the basin eventually leads to instability, and the broad plate ruptures

The San Andreas Fault

The edges of most plates are underwater and are therefore studied by indirect methods of observation and sampling such as those described in the feature, "Probing the Sea Floor." An exception is a spectacular exposure of a complex fault system known as the San Andreas fault, which slices through the countryside of western and southern California (**Figure B3–1a**). Aerial views of the landscape that borders the San Andreas fault show a linear topography (**Figure B3–2a**) underlain by fractured crustal rocks that have been forced upward into craggy mountains or downward into splintered valleys. Earthquakes in this region (**Table B3–1**) are frequent and powerful, as rocks on either side of the fault alternately grab and slip past one another, often with dire consequences for people and their structures (**Figure B3–2b**). The reason for this tectonic activity is now well understood, because this 1,300 kilometer-long (~806 miles) fault system represents a plate boundary, a transform fault that separates the Pacific plate from the North America plate. Remarkably, it runs right through southern California.

The San Andreas fault is a long transform fault that joins two spreading ridges, one located in the Gulf of California, the other off the northwestern United States and southwestern Canada (Figure B3–1a). Examining maps of plate boundaries (see Figure 3–7a) one gains the impression that the breaks are clean, the sort that one could perhaps straddle, each foot placed firmly on a different plate. These maps, in a sense, are cartoons, meant to suggest, rather than precisely delineate, the edges of the plates. The San Andreas fault is not a single fracture, but a complex system of intertwined breaks that have splintered the rocks that outcrop on either side of the boundary. In fact, both ends of the transform boundary on land splay out, or bifurcate, into a branching network of faults (**Figure B3–1b**). As along any transform boundary, the dominant motion is strike-slip, meaning that the crust is being sheared horizontally, that is, parallel to the fault trace. Irregularities in the trace of the transform fault can, however, lead to localized compression (pushing together) or tension (pulling apart). This happens at a number of spots along the San Andreas fault, where the geometry of the fault trace leads either to compression and transverse topography, such as the San Gabriel Mountains near Los Angeles, or to tension and "pull-apart" basins (**Figure B3–1c**), such as Death Valley in east central California.

In an effort to gauge the probability of earthquakes in the region, new instruments have been developed that measure seismicity and small changes in the ground's elevation and in its horizontal displacement. The slippage rate, which is not uniform along the fault's length, is estimated to be between 1 and 10 centimeters (~0.5 and 4 inches) per year. This rate and the direction of crustal displacement imply that southern California and Baja are moving to the northwest relative to North America and, eventually (some 15 million years into the future), will become detached and form a large island off California (Figure B3–1a).

Visit www.jbpub.com/oceanlink for more information.

TABLE **B3-1**

Major Earthquakes along the San Andreas Fault

Location	Year	Magnitude
Santa Barbara	1812	7
San Francisco area	1838	7
Fort Tejon	1857	8.3
Hayward	1868	7
San Francisco	1906	8.3
Imperial Valley	1940	7.1
Kern County	1952	7.7
San Fernando Valley	1971	6.5
Santa Cruz Mountains	1989	7.1

Source: Adapted from E. J. Tarbuck and F. K. Lutgens, *The Earth* (New York: Macmillan,1993).

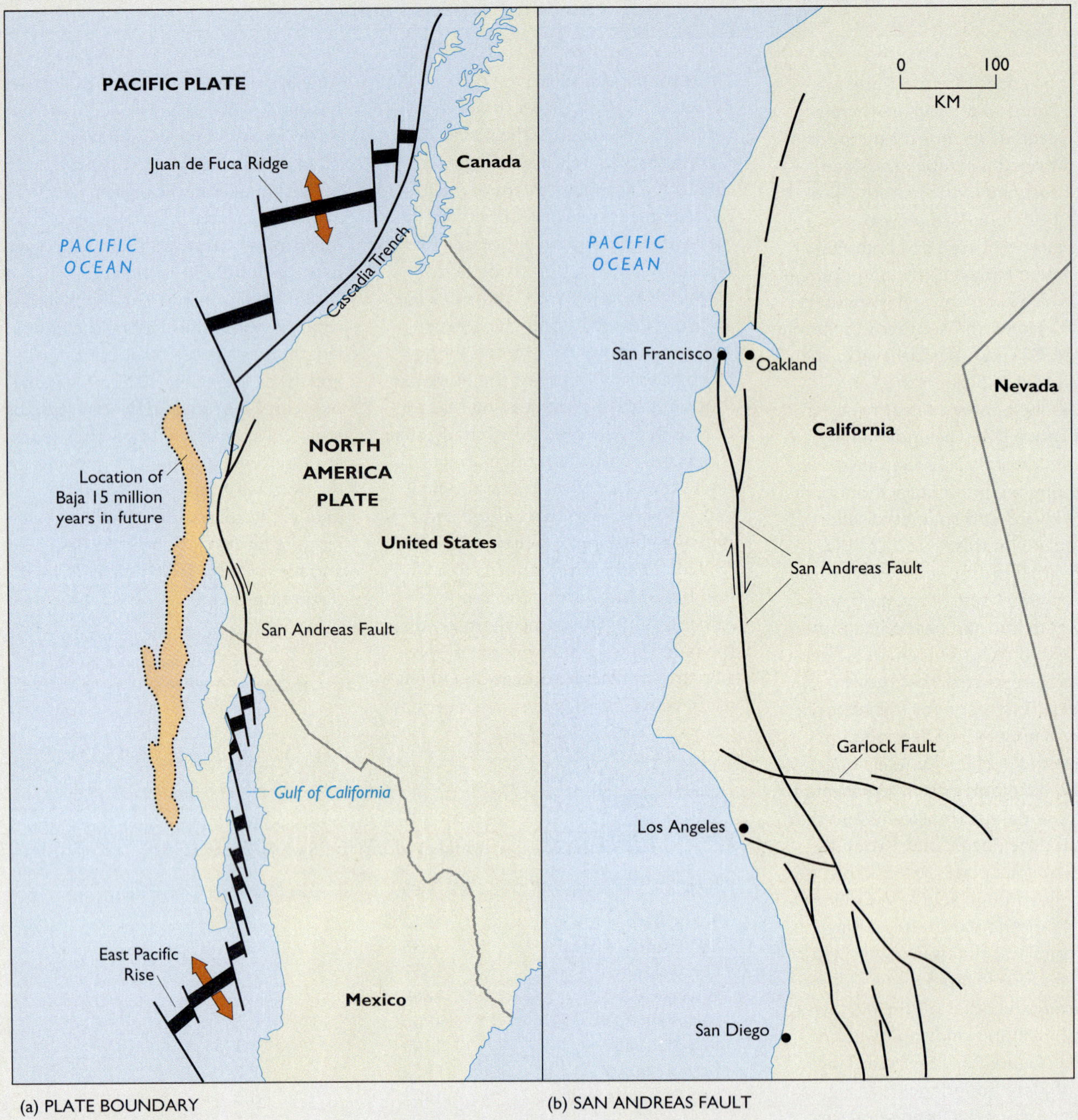

FIGURE B3-1

The San Andreas fault. (a) The Pacific and North America plates are separated from each other by the San Andreas fault. (b) The San Andreas fault is a transform fault that slices through western California. (c) Plate slippage along the irregular trace of the San Andreas fault produces localized zones of compression and tension, which create transverse mountains and "pull-apart" basins respectively.

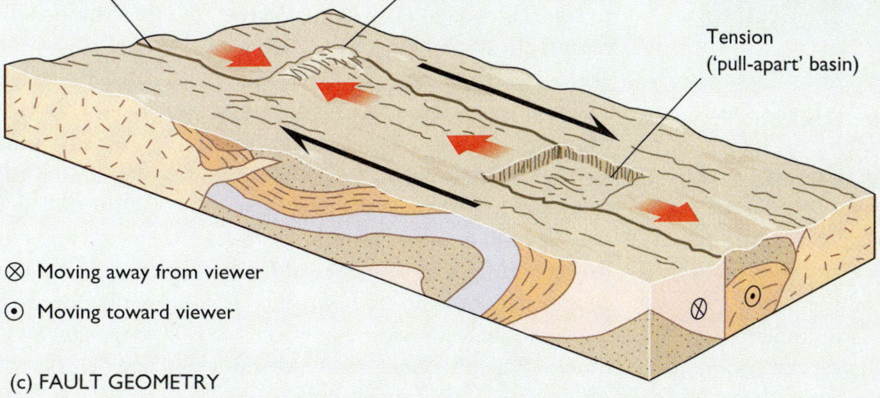

(c) FAULT GEOMETRY

(a) SAN ANDREAS FAULT

(b) DAMAGE IN OAKLAND, CALIFORNIA, FROM 1989 LOMA PRIETA EARTHQUAKE CENTERED NEAR SANTA CRUZ, CALIFORNIA

FIGURE **B3-2**

Plate slippage along the San Andreas fault. (a) Aerial photograph of a segment of the San Andreas fault 100 miles north of Los Angeles. (b) Tragic damage in Oakland, California, from the 1989 Loma Prieta earthquake.

THE PROCESS OF SCIENCE

The Scientific Process: Sea-Floor Spreading

An essential part of the scientific method is the falsification of hypotheses. This means that one should be able to test any valid scientific hypothesis in such a way that it can be disproved by the test results. Let's examine an actual test of a hypothesis. In 1963 the British geologists Fred Vine and Drummond Matthews proposed that the symmetrical magnetic anomaly stripes that lie parallel to midocean ridges in all the oceans were the result of sea-floor spreading (see Figure 3–5). If correct, the hypothesis of sea-floor spreading had revolutionary implications for interpreting the geological development of ocean basins. But how exactly to test this radical idea? Actually, this turned out to be quite easy. Marine geologists reasoned that, if all of the sea floor were created along the crests of spreading ocean ridges, as proposed by Vine and Matthews, then it stands to reason that the age of the ocean crust should increase with the distance from the ridge axis. Also, the older the basaltic crust of the sea floor, the thicker must be its sediment cover. In other words, the Vine-Matthews model of sea-floor spreading predicts clearly that the crust should get older and be buried by a thicker blanket of sediment with increasing distance from the ridge crest. These were testable predictions.

To make these observations, geologists on the drilling ship *Glomar Challenger* (see Figure 1–10a) could, for the first time, drill anywhere in the ocean in thousands of meters of water. It was relatively simple to drill through the sediment layers to the underlying basaltic crust on the flanks of the ocean ridges. The total thickness of the sediment cover at the drilling site could thus be determined and paleontologists could obtain samples of the remains of microfossils embedded in the sediment that was deposited directly on the basalt. At any particular site the fossils could then be used to establish the age of the oldest layer of sediment, which would be very close to the age of the basalt directly beneath it. The data obtained from numerous drill sites indicated that both the age of the oldest sediment and the total thickness of the sediment cover

---

where the lithosphere is old and is supporting a heavy load of sediment. This tends to occur at continental margins because of the tremendous pile of sediment that accumulates there. The process of subduction begins where one side of the now-fragmented plate overrides the other. The ocean basin then enters a *declining* stage as the ocean lithosphere, and ultimately the spreading ocean ridge itself, are subducted and disappear from the face of the Earth, a situation that is currently happening to the East Pacific Rise in the Pacific Ocean (see boxed feature, "The San Andreas Fault"). As the basin continues to close up, the *terminal* stage is reached. Continents and subduction zones on either side of the basin collide, crushing and uplifting the sedimentary deposits of the basin into a young mountain belt of folded and faulted marine sedimentary rocks. Finally, the two colliding continental masses become *sutured* (fused) tightly together, and the sedimentary deposits and the ocean crust are buckled in a viselike grip and forced upward into a majestic, towering mountain range. A fine example of this suturing stage is the Himalayan range, where sedimentary rocks with marine fossils attest to a time when they were submerged in seawater in an ocean basin that once existed between India and mainland Asia (**Figure 3–11**). Remarkably, the top of the world, Mount Everest, was once an ocean basin at the very bottom of the world.

A SUMMARY OF GLOBAL PLATE TECTONICS

We've covered a lot of ground, so to speak, in this chapter. Let's take a moment to reflect about the chapter's main points; these may have become lost among the details that were needed to flesh out the concept of global plate tectonics. The main idea is that many features of the Earth's surface—its volcanoes, earthquakes, ocean ridges, deep-sea trenches, mountain belts, magnetic anomaly stripes—can be understood as the result of the interactions between lithospheric plates. Some of these plates are moving apart at midocean ridges. This process, called sea-floor spreading, is occurring in all the ocean basins increased systematically with the increasing distance from the ridge axis. The conclusion was inescapable: the hypothesis of sea-floor spreading that Vine and Matthews proposed was not falsified by the test. This meant that sea-floor spreading offered a legitimate new model for interpreting the geologic history of the Earth's ocean basins.

The map in **Figure B3–3** shows the global age pattern of the oceanic crust determined from a synthesis of all the latest research findings. The youngest oceanic crust (red color) is associated everywhere with the crest of the spreading ridges. The age of the basalt increases systematically with increasing distance from the ridge crest, with the oldest crust (blue color) located everywhere far down the flanks of the spreading ridges.

FIGURE **B3-3**

Age of the ocean crust. Age determination indicates that the basaltic crust gets older with increasing distance from the crest of spreading ocean ridges.

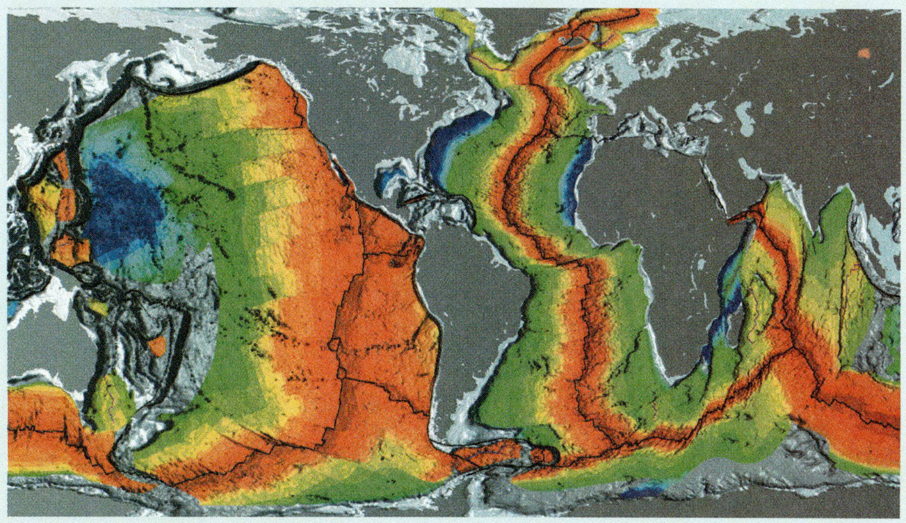

and results in the formation of new sea floor and the drift of continents. Other pairs of lithospheric plates are actively colliding. One overriding plate forces another downward into the asthenosphere, where it melts to produce andesite lava. This lava feeds volcanoes on the surface. The process, called subduction, occurs principally around the edges of the Pacific Ocean. Over geologic time, the quantity of new lithosphere produced at spreading ridges has equaled the quantity that has been consumed at subduction zones. Plate boundaries where the lithosphere is conserved (neither created nor destroyed) are called transform faults. These immense fractures separate two plates that are sliding laterally past each other.

The granite of the continents is imbedded in the lithospheric plates and is swept passively across the Earth's surface as the ocean floors spread and grow in size. When continents on two separate plates meet at a subduction zone, they collide and squeeze the intervening marine sediments tightly, lifting them skyward to create a high, folded mountain belt. This very process is presently uplifting the Himalayas, as India collides with the Asian mainland (see Figure 3–11).

This is the basic theory of global plate tectonics, a magnificent and unexpected discovery of modern oceanography. With this background, we can move ahead to the study of marine sediments, examining the origin of these materials that collect on the sea bottom and their history of transport by the moving lithospheric plates.

3-4

Future Discoveries

A great deal of effort is being directed at understanding how undersea eruptions have affected ocean processes and marine life. Mantle plumes like the one that created the Hawaiian Islands may generate megaplumes,

The Red Sea

The Red Sea (**Figure B3–4a**) is a roughly rectangular basin about 1,900 kilometers (~1,178 miles) long and about 300 kilometers (~186 miles) wide (**Figure B3–4b**). Much of the sea bottom is quite shallow, with an average water depth of 490 meters (~1,617 feet) and a maximum water depth of 2,850 meters (~9,405 feet). Running along the center of the Red Sea is a narrow trough with an average depth of about 1,000 meters (3,300 feet). Basalt—new ocean crust—is being injected into this deep axial trough as Arabia drifts away from Africa. In effect, the Red Sea is a miniature ocean (**Figure B3–4c**), a classic juvenile ocean basin that is slowly widening as a result of

(a) SATELLITE IMAGE OF THE RED SEA REGION

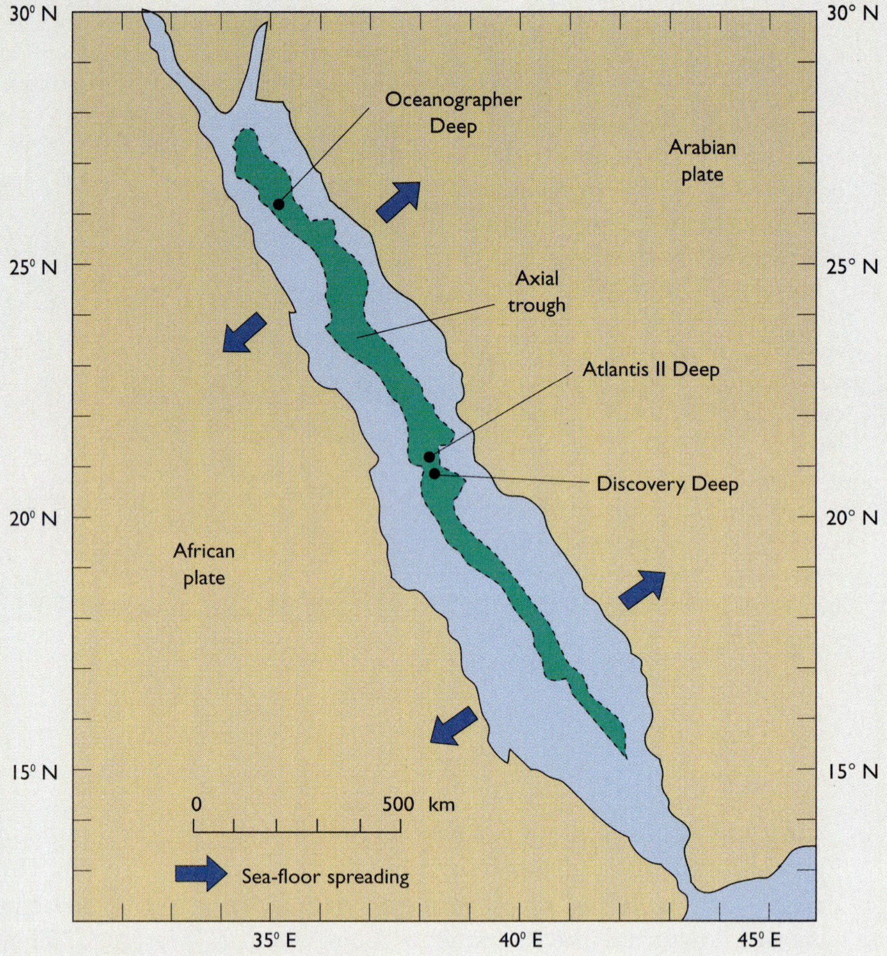

(b) THE RED SEA SPREADING CENTER

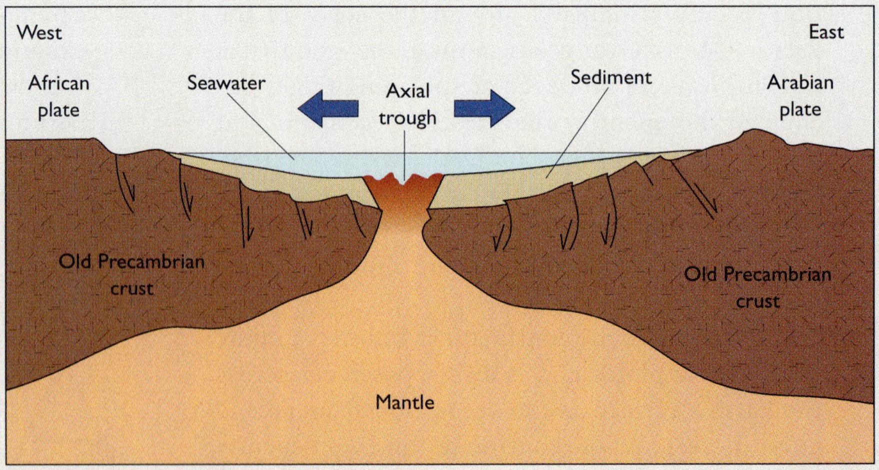

(c) GEOLOGIC (WEST–EAST) CROSS SECTION

FIGURE **B3-4**

The Red Sea. (a) This satellite image shows the terrain of the Red Sea region. (b) The Red Sea is a juvenile ocean basin that has just recently opened up as the Arabian plate separates from the African plate. (c) New oceanic crust is forming in the axial trough of the Red Sea by the process of sea-floor spreading. [Adapted from D. A. Ross, *Mineral Resources Bulletin* 22 (1977): 1–14.]

sea-floor spreading. This process was responsible for the opening of the Atlantic Ocean, and this is what the Atlantic basin must have looked like in its early development as Africa and Europe moved away from North and South America.

It appears as if the Red Sea basin began to develop some 20 million to 30 million years ago as the granitic crust of East Africa and Arabia was stretched until it broke apart along a system of normal faults. These large faults have splintered the thick granitic crust into large blocks (see Figure B3-4c). The sediments that blanket the sea bottom include salt deposits up to 7 kilometers (~4.3 miles) thick. Their presence indicates that much of the ocean dried up periodically as its water was evaporated and salt deposits were laid down on its bottom. To imagine what this was like, fill a glass with seawater and leave it in the sun for a few days. What will happen, of course, is that the water will disappear because of evaporation, and the bottom of the glass will be encrusted with salt.

Not only is there salt on the sea floor, but the water itself, which fills the deeps of the axial trough—such as the Atlantis II deep (**Figure B3–5**), the Discovery deep, and the Oceanographer deep—is unusually salty. It is so much saltier than normal seawater that it is referred to as **brine.** This brine is hot (50 to 60°C) and filled with dissolved metals. The source of the unusual salt and metal content of the water in these deeps is the flow of groundwater (subterranean water) through fractures in the underlying rocks. This briny groundwater is heated as it flows through the hot crust, becoming corrosive and leaching metals from the basalt rocks. The heated, high-salinity, metalliferous water is discharged along fractures and faults on the bottom of the deeps, where it is trapped because of its high density. As the levels of dissolved metals (iron, manganese, copper, silver, lead, and zinc) build up, many of them are precipitated as sulfide deposits that impart bright colors to the sediment.

Geochemical surveys indicate that the metal deposits of the Atlantis II deep are sufficiently concentrated to be exploited commercially. Several field tests indicate that it is feasible to mine this resource. High-pressure water jets (large, powerful hoses in effect) lowered from a vessel could convert the bottom sediment into a mud slurry, which would then be pumped to the surface at a rate estimated to be about 200,000 tons each day! This enormous volume of slurry would have to be processed aboard the mining vessel while at sea, because it would be too expensive to transport it to land. Unfortunately, once processed, the residue would become a major waste-disposal problem, because it contains highly toxic heavy metals. Engineers have developed a processing technique whereby only 1 percent of the metal concentrate would be transported to a smelter on land. The remaining 99 percent of the slurry would be diluted with seawater and treated with chemicals before being discharged into water deeper than 1,000 meters (~3,300 feet). Marine life is sparse at those depths in the Red Sea so the engineers reason that the effect of the metal toxins on the ecosystem of the area would be minimized.

Visit www.jbpub.com/oceanlink for more information.

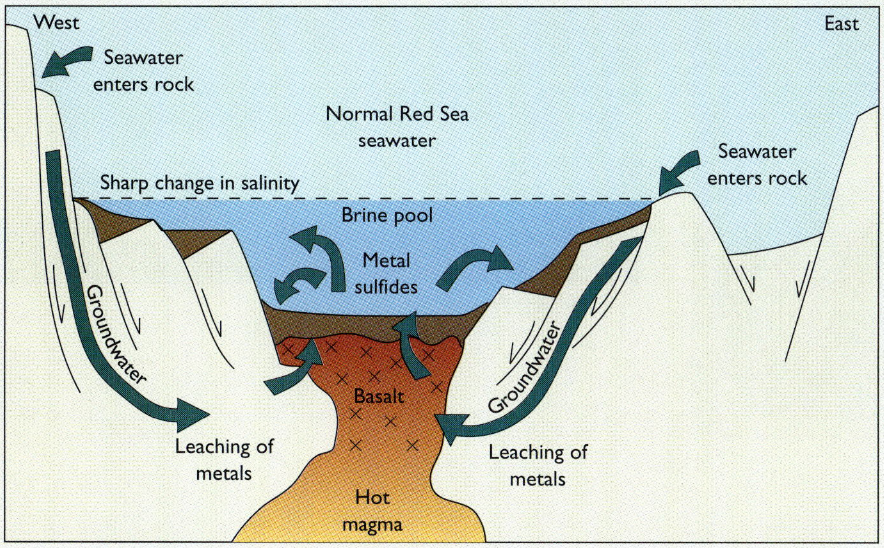

FIGURE **B3-5**

The Atlantis II deep. Groundwater flowing through fractures in the basaltic crust is acidic and corrosive. These hot salty fluids leach out metals from the rocks. When they seep out of fractures, the very dense water is trapped in the deeps. [Adapted from H. Backer, *Erzmetall* 26 (1973): 544–55.]

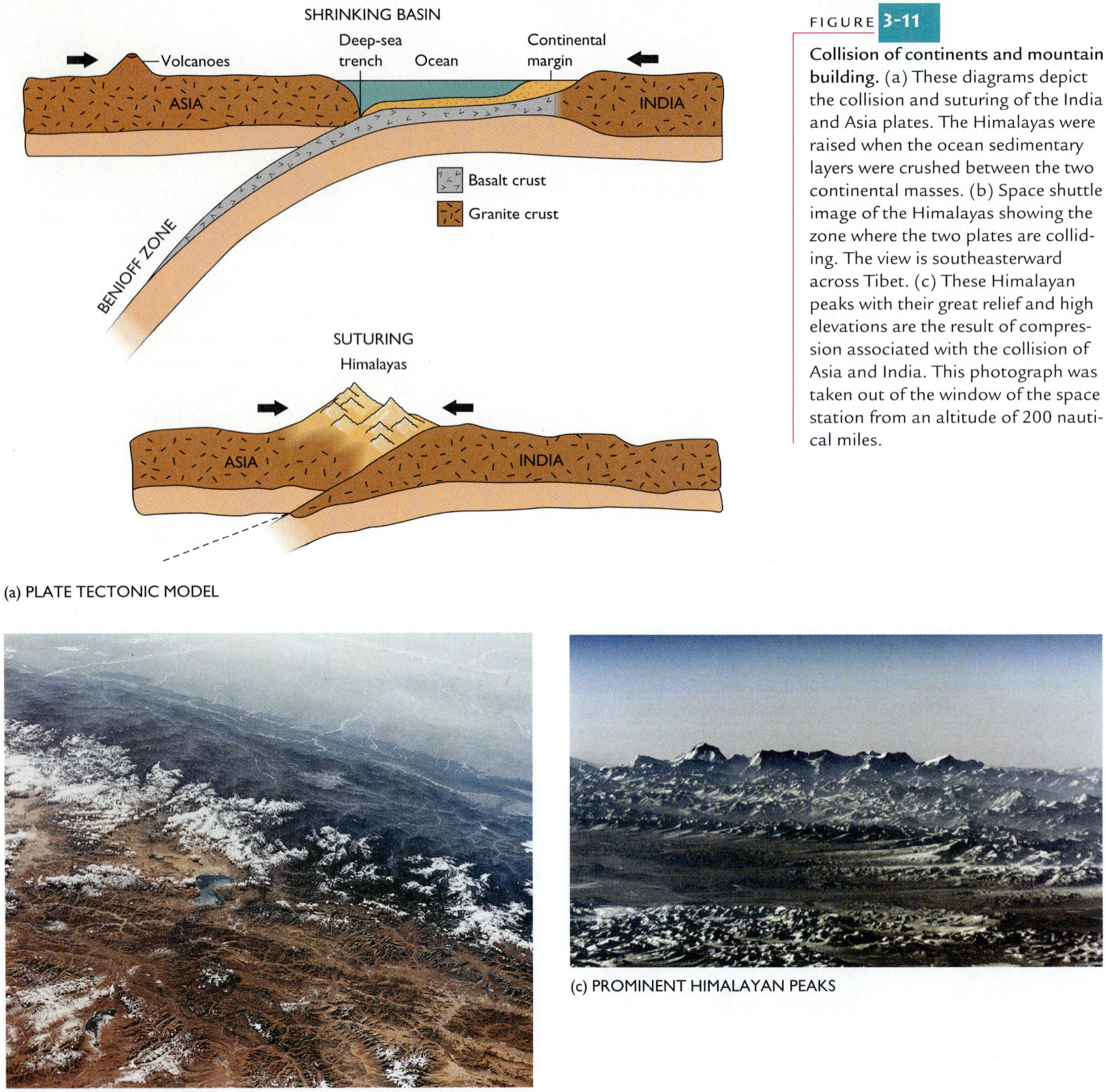

(a) PLATE TECTONIC MODEL

(b) SATELLITE IMAGE OF HIMALAYAS

(c) PROMINENT HIMALAYAN PEAKS

FIGURE 3-11

Collision of continents and mountain building. (a) These diagrams depict the collision and suturing of the India and Asia plates. The Himalayas were raised when the ocean sedimentary layers were crushed between the two continental masses. (b) Space shuttle image of the Himalayas showing the zone where the two plates are colliding. The view is southeasterward across Tibet. (c) These Himalayan peaks with their great relief and high elevations are the result of compression associated with the collision of Asia and India. This photograph was taken out of the window of the space station from an altitude of 200 nautical miles.

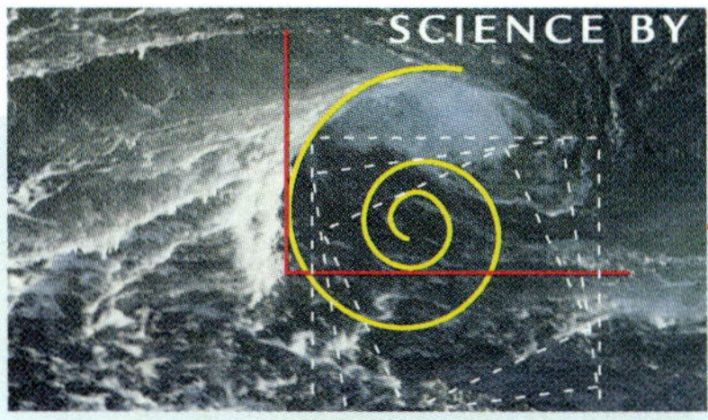

SCIENCE BY NUMBERS

Sea-Floor Spreading Rates

Lithospheric plates are moving apart at midocean ridges because of sea-floor spreading. Oceanic crust, once created, will slowly move away from the axis of the ridge. This means that the sea floor will have a speed. Speed is distance (d) traveled per unit of time (t), or

$$\text{speed} = d/t.$$

The sea floor is moving slowly, such that its speed cannot be measured directly with a stopwatch. Yet we can easily determine its speed (its spreading rate) by noting the age of the sea floor at any distance from the ridge. The older the sea floor is, the farther it will be from the ridge axis. Also, the faster the rate of sea-floor spreading, the farther the sea floor of a particular age will be from the ridge axis. Let's calculate a sea-floor spreading rate.

First, assume that we obtain a piece of basalt 400 kilometers from the ridge axis and determine its age to be 10 million years old. This implies that this rock took 10 million years to travel a distance of 400 kilometers. The number 400 can be expressed as 4×10^2 (i.e., 4 times 100). The number 10 million can be expressed as 10^7 (i.e., 10 times itself seven times). If this notation is unclear, review the Math Box called "Powers of 10" in Chapter 2. The rate of the sea-floor spreading is

$$\text{speed} = \text{spreading rate} = d/t = 4 \times 10^2 \text{ km}/10^7 \text{ yr}.$$

When you divide powers of 10, you merely subtract the exponents. So,

$$4 \times 10^2 \text{ km}/10^7 \text{ years} = 4 \times 10^{(2-7)} \text{ km/yr} = 4 \times 10^{-5} \text{ km/yr}$$

is the rate of sea-floor spreading for this ridge.

Let's now convert this rate to centimeters per year (cm/yr), which would make more intuitive sense to us. The trick, as discussed in the Chapter 1 Math Box "Conversions," is to keep tab of the units.

$$(4 \times 10^{-5} \cancel{\text{km}}/\text{yr})\ (10^3 \cancel{\text{m}/\text{km}})\ (10^2 \text{ cm}/\cancel{\text{m}}) = (4 \times 10^{-5})\ (10^3)\ (10^2) \text{ cm/yr}.$$

When you multiply powers of ten, you merely add their exponents. So,

$$(4 \times 10^{-5})\ (10^3)\ (10^2) \text{ cm/yr} = 4 \times 10^{(-5+3+2)} \text{ cm/yr} = 4 \times 10^0 \text{ cm/yr}.$$

Because any number raised to a power of zero is 1,

$$4 \times 10^0 \text{ cm/yr} = 4 \text{ cm/yr}.$$

This is the spreading rate for that side or flank of the midocean ridge.

immense twisting, swirling "mushroom clouds" of very hot water that are tens of kilometers (>6.2 miles) broad. Megaplumes carry gases such as carbon dioxide and methane and countless microbes as they spin and drift about in the ocean for months, moving hundreds of kilometers away from their site of origin. Some biologists believe that life itself may have begun at the bottom of the sea in sea-floor geysers located at spreading ridge crests. These hydrothermal vents may have been sites where raw inorganic materials were miraculously shaped by chemical reactions into simple organic compounds that ultimately led to the origin of life some four billion years ago. So the study of megaplumes and hydrothermal systems may lead to new theories about the origin of life and its dispersal throughout the world's oceans.

Other exciting new research in marine geology involves probing the Earth's interior to determine how the planet works. For example, geophysicists have uncovered old, cold slabs of subducted lithosphere in the Earth's upper mantle; others may, perhaps, be deep in the lower mantle as well. These "fossil" slabs are surviving remnants of plate tectonic events that occurred during the Earth's ancient geologic past. Understanding how they got there and what happens to them over time will give geologists a clearer and more complete view of the Earth's history and geochemical evolution.

STUDY GUIDE

KEY CONCEPTS

1. Geologic and paleontological evidence supports the existence in the geologic past of an immense landmass known as *Pangaea*. Beginning about 150 million years ago, Pangaea broke apart, and the pieces drifted across the Earth's surface to form today's continents. This process is called *continental drift*.

2. Studies of the magnetic properties of the sea floor reveal that the crests of the midocean ridges are sites where new oceanic crust is forming. The divergence of the basaltic crust away from the crestline of the ridge—a process called *sea-floor spreading*—leads to the expansion of ocean basins and to the drift of the continents. The repeated flipping back and forth of the geomagnetic poles causes basalts ejected at spreading ocean ridge crests to be normally and reversely magnetized, giving rise to a symmetric pattern of *magnetic anomaly stripes* that run parallel to the ridge axis. These magnetic anomaly stripes can be used to calculate the rate of sea-floor spreading.

3. Basaltic crust is being destroyed at *subduction zones,* which are associated with volcanic arcs and deep-sea trenches. Here, large slabs of rock are being compressed as they converge and override one another, producing an inclined plane of earthquakes called the *Benioff zone*.

4. The global distribution of earthquakes indicates the Earth's surface is broken into a mosaic of irregularly shaped plates. Each plate, which is 100 to 150 kilometers (~62 to 93 miles) thick, consists of the entire crust and the uppermost mantle, which combined constitute the *lithosphere.* The lithospheric plates are cold, rigid, and brittle, and overlie a hot, ductile (plasticlike) layer of the mantle, the *asthenosphere.*

5. Lithospheric plates have three kinds of boundaries: *spreading ridges* that are under tension, where new lithosphere is formed; *subduction zones* that are under compression, where lithosphere is destroyed; and *transform faults*, where lithosphere is preserved as plates slip laterally (strike-slip motion) past one another.

6. The vast majority of volcanoes are formed at plate boundaries. Many lie along subduction boundaries where the lithosphere has melted as it was forced downward into the Earth's hot interior by an overriding lithospheric plate. Other volcanoes form along the crests of spreading ocean ridges as well, as lithospheric plates diverge and lava spills out of fractures onto the sea floor. In contrast, some linear chains of volcanoes form on the plate interior, far from its edges, as the lithosphere drifts slowly over a *mantle plume*, or "hot spot," in the underlying asthenosphere. The mantle plume feeds lava to the plate's surface and creates a long, linear chain of volcanoes that mark the track of the plate over the mantle plume.

7. Ocean basins undergo a regular evolutionary history related to plate tectonics. A narrow, embryonic basin forms between the splintered pieces of a continent; expands into a broad, mature ocean basin by the mechanism of sea-floor spreading; declines into old age as the basin is consumed by subduction; and disappears as colliding continental masses become tightly sutured together and form an immense mountain belt on land.

QUESTIONS

REVIEW OF BASIC CONCEPTS

1. Explain the concepts of continental drift, sea-floor spreading, and global plate tectonics.

2. What exactly is a magnetic anomaly, and why does it appear as a stripe on the sea floor?

3. How do plate motions differ among a spreading ridge, a transform fault, and a subduction zone? Which of the three is characterized by the strongest earthquakes, the deepest earthquakes? Why?

4. Draw a detailed cross section of the Earth's outer layers, showing the relationships among deep-sea trenches, island arcs, Benioff zones, andesite volcanism, the lithosphere, the ocean crust, the asthenosphere, and the Moho. (You may have to refer back to Chapter 2.)

CRITICAL-THINKING ESSAYS

1. Why do magnetic anomaly stripes of similar age have the same magnetic polarity, regardless of where they are discovered in the ocean?

2. It is unlikely that magnetic anomaly stripes older than about 200 million years will be found anywhere in the oceans. Why?

3. In which ocean is the oldest oceanic crust likely to be found? Why?

4. Assume that you discover a series of large submarine volcanoes on the deep-ocean floor of the Pacific. How would you determine whether or not these volcanoes had been created by a mantle plume?

5. Cite a variety of evidence that supports the notion of (a) continental drift, (b) sea-floor spreading, (c) subduction.

DISCOVERING WITH NUMBERS:

1. Convert 33 kilometers into centimeters: 4.1×10^6 cm into kilometers (see Appendix II).

2. Assume that magnetic anomaly C in Figure 3–3 is 650,000 years old. Calculate a spreading rate for the sea floor. Now, using the calculated sea-floor spreading rate, estimate the age of magnetic anomaly B in Figure 3–3.

3. Assume that you are conducting geophysical work on the flank of a spreading ocean ridge that trends directly north-south. As the captain steers the vessel to the east, your magnetometer measures a series of magnetic highs and lows. A strong, broad magnetic high is positioned directly over the ridge crest. It is followed to the east by a magnetic low of modest width and, at 45 kilometers from the ridge crest, a narrow but prominent magnetic high. Determine the age of this latter magnetic high by consulting Figure 3–4c, and then determine the spreading rate in centimeters per year for this midocean ridge.

4. In Problem 3 above, how many kilometers from the ridge crest would you have to sail to the west in order to be positioned over ocean crust that is 15 million years old?

5. Consult Figure 3–9a. Given that 10 degrees of latitude equals about 1,110 kilometers, calculate an approximate rate of sea-floor spreading in centimeters per year for the North Pacific plate, assuming that mantle plumes do not drift over time. Does the rate of spreading change over time?

KEY WORDS

andesite
asthenosphere
Benioff zone
compression
continental drift
lithospheric plate
lithosphere
magnetic anomaly stripe
mantle plume
normal fault
normal polarity
paleomagnetism
paleontology
Pangaea
plate tectonics
reverse polarity
rift valley
sea-floor spreading
seismicity
strike-slip fault
subduction zone
tension
thermal convection
transform fault

SELECTED READINGS

Anderson, R. N. 1986. *Marine Geology: A Planet Earth Perspective*. New York: John Wiley.

Ballard, R. D. 1975. Dive into the great rift. *National Geographic.* 147 (5): 604–15.

Ballard, R. D. 1983. *Exploring Our Living Planet.* Washington, D.C.: National Geographic Society.

Bonatti, E. 1987. The rifting of continents. *Scientific American* 256 (3): 96–103.

Bonatti, E. 1994. The Earth's mantle below the ocean. *Scientific American* 270 (3): 44–51.

Bonatti, E., and Crane, K. 1984. Oceanic fracture zones. *Scientific American* 250 (5): 40–51.

Courtillot, V., and Vink, G. E. 1983. How continents break up. *Scientific American* 249 (1): 42–49.

Dalgiel, I. W. D. 1995. Earth before Pangaea, *Scientific American* 271 (1): 58–63.

Dewey, J. R. 1972. Plate tectonics. *Scientific American* 226 (5): 56–68.

Dietz, R. S. 1971. Those shifting continents. *Sea Frontiers* 17 (4): 204–12.

Dietz, R. S., and Holden, J. C. 1970. The breakup of Pangaea. *Scientific American* 223 (4): 30–41.

Francheteau, J. 1983. The oceanic crust. *Scientific American* 249 (3): 114–29.

Frohlich, C. 1989. Deep earthquakes. *Scientific American* 260 (1): 48–55.

Fryer, P. 1992. Mud volcanoes of the Marianas. *Scientific American* 266 (2): 46–52.

Gurnis, M. 20021 Sculpting the Earth from inside out. *Scientific American* 284 (3): 40–47.

Hallam, A. 1975. Alfred Wegener and the hypothesis of continental drift. *Scientific American* 232 (2): 88–97.

Heirtzler, J. R. 1968. Sea-floor spreading. *Scientific American* 219 (6): 60–70.

Heirtzler, J. R., and Bryan, W. B. 1975. The floor of the Mid-Atlantic Rift. *Scientific American* 233 (2): 78–91.

Hekinian, R. 1984. Undersea volcanoes. *Scientific American* 251 (1): 46–55.

Hoffman, K. A. 1988. Ancient magnetic reversals: clues to the geodynamo. *Scientific American* 258 (5): 76–83.

Hyndman, R. D. 1995. Great earthquakes of the Pacific Northwest. *Scientific American* 276 (6): 68–75.

Lawrence, D. M. 2002. *Upheavals from the Abyss.* New Brunswick, N.J.: Rutgers University Press.

Larson, R. I. 1995. The mid-Cretaceous superplume episode. *Scientific American* 272 (2): 82–89,

Lonsdale, P., and Small, C. 1991–92. Ridges and rises: a global view. *Oceanus* 34 (4): 26–35.

Macdonald, J. B., and Fox, P. J. 1990. The mid-ocean ridge. *Scientific American* 262 (6): 72–79.

Macdonal, K. C., and Lyendyk, B. P. 1981. The crest of the East Pacific Rise. *Scientific American* 244 (5): 100–16.

Molnar, P. 1986. The structure of mountain ranges. *Scientific American* 255 (1): 70–79.

Molnar, P. and Tapponnier. P. 1977. The collision between India and Eurasia. *Scientific American* 236 (4): 30–41.

Morgan, W. J., and Vogt, P. R. 1986. The earth's hot spots. *Scientific American* 252 (4): 50–57.

Murphy, J. B., and Nance, R. D. 1992. Mountain belts and the supercontinent cycle. *Scientific American* 266 (4): 84–91.

Nance, R. D., Worsley, T. R., Moody, J. B. 1988. The supercontinent cycle. *Scientific American* 259 (1): 72–79.

Oceanus. 1991–1992. Midocean ridges. Special issue 34 (4).

Oceanus. 1993–1994. 25 years of ocean drilling. Special issue 36 (4).

Oceanus. 1998. The Mid-Ocean Ridge: Part I. Special issue 41 (1): 1–37.

Open University Course Team. 1989. *The Ocean Basins: Their Structure and Evolution.* New York: Pergamon Press.

Powell, C. C. 1991. Peering inward. *Scientific American* 264 (6): 100–11.

Sclater, J. G., and Tapscott, C. 1979. The history of the Atlantic. *Scientific American* 240 (6): 156–74.

Siever, R. 1983. The dynamic earth. *Scientific American* 249 (3): 46–55.

Tokosoz, M. N. 1975. The subduction of the lithosphere. *Scientific American* 233 (5): 88–98.

Toomey, D. R., 1991–92. Tomographic imaging of spreading centers. *Oceanus* 34 (4): 92–99.

TOOLS FOR LEARNING

Tools for Learning is an on-line review area located at this book's web site OceanLink (**www.jbpub.com/oceanlink**). The review area provides a variety of activities designed to help you study for your class. You will find chapter outlines, review questions, hints for some of the book's math questions (identified by the math icon), web research tips for selected Critical Thinking Essay questions, key term reviews, and figure labeling exercises.

4

Marine Sedimentation

Nothing under heaven is softer or more yielding than water, but when it attacks things hard and resistant there is not one of them that can prevail.

—*Tao Te Ching, sixth century* B.C.

web navigator

critical thinking on the web

math tutor on the web

www.jbpub.com/oceanlink

PREVIEW

MUCH OF THE OCEAN FLOOR is covered by layers of sediment; each has a story to tell about the geologic past. These sedimentary layers contain a great variety of particles. Some of these grains were swept to the ocean bottom from land, or dropped down from outer space, or were produced by a variety of biological or chemical processes. A careful study of the properties of these sedimentary deposits by trained geologists yields valuable clues about the Earth's plate-tectonic history, the evolution of marine life, the chronicle of past climates, variations in the flow pattern of water currents, impacts of meteorites, eruptions of submarine volcanoes, fluctuations of sea level, occurrences of mass extinction, and many more events. The epic stories can be read from the record that is preserved in the vast sedimentary accumulations on the sea bottom.

Water flowing incessantly downhill wears down the land's bedrock and then transports the bits of rock particles downstream. Most of this eroded debris eventually finds its way to the oceans, where it is dispersed and deposited on the sea bottom. At the same time, countless small floating microorganisms—single-celled plants and animals—that dwell in the surface waters of the oceans die and their hard parts either dissolve or slowly settle and accumulate on the ocean floor.

The central topics of this chapter are the nature and significance of the tremendous amount of sediment that covers the floor of the deep sea and the submerged edges of the continents. As the oceans spread apart and the continents are dragged along, water wears down the bedrock of the land. We will examine the character and history of sedimentary material—rock debris and fossil remains—that are slowly being deposited in the ocean basins. These beds of sediment, when sampled and examined carefully, reveal clues about the chemistry and biology of ocean water of the distant past, the geologic history of ocean basins, and the nature of past climates.

4-1

Sediment in the Sea

Sediment is produced by the weathering (the chemical and mechanical breaking down) of rock such as granite or basalt into particles that are then moved by air, water, or ice. Sediment can also form by the accumulation of shells of dead organisms. Therefore, sediment can consist of either mineral or fossil particles, and both types can be found in many places on the bottom of the sea. Most erosion of rock occurs on land and most deposition of sediment occurs in the ocean. The net effect of these two processes—erosion and deposition—would be to even out the Earth's surface, cutting away its high points (the landmasses) and filling in its low points (the oceans) if it were not for plate tectonics. The collision of continents along subduction zones squeezes and deforms the layers of marine sediment that have accumulated between landmasses, raising long, linear mountain belts. The mountain peaks rise high above sea level until erosion wears them away, and the weathered debris is swept into the ocean and accumulates as sediment. This results over geologic time in a grand tectonic cycle of raising and leveling mountains and deepening and filling ocean basins. Even on the land, marine sedimentary rocks are common, because seas have regularly invaded the land in the geologic past. These marine sedimentary rocks make up more than 50 percent of the outcrops on land and record the existence of seas that dried up long ago.

Before examining some of the factors that determine the patterns and depositional rates of marine sediment, we need to be able to distinguish between different types of sediment. So let's consider two simple schemes that geologists use to do this.

CLASSIFICATION OF MARINE SEDIMENT

The first step in classifying anything is to establish criteria for defining categories. Sediments can be subdivided on the basis of the size of the particles (**grain size**) or on the basis of their mode of formation. In the first case, the classification depends on a measurement of particle size; in the second case, the classification requires an interpretation of the origin of the deposit. Both classifications are useful and are widely employed by geological oceanographers. Let's examine both of them more carefully.

The size of particles produced from the breakdown of rock ranges from enormous boulders to tiny grains of microscopic clay or even finer particles called **colloids**. From the largest to the smallest particles common in sediment, there are gravel, sand, silt, and clay (**Table 4–1**). We can ignore the even finer colloids because they are unimportant as sediment. Silt and clay particles are typically mixed together and form a deposit of **mud**. The most common sedimentary deposits in the ocean are mud and sand; gravel is very rare in the sea.

TABLE 4-1

Wentworth grain-size scale.

Sediment	Type	Diameter (mm)
Gravel	Boulder	>256.0
	Cobble	65.0–256.0
	Pebble	4.0–64.0
	Granule	2.0–4.0
Sand	Very coarse	1.0–2.0
	Coarse	0.50–1.0
	Medium	0.25–0.50
	Fine	0.125–0.25
	Very fine	0.0625–0.125
Mud: Silt		0.0039–0.0625
Mud: Clay		0.0002–0.0039
Colloid		<0.0002

Source: Adapted from C. K. Wentworth, A scale of grade and class terms for clastic sediments, *Journal of Geology* 30 (1922): 377–92.

Now let's classify sediments by the way that they form. Using the broadest scheme possible, we can subdivide sediment into five categories, presented here with their general characteristics:

> ***Terrigenous sediment:*** *fine and coarse grains that are produced by the weathering and erosion of rocks on land; typically sands and mud.*
>
> ***Biogenic sediment:*** *fine and coarse grains that are derived from the hard parts of organisms, such as shells and skeletal debris: typically lime (composed of calcium carbonate) and siliceous (composed of silica) muds.*
>
> ***Authigenic sediment:*** *particles that are precipitated by chemical or biochemical reactions in seawater near the sea floor; manganese and phosphate nodules are examples.*
>
> ***Volcanogenic sediment:*** *particles that are ejected from volcanoes; ash is an example.*
>
> ***Cosmogenous sediment:*** *very tiny grains that originate from outer space and tend to be mixed into terrigenous and biogenic sediment.*

Notice how the two classification schemes—one based on grain size, the other on sediment origin—are intertwined with each other. For example, sand and mud, which are separated on the basis of grain size, can be terrigenous, biogenic, cosmogenous, authigenic, or volcanogenic, depending on their origin. Now that we have a common vocabulary for distinguishing types of sediment, let's consider how exactly the grains are transported and deposited on the sea floor to form those various types.

FACTORS THAT CONTROL SEDIMENTATION

Two of the most important factors that determine the nature of a sediment deposit are particle-size distribution and energy conditions at the site of deposition. Because these two factors interact to produce the properties of a sediment deposit, a geologist is able to deduce what they were at the time of deposition by examining those properties of even an ancient sediment.

Terrigenous sediment is a collection of rock and mineral fragments with compositions that are directly related to their source rocks. For example, if fragments of granite are present, then the source area that yielded the grains contained granitic rocks. If limestone and shale fragments abound in the deposit, then the source rocks were composed of sedimentary rocks, including limestones and shales. Rivers, glaciers, and wind transport these particles out of the source area, a process termed **erosion.** Much of this eroded sediment eventually reaches the ocean, where currents disperse them even more before they are deposited as layers of sediment on the sea bottom.

Probing the Sea Floor

Various simple but durable devices are available for collecting sediment samples from the ocean floor, even from the deepest, most inaccessible parts of the sea. A long-established technique is scraping the ocean bottom with a **dredge**—a rigid metal frame to which is attached a sampling bag made of chain or tough netting (**Figure B4–1a**). Dredges are suitable for obtaining large, bulk samples of either rock or sediment. As they are dragged, however, they bite the bottom indiscriminately and mix samples together in the sampling bag. Also fine sediment such as mud tends to be washed out of the sample. Because of these effects, oceanographers employ dredges almost exclusively to collect hard rock rather than soft sediment. Less disturbed samples of mud and sand are collected by **grab samplers**—spring-loaded metal jaws that take a bite out of the bottom and close tightly around the sediment sample (**Figures B4–1b** and **1c**).

Dredges and grab samplers merely sample the surface veneer of sediment. Deeper penetration of soft sediment is accomplished by a **gravity corer.** This hollow metal tube, known as a **core barrel,** is pushed into the sediment by the force of gravity. The corer is lowered to the bottom, where the heavy weight at the top of the device drives the barrel into the sediment (**Figure B 4–2a**). A plastic liner that has been inserted into the core barrel allows oceanographers to extract the sediment core intact from the sampler and also serves as a temporary storage container. Gravity corers are capable of taking cores of between 1 and 2 meters (~3 and 6 feet) long, depending on the properties of the sediment. Sediment cores longer than 20 meters (~66 feet) are routinely obtained by **piston corers** (**Figure B4–2b**). This type of corer has a piston that slides up the core barrel as it penetrates the bottom. The action of the piston extrudes water from the core barrel, allowing the sediment core to enter the liner with minimal disturbance and compaction. Once the core is on deck, the plastic liner with its sample of sediment is extruded from the core barrel and taken to a laboratory for detailed examination (**Figure B4–3**). Geologists carefully study the layering and composition of the sediment particles to determine the geological history of the Earth.

At present, the best technique for sampling the ocean bottom is **platform drilling,** which was first developed by petroleum engineers on land and is now adapted to the ocean, even the deep ocean. The procedure is very expensive, but the scientific results are priceless. Marine geologists not only recover cores of sediment more than 1 kilometer (~0.6 miles) in length, but also they can drill into the hard rock of the crust beneath the sedimentary layers. The *Glomar Challenger*—the 122-meter-long (~403 feet) vessel that completed an illustrious international career of drilling, even in the remotest regions of the oceans, collecting hundreds of kilometers of core sample—has been retired. Its successor, the larger *Joides Resolution,* has a much more efficient and deeper drilling capacity than the *Challenger* (see Figure 1–10 in Chapter 1).

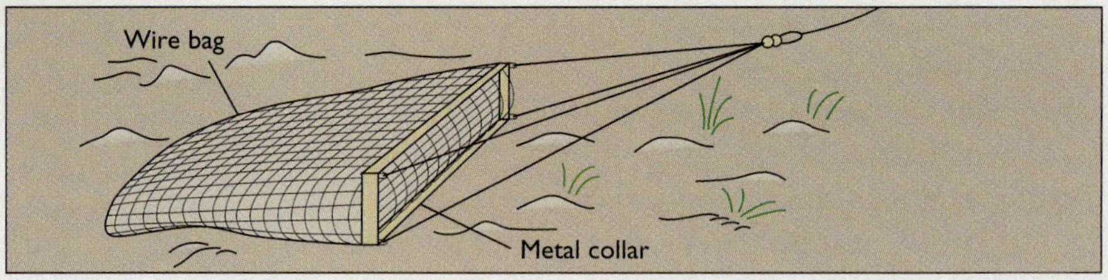

(a) BOTTOM DREDGE

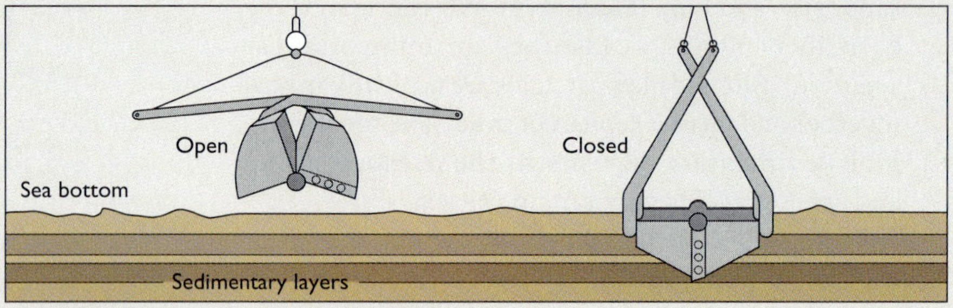

(b) GRAB SAMPLER

(c) GRAB SAMPLER

FIGURE **B4-1**

Gear for sampling sediment. (a) Hard rock can be sampled with the durable bottom dredge. (b) Surface sediments are collected by grab samplers that take a "bite" out of the sea bottom. (c) A grab sampler back up from the sea bottom.

Visit www.jbpub.com/oceanlink for more information.

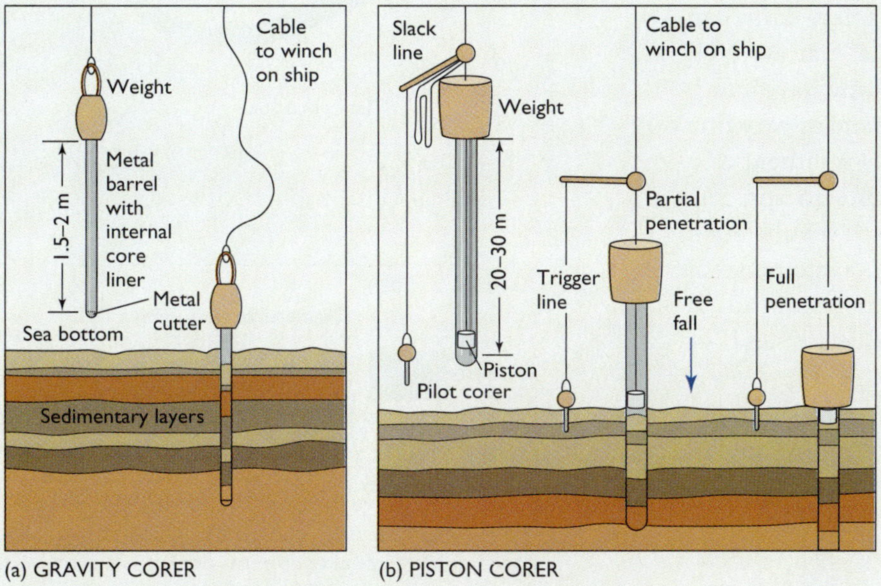

FIGURE **B4-2**

Corers. (a) A heavy weight drives the core barrel into the sediment. (b) A piston corer can take a much longer core than can a gravity corer because of a piston in the core barrel.

FIGURE **B4-3**

Sediment cores. A scientist on board the CSS *Hudson* is examining the particle content of each layer in a sediment core to interpret the Earth's climatic history.

If the rock eroded slowly, then sediment was supplied at a low rate and likely was reworked by currents before it was buried by ever younger deposits. A slow rate of sedimentation generally gives more opportunity for water to sort the grains according to size, shape, and density. This can result in deposits of mud and sand that are well sorted (a small range of grain sizes) and have a uniform appearance, such as a coarse sand or very fine sand or silt. Rapid erosion supplies sediment at a high rate. Currents have little time to sort the grains before they are buried. As a result, such deposits tend to be poorly sorted (a large range of grain sizes) and heterogeneous (nonuniform in appearance, such as gravel mixed with sand, or sand mixed with mud).

In most cases, a clear relationship exists between the average grain size and the strength (energy) of bottom currents at the time that sediment is deposited. Stated simply, the average particle size of a deposit is proportional to the energy level present at the time of deposition. Under high-energy conditions, water is swift and turbulent, keeping fine grains in suspension and resuspending those fine particles that momentarily settled to the ocean floor. This constant agitation of the sea bottom separates small grains and transports them into quieter water, which typically is deeper than turbulent water. Thus, a coarse sand (rather than medium or fine) is deposited under high-energy conditions. Low-energy environments, where currents are weak and water is quiet, do not receive a supply of coarse grains, because the weak currents cannot transport them to these sites. Here muds typically accumulate. Consequently, the average grain size of a deposit of sediment serves as a good measure of the energy of the environment at the time of deposition. Fine-grained sediments denote low-energy conditions; coarse sediments, high-energy conditions.

In laboratory experiments, geologists have worked out the relationship between grain size and current velocity that specifies whether sediment of a particular size will be eroded, transported, or deposited under a given set of energy conditions. The general results of these experiments are summarized in the Hjulström diagram (**Figure 4–1**). This chart plots the average current velocity (the *y*-axis) against the particle diameter (the *x*-axis). Two curves divide the area of the plot into three distinct fields; these are labeled erosion, transportation, and deposition. The top curve specifies the speed that a current must have in order to erode grains of a particular size from the bottom. Note that, for sand and gravel, the larger the particle, the stronger the current must be to erode the material. Surprisingly, though, higher current velocities are needed to erode tiny clay and silt particles than to erode fine sand! At first thought, this may not appear to make sense. Part of the explanation lies in the fact that mud—a mixture of clay and silt—is more cohesive and, hence, "stickier" than sand, and so a stronger current is needed to dislodge a mud particle from the bottom than a fine grain of sand. However, if you examine the lower curve on the diagram, it shows that once eroded and moving, a clay particle, but not a fine particle of sand, can be

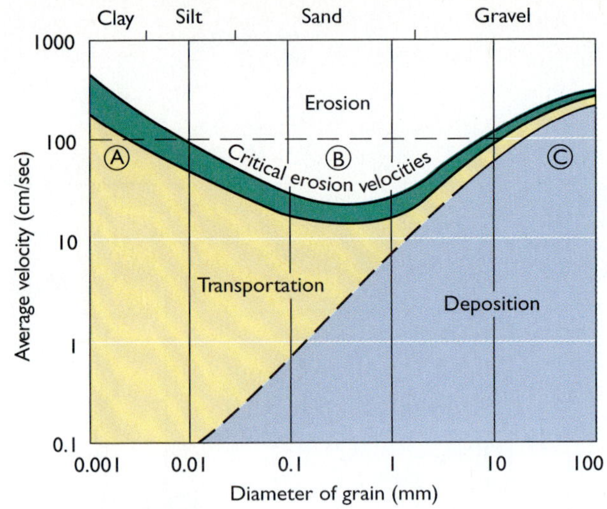

FIGURE **4-1**

Hjulström's diagram. This plot shows the average current velocities necessary for the erosion, transportation, and deposition of sediment particles of different sizes. The broad upper curve indicates that strong currents are needed to erode mud (clay and silt), coarse sand, and gravel, whereas weaker currents are sufficient to erode fine to medium sand. The lower curve indicates that once sediment is moving, the current velocity necessary to transport grains varies directly with particle size. For a given grain diameter, deposition occurs when the current speed is less than some critical value defined on the graph by the lower line. The letters A, B, and C near the top of the chart are discussed in the text. [Adapted from A. Sunborg, The River Klavaren, a study of fluvial processes, *Geografiska Annaler* 38 (1956): 125–316.]

transported by a weak current. This is because the settling rate of suspended particles also varies with their diameter. Small particles settle more slowly than do large grains and so are kept in suspension longer and are easily moved about by weak currents.

Let's review these important ideas before proceeding. Look at the top of the Hjulström diagram. The top line represents a current velocity of 1,000 centimeters per second (~22 mph). Note that this strong bottom current will erode all grain sizes ranging from clay-size to gravel-size material. Now examine the 100 cm/sec line (~2 mph). Under this energy condition, clay (Point A) will be transported but not eroded, grains ranging in size from silt to fine gravel (Point B) will be eroded, and coarse gravel (Point C) will be deposited. A bottom current flowing at 10 centimeters per second (~0.2 mph) is too weak to erode any sediment at all. Yet these current speeds are capable of transporting clay, silt, and sand, provided they are already in suspension, but not gravel, which will be deposited.

4-2

Sedimentation in the Ocean

Recall how water depth varies with distance from the continent. Proceeding seaward, the water of the continental shelf is shallow, rarely deeper than 150 meters (~495 feet). The shelf gives way to the continental slope. The water gets progressively deeper with distance from shore until in the deep sea, water depths average about 4,000 meters (13,200 feet). Away from a continent, two things happen: the water gets deeper and the sea bottom is farther from the source of terrigenous sediment, which is mainly sand and mud eroded from the land and supplied by rivers. On the basis of water depth, the ocean can be subdivided into two major areas of sedimentation. The continental shelf is shallow and close to sources of sediment from the land, and the deep sea is deep and far from river supplies of terrigenous sediment. Sedimentation processes are quite different in these two areas, and therefore each is examined separately.

SHELF SEDIMENTATION

A continental shelf is a relatively broad, essentially flat platform 70 to 100 kilometers (~43 to 62 miles) wide that represents the submerged edge of a continent. Water depths on the shelf are shallow, varying from zero at the shoreline to 120 to 150 meters (~396 to 495 feet) at the shelf break, where the gradient of the sea bottom steepens to about 4 degrees and marks the beginning of the continental slope. The sea floor of the shelf is nearly horizontal, with a regional slope that rarely exceeds 1 degree. Energy for eroding and transporting sediment grains is provided by the tides and wind-generated waves and currents.

Over most continental shelves, waves seem to be the dominant process affecting the sea bottom. Drawing from our swimming experience at the beach, we know that large waves contain more energy than do small waves. Diving beneath an unbroken wave, we will notice that the water becomes increasingly calmer with depth. In fact, if we go deep enough, the water will be relatively calm despite the surface agitation by waves. The bigger the waves, the deeper we must dive in order to escape the wave motion. We can infer that *bottom energy* induced by surface-water waves must diminish with distance offshore, because water depths increase seaward. This means that, if you began from the shore and walked seaward on the shelf bottom into deeper water, you first would be pushed around by the wave energy. As the bottom got deeper, you would feel less and less of an effect from the waves, until at some depth the water would be quite still.

The shoreline is affected by breaking waves, and these high-energy conditions suspend and remove all the fine sediment and allow mainly medium and coarse sand and gravel to be deposited on the beach and in the nearshore zone. Seaward from the nearshore zone, bottom energy induced by waves decreases because of increasing water depths. This decrease of bottom energy with water depth results in a systematic decrease in grain size with distance offshore. The beach, which is composed of coarse to medium sand and gravel, grades into fine sand farther offshore and, moving seaward, into muddy sand (sand with some mud), sandy mud (mud with some sand), and finally mud far offshore (**Figure 4–2a**).

There would be a decrease in grain size as the water depth over the shelf increased if sea level remained relatively fixed over time. Yet we know that because of glaciation and deglaciation in the recent geologic past, sea level has gone up and down. Sometimes it flooded the shelf as it does at present, and at other times it exposed the shelf so that it became covered with grasslands and forests rather than seawater. Sea level about 15,000 years ago was 130 meters (~429 feet) lower than it now is (**Figure 4–2b**). At that time the shoreline was located several hundred kilometers farther seaward than it is at present, virtually at the shelf break (**Figure 4–3a**). Consequently, the beaches were displaced some 150 kilometers (~93 miles) seaward of New York City (see Figure 4-3a) and about 300 kilometers (~186 miles) southeast of Texas. Since that time, glaciers have been melting, causing the seas to rise and flood the continental shelf. As the shoreline has advanced over the land, coarse sand and even gravel have been deposited on the outer shelf because of the shallow water depths that existed there in the past. Sea level is still rising worldwide, and much more flooding of land will occur if the ice sheets continue to melt.

So it seems that our original premise that grain size varies systematically with water depth is a valid one. Shallow waters over the inner shelf tend to be high-energy environments where coarse sed-

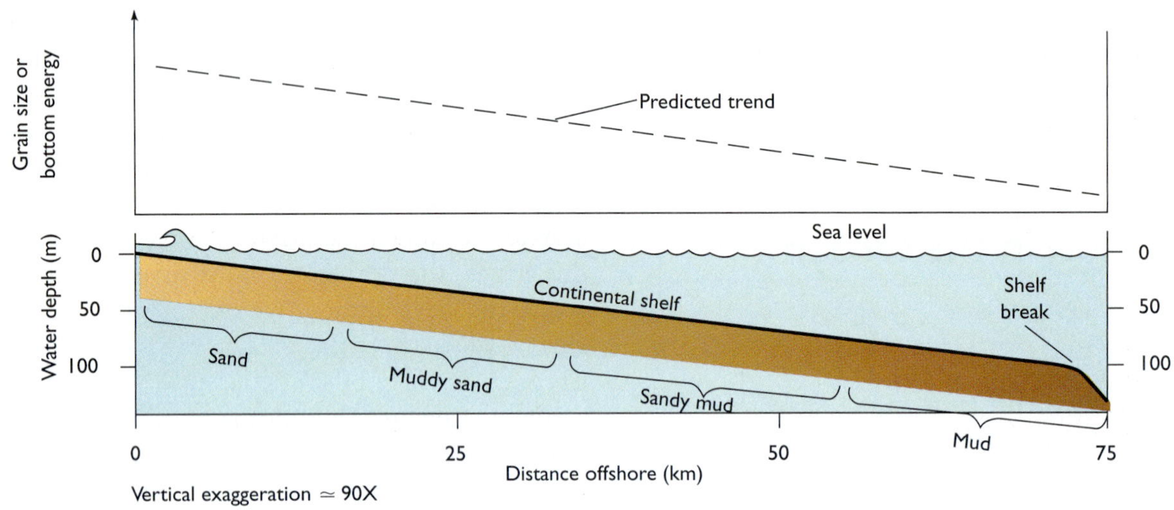

(a) MODEL PREDICTION OF SHELF SEDIMENTS

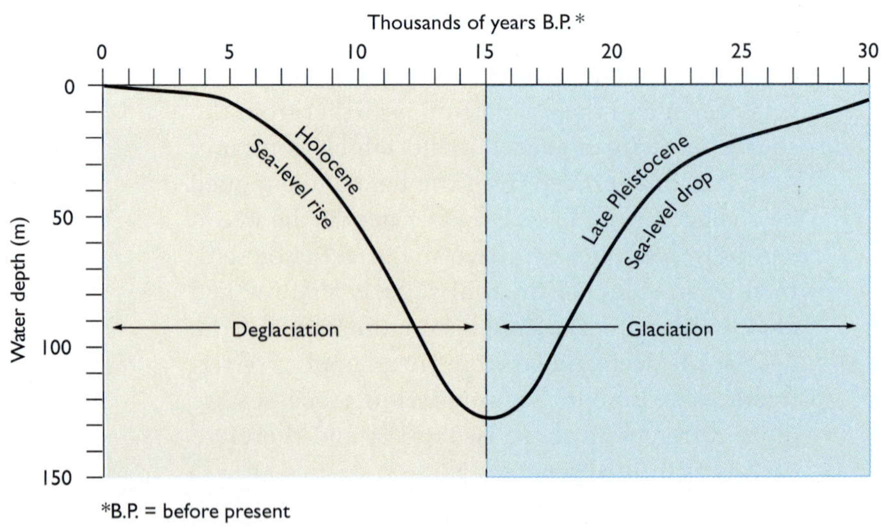

(b) POSITION OF SEA LEVEL FOR PAST 30,000 YEARS

FIGURE **4-2**

Shelf sedimentation. (a) Assuming that bottom energy on the continental shelf is inversely proportional to water depth and, hence, distance offshore, the sedimentary cover should grade systematically from coarse sands and gravel onshore to mud at the shelf break. (b) This curve shows that sea level was 130 to 140 meters below its present level some 15,000 years ago. Since then, sea level has risen to its current position, an event known as Holocene sea-level rise. [Figure 4-2b is adapted from K. O. Emery, The continental shelves, *Scientific American* 221, no. 3 (1969): 106–22.]

iment accumulates, whereas deep waters over the outer shelf tend to be low-energy environments where fine sediment is deposited. However, we must take into account fluctuations of sea level that cause the water depths at any point on the continental shelf to vary as a function of time. This explains why coarse sediment (sand and even gravel) blankets the outer shelf where water depths are deep and the bottom is quiet. These coarse-grained deposits on the outer two-thirds of the shelf are not in equilibrium with the low-energy conditions that exist there at the moment. Such material is termed **relict sediment** because it accumulated at an earlier time and under very different depositional conditions. By contrast, the coarse to fine sands that cover the inner third of the continental shelf are modern sediments that are in equilibrium with the bottom energy conditions. There has not yet been sufficient time for fine sediment to bypass this inner band of modern sand and be deposited farther offshore and cover the relict sediment (**Figure 4-3b**).

WORLDWIDE DISTRIBUTION OF SHELF SEDIMENTS

Many surveys of the distribution of sediment types of the continental shelf have been conducted

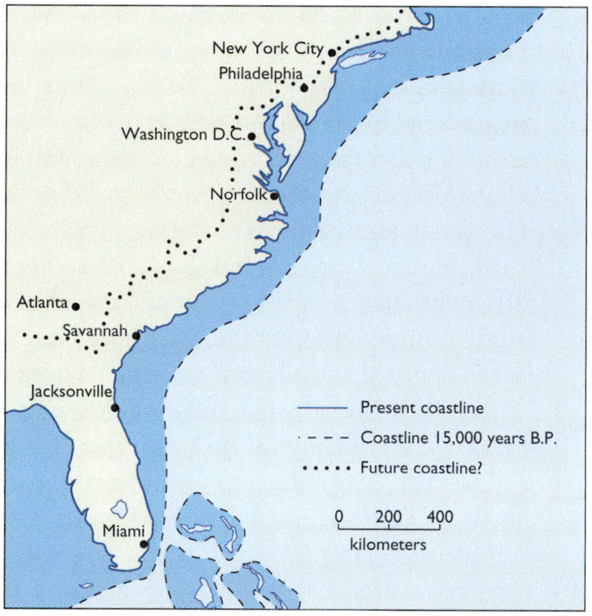

(a) COASTLINES PAST AND FUTURE

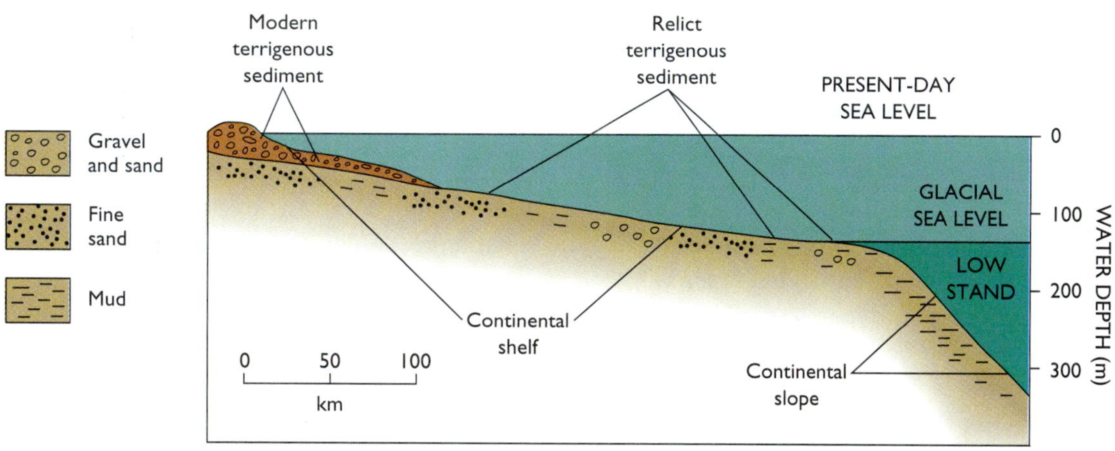

(b) RELICT SEDIMENT

FIGURE **4-3**

Shelf sedimentation. (a) About 15,000 years ago, the shorelines were displaced seaward by more than 100 kilometers, as sea level dropped due to widespread glaciation. Presently sea level is rising worldwide. If the world's ice sheets continue to melt, the oceans will flood many of the world's major cities in the future, as shown by the dotted line on the map for the East Coast of the United States. [Figure 4-3a is Adapted from K. O. Emery, The continental shelves, *Scientific American* 221, no. 3 (1969): 106–22.] (b) This cross-section shows that modern terrigenous sediment forms an apron over the inner continental shelf and that most deposits are relict in nature, having been deposited during one or more previous low stands of sea level.

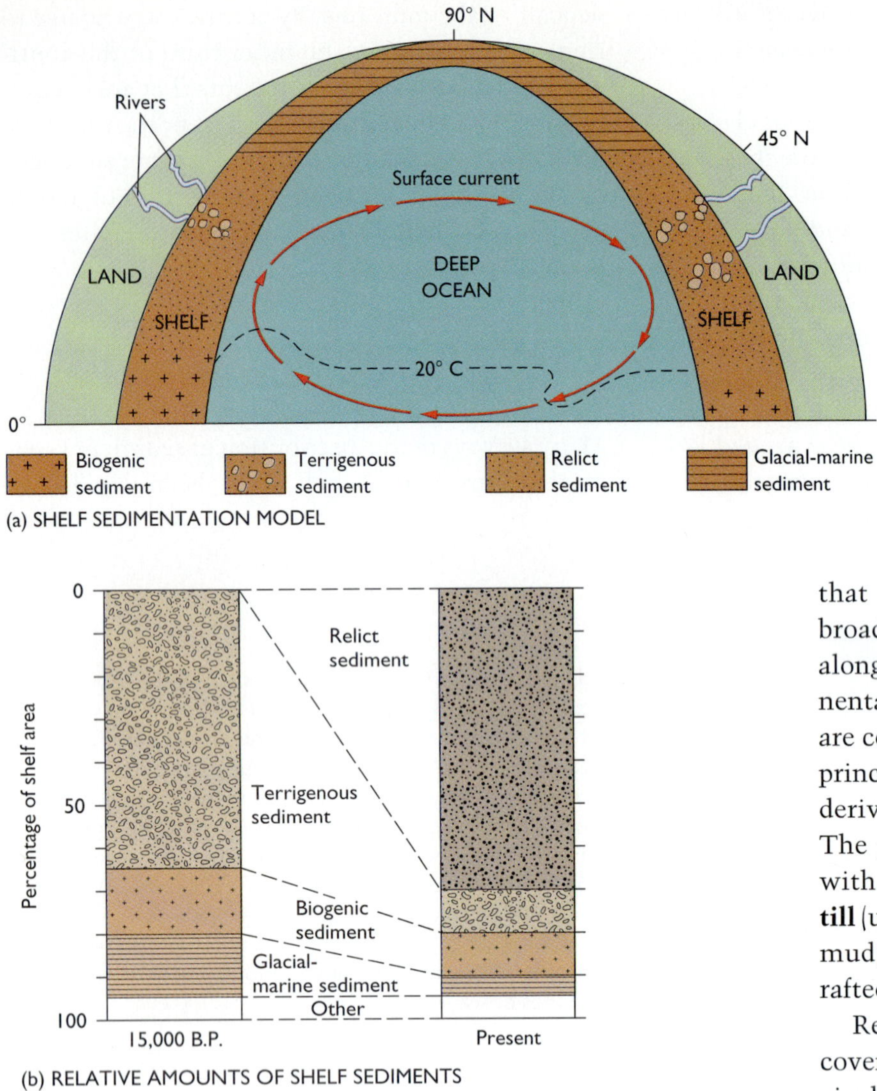

FIGURE 4-4

Distribution of shelf deposits. (a) To a first approximation, sediments on the continental shelves of the world vary with latitude. Biogenic sediment dominates the low latitudes, terrigenous sediment the middle latitudes, and glacial-marine sediment the polar regions. (b) About 60 percent of the sediment that blankets the world's continental shelves is relict in nature, meaning it was deposited in the past under conditions that are no longer evident. Relict sediments are not in equilibrium with the present-day shelf environments. [Adapted from K.O. Emery, The continental shelves, *Scientific American* 221, no. 3 (1969): 106–22.]

worldwide. What emerges from these studies is a regular pattern of sediment types that vary with latitude (**Figure 4–4a**) and that depend on climate. A broad band of biogenic sediment straddles the equator and extends into the subtropics. These deposits include coral reefs and accumulations of grain fragments composed principally of calcium carbonate ($CaCO_3$) derived from the hard parts of organisms such as clams, snails, sand dollars, coral, and calcareous algae. This band is broader along the western than the eastern edges of the oceans. Warm, westward flowing equatorial currents diverge from the equator and move poleward along the western sides of all the ocean basins. In contrast, the eastern sides of the ocean basins are bathed by cold currents that originate from high latitudes (see Figure 4–4a). Therefore, coral reefs that require warm water for growth occupy a broader part of the shelf along the western than along the eastern edges of ocean basins. The continental shelves of the middle or temperate latitudes are covered by river-supplied terrigenous deposits, principally sand-sized grains of quartz and feldspar derived from the weathering of granite on land. The polar shelves, not surprisingly, are littered with poorly sorted glacial deposits, either glacial **till** (unsorted deposits of boulders, gravel, sand, and mud) dumped there directly by glaciers or ice-rafted debris dropped from melting icebergs.

Recall, however, that most of the sedimentary cover of the continental shelves is not of recent origin, but is relict in nature. This material is not currently in equilibrium with the present-day water depths (or energy levels). Rather, it was deposited long ago when shorelines were displaced seaward of their present positions. The recent **Holocene sea-level rise** (see Figure 4–2b) has been so rapid in a geologic sense that shelf sediments have not had sufficient time to regain equilibrium with the new, deeper water conditions. No more than 30 to 40 percent of the sediment on the surface of the continental shelf is recent (modern) in origin, and this "new" sediment is confined largely to the inner shelf. The remaining 60 to 70 percent is relict in character (**Figure 4–4b**).

GEOLOGIC CONTROLS OF CONTINENTAL SHELF SEDIMENTATION

In order to understand the geology of continental shelves, it is necessary to recognize the importance of time in their development. Asking which factors distribute sediment across the shelf from day to day is very different from posing the same question for longer time spans, such as millennia to millennia (thousands of years) or geologic period to geologic period (tens of millions of years). An oceanographer studying sedimentation on the Gulf Coast shelf of Texas during the past year will draw conclusions far different from those of a geologist investigating the million-year-long sedimentary record of the same shelf. To clarify this very important point, let's assume that this part of the Gulf Coast is ravaged by hurricanes on the average of once every fifty years or so. Chances are very good that no hurricanes would have occurred during the brief one-year field survey. About 20,000 hurricanes would, however, have swept through the area in the past million years. Thus, what is perceived as a rare storm event when measured against a human lifetime becomes a regular, ongoing geologic process of critical importance to an understanding of the million-year-long sedimentary history of the area. Given this distinction, we will consider the geologic development of continental shelves from the perspective of three very different time frames: a thousand years to the present day, a million (10^6) to a thousand (10^3) years, and 100 million (10^8) to a million (10^6) years. (Before continuing, turn to Appendix III and take a minute to study the geologic time scale.)

■ 10^3 TO 0 YEARS: The comparatively brief time interval of a thousand years to the present day (10^3 to zero years) focuses on the "day-to-day" sedimentation processes on the continental shelves. Ocean surveys conducted during the past few decades probably are quite representative of conditions in general for the past thousand years. The climate has been relatively stable, and sea level has risen slightly—about a meter (~1 yard) or so—during this time span. Numerous ocean studies worldwide indicate that various currents move sediment across the shelf. The principal currents are generated by winds and tides.

Let's consider *wind effects* first. As the wind blows across the ocean's surface, it creates waves and currents. Waves grow in size and acquire more energy as wind velocity intensifies. Under calm winds, waves are small and have little effect on the shelf bottom. Storm waves, in contrast, are large and packed with energy, and their motion can extend downward to the very deepest parts of the shelf. As you will learn in Chapter 7, the water motion at the sea bottom induced by waves is back and forth, so that in theory there is no net movement of sediment grains. What waves do, however, is to raise sediment particles off the bottom momentarily. Once in suspension, the particles are moved by other currents—even weak ones that, acting alone, could not erode the particle from the sea floor. For example, along the coasts of Washington and Oregon, marine geologists estimate that sediment on the continental shelf at a water depth of 80 meters (~264 feet) is moved by wave action about 10 percent of the year, mostly during winter storms. Sediment on the outer New England shelf also is affected largely during winter storms called "nor'easters."

The drag of the wind on the water surface, if persistent, also produces currents that deepen with time. Because these currents are weather related, the direction and strength of their flow vary markedly from day to day. Under fair-weather conditions, wind-generated currents are weak and exert little influence on the shelf bottom. Currents raised by powerful gale winds, however, can attain impressive speeds and move even coarse sand on the shelf bottom. For example, wind-generated currents moving faster than 80 centimeters per second (~1.7 mph) transport silt and sand (see Figure 4–1) across the Washington-Oregon shelf.

Now let's consider *tidal currents*. The water level on continental shelves fluctuates daily in response to the tides. Locally, the strength of the tidal currents depends on the tidal range (the vertical distance between high and low tides), the slope and roughness of the sea bottom, and the configuration of the shoreline. Commonly, tidal currents by themselves are too weak to rework bottom sediment. Some shelves, however, are dominated by tidal *scour* (sediment-bed erosion), especially those that are affected by a large tidal range and have tidal currents that are constricted by land or by a shallow bottom. The effect is something like what happens when you place your thumb over the spigot of a hose that is dribbling water. The constriction of the flow by your thumb causes it to shoot out strongly. A prime example of this effect is the shallow shelf around the shores of England, France, and Ireland. Here the speed of tidal currents regularly exceeds

Dust Storms

In the Caribbean Sea, you can walk the deck of your boat anchored off the Virgin Islands and have your bare feet get covered with red African soil. It's hard to imagine that these small particles of sediment that have rained down on your vessel during the calm night originated from a massive dust storm that occurred a week before in the Sahara Desert of North Africa. Satellite images (**Figure B4-4a**) reveal that dust plumes intermittently cover enormous tracts of the North Atlantic Ocean. The photographs clearly show that these vast murky clouds originate in northern Africa, the bigger dust billows extending far to the west and reaching the southeastern United States, the Caribbean Sea, and even the faraway Amazon rainforests. Worldwide, scientists estimate that perhaps as much as two billion metric tons of dust are blown into the air every year. Surprisingly, microorganisms and chemical toxins that adhere to these small particles are carried aloft as well and are dispersed far downwind from their points of origin. This flux of airborne particles indicates how the land, the ocean, the atmosphere, and, as we shall see, human activity are linked in a surprising variety of ways.

Actually, dust storms have been an ongoing presence in the region for millions of years, as indicated by the content of desert sand mixed into the pelagic mud that has been cored from the deep-sea bed of the North Atlantic Ocean. Recent investigations indicate that about 500 million metric tons of airborne dust advected from North Africa settles into the North Atlantic Ocean each year. During the last few decades, however, the number, size, and concentration of particles in these dust plumes have increased dramatically because of the agricultural and industrial impact of human beings living in northern Africa. Here, particularly in the Sahel region, which borders the southern fringe of the Sahara Desert, intensive farming to feed hungry and starving people combined with prolonged droughts that began in the 1960s has allowed desert winds to erode soil, microbes, pesticides, plastics, and other natural and industrial chemicals from northern Africa and sweep them across the Atlantic Ocean. Some of this fine particulate debris settles on boats that cruise the Caribbean Sea. Interestingly, scientists are just now discovering that there are grave ecological consequences associated with this massive influx of airborne dust. What follows are brief descriptions of some of these potential ecological effects.

■ NUTRIENT INPUT TO RAINFORESTS

Biologists have documented the fact that a regular airborne supply of essential nutrients, some thirteen million metric tons per year derived from African soils, drops onto the upper canopy of the Amazon rainforests. Rainwater washes the nutrients off the leaves onto the soil for uptake by the shallow tree roots. The fallout of these wind-blown fertilizers, in part, helps explain the incredible productivity and diversity of the plant communities of the Amazonian rainforests, despite the sterility of their leached soils.

■ DECLINE OF CARIBBEAN CORAL REEFS

Dust plumes from northern Africa inject nutrients and other vital chemicals into the clear blue waters of the Caribbean Sea. This promotes the growth of seaweeds, which overgrow and suffocate the coral polyps. Also, 90 percent of the Caribbean-wide population of the sea urchin *Diadema anitllarum* was recently killed by a disease allegedly transmitted by pathogens carried in dust plumes from North Africa. These sea urchins are voracious grazers, and they kept the coral clean of algae. Their collapse has allowed an uncontrolled growth of algae, which has overwhelmed and killed many corals of the Caribbean Sea.

■ RED TIDE OUTBREAKS

The wind-blown dust reaching the western Atlantic Ocean from Africa is iron rich and its fallout can change dramatically the chemistry of seawater.

100 centimeters per second (~2.2 mph) and locally exceeds 500 (~11 mph). An inspection of the Hjulström curve (see Figure 4–1) shows that such currents are capable of transporting sediment as coarse as gravel.

In contrast to wave-dominated shelves, which are static most of the year because of fair weather, much of the sediment of tide-dominated shelves is mobilized each day by bottom currents. However, the greatest volumes of sand are still moved during storm activity. Under storm conditions, the tide- and wave-generated bottom currents act together, causing an enormous drift of sand and gravel along the sea bottom. It has been estimated that the transport rate of sediment on the tide-dominated shelf of England is increased substantially during a storm, perhaps by as much as a factor of 20.

For example, the dissolved iron content of the waters of the Florida Keys has, on occasion, increased by as much as 300 percent following a dust-plume event. This infusion of iron has caused a large plankton bloom of bacteria, which in turn raised the nitrogen levels in the shallow water of the Florida Keys. The elevated nitrogen concentrations, marine biologists suspect, triggered a red tide (**Figure B4-4b**) that devastated local ecosystems, killing millions of fish and a hundred or so manatees, as well as making many people ill.

MICROBE TRANSPORT

An astonishing discovery is that a large variety of microorganisms can be wafted across the Atlantic Ocean in dust plumes. The soils of the Sahel region of Africa have a rich assortment of bacteria, viruses, fungi, and other microbial types. Some are pathogens and they may have infected and killed sea fans in the coral reefs of Florida and the Caribbean region. Recent kills of orange groves in Florida by infectious diseases likewise may be the result of the wind-blown transport of viruses from northern Africa.

HUMAN HEALTH EFFECTS

The use of pesticides and herbicides and the burning of garbage with its load of plastics are releasing many toxins into the air. These chemicals get incorporated into the dust plumes that rain down in the southeastern United States and the Caribbean region. High levels of mercury, radioactive lead, and beryllium-7 have been measured in the fallout from the dust plumes emanating from northern Africa. The long-term health consequences for people of these chemical toxins are just beginning to be investigated. Even the dust itself is a health hazard, as indicated by the dramatic increase in cases of asthma attacks in Barbados when the air is choked with African dust.

Visit www.jbpub.com/oceanlink for more information.

FIGURE **B4-4**

(a) This satellite image reveals the presence of dust plumes stretching from Africa far into the Atlantic Ocean. (b) The reddish tints in this satellite image reveal a red tide (the dark reddish color) along Florida's Gulf coastline. The fallout of wind-borne, iron-rich dust that orginated in North Africa triggers some of these events.

(a)

(b)

■ 10^6 TO 10^3 YEARS: By geologic standards, a million (10^6) years is a mere fleeting moment of time. What occurs over the course of a million years usually has little consequence for the overall evolutionary history of a continental margin, which can stretch over hundreds of millions of years. An exception to this way of thinking, however, is the record of glaciation and deglaciation events over the past few million years. These are so current that their record has not yet been destroyed by more recent events and driven into geologic obscurity. This is like trying to remember details of your life as a six-year-old child when you are sixty years old. There is not much that you can recall about those times. Yet when you turned seven, abundant details of the previous year were fresh in your mind, not yet having been obliterated by living a long life.

The last million years (the **Pleistocene epoch** or Ice Age; see Appendix III) have been dominated by glaciation. During this time ice sheets affected the surface morphology and the sediment composition of the world's continental shelves by controlling the level of the oceans. Sea level worldwide went up and down in direct response to the expansion and contraction of ice caps. With the onset of glaciation, snow was stored on land and converted to glacial ice. This caused sea level to drop about 100 to 150 meters (~330 to 495 feet) below its present level, and the shelf surface became exposed land. During the warm interglacial intervals of the Pleistocene epoch, ice sheets melted, causing sea level to rise worldwide and flood the continental shelves. The Earth's temperature trends for the past thousand years of the current interglacial stage are examined in the featured box "Climate Variability and Change."

During the low stands of sea level, which were times when water was stored as ice on land, sedimentation on the shelves was altered in many ways. Some of the more important modifications follow.

1. Ice caps and glaciers extended onto the shelf proper, particularly in the high and middle latitudes. Here, moving ice scoured, abraded, and plucked the shelf bottom. In some cases, it smoothed out the tops of hills; in others, it gouged out holes and created considerable topographic relief. Also, glaciers dumped debris on the shelf floor, forming blankets and ridges of glacial till.

2. As sea level dropped, rivers extended their channels across the continental shelves, cutting into the marine deposits. This resulted in the deposition of river, sands and mud over marine deposits. During the height of glaciation, river deltas formed at the shelf break, which was the location of the new shoreline at that time. Eventually, the exposed shelf surface became vegetated, supporting forests and grasslands, and was populated by terrestrial animals.

3. Deposition and erosion rates increased in the deep-sea environments that are adjacent to the continental shelves. During low stands of sea level, rivers dumped their sediment loads near the shelf break and on the upper continental slope. Many of these deposits were unstable and slumped, producing sediment-laden bottom currents, which flowed downslope into the sea because they were denser than the surrounding water. These currents, in particular, helped to carve out or deepen submarine canyons. Massive amounts of sediment were funneled through these submarine canyons onto the deep-sea floor. Many of these canyons are now inactive, because the shoreline and rivers with their large supply of sediment were displaced landward away from the submarine canyons as sea level rose.

4. Coral reefs were killed on a massive scale. Reef banks that flourished during high stands of sea level were stranded high and dry once sea level dropped with the onset of a period of glaciation. This led to widespread death of coral reefs. New coral banks were established seaward of the old reef mass, or new coral growth occurred along the lower fringes of the older, now dead reef bank. When sea level rose during the warm interglacial periods, the reefs responded by growing upward and keeping pace with the rise of sea level.

- 10^8 TO 10^6 YEARS: The span of 100 million to 1 million years (10^8 to 10^6) is enormous and stretches far back into the geologic past to the age of the megacontinent Pangaea. The central and western parts of North America would have been unrecognizable to us today, as much of them were covered by warm, shallow seas, with a long chain of volcanic islands marking a complex subduction zone (**Figure 4–5a**). Recall that subduction zones are compressional boundaries (areas of "squeezing" together) where one plate is forced to dive beneath another plate. What is now the eastern edge of North America was a zone of tension (pulling apart), marked by fault-bounded rift valleys. At that time, North America had separated from Pangaea (**Figure 4–5b**), its eastern edge having just broken off from Africa, Europe, and Greenland. As Africa, Europe, and Greenland were moving eastward relative to North America because of seafloor spreading, a gulf was created between them and North America that widened with time into a young Atlantic Ocean. Thus, the North American continent had developed two distinct continental edges: a tectonic (active) Pacific-type of continental margin to the west, and a nontectonic (passive) Atlantic-type of continental margin to the east (**Figures 4–5c** and **4–5d**).

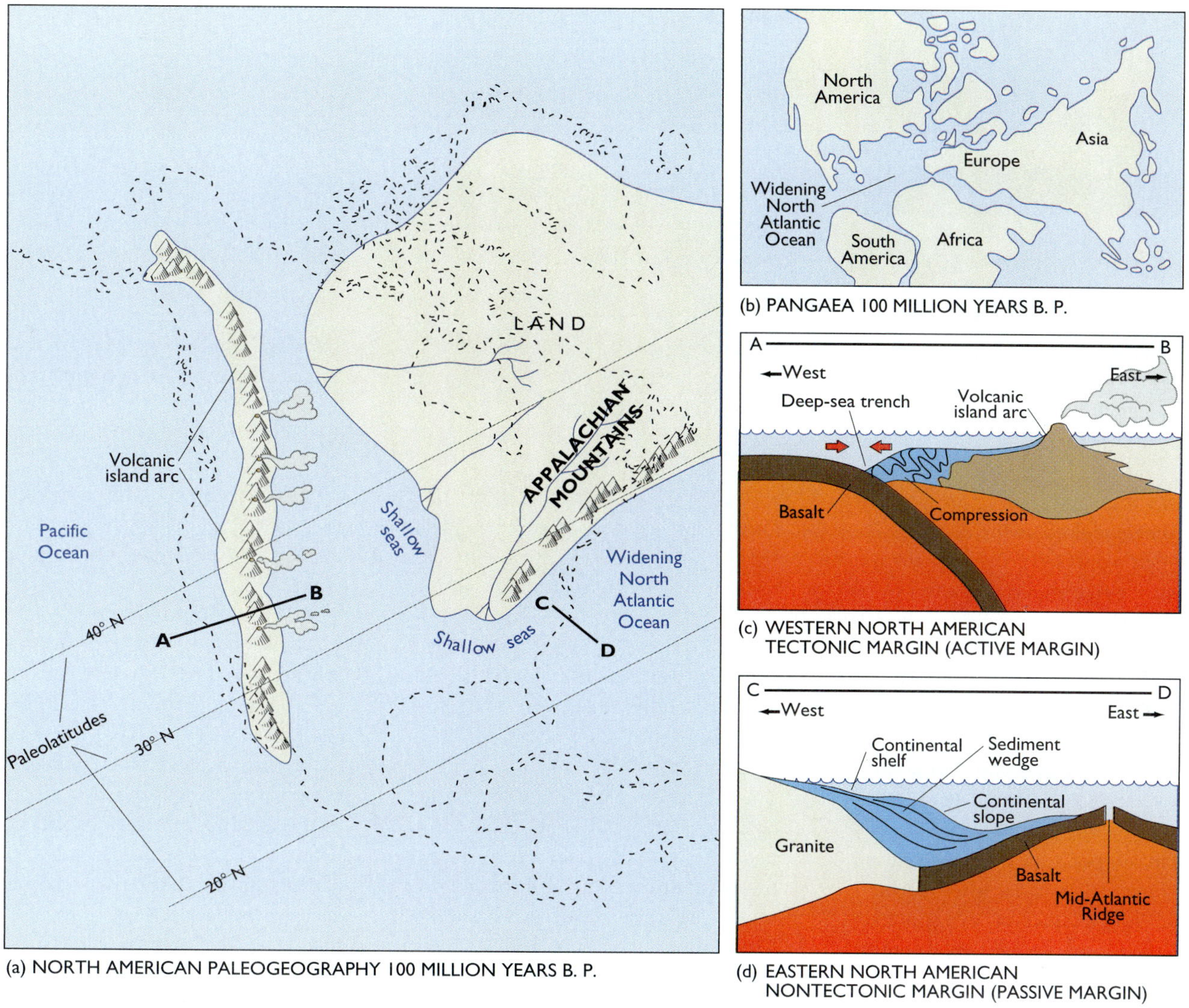

FIGURE 4-5

Paleogeography of North America. (a) During the Cretaceous period (100 million years ago), the eastern edge of North America was undergoing rifting and sinking as the newly formed Atlantic Ocean was expanding. Most of central and western North America was covered by shallow seas, except for a prominent chain of volcanoes that marked a subduction zone where two plates were colliding. [Adapted from R. H. Dott, Jr., and R. L. Batten, *Evolution of the Earth* (New York: McGraw-Hill, 1981), 402.] (b) The relative position of landmasses during the Cretaceous period indicates that the juvenile North Atlantic Ocean was widening slowly during that time. (c) Western North America was a subduction boundary, with the Pacific plate sliding beneath North America. (d) Along the eastern edge of North America, sand and mud eroded off the Appalachian Mountains were deposited to form a thick and wide continental margin.

Passive **Atlantic-type margins** are characterized by a long history of sedimentation. The sea floor at the edge of the continent sinks so gradually that the buildup of sediment keeps pace with the subsidence (sinking), and the shelf bottom remains shallow. Tectonic processes are important mostly during the earliest stage of evolution of passive continental margins. This is the time when continents are being ripped apart by plate tectonics (**Figure 4–6a**) and the crust at the new edge of the continent has been fractured, heated, and intruded by basalt rising along the newly formed spreading ridge. This early tectonic phase is comparatively short-lived, however, because the continents spread away from the active ridge. What follows is a long period of sedimentation. Terrigenous sediment eroded from the continent accumulates on the shelf, causing the continental margin to widen

THE PROCESS OF SCIENCE

Climate Variability and Change

The Pleistocene Epoch consisted of a series of Ice Ages, each lasting about a hundred thousand years, when cold temperatures resulted in the buildup of huge ice masses on land. The Ice Ages were separated from one another by interglacial stages lasting between ten and twenty thousand years during which ice sheets melted as the climate warmed worldwide. Because the most recent Ice Age ended some ten to fifteen thousand years ago, it is reasonable to suspect that the Earth may be poised at the brink of another Ice Age. Can science look into the future and test this possibility?

One approach at answering the question is to determine as accurately as possible the pattern of average global temperature over the past millennium and consider whether the trend can be justifiably extrapolated into the future. Accurate temperature measurements with thermometers go back only to the beginning of the twentieth century. Surface temperatures from earlier times are reconstructed from proxy evidence, such as tree-line positions, tree rings, the growth of coral reefs, and variations in lake and ocean sediments. As expected, the margin of error for establishing an average global temperature for the Earth increases the further back in time one goes.

Over the past thousand years, the Earth's surface temperature has varied significantly. There is a strong indication that an irregular cooling trend was underway from A.D. 1000 to about 1850 (**Figure B4–5**) although with wide geographic variability. This cooling trend ended the warmer "Medieval Little Optimum" that occurred between A.D. 900 and 1100. That was a time of sea exploration and Norsemen colonized Greenland and Iceland because the Arctic pack ice lay far to the north and no longer impeded ship travel. Grain was grown in Iceland and Greenland at that time, and the local fisheries flourished. Beginning in the thirteenth and fourteenth centuries, temperatures began to drop globally, initiating the "Little Ice Age" (see Figure B4–5). Mountain glaciers advanced, the cover of sea ice in the North Atlantic expanded, snowfall increased, the snow cover lasted longer, and rivers froze during the winter as never before. The Norse settlements in Iceland and Greenland collapsed as grain harvests and fisheries failed, and expansive pack ice in the North Atlantic Ocean prevented sea voyages between those colonies and Europe.

This nine hundred-year-long cooling suddenly ended in the early twentieth century; surface Earth temperatures have been rising sharply since then (see Figure B4–5). The general consensus among climatologists is that this temperature upswing is mostly a response to the buildup in the atmosphere of heat-trapping greenhouse gases such as carbon dioxide that have been released as a byproduct of the burning of fossil fuels. The carbon dioxide loading of the atmosphere by humans is expected to

and its deposits to thicken (**Figures 4–6b** and **4–6c**). With continued drifting away from the spreading ocean ridge, the crust and the underlying mantle at the passive edge of the continent cool and contract, and are weighed down by the sediment load. The edge of the continent sinks continually, providing ample room for the incoming loads of river sediment. The sand fraction of these river-supplied sediments is reworked by waves and currents and dispersed across the shelf. But the fine-mud load is kept in suspension and is moved seaward, where it may settle to the floor of the continental slope or bypass the continental margin entirely and eventually reach the abyssal depths of the deep-ocean floor. Study Figure 4–6 carefully; it demonstrates the development of the Atlantic continental margin of North America. The result of this long history of deposition is a broad, smooth continental shelf.

Now, let's contrast the Atlantic-type margin to the Pacific-type. Active **Pacific-type margins** not only receive sediment eroded from the nearby land, but also are affected by strong deformational forces. Subduction zones, you will recall, are boundaries where plates converge and are compressed. These strong tectonic forces squeeze the beds of sediment between the colliding plates, folding and faulting the sedimentary layers (**Figure 4–7a**). Also, the sedimentary layers and basalt are scraped off the top of continue well into the twenty-first century. This suggests that the Earth's surface temperatures are going to continue to climb into the foreseeable future. At present, the global warming rate is estimated to be about 1°C every forty years, a rate that is much faster than it has probably been at any time in the past. So, rather than a prospect of chilly temperatures and the onset of an Ice Age, it appears as if the Earth will become warmer than it has been for several million years. Let it be noted, however, that there have been geologic periods in the past when global temperatures were much higher than those expected during the twenty-first century.

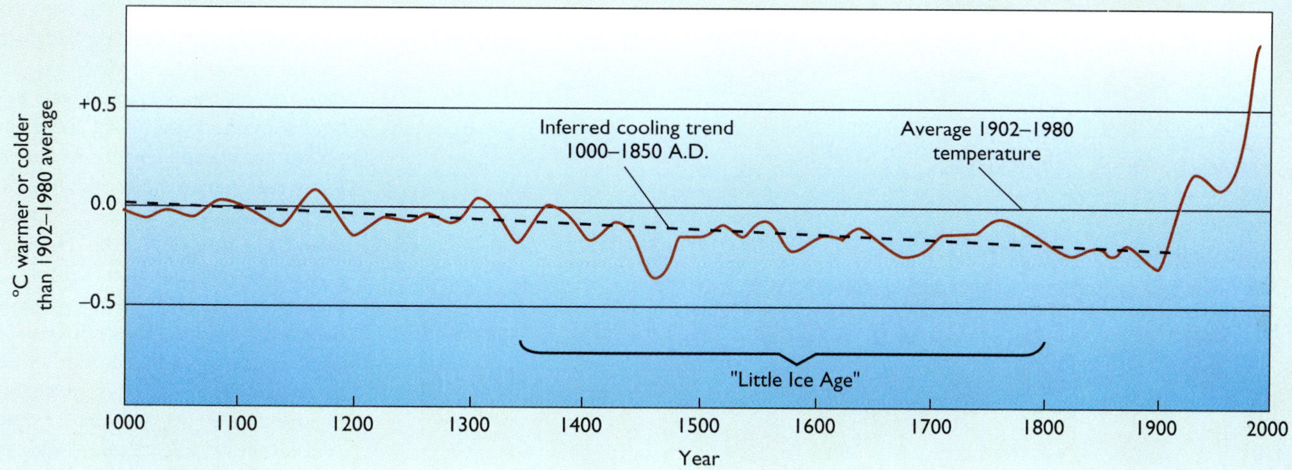

FIGURE **B4-5**

Earth's surface temperature. Since 1000 A.D., the average global temperature has been dropping slowly. This trend was reversed in the early twentieth century as global surface temperatures began to rise sharply. Temperatures continue to rise, presumably due to atmospheric loading of heat-trapping gases such as carbon dioxide. [Adapted from M. E. Mann, R. S. Bradley, and M. K. Hughes, *Geophysical Research Letters*, 15 March, 1999.]

the plate that is being forced downward by the upper plate. (This is roughly analogous to passing your hand across the top of a cake covered with frosting; as you do this, the icing piles up against your hand.) This tectonic activity creates an **accretionary prism** (**Figure 4–7b**)—a compressional zone situated between the deep-sea trench and the volcanic arc that widens with time as sediments are continually deformed and plastered to its seaward side by the subduction of plates. Along the nearby active volcanoes, sedimentary debris is derived from the erosion of the volcanic flows and deposited mostly underwater around the flanks of the volcanoes. Earthquakes abound here as well, and these trigger underwater landslides and slumps, which move large volumes of sedimentary debris to the deep-sea trench. Eventually, these materials get crushed against and are added to the accretionary prism. The result is a continental shelf that tends to be narrow and to have an irregular surface.

CARBONATE SHELVES

Because continents are drained by large river systems, a large supply of terrigenous debris is deposited onto continental shelves. Therefore, most shelves are covered by sand and mud composed of quartz and feldspar, the two principal minerals

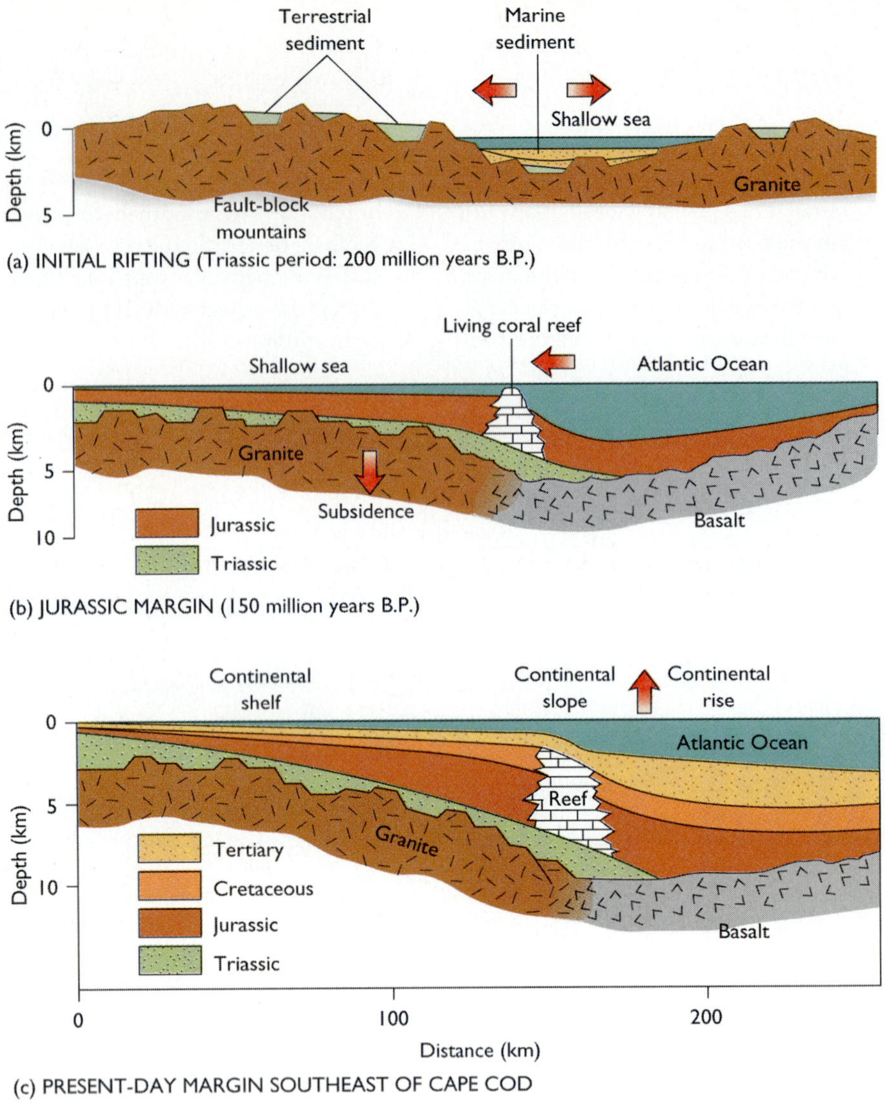

FIGURE 4-6

Development of a passive Atlantic-type margin. (a) The Atlantic continental margin of North America began when tension broke the crust into a series of fault-bounded ridges and basins. The rift basins were first filled in with river deposits and then with shallow marine deposits as seawater flooded the juvenile Atlantic basin. (b) Fifty million years after the initial rifting, subsidence of the granitic crust and deposition led to the development of a thick cover of marine sedimentary deposits. Far offshore, a thriving coral reef grew upward, keeping pace with crustal subsidence. (c) The present-day continental margin off New England shows a thick, broad sedimentary prism that completely buries the offshore reef, which died sometime during the Cretaceous period. [Adapted from J. A. Grow, R. E. Matlick, and J. S. Schlee, Multichannel seismic depth sections and interval velocities over outer continental slope between Cape Hatteras and Cape Cod in *Geological and Geophysical Investigations of Continental Margins,* Memoir 29, J. R. Watkins et al., eds. (Tulsa, Okla.: American Association of Petroleum Geologists, 1979).]

that comprise the granitic rock of the continents. Where the supply of river sediment is low and the waters of the shelf are warm, there can be a buildup of calcium carbonate ($CaCO_3$) material composed of the shells of organisms. These shallow areas are called *carbonate shelves* or *platforms*. At present, carbonate sediment covers relatively few continental shelves of the world. Modern examples are located in tropical and subtropical oceans, such as those off southern Florida, the Bahamas, the Yucatán Peninsula of Mexico and nearby Central America, and northern Australia (**Figure 4–8**). The shelf water in these regions is clear, warm, shallow, and sunlit—conditions essential for the growth of abundant carbonate-secreting organisms. Moreover, if carbonate sediment is to accumulate, the input of terrigenous sand and mud must be minimal, because muddy water not only interferes with the growth of reef organisms, but also dilutes the carbonate contribution by organisms to the bottom sediments. Therefore, carbonate shelves are located away from large rivers with their supply of terrigenous sediment. Carbonate deposits also accumulate around the shallow edges of some islands, forming carbonate platforms; a fine example is the Bahama platform located east of southern Florida. The presence of thick and extensive sequences of carbonate rocks, such as **limestone** (cemented calcium-carbonate mud and sand), in the Rocky Mountains, the Alps, the Himalayas, the Andes, and other areas indicates that carbonate deposition was more widespread during certain geologic periods of the past than it is today. This shows how changeable environments are over long peri-

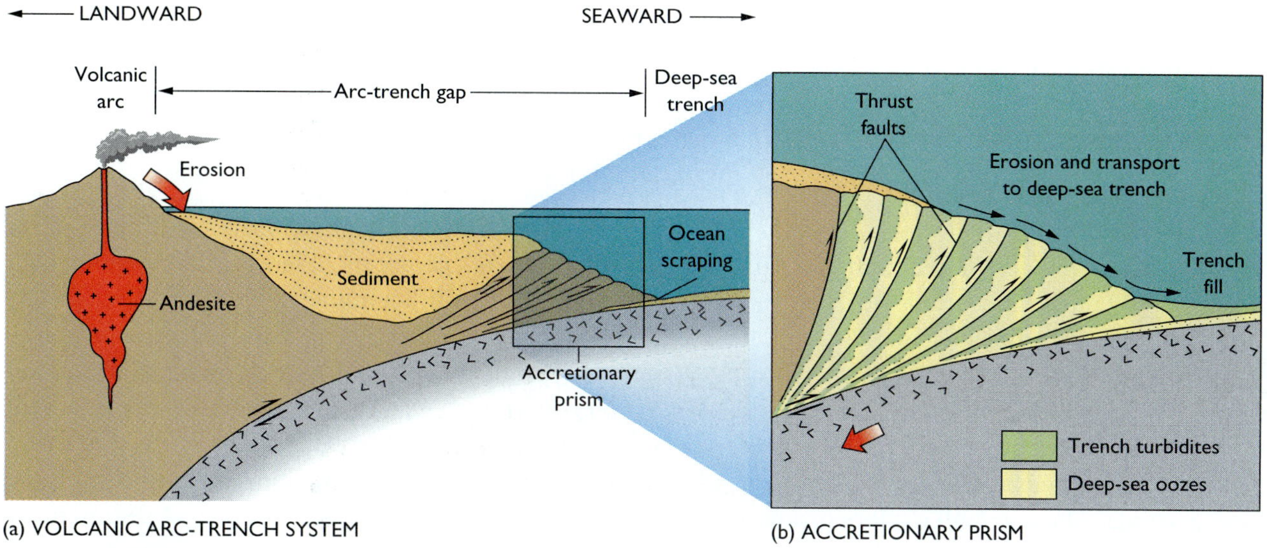

FIGURE **4-7**

Subduction tectonics and sedimentation. (a) A subduction boundary includes three principal zones: the volcanic arc, the arc-trench gap (consisting of a sediment basin and an accretionary prism), and a deep-sea trench. Sediment is added to the arc-trench in two ways: (1) debris erodes from the volcanic arc slumps or is dumped by bottom currents into the ocean basin; or (2) deep-sea deposits and trench sediments are scraped off the descending plate and plastered onto the accretionary prism. (b) The accretionary prism widens with time as slices of trench turbidites and deep-sea oozes are added systematically to its base. [Adapted from D. R. Seely, P. R. Vail, and G. G. Walton, Trench slope model in *The Geology of Continental Margins*, C. A. Burke and C. L. Drake, eds. (New York: Springer-Verlag, 1974).]

FIGURE **4-8**

Distribution of carbonate shelves. Carbonate shelves are confined to tropical and subtropical settings, where the water is shallow, warm, and clear. Such sunlit environmental conditions foster the vast growth of carbonate-secreting organisms. The accumulation of carbonate sediment on the continental shelves occurs where the input of river sediment is negligible.

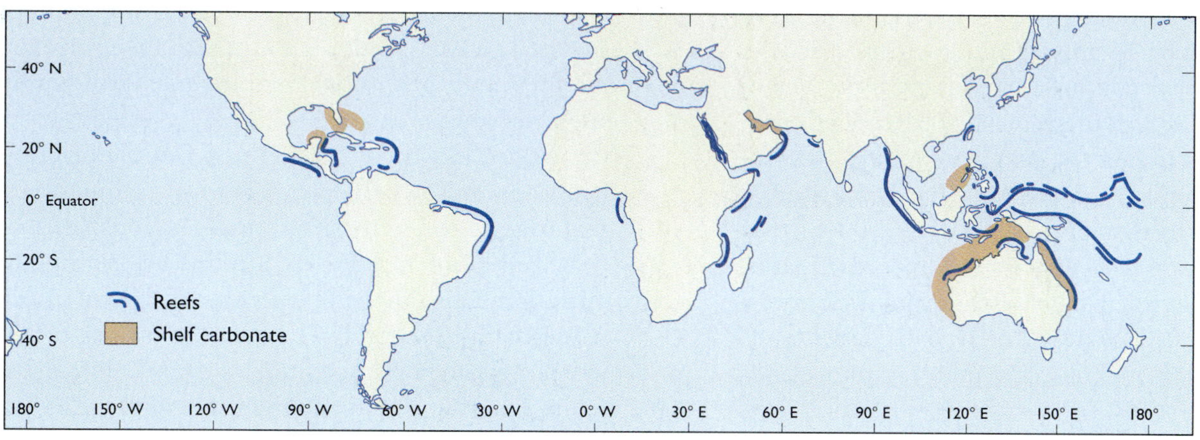

ods of time and how, in the geologic past, seas have repeatedly drowned the land and then retreated.

DEEP-SEA SEDIMENTATION

There are two main sources of sediment for the deep-ocean floor: (1) terrigenous mud and sand that bypass the shallow continental shelf and (2) the hard parts of surface-water microorganisms that settle to the deep-sea bottom. We will begin by examining a few details of these sediment sources and then talk about the way that sediment is dispersed along the floor of the deep sea. Then we will explore in some depth (pun intended!) how deep-sea sedimentation patterns are influenced by plate tectonics.

SOURCES OF SEDIMENT TO THE DEEP SEA

Sediment that settles to the bottom of the deep sea is derived from either external or internal sources (**Figure 4-9a**). External sources are the terrigenous rocks of the land. Weathering, the chemical and mechanical disintegration of rock at or near the Earth's surface, breaks down the bedrock of the land into small particles—mainly sand and mud—that are transported to the oceans by rivers and winds. The major sources of terrigenous sediment in the oceans are rivers that drain large mountain belts, such as the Himalayas of Asia (**Figure 4–9b**). Here, steep, swift, and powerful rivers disgorge large quantities of sand and mud to the ocean. Internal sources of sediment furnish material that is produced largely by organisms and, to a lesser degree, by geochemical and biochemical precipitation of solids, such as ferromanganese nodules (hard pebbles enriched in metals). As a general rule, the proportion of deep-sea sediment derived from external sources (the terrigenous material) relative to that derived from internal sources (the biogenic material) decreases with distance offshore. In other words, the farther from the river supply, the greater tends to be the fraction of biogenic material in deep-sea deposits.

Next, let's examine the specific processes responsible for dispersing sediment to the floor of the deep sea. Once we understand these processes, we will be in a position to discuss patterns of sediment distribution in the open ocean far from the influence of land.

SEDIMENTATION PROCESSES IN THE DEEP SEA

A simple classification of deep-sea deposits uses three broad categories based on the mode of sedimentation. **Bulk emplacement** is the means by which large quantities of sediment are transported to the deep-sea floor as a mass rather than as individual grains. The processes of bulk emplacement are induced by gravity: material resting high up on a slope moves downward and comes to rest on the sea floor of the deep sea. All types of sedimentary debris—terrigenous and biogenic, both fine- and coarse-grained—can be swept seaward and dispersed across the deep-ocean floor by bulk emplacement. In contrast, **pelagic sediment** is the fine-grained fallout of terrigenous and biogenic material that settles through the water column, particle by particle, much as snowflakes fall out of the sky and accumulate as a snow cover on land. **Authigenic sediment,** as you may recall, consists largely of biochemical precipitates that form in situ (in place); that is, they originate at the site of deposition by geochemical and biochemical reactions. Let's examine each type of sediment in more detail.

■ BULK EMPLACEMENT: Terrigenous debris supplied by rivers enters the ocean along its edges, where most of it is deposited at the shoreline and the inner continental shelf. However, during low stands of sea level, rivers extend their channels seaward and drop sediment on the outer continental shelf and upper continental slope. Buildup of such sediment can cause local instability and slope failure, which, under the influence of gravity, leads to downslope transport of sedimentary material as coherent slump masses, loose debris flows, or fluid mudflows. **Slumps** are sediment piles that slide downslope intact, with little internal deformation of the moving mass. In other words, bedding in the sediment pile is disturbed by folding, but is preserved when the whole mass of the deposit slides downhill as a sedimentary package. Examples of large slump masses are provided by seismic-reflection profiles (see the boxed feature, "Probing the Sea Floor"), which show large

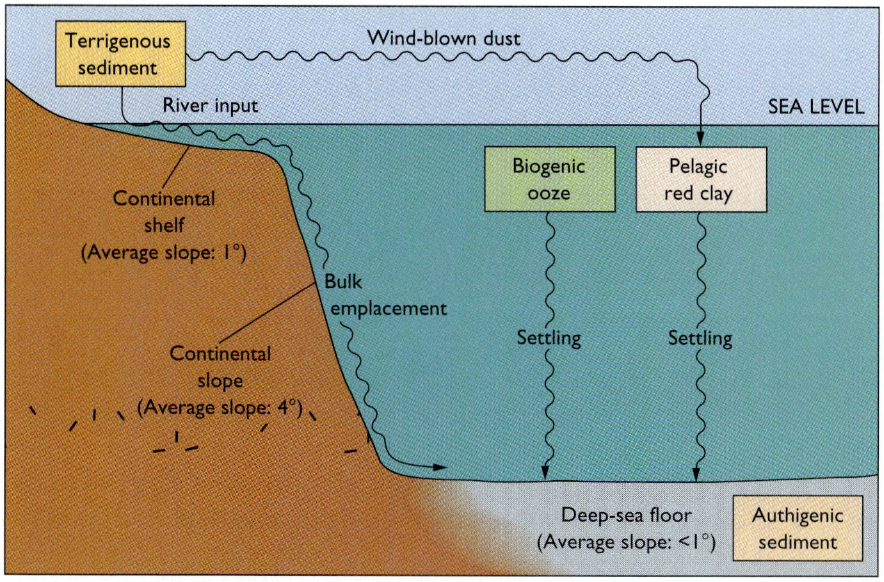

(a) SEDIMENTATION IN THE DEEP SEA

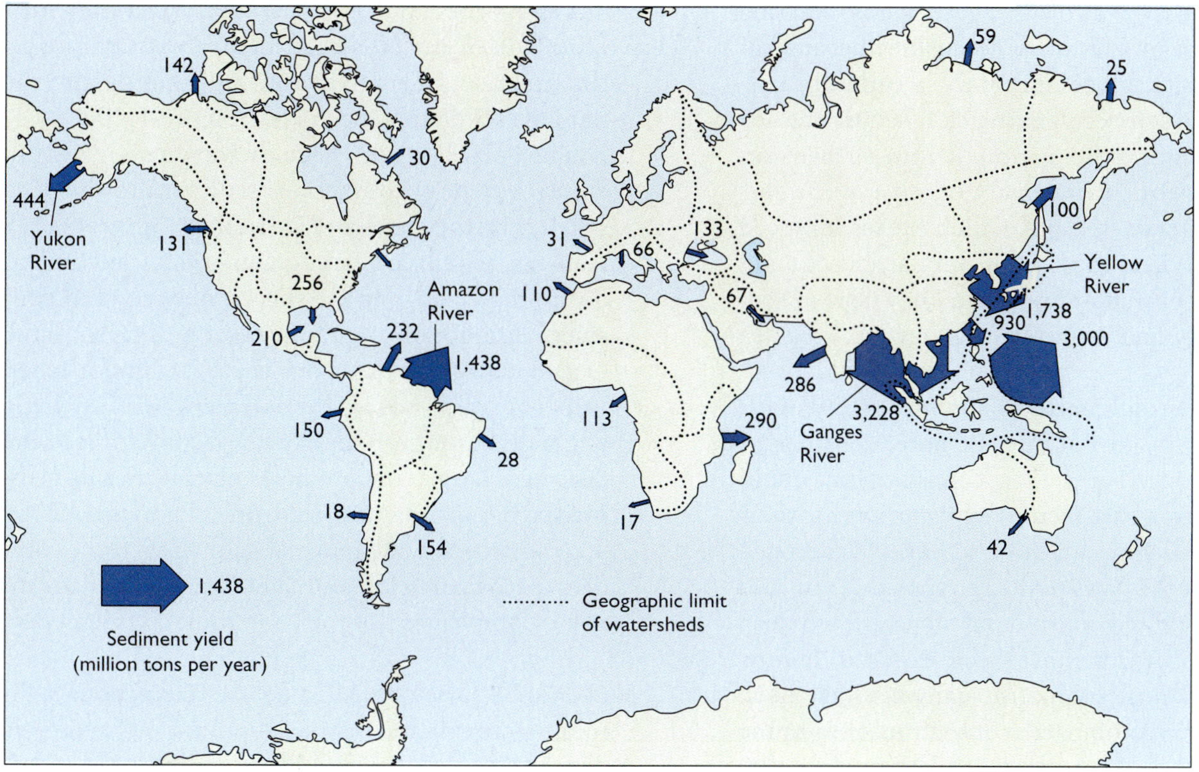

(b) RIVER INPUT OF SILT TO OCEANS

FIGURE 4-9

Deep-sea sedimentation. (a) Terrigenous sediment is derived from the breakdown of rocks on land and is supplied to the deep-sea by bulk emplacement. Pelagic sediment consists of oozes and red clay. The former is derived from biological production, the latter from erosion of rocks on land and transported as mud by rivers or by wind from land deserts. Authigenic sediments are chemical precipitates from seawater. (b) The influx of terrigenous debris to the ocean is quite variable. The vast bulk of it originates from the rivers that drain Asia. Another important source of river silt is the Amazon River of South America. [Adapted from J. D. Milliman and R. H. Meade, World-wide delivery of river sediment to the oceans, *Journal of Geology* 91 (1983): 1–21.]

and small masses of crumpled sediment that must have slid downslope toward the deep sea (**Figure 4–10a**). Debris flows and mudflows are **slurries**, a mixture of water and sediment that can sweep even large boulders downslope. Debris flows are mixtures of rock, sand, and mud; mudflows, of silt and clay. Unlike slumps, slurries destroy any previous bedding that may have existed in the deposits before they were disturbed.

Turbidity currents are important agents of transport to the deep sea. These powerful bottom currents are sediment-laden slurries that, under the influence of gravity, move rapidly downslope as turbulent underflows that push aside less dense water (**Figure 4–10b**). The slurry is created when sediment from the sea bottom is suspended and mixed with water. Imagine a large slump mass sliding downslope. The motion of the slump causes mud and sand to go into water suspension, and this slurry, too, begins to move downslope because of its high density. Once underway, a turbidity current becomes self-accelerating as it scours the sea bottom, placing more sediment into suspension and increasing its own density relative to the surrounding water even more, which causes its speed to increase. Where the sea bottom flattens out and the turbidity currents slow down, they deposit sediment rapidly and intermittently on the deep-sea bottom.

Large, deep submarine canyons, many with steep shoulders, are cut into the outer continental shelf and slope. For a long time, nobody could come up with a satisfactory explanation for their origin. The canyons appeared to have been eroded by rivers, but the parts of the canyons that cut into the outer shelf and continental slope are too deep in the ocean ever to have been exposed. It now appears that most submarine canyons have been excavated by a combination of sediment slumping and turbidity currents that have deepened a gully, or depression, on the sea bottom. Submarine canyons serve as chutes for funneling large quantities of terrigenous mud and sand from the shelf and slope into the deep sea by turbidity-current transport (**Figure 4–10c**).

Apparently, most submarine canyons are active when sea level is low, as this allows rivers to form deltas at the heads of canyons. Much of this deltaic sediment is unstable and slumps, and this leads to the formation of turbidity currents, which then flow down the canyon axis. At the base of the continental slope, where the mouth of the submarine canyon opens onto the continental rise and the abyssal floor, the turbidity current is no longer channeled (confined by the shoulders of a canyon) and it decelerates because the bottom slope flattens out and spreads the flow. As the current slows down, grains are dropped out of suspension according to size—the largest first, the finest last—producing **graded bedding** (**Figure 4–10d**). The beds of sediment laid down by turbidity currents are called **turbidites**. These graded beds accumulate one on top of another, each representing a distinct turbidity-current flow. Turbidites abound at the mouth of submarine canyons, where they form thick sequences of cone-shaped deposits known as **deep-sea fans**. They are similar to river deltas and to alluvial fans in desert environments (**Figure 4–11**). As the fan deposits grow in size, they may unite with other fans near them, creating the continental rise, and extend seaward, grading into the flat abyssal plain deposits (see Figure 4–10c).

The polar latitudes have their own style of bulk emplacement of terrigenous sediment to the deep sea. It is by the process of **ice rafting** (**Figure 4–12a**). Icebergs, which are large fragments of ice broken off from glaciers and ice sheets, may contain considerable amounts of sediment scraped off the land. Ocean currents transport these sediment-laden icebergs (rafts) away from land where, melting gradually, they drop their load of sediment into the deep sea. Ice rafting produces **glacial-marine sediments** (**Figure 4–12b**) that are characterized by poor sorting (a wide range of gain sizes, from boulders to clay particles) and a heterogeneous (nonuniform) composition of rock and mineral fragments.

■ PELAGIC SEDIMENT: Most coarse terrigenous sediment—gravel and sand—supplied by rivers is deposited along the shoreline or the inner continental shelf. An exception previously discussed is the graded turbidite found on the deep-sea floor near the mouths of submarine canyons. Most of the deep-sea bottom, however, is blanketed by fine-grained mud composed of clay-sized and silt-sized particles that have settled slowly out of suspension in quiet, deep water far from the influence of land. These pelagic muds may be either inorganic or biogenic (organic) in origin; sometimes both types are mixed together. The former

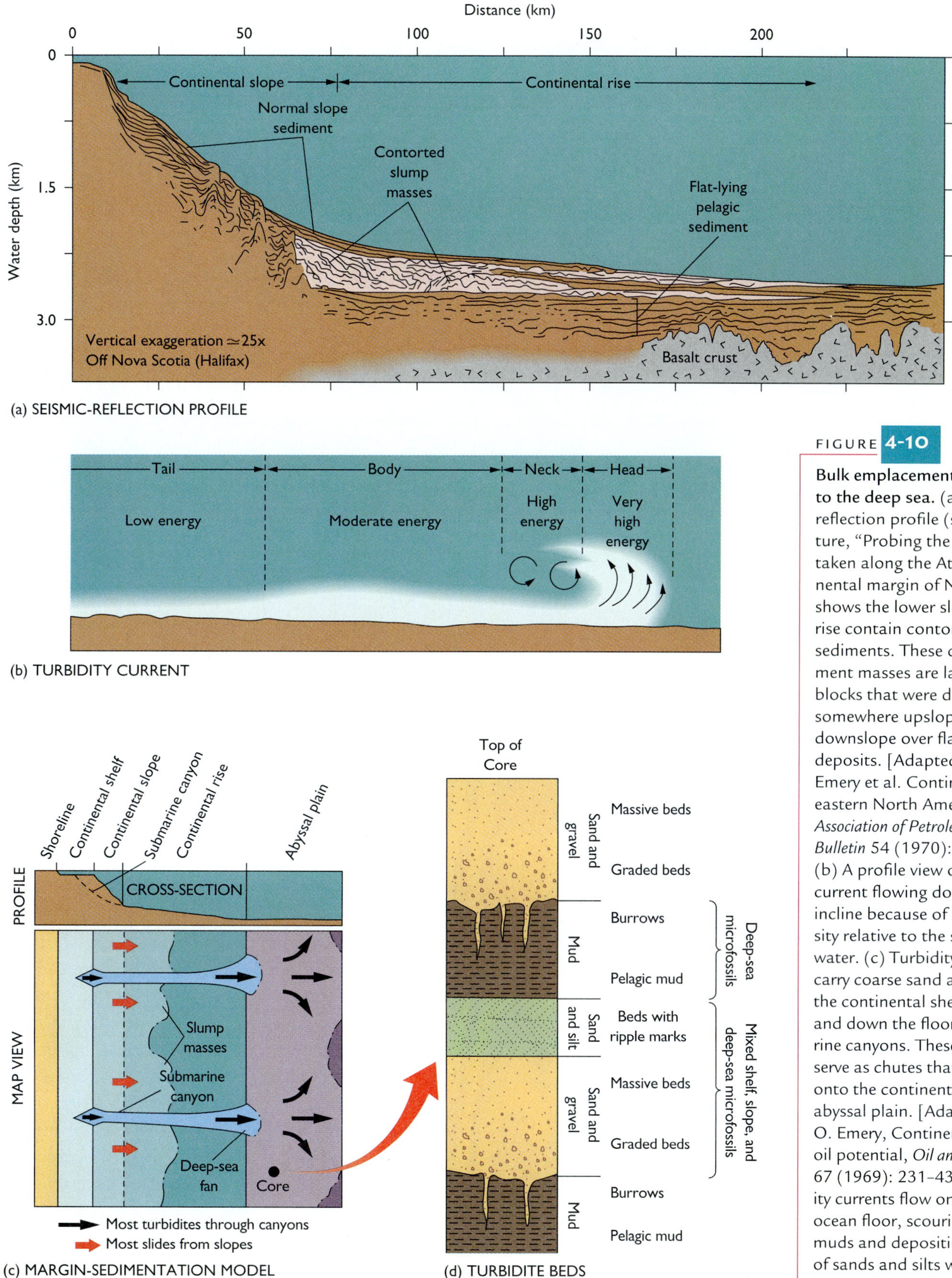

FIGURE **4-10**

Bulk emplacement of sediment to the deep sea. (a) A seismic-reflection profile (see boxed feature, "Probing the Sea Floor") taken along the Atlantic continental margin of Nova Scotia shows the lower slope and upper rise contain contorted, chaotic sediments. These deformed sediment masses are large slump blocks that were derived from somewhere upslope and slid downslope over flat-lying pelagic deposits. [Adapted from K. O. Emery et al. Continental rise off eastern North America, *American Association of Petroleum Geologists Bulletin* 54 (1970): 44–108.] (b) A profile view of a turbidity current flowing down a slight incline because of its high density relative to the surrounding water. (c) Turbidity currents can carry coarse sand and gravel past the continental shelf and slope and down the floors of submarine canyons. These canyons serve as chutes that funnel sand onto the continental rise and the abyssal plain. [Adapted from K. O. Emery, Continental rises and oil potential, *Oil and Gas Journal* 67 (1969): 231–43.] (d) Turbidity currents flow onto the deep-ocean floor, scouring the pelagic muds and depositing a sequence of sands and silts with characteristic graded bedding.

FIGURE **4-11**

Desert alluvial fan. Deep-sea fans resemble the alluvial fans that form at the mouths of steep canyons on land such as this one in the Anza-Borrego Desert State Park, California.

are mostly *red clays,* the latter are oozes. We will discuss both in detail.

The inorganic type of pelagic deposit is *red clay*, extremely fine-grained particles that typically have a brownish rather than reddish color, despite their name. In fact, many oceanographers simply call this **brown clay** or **pelagic clay**. These deposits are composed of various clay minerals, such as kaolinite and chlorite, and silt-sized and clay-sized grains of quartz and feldspar. Their color is the result of iron-bearing minerals that have been oxidized (producing, essentially, rust) by the oxygenated deep water. The precise origin of red clay is uncertain. It appears to be derived from several sources, including the weathering of granitic and volcanic rocks, wind transport of dust from land, the fallout from space of extraterrestrial dust, and perhaps even the chemical precipitation of clays from seawater itself.

There is little doubt, however, that certain clay minerals in red clay deposits originate by weathering of granitic rocks on land and are transported to the ocean in a variety of ways. Some are supplied by rivers, others are blown by the wind to the ocean, and still others are rafted by ice. To a large degree, climate in the source area controls the kind of clay minerals that form by weathering. For example, the warm, moist, acidic soils of the tropics favor the formation of kaolinite and the destruction of chlorite, two common species of clay. Given this relationship, the pelagic deposits of the low latitudes—the tropics and subtropics—are rich in kaolinite and poor in chlorite. This is indicated on the map of **Figure 4–13** by a kaolinite/chlorite ratio that is very high. Yet, you may ask, what exactly is a kaolinite/chlorite ratio? Actually, it sounds technical, but it's quite a simple concept. A ratio of 1 indicates that the amount of the two clay minerals is identical, because if you divide any number by itself you get a value of 1. A kaolinite/chlorite ratio greater than 1 signifies that there is more kaolinite than chlorite, so that you're dividing a number by a smaller value. For example, a kaolinite/chlorite ratio of 10 indicates that there is ten times more kaolinite than chlorite in the sediment sample. So if there are 10 grams of kaolinite, there must be 1 gram of chlorite. To test yourself, answer this: What does a kaolinite/chlorite ratio of less than 1 denote? Now go to the map of Figure 4–13 and note that the highest kaolinite/chlorite ratios occur in the low latitudes, the lowest ratios in the high latitudes. This indicates that kaolinite predominates in the tropics and subtropics and chlorite in the temperate and subpolar latitudes. Also note that there are two "tongues" of pelagic sediment where the kaolinite/chlorite ratios are greater than 10. One of these is located off the

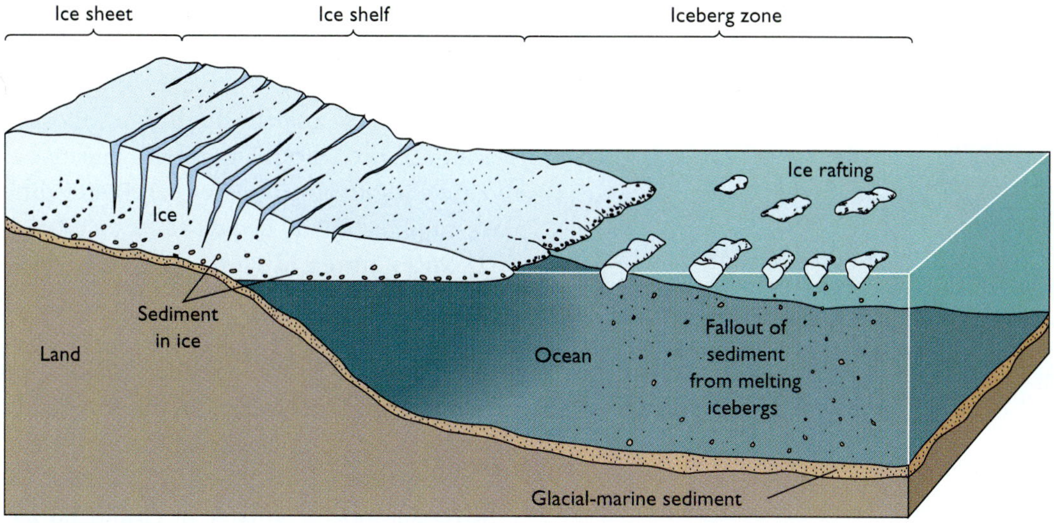

(a) ICE RAFTING

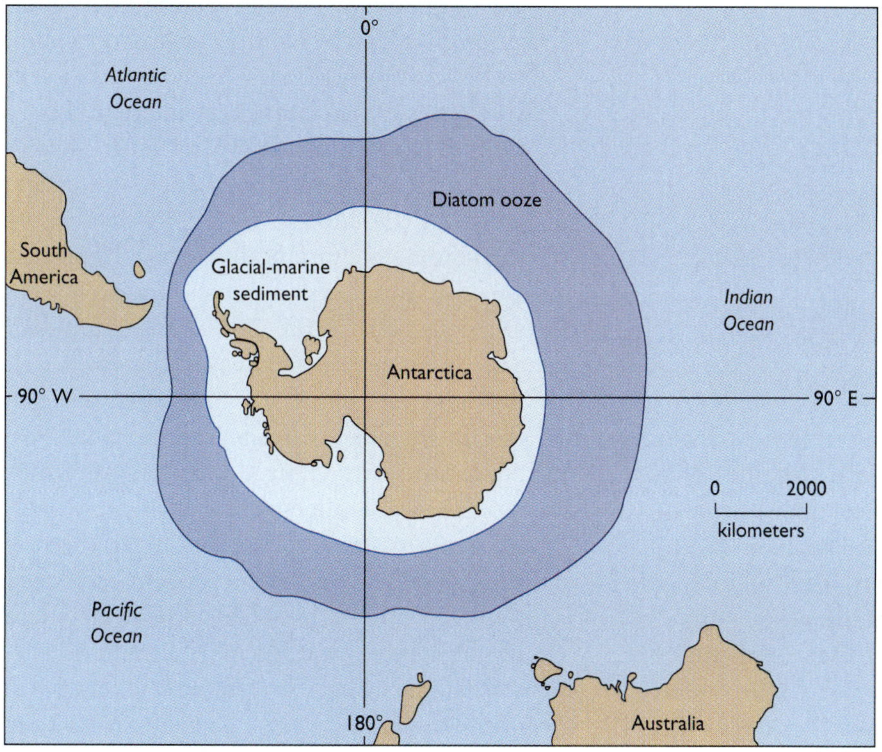

(b) DEEP-SEA DEPOSITS AROUND ANTARCTICA

FIGURE 4-12

The formation of glacial-marine sediments. (a) Glaciers extend into the ocean, where they form floating ice shelves that disintegrate into icebergs. The icebergs release their sediment load as they melt—a transport process termed ice rafting. (b) Ice rafting has created a 1,000-kilometer-broad band of glacial-marine sediment that surrounds Antarctica. [Adapted from J. D. Hays, *Progress in Oceanography*, M. Sears, ed. (New York: Pergammon Press, 1967).]

Amazon River of South America and reflects the large quantity of kaolinite supplied to the deep sea by this very large river. The other is located offshore of the western Sahara Desert of Africa, where strong trade winds blow desert sand and dust as far westward as Barbados in the eastern Caribbean Sea—a distance of over 6,000 kilometers (~3,725 miles)! Much of this windblown material is kaolinite, and it drops out of the air and settles to the bottom of the deep sea, helping to produce a band of deep-sea mud.

By definition, a **biogenic ooze** consists of 30 percent or more of the skeletal debris of microscopic organisms, most of which live in water far above the deep-sea floor, within a few hundred meters (a few hundred yards) of the ocean surface. The remaining 70 percent or less of the nonskeletal particles in oozes consists typically of inorganic mud

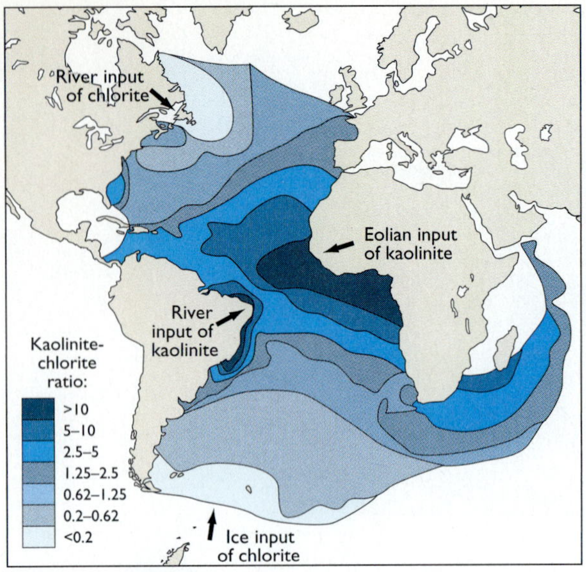

FIGURE **4-13**

Clays in deep-sea muds. The kaolinite-chorite ratio helps identify the source regions for much of these clay minerals, because the quantity of a particular clay type decreases with distance from its source region. Kaolinite-chlorite ratios in pelagic deposits of the Atlantic Ocean reveal a distribution that is a function of latitude. Kaolinite abounds in the tropical latitudes; chlorite, in the polar latitudes. [Adapted from P. E. Biscaye, Mineralogy and Sedimentation of recent deep-sea clay in the Atlantic Ocean and adjacent seas and oceans. *Geological Society of America Bulletin* 76 (1965): 810.]

particles. Biogenic deposits are divided into two major types according to their chemical composition: calcareous ($CaCO_3$) oozes and siliceous (SiO_2) oozes. **Calcareous oozes** are composed mainly of the tiny shells of **zooplankton** (floating single-celled animals), such as **foraminifera** and **pteropods**, and **phytoplankton** (floating single-celled plants), such as **coccolithophores** (**Figure 4–14**). Although abundances vary from area to area, all of these organisms are widely distributed in the surface water of the world's oceans. After the organisms die, their tiny hard parts dissolve or are incorporated into fecal pellets (the waste products of invertebrates) that settle slowly through the water column and eventually accumulate on the deep-sea floor. These shells may be dissolved, however, because cold bottom water tends to be slightly acidic (as will be explained in Chapter 5), and acid readily dissolves calcium carbonate. Oceanographers have defined the **carbonate compensation depth** (CCD) as the ocean level below which the preservation of $CaCO_3$ shells in surface sediments is negligible. The CCD depends on the rate of supply of carbonate and the acidity, temperature, and pressure of the water. Therefore, because the supply and dissolution rates of carbonate differ from place to place in the ocean, the exact depth of the CCD varies quite a bit; it tends to lie between 4 and 5 kilometers (~2.5 and 3.0 miles) below the sea surface. Rarely does carbonate ooze accumulate on ocean floor that is deeper than 5 kilometers (~3 miles). The CCD is an important chemical zone in the ocean that strongly controls the distribution of calcareous oozes.

Siliceous oozes consist of the remains of **diatoms** (floating, single-celled plants) and **radiolaria** (floating, single-celled animals) (see Figure 4–14). These organisms, which secrete hard parts made of silica (SiO_2), grow rapidly and are most abundant in water rich in nutrients. Silica is dissolved at very slow rates in seawater everywhere in the water column and tends to accumulate on the deep-sea bottom in areas where there is high biological productivity in the surface water. Two such regions of high biological production are sections of the polar and equatorial oceans; each produces a blanket of siliceous ooze on the underlying sea bottom. Typically, carbonate oozes accumulate on sea floor that is shallower than the CCD; silica oozes accumulate in the deep water below the CCD. Carbonate oozes are between three and nine times more abundant than are siliceous oozes in the world's ocean basins (see **Table 4–2**). This means that much of the ocean floor of the deep sea in each ocean basin lies well above the CCD; otherwise there would not be so much carbonate sediment on the floor of the deep sea. Table 4–2 shows the relative amounts of these various pelagic deposits. Note that 85 percent consist of either red clay (38 percent) and foraminiferal ooze (47 percent).

■ AUTHIGENIC DEPOSITS: Authigenic deposits are chemical precipitates that form in place within an ocean basin. **Ferromanganese nodules** are one of the best-known examples of such material. These nodules (irregular to sphere-shaped masses) consist of concentric layers of various metal oxides (**Figure 4–15a**) that have precipitated around nuclei such as grains of sand, gravel, and even sharks' teeth. In addition to oxides of iron and manganese, which

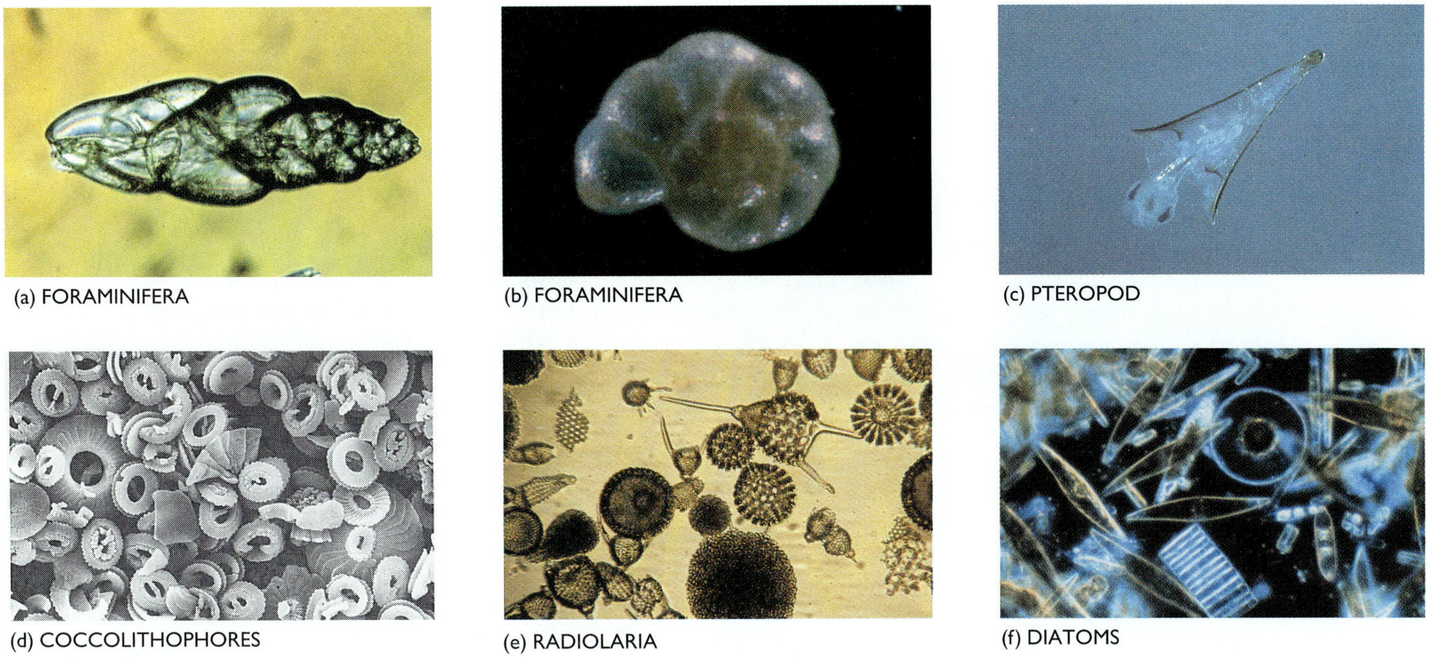

FIGURE 4-14
Common microfossils in biogenic oozes.

TABLE 4-2
Distribution of pelagic sediment

Type	Composition	Atlantic (%)	Pacific (%)	Indian (%)	Global (%)
Foraminiferal ooze	Carbonate	65	36	54	47
Pteropod ooze	Carbonate	2	0.1	—	0.5
Diatom ooze	Silica	7	10	20	12
Radiolarian ooze	Silica	—	5	0.5	3
Red clay	Aluminum silicate	26	49	25	38

Source: Adapted from W. H. Berger, Biogenous deep sea sediments: production, preservation and interpretation in *Chemical Oceanography,* vol. 5, J.P. Riley and R. Chester, eds. (New York: Academic Press, 1976), 265–388; and J. Kennett, *Marine Geology* (Englewood Cliffs, N.J.: Prentice-Hall, 1982).

average about 20 to 30 percent by weight, ferromanganese nodules contain a variety of other metals, including nickel, copper, zinc, cobalt, and lead, making them a potentially valuable economic resource. It appears that the iron and manganese are derived from several sources, notably the chemical weathering of basalt, river-supplied compounds that are dissolved in water, and hot-water seepage (**hydrothermal**) from vents on the crest of spreading ocean ridges.

The exact origin of these nodules is controversial. Some of the metals seem to have been concentrated in nodule layers by the activity of bacteria and foraminifera, which extract trace elements from the feces of microorganisms and transform them biochemically to a solid deposit on the nodule. Some geochemists, however, favor the theory of chemical precipitation directly from seawater or perhaps from hydrothermal submarine solutions. Whatever their origin, nodules clearly grow at the water-sediment interface (the contact zone between the water and the sediment). Also, they probably are rolled around, as indicated by their spherical shape. Because currents are usually weak in these areas, organisms

(a) PACIFIC NODULE

FIGURE 4-15

Ferromanganese nodules. (a) This photograph of a sectioned ferromanganese nodule reveals the concentric layering that typifies these authigenic deposits. (b) The global distribution of nodules shows that they are particularly abundant in the Pacific and South Atlantic Oceans. [Adapted from D. S. Cronan, Deep-sea nodules: distribution and geochemistry in *Marine Manganese Deposits*, G. P. Glasby, ed. (New York: Elsevier Science Publishing, 1977).]

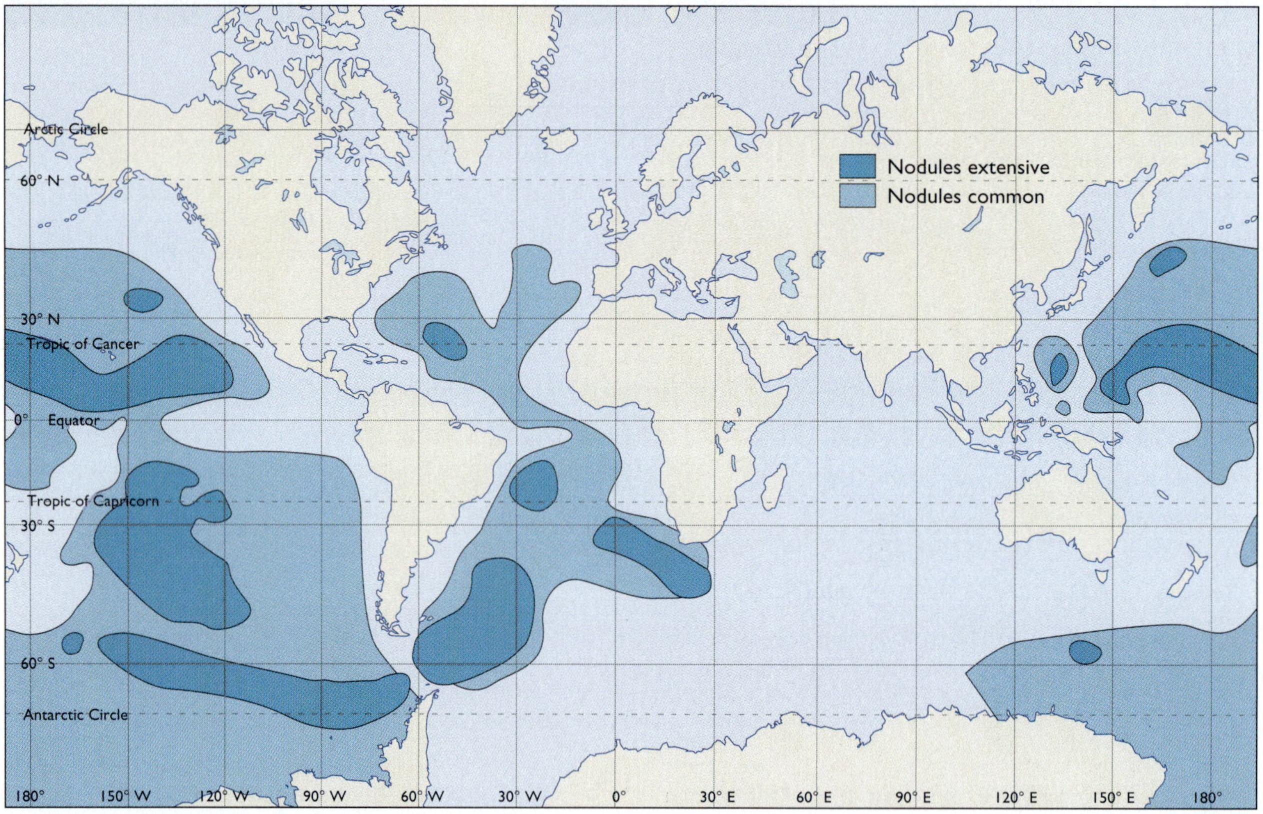

(b) GLOBAL DISTRIBUTION OF FERROMANGANESE NODULES

burrowing in the sediment are believed to be the cause of nodules slowly shifting about over the sea bottom. Nodules grow very slowly, usually at rates of between 1 to 4 millimeters (~0.039 to 0.15 inches) per million years. Many reach the size of gravel and even baseballs. Although they can be found just about anywhere in the deep sea, they are particularly abundant in parts of the North and South Pacific and in the South Atlantic (**Figure 4–15b**).

Phosphorite, another mineral deposit of possible economic importance, is composed of up to 30 percent P_2O_5 by weight. Unlike the ferromanganese nodules of the deep-ocean floor, phosphorite nodules generally are restricted to the continental shelf and upper continental slope, where cold water rich in nutrients typically upwells (moves upward) to the surface. This upwelling water creates considerable biological productivity, and large quantities of matter rich in organic phosphate settle to the ocean floor. Once buried in sediment, the unoxidized organic detritus is eventually transformed into phosphorite, which can grow as nodules at a rate of 1 to 10 millimeters (~0.04 to 0.39 inches) per thousand years.

Other important chemical precipitates in the deep sea are the metal sulfide deposits discovered around hydrothermal vents of actively spreading ocean-ridge crests. The hot water that escapes from these volcanic vents is loaded with dissolved metals that are precipitated on the adjacent ocean floor as cooling and chemical adjustment take place.

GLOBAL DISTRIBUTION OF DEEP-SEA SEDIMENTS

We will now survey the distribution of sediments across the sea floor and examine their variation with depth below the sea bottom. Once sediment is buried, the grains are compacted, cemented, and slowly transformed into sedimentary rock. Sand becomes **sandstone**; mud becomes either **shale**, if composed of clay minerals, or limestone, if composed of carbonate ooze.

SURFACE DEPOSITS

Figure 4–16a is a map of the sedimentary deposits that blanket the deep-sea floor. The clear distribution patterns that are evident reflect the source of these various materials. The continents are the principal suppliers of terrigenous debris. The bulk of this material that is supplied to the ocean by rivers is deposited on the continental margins. A small fraction of the terrigenous sediment bypasses the margins and is dispersed into the deep sea by downslope slumping, debris and mudslides, and turbidity currents. These processes have provided terrigenous sediment to the abyssal plains of the North Atlantic Ocean and of the Indian Ocean on either side of India (see Figure 4–16a). Few such deposits are evident in the Pacific Ocean because this basin is surrounded by deep-sea trenches—deep holes that trap any terrigenous sediment, such as turbidites, that manages to bypass the continental shelf. In the polar seas, ice rafting introduces terrigenous sediment to the deep sea. These glacial-marine deposits derived from melting icebergs are evident north of Iceland and in a broad band that circles the Antarctic continent (see Figures 4–12b and 4–16a).

Toward the center of all the ocean basins, far away from continental inputs of terrigenous debris, the sea floor is blanketed by pelagic deposits. Areas that have high biological productivity support large populations of planktonic organisms that contribute large quantities of siliceous and calcareous shells to the deep-sea bottom. The fertile polar seas favor the formation of diatom oozes, evident especially in the northern Pacific Ocean and off Antarctica seaward of the band of glacial-marine deposits (see Figure 4–12b). The high biological productivity of the equator produces a band of siliceous ooze in this region as well.

Calcareous oozes accumulate in water depths above the CCD and cover bathymetric highs, such as the crest and flanks of the spreading ocean ridges, the tops of seamounts, and the shallow, broad plateaus of the southwestern Pacific Ocean (see Figure 4–16a). Pelagic clay, which accumulates very slowly in the deep sea, forms in quiet environments far from other sources of sediment, such as terrigenous debris and biogenic oozes, and in deep water where calcareous particles are dissolved. Therefore, clay deposits are usually found in the deepest parts of the ocean basins, below the CCD and away from continents and areas of high surface productivity. Actually, clay particles are deposited in most areas of the deepest sea, but at such slow rates that they are diluted by the abundance of other sedimentary components. Clay deposits are most extensive in the Pacific Ocean. This is an ocean basin that is geologically old and deep (below the CCD) and has extensive areas of low biological productivity. Hence, siliceous and carbonate oozes are uncommon and do not dilute the red or brown pelagic clays that accumulate in much of the Pacific.

Rates of sedimentation in the ocean basins vary with the composition of the sedimentary material (**Figure 4–16b**). The highest rates in the deep sea exceed 5 centimeters (~2 inches) per 1,000 years and are associated with the thick terrigenous deposits of the continental margins. Biogenic oozes accumulate in the deep sea at rates of 1 to 3 centime-

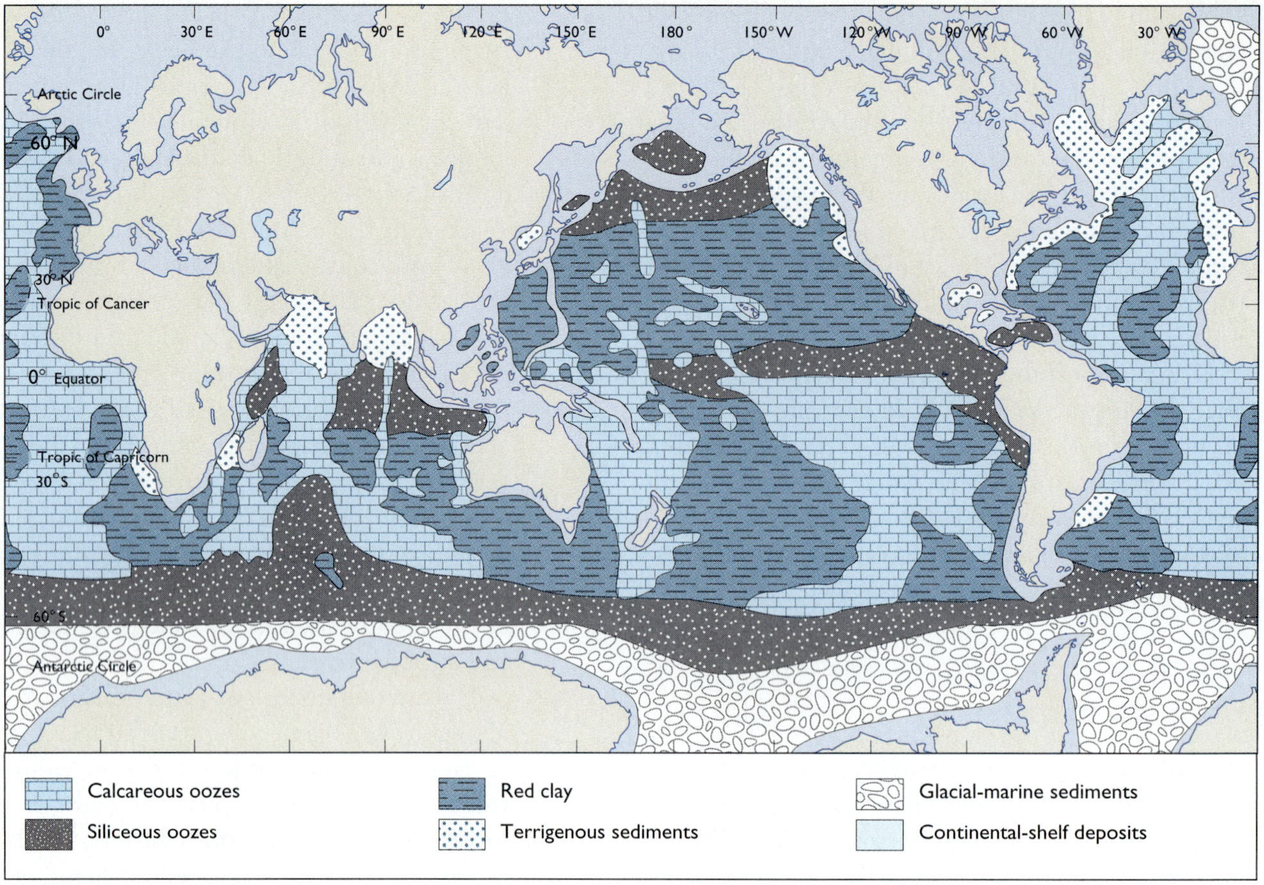

(a) DEEP-SEA SEDIMENT DISTRIBUTION

FIGURE **4-16**

Global deep-sea deposits. (a) Deep-sea deposits vary worldwide with distance from land and water depth. [Adapted from T. A. Davies and D.S. Gorsline, Oceanic sediments and sedimentary properties in *Chemical Oceanography*, J. P. Riley and R. Chester, eds. (Orlando, Fla.: Academic Press, 1976).] (b) A comparison map shows that terrigenous deposits have the highest sedimentation rates and red clays have the lowest. Oozes have sedimentation rates that lie between these two extremes. [Adapted from B. C. Heezen and C. D. Hollister, *The Face of the Deep* (New York: Oxford University Press, 1971).]

ters (~0.4 to 1.2 inches) per 1,000 years. The slowest depositional rates, which are less than 1 centimeter (~0.4 inch) per 1,000 years are associated with the pelagic red clays that lie in the remotest parts of the ocean basins at depths well below the CCD.

DEEP-SEA STRATIGRAPHY

Sea-floor spreading means that new ocean floor is created at the ocean ridge crests and then spreads away. During its long journey away from the axis of the Mid-Atlantic Ridge, the ocean floor of the North Atlantic basin has been receiving a steady supply of sediment at a rate of between 1 and 5 centimeters (~0.4 and 2 inches) per 1,000 years. Near the ridge crest, the basalt sea floor is bare, having just been formed. With distance down the flank of the ridge, the sediment cover thickens. In fact, there is a good correlation between the age of the basalt sea floor and the thickness of the sedimentary cover (**Figure 4–17**). At the very top of the ridge—far from sources of terrigenous sediment and well above the CCD—foraminiferal (foram for short) oozes accumulate and cover the basalt rock. With time, these oozes are buried and solidify into limestone. The biogenic deposits thicken with distance from the ridge axis because the underlying sea floor becomes older away from the ridge crest

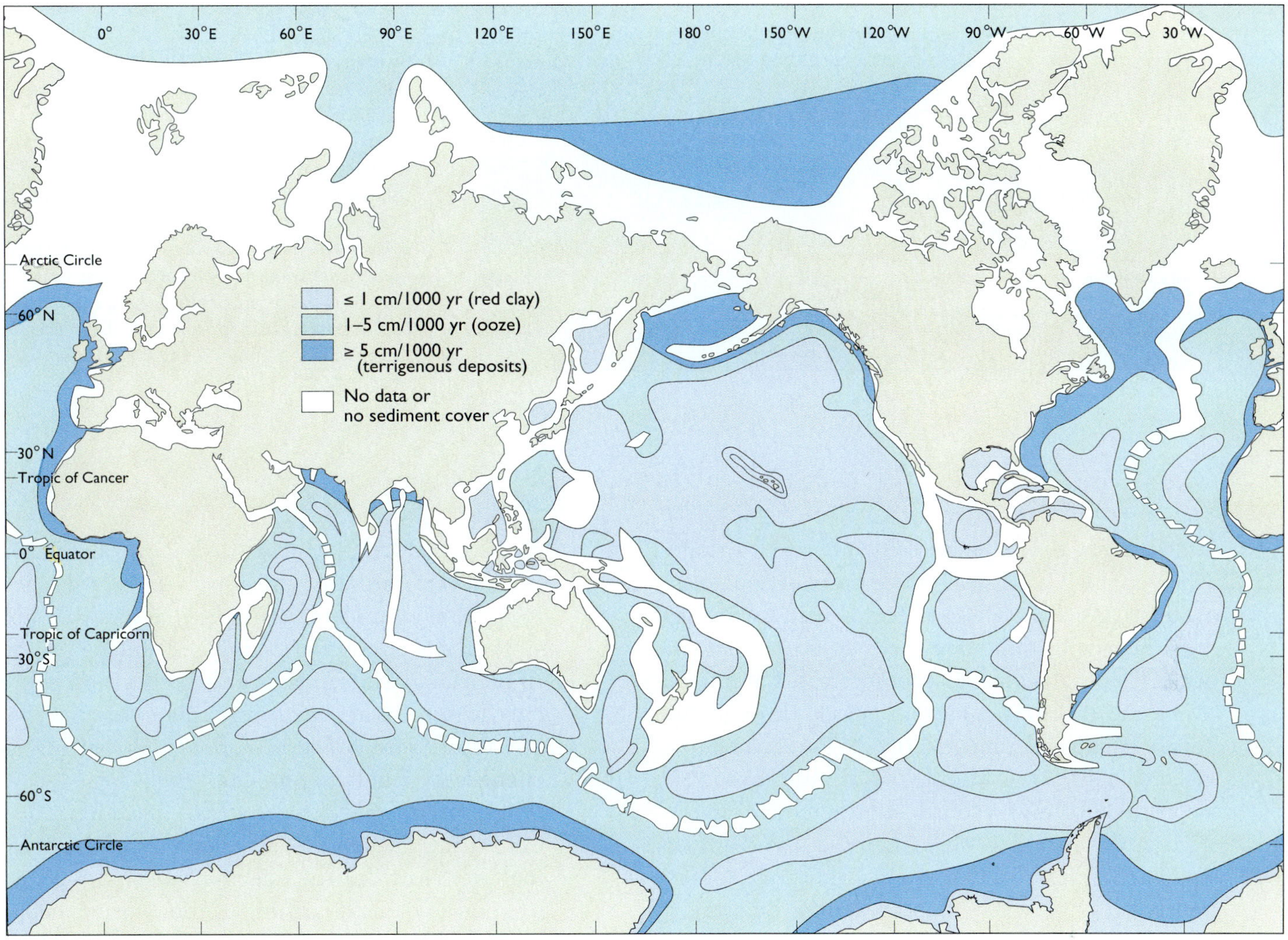

(b) SEDIMENTATION RATES

and has been subjected to a longer history of pelagic sedimentation. As the plate spreads and it cools and subsides below the CCD, the foram shells that reach the sea floor dissolve and only pelagic mud accumulates, burying and protecting the older limestones. Consequently, a two-layered sequence develops, consisting of a top layer of mud and an older bottom layer of limestone (see Figure 4–17). Oceanographers refer to this as a simple layer-cake **stratigraphy** (stratified or bedded rocks).

Figure 4–18 is an interpretation of the deep-sea stratigraphy of the central Pacific Ocean. Notice that it is more complex than the simple two-layered stratigraphy of the Atlantic Ocean. This is because of the orientation of the East Pacific Rise (the active spreading ridge of the Pacific Ocean). Unlike the Atlantic plates, which are spreading east-west, parallel to lines of latitude, the Pacific sea floor is spreading west-northwest and east-southeast. This means that the sea floor of the Pacific is crossing climatic belts as it spreads, producing a more complex stratigraphy as sediment accumulates. Note that the two deepest sedimentary units that lie just above the basalt crust are identical to the deposits of the Atlantic basin (see Figure 4–18b). A thick layer of basal limestone that accumulated on the crest and upper flank of the East Pacific Rise above the CCD is overlain by shale (mud compacted into rock) produced by the settling of mud particles on the lower flank of the ridge below the CCD. Because of sea-floor spreading, the Pacific plate crosses the equator, a fertile area that produces a great deal of biogenic ooze (Figure 4–18c). As a consequence of this passage across the equatorial zone, another

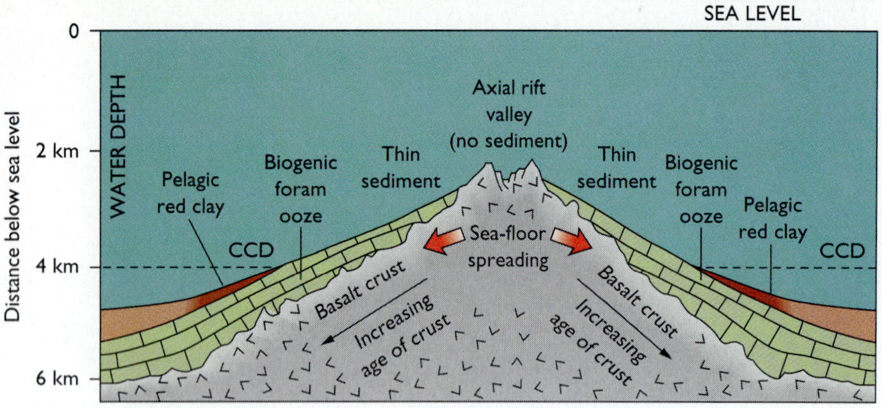

FIGURE 4-17

Stratigraphy of the Atlantic basin. The deposits that cover the basalt crust form a "layer cake" stratigraphy: a thick layer of basal ooze, formed above the CCD (carbonate compensation depth), is overlain by a thinner mud layer that accumulates below the CCD. Note that there is no sediment at the crest of the midocean ridge and that the sediment cover thins away from the crest. This pattern is the result of sea-floor spreading. As the basalt crust spreads away from the ridge crest, it acquires an ever thicker cover of sediment.

limestone layer was produced that buried the shale. The very high biological productivity of equatorial waters causes the CCD to be lowered to the sea floor, allowing carbonate shells to accumulate at water depths below 4 kilometers (~2.5 miles). Once the Pacific sea bottom passed through the fertile equatorial zone and the CCD returned to a normal depth of 4 kilometers, a mud layer was deposited, forming a fourth stratigraphic unit (see Figures 4–18b and 4–18c). Finally, as the sea floor approaches the trench of a subduction zone, it may receive a supply of volcanic debris, creating a localized fifth layer of volcanogenic sediment that overtops the older sedimentary layers.

Tectonics at subduction zones also influence the composition and stratigraphy of deep-sea deposits. One such example is described in the boxed feature, "The Drying Up of the Mediterranean Sea."

Future Discoveries

Sediment accumulates on the sea bottom and its layers represent a remarkable historical record of the geologic past. Exciting studies of sediment cores taken from the sea bottom all over the world are underway to document variations in climate and fluctuations in sea level extending back several hundred million years into the geologic past. By understanding the factors that possibly induced climate changes during earlier times, scientists will be in a better position to predict future climate and to anticipate the effects on life forms and processes that such worldwide changes are likely to produce.

Important questions that can be addressed by examining the sedimnetary deposits of the ocean include: How did the ocean-atmosphere system respond to global warming events of the geologic past? How sensitive was the marine biota to extreme climatic changes in the past? After global catastrophic events, how long did it take marine plants and animals to rebound and why did some groups recover more quickly than others?

Other marine geologists are testing the theory that some mass extinctions in the geologic past, such as the disappearance of dinosaurs 65 million years ago, may have been the result of giant meteorites striking the Earth. Scientists are searching the sedimentary record of the sea bottom for evidence of meteorite collisions with the Earth and looking for the impact craters themselves. These important findings will undoubtedly change our views of the Earth's history and the evolution of its marine and terrestrial biota.

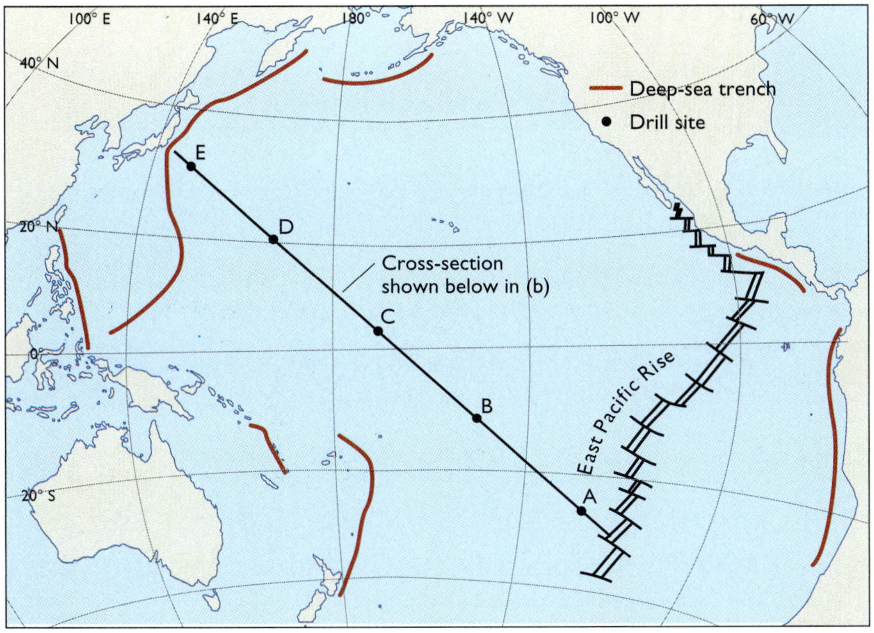

(a) PACIFIC OCEAN

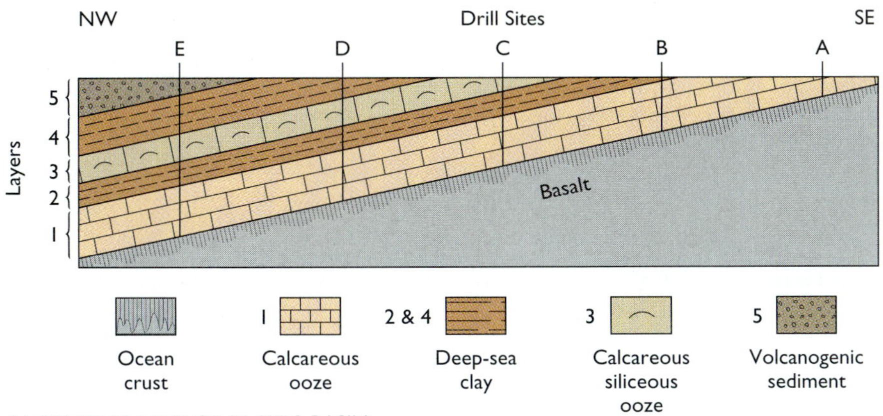

(b) STRATIGRAPHY OF PACIFIC BASIN

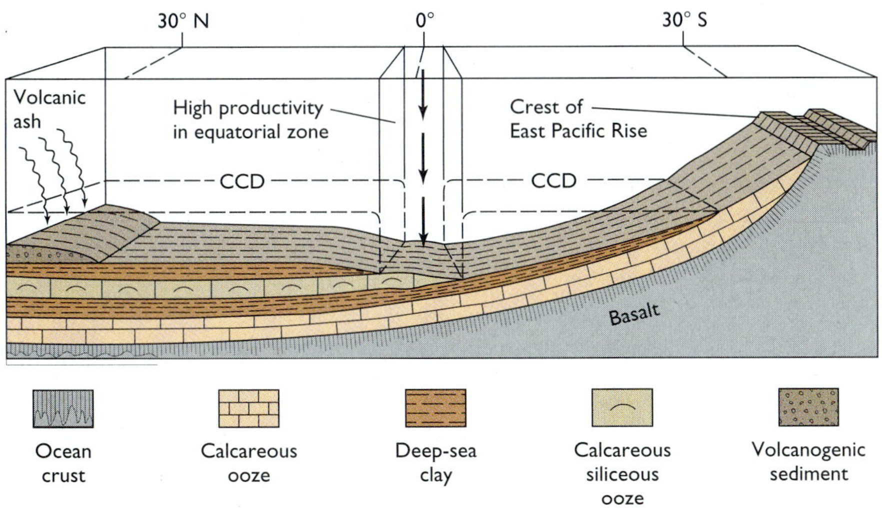

(c) MODEL TO ACCOUNT FOR PACIFIC STRATIGRAPHY

FIGURE **4-18**

Stratigraphy of the Pacific basin. (a) This generalized map of the Pacific basin shows the location of the stratigraphic cross section in part b. (b) Because the floor of the Pacific Ocean is spreading to the northwest across lines of latitudes, the stratigraphy is more complex than the two-layered stratigraphy of the Atlantic (see Figure 4-17). The letters are keyed to the drill sites shown in part a. [Adapted from B. C. Heezen et al., Diachronous deposits: a knematic interpretation of the post jurassic sedimentary sequence on the Pacific plate, *Nature* 241 (1973): 25–32.] (c) A basal layer of limestone forms on the flank of the ridge above the CCD. This limestone is topped by deep-sea mud when the ocean floor sinks below the CCD, creating the simple two-layered stratigraphy evident in the Atlantic Ocean. As the ocean floor moves across the fertile equatorial belt, a second layer of limestone, rich in calcium-carbonate and siliceous fossils, accumulates and then becomes covered by a second deposit of deep-sea mud north of the equator. Finally, a fifth layer of volcanogenic debris accumulates as the spreading sea floor approaches the volcanic arc of a subduction zone. [Adapted from B. C. Heezen and I. D. MacGregor, The evolution of the Pacific," *Scientific American* 229 (1973):102–12.]

The Drying Up of the Mediterranean Sea

The Romans coined the term "Mediterranean" because, to them, the region with its land-locked sea represented the very middle of the earth. The Mediterranean Sea is surrounded by three continents (Africa, Asia, and Europe) and is connected to the Atlantic Ocean by a narrow seaway, the Strait of Gibraltar (**Figure B4–6**). The northern edge of the Mediterranean Sea is dotted with many volcanoes and underlain by rock that has been squeezed tightly. These features, together with the abundant earthquakes in the region, indicate that this is a subduction zone along which two lithospheric plates are colliding and crushing each other. Africa, as it has been doing for hundreds of millions of years, is drifting slowly northward and encroaching on Europe and, in the process, is folding and uplifting marine sedimentary beds to form the Alps.

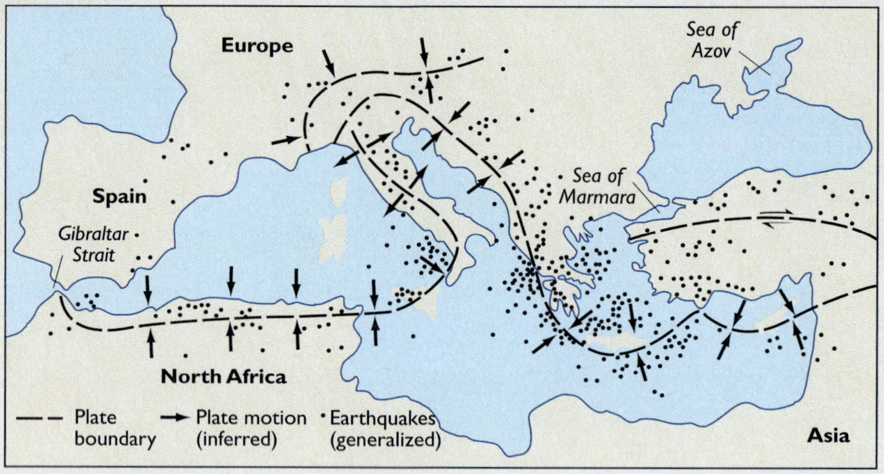

FIGURE **B4-6**

Plate tectonics in the Mediterranean region. The Mediterranean area is a tectonically active subduction zone where plates are colliding. The Mediterranean and Black Seas are the remnants of a much larger ocean that is slowly contracting as Africa drifts northward into Europe.

In 1972 deep-sea drilling in the western Mediterranean Sea by the *Glomar Challenger* uncovered unusual sedimentary deposits of Miocene age (5 to 25 million years old). Specifically, these deposits were anhydrite and stromatolites. **Anhydrite** (calcium sulfate) forms in hot, arid desert regions where saline water in the ground, but close to the surface, evaporates and causes calcium sulfate to precipitate out of solution. How can evaporites that form on dry desert land be part of the thick sedimentary deposits that lie on the bottom of the Mediterranean Sea? This is a very intriguing question indeed.

Stromatolites are also unrelated to deep-sea environments. These are crinkled laminations of carbonate mud. Today, stromatolite deposits are forming in the broad, intertidal mud flats of the Bahama Islands off Florida and in certain salty bays in Australia. There, dense mats of algae grow on the bottom and trap mud. The distinctive, crinkled structure of the mud laminations results from the irregular surface of the algal mat. The important point to remember is that stromatolites indicate shallow-water—not deep-water—deposition.

To Kenneth Hsü, the chief scientist aboard the *Glomar Challenger,* the conclusion was inescapable. Miocene deposits now lying on the deep-sea bottom of the Mediterranean Sea must have accumulated in desert environments (anhydrite) and in salty, shallow seas (stromatolites). This seemed preposterous at the time, because the sedimentary units that immediately overlie and underlie these Miocene deposits are unmistakably deep-sea carbonate oozes. The sequence implies that the floor of the Mediterranean Sea must have been shoved up thousands of meters during the Miocene epoch, drained of its seawater and, shortly thereafter, must have dropped several thousand meters to abyssal depths—as if the basin were a veritable tectonic yo-yo (**Figure B4–7a**). Yet there is another way to interpret this story. The crust may not have moved up or down at all. Rather, the Mediterranean Sea may have dried out during the Miocene and then refilled quickly with seawater. In this case, the only fact for researchers to explain is how the Mediterranean Sea could have been drained of its water (**Figure B4–7b**)! This question is less difficult to answer than one might expect.

Only two conditions are necessary for the Mediterranean Sea to dry out entirely. First, the inflow of seawater from the Atlantic Ocean through the Strait of Gibraltar must be blocked.

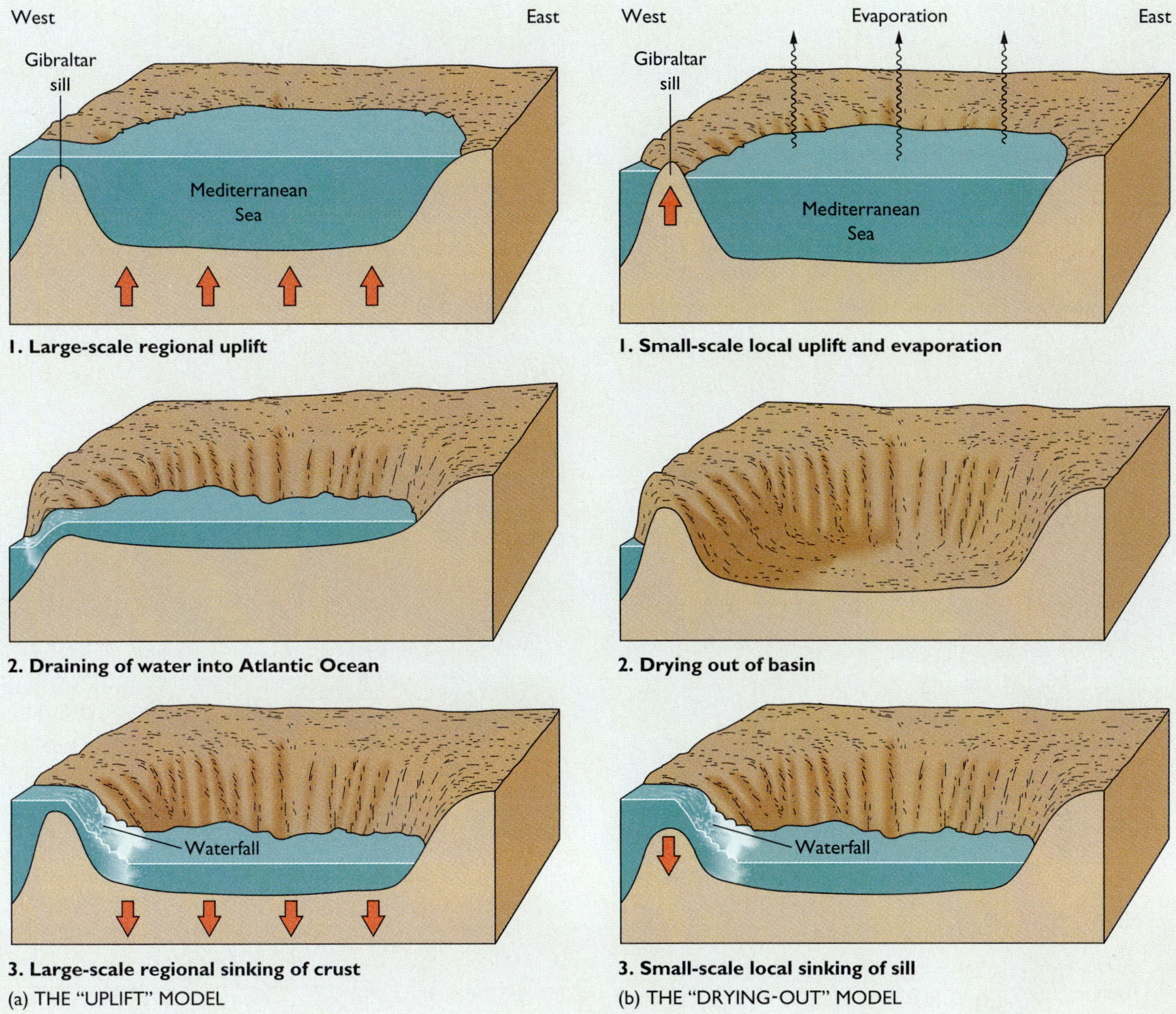

FIGURE **B4-7**

Models for emptying the Mediterranean Sea. Deep-sea drilling in the Mediterranean Sea has uncovered anhydrite and stromatolites of Miocene age. These are indicators of shallow, high-salinity conditions similar to those that exist in the present-day Arabian Gulf. These shallow-water deposits are overlain and underlain by deep-sea oozes, a peculiar stratigraphy that can be explained in two ways: (a) According to the "uplift" model, the sea bottom was thrust up thousands of meters and drained of seawater and then quickly plunged downward to abyssal depths and refilled with seawater. (b) According to the "drying-out" model, the Mediterranean Sea dried up by the evaporation of all of its water when it became separated for a while from the Atlantic Ocean and then refilled with seawater when the connection to the Atlantic was reestablished.

It turns out that this is easy to do, because the strait is a rock sill with its top only 100 meters (~330 feet) deep. Second, the climate needs to be arid to evaporate all the water trapped in the basin. The rock record indicates that the climate was hot and dry in the Mediterranean region during the Miocene epoch. Hsü calculated that it would have taken about a thousand years to evaporate all of the water trapped in the basin, once the sill had been raised and cut off the inflow of Atlantic seawater. This means that a sea would have been converted into a desert and back into a sea during a single millennium!

When the western Mediterranean had dried out , the sea floor would have been converted into a desert terrain with isolated salty lakes (**Figure B4–8a**). The deeper sea floor of the eastern Mediterranean at that time was covered by a system of large lakes, similar to the Great Salt Lake in Utah, fed by spillover from the Black Sea to the east. Eventually, the Mediterranean basin refilled in a rather catastrophic fashion. Atlantic seawater cascaded down the Gibraltar Strait once its sill had dropped a bit, creating a monumental waterfall (**Figure B4–8b**). The quantity of water that poured over the sill exceeded the discharge of Niagara Falls by a factor of 1,000 according to Hsü's calculations. At that rate, the Mediterranean basin would have entirely filled with Atlantic seawater in a mere 100 years!

Although this model appeals to our imaginations, it must be noted that some geologists believe that the evidence Hsü cites to support his model can be interpreted in other, less dramatic ways. The debates continue.

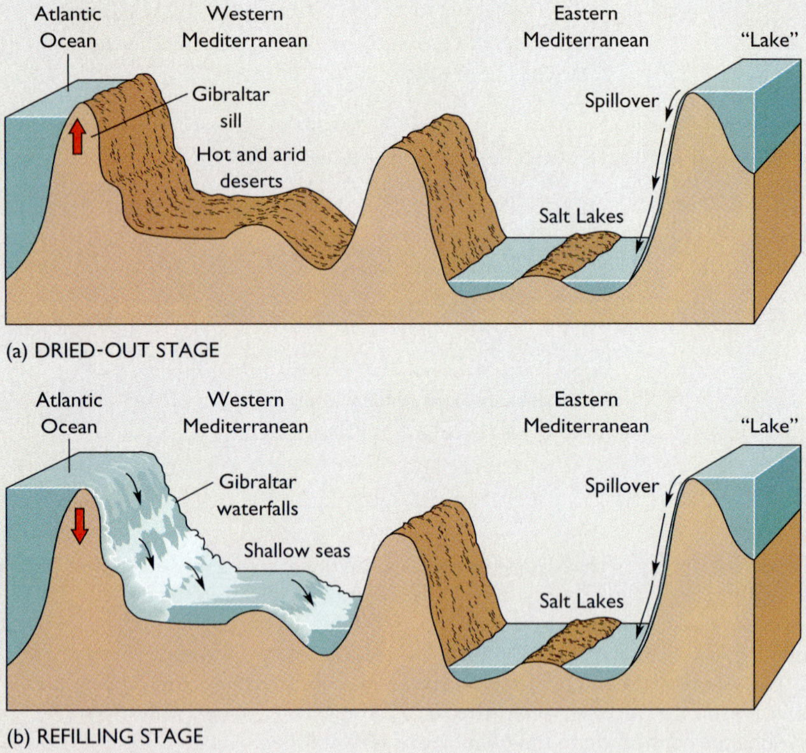

FIGURE **B4-8**

The Mediterranean region during Miocene time. (a) During the Miocene, the Mediterranean Sea apparently dried up, creating deserts and large lakes of high-salinity water. (b) Subsequently, seawater from the Atlantic Ocean cascaded down the rock face of Gibraltar sill, refilling the entire Mediterranean basin. [Adapted from M. B. Cita and W. B. G. Ryan, Time scale and general synthesis, in *Initial Reports of the Deep-Sea Drilling Project,* vol. 13 (Washington, D.C.: U.S. Government Printing Office, 1973), 1405–15.]

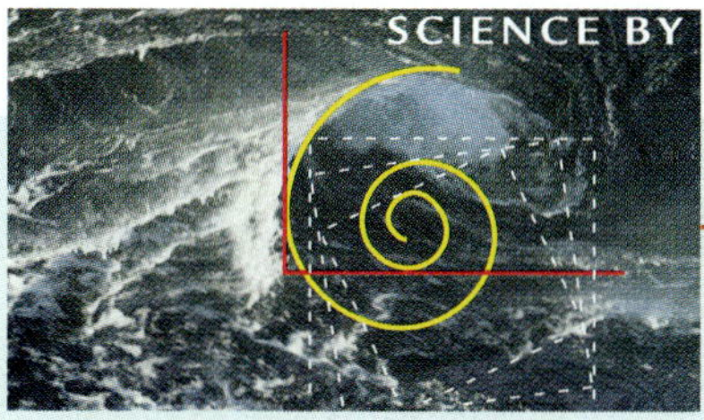

SCIENCE BY NUMBERS

Sedimentation Rates

Let's assume that sediment is settling on a sea bed at an average rate of 5 centimeters per thousand years (5 cm/10^3 yr). Now, let's assume that this sea floor has been spreading at a rate of 4 centimeters per year. How thick will the sedimentary cover be for sea floor that is 400 kilometers down the flank of this spreading ridge? If we think a bit about this problem, we discover that to solve it all we need to know is the age of the sea floor that is located 400 kilometers from the ridge axis. The thickness of the sediment cover will be equal to the sedimentation rate times the age of the sea floor. Obviously, the older the sea bed, the thicker must be its sediment cover for a given rate of deposition.

The first step is determining the age of the sea floor that is located 400 kilometers from the axis of a ridge that is spreading at a rate of 4 centimeters per year. This is similar to determining the time that has elapsed since you left home in your car, traveling at 50 mph for 100 miles. The answer is 2 hours, which you determine by dividing 100 miles (the distance traveled) by 50 miles per hour (the speed of the automobile). We do exactly the same calculation for the sea-bed problem. We divide 400 kilometers (the distance traveled) by 4 centimeters per year (the speed of sea-floor spreading). In order to do that, of course, we need the same distance units. So, we must first convert kilometers into centimeters. Let's do that.

$$400 \text{ km} = 4 \times 10^2 \text{ km},$$

and

$$(4 \times 10^2 \cancel{\text{km}})\,(10^3 \cancel{\text{m}}/\cancel{\text{km}})\,(10^2 \text{ cm}/\cancel{\text{m}}) = 4 \times 10^{(2+3+2)} \text{ cm} = 4 \times 10^7 \text{ cm}.$$

Now we divide the distance by the speed of sea-floor spreading. (Recall the car-travel example.)

$$4 \times 10^7 \cancel{\text{cm}}/4 \cancel{\text{cm}}/\text{yr} = 10^7 \text{ yr}.$$

So, the age of the sea floor 400 kilometers away from the ridge axis is 10 million years.

The next step is determining how much sediment would have built up in 10 million years if the average sedimentation rate is 5 centimeters per 10^3 years. This is a simple matter. We multiply the sedimentation rate by the time over which it operates. Think about it this way. If you're building up a brick wall at 1 foot per hour, how high will the wall be after 3 hours? Obviously, 3 feet, which you determined by multiplying 1 foot per hour by 3 hours. We'll do the same for our sediment-thickness problem.

$$(5 \text{ cm}/10^3 \cancel{\text{yr}})\,(10^7 \cancel{\text{yr}}) = 5 \times 10^{(7-3)} \text{ cm} = 5 \times 10^4 \text{ cm} = 50{,}000 \text{ cm}.$$

Now we convert centimeters into kilometers as follows:

$$(5 \times 10^4 \cancel{\text{cm}})\,(1 \cancel{\text{m}}/10^2 \cancel{\text{cm}})\,(1 \text{ km}/10^3 \cancel{\text{m}}) = 5 \times 10^{(4-2-3)} \text{ km} = 5 \times 10^{-1} \text{ km}.$$

The sediment thickness 400 kilometers from this ridge is

$$5 \times 10^{-1} \text{ km} = 0.5 \text{ km} = 500 \text{ m}.$$

STUDY GUIDE

KEY CONCEPTS

1. Deposition of sediment is affected by the type, size, chemical composition, and quantity of the sediment supply; by the energy conditions at the site of deposition; and by fluctuations of sea level. Typically, coarse sediments (sand) signify high-energy conditions, fine sediments (mud) signify low-energy conditions.

2. Shelf sediments vary in composition with latitude. In the tropics and subtropics, they consist of *biogenic* deposits composed of the hard calcium carbonate remains of organisms. Shelf sediments of the midlatitudes consist mainly of river-supplied quartz and feldspar sands, *terrigenous sediment,* and those of the high latitudes of unsorted glacial debris.

3. The bulk of the sediment of continental shelves is *relict,* having been deposited during the latest glacial event of the Pleistocene epoch when sea level was much lower than it is at present. Relict sediments are not in equilibrium with the present-day water depths and bottom-energy conditions of the shelves, but, rather, reflect an old pattern of deposition.

4. Over time intervals of more than a million years, the evolution of continental shelves depends on the plate-tectonic setting. *Atlantic-type margins* develop along the edges of continents that are located far from the tectonic effects of plate boundaries and have a long history of passive sedimentation. In contrast, *Pacific-type margins* are dominated by a subduction zone and are built up of highly deformed sedimentary beds that may be interlayered with lavas and volcanogenic sediment.

5. During the past million years, continental shelves have been periodically uncovered and covered by the ocean as glaciers advanced and retreated on land. Low stands of sea level resulted in glaciers and rivers cutting into the exposed shelf surface, and *slumping* on the continental slope and the flow of *turbidity currents* down submarine canyons increased the amount of sand and mud transported to the deep sea. Also, exposure caused by the drop in sea level during glacial periods caused widespread killing of coral reefs in the low latitudes.

6. Modern deposits on shelves have been influenced by waves and tidal currents. Most of the coarse sediment of continental shelves is moved during storms, when the combination of large waves and tidal currents create high-energy conditions along the bottom.

7. *Terrigenous sediments* are derived from the weathering and erosion of rocks on land. This material blankets the continental shelves, largely as relict material, and is transported to the deep sea by *ice rafting* and by *bulk emplacement* (slumping, debris flows, and turbidity currents).

8. *Pelagic sediments* are fine-grained materials that have settled out of the water, particle by particle. *Red clay* is largely derived from the weathering of rocks on land. *Biogenic sediments,* which are composed of the skeletal remains of microscopic organisms, are divided into *calcareous oozes* and *siliceous oozes* based on the chemical composition of the dominant fossils.

9. *Authigenic sediments* (such as *ferromanganese nodules*) are geochemical and biochemical deposits that form at the sea bottom and are not transported to the site of deposition.

10. In the deep sea, terrigenous deposits tend to collect near the mouths of submarine canyons and are brought there by turbidity currents that flow down the canyons. Siliceous oozes form distinct bands of deposits at the equator and at the polar latitudes. Carbonate oozes accumulate in the center of ocean basins in water depths above the *carbonate compensation depth* (CCD). Red clays are deposited in the most remote and deepest regions of the sea below the CCD, and away from areas having high biological productivity.

11. Because the crests of ocean-spreading ridges lie above the CCD, the first sediment to cover the basalt crust is carbonate ooze, which thickens with distance down the flank of the ridge. Once the sea floor of a spreading ridge sinks below the CCD, pelagic clays accumulate and form a second sedimentary layer that covers the basal ooze. If the spreading ocean floor crosses the equator, a third layer of ooze is produced by the tremendous fallout of biogenic debris from above. Then, beyond the zone of high surface productivity, a fourth layer composed of mud accumulates.

KEY WORDS

accretionary prism
Atlantic-type margin
authigenic sediment
biogenic sediment
bulk emplacement
calcareous ooze
carbonate compensation depth (CCD)
diatoms
foraminifera
glacial-marine sediment
graded bedding
ice rafting
Pacific-type margin
pelagic sediment
radiolaria
red (or brown, or pelagic) clay
relict sediment
siliceous ooze
slumps
stratigraphy
terrigenous sediment
till
turbidity current

QUESTIONS

REVIEW OF BASIC CONCEPTS

1. What is the difference between biogenic and terrigenous sediment?

2. Cite evidence that clearly indicates that sea level has fluctuated widely in the geologic past.

3. What exactly is relict sediment, and why is this the dominant type of material that blankets the continental shelves of the world?

4. What distinguishes an Atlantic-type from a Pacific-type continental margin?

5. Distinguish clearly between bulk emplacement and pelagic sedimentation as processes that supply sediment to the deep sea.

6. How would you distinguish turbidites from glacial-marine sediment, diatom ooze from red clay, and diatom ooze from foram ooze?

7. What is the CCD, and how does it affect deep-sea sedimentation?

CRITICAL-THINKING ESSAYS

1. Is there relict sediment on the bottom of the deep sea? Why or why not? (Hint: think carefully about the definition of the term.)
2. If you were sampling biogenic ooze from water depths of about 5.5 kilometers, would it be composed of silica or calcium carbonate? Explain. Would you be most likely to sample this ooze from the deep-sea bottom of the low, middle, or high latitudes? Why?
3. Examine the maps in Figures 4–16a and 4–16b, and account for the distribution and accumulation rates of bottom sediments in the Indian Ocean.
4. How would you determine whether or not a section of sea floor and its sediments had drifted across the equator?
5. Examine Table 4–2. Why is the dominant pelagic sediment in the Pacific Ocean composed of red clay and the dominant pelagic sediment in the Atlantic Ocean composed of foraminiferal ooze?

DISCOVERING WITH NUMBERS

1. Consult Figure 4–1. Estimate the current velocity required to erode gravel and to erode clay. Given the great difference in their size, why are their erosion velocities similar?
2. Consult Figure 4–2b. Calculate the rate of sea level rise in meters per year and in meters per 10^3 years during the past 5,000 years, and between 5,000 and 10,000 years ago.
3. Considering your response in Question 2 above, estimate in years when a city that lies 2 meters above present-day sea level is likely to be flooded. What critical assumption must you make in order to make the estimate?
4. Assuming a sedimentation rate of 5 centimeters per 1,000 years and a sea-floor spreading rate of 5 centimeters per year, calculate how far you would have to travel from the crest of a spreading ridge to encounter sediment that is 100 meters thick.
5. Assume that you go to the spot calculated in Question 4 above and you discover that the sediment is not 100 meters thick here, but is 185 meters thick! Assume that the spreading rate in Question 4 is correct and recalculate the sedimentation rate. Now assume that the sedimentation rate in Question 4 is correct and recalculate the spreading rate.

SELECTED READINGS

Droxler, A. 1993–94. Shallow carbonates drilled by DSDP and ODP. *Oceanus* 36 (4): 111–15.

Dudley, W. 1982. The secret of the chalk. *Sea Frontiers* 28 (6): 344–49.

Heath, G. R. 1982. Manganese nodules: Unanswered questions. *Oceanus* 25 (3): 37–41.

Heezen, B. C., and MacGregor, I. D. 1973. The evolution of the Pacific. *Scientific American* 229 (5): 102–12.

Hollister, C. D., Flood, R. D., and McCave, I. N. 1978. Plastering and decorating the North Atlantic. *Oceanus* 21 (1): 5–13.

Hsü, K. J. 1983. *The Mediterranean Was a Desert: A Voyage of Discovery*. Princeton, N.J.: Princeton University Press.

Keigwan, L. D. 1996. Sedimentary record yields several centuries of data. *Oceanus* 39 (2): 16–18.

McGregor, B. A. 1984. The submerged continental margin. *American Scientist* 72 (3): 275–81.

Normark, W. R., and Piper, D. J. W. 1993–94. Turbidite sedimentation. *Oceanus* 36 (4): 107–10.

Oceanus. 1992. Marine geology and geophysics. Special issue 35 (4).

Oceanus. 1993–94. 25 years of-ocean drilling. Special issue 36 (4).

Open University Course Team. 1989. *Ocean Chemistry and Deep-Sea Sediments.* New York: Pergamon Pergamon Press.

Rea, D. K. 1993/94. Terrigenous sediments in the pelagic realm. *Oceanus* 36 (4): 103–06.

Schlee, J. S., Folger, D. W., Dillon, W. P., Klitgord, K. D., and Grow, J. A. 1979. The continental margins of the western North Atlantic. *Oceanus* 22 (3): 40–7.

Sobey, E. 1982. What is sea level? *Sea Frontiers* 28 (3): 136–42.

Spencer, D. W., Honjo, S., and Brewer, P. G. 1978. Particles and particle fluxes in the ocean. *Oceanus* 21 (1): 20–6.

Stanley, D. J. 1990. Med desert theory is drying up. *Oceanus* 33 (1): 14–23.

Wanless, H. R. 1989. The inundation of our coastlines: past, present and future with a focus on south Florida. *Sea Frontiers* 35 (5): 264–71.

TOOLS FOR LEARNING

Tools for Learning is an on-line review area located at this book's web site OceanLink (**www.jbpub.com/oceanlink**). The review area provides a variety of activities designed to help you study for your class. You will find chapter outlines, review questions, hints for some of the book's math questions (identified by the math icon), web research tips for selected Critical Thinking Essay questions, key term reviews, and figure labeling exercises.

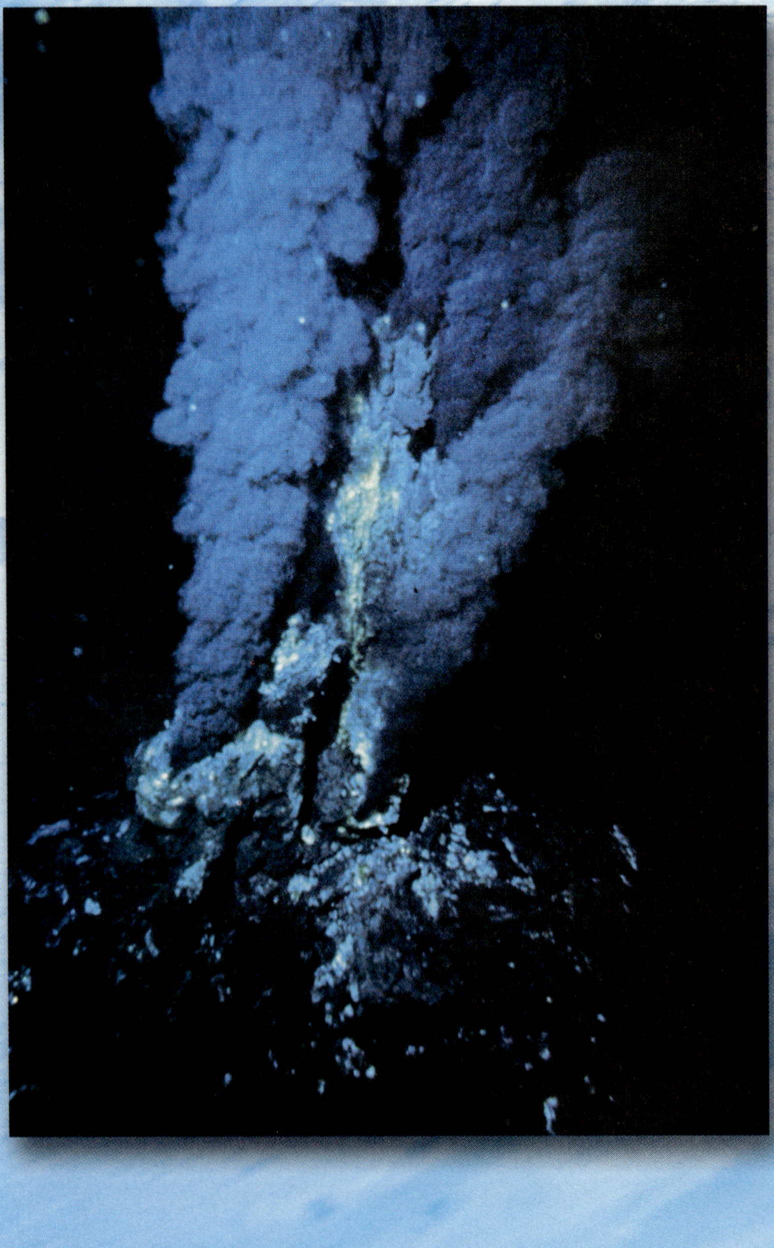

5

The Properties of Seawater

Water is just not water. It is a vital substance, the earth's blood, something of rich and infinite variety.

—*L. Watson,* The Water Planet, *1988*

PREVIEW

WATER IS EVERYWHERE. Its presence on our planet is crucial for the existence of life. "Biology," said Berg, "is wet and dynamic."

What exactly are the chemical and physical properties of water generally and seawater specifically? How variable are these properties over time and space? How do chemical elements enter the oceans, and—once there—how do they interact with other substances, both living and nonliving? A discussion of these questions forms the basis of this chapter. Here, we will examine the chemical and physical nature of seawater and lay the foundation for later discussions of ocean circulation and marine life. The concluding section of the chapter integrates these ideas into the global water cycle and examines the ocean as a complex and dynamic biogeochemical system.

web navigator

critical thinking on the web

math tutor on the web

www.jbpub.com/oceanlink

One of the most obvious properties of seawater is its salty taste. If you believe that its taste is due to dissolved salt, you are correct, and this is what makes the water of the oceans different from the water of lakes and rivers. Seawater is made up mainly of liquid water (about 96.5 percent by weight) in which chloride (Cl) and sodium (Na) are the dominant dissolved chemicals. The common table salt you use to flavor your food is composed of precisely the same elements. Before continuing our discussion about the chemistry of water, we must, however, become familiar with a few basic chemical and physical concepts. So let's begin.

5-1

Basic Chemical Notions

All matter is composed of "building blocks" termed **atoms**. An atom is the smallest unit of a substance that retains all of its chemical properties. For example, a single atom of hydrogen possesses all of the chemical characteristics of a large collection of hydrogen atoms. If you tried to divide a single hydrogen atom into simpler units, you could do so, but these bits of matter would no longer display the properties of a hydrogen atom. Atoms that are chemically bonded to one another comprise a **molecule**. Or, if you prefer, a molecule is a chemical substance that can be separated into distinct atoms. Sodium chloride (NaCl) is a molecule of salt that can be separated into a positively charged atom of Na and a negatively charged atom of Cl.

The internal structure of any atom consists of elementary particles that possess mass and electric charge. The center of an atom, the **nucleus**, is composed of two distinct kinds of particles that contain essentially all of its mass: **protons** with a positive electric charge and **neutrons** with no electrical charge (**Figure 5–1a**). Surrounding the nucleus are collections of orbiting **electrons** that have little mass and carry a negative electrical charge. The electron orbits are not randomly distributed around the nucleus, but are confined to discrete levels termed **electron shells**. All elements (except for one type of hydrogen that does not have neutrons) contain protons, neutrons, and electrons, and differ from one another because of the number and structural arrangement of these fundamental subatomic particles. A stable atom of an element is electrically neutral, indicating that the positive charges from the protons are balanced by the negative charges from the electrons.

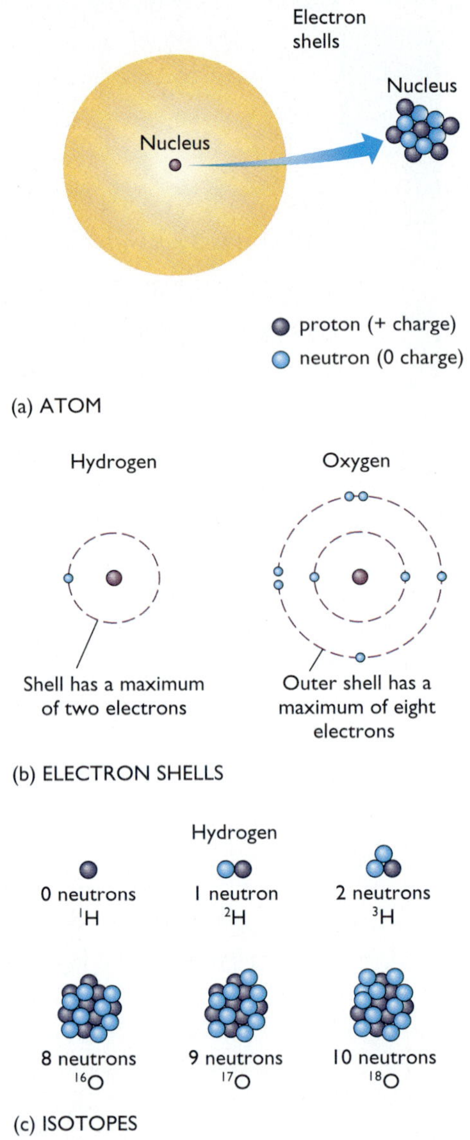

FIGURE 5-1

Atomic structure. (a) A simplified version of an atom depicts a nucleus of protons and neutrons, surrounded by shells of orbiting electrons. (b) A depiction of electrons within the shells of hydrogen (one electron) and oxygen (eight electrons) atoms. (c) The isotopes of an element have a variable number of neutrons. Hydrogen isotopes have zero, one, or two neutrons; oxygen isotopes have eight, nine, or ten neutrons.

Let's clarify this by being specific. Hydrogen (**Figure 5–1b**) possesses one proton in its nucleus; the positive charge from its single proton is neutralized by the negative charge from its one orbiting electron. Oxygen, on the other hand, contains eight protons in its nucleus, balanced electrically by eight orbiting electrons. So hydrogen atoms consist of a single proton and a single electron, and oxygen atoms of eight protons and eight electrons. If electrons are either added or removed from any single atom, the atom is no longer electrically balanced. Atoms with more electrons than protons have a net negative charge. Atoms with more protons than electrons have a net positive charge. An atom with either a positive or a negative charge is called an **ion**. For example, when NaCl dissolves in water, it separates into the ions Na^+ and Cl^-. The charge of an ion is the single most important reason for its ability to bond with other elements.

The story does not end there though. Although the number of protons is fixed for any element, the quantity of neutrons in its nucleus can vary. Because the neutron carries no electrical charge but has mass, variations in the number of neutrons change the weight of the element, but not its basic chemistry. Atoms of the same element that differ in weight due to variable numbers of neutrons are called **isotopes**. Hydrogen has three isotopes. Each isotope contains a single proton, but either zero, one, or two neutrons (**Figure 5–1c**). Oxygen similarly has three isotopes containing either eight, nine, or ten neutrons in a nucleus (see Figure 5–1c). The most abundant hydrogen isotope has zero neutrons and the most abundant oxygen isotope has eight.

5-2

Basic Physical Notions

Another important notion that we must consider in order to understand the behavior of water is the physical concept of **heat**, the property that one measures with a thermometer and that results from the physical vibrations of atoms and molecules. These physical vibrations represent energy of motion, called **kinetic energy**. The more heat in a material, the greater the agitation of its atoms and molecules.

Let's consider a specific example. A block of ice consists of an orderly, rigid arrangement of water molecules (**Figure 5–2a**) that are held firmly in place by strong electrical bonds between the molecules. Although an ice cube appears inert to the naked eye, it is not inert on an atomic scale. Its molecules are vibrating back and forth even though they are locked together into a crystalline framework. If we add heat to the ice cube, the molecules vibrate faster and move farther back and forth in the crystal. Above 0°C, the melting temperature of ice, the back-and-forth motions of the molecules are so vigorous that they exceed the strength of the electrical bonds holding the molecules in place in the crystal. At that temperature, the crystal's structure disintegrates (melts), and the solid ice becomes liquid water (**Figure 5–2b**). Liquids are loose aggregates of molecules that are in contact, but are free to move relative to one another, unlike the fixed molecules in a solid. As

FIGURE 5-2

States of matter. Water occurs in three states, which depend on temperature and pressure. (a) Solid water (ice) consists of ordered molecules that are tightly bonded to one another. (b) Liquid water consists of molecules that move relative to one another. Polymers are bits of crystalline structure that can exist in liquid water near its melting temperature. (c) Gaseous water (gas) is made up of independently moving molecules.

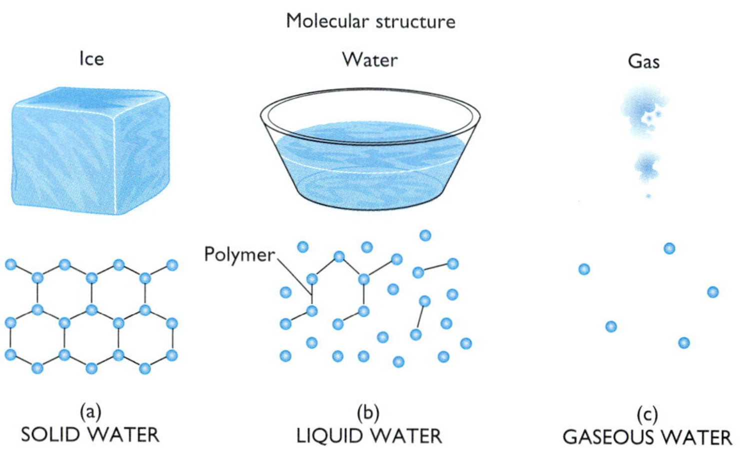

more heat is added to the liquid water, its temperature rises as its molecules vibrate more energetically. Some of these molecules contain so much kinetic energy that they escape from the liquid surface and become a gas (**Figure 5–2c**); this process is called **evaporation**. At 100°C, the boiling temperature of water, all of the molecules are highly energized and, hence, are **vaporized** (converted into **gas**). Free gas molecules move independently of one another between collisions, and are the least ordered of the three states of matter because of their very high kinetic energy.

The kinetic theory of heat has important implications for measuring temperature and for understanding the density of materials, whether they be solids, liquids, or gases. Imagine a thermometer, which is nothing more than a narrow glass tube filled with mercury. If we wish to determine the temperature of a water sample, we stick the thermometer in the liquid. What happens? The mercury in the tube rises and eventually stops. It is then an easy matter to read the temperature of the water off the thermometer's scale. But what exactly does the rise of mercury in the tube represent in a physical sense? Well, it's quite simple, provided you grasp the kinetic theory of heat. Recall that the water molecules are vibrating at a rate that depends on the water's temperature. When the thermometer is placed in the liquid, water molecules strike the tube; these collisions add energy to the molecules of the tube, so they vibrate faster. This in turn transfers kinetic energy to the atoms of mercury. The atoms of mercury begin to vibrate vigorously and collide with one another harder and harder, forcing them on the average farther apart than they were before the addition of kinetic energy from the water. Therefore, the mercury expands and rises in the tube of the thermometer, which is calibrated precisely to read temperature.

Temperature controls another fundamental property of matter called **density**—the amount of mass contained in a unit volume, expressed as grams (mass) per cubic centimeter (volume), that is, g/cm^3 or kg/m^3. The more mass that is contained in a cubic centimeter, the more dense is the material. The amount of heat contained in a substance determines how vigorously the molecules in that substance vibrate and collide with one another. The harder they vibrate, the farther apart they tend to be, which controls directly the amount of mass that is contained in a unit volume. This means that 10°C water is denser than 15°C water, and that warm air is less dense than cold air is at the same pressure.

Now that we have a solid conceptual understanding of the structure of atoms, of the kinetic theory of heat, and of temperature and density under our caps, we can examine the chemical and physical properties of water in general, and seawater in particular. Let's start by inspecting the common molecule on the Earth that we refer to simply as water.

5-3 The Water Molecule

Most people know that the chemical formula for water is H_2O. This formula means that water consists of two atoms of hydrogen (H) that are chemically bonded to one atom of oxygen (O) (**Figure 5–3a**). However, despite its simple chemical composition, water is a complex substance with truly remarkable physical properties. For example, the melting and boiling temperatures of water are much higher than expected when compared to chemically related hydrogen compounds (**Figure 5–3b**). This is fortunate; otherwise, water would be able to exist only as a gas at the temperatures that prevail at the Earth's surface. Consequently, the oceans could not have formed, and life could not have developed. In fact, H_2O is the only substance that can coexist naturally as a gas, a liquid, and a solid on the Earth's surface. Therefore, it is not surprising to discover how fundamental it is to all forms of life.

Water also has an unusually high heat capacity and tremendous solvent power. **Heat capacity** is defined as the quantity of heat required to raise the temperature of 1 gram of a substance by 1°C. More energy is required to raise the temperature of a substance with high heat capacity than one with low heat capacity. In other words, adding the same amount of heat will raise the temperature of a substance with a low heat capacity to a greater degree than one with a high heat capacity. The high heat capacity of water explains why so much energy is required to heat water.

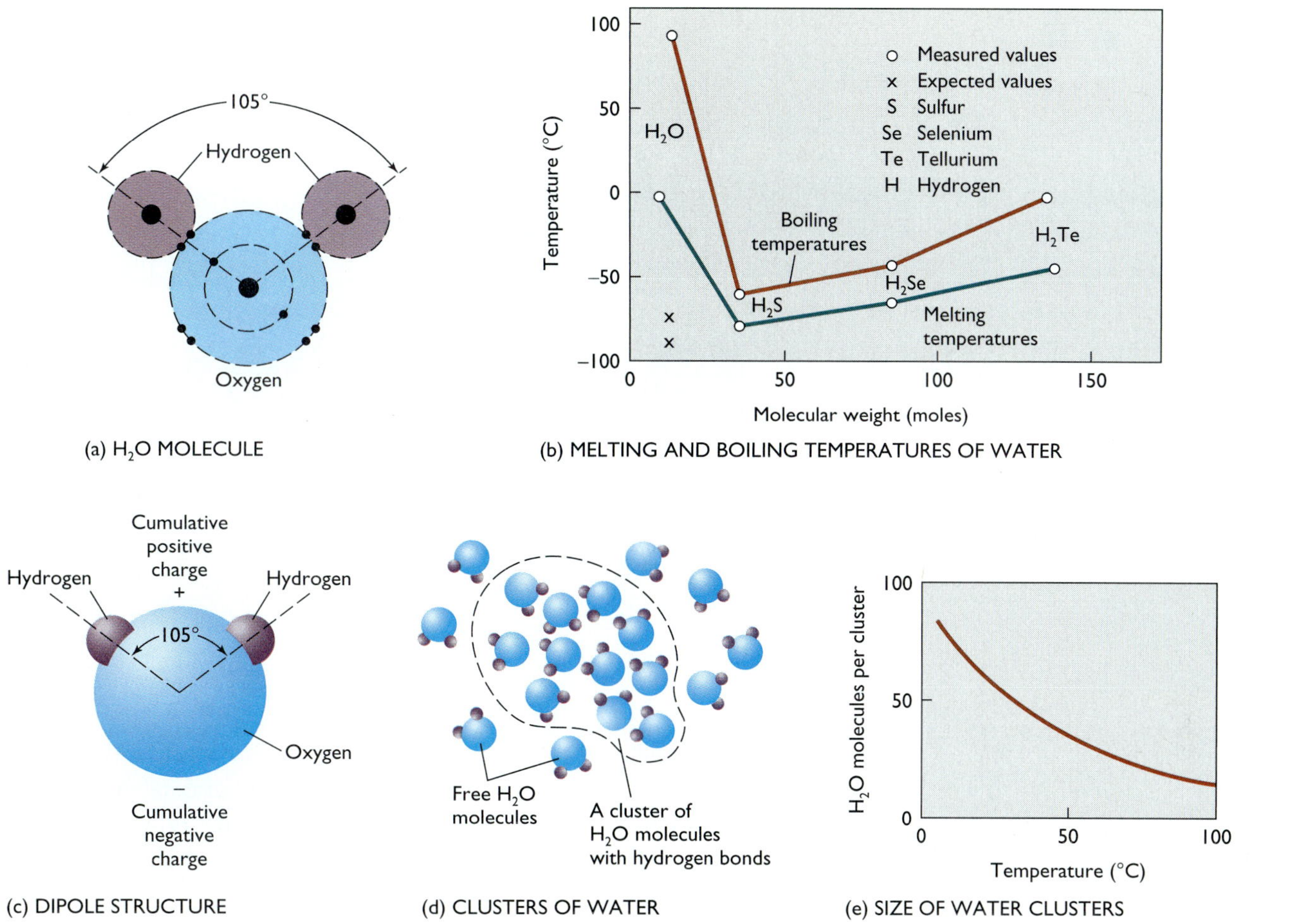

FIGURE 5-3

The water molecule. (a) The chemical bonding of two hydrogen atoms and one oxygen atom produces a water (H_20) molecule. (b) The observed (o) melting and boiling temperatures of water are much higher than theory (x) indicates. [Adapted from R. A. Home, *Marine Chemistry: The Structure of Water and the Chemistry of the Hydrosphere* (New York: John Wiley, Interscience, 1969).] (c) The two small hydrogen atoms are separated from each other at their points of attachment to the large oxygen atom by an angle of 105 degrees, so that the molecule resembles the familiar caricature of a mouse's head. This structure creates a dipolar molecule with a residual positive charge at one end and a residual negative charge at the other end. (d) A cluster of H_20 molecules with hydrogen bonds is contrasted here with free H_20 molecules. (e) The size of the clusters decreases with increasing temperature.

In addition to its unusually high heat capacity, the capability of liquid water to dissolve material, its **solvent power**, is unsurpassed by any other substance. In fact, chemists refer to water as the "universal solvent," meaning that virtually anything can be dissolved to some extent in liquid water.

Water possesses other unusual properties as well, but the case for this substance's chemical uniqueness should be clear. To account for water's singular properties we must examine the physical structure of the water molecule, H_2O. The molecule's asymmetrical shape is as important as the chemical identities of the two elements H and O. The two hydrogen atoms, rather than being attached symmetrically to either side of the oxygen atom, are separated from each other by an angle of 105 degrees for water in the liquid and gaseous states (**Figure 5–3c**). This molecular architecture resembles a mouse's "head" with hydrogen "ears." The hydrogen and oxygen atoms share electrons, and this **covalent bond** creates the water molecule H_2O. Each hydrogen atom within the molecule possesses a single positive charge; the oxygen atom, a double negative charge. As such, the H_2O molecule is electrically neutral

FIGURE 5-4

Halite (rock salt). The dipole structure of the H_20 accounts for its unsurpassed properties as a solvent. Na^+ and Cl^- ions are dislodged from the halite crystal by, respectively, the negatively and positively charged ends of the water dipole.

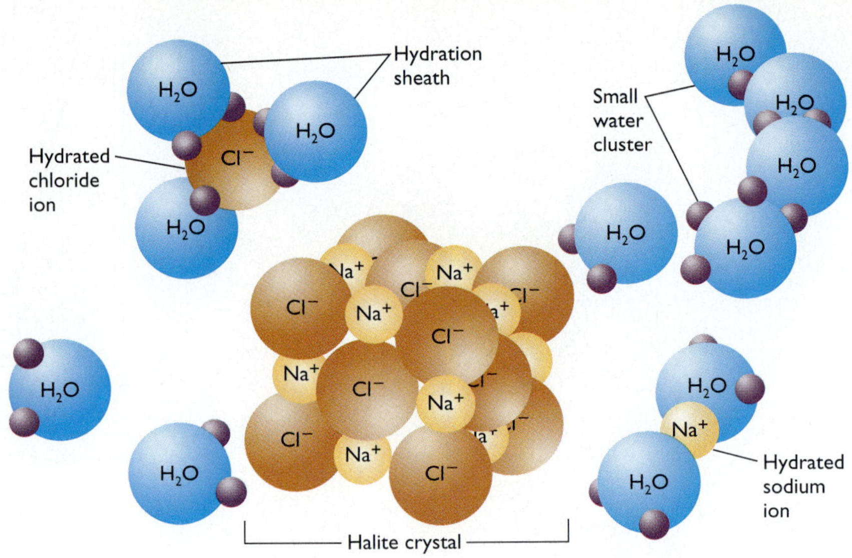

because of the balance between the positive and negative charges. However, the structural asymmetry of the molecule imposes a slight electrical imbalance, because the positive hydrogen atoms are bonded to one end of the oxygen atom and the electrons associated with that atom are concentrated on the other side. This gives rise to a **dipole** (two-pole) structure (see Figure 5–3c). Thus, there is a residual positive charge at the hydrogen end of the molecule and a residual negative charge at the oxygen end, despite the overall electrical neutrality of the water molecule. Consequently, liquid water is not merely a collection of freely moving molecules. Rather, its dipole structure causes the negative end of the molecule to be attracted to and become electrically bonded to the positive end of a nearby water molecule. This electrostatic bonding, called **hydrogen bonding**, produces irregular chains and clusters of H_2O molecules (**Figure 5–3d**). The size of water-molecule clusters decreases with increasing temperature (**Figure 5–3e**). Although hydrogen bonding is only 4 percent as strong as the covalent bonds that hold the atoms of the molecule together, it is directly responsible for many of water's extraordinary physical properties. Let's examine how this comes about.

The higher than expected melting and boiling temperatures of water (see Figure 5–3b) depend directly on the dipole structure of the H_2O molecule. More energy is required than expected to vaporize liquid water and to melt ice, because hydrogen bonds that link H_2O-molecule to H_2O-molecule must first be broken before the solid can melt and the liquid can vaporize. This is also the reason for water's high heat capacity. When heat is added to water, only a fraction of this energy is actually used to increase the vibrations of the molecules, which would be detected as a rise of temperature. Much of the added heat is used to break hydrogen bonds that link the H_2O molecules into irregular clusters. Hence, as water is heated, its temperature rises slowly relative to the amount of energy used. Conversely, when cooled, water releases more heat than expected from the decrease in its temperature.

The unusually high heat capacity of water prevents extreme variations in the temperature of the oceans and explains why the climates of coasts and islands experience less extreme temperature variations than those of land located far from the ocean or large lakes. During the summer, large bodies of water absorb solar heat, helping keep air temperatures cool. During the winter, large lakes and the ocean radiate great quantities of stored heat to the atmosphere as their water cools, warming the adjacent shoreline.

The high solvent power of liquid water likewise depends on the dipole structure of its molecule. Sodium and chloride are the most common salts

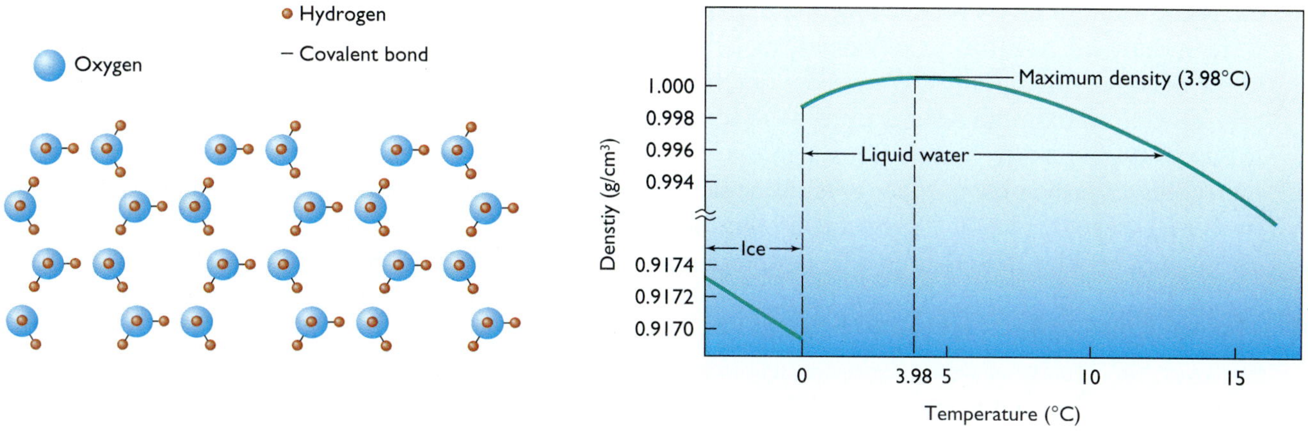

(a) HEXAGONAL CRYSTAL STRUCTURE OF ICE (b) DENSITY OF WATER

FIGURE 5-5

Properties of water. (a) The open network of the hexagonal structure of ice crystals is shown here. (b) Because of the open crystal structure of solid water, ice is less dense than liquid water. Water attains its maximum density at a temperature of 3.98°C; polymers exist in liquid water colder than this temperature.

that are dissolved in seawater. Sodium is a positively charged atom (Na^+) called a **cation**. In contrast, chloride is a negatively charged atom (Cl^-) called an **anion**. When these two ions come into contact, they are attracted to each other because of their opposing charges, and can be held together by **ionic bonds** to form halite (rock salt or common table salt). When halite crystals are put in water (**Figure 5–4**), the negatively charged end of the H_2O dipole dislodges sodium ions (Na^+) from the solid, and the positive end of the H_2O molecule tears off chloride ions (Cl^-). Dissolution (the noun form of the verb to dissolve) continues until either the entire crystal of halite is gone or the volume of water can no longer accommodate more ions of salt because it is **saturated**, meaning that there is physically no more "room" for the sodium and chloride in the water. (By way of an analogy, think about what happens when you keep adding spoonful after spoonful of sugar to your cup of coffee.) In solution, the sodium and chloride ions are surrounded by water molecules (see Figure 5–4). This keeps the cations separated from the anions, a process known as **hydration**. In other words, water acts as a solvent by preventing the chemical recombination of Na^+ and Cl^- to form the solid halite.

The density of water is yet another unusual property of this familiar substance. Ice floats on water! All other solids sink in their own liquids. It's hard to imagine a solid bar of steel floating in a vat of molten steel. Once again, water behaves in a peculiar way. Why is this so?

At the freezing point, solids crystallize from liquids because the thermal vibrations of molecules are low enough so that chemical bonding can occur and a crystal forms. The loose assortment of molecules in the liquid is reconstructed into a rigid solid. Because the molecules in the solid vibrate less than do the molecules in the liquid, they are more tightly packed and denser in the solid than in the liquid state. Therefore, solids, because of their higher density, sink in their own liquids. Water does not behave in the same way because ice molecules are arranged into an open crystal framework (**Figure 5–5a**), whereas liquid water molecules are packed into snug clusters by the hydrogen bonds between molecules (see Figure 5–3d). Also, the angle of separation between the two hydrogen atoms, which is 105 degrees in liquid and gaseous water, expands to 109.5 degrees in ice. This makes ice about 8 percent less dense than water. The H_2O molecules in ice are ordered into a porous **hexagon** (a six-sided structure) by the hydrogen bonds between oppositely charged ends of neighboring molecules (Figure 5–5a). When ice is warmed to 0°C, it begins to melt as thermal vibrations of the molecules cause the crystalline structure to break apart. Because of hydrogen bonds, the freed H_2O molecules in the liquid become more closely packed together than they were in the solid, resulting in an increase of density (more mass per unit volume). Water does not, however, reach its maximum density until it is warmed to 3.98°C (**Figure 5–5b**), because loose aggregates of molecules that resemble the open crystalline structure of ice (see Figure 5–2b), termed **polymers**, persist in water cooler than that critical temperature. Above 3.98°C, the density of water decreases with increasing temperature, as expected. Below 3.98°C, however, the density of water *decreases* as temperature *decreases* until it freezes into ice.

What we have learned so far about seawater is that it consists dominantly of H_2O, a chemical substance that can occur in a solid, liquid, and gaseous state at the temperatures and pressures that exist at the Earth's surface. Although we find water everywhere on the planet, its unusual chemical and physical properties should not be taken for granted; it is these very properties that have enabled life to appear and evolve through geologic history. We can now proceed to examine seawater's other chemical ingredients.

THE SOLUTES OF SEAWATER

Chemical analyses of samples from all over the world show that seawater consists of a small quantity of salt dissolved in water. The salt occurs as charged particles, cations and anions, that are dispersed among the molecules of liquid water. Seawater is a chemical solution. The dissolving agent, the liquid water, is the **solvent** and the dissolved substances, the salt ions, are the **solute**. Seawater also contains minor to trace amounts of dissolved metals, nutrients, gases, and organic compounds of seemingly infinite variety.

Before examining the chemistry of seawater solutions, we need to determine the amount of material that is dissolved in seawater, so that we can compare samples taken from different parts of an ocean or, for that matter, from different oceans. We can taste seawater samples and say qualitatively that this sample tastes "saltier" than this other one. But this is a rather subjective technique, and a scientist needs to know exactly and precisely how salty a parcel of water is. Oceanographers specify the concentration of a solute in seawater in units called **parts per thousand**, represented by either the abbreviation **ppt** or, as preferred by marine scientists, the symbol ‰. Although oceanographers also express salinity as a dimensionless unit in terms of PSS78—Practical Salinity Scale 1978—we will use ppt. Average or "normal" seawater has a salinity of about 35‰. This means that the dissolved salt occurs in a concentration of 35 parts per thousand (ppt). That is, the salt comprises 3.5 percent (divide 35 by 1,000, and convert it to a percentage by multiplying it by 100) of the sample, the rest (96.5 percent or 965 parts per thousand) being H_2O molecules. (Notice that 965 + 35 = 1,000.) This signifies that a volume of seawater weighing 1,000 grams (or 1 kilogram) with a salinity of 35‰ contains 35 grams of solute. Obviously, a 100 gram sample of seawater with a salinity of 35‰ contains 3.5 grams of dissolved salts.

As you know, because of its high solvent power (ability to dissolve), many types of chemicals are dissolved in ocean water and most are found in minute quantities. All these solutes can be grouped into five broad categories: major constituents, nutrients, gases, trace elements, and organic compounds. A review of the general characteristics of each group follows:

MAJOR CONSTITUENTS

In terms of quantity, the primary solutes in seawater are cations and anions. By weight, chloride (Cl^-) and sodium (Na^+) together comprise more than 85.65 percent (**Table 5–1**) of all the dissolved substances in seawater. When these two ions bond chemically into a solid, they form halite and give seawater its most distinctive property—its saltiness. Surprisingly, the six most abundant ions—chloride (Cl^-), sodium (Na^+), sulfate (SO_4^{2-}), magnesium (Mg^{2+}), calcium (Ca^{2+}), and potassium (K^+)—make up over 99 percent of all of seawater's solutes. The addition of five more solutes to the list—bicarbonate (HCO_3^-), bromide (Br^-), boric acid (H_3BO_3), strontium (Sr^{2+}), and fluoride (F^-)—elevates the quantity of dissolved ingredients in seawater to 99.99 percent (see Table 5–1). This means, of course, that everything else dissolved in seawater occurs in trace amounts and collectively comprises only 0.01 percent! But what appears to be insignificant cannot be ignored, because, even in tiny quantities many of these chemicals are absolutely critical for life in the ocean. Because the concentrations of these **major constituents** in sea-

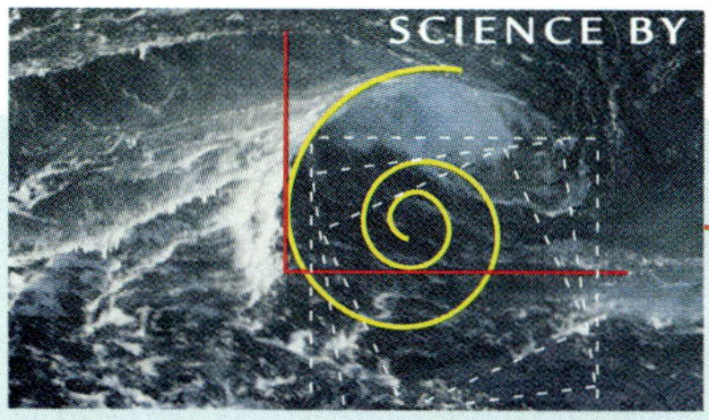

SCIENCE BY NUMBERS

Parts per Thousand

For those who are not acquainted with such measures, the expression parts per thousand (ppt) may seem mysterious. This measure of concentration is, actually, quite simple to understand, provided you spend a few moments learning and thinking about its meaning.

Let's begin by measuring the concentration of red marbles in a large vat filled with red, blue, yellow, and green marbles. A direct way to do this is to take a random collection of marbles in a bucket and count the marbles of each color that you sample. You do this until you have counted 1,000 marbles and you get the following results:

red marbles	112
blue marbles	464
green marbles	416
yellow marbles	8
Total	1,000 marbles

A convenient way to report these results is in parts per thousand. In other words, the red marbles appear in a concentration of 112 parts per thousand (ppt), meaning that, out of one thousand marbles, 112 are colored red. The concentrations of the others are: blue marbles 464 ppt, green marbles 416 ppt, and yellow marbles 8 ppt. It is that simple.

The conversion of ppt into a percentage is as straightforward. You simply divide the concentration of each colored marble by 1,000 and multiply the result by 100 to get the percentage value. Let's do this for the red and yellow marbles in our example.

Red marbles: $(464/1{,}000)\,(100)\% = (464/10^3)\,(10^2)\% = 464\times 10^{(2-3)}\%$
$= 464 \times 10^{-1}\% = 46.4\%$

Yellow marbles: $(8/1{,}000)\,(100)\% = (8/10^3)\,(10^2)\% = 8 \times 10^{(2-3)}\%$
$= 8 \times 10^{-1}\% = 0.8\%$

This means that 464 ppt is equivalent to 46.4% and 8 ppt to 0.8%. A salinity of 35‰ (ppt) means that the salt by weight comprises 3.5% of the seawater sample. This, in turn, implies that 97.5% of the weight is made up of water molecules in the sample. Can you provide the calculations to bear this out?

TABLE 5-1

Major solutes in seawater

Salt Ion	Ions in Seawater* (‰°)	Ions by Weight (%)	Cumulative (%)
Chloride (Cl^-)	18.980	55.04	55.04
Sodium (Na^+)	10.556	30.61	85.65
Sulfate (SO_4^{2-})	2.649	7.68	93.33
Magnesium (Mg^{2+})	1.272	3.69	97.02
Calcium (Ca^{2+})	0.400	1.16	98.18
Potassium (K^+)	0.380	1.10	99.28
Bicarbonate (HCO_3^-)	0.140	0.41	99.69
Bromide (Br^-)	0.065	0.19	99.88
Boric acid (H_3BO_3)	0.026	0.07	99.95
Strontium (Sr^{2+})	0.013	0.04	99.99
Floride (F^-)	0.001	0.00	99.99
Total	34.482	99.99	99.99

*The gram weight of ions per 1 kg of seawater, or g/kg.

Source: Adapted from H. U. Sverdrup, M. W. Johnson, and R. H. Fleming, *The Oceans* (Englewood Cliffs, N.J.: Prentice-Hall, 1942).

TABLE 5-2

Near-surface nutrient concentrations in seawater

Nutrient Element	Concentration (ppm)*
Phosphorus (P)	0.07
Nitrogen (N)	0.5
Silicon (Si)	3

*ppm = parts per million

water vary little over time at most localities, they are described as **conservative ions** of the ocean.

NUTRIENTS

Nutrients are essential for plant growth, as anybody who has fertilized a lawn or garden knows. All plants, including those that live in the ocean, convert nutrients into food (organic compounds such as sugar) by photosynthesis. Nutrients in seawater are compounds that consist primarily of nitrogen (N), phosphorous (P), and silicon (Si). Representative concentrations of these nutrients in the ocean are listed in **Table 5–2**; note that the concentrations are specified in **parts per million (ppm)**. Most plants cannot use elemental nitrogen and phosphorus and so satisfy their nutrient needs by absorbing phosphate (PO_4^{3-}) and nitrate (NO_3^-). Silicon is used by important groups of microscopic plants (diatoms) and animals (radiolaria) to precipitate silica (SiO_2) shells around their fragile cells. Because of biological uptake and release, the concentrations of nutrients in seawater, as on land, vary from place to place and over time at any one place. Hence, oceanographers refer to these substances as **nonconservative ions** of seawater, signifying that levels of these substances are not constant in water, but vary over time and from place to place.

GASES

Listed in order of decreasing abundance (**Table 5–3**), gases in seawater include nitrogen (N_2), oxygen (O_2), carbon dioxide (CO_2), hydrogen (H_2), and the noble gases argon (Ar), neon (Ne), and helium (He). Nitrogen and the three noble gases are **inert** (unreactive) and rarely involved directly in plant photosynthesis. In contrast, levels of dissolved O_2 and CO_2 are greatly influenced by photosynthesis and respiration of organisms. Therefore, they vary greatly in space and time depending on the activities of plants and animals and are regarded as nonconservative.

TRACE ELEMENTS

Trace elements are all chemical ingredients that occur in minute (trace) quantities in the oceans. Most trace elements, such as manganese (Mn), lead (Pb), mercury (Hg), gold (Au), iodine (I), and iron (Fe), occur in concentrations of less than 1 ppm (part per million) (**Table 5–4**). Many occur in quantities of less than 1 **part per billion (ppb)** and even at 1 **part per trillion**. These low concentrations make certain trace elements difficult and sometimes even impossible to detect in seawater. However, despite their extremely low concentrations, trace elements can be critically important for marine organisms, either by helping to promote life or by retarding or killing life (**toxicity**).

ORGANIC COMPOUNDS

Organic compounds are large, complex molecules produced by organisms. They include substances such as lipids (fats), proteins, carbohydrates, hormones, and vitamins. Typically, they occur in low

TABLE 5-3

Quantities of gas in air and seawater

Gas	In Dry Air (%)	In Surface Ocean Water (%)	Water-Air Ratio
Nitrogen (N_2)	78.03	47.5	0.6
Oxygen (O_2)	20.99	36.0	1.7
Carbon dioxide (CO_2)	0.03	15.1	503.3
Argon (Ar), hydrogen (H_2), neon (Ne), and helium (He)	0.95	1.4	1.5

TABLE 5-4

Examples of trace elements in seawater

Trace Element	Concentration (ppb)*
Lithium (Li)	170
Iodine (I)	60
Molybdenum (Mo)	10
Zinc (Zn)	10
Iron (Fe)	10
Aluminum (Al)	10
Copper (Cu)	3
Manganese (Mn)	2
Cobalt (Co)	0.1
Lead (Pb)	0.03
Mercury (Hg)	0.03
Gold (Au)	0.004

*ppb = parts per billion

concentrations and are produced by **metabolic** (physical and chemical processes in the cell of an organism that produce living matter) and decay processes of organisms. For example, vitamin complexes are vital for promoting the growth of bacteria, plants, and animals, as shown by the control that thiamine and vitamin B_{12} have on the growth rate, size, and number of microscopic plants grown in laboratory experiments.

Now that we have a general understanding of the chemical makeup of seawater—a solution of mainly water with some salts, and tiny quantities of nutrients, gases, trace elements, and organic compounds—we can proceed to examine the nature of salinity and its effect on the properties of water, as well as the factors that control the saltiness of the ocean. In other words, why are the oceans salty, why are there variations in the salinity of the oceans, both on its surface and within its depths, and how does dissolved salt affect the physical properties of water? Answers to such questions are easy to grasp, provided that you understand the concepts just introduced. Also, the boxed feature, "Chemical Techniques," reviews a few of the methods that chemists employ to measure the properties of seawater while at sea and in the laboratory.

5-4

Salinity

A simple way to determine salinity is to evaporate water from a container of seawater and then compare the weight of the solid residue left behind in the bottom of the container—the salts—to the weight of the original sample of seawater. Unfortunately, the method is neither precise nor accurate, because salt crystals hold on to variable amounts of H_2O molecules, and that affects the weight of the salt residue. In order to compare accurately salinity data gathered from many parts of the ocean and measured in many different laboratories and ships, chemists have adopted a standardized and what seems to the non-chemists to be a rather cumbersome definition of **salinity**: the total mass expressed in grams of all the substances dissolved in 1 kilogram of seawater, when all the carbonate has been converted to oxide, all the bromine and iodine have been replaced by chlorine, and all organic compounds have been oxidized at a temperature of 480°C. Because we are not chemical oceanographers, we can simplify the definition of salinity as follows to suit our more general purpose: the total weight in grams of dissolved salts in 1 kilogram of seawater expressed as ‰ (parts per thousand).

PRINCIPLE OF CONSTANT PROPORTION

Salinity determinations from the world's oceans have revealed an important, unexpected finding. Although salinity varies quite a bit because of differences in the total amount of dissolved salts, the relative proportions of the major constituents are constant. In other words, the ratio of any two major constituents dissolved in seawater, such as Na^+/K^+ or Cl^-/SO_4^{2-}, is a fixed value, whether the salinity is 25, 30, 35‰, or whatever. To put it in more familiar terms, let's imagine that the ratio of females to males in a population is ¼ (1 female for every 4 males) and that this ratio never changes regardless of population size. This means that the total number of people in the population can vary, but the *relative proportion* of females to males does not

Chemical Techniques

In trying to characterize and explain the chemical properties of seawater, marine chemists find it critically important to collect sufficient and appropriate seawater samples, prevent their chemical contamination, determine sampling depths, and use accurate and precise analytical procedures.

SAMPLE COLLECTION

Seawater samples for chemical analysis are collected in metallic or plastic cylindrical bottles. The metallic bottles have interior liners composed of inert plastic to prevent contamination of the sample by metal ions. One such sampling device, the Niskin bottle, has valves on both ends that are opened and attached to a cable (**Figure B5–1**). Typically, several open bottles are attached at predetermined positions on the cable. After the bottles are lowered, a weight, known as a *messenger,* is fastened to the cable and released. When the messenger strikes the first water bottle, it causes it to close tightly, trapping a sample of seawater. A messenger attached to a clamp beneath the first bottle is then released, dropping and triggering the next bottle below. The procedure is repeated until all the bottles on the cable have been closed. Depending on the nature of the study, chemists usually collect seawater volumes of between 1 and 3 liters (~1.06 and 3.17 quarts) with these bottle samplers.

A more elaborate sample-bottle configuration is known as the *rosette cluster*. It consists of a rigid frame that holds a number of collection bottles upright, arranged in a circular pattern (**Figure B5–2**). The bottles can be set to open and close automatically, or a technician can trigger any of the bottles electronically from a shipboard console.

FIGURE **B5-1**

Niskin bottles. Open Niskin bottles are attached to a cable and lowered to water depths where seawater samples are to be obtained for chemical analysis. A metal messenger "trips" each bottle on the cable individually, causing it to fill with water and close securely.

FIGURE **B5-2**

A rosette cluster. Water collecting bottles are arranged around a rigid, circular frame in a rosette pattern. Technicians are able to close the bottles individually as the array is lowered or raised through the water column.

SAMPLING DEPTHS

Chemists must establish the exact sampling depth for each bottle. Otherwise, the analytical work, no matter how accurate, is of limited use in determining the exact chemical structure of the water column. A common technique is to measure the length of the cable between the ocean surface and the depth at which the bottle was triggered by the messenger. However, the cable rarely hangs straight down, because of the drift of the ship relative to the bottles on the cable. Depth corrections are applied by measuring the angle of the cable and by noting the difference between the temperature readings on the pressure-protected and unprotected thermometers mounted on the sampling bottles. (Temperature discrepancies are indicators of water pressure, which is a function of water depth.) When near-bottom water samples are collected, it is customary to attach a *pinger* (a pulsing sound source) to the free end of the cable. Sound signals reflected off the sea floor and transmitted to the ship are used to determine the distance between the pinger and the bottom to within a meter or so.

ANALYTICAL PROCEDURES

Analytical procedures reveal temperature and salinity of water. Reversing thermometers, which are fastened to water-sampling bottles, are used to measure the temperature of water *in situ*. Better precision (up to 0.0001 °C) is obtained by using temperature-sensitive materials, such as quartz crystals, which vibrate at frequencies that depend on temperature. These signals are transmitted electronically to the ship. This allows the temperature of the water to be monitored continuously as the instrument is lowered.

Because the composition of seawater is constant, chemists traditionally have determined water *salinity* by chemical titration—the process of standardizing silver nitrate against a normal seawater sample of known chemical composition. The electrical conductivity of seawater, which is proportional to the total concentration of dissolved ions, is now used routinely to determine salinity rapidly. The salinometer compares the electrical conductivity of an unknown sample with that of a known, standard sample of seawater, and converts the difference into a salinity value after correcting for temperature effects. An important instrument called the CTD (conductivity, temperature, depth) consists of a salinometer, an electronic thermometer, and a pressure sensor. As it is lowered through the water column, the CTD transmits electronic signals to the ship, where they are stored in a shipboard computer for analysis later.

change. In other words, the ratio of females to males is constant and is independent of population size. Just so, the ratio of any two major salt constituents in ocean water is constant and is independent of salinity.

This important discovery, made during the *Glomar Challenger* expedition (see Figure 1–10 in Chapter 1), is termed the **principle of constant proportion** or **constant composition**, and was a major breakthrough in determining salinity of seawater in a rapid, accurate, and economical manner. In theory, all that need be done to quantify salinity is to measure the amount of only a single major ion dissolved in a sample of seawater, because all the other major constituents listed in Table 5–1 occur in fixed amounts relative to that ion. Chemists chose to measure Cl^- for determining the salinity of seawater, because it is the most abundant solute in seawater and its concentration is easily determined. Actually, elements in the halogen family, which include chlorine, bromine, iodine, and fluorine, are difficult to distinguish analytically from one another. Therefore, chemists determine, not merely the Cl^- content, but the **chlorinity**, that is, the total quantity of halogens dissolved in water, expressed as g/kg (‰). It is then a simple matter to convert chlorinity to salinity by the formula

$$\text{salinity (‰)} = 1.80655 \times \text{chlorinity (‰)}.$$

Some representative chlorinity values and their corresponding salinities appear in **Table 5–5.**

Today, oceanographers rely on a variety of methods, including the electrical conductivity of seawater, to make routine determinations of salinity.

TABLE 5-5

Chlorinity and salinity values

Chlorinity (%°)	Salinity (‰°)
5	9.03
10	18.07
15	27.10
20	36.13
25	45.16

The electrical conductivity of a solution is its ability to transmit an electrical current directly proportional to the total ion content of the water at a given temperature, which, of course, is its salinity. A **salinometer** indirectly measures salinity by measuring the electrical conductivity of seawater. Its calibrations allow oceanographers to determine salinity directly and quickly by inserting an electrical probe into the water, a very simple matter compared with the rather laborious chemical determination of chlorinity.

FACTORS THAT REGULATE THE SALINITY OF SEAWATER

There is a lot of dissolved salt in the ocean. An important question is, what supplied all this material to the sea? The answer is straightforward (**Table 5–6**). Rivers disgorge huge volumes of freshwater into the ocean every year. Chemical analyses of water samples from rivers all over the world indicate that they contain a variety of dissolved substances in concentrations of ppm (**Table 5–7**). Rivers have a dissolved load of chemicals because of the chemical weathering of rocks on the land. These rocks are made up of an assemblage of minerals composed predominantly of the elements silicon, aluminum, and oxygen. Acidic water breaks down these rocks into their component elements. When carbon dioxide (CO_2) is dissolved in water, it reacts with H_2O molecules to produce H_2CO_3, a weak acid called **carbonic acid**. In turn this acid

TABLE 5-6

Sources and sinks of some seawater components

Chemical Component	Sources	Sinks
Chloride (Cl^-)	Volcanoes River influx	Evaporative deposition as NaCl (rock salt) Net air transfer Pore-water burial
Sodium (Na^+)	River influx	Evaporative deposition as NaCl (rock salt) Net air transfer Cation exchange with clays Basalt-seawater reactions Pore-water burial
Potassium (K^+)	River influx	Uptake by clays
	Volcanic-seawater reactions (high temperature)	Volcanic-seawater reactions (low temperature)
Calcium (Ca^{2+})	River influx	Biogenic secretion of shells
	Volcanic-seawater reactions Cation exchange	Evaporitic deposition of gypsum ($CaSO_4 \cdot 2H_2O$) Precipitation as calcite
Silica (H_4SiO_4)	River influx	Biogenic secretion of shells
	Basalt-seawater reactions	
Phosphorus (HPO_4^{2-}, PO_4^{3-}, $H_2PO_4^-$, organic P)	River influx Rainfall and dry fallout	Burial as organic P Adsorption on volcanic ferric oxides Formation of phosphorite rock

Source: Adapted from E. K. Berner and R. A. Berner, *The Global Water Cycle* (Englewood Cliffs, N.J.: Prentice-Hall, 1987).

TABLE 5-7

Dissolved substances in river water

Substance	Concentration (ppm)	Concentration (%)
Bicarbonate/carbonate (HCO_3^-/CO_3^{2-})	58.8	48.7
Calcium (Ca^{2+})	15.0	12.4
Silica (SiO_2)	13.1	10.8
Sulfate (SO_4^{2-})	1.2	9.3
Chloride (Cl^-)	7.8	6.5
Sodium (Na^+)	6.3	5.2
Magnesium (Mg^{2+})	4.1	3.4
Potassium (K^+)	2.3	1.9
Nitrate (NO_3^-)	1.0	0.8
Iron aluminum oxide [$(Fe, Al)_2O_3$]	0.9	0.8
Remainder	0.3	0.3

Source: D. A. Livingstone, Chemical composition of rivers and lakes, U.S. Geological Survey, Professional Paper 440-G (U.S. Government Printing Office, 1963).

separates into hydrogen (H^+) and **bicarbonate** (HCO_3^-) ions. The specific chemical reactions are reversible, are quite simple and are represented by

$$H_2O + CO_2 \rightleftharpoons H_2CO_3 \rightleftharpoons H^+ + HCO_3^-.$$

Notice that the reaction yields free ions of H^+, which because of their small size and high chemical reactivity, replace cations such as Na^+ and K^+ that are bound to minerals in rocks. The amount of H^+ is a measure of the acidity of the water. This process—the bathing of rocks in acidic water—slowly weathers minerals, releasing ions, which go into solution and become part of a river's dissolved chemical load (Table 5–7).

Let's firm up our understanding of weathering by examining the chemical breakdown of an actual mineral, **orthoclase** ($KAlSi_3O_8$), the common potassium-bearing feldspar of granite. The chemical reaction is

$$\underset{\text{(ORTHOCLASE)}}{2KAlSi_3O_8} + 2H^+ + H_2O$$

$$\rightarrow 2K^+ + \underset{\text{(KAOLINITE)}}{Al_2Si_2O_5(OH)_4} + \underset{\text{(DISSOLVED SILICA)}}{4SiO_2.}$$

This formula indicates that the mineral orthoclase in the presence of acidic water (H^+) is broken down into potassium ions (K^+), silica (SiO_2), and aluminum silicates. The latter compound bonds with H_2O, forming another mineral, the clay **kaolinite**. Rivers transport these materials to the ocean in two distinct states: as a **dissolved load** (K^+ and SiO_2) and as a **suspended load** (particles of kaolinite). The dissolved K^+ and SiO_2 contribute to the salinity of seawater; the clay accumulates as sediment on the ocean floor (see Figure 4–16a in Chapter 4). So what was a solid mineral in rocks on land becomes, by the processes of chemical weathering and transport, dissolved salts in the ocean and mud on the sea bottom.

How much mass is actually added to the ocean by the river influx of dissolved matter? The first response by most people is that the amount could not be much, because the concentration of solutes in rivers is low (that's why freshwater doesn't taste salty), and it's dissolved, so you can't see it (how can anything that can't be seen amount to much?). It turns out that the annual river input of material in solution to the oceans is somewhere between 2.5×10^{15} and 4×10^{15} grams. True, 10^{15} (10 multiplied by itself fifteen times) seems to be quite a big number. But how big is it, really? This rate of influx, 10^{15} grams per year, is about equivalent to the mass of mud supplied each year to the oceans by the rivers of North America, South America, Africa, and Europe combined! Although dissolved material is invisible to the naked eye, it represents a major annual input of mass to the oceans, all of it derived by the chemical weathering of rocks on land. In addition to the supply from rivers, the Earth is degassing. This means that volcanoes on the crests of spreading ocean ridges and in the volcanic arcs of subduction zones spew out large quantities of cations (including Ca^{2+} and K^+) and anions (including SO_4^{2-} and Cl^-) into the water column, although the exact amount of this input has yet to be determined reliably.

The fossil record and sedimentary rocks themselves indicate that oceans have existed on the Earth for at least as long as 3.4 billion years. Geochemical data indicate that the salinity of the oceans has changed little over the past 1.5 billion years. This constant ocean salinity despite the tremendous annual supply of dissolved chemicals to the oceans by rivers can only mean that on average a similar quantity of salt must be removed from the oceans each year. Otherwise, the salinity of the world's oceans would have increased over geologic time. This balance between inputs and outputs of salt to the ocean is called a **steady-state equilibrium**.

Oceanographers refer to inputs of material as **sources** and their outputs as **sinks**. We've already identified several of the principal sources of the salt ions to the ocean. Let's now examine the principal sinks of dissolved salt in the ocean (see Table 5–6). The removal of salt occurs by both inorganic and organic processes. Evaporation is an excellent example of an inorganic process and, as discussed in the boxed feature, "Desalination," is a technique for producing drinkable water from seawater. In arid climates, evaporation rates are high. Evaporation removes water from the ocean, but not the dissolved salt ions. This indicates that, with time, the concentration of salt will rise by the evaporation of water molecules, creating a **brine**, or a very salty solution. The Dead Sea and the Great Salt Lake, both in arid settings, are fine examples of this very process. As more and more water molecules are evaporated from the water, the solution eventually becomes saturated, which means that the solution is holding as much material in a dissolved state as it can for the temperature and pressure conditions of the water. The removal of more water creates a **supersaturated** solution (a solution containing a quantity of dissolved ions that exceeds the theoretical saturation value). This leads to the precipitation of **evaporite** minerals from seawater, notably halite (NaCl) and gypsum ($CaSO_4 \cdot H_2O$). Precipitation of evaporite minerals from seawater represents a sink, because dissolved salt ions are being removed from the ocean to form sedimentary deposits on the seafloor.

Wind also blows onshore a large amount of sea spray, which on evaporation, forms a coating of salt on land—as anybody who wears glasses and lives by the seashore can attest. In addition, freshly extruded basalt lavas on the ocean floor are quite reactive and extract dissolved ions, such as Mg^{2+} and SO_4^{2-}, from the seawater that comes in contact with the hot lava. Finally, **adsorption** (the "sticking" of ions to a surface) of cations like K^+ and Mg^{2+} by certain clay minerals in the ocean and the formation of authigenic minerals, such as ferromanganese nodules (see Chapter 4), remove a large, unknown quantity of ions from the sea. All of these represent chemical sinks for ions dissolved in seawater.

Organisms help maintain the steady-state equilibrium of the ocean's salinity as well. We already know that diatoms have silica shells and forams carbonate shells that are precipitated from the uptake of Si^{4+} and Ca^{2+}, respectively, from seawater. Once these organisms die, their hard parts may settle to the sea bottom to form deep-sea oozes (see Figure 4–16a in Chapter 4). Also, many species of animals extract certain chemical substances that are dissolved in seawater. Some of these chemicals are concentrated in fecal pellets that sink to the ocean floor and become incorporated in sediment.

Let's synthesize all that we have learned about steady-state equilibrium, sources and sinks of salt ions, and geologic cycles. The salt ions dissolved in

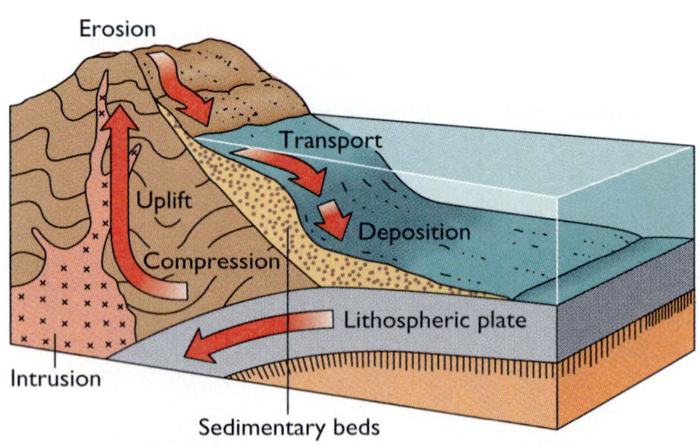

FIGURE 5-6

Sedimentary cycle. Over geologic time, mountains are leveled by rivers. The weathering products are dispersed into the ocean and collect on the sea bottom, forming sedimentary beds. Eventually, these accumulations of sediment are deformed and uplifted into mountain ranges by plate tectonics. Then a new cycle of river erosion begins.

ocean water are derived largely from the weathering and erosion of rocks on land. Because the average chemical composition of seawater has remained remarkably stable (in a steady state) over geologic time, the inputs of salt must be balanced by the outputs. The river-supplied ions remain in ocean water for a long time, but eventually are extracted by inorganic and organic processes and become part of the ocean's sedimentary record. These sediments, as they are buried, become cemented into sedimentary rocks and eventually are subducted along the colliding boundaries of lithospheric plates. Some of these sedimentary units are melted and intruded as igneous rocks, others are crumpled and raised into large mountain belts, where chemical attack by acidic water once again releases these ions to the sea. In effect, this is a grand sedimentary cycle (**Figure 5–6**). Elements bonded into solid minerals in rocks are put into solution and transported to the ocean by rivers, where organic and inorganic processes cause them to be precipitated into solids (such as halite, and silica and carbonate shells) and reincorporated into sedimentary rocks that become uplifted by tectonic processes and weathered once again, repeating this grand cycle.

If rivers are the primary source of salt ions to the oceans, why aren't the ionic compositions of freshwater and seawater similar? They clearly are not, as a comparison of Tables 5–1 and 5–7 shows. Shouldn't they be, if one is supplying material directly to the other? The difference in the relative composition of solutes in seawater and river water is a result of the **residence time** of ions in the ocean, which is simply the average length of time that an ion remains in solution there. It ranges between 2.6×10^8 years for sodium and 1.5×10^2 years for aluminum (**Table 5–8**). This is no different from saying that your "residence time" in your bedroom, asleep in bed, is eight hours a day. Note that the two most abundant components of seawater (Na^+ and Cl^-) have long residence times, on the order of hundreds of million years. Their persistence in a dissolved state in ocean water reflects their low geochemical and biochemical reactivity; in other words, they are essentially inert. (To carry our analogy one step further, an inert, listless person would tend to spend more time in bed and have a longer residence time in the bedroom than would an active, alert person.) By contrast, many of the principal ions in river water are characterized by short residence times in the oceans because they are much more reactive or are important for biological cycles. For example, many marine organisms require dissolved Ca^{2+} to secrete their carbonate ($CaCO_3$) shells. These calcium ions are in constant demand by the marine biota, so calcium has a relatively low residence time of 8×10^6 years.

Long residence times also help to explain the principle of constant proportions. Water is stirred and mixed by ocean currents, much as stirring a pot of soup with a spoon creates eddies and swirls (turbulence) that mix the ingredients. Studies of currents indicate that mixing rates in the oceans are on the order of a thousand (10^3) years or less. This rate is much lower than are the residence times of the major ions of seawater, which range from millions (10^6) to hundreds of millions (10^8) of years (see Table 5–8). Thus, rapid mixing and very long residence times of salt ions in seawater assure that these substances are distributed uniformly throughout the oceans. A more familiar way to think about this is to imagine yourself making a cake. Slowly adding dye along the edge of a bowl (this is equivalent to a river supplying ions at the edge of an ocean basin) to a cake batter that is being rapidly stirred by an electric beater (this is equivalent to mixing by ocean currents) quickly distributes the dye molecules evenly throughout the batter, so that its color (this is equivalent to the ocean's salinity) is uniform (**Figure 5–7**).

TABLE **5-8** Residence in ocean waters

Substance	Residence Time (× 10^6 yr)
Chloride (Cl^-)	∞
Sodium (Na^+)	260
Lithium (Li^+)	20
Strontium (Sr^{2+})	19
Potassium (K^+)	11
Calcium (Ca^{2+})	8
Zinc (Zn^{2+})	0.18
Barium (Ba^{2+})	0.084
Cobalt (Co^{2+})	0.018
Chromium (Cr)	0.00035
Aluminum (Al)	0.00015

Source: Adapted from C.K. Wentworth, A scale of grade and class terms for clastic sediments, *Journal of Geology* 30, (1922): 377–92.

FIGURE 5-7

Rapid mixing spreads dye evenly throughout the cake batter.

EFFECTS OF SALINITY ON THE PROPERTIES OF WATER

In the previous sections, we learned a great deal about the structure of the water molecule and the salinity of the ocean. A few other important properties of water are reviewed in the boxed feature, "Other Physical Properties of Water." Now we are ready to examine the effect dissolved ions have on the physical properties of water. As you might guess, the addition of salt modifies some of the properties of water in a number of significant ways. Most of these changes come about because the ions are hydrated (see Figure 5–4), which modifies the chemical behavior of the H_2O molecules in the solution. Let me explain by examining some specific water properties as they are affected by solutes.

FREEZING POINT

Pure freshwater freezes at 0°C. The addition of salt to the water lowers its freezing point. For example,

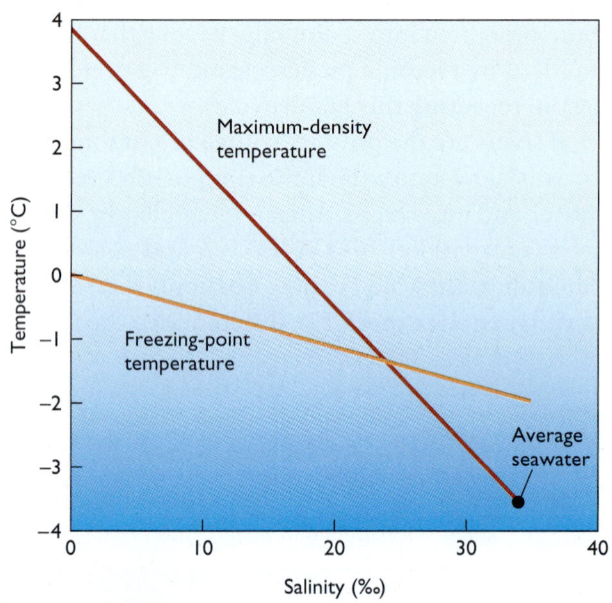

FIGURE 5-8

Effects of salinity on the maximum-density and freezing-point temperatures of seawater. The addition of dissolved ions to water lowers the initial freezing point temperature of the solution because the hydrated ions interfere with the rearrangement of the H_2O molecules into an ordered ice structure. Also, an increase in salinity depresses the maximum-density temperature of seawater. Consequently, the temperature of maximum-density of average seawater is well below its freezing point. This means that seawater of average salinity will freeze before it will sink.

seawater with a salinity of 35‰ freezes at a temperature of –1.91°C (**Figure 5–8**). The reason that the freezing temperature of seawater is depressed relative to that of freshwater is quite simple. The hydrated salt ions "hold on" to individual H_2O molecules, interfering with their rearrangement into an ordered ice crystal.

DENSITY

Because solutes have a greater atomic mass than do H_2O molecules, the density of water increases with salinity. This means that freshwater floats on salt water. For salinities >24.7‰ (see Figure 5–8), the temperature of maximum density (3.98°C for freshwater; see Figure 5–5b) is below the freezing point (0°C).

VAPOR PRESSURE

The vapor pressure on a liquid surface is the pressure exerted by its own vapor. When a liquid such as water is placed in a closed container, some of the molecules will vaporize, decreasing the amount of liquid. Once equilibrium is reached, the vapor pressure is the pressure exerted by the molecules in the vapor phase. As the salinity of water increases, vapor pressure drops; this means that freshwater evaporates at a faster rate than seawater does. The depression of vapor pressure in seawater is directly related to the total number of solute molecules. This effect is a consequence of the hydrated ions, which "hold on" to the water molecules, making their vaporization more difficult.

5-5 Chemical and Physical Structure of the Oceans

Oceanographers have acquired a great deal of information about variations in the fundamental properties of ocean water, including the regional distribution of temperature, salinity, and density along the sea surface and throughout the ocean's depths. This section summarizes and explains the general distribution of these three important parameters of ocean water. This will provide the background necessary to understand variations in the ocean's content of gases, the important topic of the following section.

TEMPERATURE

Latitude (Appendix IV) exerts a strong control on the surface temperature of the ocean, because the amount of **insolation** (the solar energy striking the Earth's surface) decreases poleward. Surface-water temperatures, therefore, are highest in the tropics and decrease with distance from the equator. **Isotherms**, imaginary contour lines that connect points of equal water temperature, generally trend east-west, parallel to the lines of latitude (**Figure 5–9**). Because the amount of insolation varies with the seasons, the surface-water temperature changes with time as well. The intense sunlight in the tropics and subtropics produces a broad band of water with a temperature higher than 25°C that shifts north or south depending on the season (see Figure 5–9). Water in the polar oceans is very cold, much of it being near or even below 0°C. Seasonal shifts of the isotherms are minor in the ocean off Antarctica, but are significant in the North Pacific and North Atlantic Oceans.

Ocean currents that flow around the periphery of each ocean affect the distribution of surface-water temperature. For example, the strips of

>25°C water in the tropical Atlantic and Pacific Oceans are much broader at their western than at their eastern margins. This distortion is produced by currents that move warm water poleward along the western side of oceans and cold water equatorward along their eastern sides (see Figure 5–9).

Investigations of water temperature with water depth reveal that the oceans in the middle and lower latitudes have a layered thermal structure (**Figure 5–10**). A layer of warm water, several hundred to a thousand meters thick, floats over colder, denser water that fills up the rest of the basin. The two water masses are separated from each other by a band of water that has a sharp temperature gradient (meaning that temperature changes rapidly with depth); it is called the **thermocline** (see Figure 5–10). The thermocline is a permanent hydrographic feature of temperate and tropical oceans, and ranges in water depth between 200 and 1,000 meters (~660 to 3300 feet). Surprisingly, in terms of volume, the most typical water of the tropics is not the warm, thin surface-water layer so familiar to sailors and tourists, but the deeper water below the thermocline, where temperatures are near freezing even at the equator!

Because of solar heating during the summer, a seasonal thermocline appears at a depth of between 40 and 100 meters (~132 and 330 feet) in the oceans

FIGURE 5-9

Sea-surface temperatures. Because water is heated by the sun, and solar radiation decreases with distance from the equator, sea-surface temperatures vary directly with latitude. Large-scale ocean circulation transports warm water poleward at the western edges of the ocean bains and cold water equatorward at their eastern peripheries. (a) Global sea-surface temperature (SST) is shown for August. (b) Global SST is shown for February. [Adapted from H. U. Sverdrup, M. W. Johnson, and R. H. Fleming, *The Oceans* (Englewood Cliffs, N.J.: Prentice-Hall, 1942).]

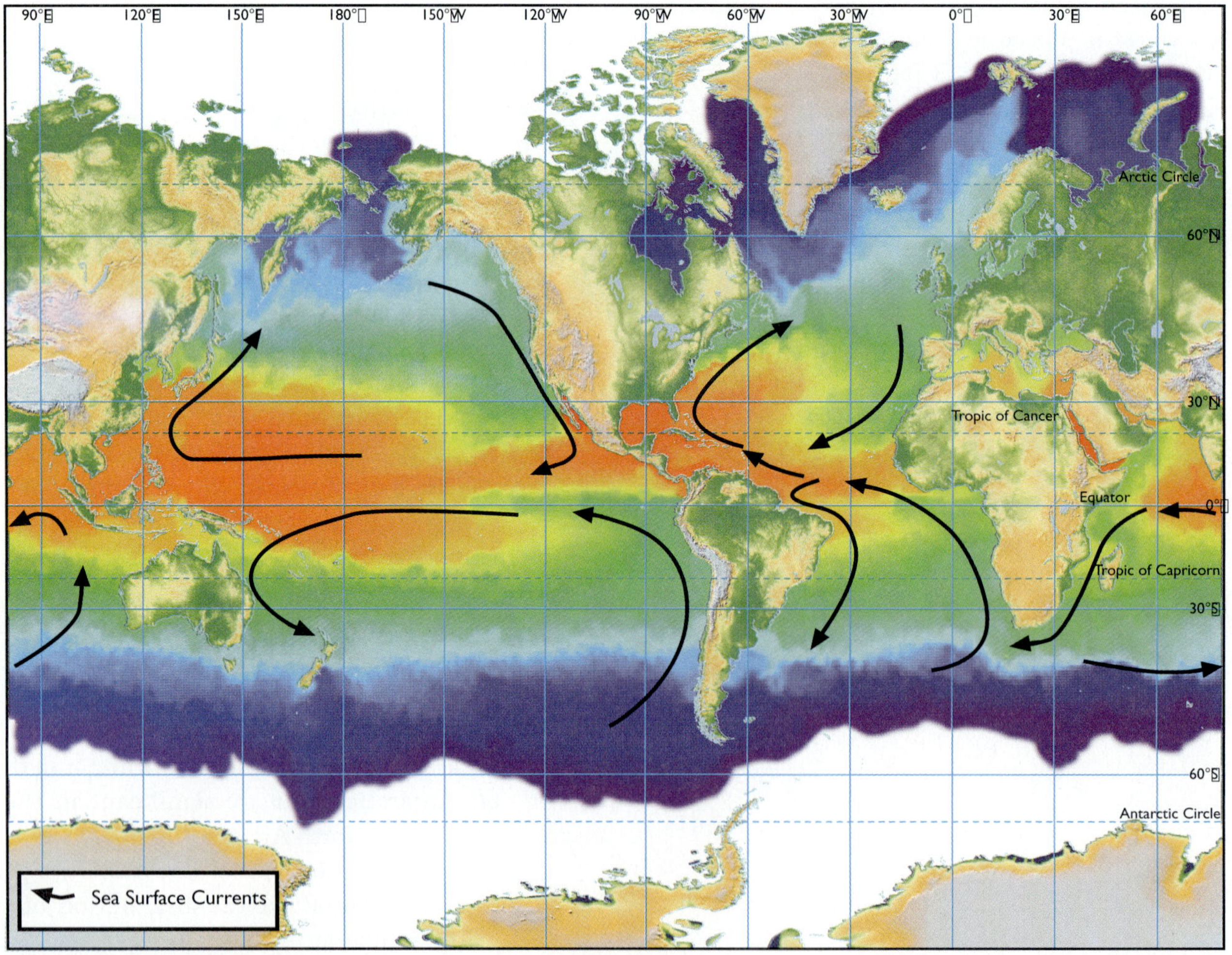

(a) SEA-SURFACE TEMPERATURE IN AUGUST

of the mid-latitudes (**Figure 5–11**). The daily thermoclines that occur at very shallow water depths (<12 meters; ~39 feet) are diurnal. Unlike the surface water, where temperature changes with the hour, the day, and the season, water below the permanent thermocline maintains a low, stable temperature over time, averaging <4°C for most of the ocean (see Figure 5–10).

SALINITY

The salinity of the ocean's surface water, although more complicated in detail than the pattern of the sea-surface temperature, also shows a clear latitudinal dependence (**Figure 5–12a**). The highest salinity values occur between 20 and 30 degrees north and south latitude and decrease toward the equator and the poles. Salinity variations are caused by the addition or removal of H_2O molecules from seawater. Processes that remove H_2O molecules from seawater include evaporation and the formation of ice. Precipitation (rain, snow, sleet), river runoff, and ice melting add H_2O molecules, diluting the salinity of seawater and reducing its density. Because these processes are dependent to a large degree on climate, and climate varies with latitude, the salinity of surface seawater varies directly with latitude.

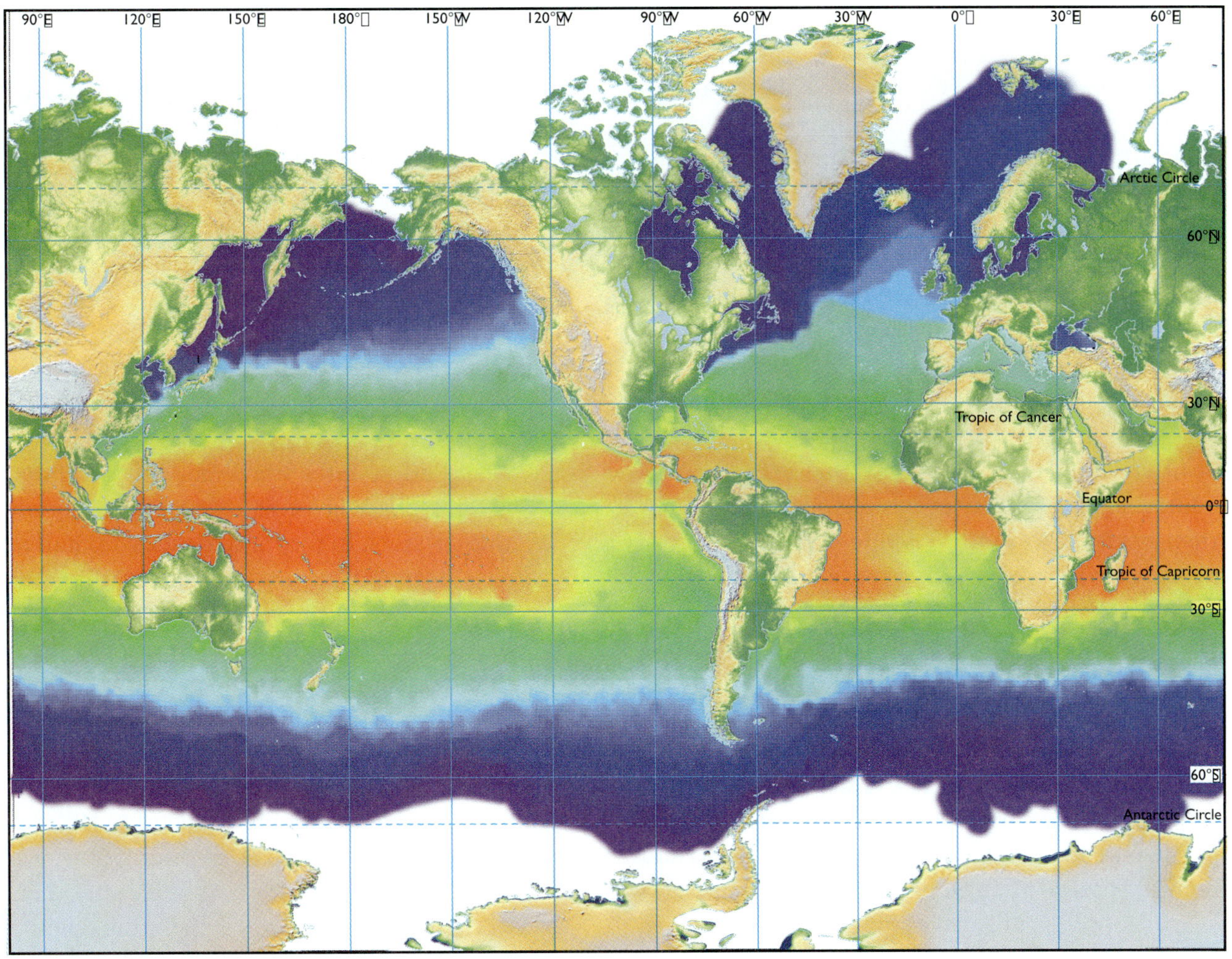

(b) SEA-SURFACE TEMPERATURE IN FEBRUARY

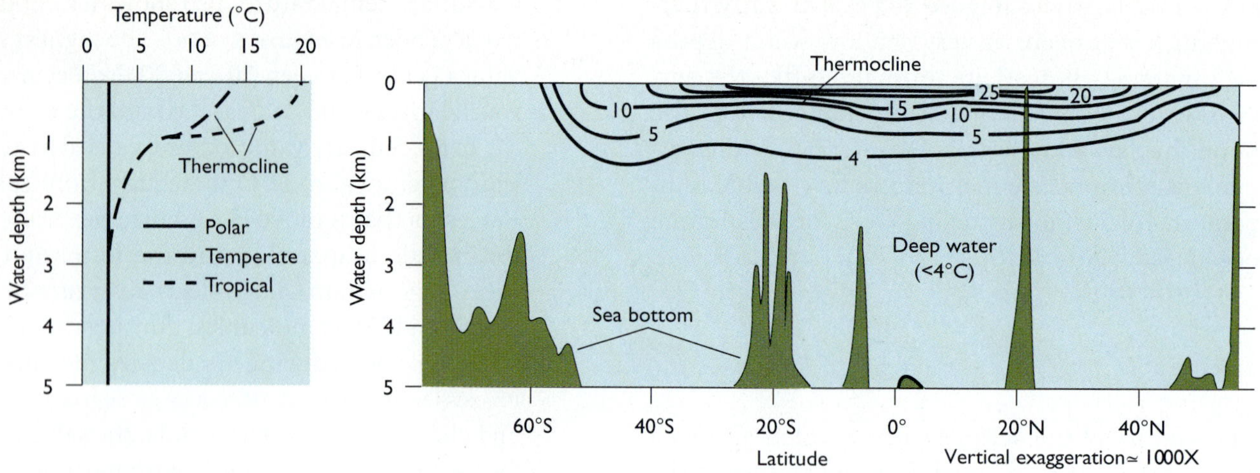

(a) TEMPERATURE PROFILES (b) TEMPERATURE DISTRIBUTION (°C) IN CENTRAL PACIFIC OCEAN

FIGURE **5-10**

Temperature profiles. (a) These vertical profiles depict variations of temperature with water depth. Note the prominent thermocline that separates cold, deep water from warm surface water. (b) A longitudinal profile of the average temperature distribution in the Pacific Ocean indicates that the bulk of all ocean water is colder than 4° Celsius. [Adapted from J. L. Reid, Jr., *Intermediate Waters of the Pacific Ocean* (Baltimore: Johns Hopkins Press, 1965).]

Figure 5–12b shows exactly how ocean salinity changes with latitude. Maximum salinity values occur in the subtropics and minimum values near the equator and the polar regions. The dashed line in this figure is a plot of the latitudinal difference between evaporation and precipitation—that is, evaporation (removal of H_2O) minus precipitation (addition of H_2O). To calculate this value for any latitude, the total amount of evaporation for one year is subtracted from the total precipitation for that same year. If the result of the calculation is a positive value, then evaporation has exceeded precipitation, and salinity in this region will be higher than normal. A negative value denotes the converse, that is, precipitation exceeds evaporation, and salinity here is lower than normal. The rate of evaporation depends strongly, thought not exclusively, on temperature and, hence, varies directly with latitude, being very low in the polar and subpolar regions, rising in the middle latitudes, and being highest in the tropics and subtropics. By contrast, rainfall maxima occur in the tropics and temperate latitudes, and minima in the subtropics and high latitudes. Therefore, during the course of a year, there is a net excess of precipitation over evaporation in the tropics and temperate latitudes, and an excess of evaporation over precipitation between 20 and 35 degrees latitude, where, not surprisingly, most of the major land deserts occur.

Notice in Figure 5–12b that the two curves representing evaporation minus precipitation and sea-surface salinity are quite similar in shape. This suggests that the major control on the surface salinity of the world's oceans is the relative effect of evaporation and precipitation—in other words, of climate. The maximum salinity levels in the subtropical oceans are produced by a strong excess of evaporation over precipitation. At the equator, evaporation rates are high, but rainfall is even greater, leading to the lower surface salinity in these waters. In the midlatitudes, rainfall rates surpass evaporation rates to an even greater degree than they do in the tropics, which reduces salinity to less than 34‰. Surface salinity in the polar seas fluctuates as ice forms and melts with the changing seasons. The water of the coasts and continental shelves is diluted by the freshwater of rivers. A case in point is the Amazon River, which injects more than 5×10^{12} cubic meters ($\sim 1.8 \times 10^{14}$ cubic feet) of freshwater into the western South Atlantic Ocean each year, lowering the salinity there.

As was the case for temperature, sharp salinity gradients characterize the water column of the world's oceans (**Figure 5–13a**). These gradients are

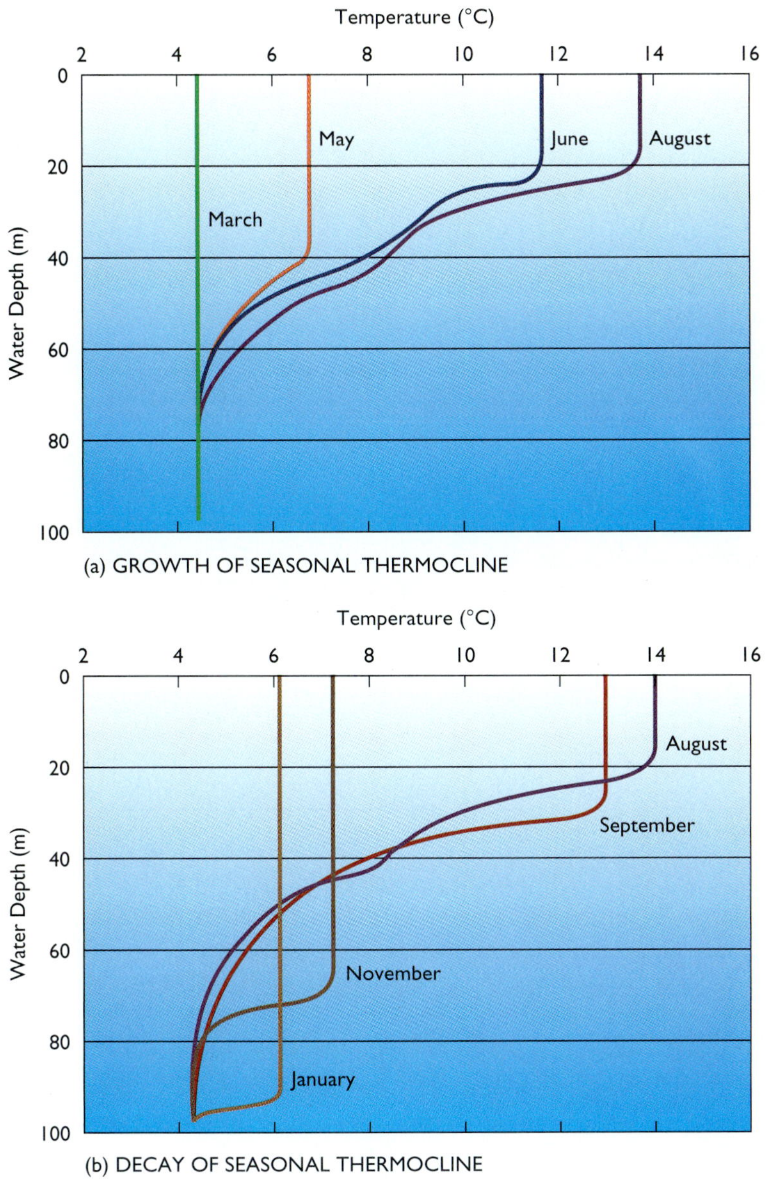

FIGURE 5-11

Seasonal thermocline. (a) Beginning in March, when there is no vertical change in the temperature of the upper water column, the thermocline develops; it is most pronounced in August. (b) After August, the thermocline weakens with the onset of winter and disappears by March (see part a). [Adapted from Open University Course Team, *Seawater: Its Composition, Properties, and Behavior* (New York: Pergamon Press, 1989).]

termed **haloclines** and, like temperature, represent boundary zones between distinct water masses. A north-south longitudinal profile of salinity in the western Atlantic Ocean reveals a well-developed layering of water masses (**Figure 5–13b**). **Water stratification** (layering) is evident between 40°N and 40°S latitudes, where a lens of high-salinity (35‰) surface water is separated from less saline water below by a sharp halocline. Also, two tongues of water with salinities of <34.74‰ extend northward from the Antarctic region: one at about a 1,000-meter (~3,300-feet) depth, the other along the deep-sea bottom. These tongues indicate that their water originated at the sea surface in the south polar seas and are separated from each other by a water mass that flowed southward out of the North Atlantic Ocean (see Figure 5–13b). What is truly remarkable about this North Atlantic water below a depth of 2 kilometers (~1.2 miles) is its uniform salinity of ~34.9‰. This condition reflects the fact that once water sinks, it is no longer in contact with the atmosphere, where precipitation and evaporation alter the salinity. Hence, the salinity of deep water remains relatively stable (unchanging) over time, although very slow mixing processes not well understood do eventually change the salinity of water masses.

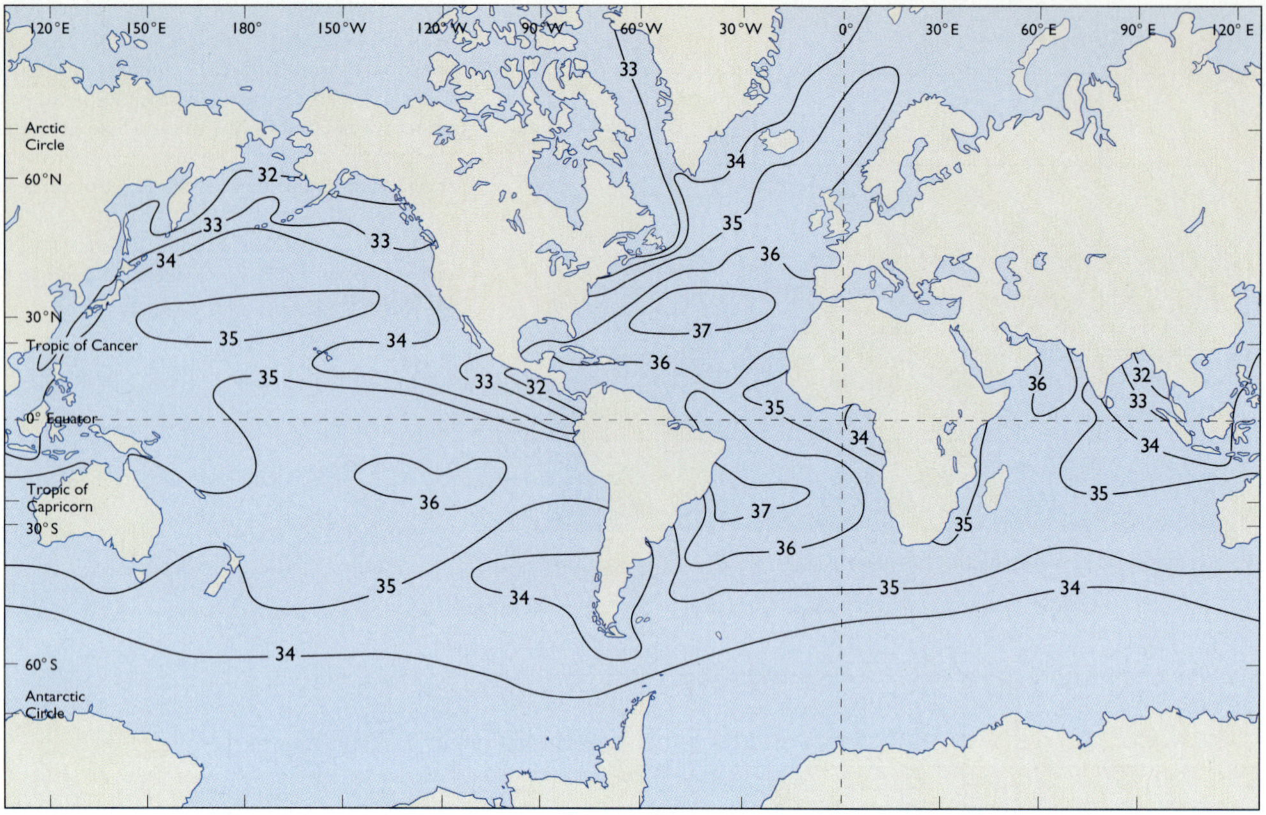

(a) SEA-SURFACE SALINITY (0/oo) IN AUGUST

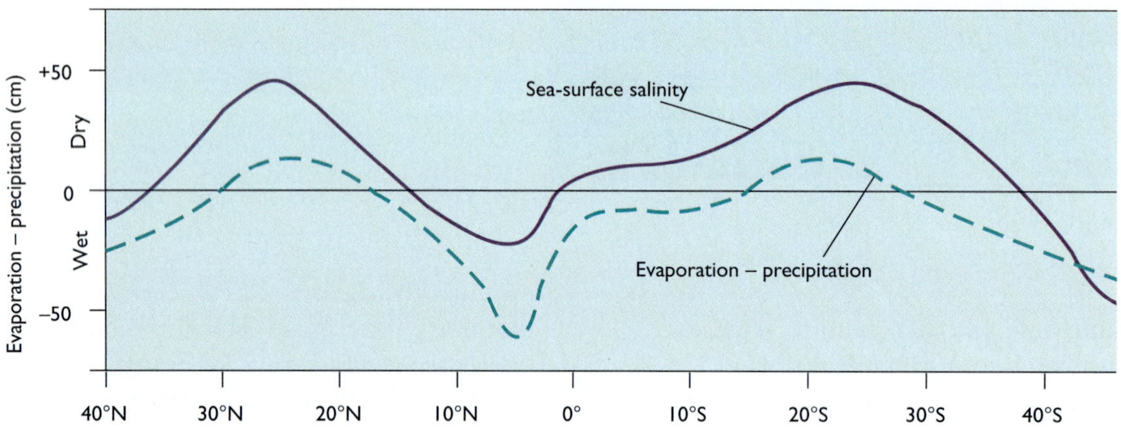

(b) LATITUDINAL VARIATIONS IN SALINITY AND "DRYNESS"

FIGURE **5-12**

Salinity variations. (a) The surface salinity of the world's ocean in parts per thousand (‰) during August shows a regular pattern that depends on latitude, with maximum amounts in the center of each ocean and minimal amounts at the equator and polar regions. [Adapted from H. U. Sverdrup, M. W. Johnson, and R. H. Fleming, *The Oceans* (Englewood Cliffs, N.J.:Prentice-Hall, 1942).] (b) A comparison of evaporation and precipitation rates accounts for the maximum salinity levels in the subtropics and the minimum salinity levels near the equator and the polar regions. This profile is a global average.

DENSITY

Density, the amount of mass per unit volume (g/cm^3), depends on the temperature, salinity, and pressure of seawater. Because water is essentially incompressible, pressure affects density only in the deepest parts of the ocean and can be ignored for our purposes. It should, however, be noted that, if water were absolutely incompressible, sea level would be 50 meters (~165 feet) higher than it presently is! Water density controls the vertical structure of the water column, with the more dense water underlying the less dense water. Basically, the density of water is increased by dropping its temperature and raising its salinity. This means that cold, saline water is more dense than warm freshwater is. The conversion of temperature and salinity data to water density can be done simply by consulting either tables or graphs (**Figure 5–14a**).

It is possible to have high-salinity water overlie low-salinity water, provided that the temperature of the surface water is much higher than the temperature of the deep water. Thus warm water with high salinity levels can overlie cold water with low salinity—a water-mass arrangement that is common in the ocean of the low and middle latitudes (see Figure 5–13a), where high evaporation rates increase the salt content of surface water relative to subsurface water.

If vertical gradients of water temperature (thermoclines) and salinity (haloclines) exist in the ocean, then it stands to reason that density will show a similar gradient, because it depends directly on variations of temperature and salinity. A sharp density gradient with depth is called a **pycnocline** (**Figure 5–14b**). Density stratification of the ocean imposes a three-layered structure on the water column: a surface layer, a pycnocline layer, and a deep layer (**Figure 5–14c**).

SURFACE LAYER

This topmost layer, the **surface layer**, is thin, averaging about 100 meters (~330 feet) in thickness, and represents about 2 percent of the ocean's volume. Being at the surface, it is the least dense water of the water column, largely because of its warm temperatures. Because it is in contact with the atmosphere, its water is affected by weather and climate, which cause diurnal, seasonal, and annual fluctuations of salinity and water temperature. Light penetrates to the bottom of this zone, allowing plant photosynthesis to occur wherever nutrient levels are adequate. In the polar regions, cooling of the surface water produces dense water that sinks. This sinking process prevents the formation of a pycnocline in the oceans of the high latitudes.

PYCNOCLINE LAYER

The boundary zone between the surface-water and deep-water layers is the **pycnocline layer**, where a sharp density gradient exists. It is not really a distinct water mass, but a transition zone where the surface-water layer grades into the deep-water layer. Water in this transition zone amounts to about 18 percent of the ocean's volume. In the low latitudes, the pycnocline corresponds to the permanent thermocline, which is created by the strong and persistent heating of the water by the tropical sun. In the midlatitudes, the pycnocline weakens and coincides with the halocline, which is produced by the abundant rainfall that dilutes the salinity of the water in the surface layer (see Figure 5–12b).

DEEP LAYER

About 80% of the total volume of the oceans belong to the **deep layer**. The bulk of this deep water originates in the high latitudes, where it is cooled while in contact with the frigid polar atmosphere. This cold (<4°C) polar water sinks to the ocean bottom because of its high density, and flows slowly equatorward, supplying the depths of the ocean with water that was once near the ocean's surface. Chapter 6 provides a thorough review of the origin and flow patterns of these deep-water masses.

We can now discuss the nature and distribution of gases dissolved in the ocean. As you will see, gas concentrations depend a great deal on the density structure of the water column.

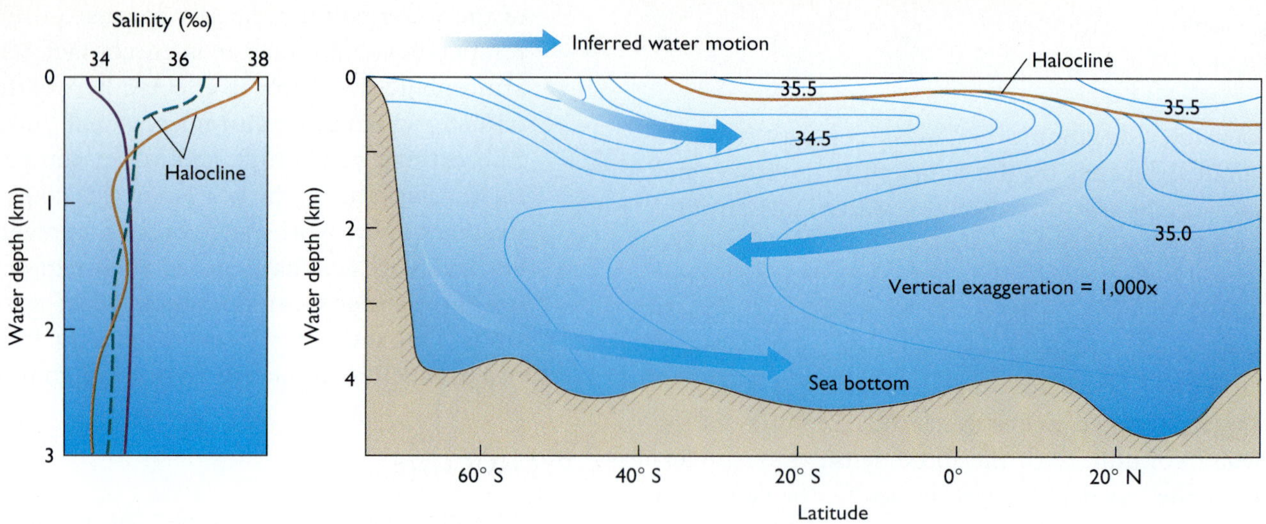

(a) SALINITY PROFILES (b) SALINITY DISTRIBUTION (‰) IN THE WESTERN ATLANTIC OCEAN

FIGURE **5-13**

Salinity profiles. (a) Vertical profiles of salinity show that sea-surface water may be more or less saline than the water below it. Note the prominent haloclines. (b) Isohalines (lines connecting points of equal salinity) in a longitudinal profile of the western Atlantic Ocean reveal distinct water-mass stratification and a prominent halocline. Below a 2-kilometer depth, the water has a remarkably uniform salinity, ranging between 34.7 and 34.9 ‰. [Adapted from D. Tolmazin, *Elements of Dynamic Oceanography* (Winchester, Mass.: Allen and Unwin, 1985).]

5-6

Gases in Seawater

To understand better the levels and distribution of gases in seawater, we need a thorough grasp of certain chemical concepts. The first is **saturation value**, which refers to the amount of gas at equilibrium that can be dissolved by a volume of water at a specific salinity, temperature, and pressure. The higher the saturation value is for a gas, the greater is its **solubility**, which is the property of being dissolved and going into solution. The solubility of gases increases with a drop in either water temperature or salinity, and a rise in pressure (**Figure 5–15a**). This means that cold, brackish (slightly salty) water can dissolve more gas than can warm, saline (salty) water at the same pressure. If a parcel of water is saturated with a gas, a change of water temperature or salinity will cause the quantity of gas to be below or above the water's new saturation value, conditions called **undersaturation** and **supersaturation**, respectively. Water that is undersaturated can dissolve more gas, whereas water that is supersaturated can release gas from solution (bubbles may form under the right conditions).

At this point, it is necessary to specify the types of units that chemists use to characterize the quantity of dissolved gases in water. We will use a volume measure expressed as milliliters—that is, a thousandth (10^{-3}) of a liter—of gas dissolved in 1 liter of water, expressed as ml gas/l. Those of you who have taken a course in chemistry realize that the amount of gas in a 1-milliliter volume depends on the temperature and pressure of the system. Therefore, the concentration of a gas dissolved in 1 liter of water is specified for a standard temperature of 0°C and a standard pressure of 1 atmosphere (equivalent to the pressure exerted by the weight of the atmosphere, which is the pressure that we feel at sea level).

The processes that produce, consume, and regulate gas concentrations in the ocean are summarized in **Table 5–9** and depicted in **Figure 5–15b**. Enormous quantities of gases are exchanged between the ocean and the atmosphere near the sea surface (see boxed feature, "The Sea-Surface Microlayer"). Ordinarily, surface seawater is near saturation with respect to the common atmospheric gases (O_2, N_2, CO_2). However, gases diffuse across the air-sea interface, as the temperature and salinity of the water change or as organisms produce or consume gases, creating supersaturated or undersaturated conditions. Also, breaking waves, particularly those associated with large storms, drive air bubbles downward into the water, where water pressure dissolves some of the bubbles before they can ascend back to the surface. In fact, this very process—the passage of air bubbles through water—is used to aerate water in your fish aquarium. Oxygen diffuses out of air bubbles into the aquarium water where fish can then use it to "breathe."

The primary regulator of gas concentrations in subsurface water is the activity of organisms (see Table 5–9). When light and nutrient conditions are adequate, plants photosynthesize. This complex

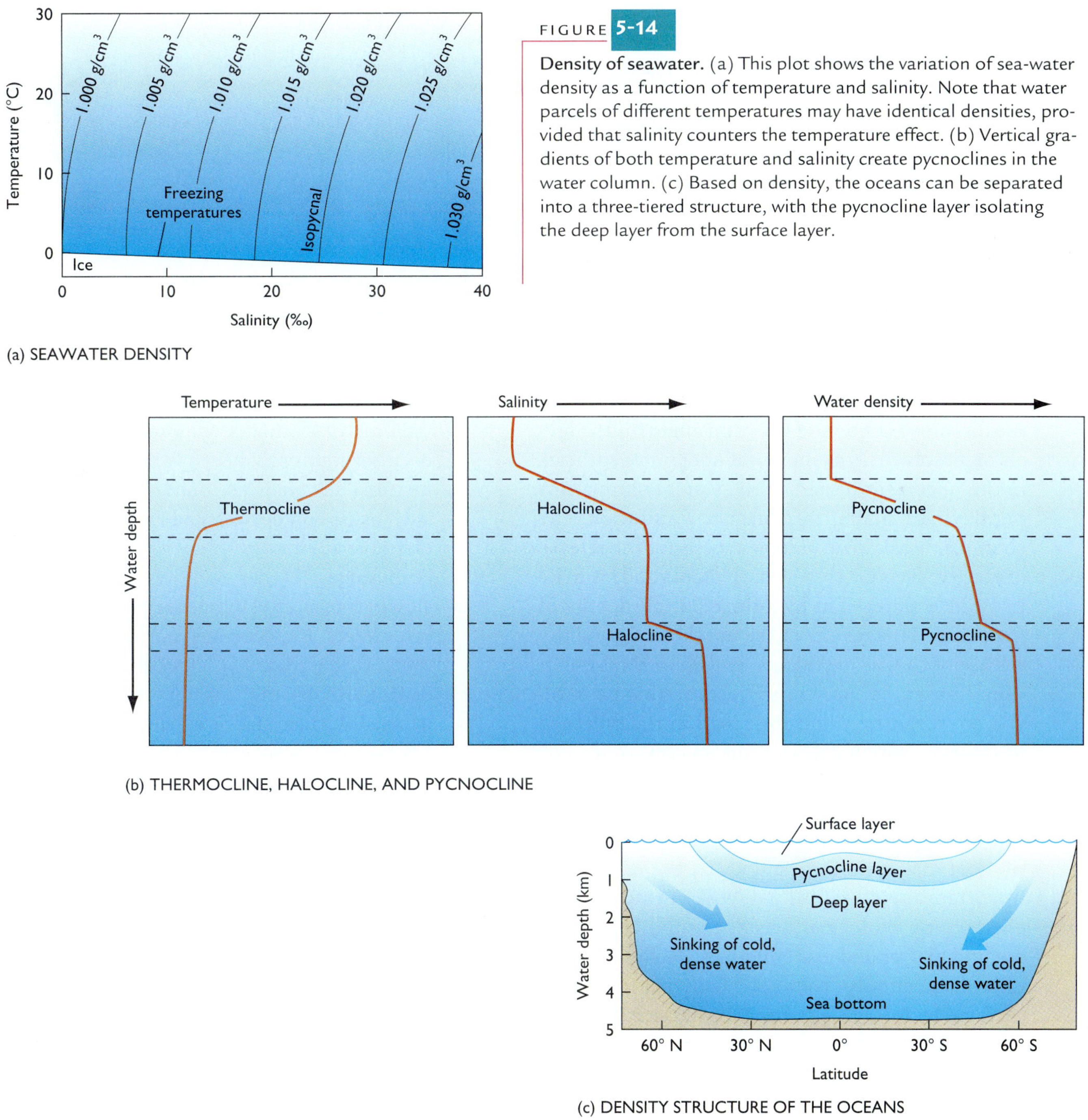

(a) SEAWATER DENSITY

(b) THERMOCLINE, HALOCLINE, AND PYCNOCLINE

(c) DENSITY STRUCTURE OF THE OCEANS

FIGURE 5-14

Density of seawater. (a) This plot shows the variation of sea-water density as a function of temperature and salinity. Note that water parcels of different temperatures may have identical densities, provided that salinity counters the temperature effect. (b) Vertical gradients of both temperature and salinity create pycnoclines in the water column. (c) Based on density, the oceans can be separated into a three-tiered structure, with the pycnocline layer isolating the deep layer from the surface layer.

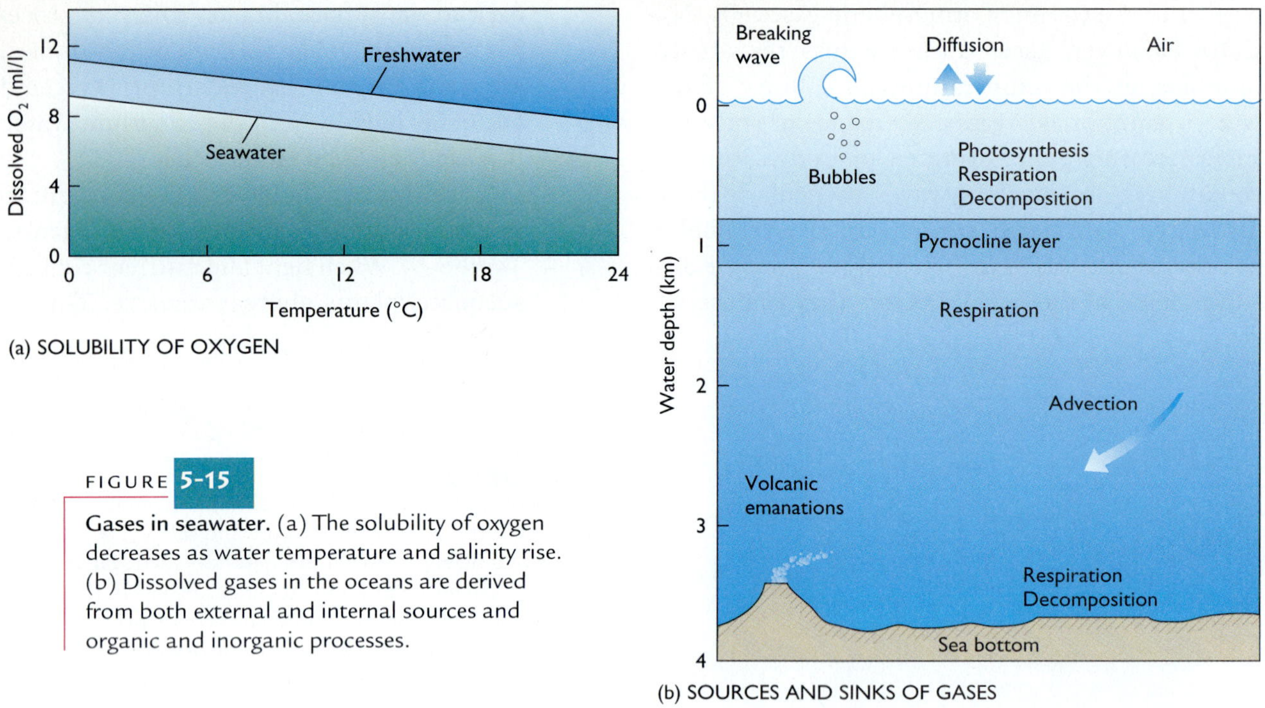

FIGURE 5-15

Gases in seawater. (a) The solubility of oxygen decreases as water temperature and salinity rise. (b) Dissolved gases in the oceans are derived from both external and internal sources and organic and inorganic processes.

biochemical process of **photosynthesis** converts water (H_2O) and carbon dioxide (CO_2) into organic matter and liberates oxygen (O_2) as a product of the reactions. Thus, plants simultaneously reduce the dissolved content of CO_2 and augment the levels of dissolved O_2 as they conduct photosynthesis in the upper, sunlit part of the water column during daylight hours. In contrast, **respiration**, the chemical breakdown of food in cells for the release of energy, is conducted by both plants and animals in surface water, and by animals at all depths of the ocean. This results in the uptake of O_2 and the release of CO_2 as organisms oxidize food for nutritional purposes. As you sit reading this passage, you are respiring, "burning" food in the presence of oxygen, which releases the energy that your body requires to live. Consequently, these life-sustaining processes—photosynthesis and respiration—have a profound impact on the concentrations of dissolved CO_2 and O_2 in the oceans, an important topic that we will cover in subsequent chapters on biological oceanography. Last, dead organic matter and excrement are decomposed by microbes, chiefly bacteria. Enzymes that are secreted by microbes chemically "attack" organic matter and break it down into simpler chemical compounds by the process of **oxidation**, a chemical reaction that consumes O_2 and produces CO_2 and other gases.

Gases other than oxygen and carbon dioxide are also released and taken up by physical and biological processes. For example, the decay of unstable radioactive elements contained in the minerals of deep-sea muds produces a variety of gases, including helium (He), radon (Rn), and argon (Ar). Some gases, such as carbon dioxide and helium, are spewed out of active submarine volcanoes on spreading mid-ocean ridges (**Figure 5–16**). These gases are then slowly transported hundreds, perhaps thousands, of kilometers by deep-sea currents before mixing dilutes them to such a degree that they are no longer distinguishable in the water column. Such gases can be used to trace the flow path of currents near the sea bottom, even very sluggish ones. Also, ocean basins filled with **anoxic** water (water without dissolved oxygen) are not populated by normal communities of plants and animals, because these organisms cannot survive without O_2. However, specialized **anaerobic** bacteria, which live without free oxygen, inhabit such waters. They use the oxygen atoms that are chemically bonded to sulfate ions (SO_4^{2-}) for conducting their metabolic processes, which yields the toxic gas hydrogen sulfide (H_2S) as a reaction by-product. Although anaerobic conditions rarely develop in the well-mixed, and hence well-ventilated, water of the open ocean, they do occur in some restricted basins where circulation is sluggish and the supply of oxygen is limited. For example, water trapped in the small basins that are part of the continental shelf of southern California commonly is anoxic because of sluggish bottom currents and poor mixing.

In order to firm up our understanding of the gas chemistry of the oceans, we will consider in the next two sections the sources, sinks, and distribution of O_2 and CO_2—two gases vital for the ocean's biota. Both are considered to be nonconservative substances in seawater because their concentrations vary greatly over short time and distance scales.

TABLE 5-9

Summary of factors that regulate the concentration of gases in water

Factors	Effects
Wave and current turbulence	Increases the exchange of seawater gases with the atmosphere.
Difference in gas concentration	Gases diffuse across the air-sea interface from high to low areas of concentration until chemical equilibrium is attained.
Temperature	A drop in water temperature increases the solubility of gases.
Salinity	A rise in salinity decreases the solubility of gases.
Pressure	A rise in pressure increases the solubility of gases.
Photosynthesis	Increases concentration of O_2; decreases concentration of CO_2.
Respiration	Increases concentration of CO_2; decreases concentration of O_2.
Decomposition	Increases concentration of CO_2; decreases concentration of O_2.
pH	Controls the relative concentrations of the various species of CO_2 in water (H_2CO_3, HCO_3^-, CO_3^{2-}).

Source: Adapted from H. S. Parker, *Exploring the Oceans* (Englewood Cliffs, N.J.: Prentice-Hall, 1985).

OXYGEN

The vertical distribution of O_2 in the ocean of the low and middle latitudes shows a distinct pattern (**Figure 5–17a**). A warm surface-water layer with a high content of dissolved O_2 is separated from cold, relatively well-aerated deep water by a distinct oxygen-minimum layer at about 150 to 1,500 meters (~495 to 4,950 feet) below the sea surface. This vertical oxygen profile reflects inputs and outputs of gases by a variety of processes. Let's examine them.

There are two principal sources of O_2 for the oceans—gas diffusion across the air-sea interface and plant photosynthesis. Both of these processes are limited to the uppermost levels of the water column, and account for the high concentrations of dissolved O_2, typically in excess of 5 milliliters per liter, in the surface-water layer. The O_2-minimum zone coincides with the pycnocline layer; this implies a connection between the two. Apparently, the sharp density gradient in this intermediate water layer slows down the settling rate of organic detritus falling from above, so that dead organic matter tends to accumulate and form a rich food source in the pycnocline. (This is somewhat similar to cars that are on a large highway entering a city area. Far from the city, cars are widely spaced and moving rapidly. Once in the city limits, however, the average speed of the vehicles slows down and they bunch up, sometimes bumper to bumper.) Organisms of all kinds are drawn to the pycnocline layer because of the ample food supply, and, as they respire, they deplete the O_2 content of the water. Also, bacteria there decompose the abundant dead organic matter, which further reduces the dissolved O_2. The sharp pycnocline indicates that the water column is stable and that vertical mixing of water is minimal at these intermediate water depths. Thus, the high demand for dissolved O_2 by animals and bacteria that are feeding on the dead and living organic matter, combined with a relatively slow rate of water mixing and, hence, of O_2 replenishment, produces the O_2-minimum layer with O_2 concentrations of <2 milliliters per liter (see Figure 5–17a).

Below the pycnocline, water is sparsely populated because food is scarce. The biological demand for dissolved O_2 is low here, and O_2 concentrations rise with depth to between 3 and 5 milliliters per liter. Because oxygen production is restricted to surface water, the dissolved O_2 found in this deep layer must have been derived from shallow depths. Temperature and salinity data (see Figure 5–13b) indicate that deep water in all the ocean basins is

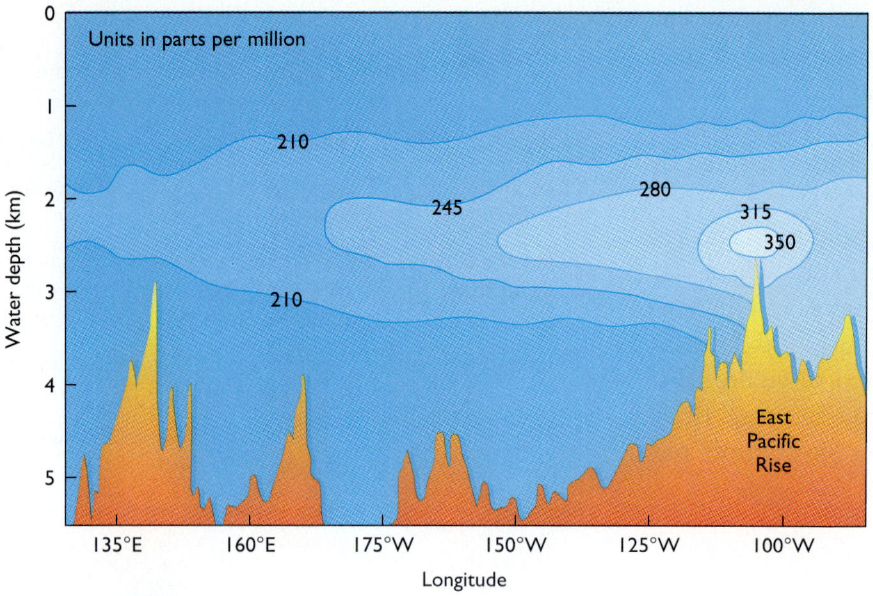

FIGURE 5-16

Helium-3. A large plume of water enriched in helium-3 extends westward from the East Pacific Rise, tracing the flow of water at mid-depth in the Pacific Ocean. Volcanic emissions are clearly the source of this gas, which originated deep in the Earth's mantle. [Adapted from W. J. Jenkins, Tracers in oceanography, *Oceanus* 35, no. 1 (1992): 47–56.]

derived from the polar regions. In the high latitudes, surface water is cooled, which raises its gas-saturation values. These cold, O_2-rich water masses sink because of their high densities. They flow equatorward, ventilating the depths of all the ocean basins (**Figure 5–17b**). Although this **advection** process (the horizontal and vertical movement of a fluid) is slow, it occurs at a rate that satisfactorily meets the respiration requirements of the scanty populations of deep-sea fauna.

CARBON DIOXIDE

Carbon dioxide (CO_2) is a gas that is actively involved in photosynthesis and respiration. Plants cannot produce food and survive in its absence. Animals release CO_2 as they break down food for energy by the chemical process of respiration. Also, CO_2 regulates the acidity of seawater. Let's examine this latter chemical property of seawater.

The degree to which water provides a suitable habitat for marine biota is determined, in part, by the concentration of dissolved hydrogen ions (H^+), a measure termed the **pH** of water. In pure water (water containing nothing but H_2O molecules) at 25°C, a very tiny fraction of the H_2O molecules, about 10^{-7}, or one in 10 million (10^7), spontaneously dissociates (breaks apart) into a hydroxyl (OH^-) and a hydrogen (H^+) ion. The free hydrogen ions are what control the acidity of water. The formal definition of pH is

$$\text{pH} = -\log_{10}(\text{H}^+).$$

At first glance, this formula looks quite difficult to grasp, but it need not be. Remember that the pH of water is, to a first approximation, merely a measure of the concentration of the H^+ ion. The parentheses in the preceding equation can be equated to the concentration of H^+, although technically they refer to the chemical activity of this ion, which is an indirect function of its concentration level. Now, if we go back to our example of pure water, measurements indicate that 10^{-7} H_2O molecules (one in 10 million) are separated into their ionic components. If we substitute 10^{-7}, the concentration of H^+, into the formula and solve for its negative log, we get a pH value of 7 (**Figure 5–18**). The log to the base ten of 10^{-7} is the exponent –7; hence, the negative log of 10^{-7} is –(–7) or 7. Water with a pH of 7 is, by definition, a **neutral solution** consisting of equal parts (10^{-7}) of OH^- and H^+ (see Figure 5–18).

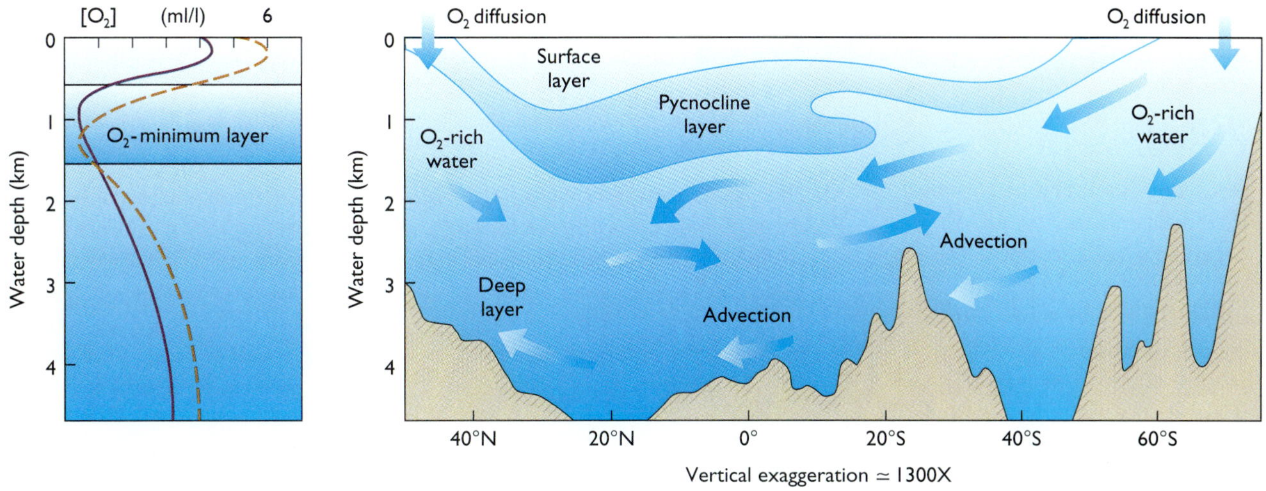

(a) VERTICAL O_2 PROFILES IN THE ATLANTIC OCEAN

(b) O_2-ADVECTION PATTERN IN THE ATLANTIC OCEAN

FIGURE **5-17**

Dissolved oxygen. (a) Vertical profiles of dissolved oxygen reveal a distinct pattern consisting of well-ventilated surface and deep waters that are separated by an O_2-minimum zone in the pycnocline layer. Apparently, the accumulation of dead organic matter in the pycnocline layer produces a high demand for O_2 because of respiration and decomposition. (b) The deep water of the ocean is aerated by the advection of cold, dense, O_2-rich polar water.

Raising the amount of H^+ to 10^{-6}—remember that 10^{-7}, one H^+ ion in 10 million of the H_2O molecules, is a much smaller value than 10^{-6}, which is one H^+ ion in a million H_2O molecules—lowers the pH to 6, a solution that is no longer neutral but **acidic** (see Figure 5–18). The log to the base 10 of 10^{-6} is the exponent –6; hence, the negative log of 10^{-6} is –(–6) or 6. Low concentrations of H^+ that impart a higher pH than the neutral level of 7 create a **basic** solution, in which the amount of OH^- surpasses the amount of H^+ (see Figure 5–18).

It's important that you understand the preceding discussion. Let's review what we've learned so far about pH. A neutral solution has a pH of 7, which means that the H^+ concentration is 10^{-7}. A basic solution has a pH value that is greater than 7 (remember that this means low concentrations of H^+) and an acidic solution has a pH value that is lower than 7 (high H^+ concentrations). So the pH scale, to a first approximation, is nothing more than a measure of the content of H^+ in the water. Because the H^+ ion is so reactive with other compounds, acid water (water with a pH of less than 7 and, hence, a relatively high concentration of H^+) is a powerful weathering agent, able to weather rocks chemically. Also, plants and animals living in the ocean are affected by the content of H^+ in the water they inhabit, because many metabolic activities are regulated by the seawater's pH. The question remaining to be answered is, what exactly determines the amount of H^+ in the ocean?

The pH of water is directly linked to the CO_2 system. When CO_2 is added to water, most of it is rapidly converted into carbonic acid (H_2CO_3) as it bonds with water molecules (hydration). This acid then rapidly dissociates into bicarbonate (HCO_3^-) and carbonate (CO_3^{2-}) ions, which yields H^+ ions, making the water acidic. The specifics of the chemical reaction are

$$CO_2 + H_2O \rightarrow H_2CO_3 \rightarrow HCO_3^- + H^+ \\ \rightarrow CO_3^{2-} + 2H^+.$$

This formula summarizes each step of the reaction when CO_2 is dissolved in water. Notice that the free ions of H^+ lower the pH of the water. Therefore, we conclude that the addition of dissolved CO_2 tends to lower the pH of water. At equilibrium, the proportion of each of the carbon compounds (H_2CO_3, HCO_3^-, and CO_3^{2-}) depends on the pH of the water. At the pH of normal seawater (7.8 to 8.2), about 80 percent of the carbon compounds occur as HCO_3^- (**Figure 5–19a**). A change in the

Desalination

Freshwater is a precious commodity that is essential for all living organisms. Its distribution on the Earth is variable. In arid regions, the amount of freshwater is very limited, making **desalination**, the production of drinkable water from seawater, very important. Unfortunately, desalination requires a great deal of energy, which means that it is an expensive process.

If seawater consists overwhelmingly of water (about 96.5 percent on the average), why is it so expensive to purify it into drinkable freshwater? The answer is related to the dipole structure of H_2O molecules, which results in tightly bonded clusters that must be disrupted for desalination to occur.

Desalination relies on five general techniques:

1. *Distillation*. The evaporation of seawater produces water vapor, which is then cooled to form a liquid-water condensate.
2. *Freezing*. As seawater begins to freeze, the H_2O molecules arrange themselves into a crystal structure, leaving behind the salt ions in a highly concentrated brine. The pure ice must then be separated from the brine.
3. *Reverse osmosis*. Seawater, under an applied pressure, is forced through a semipermeable membrane. Only the H_2O molecules are able to pass through the membrane, which separates them from the salt ions.
4. *Electrodialysis*. Electrically charged, semipermeable membranes draw salt ions out of the seawater solution, leaving behind freshwater. This technique is most effective for brackish (low-salinity) water.
5. *Salt absorption*. Chemically active resins or charcoals are used to draw off the dissolved salt ions from seawater, producing freshwater.

All of these techniques are currently in use somewhere in the world (**Figure B5–3**). Perhaps the simplest, least expensive, and most widely employed one is distillation. In Kuwait, Saudi Arabia, Israel, Greece, Pakistan, Australia, India, and Chile, as well as in the United States in Texas, southern California, and the Florida Keys, freshwater is obtained by distillation. A large saltwater pond is covered with plastic, glass, or even canvas. Solar radiation heats the enclosed volume of saltwater. Evaporation produces water vapor, which rises, cools, and condenses on the covering fabric of the structure (**Figure B5–4**). This condensed water then trickles down the underside of the sloping roof, and is collected in large receptacles. In Kuwait, the heat of the sun combined with waste heat generated by petroleum-fired power plants greatly increases evaporation and the production rate of freshwater at little additional cost.

Visit www.jbpub.com/oceanlink for more information.

FIGURE **5-18**

The pH scale. The concentration of H^+ in water is specified by pH. A pH of 7 denotes that the H^+ concentration is 10^{-7}, or one part in 10 million. Solutions with low pH values (high concentrations of H^+) are acidic. Solutions with high pH values (low concentrations of H^+) are basic.

	pH	$[H^+]$	$[OH^-]$	
Acidic Solutions	1	10^{-1}	10^{-13}	Hydrochloric acid
	2	10^{-2}	10^{-12}	Lime juice
	3	10^{-3}	10^{-11}	Acetic acid
	4	10^{-4}	10^{-10}	Tomato juice
	5	10^{-5}	10^{-9}	Black coffee
	6	10^{-6}	10^{-8}	Milk
Neutral	7	10^{-7}	10^{-7}	Pure water
Basic Solutions	8	10^{-8}	10^{-6}	Seawater
	9	10^{-9}	10^{-5}	Borax solution
	10	10^{-10}	10^{-4}	
	11	10^{-11}	10^{-3}	Milk of magnesia
	12	10^{-12}	10^{-2}	Household ammonia
	13	10^{-13}	10^{-1}	Lye
	14	10^{-14}	10^{-0}	Sodium hydroxide

FIGURE **B5-3**

Reverse osmosis. At the desalination facility on Culebra Island, Puerto Rico, the seawater is forced through a series of membranes in each white tube. By the time the seawater is forced through the last tube and reaches the tank in the back, it is fresh enough to drink

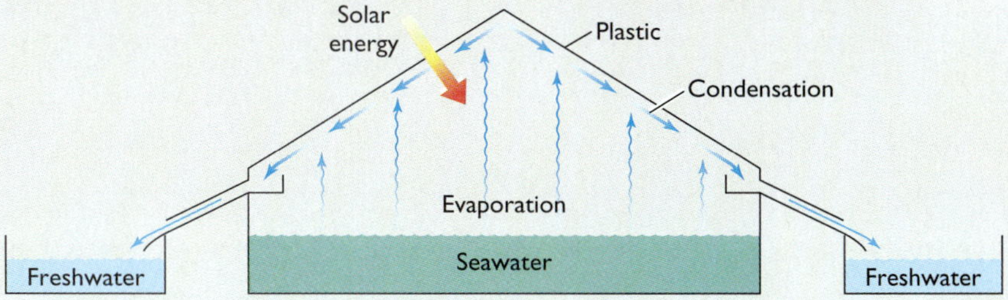

FIGURE **B5-4**

Distillation of seawater. A widely used technique for desalination (taking salt out of seawater) involves the evaporation of seawater from a pool, the condensation of freshwater on a covering fabric, and its collection in a storage vat.

amount of any of these carbon substances disrupts the balance of the CO_2 system, and reactions occur to reestablish equilibrium. For example, the removal of CO_2 from water by plant photosynthesis (**Figure 5–19b**) or by solar heating, which reduces its saturation value and causes bubbles to form, initiates a series of chemical responses that shift the reactions specified in the formula to the left. This results in the production of CO_2. Respiration, in contrast, releases CO_2 into the water (see Figure 5–19b). This process causes the reactions in the formula to move to the right, which increases the concentrations of the other chemical species at the expense of CO_2.

Organisms are sensitive to pH, quite a few extremely so. Fortunately, the many rapid chemical reactions within the CO_2 system (Figure 5–19) prevent large fluctuations in the pH levels of the world's oceans. Seawater is essentially a stable solution with a pH that rarely ranges below 7.5 or above 8.5. This condition is described as **buffered**, meaning that the mixture of compounds and the nature of the reactions are such that the pH of the solution is hardly affected despite an input or output of H^+. Let me explain. Increasing the level of H^+, which should lower the pH (make the water more acidic), causes the reactions (see **Figure 5–19c**) to shift to the left; as a consequence, H^+ is removed as it complexes with HCO_3^- to form H_2CO_3, buffering the solution (keeping the pH near its original value, **Figure 5–19d**). Conversely, reducing the level of H^+(making the solution more basic by raising the pH) reverses the reactions. There is a rapid release of H^+ into solution, as H_2CO_3 dissociates to HCO_3^-

FIGURE 5-19

Carbon species and the CO_2 system in the oceans. (a) Carbon dissolved in water occurs in the form of carbon dioxide (CO_2), bicarbonate (HCO_3^-), and carbonate (CO_3^{2-}). At equilibrium, the relative proportion of these carbon species depends on the pH of the water. Basic solutions are dominated by CO_3^{2-}; acidic solutions, by CO_2. At the pH of normal seawater, HCO_3^- makes up about 80% of the carbon species. (b) The CO_2 system is also involved in biological processes, notably photosynthesis, which removes dissolved CO_2 from the water, and respiration, which liberates CO_2 into the water. Hence, the CO_2-carbonate cycle influences and is influenced by both chemical and life processes. (c) Carbon dioxide diffuses from the atmosphere into the oceans, where it complexes with water to form weak carbonic acid (H_2CO_3) and hydrogen (H^+) ions. Some of the bicarbonate ions dissociate into carbonate (CO_3^{2-}) ions, which may complex with calcium (Ca^{2+}) to form calcium carbonate ($CaCO_3$). (d) The pH of seawater (7.8) is buffered by the CO_2- carbonate cycle. If the pH rises, H_2CO_3 dissociates and yields H^+, reversing the trend. If the pH drops, H^+ complexes with HCO_3^-, causing the pH to rise again.

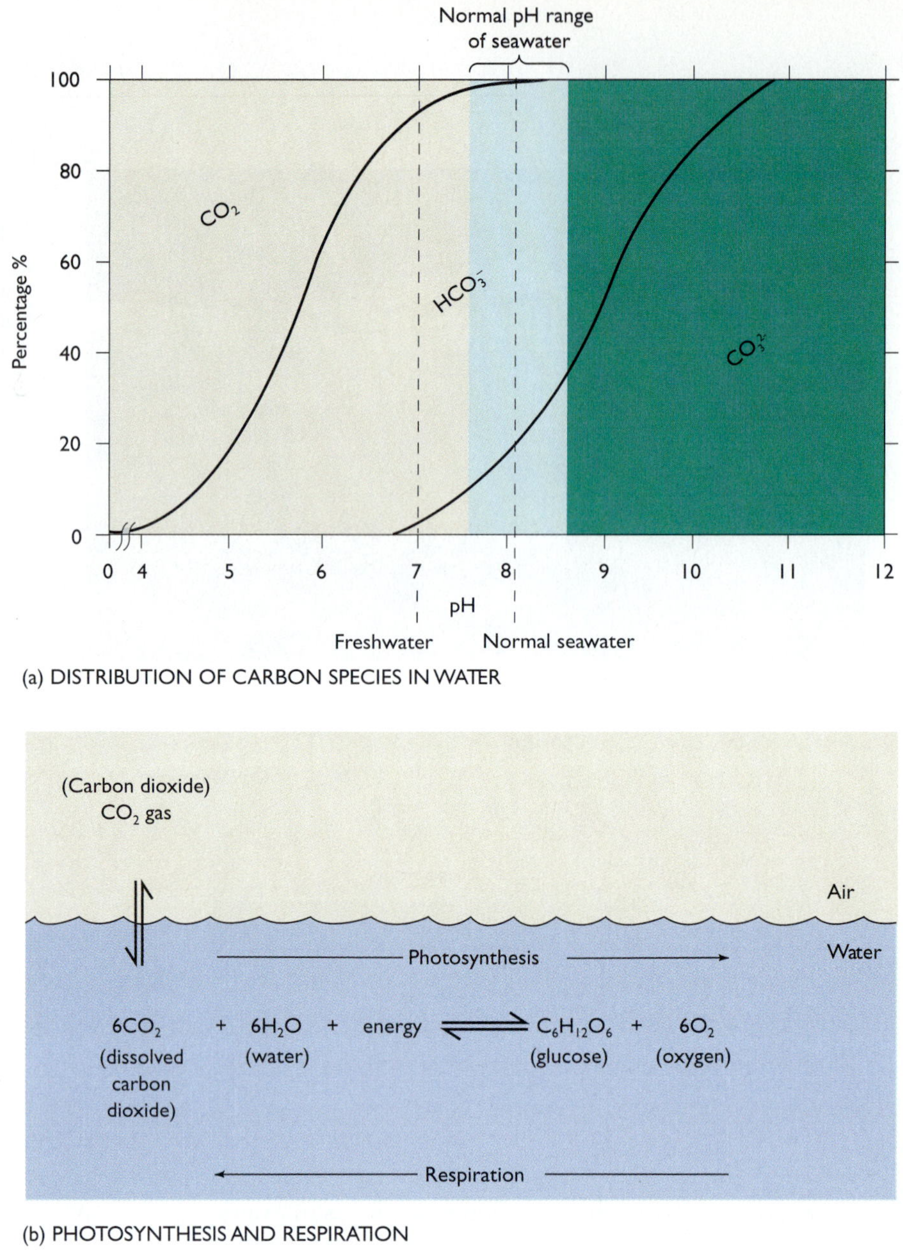

and releases H^+ (see Figure 5–19d). The end result is that the pH is stable despite changes in the relative amounts of the carbon species in the CO_2 system; the system is basically self-regulating or buffered.

With this background, we can now explain the dissolution of $CaCO_3$ shells in cold, deep water, but not in warm, shallow water, and the ocean's carbonate compensation depth (CCD) (see Figure 4–17 in Chapter 4). Recall that the pH level of water is inversely proportional to the concentration of dissolved CO_2. This means that the higher the CO_2 content of the water is, the lower the pH (or, if you prefer, the greater the content of H^+). Cold water has a higher gas-saturation value than does warm water because of its low temperature (see Figure 5–15a). Also, the saturation value of gases in water increases with increasing pressure and therefore water depth. This means simply that the cold, dense water that fills the ocean depths contains high levels of dissolved gases like CO_2 (see Figure 5–17b). Its high CO_2 concentration lowers the pH, making the water acidic and dissolving $CaCO_3$ shells that sink to the deep-sea floor. Shallow water, on the other hand, is rela-

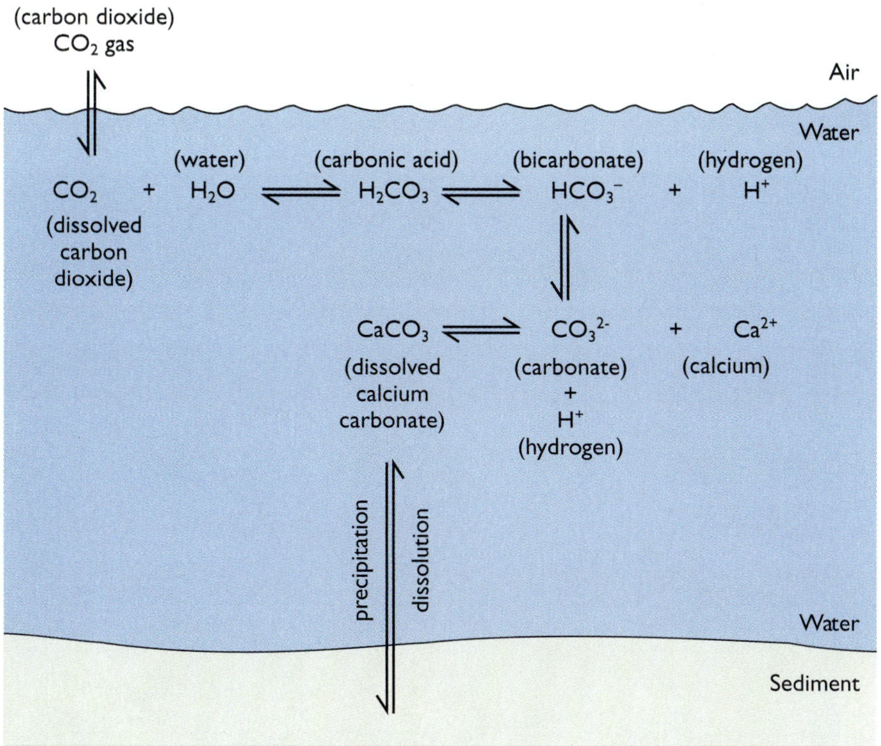

(c) THE CO_2 SYSTEM

Seawater too basic:	$H_2CO_3 \longrightarrow HCO_3^- + H^+$	pH drops
Seawater too acidic:	$HCO_3^- + H^+ \longrightarrow H_2CO_3$	pH rises

(d) CARBONATE BUFFER

tively warm and has a lower concentration of CO_2 at saturation than does the deep, cold water. This raises the water's pH level, which releases carbonate ions (CO_3^{2-}) that chemically bond to the abundant calcium ions (Ca^{2+}) and precipitate $CaCO_3$. Therefore, as discussed in Chapter 4, carbonate oozes tend to accumulate in relatively shallow water of the deep sea above the CCD, because water below the CCD is cold and acidic, and dissolves shells composed of $CaCO_3$.

5-7

The Ocean as a Physical Chemical System

For teaching purposes, in Chapter 2 we treated the oceans separately, as if each were a self-contained basin of water, separated cleanly from the others. But such is not the case. Water

Other Physical Properties of Water

In the body of the chapter, we discussed the unusual nature of water as a chemical compound. Its unique and remarkable properties are extensive. Here we examine a few more of these properties, including the formation of sea ice and the transmission of light and sound in water.

FORMATION OF SEA ICE

When water is chilled below its freezing temperature, it solidifies and crystals of ice form and grow. The very same process happens to seawater, except that seawater freezes at a lower temperature than freshwater does (see Figure 5-8). Once ice crystals begin to form, they impart a dull, cloudy appearance to the water as they grow into needles and platelets several centimeters long. When these solids become abundant in the water, a slush forms. The dissolved ions of salt are not, however, incorporated into the structure of the crystals, that is, the ice. Their exclusion from the solid ice raises the salinity of the remaining water and, as a consequence, depresses its freezing temperature. As more and more sea ice forms, some seawater becomes trapped in the thickening ice mass; the quantity depends on the rate of ice formation. Slow freezing rates increase the chances that trapped brines will escape out of the ice. Aging of the sea ice also leads to the slow drainage of the trapped brine, so that, in time, the mass of ice consists mostly of freshwater and is drinkable.

Once a continuous sheet of sea ice is formed, wind and waves break the cover into pieces, creating **pancake ice** (**Figure B5–5a**). These pieces of ice drift about and unite into rather flexible masses. Some of them collide and buckle, raising hummocks called **pressure ridges** (**Figure B5–5b**). Ice masses thicken with time, as seawater

(a) PANCAKE ICE

(b) PRESSURE RIDGES

FIGURE **B5-5**

Sea ice. (a) Wind and waves can break newly formed sea ice into irregular masses, or pancake ice. (b) The drift and collision of ice masses raise large pressure ridges.

freezes to their bottom and snow freezes to their surface. In the polar latitudes, an ice mass about 2 meters (~6.5 feet) thick typically will form during the winter. A continuous ice cover may be broken into small and large **floes**, which keep separating and freezing together again in response to wind, waves, and currents. Floes should not be confused with **bergs**, which are large, very thick masses of floating ice that became detached from a glacier near a shoreline and are swept out to sea.

LIGHT TRANSMISSION

Life—with a few exceptions—depends directly or indirectly on energy from sunlight. As do land plants, marine plants use green chlorophyll and a few other pigments to capture the visible light from the sun. As solar radiation strikes the ocean, a large fraction of it is reflected from the sea surface back into the atmosphere. The amount of reflected light depends on the height of the sun above the horizon and the smoothness of the water surface. What isn't reflected enters the water and is absorbed by water molecules. About 65 percent of the visible light is absorbed within 1 meter (~3.3 feet) of the sea surface in shelf and open-ocean water. This absorbed energy manifests itself as heat, elevating the temperature of the surface water. The longer wavelengths of visible light, the reds and yellows, are absorbed by water more readily than are the shorter wavelengths, the greens and blues (**Figure B5–6a**). This property of water—the selective absorption of certain wavelengths of light—combined with the scattering of the visible light accounts for the blue color of the open ocean. In very clear water, not even 1 percent of the light that enters the ocean reaches a water depth of

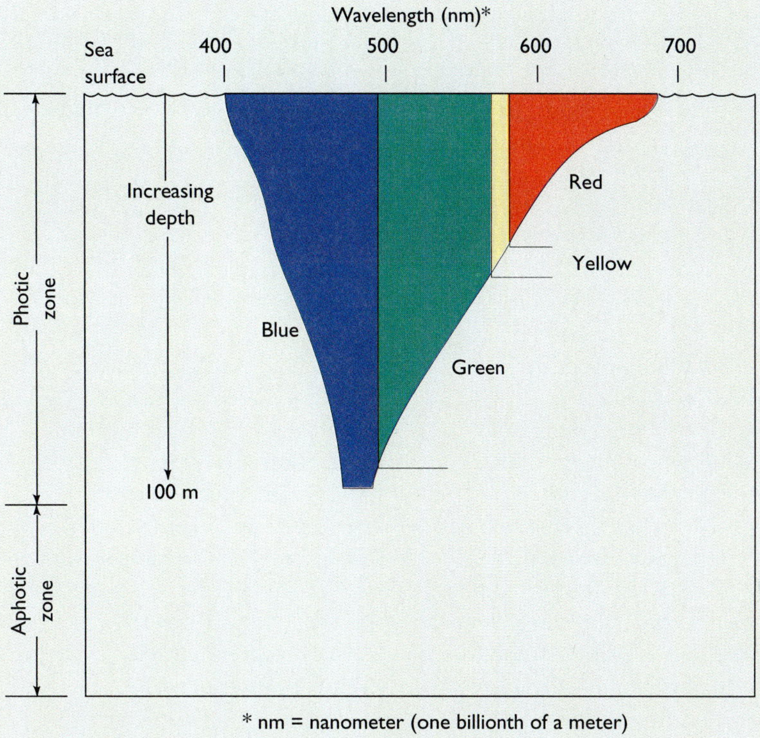

(a) LIGHT ABSORPTION IN THE OPEN OCEAN

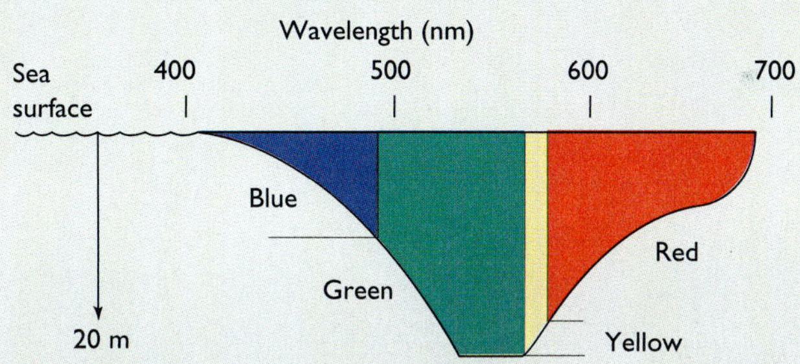

(b) LIGHT ABSORPTION IN NEARSHORE WATERS

FIGURE **B5-6**

Light absorption. (a) The selective absorption of visible light by water permits the shorter blue and green wavelengths to penetrate much deeper than the longer red and yellow wavelengths. This effect imparts blue and green colors to the sea. No more than 1 percent of all sunlight penetrates to a water depth of 100 meters. This level separates the photic zone (where plants can photosynthesize) from the aphotic zone (where plants cannot photosynthesize). (b) Inshore water contains a high sediment load, which limits light penetration to a depth of no more than 20 meters below the sea surface. Note that the yellow and green wavelengths extend the deepest here and impart the green and yellow colors of coastal water. [Adapted from G. S. Levine, Vision underwater, *Oceanus* 23 (1980): 19–26.]

100 meters (~330 feet), and it is entirely in the blue part of the visible spectrum. Inshore waters tend to have high loads of suspended solids, both organic and inorganic. Here, light may not penetrate this turbid water any deeper than 20 meters (~66 feet) (sometimes less than 1 meter!) below the surface (**Figure B5–6b**). The color of nearshore water typically is green and yellow because suspended particles reflect these wavelengths (**Figure B5–7**).

Because light intensity diminishes with depth, it is useful to divide the water column into vertical domains (see Figure B5-6a). In the sunlit upper layer, termed the **photic zone**, plants receive adequate levels of sunlight and can photosynthesize. Below the illuminated photic zone lies the dark **aphotic zone**, where plants cannot survive. As we shall discover later in the book, a large, diverse community of animals inhabits the unlit lower portions of the ocean, despite the absence of light and of live plants.

SOUND TRANSMISSION

Sound is transmitted more effectively and rapidly in water than it is in air. In the sea, the velocity of sound averages about 1,445 meters per second (~3,237 mph), compared with about 334 meters per second (~748 mph) in air. It increases by about 1.3 meters per second (~29 mph) for every one ‰ increase in salinity, by 4.5 meters per second (~10.1 mph) for every 1°C increase in temperature, and by 1.7 meters per second (~3.8 mph) for every 100 meters (~330 feet) increase in water depth (in effect, an increase in water pressure).

Because salinity varies slightly with water depth in most areas of the ocean, the speed of sound in water is affected mostly by temperature and pressure. Although pressure increases linearly with depth, temperature does not (**Figure B5–8a**). For example, the permanent thermocline marks a pronounced temperature gradient in the water column. In the surface water above the thermocline, sound velocity increases a bit with depth. The speed of sound drops sharply to a minimum value within the thermocline itself, but sound transmission increases in the water below the thermocline because of increasing water pressure. The result of these variations is a channel of minimum sound velocity at a water depth of about 1,000 meters (~3,300 feet) (see **Figure B5–8b**)—the so-called deep-sound channel or **SOFAR channel** (SOund Fixing And Ranging). Sound generated in this channel is focused as it is refracted (bent) from above and from below toward the center of the channel, where the speed is at a minimum. Because the sound waves are not spread vertically, there is little loss of energy, and , even moderate amounts of sound energy can be focused and transmitted over great distances within the SOFAR channel. The sound in this zone is not dispersed; rather, it is trapped and confined to this relatively thin layer of the water column. Field tests have resulted in sound transmission along the SOFAR channel from Australia to Bermuda—a distance of 25,000 kilometers (~15,500 miles)!

Visit www.jbpub.com/oceanlink for more information.

FIGURE B5-7

Water color. The different color tints of the water pouring out of the Yangtzei River into the East China Sea are created by the light-reflecting properties of different concentrations of suspended mud particles.

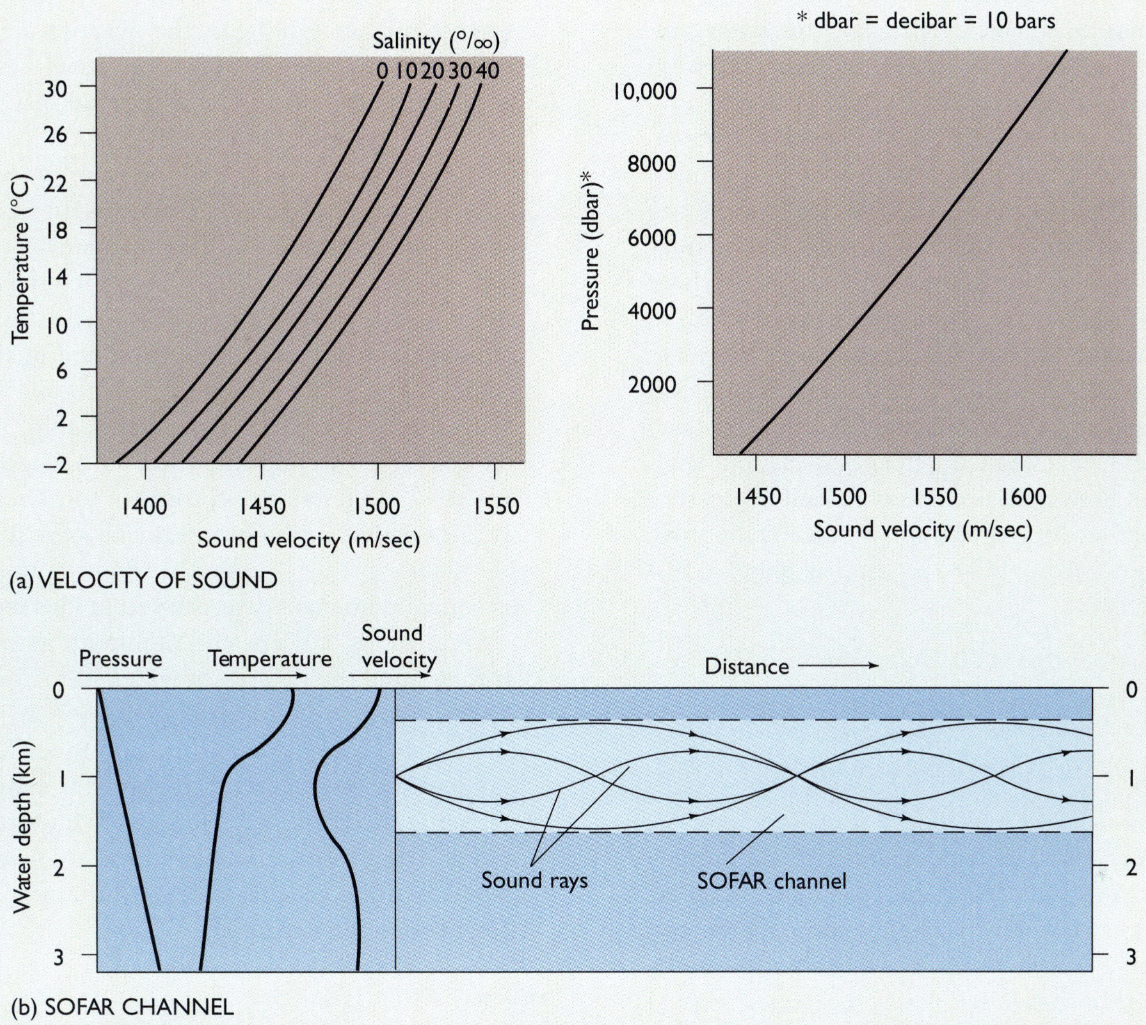

FIGURE **B5-8**

Sound in seawater. (a) The speed of sound in water increases as salinity, temperature, and pressure increase. Note that temperature and pressure changes have a more marked effect on sound speed than salinity does. [Adapted from G. Neuman and W. J. Pierson, *Principles of Physical Oceanography* (Englewood Cliffs, N.J.: Prentice-Hall, 1966).] (b) Vertical variations in temperature and pressure create a zone of minimum sound velocity at a water depth of about 1 kilometer. Consequently, sound waves are bent toward the center of the zone, the SOFAR channel, so that the trapped sound energy is focused and can be transmitted over long distances. [Adapted from L. E. Kinsler and A. R. Frey, *Fundamentals of Acoustics* (New York: John Wiley, 1962).]

does flow across geographic boundaries, both near the sea surface and at depth. H_2O molecules in seawater are not even confined to the oceans, but are passed into the atmosphere by evaporation; winds and air masses then transport this vapor, some of it across great distances. Eventually, the vapor condenses and falls as snow, sleet, or rain, often on land, whence it returns to the ocean as river runoff.

Dwelling in this watery wilderness is an incredible variety of life-forms: delicate, lacy plants, spindly-legged shrimp, swift, powerful sharks and tuna, microscopic, floating algae, and solidly anchored oysters. We marvel at the ability of these creatures to live in the ocean, but many of us fail to realize that we, as terrestrial inhabitants, live in an ocean as well. We dwell on land beneath an ocean of air that is an order of magnitude (differs by a factor of 10) deeper than the deepest seas and that, because of its weight, exerts considerable pressure on us. (For those of you who are technically inclined, atmospheric pressure at sea level is about 1,013 millibars, or

$$\sim 1{,}013 \times 10^6 \text{ dynes/cm}^2,$$

or

$$\sim 10 \text{ metric tons/m}^2.)$$

This transparent envelope of fluid—the gaseous atmosphere—is in direct contact with the surface of all the land and oceans and is a crucial element of the planet's water cycle. The atmosphere and hydrosphere, which enclose the Earth's crust, interact through a vast network of processes. Powered by the energy of the sun, water is exchanged among the oceans, the atmosphere, and the landmasses through evaporation, precipitation, river flow, groundwater (subterranean water) percolation, ocean circulation, and a host of related and intertwined processes. Some of these operate quickly; others, quite slowly. However, before we examine the workings of this immense global water cycle, we need to consider the distribution of water on the Earth.

RESERVOIRS OF WATER

Water covers more than 60 percent of the Earth's surface in the Northern Hemisphere and over 80 percent in the Southern Hemisphere. Not surprisingly, most of the water on the Earth—in fact, 97.25 percent of it by volume—is found in the oceans (**Table 5–10**). It may astonish some of you to discover that rivers and lakes, which are so familiar and so indispensable to the activities of humans, are of little significance in the Earth's overall water inventory. In fact, **groundwater**—the water below the ground surface that saturates the void space in soil, sediment, and rock—exceeds the combined volume of water in lakes, rivers, and the atmosphere by one order of magnitude (see Table 5–10). Also, the atmosphere contains a significant quantity of water, both as gas and as droplets (clouds), that far exceeds the volume of water stored in rivers. Finally, a minuscule quantity of water is stored in the biosphere, in the cells and tissues of plants and animals.

Although the values cited in Table 5–10 are accurate for the present, they do not reflect the quantities that existed in some of these reservoirs in the geologic past. During glacial ages, for example, seawater that evaporated from the oceans fell as snow on land and was converted into ice, forming enormous ice caps and mountain glaciers. So, at that time, the volume of water stored in the ice reservoir was much greater than it is at present. Obviously, water does not reside indefinitely in any one state or reservoir; rather, it is continually changing back and forth, from gas to liquid to solid, and shifting from one reservoir to another. Let's examine some specific processes that control the flux of water on the Earth.

THE GLOBAL WATER CYCLE

The exchange of water among the ocean, the land,

TABLE 5-10

Earth's water reservoirs

Reservoir	Water Volume (10^6 km^3)	Total Water (%)
Oceans	1370	97.25
Ice masses	29	2.05
Groundwater	9.5	0.68
Lakes	0.125	0.01
Atmosphere	0.013	0.001
Rivers	0.0017	0.0001
Biosphere	0.0006	0.00004
Total	1408.64	99.99

Source: Adapted from E. K. Berner and R. A. Berner, *The Global Water Cycle* (Englewood Cliffs, N.J.: Prentice-Hall, 1987).

and the atmosphere is termed the **hydrologic cycle** (**Figure 5–20**). The sun's heat evaporates surface water from the ocean and the land. (Just leave a glass of water in sunlight for a few days and see what happens!) Evaporated water enters the atmosphere as vapor and most of this returns directly to the sea as precipitation. Air currents transport the remainder of this water vapor over land, where it condenses and falls as rain or snow (**Figure 5–21**). This precipitation flows as runoff into rivers, collects temporarily in lakes, ponds, and wetlands, infiltrates the ground (groundwater) only to emerge later in rivers and lakes, or remains in solid form as snow or ice. With time, all of this water finds its way back to the oceans, either as river outflow, melting glaciers and icebergs, groundwater seepage, or evaporation and precipitation, thereby closing the hydrologic cycle. In effect, the same molecules of water are being continually recycled from reservoir to reservoir, the rates of flux and the quantity exchanged and stored depending on climate.

Over the oceans, evaporation exceeds precipitation, and the balance of water is maintained by river inflow. On all the land combined, in contrast, precipitation far exceeds evaporation, and the excess water travels to the ocean by river and groundwater flow. Estimates of these various fluxes are summarized in **Table 5–11**. These calculations assume that the total volume of water on the planet is fixed

FIGURE 5-20

Hydrologic cycle. The hydrologic cycle incorporates the major pathways for the transport and exchange of water among the various reservoirs of the earth. Basically, water circulates and changes phases (solid, liquid, gas) continually. In the gaseous state, water is supplied to the atmosphere by the evaporation of surface liquid water, much of it from the oceans, and by transpiration, the passage of water vapor through the surface of leaves. Condensation then causes the atmosphere to release this water as a liquid (rain) and as a solid (snow). These liquid and solid forms of water eventually return to the oceans by river runoff, closing the hydrologic cycle.

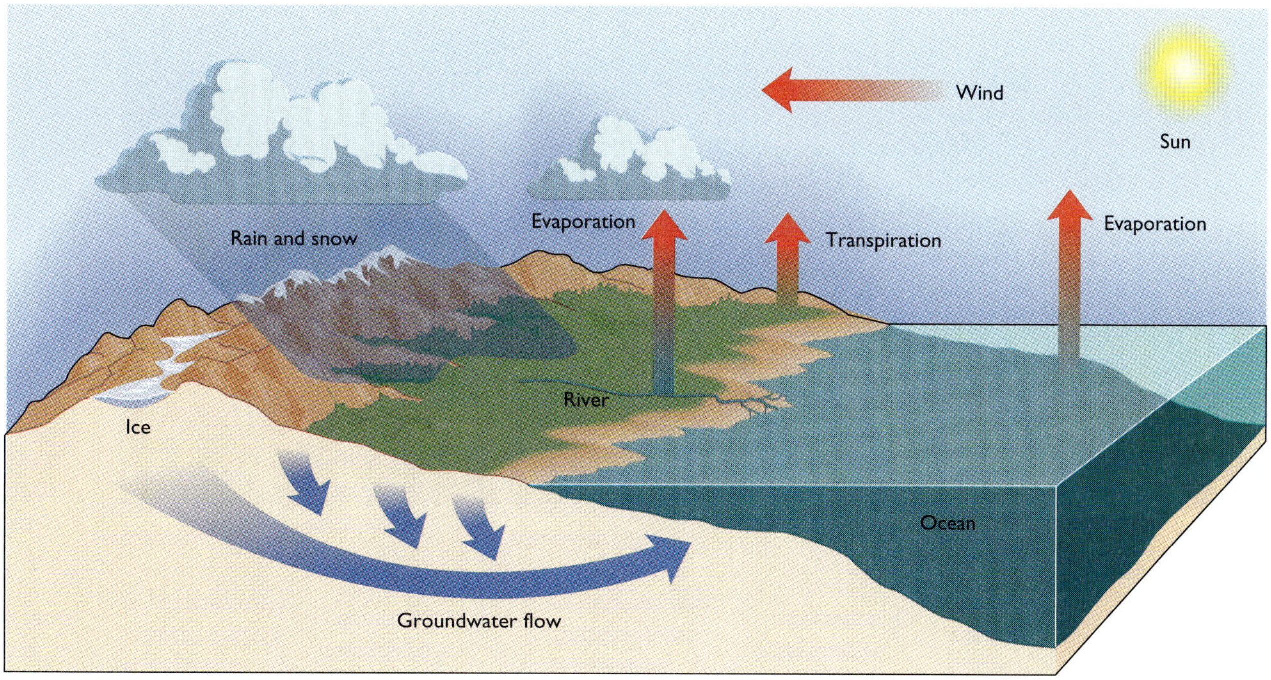

on a global scale (a perfectly reasonable supposition on a short time scale), so that precipitation and evaporation are balanced for the planet as a whole.

THE OCEAN AS A BIOGEOCHEMICAL SYSTEM

Now that we have examined the chemistry of seawater and the physical nature of the water column, we can combine them into a generalized biogeochemical system (**Figure 5–22**). Rivers supply the ocean with most of its dissolved ions, these being derived mainly from the chemical weathering of rocks on land. The driving force underlying the ocean's biogeochemical system is the sun. Solar radiation absorbed by ocean water raises its temperature and stratifies the water column. A thermocline, separating the warm surface water from the cold deep water, forms. In addition sunlight is used by plants to convert carbon dioxide and nutrients into organic compounds (food) by photosynthesis. Small marine animals that graze on the multiplying plants in turn serve as food for even larger animals. This results in the transfer of energy (radiation from the sun) and matter (nutrients from rocks) through biological systems. When marine organisms die, they sink below the permanent thermocline, taking with them nutrients (phosphorus and nitrogen compounds) that are so crucial to plant photosynthesis. Bacteria decompose this dead organic matter, freeing these substances and producing nutrient-rich water. The vertical mixing of ocean water then slowly transports this water and its dissolved load of chemicals upward, reinfusing the sunlit surface layer with nutrients that, once again, are incorporated into living plant tissue by photosynthesis. Some of these materials collect on the sea bottom as oozes, where they become buried and lithified into rock, and, by the process of plate subduction, are deformed and raised to form a mountain belt (see Figure 5–6). As these rocks are squeezed and uplifted by the tremendous forces of

FIGURE **5-21**

Recycling of water. Water evaporated from the ocean condenses as clouds that drop rain onto these mountains. River runoff and groundwater flow discharge this water into the ocean, closing the hydrologic cycle.

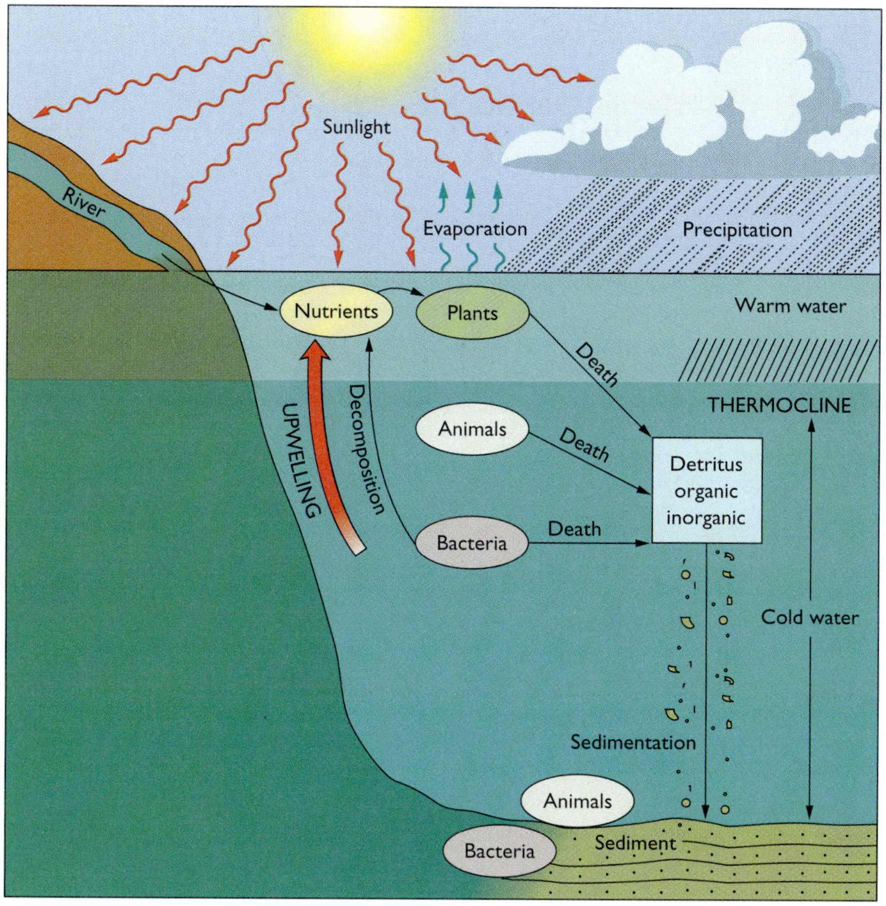

FIGURE 5-22

Biochemical recycling of matter. Inorganic nutrients are converted into food by plant photosynthesis in the surface-water layer of the ocean. Animals eat plants and one another. When plants and animals die, their organic matter settles through the water column, where it is converted into simple nutrients by bacterial decomposition. This nutrient-charged water is then returned slowly to the surface by upwelling currents, completing the biochemical recycling of key nutrients. [Adapted from D. J. Repeta, G. Eglinton, and C. Lee, Marine organic geochemistry, *Oceanus* 35, no. 1 (1992): 38–46.]

the colliding plates, they undergo chemical weathering and mechanical erosion until the chemicals are released once more and are transported by rivers to the ocean. The result is a continuous biochemical recycling of elements between living and nonliving matter, and between surface and subsurface water (see Figure 5–22). Much of the remainder of this book is devoted to examining the different physical, chemical, geologic, and biological pathways involved in this majestic planetary cycle.

TABLE 5-11

Water fluxes

Process	Water Flux (km^3/yr)
Evaporation from land	72,900
Precipitation on land	110,300
Precipitation on oceans	385,700
Evaporation from oceans	423,100
Total precipitation on Earth	496,000
Total evaporation on Earth	496,000

Source: Adapted from E. K. Berner and R. A. Berner, *The Global Water Cycle* (Englewood Cliffs, N.J.: Prentice-Hall, 1987).

The Sea-Surface Microlayer

The seawater surface is a unique interface where chemicals are exchanged between the ocean and the atmosphere. This actual surface, called the **surface microlayer**, is incredibly thin—a few hundred micrometers (~0.0039 inch)—and, although regarded as being important for the chemistry of the upper ocean, has not been examined extensively because of the difficulties of sampling such a tiny vertical section of the water column. However, it represents a critical link between the atmosphere and the ocean (**Figure B5–9**). The surface microlayer receives and transmits energy, gases, and solids and collects matter transported by winds from above and by water from below. Ferren MacIntyre (1974) aptly states, "Through the ocean's 360 million square kilometers of surface pass 70 percent of the solar energy that the earth absorbs, most of its supply of freshwater, a large fraction of the annual production of carbon dioxide and oxygen, a huge tonnage of particulate matter and unmeasured volumes of man-made pollutants." Clearly, an understanding of the chemical and physical processes that occur across the surface microlayer of the ocean is essential for an understanding of processes of global proportion.

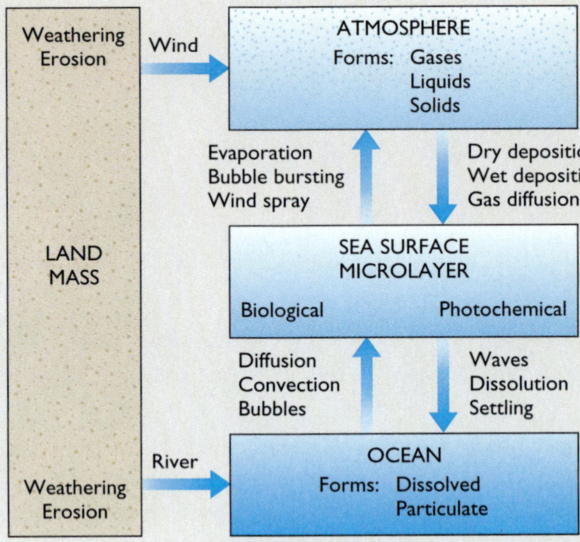

FIGURE B5-9

Box model of the sea-surface microlayer. This simple model shows how the sea-surface microlayer connects the atmosphere with the ocean. [Adapted from P. S. Liss, in *Chemical Oceanography*, J. P. Rilery and G. Skirrow, eds. (New York: Academic Press, 1975), 193–243.]

What exactly is meant by the ocean's surface? In other words, how thick is this surface? Actually, providing an answer to this question is not a simple matter. The definition of the surface microlayer is quite loose and depends on the research interests of the chemists, physicists, and biologists who study ocean phenomena at the air-sea interface. Most oceanographers would probably agree that the surface microlayer extends from 3 Angstroms (10^{-10} meters), the diameter of a water molecule, to a depth of about 3 millimeters (~0.1 inch) or so, the lower limit of nonturbulent (smooth or laminar) water flow in the absence of a wind. Using a log scale to plot a vertical profile of the ocean lets us expand the surface microlayer, so that it occupies about half the water depth on that diagram (**Figure B5–10**). Some chemists and physicists believe that this 3-millimeter-thick (~0.1-inch) surface layer may provide as many research problems as has the remainder of the ocean! Biologists refer to the organisms that inhabit the surface microlayer as the **neuston** and that thin habitat as the **neuston layer** (see Figure B5–10).

Three processes transport matter to the ocean's surface from below: molecular diffusion, convective motion, and rising air bubbles. *Diffusion* is a slow process and comes about by the random motion of molecules. *Convection* refers to the vertical circulation of water that results in the transfer of heat and both dissolved and particulate matter. The most significant process in the surface microlayer by far is, however, the ascent and bursting of *bubbles*. Bubbles created by waves and wind rise through seawater because of their buoyancy and adsorb or scavenge inorganic and organic matter in both dissolved and solid states. When the bubble bursts, a portion of the scavenged material is

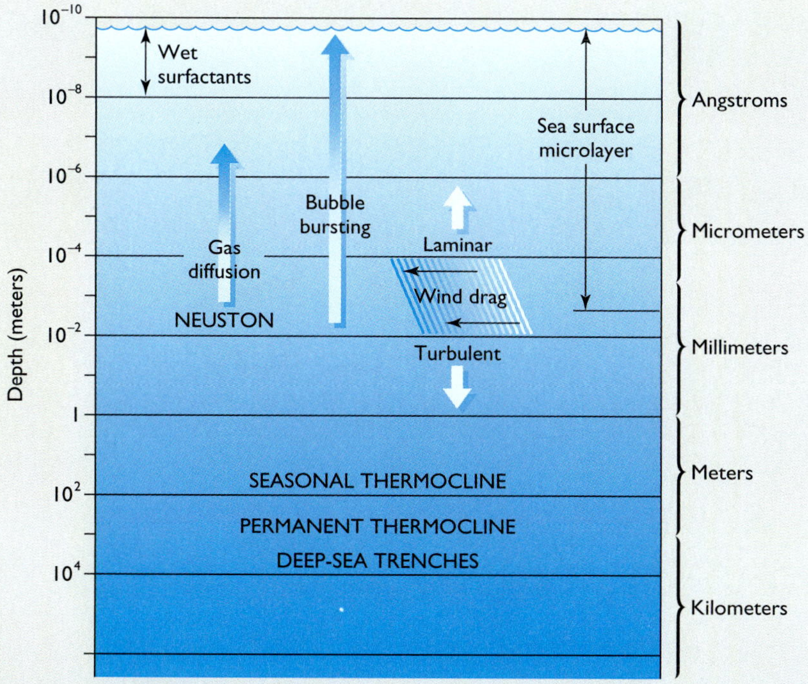

FIGURE **B5-10**

Profile of the ocean. This cross section of ocean depth is plotted on a logarithmic scale that expands the sea-surface microlayer, which actually extends no more than 3 millimeters below the sea surface, relative to the remainder of the water column.

ejected into the air, while the remainder collects in the surface microlayer. Material also is added to the sea surface from the atmosphere either as "wet" precipitation (rainfall and snowfall) or as "dry" deposition (particle settlement and gas diffusion). The result of this atmospheric and oceanic flux of material is the enrichment of both dissolved and particulate matter in the surface microlayer. The concentrated materials represent a surface coating that can reduce significantly the transfer of gases and water vapor across the air-sea interface, an effect that influences the chemistry of the lowermost atmosphere and the uppermost ocean and, quite possibly, climate in the long run.

The processes that occur within the surface microlayer can be divided into biological and **photochemical effects**. Sampling indicates that bacteria and plankton are more numerous in the surface microlayer than they are in the water immediately below. These organisms consume and produce a large variety of organic and inorganic substances and, presumably, influence the chemistry of this surface film of water. Evidence suggests that, when microorganisms living in the surface microlayer wiggle their tiny flagella (whiplike appendages), the tiny currents they induce can increase the rate of evaporation by as much as a factor of 3. Field experiments have demonstrated that microorganisms are scavenged by rising air bubbles and are ejected into the air when the bubble bursts. Ultraviolet radiation is not attenuated in the surface microlayer, and, in theory, this should lead to photochemical reactions, such as the oxidation of organic compounds. For example, it has been suggested that photochemical reactions, as well as microbial processes, are responsible for the production of much of the carbon monoxide (CO) in the surface microlayer.

Besides bacteria and plankton, the neuston includes macroscopic organisms. The snail *Janthina prolongata* is such a member, and apparently it feeds directly on the organics that collect in the surface coating. Insects that abound on land are uncommon in the ocean. However, five species of sea striders, *Halobates* (**Figure B5–11**), live on top of the surface microlayer, floating because air is trapped in their body hair. Apparently, they feed on small organisms that they encounter in the surface film.

Some marine chemists believe that human pollutants may be enriched in the surface microlayer and these are rapidly dispersed far from their point of origin. For example, in the past, DDT, a toxic chlorinated hydrocarbon, was used extensively in North America as a pesticide to enhance crop yield. Some fraction of the DDT sprayed by planes on croplands remained airborne, and winds transported this load westward, where it rained down in the North Atlantic Ocean. Scientists surmise that much of the DDT likely remained in the surface microlayer and became concentrated in neuston organisms. Wilson's petrel, allegedly the most abundant bird in the world, transported DDT to the high latitudes of the Southern Hemisphere. It accomplished this because it is a long-distance migrant that feeds in the North Atlantic and breeds on the Antarctic continent. By this mechanism, DDT was dispersed rapidly across the globe, so that even the ostensibly isolated Antarctic fauna (birds and seals) became contaminated with the exceedingly toxic pollutant.

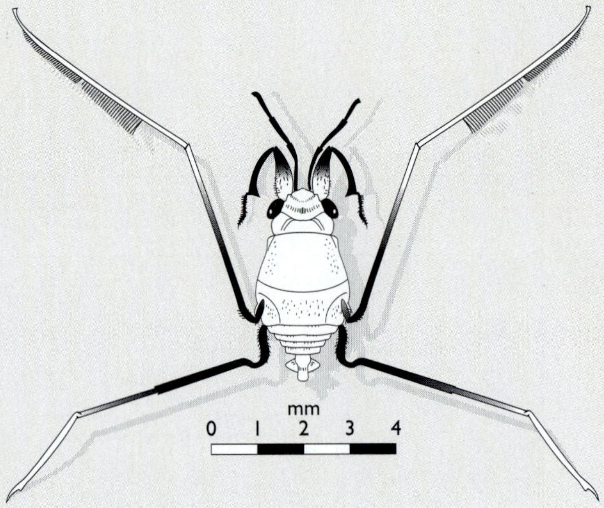

FIGURE B5-11

The sea strider *Halobates*. This is a sketch of one of the few marine species of insects that strides atop the sea-surface microlayer. [Adapted from P. M. David, *Endeavor* 24 (1965): 95–100.]

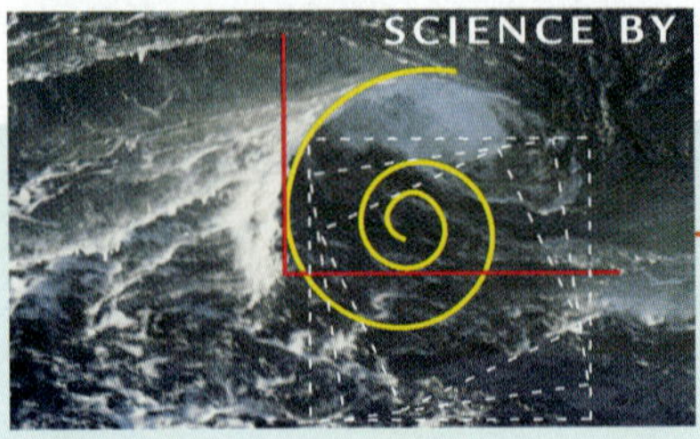

SCIENCE BY NUMBERS

Order of Magnitude

A concept used by scientists is the order of magnitude. A difference of one order of magnitude between two measurements means that they vary by a factor of 10. For example, a current that is flowing at 2.3 meters per second is one order of magnitude faster than a current with a speed of 0.6 meters per second.

Perhaps we can see this better if we express both current velocities as a power of 10 (see Math Box, "Powers of Ten"). Let's do that.

$$2.3\ \text{m/sec} = 2.3 \times 10^{0}\ \text{m/sec}$$
$$0.6\ \text{m/sec} = 6.0 \times 10^{-1}\ \text{m/sec}$$

The order of magnitude difference is 1, that is, 10^{0} is larger than 10^{-1} by a factor of 10.

If three fish are, respectively, 0.12, 0.43, and 1.88 meters in length, the first two values are of the same order of magnitude, and both are one order of magnitude smaller than the third value. Do you agree? It may help you to express each measurement of length as a power of 10.

$$0.12 = 1.20 \times 10^{-1}$$
$$0.43 = 4.30 \times 10^{-1}$$
$$1.88 = 1.88 \times 10^{0}.$$

This shows clearly that the first two measurements are one order of magnitude smaller than the third (10^{-1} versus 10^{0}), which means simply that they vary by a factor of 10.

To see if you grasp the concept, try this problem. The average depth of the ocean is 4 kilometers (4,000 meters). How many orders of magnitude is this greater than the depth of the shelf break, which averages about 130 meters? Try to solve this problem on your own before continuing.

Now let's check your answer and your reasoning. The average depth of the sea is one order of magnitude greater than the average depth of the shelf break. This can be shown to be the case by expressing each value as a power of 10:

$4{,}000 \text{ m} = 4 \times 10^3 \text{ m}$, and $130 \text{ m} = 1.3 \times 10^2$
10^3 is one order of magnitude greater than 10^2.

Deep-sea trenches average about 12 kilometers in water depth. How many orders of magnitude greater is this depth (12,000 meters) in comparison to the depth of the shelf break (130 meters)? This should now be routine for you to answer. Your reasoning might be as follows:

$12{,}000 \text{ m} = 1.2 \times 10^4$
$130 \text{ m} = 1.3 \times 10^2$

A comparison of the exponents shows that deep-sea trenches are two orders of magnitude deeper (10^4 versus 10^2) than are shelf breaks.

STUDY GUIDE

KEY CONCEPTS

1. The water molecule H_2O has a dipole structure, which produces *hydrogen bonding* and allows water molecules to collect into snug *clusters*. This dipole structure explains many of the curious properties of water, including its elevated melting and boiling temperatures, high *heat capacity,* unusual *solvent* power, and the fact that ice, the solid form of water, is less dense than liquid water.

2. The quantity of *solutes* in seawater, its *salinity,* is expressed as *parts per thousand* (‰). Average seawater has a salinity of 35‰, meaning that a sample of seawater weighing 1,000 grams contains 35 grams of dissolved salt, collectively referred to as the *major constituents* of seawater. Other solutes in the ocean include *nutrients* (such as compounds of N, P, and Si), *gases* (such as O_2 and CO_2), *trace elements* (such as metals), and *organic compounds* (such as fats, proteins, and carbohydrates).

3. Salinity can be measured simply by determining the amount of chloride (Cl^-) in the seawater sample. This is possible because the relative amounts of all the major elements are fixed, regardless of the total salt content of the seawater—a fact expressed as the *principle of constant proportion.* Today, salinity is routinely determined by such indirect techniques as *electrical conductivity.*

4. The salt content of the ocean is derived from the chemical breakdown of rocks on land and from volcanic outpourings along the crests of spreading ocean ridges. The major elements dissolved in seawater have a long *residence time* in the ocean, meaning that they remain dissolved in seawater for between tens of millions and hundreds of millions of years before they are removed from the system. Because mixing rates are on the order of tens to thousands of years, elements with long residence times, such as Na^+ and Cl^-, are distributed in a fixed ratio to one another throughout the oceans.

5. The *density* of seawater depends largely on water temperature and salinity. Water temperature is regulated by solar heating and varies directly with

latitude. Salinity depends on the relative effects of evaporation and rainfall, and like temperature, varies with latitude. Sharp vertical gradients of temperature are called *thermoclines,* of salinity are called *haloclines,* and of density are called *pycnoclines.*

6. Sources of gases in seawater include diffusion from the atmosphere and biological processes. Plant *photosynthesis* uses up CO_2 and releases O_2; plant and animal respiration consumes O_2 and releases CO_2. The production of O_2 is limited to surface water where gas can diffuse across the air-sea interface and where plants can photosynthesize. The *sinking* and *advection* (horizontal transport) of cold, dense, well-oxygenated polar water supply O_2 to the deep levels of the ocean, ventilating the deepest parts of the ocean basins.

7. The *hydrologic cycle* describes the exchange of water in gaseous, liquid, and solid states among the various water reservoirs. Water is evaporated from the ocean's surface, falls as precipitation on land, and returns to the oceans as river runoff and groundwater flow, thereby completing the hydrologic cycle.

8. Matter is recycled throughout the oceans. Through photosynthesis, plants convert inorganic CO_2 and nutrients into food. When consumed, this matter and energy are passed on to animals. When marine organisms die, they sink, and their dead tissues are decomposed by bacteria, a process that releases simple inorganic nutrients. The upwelling of this nutrient-rich water recharges the surface water with vital elements that are once again converted into food by plants, initiating a new biochemical cycle.

KEY WORDS

anaerobic
anion
anoxic
atoms
cation
conservative ions
deep layer
density
gas
groundwater
halocline
heat capacity
hydrologic cycle
ion
molecules
nonconservative ions
nutrient
parts per thousand (ppt)
pH
photosynthesis
pycnocline
residence time
salinity
thermocline
vaporization

QUESTIONS

REVIEW OF BASIC CONCEPTS

1. What is a dipolar molecule? How does it account for the singular physical properties of water?

2. Why does water attain its maximum density at 3.98°C, rather than at its freezing point of 0°C?

3. What are the major dissolved constituents of seawater? Why do they occur in fixed ratios, despite variations in the salinity of ocean water?

4. Describe the important processes that change the salinity of seawater without altering the ratio of its dissolved salts.

5. What is residence time for an element in seawater?

6. List and discuss the mechanisms for the input and output of elements in ocean water. Which of these input processes is the dominant supplier of salt to the ocean?

7. What exactly is pH? Explain how photosynthesis and respiration affect seawater pH.

CRITICAL-THINKING ESSAYS

1. Why is the amount of dissolved Na^+ in seawater constant over time whereas the content of dissolved CO_2 is not?

2. Explain why the quantity of nutrients in ocean water often varies inversely with the measured amount of dissolved O_2 in the water.

3. Can the density of an ice cube vary? Explain your reasoning. (Hint: Think about kinetic energy.)

4. Infer how the relative content of dissolved CO_2 in the water of a tidal pool will vary over a 24-hour period. Will variations in the level of dissolved CO_2 affect the pH of this tidal pool water? If so, how? If not, why not?

5. Would you expect the salinity of the oceans to change during a geologic period when many continents are colliding and major mountain belts are being raised worldwide? If so, how? If not, why not? (Hint: Think about the control that mountains exert on the content of dissolved chemicals discharged into the ocean by rivers.)

6. What are a few interconnections among global plate tectonics, the hydrologic cycle, seawater salinity, and climate?

7. What would happen to the hydrologic cycle if water could not exist in the gaseous phase on the Earth's surface? What would the implications be for the landmasses? For the oceans?

8. If global warming occurs in the next century, what will be the consequences, if any, of these elevated air temperatures for the salinity of the ocean? What is likely to happen in a qualitative sense to the volumes of water in the various reservoirs listed in Table 5–10 in the event that the Earth does undergo warming in the future?

DISCOVERING WITH NUMBERS

1. Determine the salinity of a seawater sample with a chlorinity of 10.3‰. Estimate the amount of chloride in this sample.

2. How many grams of sodium and chloride are present in a 1 kilogram of seawater with a salinity of 35‰? With a salinity of 10‰?

3. Using orders of magnitude, how many times is 1 ppt greater than 1 ppm?

4. Calculate the residence time of sodium in millions of years if the total amount of sodium in the oceans is 1.5×10^{22} grams, and it is supplied to the ocean at a rate of 2.2×10^{14} grams per year. Estimate the rate of removal of dissolved sodium from the oceans? What must you assume in order to make this estimation?

5. Calculate the annual supply rate of potassium in seawater assuming the ocean's total content of dissolved potassium is 5.21×10^{19} grams. (Hint: Remember that residence time of a substance equals the total mass of that substance in seawater divided by the rate of that substance's input to the ocean.)

6. Imagine that you're on a cruise in an area of the ocean where the difference between the annual rates of evaporation and precipitation (E - P) is zero. What would you predict the seawater salinity of the area to be in parts per thousand? (Hint: See Figure 5–12.) The following day after a night of heavy rain, you take a surface-water sample and measure a salinity of 34.55‰. Why does this measurement of salinity not agree with your estimate made the day before? (Hint: Think of the salinity of an area averaged over a long period of time as compared with its day-to-day salinity values.)

7. From the information in Table 5–11, calculate as a percentage how much more precipitation there is on land than evaporation, and how much more evaporation there is on the ocean than precipitation. Shouldn't these two percentages be the same if we have a steady-state hydrologic cycle?

8. Plot the following data for an ocean in the low latitudes:

Depth (m)	Temp. (°C)	Salinity (‰)	Oxygen (ml/l)
0	24.4	36.5	4.6
250	21.2	36.3	4.7
500	6.9	35.6	2.8
750	5.1	34.7	3.5
1,000	4.9	34.4	3.8
2,000	4.8	34.8	5.1
3,000	4.7	34.9	5.1
4,000	4.6	34.8	5.1

a. Identify the thermocline and halocline in the water column, and account for their mode of formation.

b. How many distinct water layers are evident, and what are their chemical properties?

c. Describe trends in the oxygen profile, and account for variations in the dissolved oxygen content of the water.

SELECTED READINGS

Andreal, M. O. 1986–87. The oceans as a source of biogenic gas. *Oceanus* 29 (4): 27–35.

Baker, J. A., and Henderson, D. 1981. The fluid phase of matter. *Scientific American* 245 (5): 130–39.

Berg, H. C. 1983. *Random Walks in Biology.* Princeton, N.J.: Princeton University Press.

Broecker, W. S. 1974. *Chemical Oceanography*. New York: Harcourt Brace Jovanovich.

Brown, N. 1991. The history of salinometer and CTD sensor systems. *Oceanus* 34 (1): 61–66.

Buckingham, M. J., Potter, J. R., and Epifanio, C. L. 1996. Seeing underwater with background noise. *Scientific American* 274 (2): 86–91.

Butler, J. N. 1975. Pelagic Tar. *Scientific American* 232 (6): 90–97.

Faulkner, D. J. 1979. The search for drugs from the sea. *Oceanus* 22 (2): 44–50.

Friedman, R. 1990. Salt-free water from the sea. *Sea Frontiers* 36 (3): 48–54.

Gabianelli, V. J. 1970. Water—the fluid of life. *Sea Frontiers* 16 (5): 258–70.

Gleick, P. H. 2001. Making every drop count. *Scientific American* 284 (2): 40–45.

Goldman, J. C. 1979. Chlorine in the marine environment. *Oceanus* 22 (2): 36–43.

Gordon, A. L., and Comiso, J. C. 1988. Polynyas in the Southern Ocean.*Scientific American* 258 (6): 90–97.

Gregg, M. 1973. Microstructure of the ocean. *Scientific American* 228 (2): 64–77.

Hill, M. N. 1963. *The Sea: Composition of Sea Water.* New York: John Wiley, Interscience.

Jenkins, W. J. 1978. Helium isotopes from the solid earth: Up, up, up and away. *Oceanus* 21 (3): 13–18.

Jenkins, W. J., and Smethie, W. M., Jr. 1996. Transient tracers track ocean climate signals. *Oceanus* 39 (2): 29–32.

McCarthy, J. J., Brewer, P. G., and Feldman, G. 1986–87. Global ocean flux. *Oceanus* 29 (4): 16–26.

MacIntyre, F. 1970. Why the sea is salty. *Scientific American* 223 (5): 104–15.

MacIntyre, F. 1974. The top millimeter of the ocean. *Scientific American* 230 (5): 62–69.

Newburgh, T. 1986. Sea ice and oceanographic conditions. *Oceanus* 29 (1): 24–30.

Oceanus. 1977. Sound in the sea. Special issue 20 (2).

Oceanus. 1992. Marine chemistry. Special issue 35 (1).

Open University Course Team. 1989. *Seawater: Its Composition, Properties, and Behavior.* New York: Pergamon Press.

Revelle, R. 1963. Water. *Scientific American* 209 (3): 93–108.

Riley, J. P., and Chester, R. 1971. *Introduction to Marine Chemistry.* New York: Academic Press.

Robinson, B. H. 1995. Light in the ocean's midwaters. *Scientific American* 273 (1): 60–65.

Rona, P. A., 1986. Mineral deposits from sea-floor hot springs. *Scientific American* 254(1): 84–92.

Schmitt, R. W. 1996. If rain falls on the ocean—does it make a sound? *Oceanus* 39 (2): 4–8.

Sobey, E. 1979. Ocean ice. *Sea Frontiers* 25 (2): 66–73.

Stewart, R. W. 1969. The atmosphere and the oceans. *Scientific American* 221 (3): 76–105.

Watson, L. 1988. *The Water Planet.* New York: Crown.

TOOLS FOR LEARNING

Tools for Learning is an on-line review area located at this book's web site OceanLink (**www.jbpub.com/oceanlink**). The review area provides a variety of activities designed to help you study for your class. You will find chapter outlines, review questions, hints for some of the book's math questions (identified by the math icon), web research tips for selected Critical Thinking Essay questions, key term reviews, and figure labeling exercises.

BENJAMIN FRANKLIN'S MAP OF THE GULF STREAM
LAND of the
ESKIMAUX's or LABRADOR
Newfoundland
ATLANTIC OCEAN
NORTH CAROLINA
SOUTH CAROLINA
GEORGIA
GULF of MEXICO
ATLANTIC OCEAN
A CHART of The GULF STREAM

6 Wind and Ocean Circulation

There is a river in the ocean. In the severest droughts it never fails, and in the mightiest floods it never overflows. Its banks and its bottom are of cold water, while its current is of warm. The Gulf of Mexico is its fountain, and its mouth is the Arctic Seas. It is the Gulf Stream.

—*Matthew Fontaine Maury,* The Physical Geography of the Sea, *1855*

web navigator

critical thinking on the web

math tutor on the web

www.jbpub.com/oceanlink

PREVIEW

IN THIS CHAPTER WE CONSIDER the large-scale circulation of the oceans. We first consider a general model of atmospheric circulation. Then we examine the persistent ocean currents that slowly transport large volumes of surface and subsurface water over vast horizontal distances across the globe. On a perfectly calm and flat sea, a boat with sails hanging listlessly from its spars appears to be motionless in the water, but it is actually drifting with the slow surface currents. Even deep below the boat's keel, water is moving continually throughout the ocean's depths. Our intent in this chapter is to explore the nature of these surface and deep-ocean currents and their driving forces.

In addition to these major ocean-circulation systems, small-scale water motions are apparent. These are worthy of examination as well, because they have a major effect on the local chemistry and biology of an ocean region.

Human beings live at the bottom of an ocean of air. This ocean of air—that is, of gas, the *atmosphere,* is in contact with water, the *hydrosphere,* over an area that covers more than 70 percent of the Earth. Circulation mixes the fluids (gases and liquids) of the two as they interchange mass and energy and interact in a variety of chemical and physical ways. Because weather and climate influence, and are in turn influenced by, the oceans, they are considered together in the science of oceanography. The main point of this section is to describe some of the close connections between the atmosphere and the oceans.

6-1 Atmospheric Processes

AIR PRESSURE

Air, as we know, is a mixture of gases. Dry air consists predominantly of nitrogen (78.08 percent by volume) and oxygen (20.95 percent) with minor amounts of argon, carbon dioxide, and various halogen (neon, helium, krypton, xenon) gases. Most air contains water vapor (gaseous water), a bit more than 1 percent on the average. Air with water vapor is less dense than is dry air at the same temperature and pressure. Also, when heated, air expands and becomes less dense. This means that cold air is denser than warm air is, and cold, dry air is much more dense than warm, moist air is.

Pressure is defined as a force-per-unit area. A formula for pressure is

$$P = \rho g h$$

where ρ is density, g is gravitational acceleration, and h is height of the mass. This is a familiar formula, because we worked with it extensively in Chapter 2 when we discussed the isostatic balance of rocks in the crust and upper mantle. Here we will concentrate on the effect of air density on atmospheric pressure.

At sea level, a column of air one square inch (in^2) that reaches to the top of the atmosphere weighs on the average 14.7 pounds. The pressure exerted by this air column (14.7 lb/in^2) is called the *standard atmospheric pressure,* and it is equivalent to one atmosphere. When the density of the air is lower than normal (because of an increase in water vapor content or temperature), atmospheric pressure drops. This creates a **low-pressure zone**. Conversely, when air density is higher than normal (because of a decrease in water vapor or temperature), atmospheric pressure rises, creating a **high-pressure zone**. Fluids flow from zones of high pressure to zones of low pressure, and this flow creates air currents, the *winds*.

The change in pressure across a horizontal distance is called a **pressure gradient.** The greater the change, the steeper is the pressure gradient. A steep pressure gradient results in strong winds, a shallow one in weak winds. We can compare this roughly to sledding on a snow-covered hill. The hill is a topographic gradient. The steeper the slope, the faster our sled moves downhill.

Before continuing this discussion, we should clarify the difference in the conventions for designating the flow paths of wind and water currents. *Wind direction* indicates where the wind originates from. Thus, what we call a "north wind" flows out of the north and moves to the south. In describing water-current flows, the convention is exactly the opposite: a "northerly current" flows toward, and not out of, the north. Bear this distinction in mind.

CORIOLIS DEFLECTION

The global winds blow in response to pressure gradients on the Earth. Variations in atmospheric pressure reflect the uneven distribution of solar radiation across the Earth's surface. For example, high solar radiation at the equator produces hot air, which rises because of its low density, creating a belt of low air pressure. This rising air is coupled to the convergence of surface winds and the divergence of upper-level winds (**Figure 6–1a**). In contrast, the air is very cold in polar regions because of limited solar heating. Here the cold air sinks because of its high density, creating a zone of high pressure near the north and south poles. The sinking air at the poles is associated with the divergence of surface winds and the convergence of upper-level winds (**Figure 6–1b**).

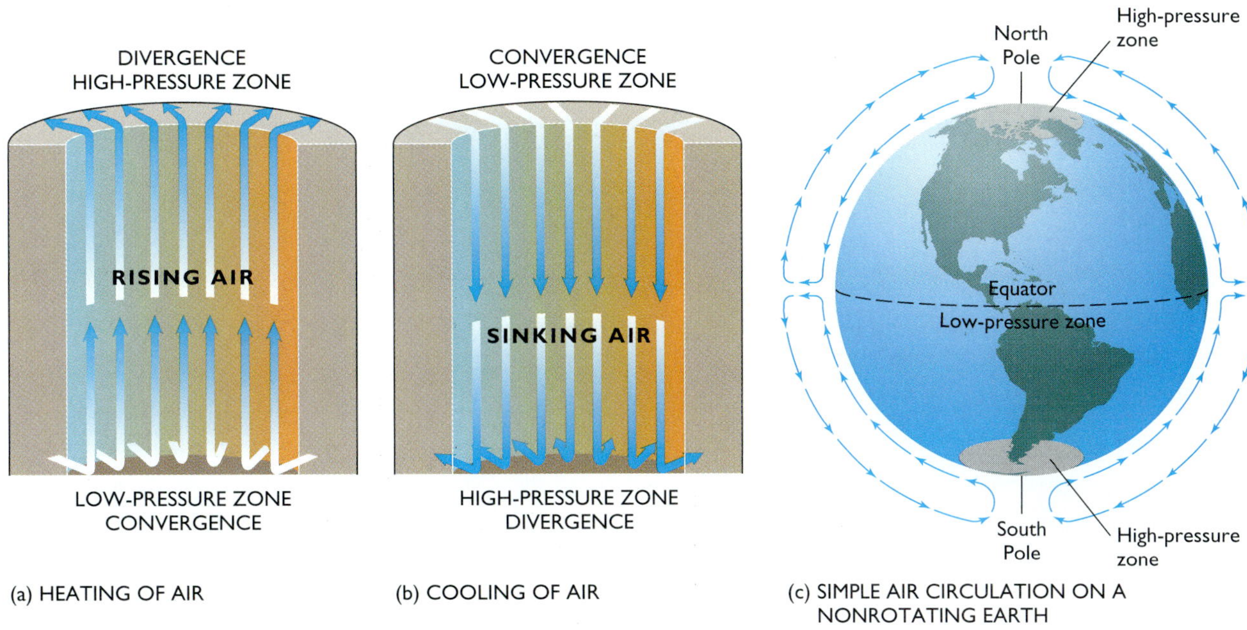

FIGURE 6-1

Air pressure. (a) Heating by the sun causes air to expand. This lowers the air density and creates a low-pressure zone of rising air at the Earth's surface. (b) Cooling causes air to contract. This increases air density and creates a high-pressure zone of sinking air at the Earth's surface. (c) The low-pressure zone at the equator and the high-pressure zone at the poles should induce a simple circulation cell in each hemisphere with surface winds directed towards the equator and directed upper-level winds towards the poles.

Fluids move in direct response to pressure gradients. Hence, we would expect that the surface winds of the Earth flow away from the north and south poles (zones of high pressure) toward the equator (a zone of low pressure) (**Figure 6–1c**). The wind patterns in this model are simple. They consist of very long circulation cells that stretch uninterrupted from the equator to the poles (see Figure 6–1c). The equator in this scheme is a zone of surface-wind convergence. The intense tropical sunshine there heats the air, and so it rises. In the atmosphere high above the equator, ascending air diverges and streams to the poles as fast, upper-level winds. In the upper atmosphere of the north and south poles, the upper winds converge, and the cold air sinks. Near the ground, the polar winds diverge and stream back to the equator. In this scheme, the surface winds in the Northern Hemisphere flow dominantly out of the north (north winds), and in the Southern Hemisphere dominantly out of the south (south winds).

This wind pattern does not agree with common experience. For example, as anybody who has spent time in the midlatitudes of the Northern Hemisphere will tell you, the dominant winds flow out of the west (the **westerlies**) and not out of the north. In fact, the winds in the model fail to describe the prevailing winds in the low and high latitudes as well. The model winds do not resemble the actual winds of the Earth because we have neglected the effect of the Earth's rotation. It turns out that the planet's rotation strongly influences the motion of fluids, both air and water. This effect is named the **Coriolis deflection** after its discoverer, the French engineer and mathematician Gaspard Gustave de Coriolis (1792–1843).

A way of beginning to understand this odd effect is to imagine yourself and a friend riding a pair of horses on a merry-go-round (**Figure 6–2a**). You decide to toss a ball to your friend, who is sitting astride a wooden horse mounted on the outside edge of the merry-go-round. To your astonishment, the ball curves sharply and your friend is unable to catch it. You've never been able to throw a curve ball. What happened, exactly? Well, the path of the ball was actually straight, as was witnessed by several

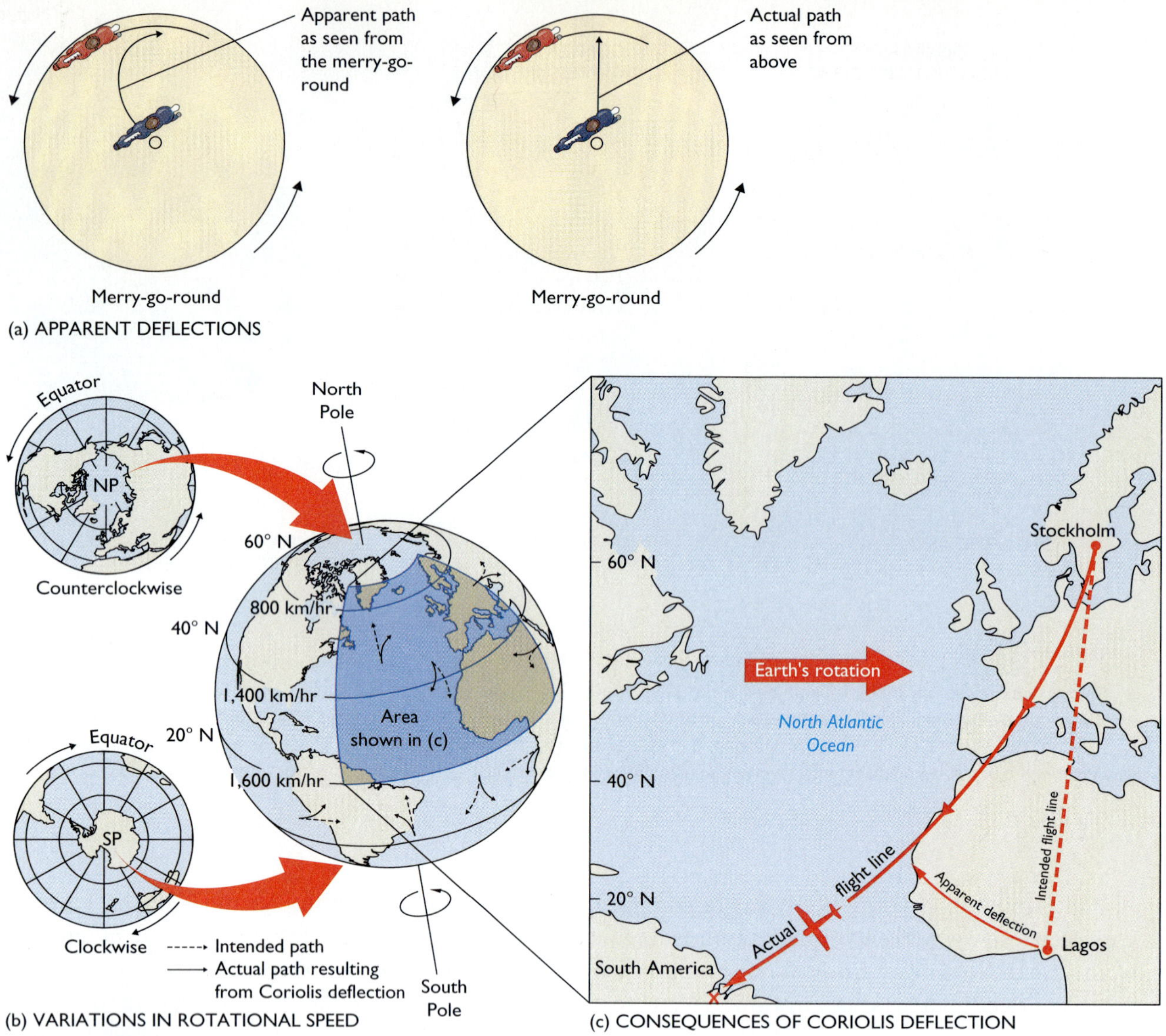

FIGURE **6-2**

Coriolis deflection. (a) The apparent and actual paths of a ball thrown between two people riding horses on a merry-go-round. (b)The rotational speed of the earth varies with latitude, so that matter not rigidly attached to the planet undergoes an apparent deflection known as the Coriolis effect. Objects that move relative to the planet appear to drift to the right of their intended path in the Northern Hemisphere and to the left of their intended course in the Southern Hemisphere. (c) If corrections are not made for Coriolis deflection, a plane departing from Stockholm, Sweden, on a six-hour flight to Lagos, Nigeria, will drift 4,800 kilometers westward (to the right) and be forced to land in South America rather than in Africa.

people riding high above the merry-go-round in an anchored hot-air balloon. But from the point of view of yourself and your friend riding the horses on the merry-go-round, the ball curved sharply. To put it another way, the deflection was not real, but was an apparent one, created by your frame of reference. Keep this image in mind as we examine Coriolis deflection on a rotating Earth.

If we think about the Earth spinning on its axis, completing a full rotation every 24 hours, it is obvious that, over the course of a day, people living near the equator describe a circle with a larger circumference than do individuals living in the middle and high latitudes (**Figure 6–2b**). Obviously, the velocity of rotation increases from the poles toward the equator, in the same way that a horse on the outside of a merry-go-round must be rotating faster than one on the inside.

What are the implications of differential rotation for objects moving relative to the Earth? Let's consider a concrete, but simplified, example. A jetliner in Stockholm, Sweden (60°N latitude) preparing for takeoff is moving to the east with the planet at a velocity of 800 kilometers per hour (~497 mph) (**Figure 6–2c**). Its destination, Lagos, Nigeria, in Africa, located almost due south near the equator, is rotating eastward at 1,600 kilometers per hour (~994 mph). Once it is airborne and traveling southward, the aircraft still has a rotational velocity of 800 kilometers per hour (~497 mph), but it is flying over land closer to the equator that is rotating eastward at a faster and ever increasing rate. If the flight lasts six hours and the pilots do not alter their course, the jet will miss the Lagos airport by 4,800 kilometers (~2,980 miles)—that is, 6 hours × 800 kilometers per hour (the difference in the rotational velocity between Stockholm and Lagos). To their dismay, they will land somewhere in South America rather than in Africa (see Figure 6–2c)! To the pilots, as well as to observers on the ground tracking the jet's course, the plane appears to be curving to the right relative to the land. Remember the "curve" that you threw to your friend on the merry-go-round. In actuality the deflection of the jet is an illusion, and the aircraft is flying a perfectly straight course, as would be evident to astronauts in a space station looking down on the Atlantic Ocean and monitoring the aircraft's flight.

Conversely, a jet headed for Sweden from Lagos, if it does not alter its course, will land 4,800 kilometers (~2,980 miles) to the east of Stockholm and, as before, will appear to be veering to the right. During its northern journey, the plane flies over land that is rotating eastward at a slower and ever decreasing rate compared to that of the jet. This is equivalent to your friend who is sitting on a horse on the outside of the merry-go-round tossing the ball to someone astride a horse nearer the center of the merry-go-round.

It is important to note that the plane, whether flying north or south in the Northern Hemisphere, appears to veer to the right. Even travel due east or west in the Northern Hemisphere will result in an apparent deflection to the right of the intended course. This is the result of centrifugal acceleration. When a car turns sharply to the left, passengers continue to travel in a straight line because of inertia, but appear to move to the right relative to the car. Water at rest has the same centrifugal acceleration as the solid crust beneath it. But, if it moves to the east in the same direction that the Earth rotates, the centrifugal "force" increases and it drifts to the right, toward the equator. If the current moves west against the Earth's rotation, the centrifugal "force" decreases and it drifts to the right, toward the pole.

The point is this: any object moving relative to the Earth, be it airplane, water, wind, artillery shell, or whatever, will be deflected to the right of its course anywhere in the Northern Hemisphere. In the Southern Hemisphere, the Coriolis deflection is to the left, rather than to the right. This is caused by the clockwise spin of the Southern Hemisphere relative to the south pole (see Figure 6–2b). Look down on the Earth from the north pole and you will see the planet rotating counterclockwise about its axis. Now take the globe and turn it over so that the south pole is in view. The Earth's west-to-east spin will then be clockwise, and the apparent deflection of moving objects relative to the ground will be to the left. Test this by making the merry-go-round in Figure 6–2a rotate clockwise instead of counterclockwise and note the apparent deflection of a tossed ball.

To summarize: *moving objects relative to the ground are deflected to the right in the Northern Hemisphere and to the left in the Southern Hemisphere.* The amount of deflection that is induced by the Coriolis effect depends on the *speed* of the moving object and its *location* on the Earth. The tendency of a wind to curve because of Coriolis deflection increases with its speed and with distance from the equator. This means that polar winds (in high latitudes) deviate much more relative to the Earth's surface than do tropical winds (in low latitudes) moving at the same speed. Also, fast winds veer more sharply than slow winds in either polar or tropical oceans. At the equator, there is in fact no Coriolis effect and hence no apparent deflection of wind.

GENERAL WIND CIRCULATION

The global wind pattern reflects the uneven distribution of solar heat on the rotating Earth. Heating at the equator causes air to warm. This hot air expands, which creates a band of low pressure at the surface (see Figure 6–1a). As this warm air rises, it cools by expansion (air pressure is lower with increasing altitude) and loses moisture by precipitation. As the rising air becomes colder and drier, its density increases. The result is a high-pressure zone (see Figure 6–1a) located high above the equatorial zone, from which winds aloft move poleward. Near 30°N and 30°S, the air in the upper atmosphere becomes so dense that it sinks back toward the Earth's surface (**Figure 6–3a**). Much of this descending air returns to the equator, becoming warmer and picking up moisture on its way. As it travels equatorward in both hemispheres, the air undergoes deflection because of the Coriolis effect. Coriolis deflection is to the right in the Northern Hemisphere. This causes the surface winds to blow out of the northeast rather than out of the north, creating the familiar northeast **trade winds**. Conversely, Coriolis deflection is to the left in the Southern Hemisphere, causing the surface winds to blow out of the southeast rather than out of the south, which produces the southeast trade winds. The trade winds converge on the belt of equatorial low pressure, known as the *doldrums*, where they rise and begin the cycle once again (see Figure 6–3a).

This wind-flow pattern, which connects surface winds and upper-level winds with ascending air at the equator and descending air at the horse latitudes of 30°N and 30°S, describes a pair of circulation cells. They are called **Hadley cells** (see Figure 6–3a), named after George Hadley (1685–1768), the British meteorologist who first described them.

FIGURE 6-3

Global wind circulation. (a) Unequal heating of the Earth's surface and the Coriolis deflection cause a zonal wind system to develop, arranged in three circulation cells—Hadley, Ferrel, and polar cells. (b) The plotting of prevailing winds on an air-pressure map of the world reveals that winds flow from high-pressure zones to low-pressure zones at an angle to the regional pressure gradients.

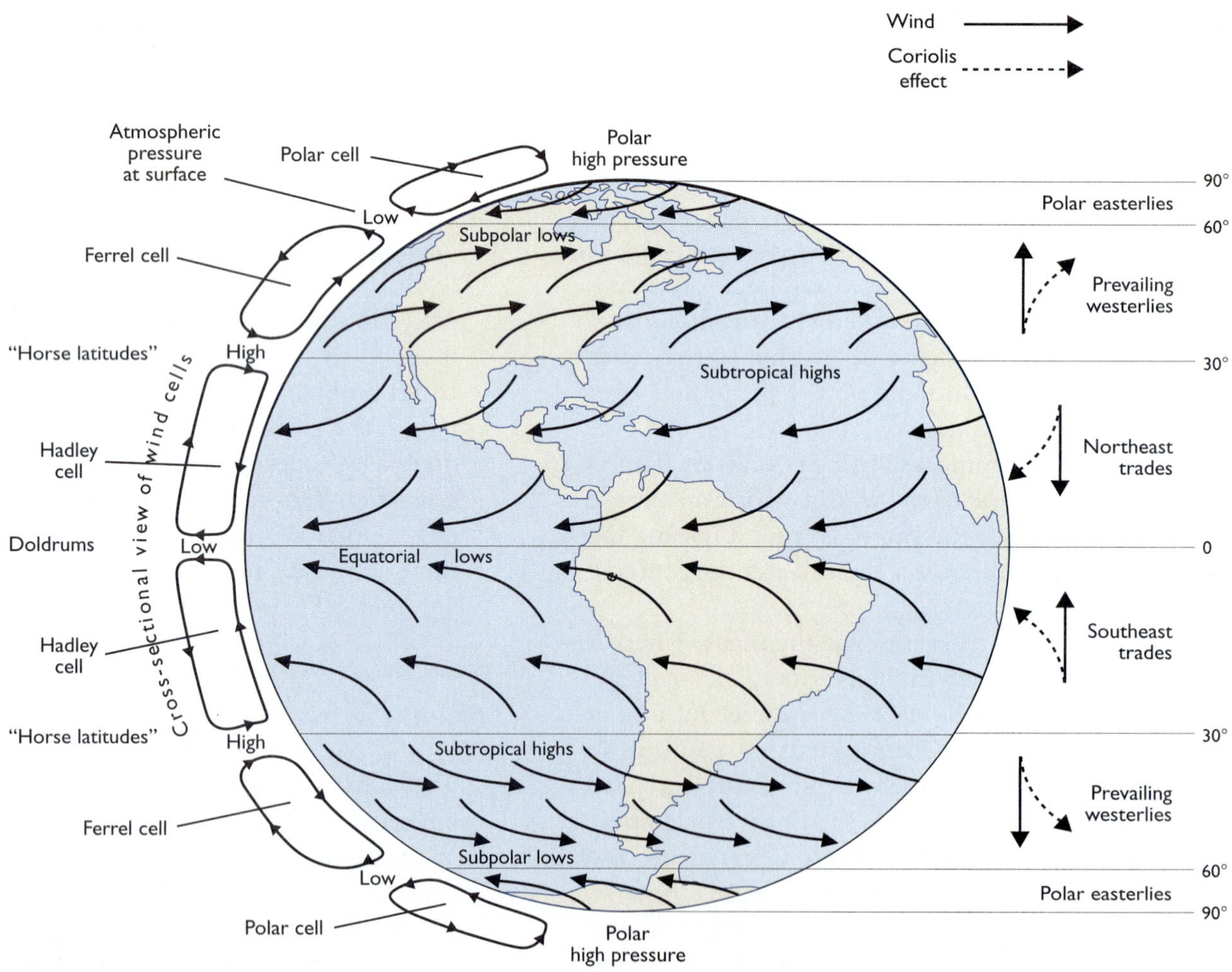

(a) GLOBAL WIND PATTERN

Other wind cells, called **Ferrel cells** in honor of the American meteorologist William Ferrel (1817–1891), dominate air circulation in the midlatitudes of both hemispheres (see Figure 6–3a). They display complex flow patterns, but we can generalize and simplify them. At 30°N and 30°S latitudes, some of the descending air travels poleward. These surface winds are deflected by the Coriolis effect, and become the westerlies of both hemispheres. They travel poleward, until at about 50°N and 50°S latitudes they encounter cold, dense air streaming from the polar regions. This zone of air mass convergence is called the *polar front*. It is a zone of convergence that forces air aloft, where it diverges, some traveling equatorward and some poleward. The former winds eventually link up with the descending air currents at 30°N and 30°S latitudes, which completes the circulation of the Ferrel cells. The upper air streaming poleward in both latitudes cools, becomes dense, and sinks back to the Earth's surface at the poles. There the cold, dense air flows equatorward, and the Coriolis deflection creates the polar easterlies of both hemispheres. This high-latitude circulation cell is known as a **polar cell** (see Figure 6–3a).

Spend some time studying Figure 6–3a, which summarizes all of this information. Note that the global winds flow in a regular pattern that varies with latitude. The circulation of winds in both hemispheres is arranged in three cells—Hadley, Ferrel, and polar cells. Each of these girdles the Earth, and each has a three-dimensional shape that resembles a large doughnut. **Figure 6–3b** is a map of the global winds, showing their relationship to belts of high and low pressure. Note that the winds flow down the pressure gradients, away from zones of high pressure toward zones of low pressure. Also, note the deflection of the winds due to the Coriolis effect. These latitudinal wind belts, which shift north and south, greatly influence the large-scale circulation of surface water in the ocean, the topic of the following section.

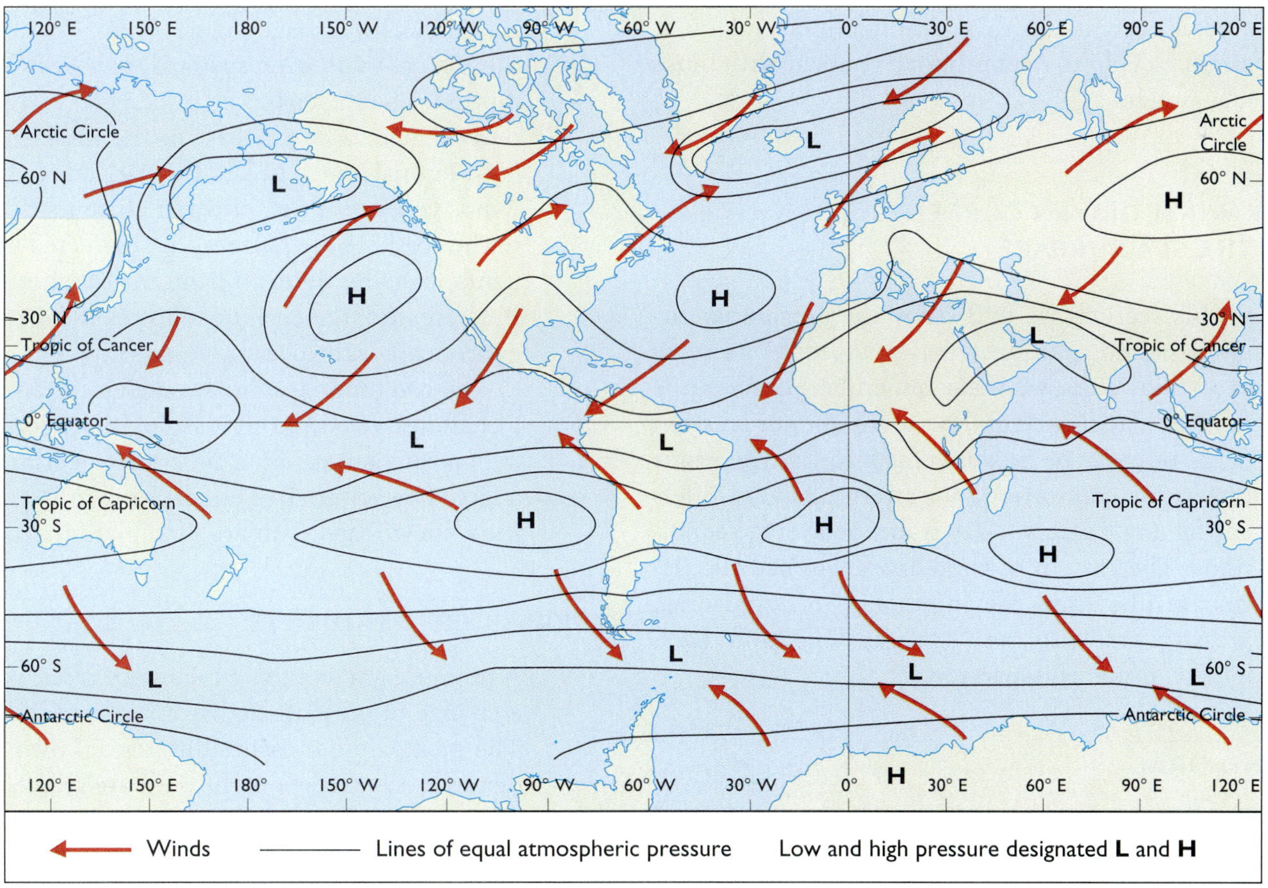

(b) AIR PRESSURE AND PREVAILING WINDS

6-2

Surface Ocean Currents

Ocean currents are divided into two types of flow, according to the factors that power them. Most surface currents are driven by the wind. Subsurface currents are density-driven. This means that dense water under the effect of gravity sinks and thus displaces less dense water. Although subsurface water is not affected by winds, it is influenced strongly by climate over the long term, because climate controls the salinity and temperature of water masses and hence their density.

The wind-driven surface currents affect only about 10 percent of the ocean's volume, yet most of this chapter is devoted to them, because people are better acquainted with the sea's surface than with its depths. Sailors propel their boats through their currents. Much less is known about water movements in the oceans' depths. The boxed feature, "Current-Measuring Techniques," describes the instruments that oceanographers use to measure surface currents. At this point in our discussion, it might be helpful to familiarize yourself with those instruments.

THE WIND-DRIVEN CURRENTS OF THE SEA SURFACE

In this section, we will study the prevailing currents that flow, *on the average,* "steadily" for years at a time. What we must remember, however, is that the actual current flow at any surface location of the ocean may, because of day-to-day variations in weather and water-flow patterns, be very different on any particular day from the average conditions. The pattern of wind-driven surface circulation results from the interaction of wind drag, pressure gradients, and Coriolis deflection. Each will be discussed separately.

WIND DRAG

Wind is moving air. As air molecules are dragged across the sea surface in a wind, they collide with water molecules at the ocean's surface. The energy transfer by frictional drag, if prolonged, raises waves and generates currents. The fact that still water is set in motion by wind implies that momentum associated with the moving air molecules is transmitted to the water molecules, setting them in motion. Careful experiments and field measurements show that the actual speed of the resulting water current is a tiny fraction of the wind speed, because the transfer of energy from the air to water is an inefficient process. You can easily estimate the speed of a current. It will be roughly 3 to 4 percent of the speed of the generating wind. This means that a wind blowing at 50 kilometers per hour (~31 mph) will produce a water current flowing at a speed of about 1.5 to 2.0 kilometers per hour (~0.9 to 1.2 mph).

What does all of this imply about the ocean's currents? Well, surface winds blow in a regular pattern (Figure 6–3a), in response to (1) differential heating of air across the Earth's surface and (2) the Coriolis deflection. The net effect of these interactions is **zonal wind flow** (the movement of air parallel or near-parallel to lines of latitude). This produces the trade winds of the subtropics with their strong easterly component and the westerlies of the midlatitudes (see Figure 6–3b). Wind drag by these large-scale wind systems sets ocean water in motion. The westerlies produce a belt of water currents that flow to the east in the midlatitudes of both hemispheres. In the low latitudes, the trade winds generate a pair of water currents that move to the west. These currents are deflected by continents, causing them to bend into each other and thus create large current loops called **circulation gyres** in all of the oceans (**Figure 6–4**).

You can roughly simulate such a current system by having two people blow strongly and persistently in opposite, but parallel, directions on either side of a large bowl of soup. The resulting flow pattern of the soup will approximate a circulation gyre.

PRESSURE GRADIENTS

A pressure gradient, as explained earlier, is merely a change of pressure across a horizontal distance. The greater the pressure differential over a given distance, the steeper is the pressure gradient. Near the sea surface, pressure gradients arise as a consequence of horizontal variations in the height of the water surface. Water that is piled up in a mound

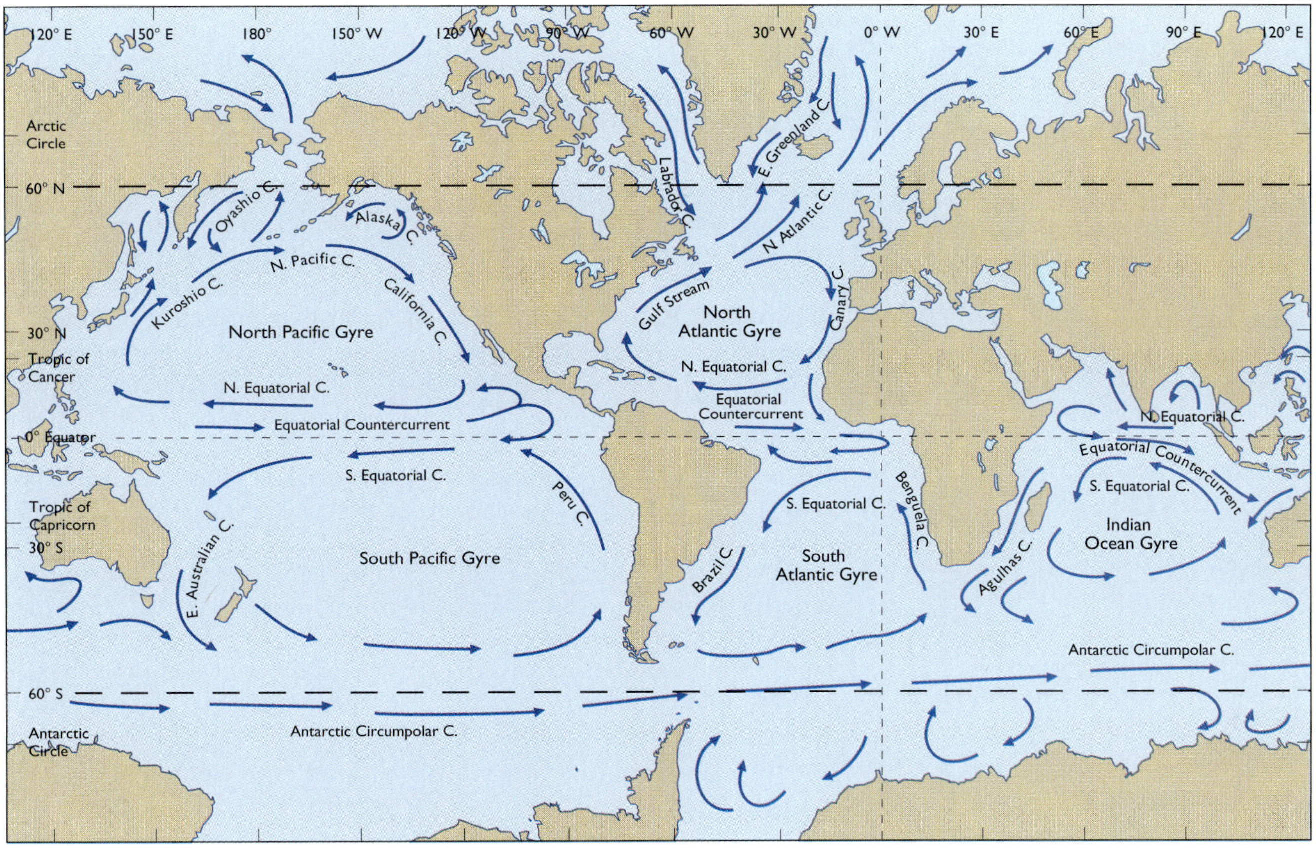

FIGURE **6-4**

Surface ocean currents. Global wind-driven ocean circulation consists of gyres that rotate clockwise in the Northern Hemisphere and counterclockwise in the Southern Hemisphere. Compare the flow direction of the water currents with the zonal wind patterns shown in Figure 6-3. In the above map, C. stands for current.

creates a zone of high pressure because of an increase in the height of the water column ($P = \rho gh$, where h is water height). Water responds by flowing down the pressure gradient. The steeper the pressure gradient, the faster is the flow of water, in the same way that a ball will roll down a steep slope faster than it will down a gentle slope.

What isn't obvious, though, is how can water be piled into a mound? With the exception of waves, water surfaces are always flat, aren't they? This is an intriguing question. After all, common experience tells us that water poured from a glass quickly flattens out to a level surface under the force of gravity. However, if we imagine pouring water from a glass with an infinite capacity , then the water streaming out of this hypothetical glass will sustain a mound of water indefinitely beneath the glass. Water spreads out from the mound, but it is replaced immediately by water pouring out of the glass, sustaining the mound and, hence, the pressure gradient. This mound of water will cause the water to continually flow outward in a radial pattern in response to a permanent horizontal pressure gradient. Later in this section, you will discover that mounds of water in the real ocean are created by converging currents.

Most people imagine that the sea surface when undisturbed by waves is flat. It is not! When examined carefully, it reveals a definite topography. It is literally warped into broad mounds and depressions, so that if you could walk across its surface, you would be moving up "hills" and down "valleys" across its uneven watery surface. True, these are not towering "hills" of water: the difference in elevation between the top of the water "hill" and the bottom of a water "valley" is

Current-Measuring Techniques

There are two principal ways to measure water currents directly. One, the Eulerian method, named after the Swiss mathematician Leonhard Euler (1707–1783), measures the current with a meter fixed to the ocean floor. The other, the Lagrangian method, named after the Italian mathematician Joseph Lagrange (1736–1813), uses a neutrally buoyant float that drifts with the water. Both types of measurements are discussed here.

EULERIAN METHOD

Eulerian current meters are mounted on buoy systems that are attached to cables anchored to the sea bottom. After being deployed from the ship, the current-meter system is left in place for a predetermined amount of time—days, weeks, months—that depends on the research objectives. Several current meters may be attached to the same cable (**Figure B6–1a**). The current meters become oriented into the water current by means of a vane, in the same way that moving air orients a wind vane. Current speed is measured by an impeller, essentially a propeller that is rotated by the force of the current; the faster the current, the faster is the rotational rate of the impeller. These measurements of current direction and speed are recorded directly into computer-chip memory. The current meter is later retrieved by the use of a sound signal, which activates an acoustic link that disengages the cable and the instrument package from the anchor. The instrument system with its valuable data set then floats to the sea surface, where it is located acoustically and retrieved.

A new but expensive meter—the Doppler Acoustic Current Meter—has been developed to measure currents at several depths. The meter can either be mounted on the keel of a research vessel or deployed on the sea bottom (**Figure B6–1b**). It relies on

FIGURE **B6-1**

Eulerian current meters. (a) An array of current meters is secured to a cable anchored to the sea bottom. (b) Narrow beam pulses of sound can be used to measure the direction and speed of currents at different depths using the Doppler shift of tiny particles suspended in the water.

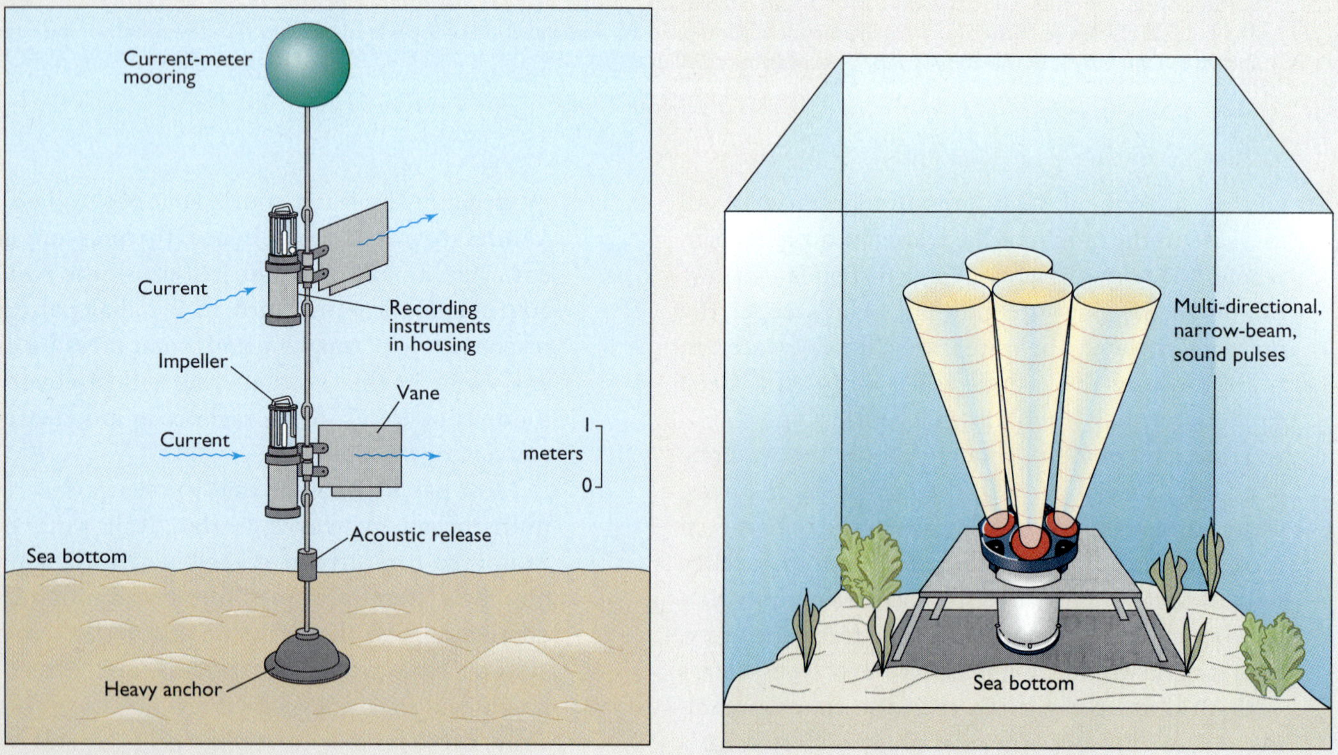

(a) TWO IMPELLER METERS

(b) DOPPLER ACOUSTIC CURRENT METER

the Doppler shift, which most people recognize as the change in the frequency or pitch of the sound that a train makes as it rushes by an observer. An approaching train has an increasing pitch, a departing train a decreasing pitch. Basically, narrow sound pulses are emitted by the current meter in different directions; these pulses are reflected from tiny particles suspended in the water that lie within about 100 meters (~328 feet) of the current meter. Particles drifting away or toward the meter cause a characteristic Doppler shift in the reflected echo that allows the direction and speed of a current to be inferred.

LAGRANGIAN METHOD

This technique involves the release of drifters, drogues, and floats that are set loose at the surface or at predetermined depths and tracked acoustically (**Figure B6–2**). They float with the moving water, allowing scientists to determine the direction and speed of the current. When plotted (**Figure B6–3**), the flow paths of the water describe intricate patterns; many of them are even looped, creating circles, ovals, and figure eights! Because they resemble strands of cooked pasta, they are referred to as *spaghetti diagrams*.

SATELLITE OCEANOGRAPHY

Satellite oceanography—remote sensing of the sea by satellite-mounted instrument packages—is providing valuable data that supplement shipborne surveys. The imagery from space covers a large expanse of ocean and allows scientists, as never before, to get measurements over a large area at an instant in time. Furthermore, satellites can make several passes over the same area in the course of a day and document in a systematic way time changes as well.

Today satellites are equipped with a variety of sensors, including radar emitters and detectors, laser reflectors, altimeters (a device to measure altitude), and cameras that detect visible light, infrared (heat), and microwave radiation. When these data are processed and interpreted, they reveal the temperature of the sea surface, sea-ice concentrations (**Figure B6–4**), water turbidity, surface wind and wave conditions, ocean currents and tides, and even the topography of the ocean floor (**Table B6–1**).

FIGURE **B6-2**

The Lagrangian method. The Lagrangian method for measuring **current** flow employs a neutrally buoyant float that is tracked acoustically by ship or satellite as the device drifts with the water current.

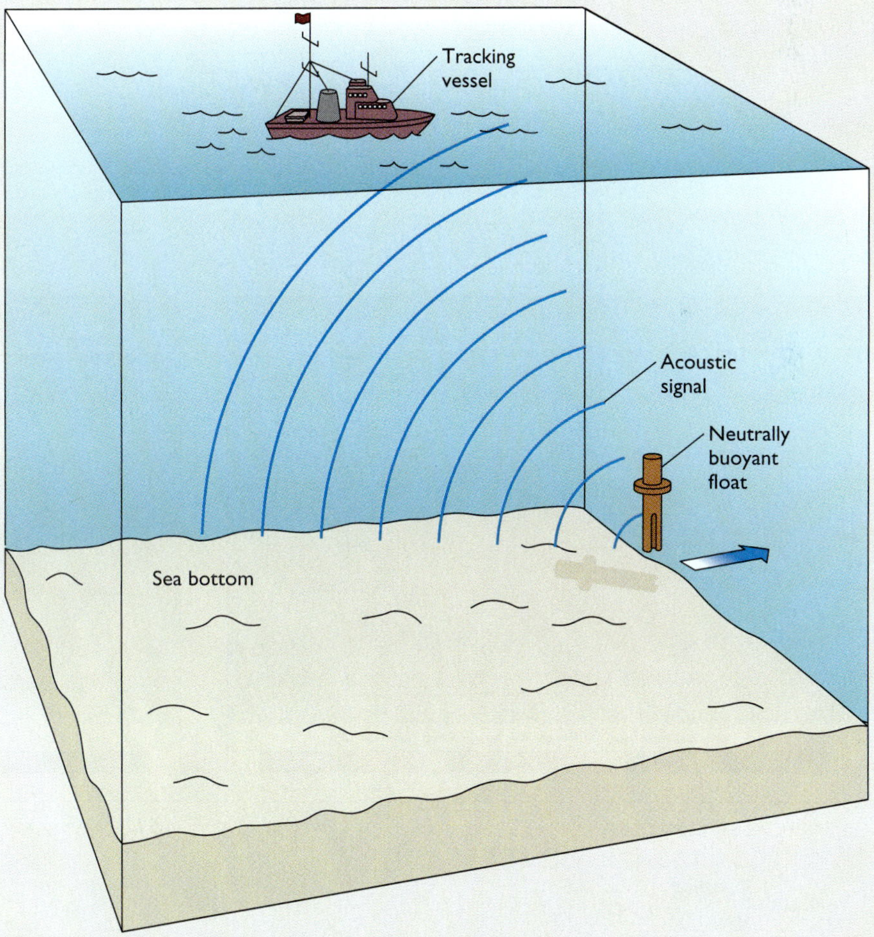

FIGURE **B6-3**

A spaghetti diagram. The plot of the drift paths of floats reveals a complicated subsurface-current pattern that is aptly called a spaghetti diagram.

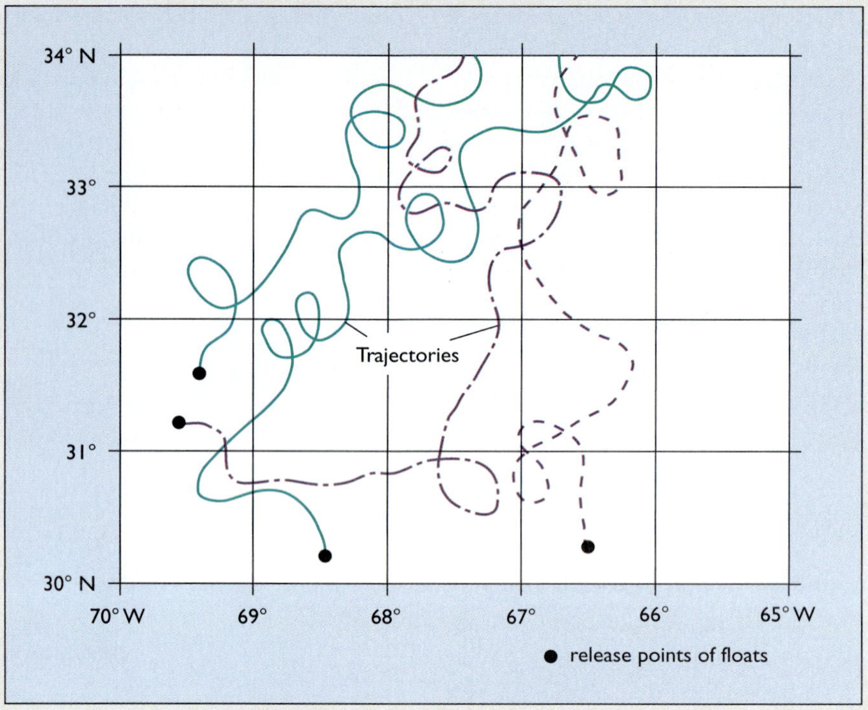

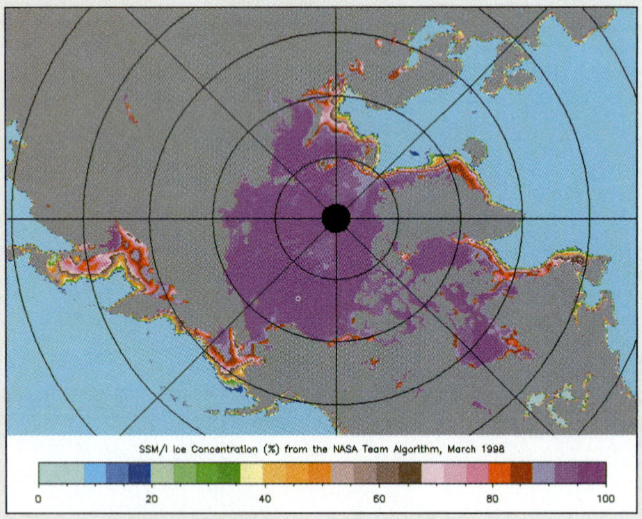

(a) MARCH SEA-ICE COVER

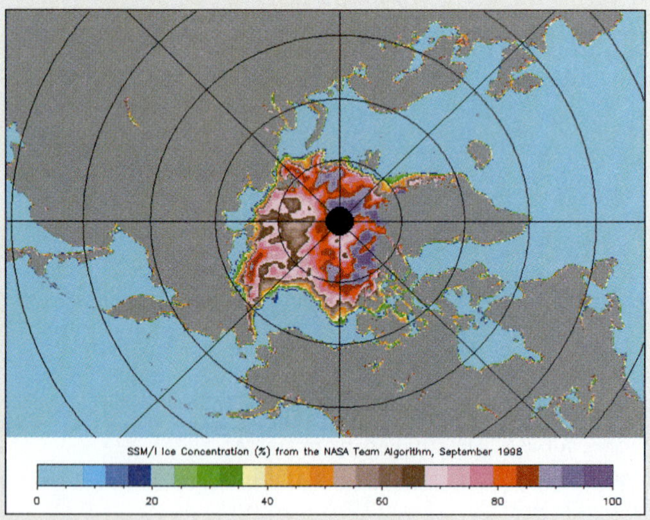

(b) AUGUST SEA-ICE COVER

FIGURE **B6-4**

Arctic sea ice. (a) The concentration of sea ice in the Arctic Ocean during March. (b) The concentration of sea ice in the Arctic Ocean during September.

TABLE **B6-1** Satellite sensors.

Sensor Type	Type of Measurement	Oceanographic Application
Visible and near infrared radiometers	Backscattered solar radiation from surface layer	Surface-water turbidity; phytoplankton concentration; sea-surface roughness
Thermal infrared radiometers	Thermal emission from sea surface; some solar reflection	Sea-surface temperature; surface heat flux
Microwave	Sea-surface microwave emission and reflected solar emission	Sea-surface temperature; surface heat flux; salinity; sea state; wind and wave conditions
Altimeter	Return time of pulse; return shape of pulse	Ocean currents and tides; significant wave height
Side-looking radar	Strength of return pulse from different directions	Surface wind speed and direction; directional wave spectra
Nadir (vertical) viewing radar	Strength of return pulse from directly below	Sea state; surface wind speed
High-resolution imaging radar	Strength of return pulse from small area with Doppler information (shift of wave frequency with distance)	Swell patterns; internal wave patterns; ocean-floor topography

Source: Adapted from I. S. Robinson, *Satellite Oceanography* (New York: John Wiley, 1985).

about 1 meter (~3.3 feet) or less. But this subtle sea-surface topography has profound effects on surface circulation.

The North Atlantic Ocean (**Figure 6–5**), for example, is dominated by a broad circulation gyre. The center of the gyre is significantly higher than its periphery, such that a boat departing from Florida and heading for Europe must first sail uphill! This rolling water topography is produced in part by the convergence and divergence of currents. Converging currents cause water to pile up, creating a water "hill"; diverging currents cause water to move apart, creating a water "valley." Because pressure varies directly with the height of the water (remember that $P = \rho gh$, where h is height), pressure gradients exist, and they cause water to flow from the "hills" of water (high-pressure zones) to the "valleys" (low-pressure zones). Actually, it doesn't work quite that simply, because we need to consider the Coriolis deflection. Let's do that now.

CORIOLIS DEFLECTION

Because of the Earth's rotation, water currents are deflected by the Coriolis effect, bending to the right in the Northern Hemisphere and to the left in the Southern Hemisphere. Consequently, water currents do not flow directly down pressure gradients, but at some angle to them. This is similar to what happens to global winds. They do not take a direct flow path from the high to the low pressure zones. Rather, they flow down at an angle to the pressure gradient (see Figure 6–3b). Their deflection, like that of water currents, depends on their speed and latitude, which control directly the magnitude of the Coriolis effect.

With these notions in mind—wind drag, pressure gradient, and Coriolis deflection—let's see how they help explain wind-driven ocean circulation. A way to proceed is first to examine a few types of surface flows. One in particular, *Ekman transport,* is critical for developing a theory of wind-driven circulation in the oceans.

FIGURE **6-5**

Currents and sea-surface topography of the North Atlantic. (a) In the North Atlantic Ocean, the westerlies drive the North Atlantic Current (NAC) and the trade winds power the North Equatorial Current (NEC). The Sargasso Sea forms the asymmetric center of the circulation gyre, which consists of a narrow, swift-flowing Gulf Stream (GS) and a broad slow-flowing Canary Current (CC). The gyre is not self-contained. Water leaves the system in the Labrador Sea and off equatorial Africa and it enters the system from the Labrador Sea and off equatorial South America. [Adapted from D. E. Ingmanson and W. J. Wallace, *Oceanography: An Introduction* (Belmont, Calif.: Wadsworth, 1985).] (b) Far from being a smooth plane, the sea surface in the North Atlantic Ocean has broad topographic highs and lows and variable topographic gradients. A steep slope faces North America and a gentle slope faces Africa. Sea level in the Sargasso Sea is about 1 meter higher at its center than along its edge. [Adapted from V. A. Burkov, *General Circulation of the World Ocean* (Leningrad: Hydrometeorizdat, 1980).]

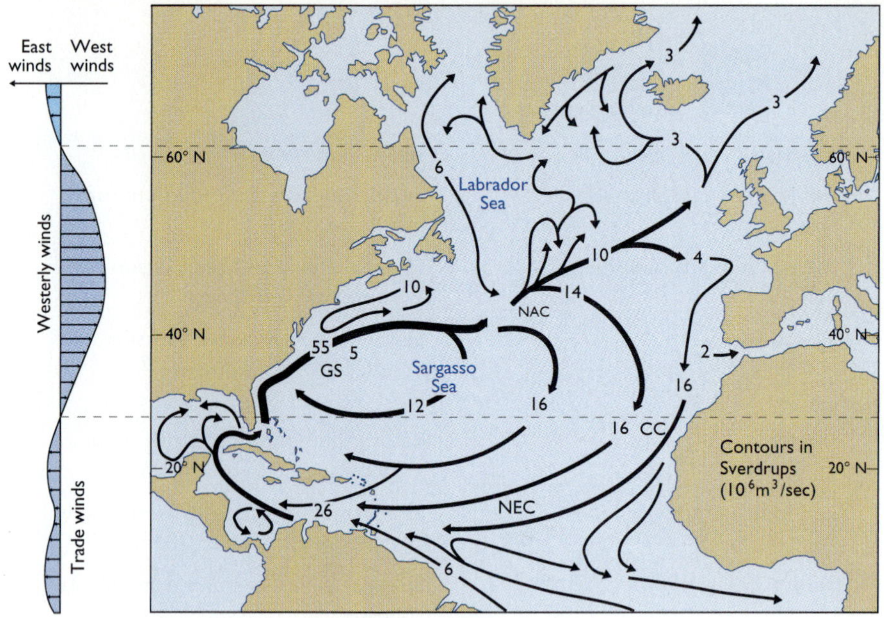

(a) FLOW RATES IN THE NORTH ATLANTIC OCEAN

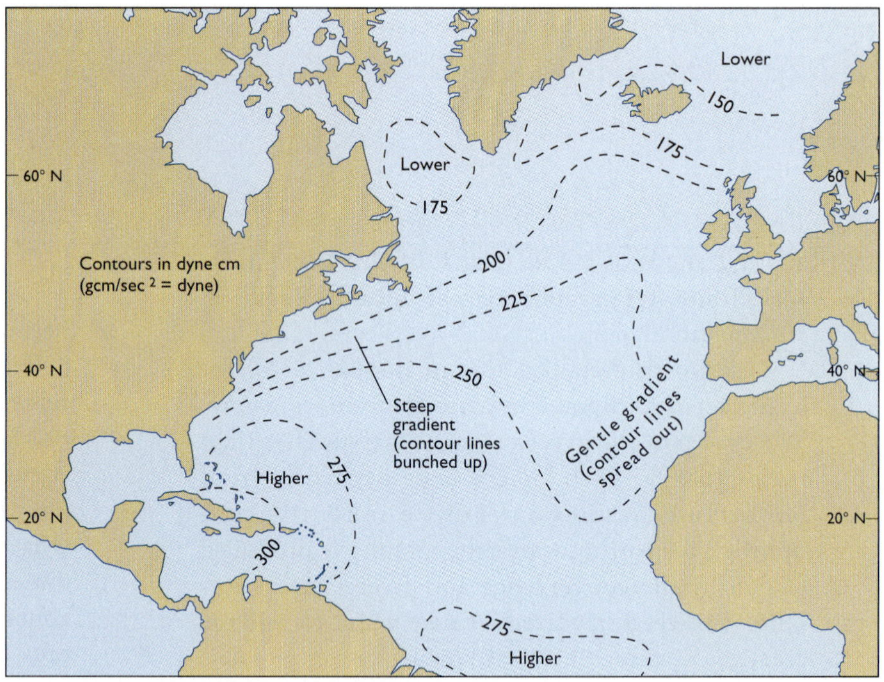

(b) DYNAMIC TOPOGRAPHY OF THE SURFACE OF THE NORTH ATLANTIC

TYPES OF SURFACE FLOWS

The collision of air molecules in wind with water molecules at the sea surface makes the latter move, generating a water current. Once this surface film of water molecules is set in motion, they exert a frictional drag on the water molecules immediately beneath them, getting these to move as well. If the wind blows persistently for a long period of time, motion is transferred downward into the water column. As this wind-driven current deepens, its speed diminishes, because of distance from the driving force, the wind. Besides its speed, the direction of the water current's flow changes with depth. This deviation results from the Coriolis deflection. In the Northern Hemisphere, the surface-water layer flows to the right of the wind direction (**Figure 6–6a**). When this topmost water layer sets the underlying layer of water in motion through frictional drag, the deeper layer moves to the right of the flow direction of the layer above it, again because of the Coriolis effect. As the current deepens with time, each successively deeper water layer is deflected to the right of the layer immediately above it. The result of this process is a spiraling current (see Figure 6–6a). The speed of this current decreases with distance below the sea surface.

This spiraling flow pattern is called the **Ekman spiral** in honor of the Scandinavian physicist, V. Walfrid Ekman, who first explained the phenomenon. In 1902 Ekman showed why water movement at the surface must deviate from the direction of the generating winds (**Figure 6–6b**). Furthermore, he accounted for the fact that each successively deeper layer moves at a lower speed and in a slightly different direction from the layer above. For example, near the base of the current, the flow rate is about 4 percent the speed of the surface current, and the

FIGURE 6-6

Ekman transport. (a) In the Northern Hemisphere, the Coriolis deflection causes a surface current to flow to the right of the generating wind. As the current deepens, each water layer is deflected slightly to the right of the layer above it, producing a spiraling current called the Ekman spiral. The speed of the current decreases with depth. Calculations indicate that the bulk of the water (the net transport) moves at an angle of 90 degrees to the right of the generating wind. (b) This map view summarizes the overall effect of Ekman transport. A persistent wind sets water adrift and generates the Ekman spiral. This results in a surface current that flows to the right of the generating wind and the net transport of water (Ekman transport) to the right and perpendicular to the direction of wind flow. In the Southern Hemisphere, Ekman transport will be to the left of the generating wind.

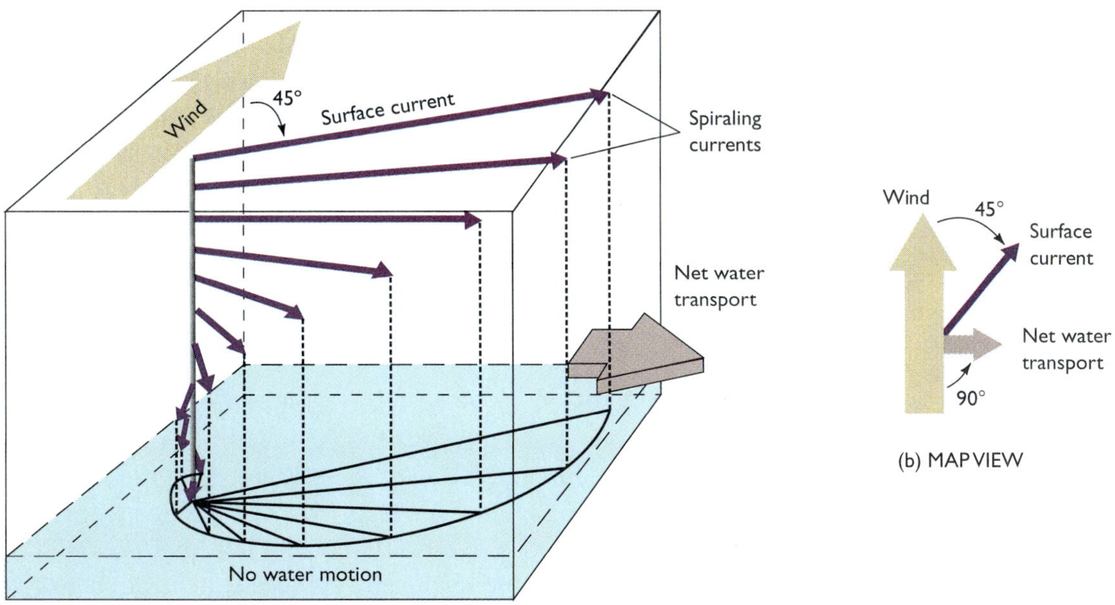

(a) EKMAN SPIRAL IN THE NORTHERN HEMISPHERE

(b) MAP VIEW

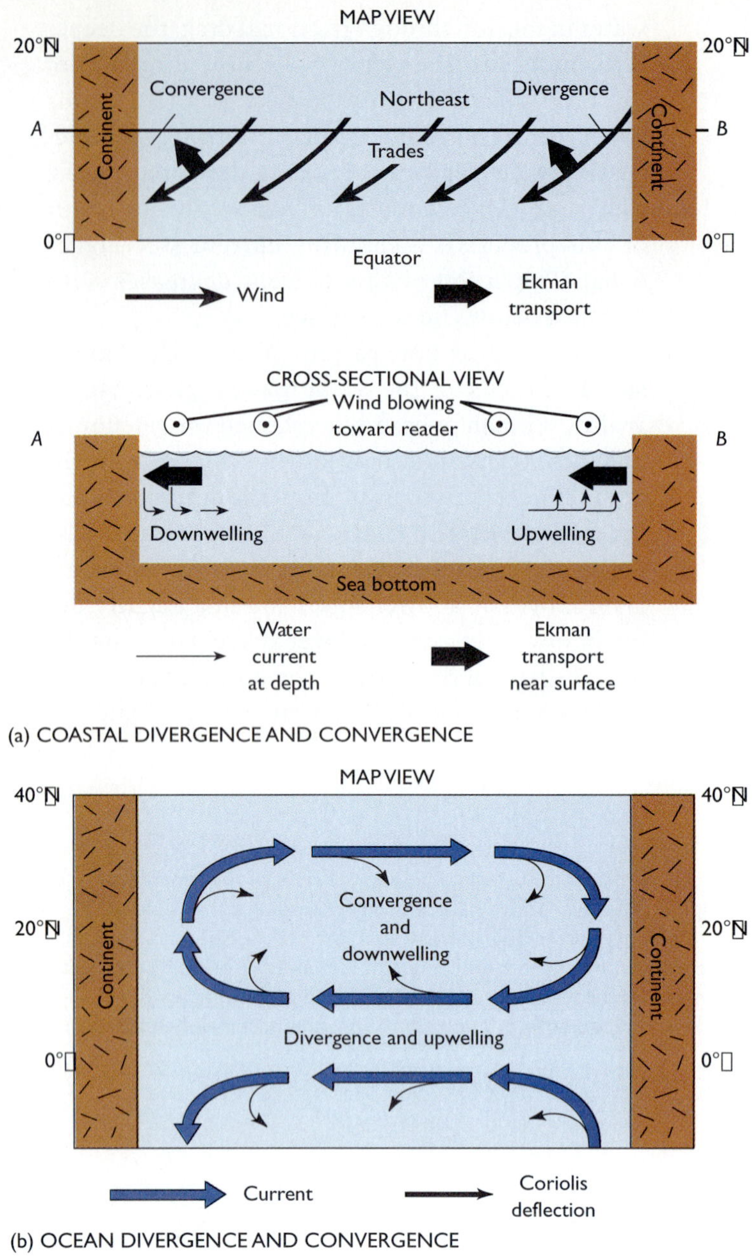

FIGURE 6-7

Convergence and divergence of water currents in the Northern Hemisphere. (a) Winds along continents produce Ekman transport, which tends to move water away from the land on the eastern side and toward the land on the western side of oceans. This results in upwelling (rising) and downwelling (sinking) of water on the eastern and western sides, respectively. (b) Coriolis deflection causes water to converge in the center of circulation gyres, which causes downwelling. Along the equator, upwelling results from current divergence, as water is deflected to the right and left in the Northern and Southern hemispheres, respectively, by Coriolis deflection.

flow direction is opposite to the surface movement because of the spiraling effect. Under the influence of a strong, persistent wind, the Ekman spiral may extend to a depth of between 100 and 200 meters (~330 and 660 feet), below which wind-induced effects are negligible. Ekman calculated that the net transport over the wind-driven spiral (appropriately called **Ekman transport**) is 90 degrees to the right of the wind in the Northern Hemisphere (see Figure 6–6b) and to the left of the wind in the Southern Hemisphere. Net transport represents the average of all the directions and speeds of the current spiral. In other words, although the current speed diminishes and the flow direction spirals with depth, the bulk of the water moves at a 90-degree angle to the generating wind.

Ekman transport has an important bearing on other types of surface water currents. For example, if the wind blows parallel to a shore in the proper direction (this depends on the hemisphere), the resultant Ekman transport moves near-surface water offshore. Water must then rise from below to compensate for the seaward surface flow, a process of vertical circulation called **upwelling** (**Figure 6–7a**). **Downwelling**, the sinking of surface water, is caused along a shoreline by winds that blow parallel to the coast; Ekman transport drives water against the land and causes surface water to sink (see Figure 6–7a).

Upwelling and downwelling are not restricted to the edges of continents; these vertical currents can occur in the open ocean, far from the influence of land. At the latitudes of about 30°N and 30°S, for example, the surface currents of the circulation gyres generated by the westerlies and trade winds converge due to the Coriolis deflection. This convergence produces two mounds of water that induce sinking, that is, downwelling (**Figure 6–7b**). In comparison, the equator is a zone of upwelling water. Here the southeast and northeast trade winds generate two westerly flowing surface currents, the North and South Equatorial Currents (see Figure 6–4). The Coriolis effect causes them to be deflected to the right (north) in the Northern Hemisphere and to the left (south) in the Southern Hemisphere (see Figure 6–7b). This divergent flow of surface water at the equator lowers the sea surface and promotes the upwelling of water from below.

In addition to the Ekman spiral, persistent winds blowing across the sea surface can induce small-scale vertical water motions as well. When the wind blows more strongly than about 3.5 meters per second (~8 mph), water flows parallel to the wind (**Figure 6–8a**). The current motion is complex and resembles a corkscrew. This flow is called **Langmuir circulation**, after its discoverer, Irving Langmuir. Each convection cell is 10 to 50 meters (~33 to 165 feet) broad, 5 to 6 meters (~17 to 20 feet) deep, and hundreds of meters (~hundreds of yards) to several kilometers (~1 mile) long. Because of rotation in opposite directions, boundaries between adjoining Langmuir cells alternate between convergent and divergent flow. Floating material, such as bubbles, oil slicks, flotsam (floating debris), and seaweed, aggregates at the zones of flow convergence, imparting a streaked appearance to the water surface (**Figure 6–8b**). Langmuir circulation, on the one hand, represents a short-term response of water to wind drag, forming several minutes to several hours after the appearance of the generating wind. Ekman flow, on the other hand, is a response of water to a wind blowing for much more than several hours and even days.

Well, now we have all the pieces that are necessary to account accurately for water flow around the large circulation gyres of the ocean. In order for you to understand what follows, it is imperative that you grasp firmly the following notions: pressure gradient, Coriolis deflection, Ekman transport, and downwelling. Be certain that you do before continuing.

A MODEL OF GEOSTROPHIC FLOW

Let's begin the analysis by considering the effect of the westerlies and the trade winds. We start with the simplest state possible, a large body of standing water (no currents) with a flat surface. As we know well, wind exerts drag on the sea surface and causes water to drift. But once set in motion, the water will veer to the right of the wind because of the Coriolis effect. In our simple model, this will produce a northerly Ekman transport of water from the drift generated by the trade winds and a southerly Ekman transport of water from the drift raised by the westerlies (**Figure 6–9a**). The result is the convergence of water in the area located

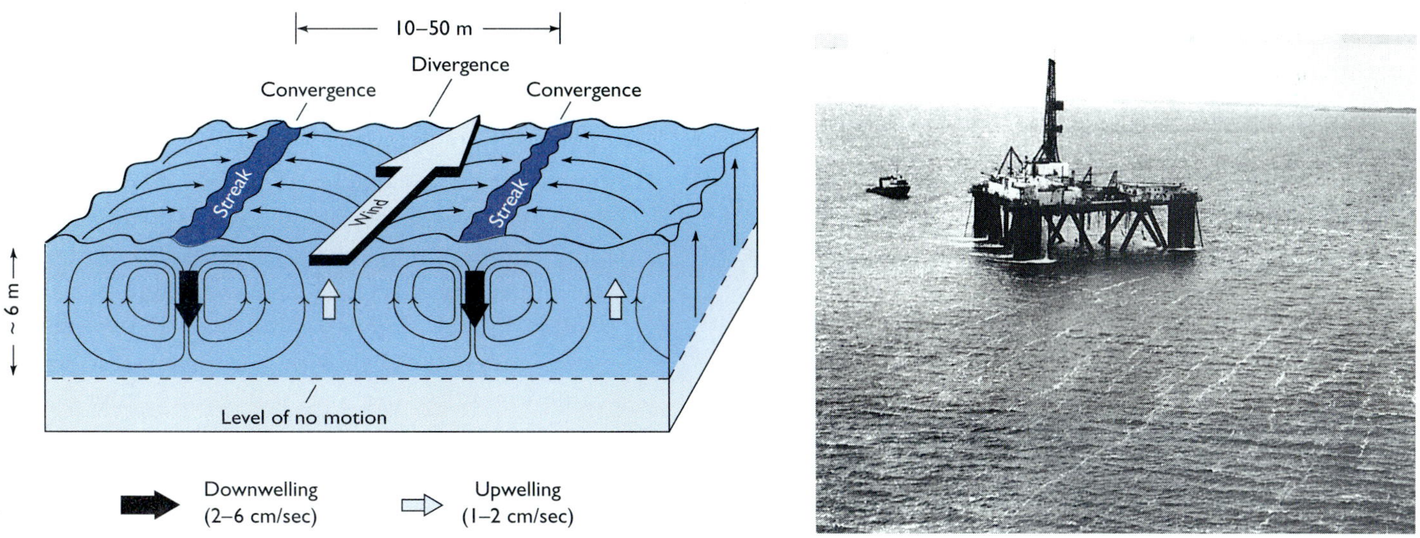

(a) LANGMUIR CELLS

(b) LANGMUIR CIRCULATION

FIGURE 6-8

Langmuir cells. (a) Langmuir circulation results from the complex interaction of a moderate, but persistent, wind with the sea surface. This produces long, narrow circulation cells in the topmost 6 meters of the water column. Adjacent cells rotate in the opposite direction, producing alternating zones of divergent and convergent currents. Convergent currents are identified by the presence of streaks on the sea surface. (b) This is a photograph of wind streaks created by the convergent motion of water associated with Langmuir.

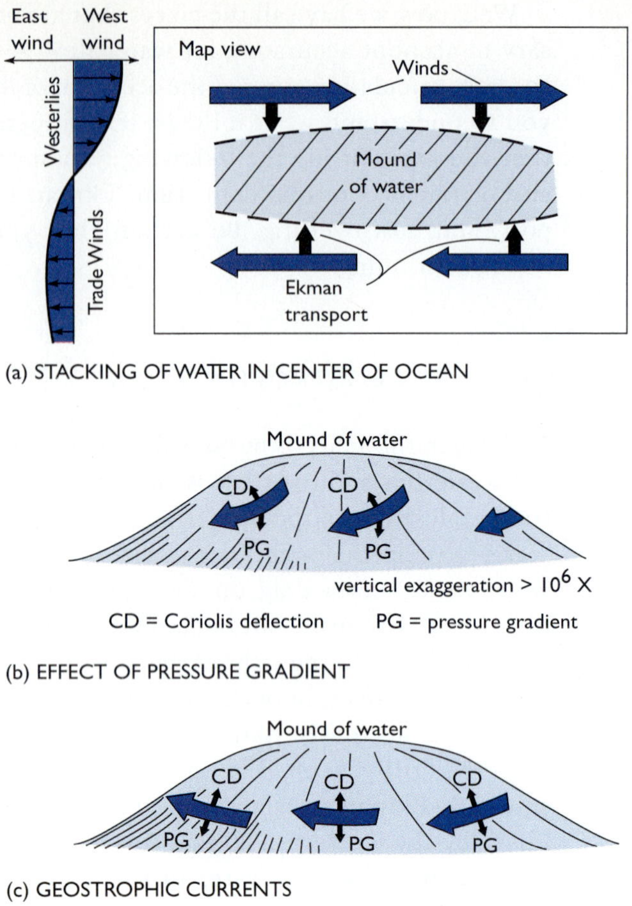

FIGURE **6-9**

Geostrophic currents. (a) In the Northern Hemisphere, the westerlies and trade winds induce Ekman transport, which causes water to flow toward the center of the ocean. The converging flow produces a mound of water—a topographic elevation of the sea surface. (b) This pile of water in the center of the ocean has slopes and, hence, pressure gradients (PG). As water begins to flow outward in response to the pressure gradient, Coriolis deflection (CD) causes the current to bend to the right. (Note that the vertical exaggeration is extreme. The mound is no higher than 1 meter but is thousands of kilometers long.) (c) Eventually, a stable flow pattern, known as a geostrophic current, establishes a balance between the pressure gradient and Coriolis deflection. This creates a circulation gyre with currents that flow clockwise around the mound of water. In the Southern Hemisphere, the geostrophic currents rotate counterclockwise because Coriolis deflection is to the left.

between the trade winds and the westerlies, and the slow buildup of a mound, a veritable "hill" of water. This water pile grows in volume and height until the pressure gradient causes water to flow downslope, away from the centerline of the mound. Because of Coriolis deflection, the water current that is flowing in response to the pressure gradient bends to the right (**Figure 6–9b**) until it has swung around and is moving uphill. At this point, the flow decelerates and the Coriolis deflection weakens, because the magnitude of the Coriolis effect depends directly on the speed of the flow. Eventually, the flow bends back downslope and accelerates in response to the pressure gradient. This renewed flow, in turn, forces the current to bend to the right once again, as the Coriolis deflection reintensifies because of the speed of the current. In theory, the intensity of this "uphill-downhill," wavy motion will become less and less with time until the pressure gradient, which wants to force the current downslope, and the Coriolis deflection, which wants to force the current upslope, are balanced. Then the system will be in a dynamic equilibrium, and the water current will be stable and constant, flowing across, rather than down, the water slope (**Figure 6–9c**). Oceanographers call the steady currents that result from a balance between a pressure gradient and Coriolis deflection **geostrophic currents**.

It is important that you grasp the fact that the ocean currents in a circulation gyre, which happen to flow parallel to the prevailing zonal winds, do so, not because of wind drag, but because of geostrophic flow—a dynamic balance between Coriolis deflection and pressure gradient. The resultant currents are indirectly driven by the zonal winds, as our simple model of geostrophic flow shows clearly (see Figure 6–9). The result is a current that seems to defy gravity, because it does not flow downslope, but alongslope. But we know that there is no defiance here, only a balance between two competing effects—the Coriolis deflection and the pressure gradient.

The landmasses are massive barriers, veritable dams of granite that interrupt the flow of the geostrophic currents. The easterly current flow in the midlatitudes is deflected equatorward by continents, and the westerly current flow at the equator is forced poleward by landmasses. The result is a closed circulation loop with a clockwise rotation in the oceans of the Northern Hemisphere, the North Pacific and North Atlantic gyres. A mirror image of this geostrophic system (the South Pacific, South Atlantic, and Indian Ocean gyres), including opposite rotation, develops in the Southern Hemisphere (see Figure 6–4), because Coriolis deflection there is to the left rather than to the right.

REFINEMENT OF THE GEOSTROPHIC-FLOW MODEL

If we examine the surface currents of the North Atlantic Ocean (see Figure 6–5a), we see that the flow pattern about the gyre is asymmetrical, with a narrow, deep, and swift current in the west, the Gulf Stream, and a broad, shallow, slow current to the east, the Canary Current. These properties of the western and eastern arms of the gyre are contrasted in **Table 6–1** and **Figure 6–10**. The powerful, swift flow associated with the western arm of all ocean gyres, irrespective of the hemisphere, is called **western-boundary intensification**. Also, note that the mound of water around which the geostrophic currents flow is not located centrally in the ocean, but is displaced to the west, and the western side of the mound of water is steeper than its eastern edge (see Figure 6–10). Our model, if it is to be accurate and useful, must account for both western-boundary intensification and the distorted shape of the mound of water.

It bears repeating that *a geostrophic current represents a steady flow that arises as a consequence of a dynamic balance between a pressure gradient and Coriolis deflection* (see Figure 6–9c). We know that the magnitude of the Coriolis deflection is directly proportional to the speed of the current. Hence, the faster the flow rate of the current, the more pronounced is the Coriolis effect, and the stronger the pressure gradient (i. e., the steeper the water slope) must be to counteract these effects and maintain a geostrophic balance with a water current flowing parallel to the water slope. Clearly then, western-boundary intensification requires that the tilt of the water slope be steeper on the western than on the eastern side of the gyre in order to achieve a dynamic balance, because Coriolis deflection is sharper on the western than the

TABLE **6-1**

Comparison of the eastern and western arms of current gyres.

Location of Current	Width (km)	Depth (km)	Speed (m/sec)	Transport (Sv)*
Eastern boundary	>1,000 km	<0.5 km	<0.3 m/sec (~10 km/day)	10–15 Sv
Western boundary	<100 km	1–2 km	>1.5 m/sec (~100 km/day)	>50 Sv

*1 Sv (Sverdrup) = 10^6 m^3/sec.

FIGURE **6-10**

Flow asymmetry around circulation gyres. In all the oceans, flow on the western side of circulation gyres is narrow, deep, and strong. This contrasts with the eastern side of the gyre, where currents are broad, shallow, and weak. Note that the volume transport of water is the same on both sides.

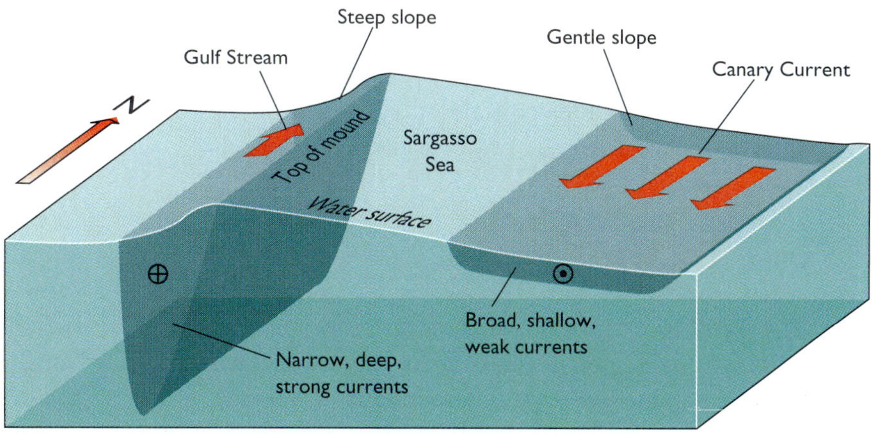

FIGURE 6-11

The Sargasso Sea. (a) The western North Atlantic Ocean is called the Sargasso Sea (see Figure 6-5a) because of the ever-present seaweed Sargassum that floats in its water. (b) Sargassum seaweed floating in the Sargasso Sea. (c) Isotherms (lines of equal water temperature) indicate that the Sargasso Sea consists of warm water that is separated from the cold slope water off North America by the northerly-flowing Gulf Stream. The upper profile shows the 1-meter-high mound of water and the location of the Gulf Stream on the steepened edge of the Sargasso Sea. [Adapted from C. O. Iselin, *Papers in Physical Oceanography and Meteorology* 4 (1936).]

(a) SARGASSUM SEAWEED

(b) THE SARGASSUM

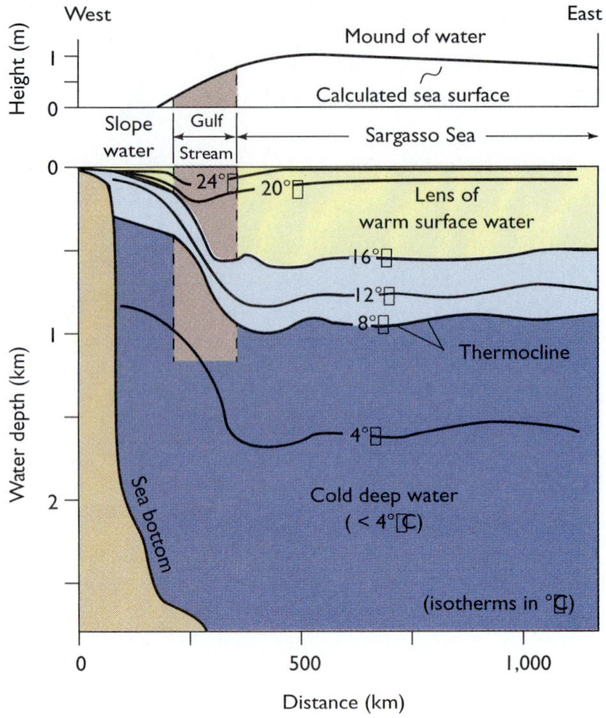

(c) TEMPERATURE SECTION ACROSS THE WESTERN SARGASSO SEA

eastern side of the gyre. This accounts nicely for the distorted shape of the mound of water, but it does not explain why western-boundary intensification exists in the first place. We must introduce a few additional ideas about current motion in order to understand the origin of western-boundary intensification.

Because the Coriolis effect increases with distance from the equator, the eastward-flowing ocean currents beneath the westerly winds in the Northern Hemisphere turn to the south much more strongly and quickly than the westward-flowing equatorial currents beneath the trade winds turn to the north. What happens as a consequence of this asymmetrical pattern of Coriolis deflection? Clearly, much more eastward-flowing water is deflected to the south into the interior of the current gyre than is deflected to the north into the gyre's center along the westward-flowing equatorial currents. This results in a very large volume of equatorial water drifting to the west and piling up against the western edges of ocean basins. This in turn necessitates accelerated geostrophic flow along the western side of the ocean gyre. This very same phenomenon—western boundary intensification—occurs along the geostrophic gyres of the Southern Hemisphere as well. Can you work out the reasoning for the South Atlantic gyre?

An accurate explanation of western-boundary intensification must include the effect of *vorticity*, which is the amount of circular rotation about a vertical axis ("whirl") that parcels of water undergo because of planetary spin and current shear (water layers with different motions that "slide" past each other). Unfortunately, the concept of vorticity is difficult to grasp without a background in physics and mathematics that is well beyond the level of an introductory textbook on oceanography. But you should at least know that vorticity controls in a very direct way western-boundary intensification.

The model of geostrophic flow just described agrees quite well with the large-scale properties of the actual ocean gyres. The classic example of such a circulation system is the **Sargasso Sea** of the North Atlantic, named after the brown seaweed, *Sargassum,* which floats over much of the water there (**Figures 6–11a** and **6–11b**). The Sargasso Sea consists of a large lens of warm water that is encircled by geostrophic currents rotating clockwise around it. The lens of buoyant, warm water is about 500 meters (~1,650 feet) thick and is separated from cold water below and to the sides by a sharp thermocline (**Figure 6–11c**). The swift-flowing Gulf Stream at the western edge of the Sargasso Sea is the result of western-boundary intensification.

There is another curious feature of the western arms of current gyres. Take, for example, the well-studied Gulf Stream of the North Atlantic Ocean. As you know, the Gulf Stream separates warm water in the center of the gyre from cold water that lies outside of the gyre (**Figure 6–12a**). Careful monitoring shows that the Gulf Stream's flow path is not always straight, but is often wavy or snakelike in appearance, analogous to the broad, curved meander loops of river channels that wind their way across floodplains. In some cases, the curvature of the meander loops in the current may become so accentuated that they pinch off from the main axis and form closed, self-contained loops. These are essentially large, swirling eddies called **rings** (**Figure 6–12b**). To the west of the Gulf Stream, the rings circulate clockwise and have a warm core of Sargasso Sea water; they are called **warm-core rings**. Rings to the east of the Gulf Stream, in contrast, spin counterclockwise and have cores filled with cold water derived from outside of the Sargasso Sea; they are called **cold-core rings**.

Considered in three dimensions, the rings are flat "wafers" of rotating water, 100 to 300 kilome-

FIGURE **6-12**

Gulf Stream meanders and rings. (a) This infrared satellite photograph reveals the wavy flow path of the Gulf Stream, which separates warm Sargasso Sea water from cold continental-slope water (see Figure 6–11c). (b) Often meander loops of the Gulf Stream become so coiled that meanders are cut off. This process generates clockwise-spinning warm-core rings and counterclockwise-spinning cold-core rings.

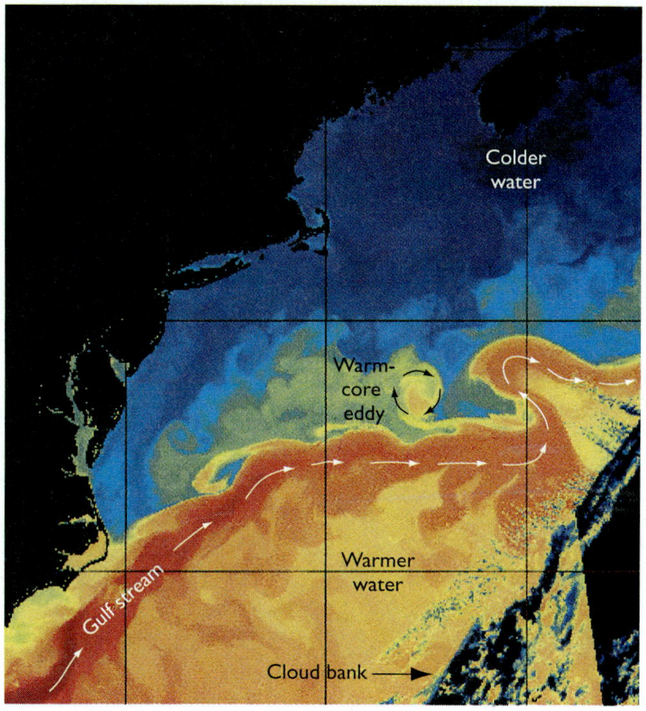

(a) "HEAT" PHOTO OF GULF STREAM

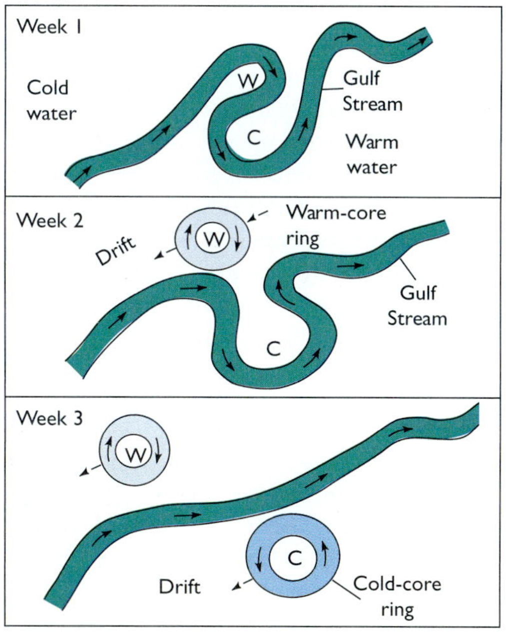

(b) FORMATION OF RINGS

Hurricanes and Typhoons

Hurricanes in the western Atlantic, which are called typhoons in the western Pacific, are one of nature's most powerful, regularly occurring phenomena. Although details of their formation are not understood completely, the general weather and ocean conditions that lead to the formation of hurricanes are clear. They form in tropical latitudes during the late summer and fall seasons. Hurricanes evolve from preexisting tropical cyclones, which are low pressure disturbances associated with wind speeds between 54.7 and 117.5 kph (39 and 73 mph) and with intense thunderstorm activity. For tropical cyclones to mature into full-fledged hurricanes with wind speeds greater than 119 kph (74 mph), the sea-surface temperature down to a water depth of about 45 m (~147.6 ft) must be warmer than 27° C (80° F) and the upper-level winds must be weak (otherwise wind shear will literally blow the developing storm system apart). If these conditions are met, chances for the tropical storm to intensify into a hurricane are good.

With time and under ideal conditions, a tropical cyclone matures by gathering energy from the warm ocean water. Basically, the air in contact with the ocean's surface gains heat and moisture and rises. As the water vapor in the rapidly ascending air condenses into storm clouds, heat is released, which energizes the spiraling weather system (**Figure B6-5**). Warm, moist air in contact with the ocean is continually drawn upward to replace the rising air, leading to an intense storm system that has strong wind convergence near the sea surface and wind divergence in the upper levels of the atmosphere. Sustaining a hurricane and its furious winds requires a regular infusion of heat and moisture from the ocean. This is why hurricanes decay, if they drift over cool water or make a landfall. However, weakened hurricanes are still quite dangerous and they can reintensify by drifting back over warm ocean water.

Once formed (**Figure B6-6**), the spiraling clouds of hurricanes and typhoons, which spin counterclockwise in the Northern Hemisphere and clockwise in the Southern Hemisphere in response to the Coriolis effect, can cover an area as wide as 485 km (~ 301 miles). The diameter of their central eye where winds are relatively light ranges from 30 to 65 km (~ 18.6 to 40.4 miles) and is surrounded by the eyewall, a 15-km (~ 9.3 mile) thick band of dense clouds with hurricane-force winds that extend upward into the sky some two to three kilometers. The complexities of the internal flow and the intricacies of the drift path of hurricanes make their speed and journey difficult to predict accurately. On the average, about 85 tropical storms develop each year across the globe, and about half of these intensify into hurricanes and typhoons.

In August 1992, Hurricane Andrew made a landfall in south Florida and a few days later in southern Louisiana. With sustained winds between 131 and 155 mph, property damage was extensive, estimated to be between 25 and 45 billion dollars. In Dade County, Florida, over 63,000 homes were com-

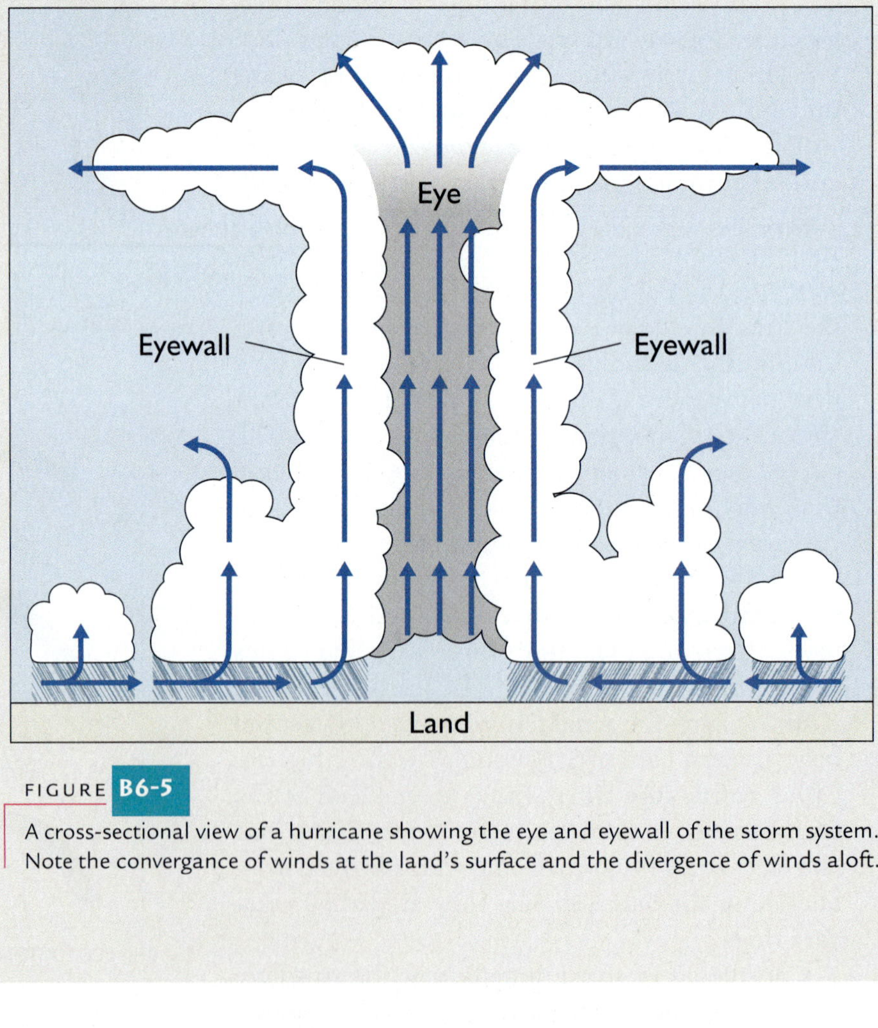

FIGURE **B6-5**

A cross-sectional view of a hurricane showing the eye and eyewall of the storm system. Note the convergance of winds at the land's surface and the divergence of winds aloft.

pletely destroyed and about 110,000 houses were seriously damaged. Coastal environments likewise were severely affected. For example, mangrove trees of south Florida were defoliated and their trunks and branches broken, resulting in the death of large tracts of old growth, especially in the northern Keys. In Louisiana, storm surge and monstrous waves destroyed 70 percent of the low-lying barrier islands, exposing and killing over 80 percent of the region's oyster reefs and devastating fragile coastal wetlands that serve as breeding and feeding grounds for many species of commercial fish.

The total property damage due to hurricanes in the 1990s worldwide exceeded the combined loss of the previous two decades. This does not, however, reflect the occurrence of more severe hurricanes and typhoons in the 1990s than earlier. Rather, the excessive damage is due to the rapid and extensive development of the shoreline for human occupation, a trend that is accelerating in the first decade of the twenty-first century

Visit www.jbpub.com/oceanlink for more information.

FIGURE **B6-6**

Hurricane Hernan. This satellite image of hurricane Hernan off Baja California (September, 2002) shows clearly the spiraling weather system. Hernan peaked as a category 5 hurricane with winds of 140 kph (154 mph)

ters (~62 to 186 miles) in diameter and 3 kilometers (~1.8 miles) or less thick, that rotate swiftly at speeds as high as 1 meter per second (~2.2 mph). Typically, about eight warm rings and ten cold rings are spawned each year from the Gulf Stream. Once formed, they persist for several months to several years, drifting to the southwest (see Figure 6–12b) at speeds of 10 kilometers (~6.2 miles) or less per day and eventually are reabsorbed back into the Gulf Stream. Similar rings have been observed in association with currents along the western arms of ocean gyres as well, such as the Kuroshio Current of the North Pacific Ocean and the Brazil Current of the South Atlantic Ocean (see Figure 6–4).

6-3

Deep-Ocean Circulation

The geostrophic currents of the ocean gyres are powered by the wind and, therefore, are confined largely to the uppermost level of the water column. **Deep water**, although not affected directly by the wind, is, however, in motion at all depths as well. These subsurface currents, collectively referred to as **thermohaline circulation**, arise from density differences between water masses produced by variations in water temperature (thermal effect) and salinity (haline effect). Understanding the nature of thermohaline circulation is a simple matter. When two water masses having different densities come into contact, the denser water mass will displace and underride the less dense water mass. If you divide a bucket with a vertical partition, fill half of it with seawater and the other half with freshwater, and then remove the partition, the dense saltwater will sink to the bottom of the pail and buoy up the less dense freshwater. This process is thermohaline circulation.

Thermohaline circulation is difficult and expensive to study, because it occurs in the subsurface of the open ocean far from land. Also, these subsurface currents tend to flow very slowly, making their speeds technically challenging to measure directly. However, despite these difficulties, physical oceanographers have successfully identified the general pattern of subsurface water motion. Keep in mind that thermohaline circulation, about which we know the least, is the dominant form of water flow in the ocean, affecting about 90 percent of the ocean's total volume of water.

WATER MASSES AND DENSITY-DRIVEN WATER FLOW

Recall from Chapter 5 that the density of water is a function of its temperature, salinity, and pressure. Because water is almost incompressible, the pressure effect on the density of seawater is negligible and, for our purposes, we can disregard it. That leaves water temperature and salinity to consider. Water gains heat from the sun and loses heat to colder air. Freshwater can be added (by precipitation, river runoff, ice melting) to or subtracted (by evaporation, ice formation) from a parcel of seawater, changing its salinity. As the water temperature rises, the density of the water decreases (an inverse relationship), and as salinity increases, so will water density (a direct relationship).

It is important to recognize that the processes altering the variables that control water density—temperature and salinity—occur near the sea surface. They depend on climate and therefore on latitude. This implies that once seawater sinks and is no longer in contact with the atmosphere, its salinity and temperature are "fixed" and remain remarkably stable over time. In fact, the temperature and salinity can be used as a tag to identify subsurface water masses, even if they move thousands of kilometers from their point of origin near or at the sea surface. Although stable, the temperature and salinity of water masses actually do change ever so slightly with time, as mixing occurs between adjoining water masses. But because they flow so slowly, the mixing process is inefficient and hence slow as well. This is fortunate, because these conservative properties of water are easily measured. They can be used to classify the deep-ocean's water masses, and to trace their lengthy flow paths across oceans.

Subsurface water masses are broadly classified into three major groupings, based on their depth of occurrence (**Table 6–2**). This means of course that the deepest water masses are the densest, so they must be very cold and/or very salty. A statistical analysis of a large number of temperature and salinity measurements of water masses has provided several important findings about the nature of seawater. First, remarkably, over 75 percent of all ocean water has a temperature of 0 to 5°C and a salinity of 34 to 35‰ (**Figure 6–13**). This indicates clearly that the ocean's depths are filled with very cold water, much colder than the ~17.5°C average temperature of the ocean's surface water. This finding, in turn, indicates that most of the water that fills the ocean must have originated in the polar latitudes, where it was chilled by losing heat to the frigid air. Global surveys of seawater chemistry and the mapping of deep-sea currents have confirmed this inference. Second, the uniformity of the temperature and salinity of subsurface seawater from ocean to ocean suggests that ocean basins are not closed, but are open systems that exchange water with one another. This has been verified by assessing the water budget of each ocean, that is, by noting water losses and gains.

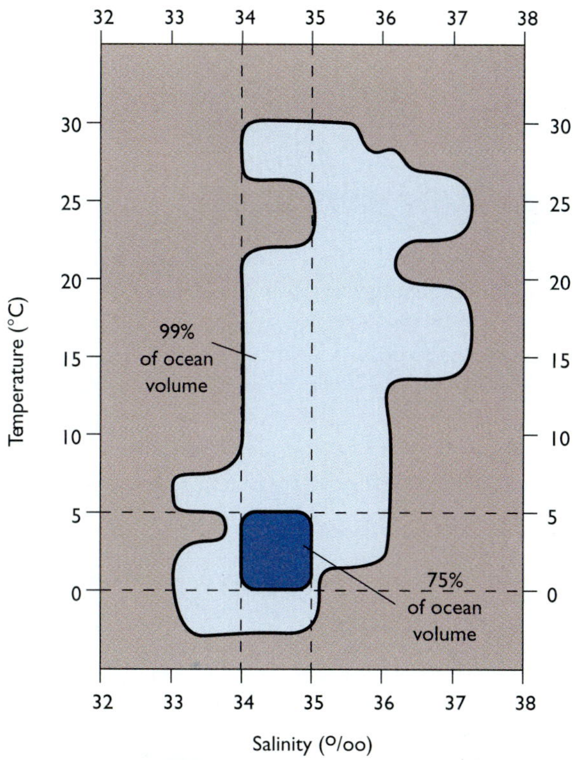

FIGURE **6-13**

Temperature and salinity range of seawater. Seawater is a remarkably uniform solution, with 75 percent of the total ocean volume having a temperature range of between zero and 5°C and a salinity range of between 34 and 35‰. [Adapted from R. B. Montgomery, *Deep Sea Research* 5 (1958): 134–48.]

TABLE **6-2**

Properties of the water masses of the global ocean.

Water-Mass Types	Approximate Depth Range (km)	Water Mass*	Temperature (°C)	Salinity (‰)
Central waters	0–1	SPCW	9–20	34.3-36.2
		NPCW	7–20	34.1-34.8
		NACW	4–20	35.0-36.8
		SACW	5–18	34.3-35.9
		SICW	6–16	34.5-35.6
Intermediate waters	1–2	NPIW	4–10	34.0-34.5
		RSIW	23	40.0
		MIW	6–11.9	35.3-36.5
		AIW	0–2	34.9
		AAIW	2.2-5	33.8-34.6
Deep and bottom waters	>2	CoW	0.6–1.3	34.7
		PSW	5–9	33.5-33.8
		NADW	3–4	34.9-35.0
		AADW	4.0	35.0
		NABW	2.5-3.1	34.9
		AABW	-0.4	34.6

*See Figure 6–15 for explanations of these abbreviations.
Source: Adapted from M. U. Sverdrup, M. W. Johnson, and R. M. Fleming, *The Oceans: Their Physics, Chemistry, and General Biology* (Englewood Cliffs, N.J.: Prentice-Hall, 1942; A. Defant, *Physical Oceanography*, vol. 1 (Oxford: Pergamon Press, 1961); and O. I. Mamaev, *Temperature-Salinity Analysis of World Ocean Waters* (Amsterdam: Elsevier, 1975).

SCIENCE BY NUMBERS

Volume Transport

To understand conceptually what volume transport is, let's begin with a garden hose. If we multiply the cross-sectional area (cm^2) of the hose by the speed of the water gushing out of it (cm/sec), we get volume transport in units of cm^3/sec. That is, $cm^2 \times cm/sec$ equals volume transport (cm^3/sec).

Imagine a large current such as the Gulf Stream of the North Atlantic Ocean. It is important to oceanographers to know how much water passes through a cross-sectional area of the Gulf Stream each second (m^3/sec). We can model this problem as if the ocean were a large garden hose. We need to know (1) the width and depth of flow of the Gulf Stream in order to calculate its cross-sectional area and (2) its average current speed. This is exactly the same information we needed to determine the volume transport of the garden hose. The Gulf Stream has an average width of about 100 kilometers, an average depth of flow below the sea surface of about 1.5 kilometers, and an average velocity of about 1.5 meters per second.

The calculation is as follows:

$$\text{Cross-sectional area} = (\text{width})(\text{depth of flow}) = (100\text{ km})(1.5\text{ km})$$
$$= 1.5 \times 10^2\text{ km}^2$$

We must now convert $1.5 \times 20^2\text{ km}^2$ into m^2, because our current measurement is in m/sec and not km/sec. We do this conversion systematically as follows:

If
$$1\text{ km} = 10^3\text{ m},$$
then
$$(1\text{ km})^2 = (10^3\text{ m})^2,$$
and
$$1\text{ km}^2 = 10^6\text{ m}^2;$$
so
$$1.5 \times 10^2\text{ km}^2 = (1.5 \times 10^2\,\cancel{\text{km}^2})(10^6\text{ m}^2/1\,\cancel{\text{km}^2}) = 1.5 \times 10^8\text{ m}^2.$$

$$\text{Volume transport} = (\text{cross-sectional area})(\text{current speed})$$
$$= (1.5 \times 10^8\text{ m}^2)(1.5\text{ m/sec}) = 2.25 \times 10^8\text{ m}^3/\text{sec}.$$

There we have it—an estimate of the volume transport of the Gulf Stream!

A GENERAL MODEL OF THERMOHALINE CIRCULATION

Longitudinal profiles of the water column in the Pacific and Atlantic Oceans (**Figure 6–14**) reveal that water temperature drops with increasing latitude and increasing depth. Water that is warmer than 10°C is confined to a thin lens that straddles the equator, and extends poleward to the 45°N and 45°S latitudes and vertically to the permanent thermocline. This water is very light, so to speak, and is mixed by the geostrophic currents of the circulation gyres that we studied in the first part of this chapter. The sun heats this surface water to such an extent that it offsets the high salinities created by the high evaporation rates that generally characterize the tropical and subtropical latitudes.

Between the poles and 45°N and 45°S latitudes, evaporation and cold air cool the surface water (see Figure 6–14), raising its density. As this polar water drifts equatorward, it encounters warmer, less dense surface water and sinks below it. The cold water descends to a depth that is appropriate to its density, sliding under less dense water and over more dense water, at which point the water mass flows horizontally rather than vertically. During this journey through the ocean depths, mixing

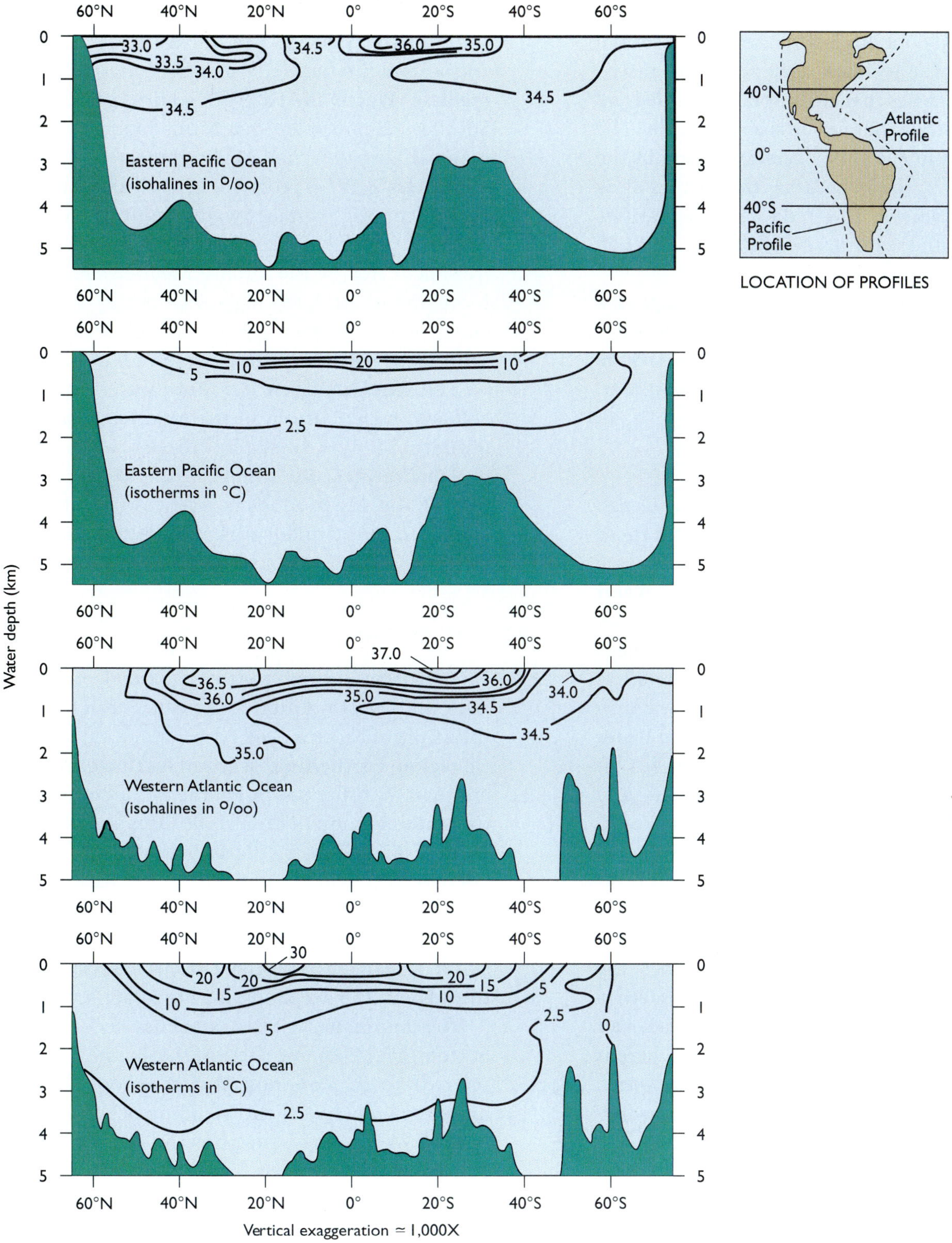

FIGURE **6-14**

Temperature and salinity profiles. These longitudinal profiles depict variations in salinity and temperature for the Atlantic and Pacific Oceans. Sharp haloclines and thermoclines are evident in the topmost kilometer of both oceans. [Adapted from A. Couper, ed., *The Times Atlas of the Oceans* (New York: Van Nostrand Reinhold, 1983).]

occurs at the boundaries of the water mass, slowly but relentlessly modifying its temperature and salinity. Eventually, this deep water is mixed upward by processes that are thus far ill defined and ultimately reaches the sea surface once again. The entire trip—from the time of its formation in the surface water of the polar seas, through its slow descent and advective flow in the deep sea, until its ascent back to the sea surface—takes roughly 1,000 years or so to complete.

With this background, we can now complete a brief survey of the distribution of the major water masses in the world's oceans. Consult **Figure 6–15** and Table 6–2 for clarification of the following text.

ATLANTIC OCEAN

The **deep** and **bottom waters** of the Atlantic Ocean are dominated by cold, dense water. Two of these water masses are created offshore Antarctica (see Figure 6–15). One, **Antarctic Bottom Water** (AABW), is produced principally at the surface of the Weddell Sea of Antarctica during the southern winter when the formation of ice leaves the water very cold and saline, creating the densest water in the global ocean. The other, **Antarctic Deep Water** (AADW), forms in the less extreme latitudes off Antarctica, and therefore is a bit less salty and slightly warmer than AABW (see Table 6–2). Because of its very high water density, AABW sinks to the bottom of the Atlantic basin and flows northward, streaming across the equator and reaching far into the North Atlantic. AADW is water that flows northward at the surface until it reaches the Antarctic Convergence (a zone analogous to a weather front but where water masses instead of air masses are in contact with one another), where it plunges beneath warmer, less dense subpolar water. Once below the surface, AADW becomes sandwiched between the denser AABW and the less dense **North Atlantic Deep Water** (NADW), which originates far to the north, near Greenland. NADW, which is moderately salty and cold, is produced by winter cooling and evaporation. It is generated at such a high rate across such a vast area of the ocean that, by volume, it is the most common type of water in the depths of the Atlantic Ocean.

Several intermediate water masses are evident in the Atlantic (see Figure 6–15). One, **Arctic Intermediate Water** (AIW), is formed in the subarctic, and as it flows south, it slides under the warm water of the Sargasso Sea, until its leading edge encounters the northerly moving **Antarctic Intermediate Water** (AAIW), its counterpart in the Southern Hemisphere that forms in a similar climatic setting. Another, **Mediterranean Intermediate Water** (MIW), is produced in the Mediterranean Sea. Incoming surface water from the North Atlantic, which is cool and has normal ocean salinities, is heated by the intense Mediterranean sun. More importantly, the high evaporation rate associated with the area's arid climate elevates the salinity of this incoming water to more than 36.5‰. This very dense but warm water sinks and collects at the bottom of the Mediterranean Sea until it spills over the shallow rock sill at the Strait of Gibraltar near the southern tip of Spain. Once free of the Mediterranean basin, MIW streams westward in the middepths of the North Atlantic Ocean as an elongated tongue of warm, salty water.

PACIFIC OCEAN

The shallow Bering Strait that extends between Asia and North America prevents much of an influx of cold bottom and deep water from the arctic region. Furthermore, the production of AADW and AAIW in the South Pacific is modest. Thus, the water column of the Pacific Ocean tends to be weakly layered, particularly below 2,000 meters (~6,600 feet), where temperature and salinity are remarkably uniform with depth. This means that density contrasts between water masses are slight, imparting a sluggish subsurface circulation throughout the deeper parts of the Pacific Ocean.

The dominant type of water that occupies the bottom of the Pacific Ocean (see Figure 6–15) is referred to as **Common Water** (CoW), a water mass that represents the blending of AABW and NADW. Apparently, the powerful Antarctic Circumpolar Current (see Figure 6–4) injects CoW that was mixed in the South Atlantic through the Indian Ocean and into the Pacific Ocean, where it sinks and flows northward along the bottom of the Pacific basin to about 10°S latitude. Above 2,000 meters (~6,600 feet), two cool, low-salinity water masses dominate the water column: Antarctic Intermediate Water (AAIW) and **North Pacific Intermediate Water** (NPIW). Both of these water masses are hybrids in that they appear to be

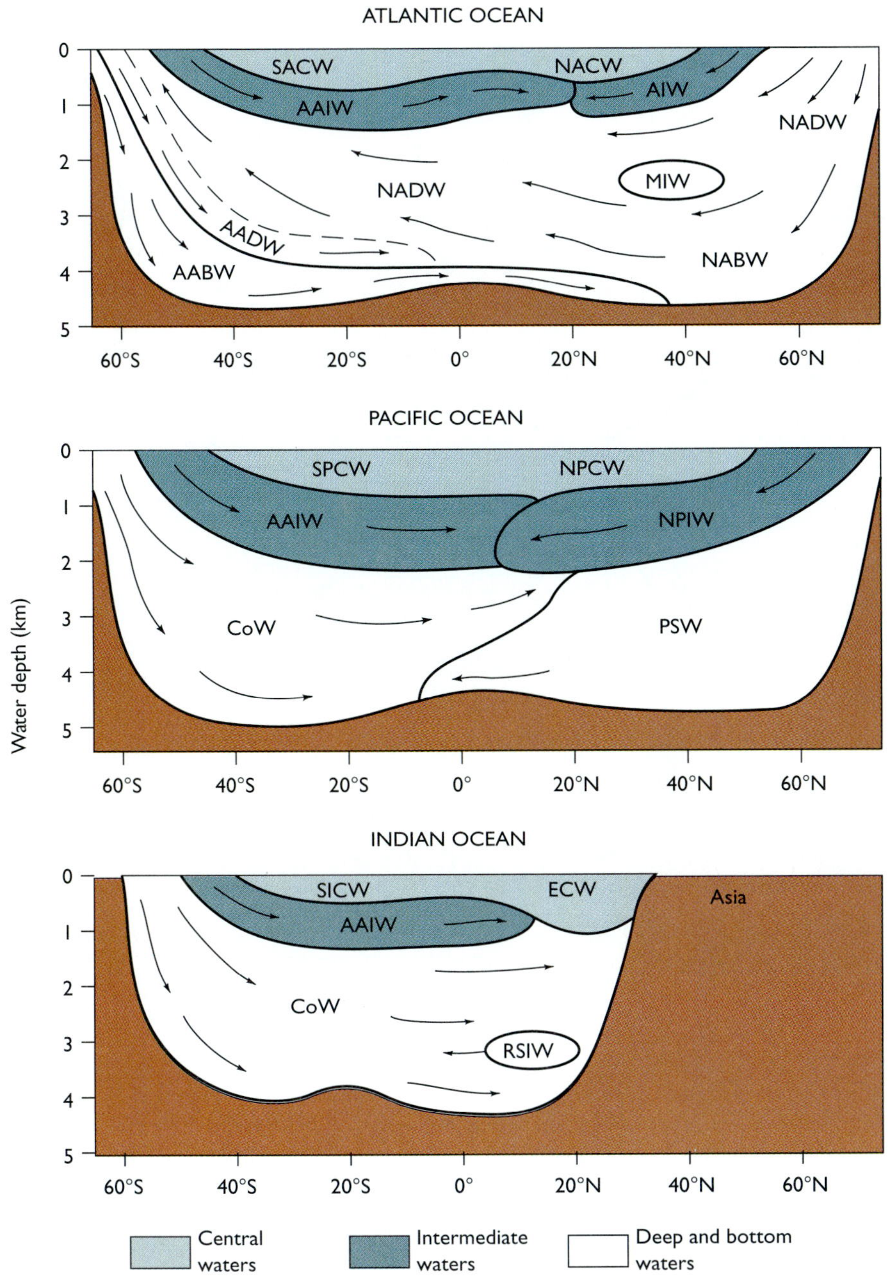

FIGURE **6-15**

Water-mass structure in the oceans. The simplified water-mass structure of the Atlantic, Pacific, and Indian Oceans is shown. Transport in the deep sea is generated largely by the formation of cold, dense water in the polar regions that then sinks and flows equatorward, displacing less dense water masses.

AABW	*Antarctic Bottom Water*
AADW	*Antarctic Deep Water*
AAIW	*Antarctic Intermediate Water*
CoW	*Common Water*
ECW	*Equatorial Central Water*
MIW	*Mediterranean Intermediate Water*
NABW	*North Atlantic Bottom Water*
NACW	*North Atlantic Central Water*
NADW	*North Atlantic Deep Water*
NPCW	*North Pacific Central Water*
NPIW	*North Pacific Intermediate Water*
PSW	*Pacific Subarctic Water*
RSIW	*Red Sea Intermediate Water*
SACW	*South Atlantic Central Water*
SICW	*South Indian Central Water*
SPCW	*South Pacific Central Water*

formed largely by complex and little-understood mixing processes in the subsurface.

INDIAN OCEAN

The Indian basin is confined essentially to the Southern Hemisphere. Consequently, deep and bottom waters derived from the arctic region are absent from the Indian Ocean, greatly simplifying its water-mass structure (see Figure 6–15). Common Water (CoW), which occupies all of its depths, is a hybrid mixture of AABW and NADW. The lower levels of CoW have properties similar to AADW; the upper levels, to NADW. AAIW is produced in modest quantities north of the Antarctic Convergence. From this point it spreads slowly northward over CoW. Also, a tongue of **Red Sea Intermediate Water** (RSIW) with a salinity of more than 40‰, penetrates into CoW at a depth of about 3,000 meters (~9,900 feet).

The preceding depictions only hint at the fact that the wide-scale exchange of water between basins is common and not the exception. **Figure 6–16a** displays the flow paths of deep thermohaline

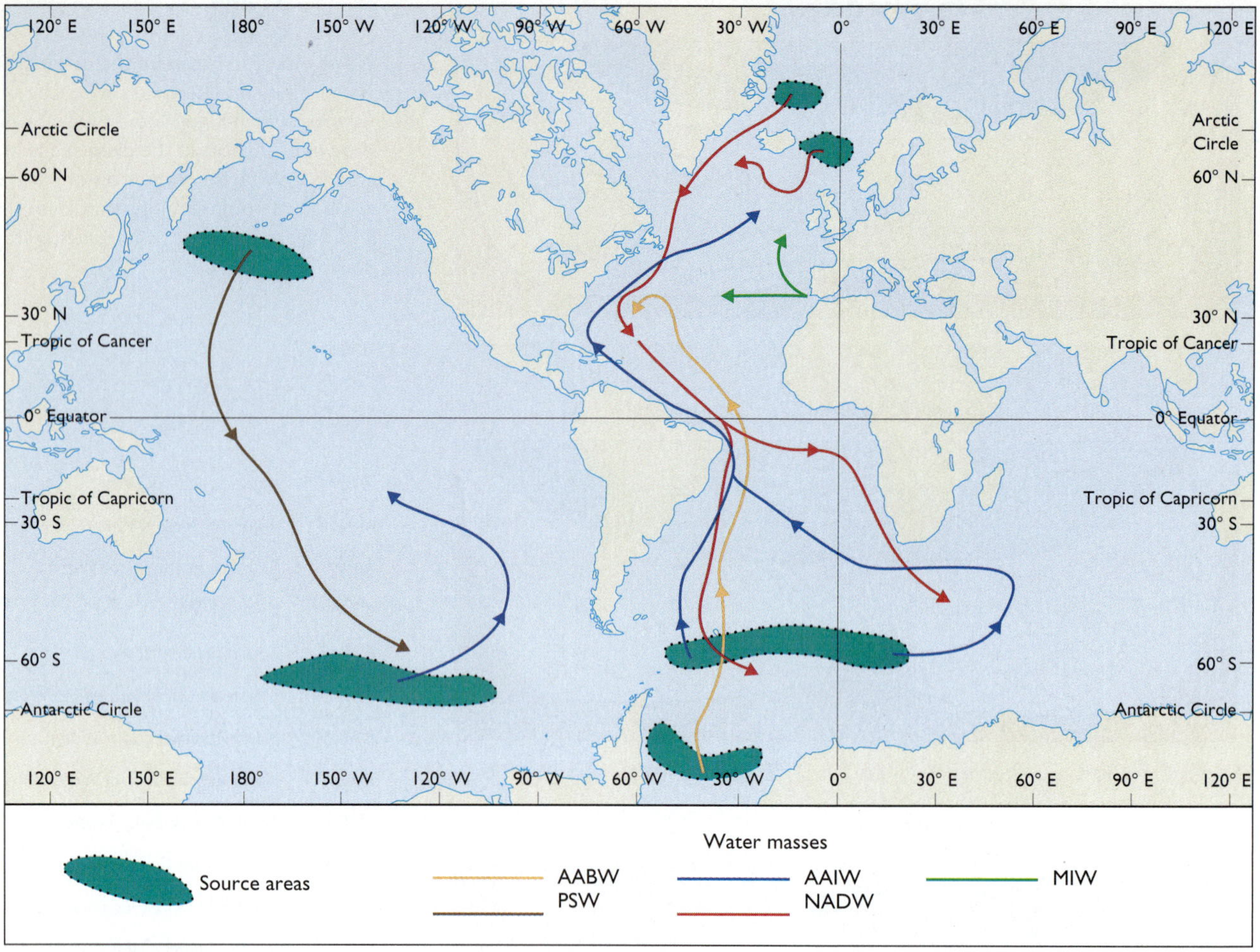

(a) THERMOHALINE CIRCULATION

FIGURE **6-16**

Large-scale circulation. (a) This map depicts the flow paths of some deep-water masses, notably Antarctic Bottom Water (AABW), Antarctic Intermediate Water (AAIW), Mediterranean Intermediate Water (MIW), North Atlantic Deep Water (NADW), and Pacific Subarctic Water (PSW). [Adapted from A. L. Gordon, *Lamont-Doherty Geological Observatory Report* 1990–91.] (b) This model portrays the exchange of surface and deep water across the ocean basins as a conveyor belt. [Adapted from M. S. McCartney, *Oceanus* 37 (1994): 5–8.]

currents, including the source regions for these water masses. This shows clearly that they form in surprisingly few areas, mostly in the polar seas in relatively shallow basins. What happens at these sites is that dense water created by a reduction in temperature and/or an increase in salinity becomes trapped by a shallow sill at the entryway to the basin. Eventually, the dense water overtops and then overflows the sill. It sinks because of its relatively high density and streams out into the water column of the nearby ocean basins (see Figure 6–16a).

Recent models connect this deep flow with surface circulation. One such model depicts water flow in the world's oceans as an immense conveyer belt (**Figure 6–16b**). It is driven largely by processes that occur in the North Atlantic Ocean. Here, warm, shallow water imported from the Pacific and Indian oceans loses heat to the atmosphere by radiation. Its density increases and, as a result, the water is forced to sink and be exported to the other basins as deep-water flow. Once back in the Indian and Pacific Oceans, the water upwells, is warmed by contact with the atmosphere, and begins a new circuit along the conveyer belt (see Figure 6–16b). This model is compatible with general observations, but it is undoubtedly highly oversimplified. With additional work and findings, the model will be refined or may be replaced by a better one.

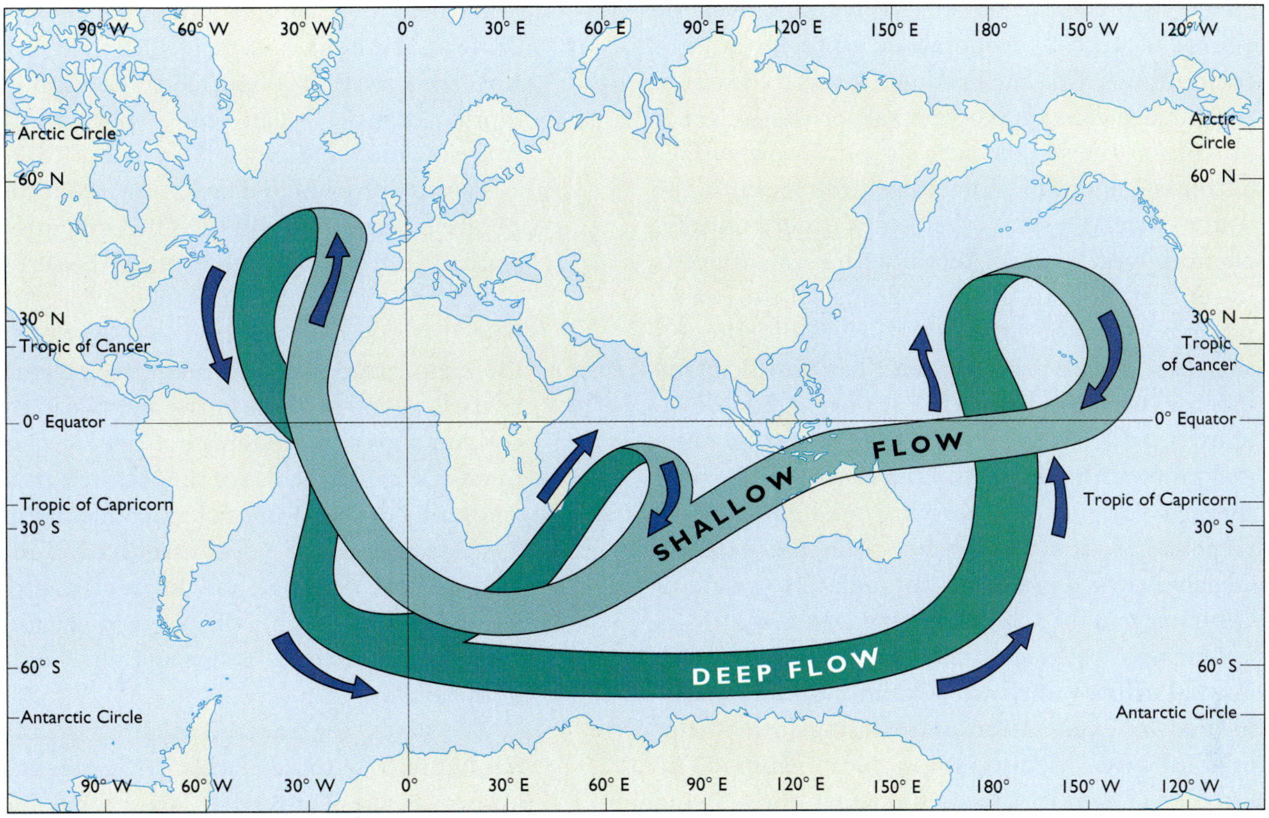

(b) SURFACE AND DEEP WATER EXCHANGE

The large-scale circulation described here has a tremendous effect on marine organisms. Where water masses sink, they carry dissolved oxygen to the deep sea. This allows life to exist at all depths. Where water masses rise, they return nutrients to the surface. This helps maintain plant productivity. Without the exchange of surface and deep water, there would be a lot less life in the oceans worldwide.

6-4

Water Flow in Semienclosed Seaways

Seas are located at the edges of the major ocean basins. Many represent large indentations into landmasses. Because they are so close to land, humans have settled along the shores and tend to be quite familiar with the oceanography of their water. Like the open ocean, seas are created by large-scale geologic processes, are filled with seawater that is mixed by surface and subsurface currents, and are populated by diverse communities of organisms. In many ways they are miniature oceans. This analogy, however, can only be taken so far. Unlike the open ocean, seas are semienclosed, and they are greatly influenced by continental climates and river drainage—conditions that impart unique oceanographic characteristics to them. By way of illustration, we will examine the circulation and water chemistry of the Mediterranean Sea and the Black Sea.

The Mediterranean Sea is noted for its distinctive subsurface-water circulation as well as for its geologic history, which is partly described in the boxed feature, "The Mediterranean Sea," in Chapter 4. Its subsurface currents are controlled by surface processes, notably evaporation, precipitation, and river runoff. The latter two dilute the salinity of seawater, the former increases it. Because the dominant climate of the Mediterranean region is hot and dry, evaporation surpasses the combined effects of precipitation and river inflow over the course of a year (evaporation > precipitation + river inflow). This means that the surface water of the Mediterranean Sea gets saltier the longer it remains at the sea surface. The prevailing surface currents generally flow eastward, making the water saltier (**Figure 6–17a**) and warmer as it travels from west to east. Actually, the sea surface is lower and tilted to the east because of the evaporative removal of water. In other words, the farther east a parcel of seawater has traveled, the more water it has lost to the atmosphere by evaporation, and this loss lowers the sea surface more and more with distance to the east. This slope of the sea surface is a pressure gradient, and in response to it, surface water from the Atlantic Ocean flows downslope through the Strait of Gibraltar and into the Mediterranean Sea.

During its easterly transit, the relatively cold, normal-salinity Atlantic Ocean water is transformed by evaporation and heating into warm, high-salinity Mediterranean Sea water. The increase of salinity offsets the effect of the rising water temperature, so that the density of Mediterranean Sea water increases eastward. At some point the dense water sinks and collects in the depths of the basin. This continues until the dense water fills the deep, overtops the shallow sill at Gibraltar, then spills back out into the Atlantic Ocean as subsurface outflow (see Figure 6–17a and the boxed feature, "Underwater Weather and Waterfalls"). Thus, the current flow through the Strait of Gibraltar reverses direction with depth. There is surface inflow in response to the pressure gradient associated with the eastward tilt of the water surface, and there is westerly subsurface outflow associated with overfilling the basin with warm, salty water. Calculations indicate that the water that flows into the Mediterranean Sea from the Atlantic Ocean remains in the basin for about 80 to 100 years before it returns back to the Atlantic.

Once free of the narrow Strait of Gibraltar, Mediterranean Sea water travelling west becomes a distinctive water mass called Mediterranean Intermediate Water (MIW; see Table 6–2). It sinks to a depth of about 1,000 meters (~3,300 feet) and then spreads out horizontally, flowing west in the depths of the North Atlantic Ocean (see Figures 6–15 and **6–17b**). Its distinctive water properties—a temperature near 13°C and a salinity range of 38.40 to 38.46‰—make it easy to identify and to trace its flow path. MIW has been detected in the western North Atlantic Ocean—thousands of kilometers from its point of origin at the Strait of Gibraltar.

Because of its high density, surface water in the Mediterranean Sea sinks. This downwelling of water does not supply nutrients to the surface, as is indicated by the small quantities of dissolved phosphate in the surface-water layer (**Figure 6–17c**). The low concentrations of nutrients keep plant productivity low, and as a result, the animals in the water column and on the basin's bottom are sparse. The intense, clear blue color of the Mediterranean Sea is testimony to the absence of much life in this water. The Mediterranean Sea is essentially a biological desert. Locally, a small amount of plant production occurs, but only during the winter and early spring, when storms stir the water and recycle nutrients from below.

The Black Sea and the Mediterranean Sea are remnants of a much larger and ancient seaway known as the Tethys Sea, which has been narrowed by the plate-tectonic collision of Africa with Europe and Asia. This continental "smash-up" is ongoing, as is indicated clearly by the numerous earthquakes that presently affect the region along a broad front that defines the boundary between the large continental plates (**Figure 6–18a**). Today, the Black Sea has a tenuous, indirect connection to the Atlantic Ocean via the Sea of Marmara and the Mediterranean Sea (see Figure 6–18a). Because the Black Sea is located in the midlatitudes, rainfall and river discharge from a large drainage basin add much more water to the basin than is removed by evaporation (evaporation < precipitation + river inflow). This is exactly opposite to the situation for the Mediterranean Sea just discussed. Consequently, the salinity of the surface water of the Black Sea is diluted to between 18 and 22‰, with the resultant lowering of its density. This surface layer of brackish water is separated from normal seawater by a sharp halocline that stratifies the water column. Also, summer heating creates a thermocline, further intensifying the layered structure of the water column (**Figure 6–18b**). These sharp salinity and temperature gradients combine to produce a pronounced density discontinuity (a pycnocline) at a depth of about 100 meters (~330 feet), which serves to isolate deep water from surface water.

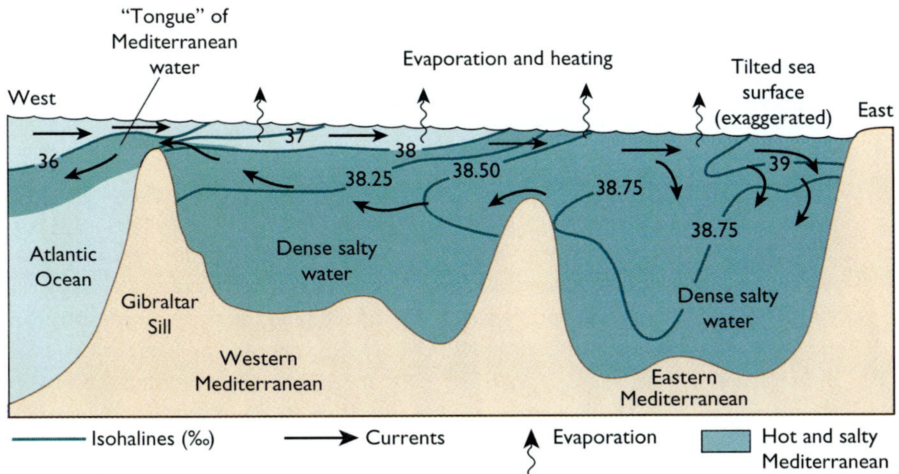

(a) CIRCULATION IN THE MEDITERRANEAN SEA

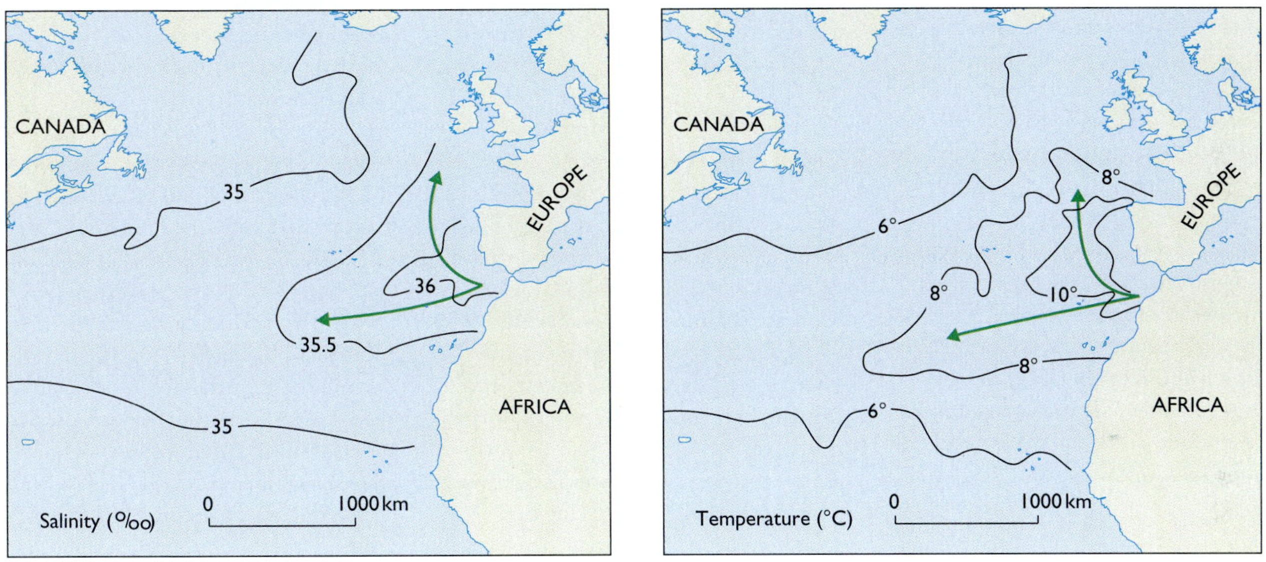

(b) SALINITY AND TEMPERATURE DATA AT 1,000-METER DEPTH

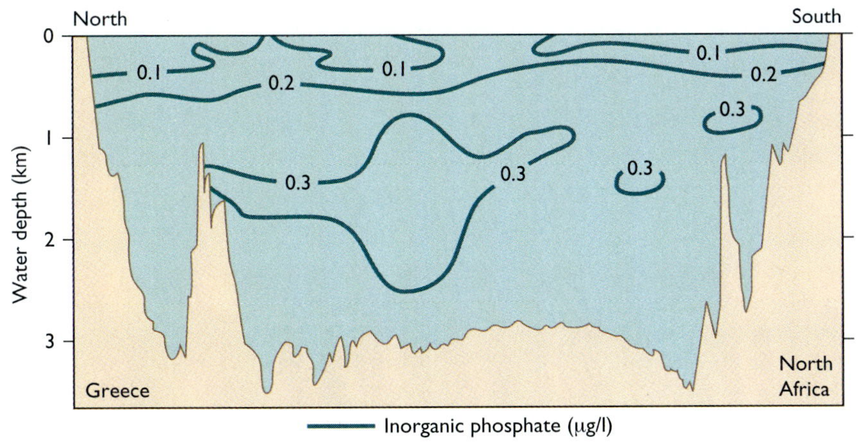

(c) PHOSPHATE CONCENTRATIONS IN THE MEDITERRANEAN SEA

FIGURE **6-17**

Circulation in the Mediterranean Sea. (a) Surface and subsurface circulation in the Mediterranean Sea occur because evaporation exceeds precipitation and river runoff combined. Warm, salty water sinks because of its high density, filling the deeps and spilling over the Gibraltar sill into the North Atlantic Ocean. This water is replaced by the surface inflow of seawater from the Atlantic Ocean. [Adapted from G. Wurst, *Journal of Geophysical Research* 55 (1961): 3261–71.]

(b) These maps trace out Mediterranean Intermediate Water (MIW) in the Atlantic Ocean using temperature and salinity data at a 1,000-meter water depth. [Adapted from Open University Course Team, *Ocean Circulation* (New York: Pergamon Press, 1989.] (c) Nutrients are transported downward and out of the photic zone by the downwelling of dense, salty water. This reduces significantly the biological productivity of the surface water. [Adapted from A. R. Miller et al., *Mediterranean Sea Atlas,* Series 3 (Woods Hole, MA: Woods Hole Oceanographic Institution, 1970).]

Underwater Weather and Waterfalls

When looking up at the sky, we notice a variety of clouds on most days. Their forms and motions reveal the direction and relative speed of air currents that are transporting uncountable numbers of water droplets. When clouds are evident at different altitudes, their drift patterns reveal that the atmosphere—this veritable ocean of air—is layered and that its winds blow at different speeds and in different directions. Much of this fluid air is moving vertically as well as horizontally and whirling in eddies as well as flowing smoothly. What we learn from these observations is that the lower atmosphere is neither homogeneous nor static. Rather, it consists of air masses that sweep across the landscape, causing changes in the weather of a region. Weather fronts represent zones where air masses abut one another. This boundary is made visible by clouds, precipitation, and winds, as the air masses to either side of the dynamic front interact with each other, producing unstable, turbulent weather.

Physical oceanographers are discovering "weather" in the ocean as well. In this case, water masses rather than air masses are in contact along fronts. Because of water's high density relative to air, the turbulent, chaotic motions associated with ocean fronts operate at slower rates and over longer time scales than they do in the atmosphere. Nevertheless, they are genuine "weather" fronts, in a watery rather than a gaseous fluid. It is not uncommon for a large mud plume to be evident far offshore of a river mouth after a flash flood on land. The edge of the plume is an **oceanic front,** where two water masses with distinct oceanographic properties—muddy freshwater and clear saltwater—are in contact. Another familiar example of an oceanic front is represented by the Gulf Stream where it abuts the water of the Sargasso Sea (see Figure 6–12a). In both cases, sharp horizontal gradients in some property of the water—temperature, salinity, density, suspended sediment load—is evident at the frontal boundary. Also, it is common to see signs of turbulent flow at the water-mass contact, such as swirling eddies splaying off a river mud plume, or cold-core and warm-core rings breaking off the Gulf Stream. Satellite imagery (see boxed feature, "Current-Measuring Techniques") enables physical oceanographers to detect these oceanic fronts, and study their dynamic changes over time. Recent work has led to some preliminary conclusions about the types and scales of ocean fronts worldwide, which are summarized in **Table B6–2**.

Vertical profiles of temperature, salinity, and density of seawater reveal conclusively that the water column of the ocean is not homogenous; rather, it reflects the presence of distinct water masses that are stacked one on top of another according to their densities. The contacts between water masses in the deep sea are near horizontal and appear as thermoclines, haloclines, and pycnoclines. Large-scale sea-floor topography can, however, interfere with the uniform flow of water masses, not unlike a tall mountain's disruption of the passage of an air mass. In such cases, spectacular "waterfalls" can occur, as dense water piles up behind and then cascades down the front of a topographic barrier, creating cataracts that are unimaginably large. Let's examine four such submarine waterfalls in the Atlantic Ocean.

Figure B6–7, representing a longitudinal cross section that runs the length of the Atlantic Ocean, shows temperature distribution of the water column and the topography of the sea bottom. Note that water colder than 0.2°C, which originated off Antarctica, is ponded behind and is spilling over the Rio Grande Rise. Because of the high density of this bottom water (AABW), it is cascading down into the Brazilian basin, with a vertical drop of more than 1,000 meters (~3,300 feet). At the equator, there exists another dam, the Ceará abyssal plain. Here AABW colder than 1.4°C is cascading into the North Atlantic basin. The highest submarine waterfall in the Atlantic Ocean is located far to the north, in the Denmark Strait. Here, 5 million cubic meters of water pour over the sill each

TABLE **B6-2** Properties of ocean fronts.

Width of fronts	10 m to 10 km
Temperature difference across fronts	1 to 6°C
Salinity difference across fronts	0.2 to 10‰
Horizontal temperature gradient	0.1 to 3.0°C/km
Horizontal salinity gradient	0.1 to 10‰/km

Source: Adapted from K. N. Fedorov, *The Physical Nature and Structure of Oceanic Fronts* (New York: Springer-Verlag, 1983).

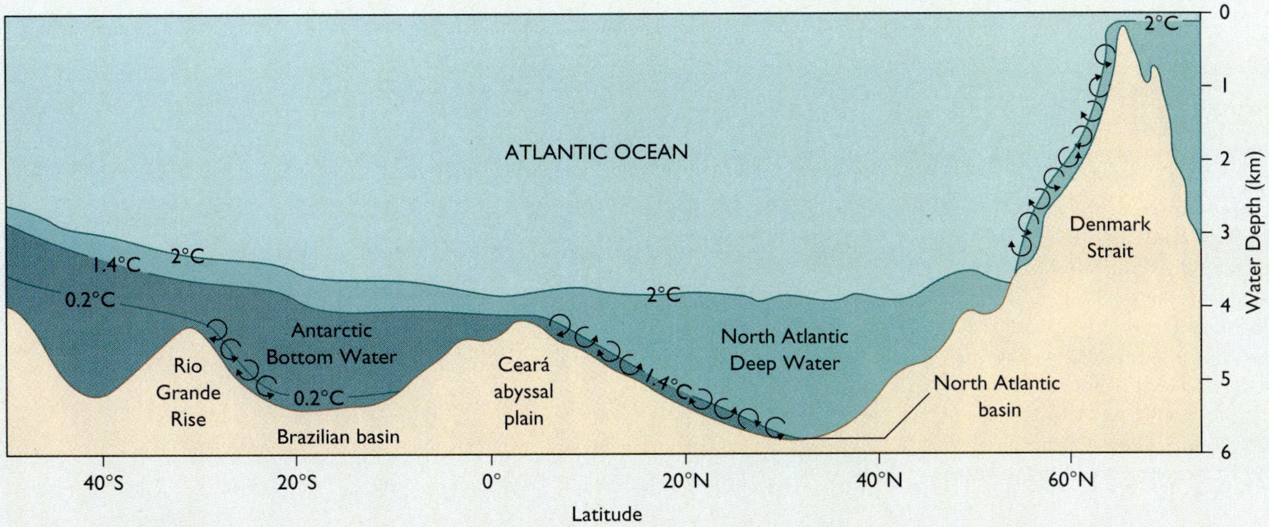

FIGURE **B6-7**

Underwater waterfalls. This representation of a longitudinal cross section through the Atlantic Ocean reveals three areas where water spills over bathymetric obstructions and forms submarine waterfalls. [Adapted from J. A. Whitehead, *Scientific American* 260, no. 2 (1989): 50–57.]

second and drops more than 3.5 kilometers (~2.2 miles), creating **North Atlantic Deep Water** (NADW), which then spreads out over the denser AABW. The dimensions of the Denmark Strait cataract dwarf those of comparable features on the continents. The Angel Falls in Venezuela are the tallest waterfalls on land, and they drop a mere 1 kilometer (~½ mile). The Guairá Falls on the Paraguay-Brazil border, which have the largest average flow rate of any waterfall on land, pour a trifling 13,000 cubic meters of water per second!

Another underwater waterfall of note in the North Atlantic occurs at the Gibraltar Strait, where high-salinity, warm Mediterranean water tumbles over the Gibraltar sill, plunging down to a water depth of about 1,000 meters (~3,300 feet) before it spreads out horizontally into the North Atlantic Ocean. This water mass, with its unique temperature-salinity signature, can be identified and traced by contouring the temperature and salinity of the water at a 1,000-meter (~3,300 foot) depth (**Figure B6–8**). Note that, in Figure B6–8,the center of this warm, saline plume of water is displaced to the north of the Gibraltar sill. Why would this be?

As expected, flow in these submarine waterfalls is exceedingly turbulent, generating eddies and whirlpools of many sizes that stir and mix water. The three-dimensional water motions in these eddies are very complex, and, if generated along fronts between water masses, can mix less dense water above with more dense water below. This effect efficiently redistributes heat and dissolved salts in the water. The mechanism for turbulent mixing, technically called *turbulent eddy diffusion,* is undoubtedly responsible for causing water masses to lose their identity as they drift away from their source areas. A case in point is the gradual dilution of Mediterranean water moving from east to west in the North Atlantic Ocean (see Figure B6–8). In addition to mixing, the cataracts may generate eddies that break off from the main water mass and drift away, transporting salt and heat from one area of the ocean to another. Warm, saline eddies spawned from Mediterranean water and called "Meddies" have been detected as far west as the Bahamas off eastern North America. Can you discern any Meddies in Figure B6–8?

Visit www.jbpub.com/oceanlink for more information.

FIGURE **B6-8**

Salinity-temperature variations at depth in the North Atlantic. (a) Temperature of the water at a depth of 1,000 meters clearly shows the distribution of Mediterranean water in the North Atlantic. (b) Salinity at a depth of 1,000 meters likewise reveals the distribution of Mediterranean water. [Adapted from Open University Course Team, *Ocean Circulation* (New York: Pergamon Press, 1989).]

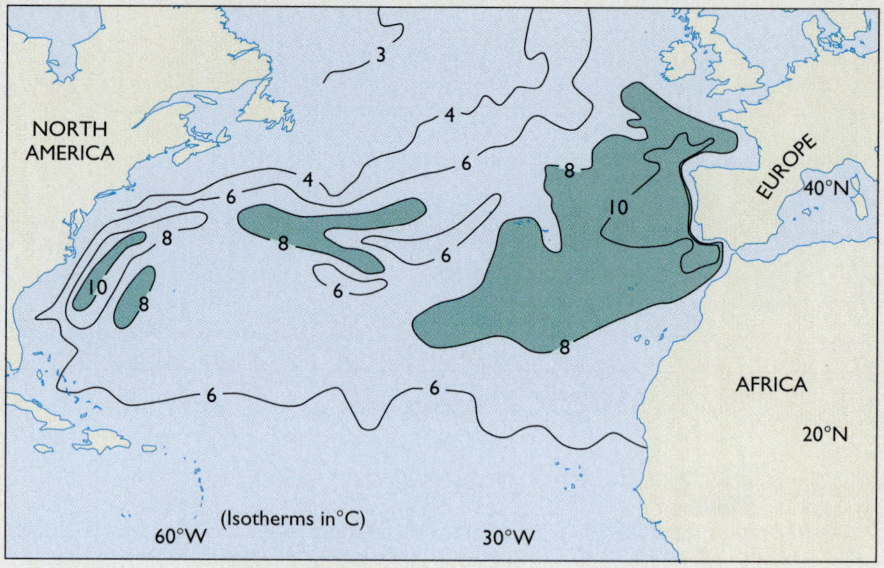

(a) TEMPERATURE AT 1,000-METER DEPTH

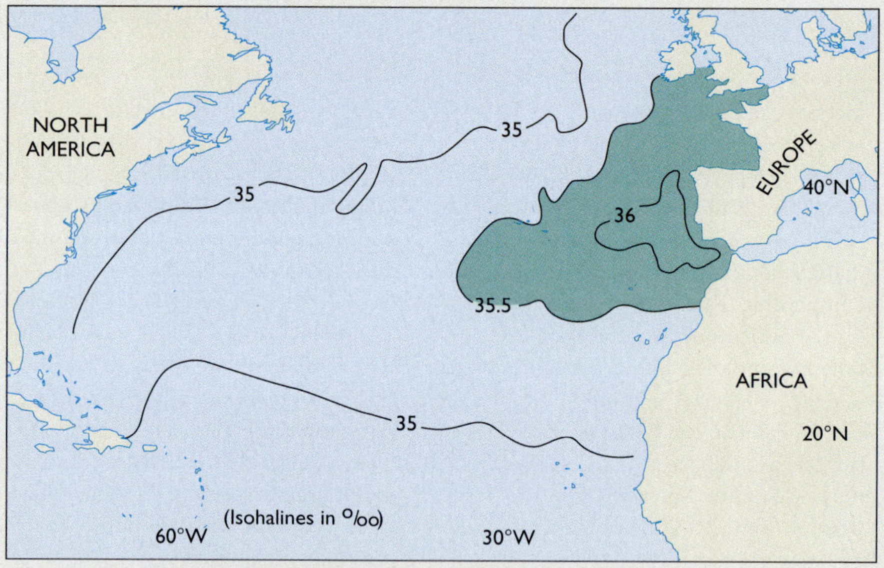

(b) SALINITY AT 1,000-METER DEPTH

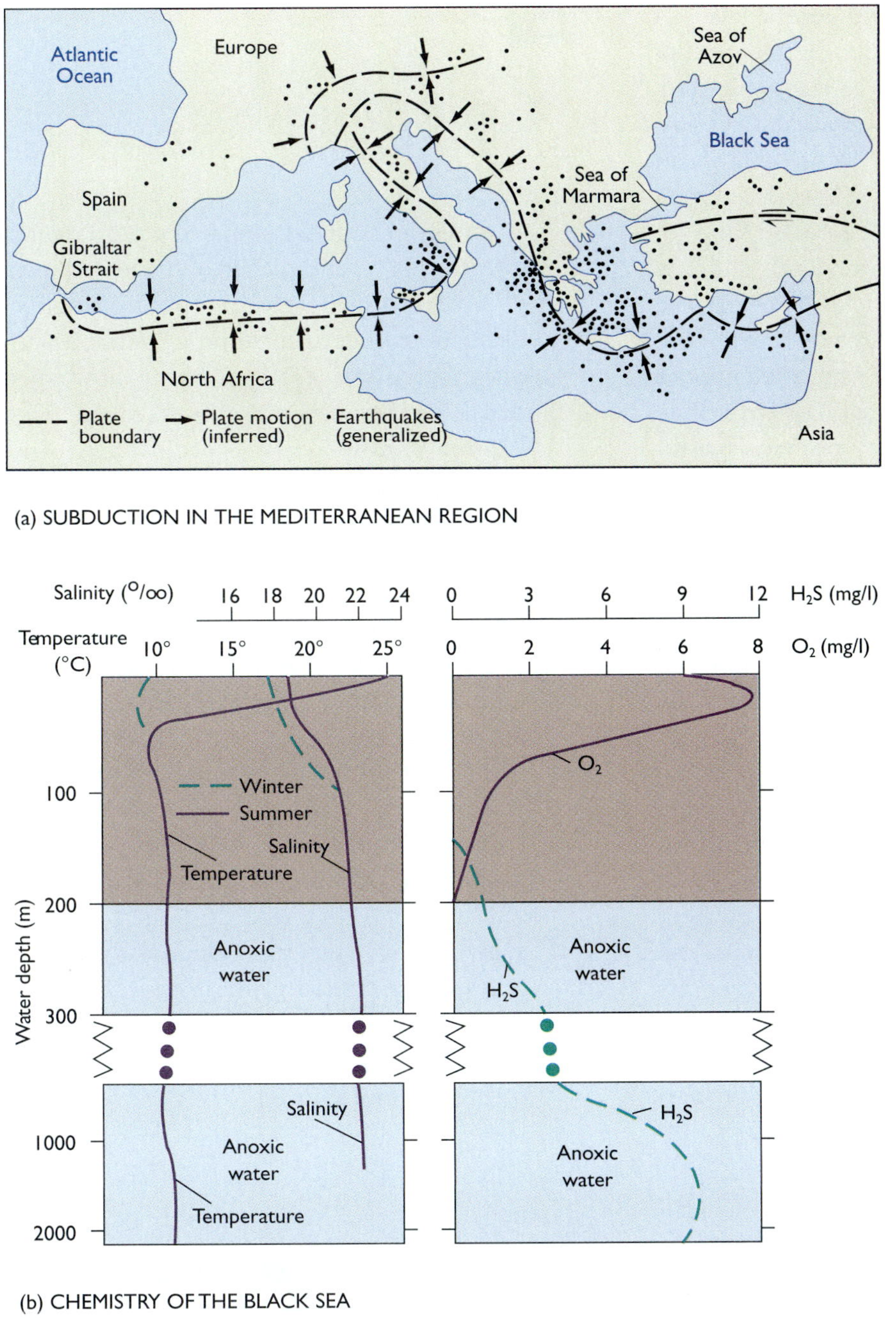

FIGURE **6-18**

The Black Sea. (a) The Mediterranean region is a broad subduction zone, as indicated by the numerous earthquakes and active volcanoes in the area. Africa is colliding with Europe. The Mediterranean Sea and the Black Sea represent the remnants of a much more extensive ocean in the geologic past. (b) Vertical profiles of temperature and salinity indicate that the water column of the Black Sea is stratified. A thin surface layer of well-oxygenated water overlies a thick layer of anoxic water with a high content of toxic hydrogen sulfide gas. [Adapted from Y. I. Sorokin, "The Black Sea" in *Estuaries and Enclosed Seas*, B. H. Ketchum, ed. (New York: Elsevier Science Publishing, 1983), 253–92.]

As is the case for the Mediterranean Sea, circulation in the Black Sea is a function of precipitation, river discharge, evaporation, and the water depth of the sill in the Sea of Marmara. The brackish, low-density surface water cannot sink, so it must be discharged from the basin across the narrow and shallow sill that connects it to the Mediterranean Sea. Otherwise, the freshwater inflow from rivers and rainfall would accumulate and cause the level of the Black Sea to rise rapidly with time, flooding the coastline. The water below the permanent halocline is essentially stagnant, trapped in the basin by the shallow sill. The turnover time for the deep water in the Black Sea is about 3,000 years, which reflects the stability of the water column.

Because buoyant, brackish water floats on top of dense, deep water, the water below the halocline is isolated from any contact with the surface. This state is clearly reflected in the distribution of dissolved gases. Surface water is charged with dissolved oxygen, produced by plants during photosynthesis. As plants die, they sink below the surface layer and are decomposed by bacteria, a process of oxidation that consumes dissolved oxygen. This constant fallout of dead plant matter and its subsequent decomposition depletes the levels of dissolved oxygen in the deep water. The sharp halocline does not allow the water to turn over and, therefore, prevents oxygen renewal at depth. Thus, the concentration of dissolved oxygen in water deeper than about 200 meters (~660 feet) is zero, a condition known as **anoxia** (see Figure 6–18b).

As oxygen disappears with water depth, a highly toxic gas—hydrogen sulfide (H_2S)—appears (see Figure 6–18b). This gas is produced by **anaerobic** (living without oxygen) bacteria, which decompose organic matter by reducing sulfate (SO_4^{2-}) to sulfide (S^{2-}), which then chemically combines with hydrogen and forms H_2S. Except for populations of anaerobic bacteria, water in the Black Sea deeper than 200 meters (~660 feet) is devoid of life. Only the surface layer is fertile; it supports a thriving community of organisms, including large stocks of fish. Although the residence time of water below the halocline is about 3,000 years, mixing within the basin occurs at the much faster rate of once every century or so. This means that water, once in the basin, resides there for thousands of years. Yet, it is actively mixed upward and downward as surface and deep water are interchanged. Apparently, the cooling of surface water during a very cold winter when combined with an unusual storm can cause overturning of the water column, and deep water is mixed upward into the surface-water layer. This overturning of the water column carries toxic H_2S gas to the surface, creating a pungent odor of "rotten eggs" and killing massive numbers of fish.

6-5

Future Discoveries

Some of the most exciting work is determining the role of both geostrophic and thermohaline circulation in global climate change. In an effort to predict future climate variability, physical oceanographers are examining how specifically ocean and atmospheric processes are coupled. For example, how do the atmosphere and ocean exchange heat, gases, and water over different time scales? What controls the growth and decay rates of sea ice, and how are these processes coupled to atmospheric warming and cooling? How are water masses in the deep ocean affected by and in turn affect climate? How will geostrophic and thermohaline circulation be affected by warmer water and air temperatures and by the inflow of freshwater from melting ice caps and glaciers due to global warming? How specifically is circulation in the tropical ocean coupled to El Niño events and their powerful effect on human and ecological systems worldwide? What role do large, swirling eddies spawned by western boundary geostrophic currents play in climate variability? These questions are numerous and fascinating, and critically important for dealing responsibly and realistically with future climate changes.

STUDY GUIDE

KEY CONCEPTS

1. Global winds in each hemisphere are arranged into large circulation cells that are caused by wind responding to pressure gradients and Coriolis deflection. These cells result in zonal wind flow, including the westerlies of the midlatitudes and the trade winds of the subtropics.
2. Ocean circulation can be divided into wind-induced surface currents, which influence about 10 percent of the total ocean's volume of water, and density-driven subsurface flow, which affects the remaining 90 percent.

3. The large wind-powered circulation *gyres* in each ocean consist of a system of *geostrophic currents* that rotate clockwise in the Northern Hemisphere and counterclockwise in the Southern Hemisphere. These currents, rather than flowing downslope, actually flow around a central mound of water parallel to the water slope. Geostrophic currents represent a dynamic balance between *Coriolis deflection* and *pressure gradient*.

4. In both hemispheres, the pattern of current flow around the gyres is asymmetrical, with narrow, deep, swift transport to the west (*western-boundary intensification*) and broad, shallow, slow drift to the east. This pattern is a response in part to the "pinching" effect of the ocean basin rotating east with the Earth and "squeezing" the western arm of the circulation gyre against the central mound of water.

5. The flow axis of western-boundary currents, such as the Gulf Stream, typically meanders like a river. At times, closed meander loops become separated from the main current and create *rings* that have either a warm-water core or a cold-water core. These rings persist for several months to several years before they are absorbed back into the main current flow.

6. Under the influence of a steadily blowing wind, an *Ekman spiral* can develop, whereby the speed of the current decreases with depth and the direction spirals around to the right in the Northern Hemisphere and to the left in the Southern Hemisphere because of Coriolis deflection. The overall result of the Ekman spiral is net transport at 90 degrees to the right of the wind in the Northern Hemisphere and to the left of the wind in the Southern Hemisphere.

7. Subsurface flow, known as *thermohaline circulation*, results from density variations produced by a difference in the temperature and salinity of water masses. When water masses converge, the more dense water sinks below and buoys up the less dense water. Most water that fills the depths of the oceans is near freezing and originated near the surface of the polar seas. Residence time for water in the deep sea is about 500 to 1,000 years.

8. Ocean basins exchange water on a regular basis. A preliminary "conveyer belt" model of this exchange suggests that warm water from the Pacific Ocean and Indian Ocean is exported to the Atlantic Ocean along the surface, where it cools and sinks, returning at depth back to the Pacific and Indian Oceans.

9. Climate exerts strong control on the water circulation in semienclosed seas. As a consequence of intense evaporation, which creates dense, salty water, the Mediterranean Sea has surface inflow and subsurface outflow over its sill. By contrast, the Black Sea, with its wet climate, has surface outflow and subsurface inflow over its sill. The water chemistry and the marine biology in each of these basins is influenced greatly by their respective circulation patterns.

KEY WORDS

anoxic
bottom water
cold-core ring
Coriolis deflection
deep water
downwelling
Ekman spiral
Ekman transport
geostrophic current
gyre
Langmuir circulation
pressure gradient
thermohaline circulation
trade winds
upwelling
warm-core ring
westerlies
western-boundary intensification

QUESTIONS

REVIEW OF BASIC CONCEPTS

1. Explain why Coriolis deflection produces an apparent rather than a real curvature in the trajectory of a body moving relative to the earth.

2. Contrast and account for the differences in speed, flow direction, width, and depth of the eastern and western arms of current gyres in both hemispheres.

3. How are upwelling and downwelling generated along coasts and in the open ocean.

4. Explain how water masses form generally, and how those generated in the high latitudes drive the thermohaline circulation of the deep sea.

5. What specifically makes Antarctic Bottom Water (AABW) the densest water in the ocean?

6. Describe the formation of Mediterranean Intermediate Water (MIW), and account for its presence in the North Atlantic Ocean.

CRITICAL-THINKING ESSAYS

1. Using diagrams, describe the relationship of circulation gyres in the open ocean of the Southern Hemisphere to (a) zonal wind patterns, (b) Coriolis deflection, (c) pressure gradients, (d) flow direction, (e) downwelling.

2. Using a series of diagrams, describe the formation of a warm-core ring spawned from the Kuroshio Current of the North Pacific Ocean.

3. Can you explain why the top of cold-core rings are concave and the top of warm-core rings are convex? (Hint: Keep geostrophic balance in mind.)

4. Why does surface water flow into the Mediterranean Sea but out of the Black Sea?

5. Why, although it is highly saline, doesn't the MIW sink to the bottom of the North Atlantic Ocean?

6. Is the Ekman spiral a geostrophic current? Why or why not?

7. Why do you suppose that AABW is confined to the western half of the Atlantic Ocean? (See Figure 6–16a.)

DISCOVERING WITH NUMBERS

1. Estimate the speed in knots and centimeters per second of the surface current of an Ekman spiral that would be generated by a 30-knot wind. (Read section entitled "Wind Drag.") Now estimate the speed of the deep current of the Ekman spiral that is flowing in the direction opposite to that of the surface current. (Read about Ekman spiral.)

2. Upwelling occurs at a rate of about 1 to 10 meters per day. Calculate the speed of that water motion in centimeters per second and in knots. (See Appendix II.)

3. Examine Figure 6–15. If water sampled from the bottom of the Pacific Ocean at the equator was "dated" at 850 years old, what is the flow rate in centimeters per second, in meters per year, and in mph of that bottom water mass? (One degree of latitude equals 60 nautical miles; consult Appendix II for conversion values.)

4. Consult Table 6–1 and Figure 6–4, and estimate the time in years it would take a water molecule to complete one circuit around the edge of the circulation gyre of the North Atlantic Ocean.

SELECTED READINGS

Armi, L. 1978. Mixing in the deep ocean—the importance of boundaries. *Oceanus* 21 (1): 14–19.

Baker, D. J., Jr. 1970.Models of ocean circulation. *Scientific American* 222 (1): 114–21.

Frye, J. 1982. The ring story. *Sea Frontiers* 28 (5): 258–67.

Gaskell, T. F. 1972. *The Gulf Stream.* New York: John Day.

Hollister, C. D., and Nowell, A. 1984. The dynamic abyss. *Scientific American* 250 (3): 42–53.

Jenkins, W. J. 1992. Tracers in oceanography. *Oceanus* 35 (1): 47–55.

Joyce, T., and Wiebe, P. H. 1983. Warm-core rings of the Gulf Stream. *Oceanus* 26 (2): 34–44.

MacLeish, W. H. 1989. Painting a portrait of the Gulf Stream from miles above—and below. *Smithsonian* 19 (12): 42–55.

Oceanus. 1976. Ocean eddies. Special issue 19 (3).

Oceanus. 1981. Oceanography from space. Special issue 23 (3).

Oceanus. 1992. Physical oceanography. Special issue 35 (2).

Oceanus. 1994. Atlantic Ocean circulation. Special issue 37 (1).

Open University Course Team. 1989. *Ocean Circulation.* New York: Pergamon Press.

Richardson, P. L. 1991. SOFAR floats give a new view of ocean eddies. *Oceanus* 34 (1): 23–31.

Richardson, P. L. 1993. Tracking ocean eddies. *American Scientist* 81 (3): 261–71.

Robinson, A. R., and Simmons, W. 1980. A new dimension in physical oceanography. *Oceanus* 23 (1): 40–51.

Spindel, R. C., and Worcester, P. F. 1990. Ocean acoustic tomography. *Scientific American* 263 (4) 94–99.

Tolmazin, D. 1985. *Elements of Dynamic Oceanography.* Winchester, Mass.: Allen & Unwin.

Toole, J. M. 1996. New data on deep-sea turbulence. *Oceanus* 39(2): 33–35.

Webster, P. J. 1981. Monsoons. *Scientific American* 245 (5): 108–19.

Whitehead, J. A. 1989. Giant ocean cataracts. *Scientific American* 260 (2): 50–57.

Whitworth, T., III, 1988. The Antarctic Circumpolar Current. *Oceanus* 31 (2): 53–58.

Wiebe, P. H. 1982. Rings of the Gulf Stream. *Scientific American* 246 (3): 60–70.

Woodard, C. 2001. *Ocean's End: Travels through Endangered Seas.* New York: Basic Books.

Wunsch, C. 1992. Observing ocean circulation from space. *Oceanus* 35 (2): 9–17.

TOOLS FOR LEARNING

Tools for Learning is an on-line review area located at this book's web site OceanLink (**www.jbpub.com/oceanlink**). The review area provides a variety of activities designed to help you study for your class. You will find chapter outlines, review questions, hints for some of the book's math questions (identified by the math icon), web research tips for selected Critical Thinking Essay questions, key term reviews, and figure labeling exercises.

7 Waves in the Ocean

On heav'nly ground they stood, and from
the shore
They view'd the vast immeasurable Abyss
Outrageous as a Sea, dark, wasteful, wilde,
Up from the bottom turn'd by furious windes
And surging waves, as Mountains to assault
Heav'n's highth, and with the Centre mix
the Pole.

—*John Milton,* Paradise Lost, *book viii*

PREVIEW

WATER IN THE OCEAN, with its content of suspended and dissolved material, is always being stirred by surface and deep currents. The surface layers of the ocean move and swirl about in response to wind stress; the deep bottom water, to differences in the density of water masses. These two styles of water motion—geostrophic currents and thermohaline circulation—stir and slowly mix the water column. We will complete our study of water movement by examining waves. Whether expressed as the regular, rolling swell of the open ocean or as monster breakers crashing on a rocky, kelp-strewn shore, they are the sea's own language.

We will begin our analysis of ocean waves by defining their properties and examining the motion of wave forms and of the water particles beneath the water surface Then we will reconstruct the life history of wind-generated waves, from their formation far out at sea to their collapse as breakers along a distant shoreline. Last, we will study wave types other than those created by the wind, including waves that are caused by earthquakes and waves that travel beneath, rather than along, the sea surface.

web navigator

critical thinking on the web

math tutor on the web

www.jbpub.com/oceanlink

One of the most memorable aspects of a beach trip is the pounding surf. When waves are small or absent, visitors are often disappointed, the excitement of the beach being markedly diminished. When the surf is high, some thrill to the experience of letting the power of a wave propel them shoreward on their surfboards. An understanding of waves and their motions should increase your enjoyment the next time you visit the beach or go for a sail in the ocean.

7-1 Properties of Ocean Waves

An ocean wave is an undulation of the sea surface, usually created by the wind. When undisturbed by wind (or some other factors, such as earthquakes), the sea surface is naturally smooth. An ocean wave represents the sea surface in regular motion, as water rises to a **wave crest** (the highest part of the wave) and sinks to a **wave trough** (the lowest part of a wave). If the seas are not too chaotic, a regular up-and-down motion is evident on the water surface affected by the passage of waves.

Waves are distinguished from one another by the following properties (**Figure 7–1a**):

Wave height: the vertical distance separating the crest from the trough. **Wavelength:** the horizontal distance between the crest of one wave and the crest of an adjacent wave. **Wave period:** the time it takes two successive crests to pass a fixed point, such as the piling of a pier.

Methods for measuring wave properties are described in the boxed feature, "Wave-Measuring Techniques."

The wave period, it turns out, is a useful way to classify waves (**Table 7–1** and **Figure 7–1b**). The smallest water waves, which have periods of less than one-tenth of a second, are called **capillary waves**. They are most readily seen on a flat, calm sea when a puff of wind abruptly disturbs the water surface, creating very tiny, short-lived wavelets. This ruffling of the water by a light breeze is called a "cat's paw." Locally generated waves, termed **chop**, have periods of about 1 second; ocean **swell**, about 10 seconds. Very long periods on the order of minutes and tens of minutes are associated with seismic sea waves, or **tsunamis**, and with the back-and-forth sloshing of water in harbors, called **seiches**. If these wave types are unfamiliar, you need not worry, because they are treated in detail in this chapter. All you need to know for the moment is the meaning of a wave period and the way it is used to classify waves. Study Figure 7–1 carefully before proceeding.

WIND GENERATION OF WAVES

A visit to the seashore on most days reveals a connection between the strength of the wind and the size of waves: the stronger the wind, the larger the waves. We need not concern ourselves here with the complex details of how wind energy is transferred to the sea surface. Instead, we will touch on some well-known ideas regarding the generation and growth of waves.

The variety and size of wind-generated waves are controlled by four principal factors: (1) wind velocity, (2) wind duration, (3) fetch, and (4) origi-

TABLE 7-1 Wave classification.

Wave	Period	Wavelength	Wave Type	Cause
Capillary wave	<0.1 sec	<2 cm	Deep to shallow	Local winds
Chop	1–10 sec	1–10 m	Deep to shallow	Local winds
Swell	10–30 sec	Up to hundreds of m	Deep or shallow	Distant storm
Seiche	10 min–10 hr	Up to hundreds of km	Shallow or intermediate	Wind, tsunami, tidal resonance
Tsunami	10–60 min	Up to hundreds of km	Shallow or intermediate	Submarine disturbance
Tide	12.4–24.8 hr	Thousands of km	Shallow	Gravitational attraction of sun and moon
Internal wave	min to hr	Up to hundreds of m	Deep to shallow	Disturbance at pycnocline

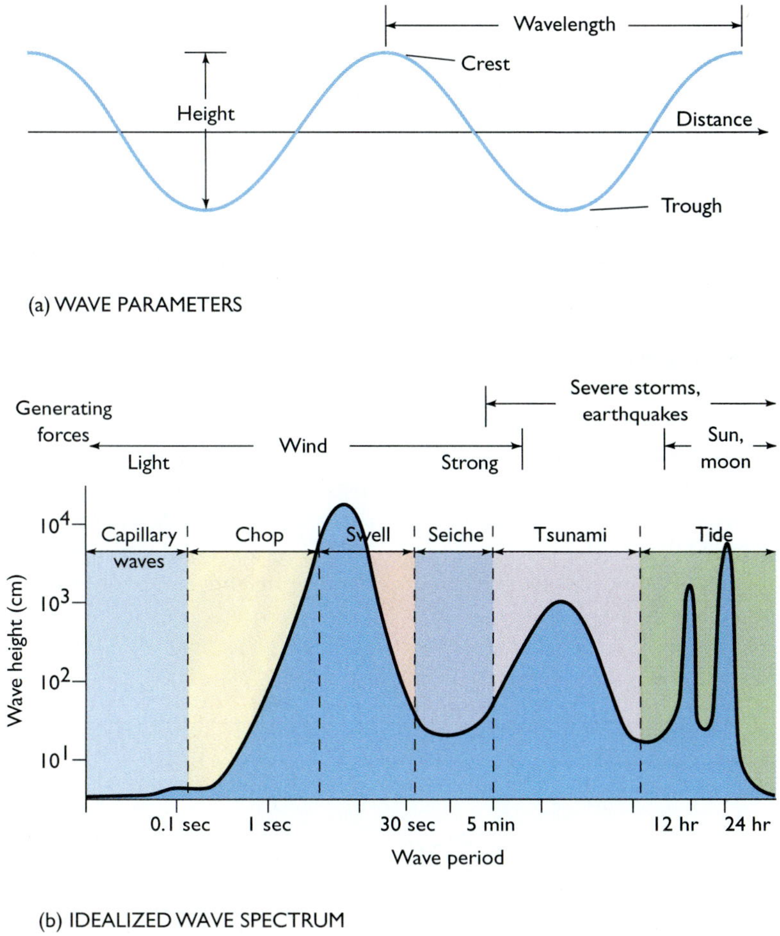

FIGURE 7-1

Waves and their properties. (a) Regular, symmetrical waves are described by their height, wavelength, and period (the time one wavelength takes to pass a fixed point). (b) Waves are classified according to their wave period. The scales of wave height and wave period are not linear, but logarithmic, being based on powers of 10. Additional information about wave properties is summarized in Table 7-1.

nal sea state. Put simply, as wind speed increases, so do the wavelength, the period, and the height of the resulting waves (**Table 7–2**)—provided the wind blows long enough and the area of water over which the wind blows, called the **fetch**, is adequate in length. These relationships are straightforward and can be confirmed during a wind storm by moving a bathtub outdoors , preferably by a large pool. As the wind strikes the water in your tub, waves will be generated as the surface is disturbed by the wind's turbulent motions. Their size, to a degree, will depend on the wind's speed. However, even if the wind is blowing persistently at 90 kilometers per hour (~56 mph), a 15-meter-high (~50 feet) wave, which theoretically could develop under these wind conditions, will not appear in your tub because of its limited fetch. But if you glance over at the pool, bigger waves will be evident there because of its longer fetch. This explains why the largest waves to develop in a pond are considerably smaller than those produced in lakes, which in turn are dwarfed by the huge waves that form in the oceans.

It makes sense that a single puff of wind blowing at 90 kilometers per hour (~56 mph) will not raise a 15-meter-high (~50 feet) wave even if the fetch is adequate. Indeed, the wind would have to blow at that speed for 42 hours before waves of such height would be evident. Also, waves present before the onset of recent winds, the original sea state, affect both the wind and the generation of the newly produced waves. Without "old seas," the sea surface is smooth. With "old seas," the sea surface is rough, and this irregular surface affects how the water and wind interact. Waves will continue to grow in size until they reach a maximum size that is determined by the wind speed and fetch. At this stage of wave growth, called a **fully developed sea**, waves can no longer grow in size under the existing wind conditions, because the energy supplied by the wind equals the energy lost by waves breaking and leaving the fetch area.

Wave-Measuring Techniques

The simplest way to measure waves is to observe the oscillations of the sea surface against a rod or pole with metric markings that is firmly imbedded in the sea bottom. Reasonable estimates of wave height and wavelength can be made with this graduated pole, and wave period can be estimated by using a watch with a second hand. The procedure is complicated, however, by the fact that waves of various sizes can pass the pole at the same time. To overcome this problem, oceanographers employ pressure sensors that are anchored to the sea floor. The passage of waves is indicated by variations in water pressure, as a crest (more water and therefore higher pressure) and then a trough (less water and therefore lower pressure) pass over the sensors (**Figure B7–1**). These pressure variations can be transmitted by cable to a shore station, where they are recorded and stored for later analysis.

Equally effective for measuring the properties of waves are certain electrical devices. One such instrument consists of pairs of wires, separated by a tiny gap, mounted every few centimeters or so on a vertical rod that is secured to the sea bottom. Immersion in seawater short-circuits the wire pairs. Therefore, the passage of waves can be recorded simply and accurately by noting over time the number of wire pairs that are short circuited and so must be under water.

All these techniques measure the passage of waves over time at a single point where the instrument is installed. Although satisfactory for certain kinds of wave studies, such as monitoring breaker heights and periods at a specific beach site, these data do not indicate how the wave characteristics

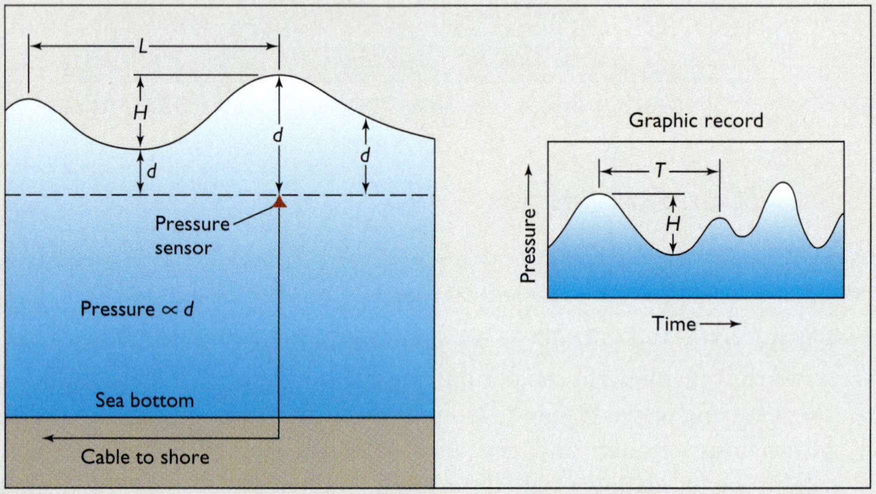

FIGURE B7-1

Wave measurements. Pressure sensors suspended below wave troughs are used to measure fundamental wave properties such as wavelength, *L*, height, *H*, and period, *T*. The height of water and, hence, hydrostatic pressure vary with the passage of waves over the sensor, decreasing beneath a wave trough and increasing beneath a wave crest.

Table 7–2 specifies a variety of predicted wave properties for fully developed seas under different wind speeds. An often used statistical wave measure included in this table is the **significant wave height**, which is the average of the highest one-third of all the waves present in an area of the sea surface. The significant wave height will always be more than the average wave height. Let's think about significant height in another, more familiar way. In your mind, compare the average height of your classmates to the average height of the tallest third of your class. Obviously, the latter, the significant height, will be considerably greater than the average height of the class members. Why might one want to know the significant height? Well, I might be a recruiter for a basketball club, and the significant height would tell me a great deal more about the taller members of a class than would the average height, wouldn't it? Likewise, big waves have much more energy than small waves do, and the taller ones are those that need to be predicted, because they are the most dangerous for ships and inflict the most damage on a shoreline. Hence, the use of the significant wave height.

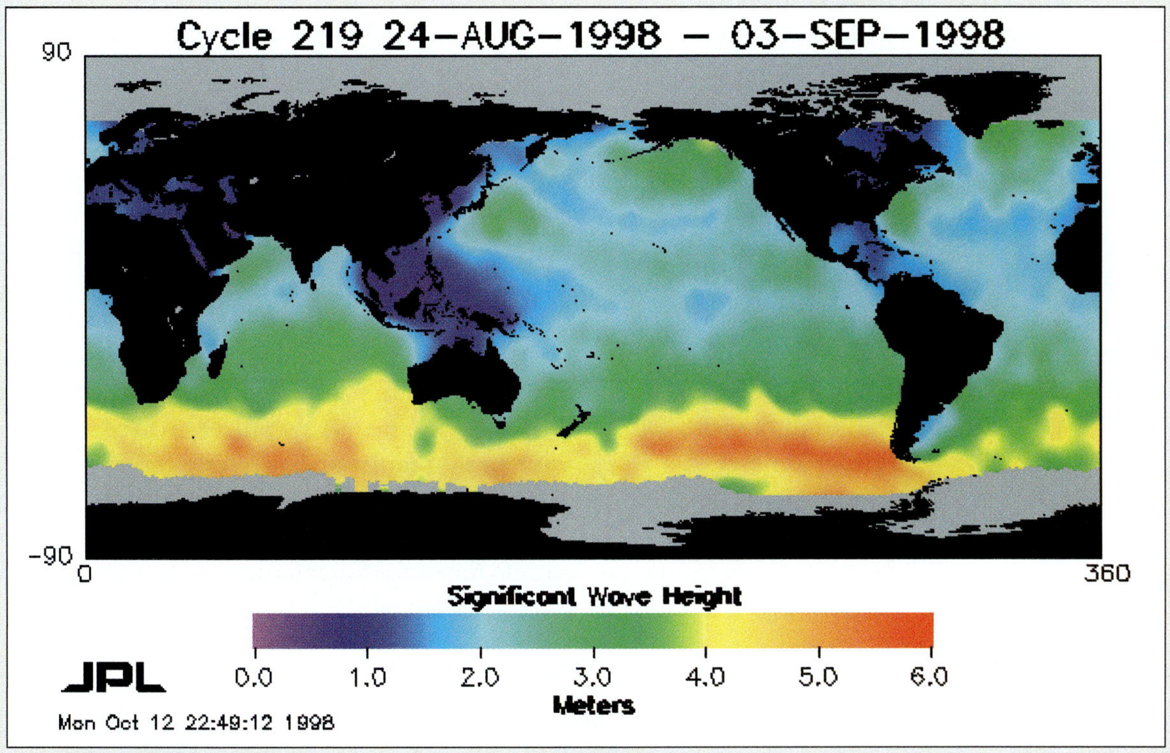

FIGURE **B7-2**

Satellite imagery. Waves can now be monitored by radar instruments mounted in satellites. The irregular sea surface created by waves reflects and distorts a pulse of radio energy, which is used as an indirect measure of wave height.

vary along the coastline. With the availability of satellites, it is now possible to conduct oceanographic studies from instruments orbiting the Earth far out in space. For example, radar pulses can be reflected from the sea surface and timed (**Figure B7–2**). In this way, the height of the satellite above the sea surface can be determined to an accuracy of about 10 centimeters (~3.9 inches). Any waves taller than this can in theory be detected and measured from an orbiting satellite.

Visit www.jbpub.com/oceanlink for more information.

TABLE **7-2**

Waves in fully developed seas.

Wind Speed (km/hr)	Average Height (m)	Average Length (m)	Average Period (sec)	Significant Wave Height (m)
20	0.33	10.6	3.2	0.5
30	0.88	22.2	4.6	1.2
40	1.8	39.7	6.2	2.5
50	3.2	61.8	7.7	4.5
60	5.1	89.2	9.9	7.1
70	7.4	121.4	10.8	10.3
80	10.3	158.6	12.4	14.3
90	13.9	201.6	13.9	19.3

Source: Adapted from H. V. Thurman, *Introductory Oceanography* (Westerville, Ohio: Charles E. Merrill, 1988).

The accuracy for predicting a significant wave height as well as other wave properties using statistical models is remarkably good. Let's consider a specific case. A fully developed sea with large waves on the order of 13 to 14 meters (~43 to 46 feet) high will arise for a wind blowing at 20 meters per second (~45 mph), provided the fetch is equal or greater than 1,500 kilometers (~930 miles) and the wind blows at that high speed for at least 40 hours. A smaller fetch or shorter wind duration will result in waves that are smaller than the theoretical maximum for a fully developed sea. In the case of a wind blowing at 20 meters per second (~45 mph), the largest waves to form, if the fetch is limited to 500 kilometers (~310 miles) or the wind duration to 22 hours, will be no higher than about 7 meters (~21 feet).

7-2

Wave Motions

Wind-generated waves are **progressive waves**, because they travel (progress) across the sea surface. If you focus on the crest of a progressive wave, you can follow its path over time. But what actually is moving? The wave form obviously is moving, but what about the water just below the sea surface? Actually, two basic kinds of motions are associated with ocean waves: the forward movement of the wave form itself and the orbital motion of water particles beneath the wave. An analogy may help you understand this better. If you tie a length of rope to a doorknob and then flick your wrist vigorously, you will cause a series of progressive waves to travel toward the door (**Figure 7–2**). The individual rope fibers will move up and down as the waves pass, but they will not move closer to the door. What does move is the energy pulses created by flicking your wrist, manifested as a series of moving (propagating) wave forms. We will now examine both of these types of wave motion, as well as a few very simple formulas for calculating the speed of waves on the water surface.

FIGURE 7-2

Wave motion. The waves created by flicking a rope represent the flow of energy and not mass.

THE MOTION OF WATER PARTICLES BENEATH WAVES

Anyone who has played in the surf is familiar with the orbital motion of water particles beneath waves. For example, floaters on an air mattress and swimmers just beyond the breakers will be pushed upward and shoreward as they rise up the front of a wave to its crest, and then will be pushed downward and seaward as they move down the back of the wave to its trough. This back-and-forth and up-and-down motion that occurs with the passage of one wave describes an orbit (**Figures 7–3a** and **7–3b**), whereby anything floating in the water, and the water particles themselves, rotate around a vertical circle. The bigger the wave, the larger is the size of the orbit. So this means that with the passage of one wave, a float and all the water particles disturbed by the wave return to the same spot where they began. The net result is that in theory there is no forward motion of mass, no matter how many waves pass through the area. Thus, wave energy, not water particles, travels across the sea surface.

If you think about it, this makes a great deal of sense, because if the water were moving forward with the wave, the hulls of ships, not to mention the bodies of swimmers, would be battered by the impact of the moving water. Also, this explains why a bay does not dry up when an offshore wind blowing for several days causes waves to move out to sea. The energy is being exported from the bay, but not the water, because water particles set in motion by waves have no net movement. Actually, the orbits of the water particles are not quite closed; they do move slightly forward in the direction of wave advance (**Figure 7–3c**). In the open ocean, this slight forward movement of water with the passage of waves, the mass transport, called *Stokes drift,* is of little consequence, except, it has been discovered, in the shallow water of the inner continental shelf. When waves arrive at a shoreline, the slow mass transport by waves moves water against the coast, causing nearshore currents in the surf zone. This important topic is discussed in Chapter 11.

The size of the orbits described by water particles that are influenced by waves drops rapidly with depth below the water surface, and is not detectable at a water depth equal to about one-half the wavelength (see Figure 7–3b). This depth is called the **wave base**. Snorkelers and scuba divers are well aware of this effect, as they can escape being pushed around by waves by swimming downward. If the wave is 10 meters long (~33 feet), they will have to dive to a depth of 5 meters (~17 feet) in order to reach the wave base, below which there is little or no wave-induced motion. A 50-meter-long (~165 feet) wave would require their diving to a water depth of 25 meters (~83 feet) in

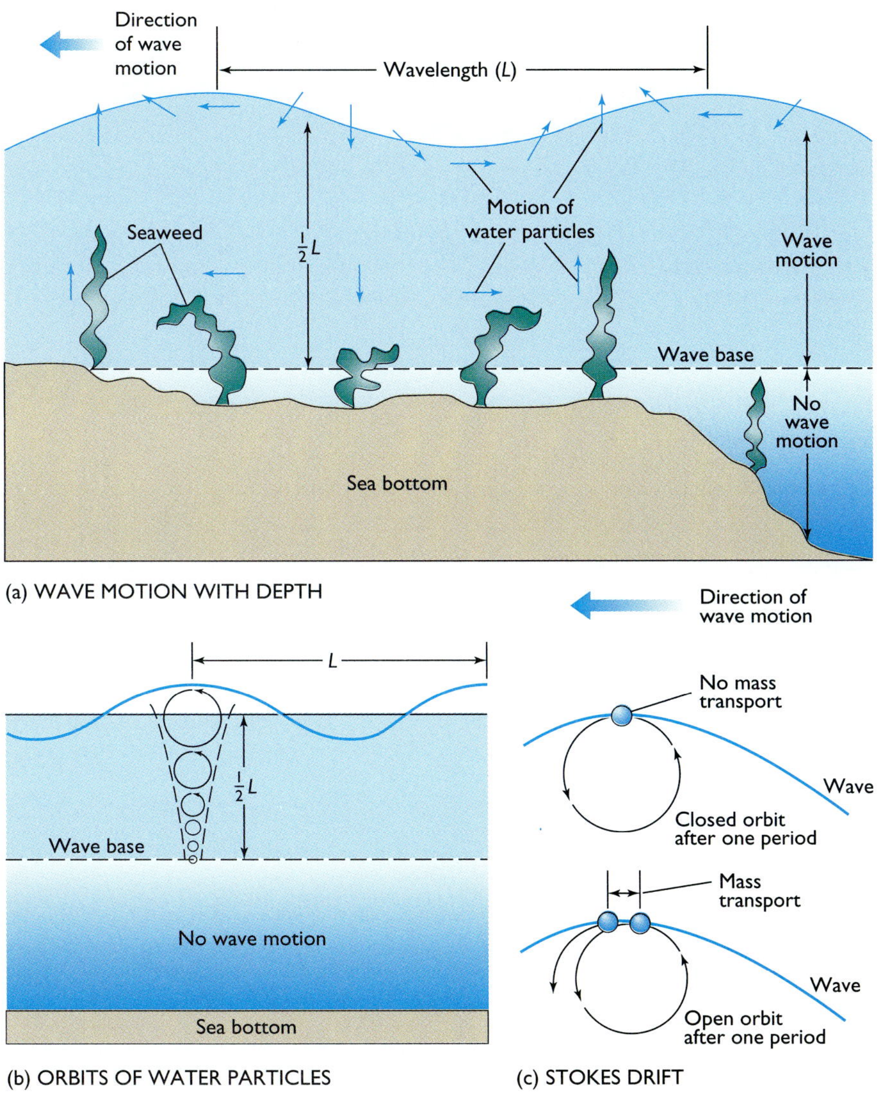

FIGURE 7-3

The motion of water particles beneath waves. (a) The arrows at the sea surface denote the motion of water particles beneath waves. With the passage of one wavelength, a water particle at the sea surface describes a circular orbit with a diameter equal to the wave's height. Water particles below the sea surface also describe orbits with the passage of each wave, as indicated by the back-and-forth motion of seaweed attached to the ocean floor. (b) The orbital diameters described by water particles beneath waves decrease rapidly with distance below the water surface. Wave-induced water motion essentially ceases at a depth (called the wave base) that is equal to one-half the wavelength. (c) The orbits described by water particles beneath waves are not closed, but slightly open. This results in a net displacement of water, called mass transport (Stokes drift), in the direction of wave advance.

SCIENCE BY NUMBERS

Wave Celerity

Let's assume that you're anchored in 9 meters of water and are fishing for bluefish. As you sit in the cockpit, you sense the regular up-and-down motion of your boat due to ocean swell. The sea is perfectly glassy except for the ocean swell. For distraction you decide to measure the time it takes for pairs of consecutive wave crests to pass your anchor line. You discover that all the measurements cluster around 10 seconds, and so you conclude, rightfully, that the swell has a 10-second period. You remember from an oceanography course you took the previous year that the period of a wave remains unchanged as the wave travels from deep to shallow water. What else can you infer about this swell? It occurs to you that you actually have the oceanography textbook below and so you get it.

As you scan the chapter on waves, you discover some simple formulas. The speed of a deep-water wave is $C = 1.56T$ m/sec, where T is the wave period. You also note that $C = 1.25\sqrt{L}$ m/sec, where L is the wavelength. So, you decide to calculate the wavelength of the swell before it became a shallow-water wave. You do this in three steps. First,

$$C = 1.56\,T\text{ m/sec} = (1.56)\,(10)\text{ m/sec} = 15.6\text{ m/sec.}$$

Second,

$$C = 1.25\sqrt{L}$$

and, therefore,

$$15.6 = 1.25\sqrt{L}.$$

Third,

$$\sqrt{L} = 15.6/1.25 = 12.5\text{m.}$$

You square both sides to get rid of the square root sign:

$$L = (12.5)^2 = 156\text{ m.}$$

Now you ask yourself, how fast is the wave traveling past your boat? It obviously is a shallow-water wave, and you quickly uncover the formula for the speed of a shallow-water wave:

$$C = 3.13\sqrt{d}\text{ m/sec,}$$

where d is water depth. Thus,

$$C = 3.13\sqrt{9}\text{ m/sec} = (3.13)\,(3)\text{ m/sec} = 9.39\text{ m/sec.}$$

You wonder what the speed of the swell will be in 4 meters and 1 meter of water. So you crank out the numbers.

$$C = 3.13\sqrt{4}\text{ m/sec} = (3.13)\,(2)\text{ m/sec} = 6.16\text{ m/sec,}$$

and

$$C = 3.13\sqrt{1}\text{ m/sec} = (3.13)\,(1)\text{ m/sec} = 3.13\text{ m/sec.}$$

order to reach its wave base. Submarines commonly escape the uncomfortable and sometimes dangerous motions of large storm waves by descending to the quiet water that exists below the wave base of even the longest surface waves.

When the sea bottom is deeper than one-half the wavelength, the orbits of the water particles are near circular (**Figure 7–4a**). Obviously, waves that are in water shallower than one-half their length will interact with the ocean floor. Water near the sea bottom cannot move in a circular orbit, only "back and forth." Here the circular orbits of the water particles become flattened into ellipses because of bottom interference (**Figure 7–4b**). Also, bottom interference changes the shape and speed of the wave and transforms it from a **deep-water wave**, where bottom interactions were absent because the wave base was above the sea floor, into an **intermediate-water wave** and then a **shallow-water wave**, where the effects from contacting the ocean floor become important (see Figure 7–4b). A deep-water wave is defined as a wave moving through water that is deeper than its wave base, which is one-half its wavelength. An intermediate-water wave travels through water depths that are between one-half and one-twentieth the wavelength, and a shallow-water wave is in water less than one-twentieth the wavelength. It is important to note that this is a relative classification, meaning that the wavelength is examined relative to the water depth. It's possible to have deep-water waves in a bathtub filled with a few inches of water and shallow-water waves in the deep ocean where water is miles deep.

MOTION OF THE WAVE FORM

It is a simple matter to calculate the speed of the wave itself. The formula is

$$C = L/T,$$

where C is **celerity** (speed), L is wavelength, and T is wave period. This means that if you simply divide the wavelength by the wave period, you have a measure of the wave's speed. Physicists use the word *celerity* instead of *speed*, because the latter applies to the motion of a mass, such as a car, or a ball, or a water particle. Since we know that there is no net displacement of water (mass) beneath a wave (remember that the water particles describe circular orbits), there technically cannot be speed. Think back to the analogy of the rope tied to a doorknob (see Figure 7–2). When you flick your wrist and create waves, what has speed is not the rope fibers (they simply go up and down), but the wave forms that represent energy traveling along the cord toward the door. Although the distinction between celerity and speed is a useful one, physical oceanographers use wave speed and not celerity in describing the rate of motion of a sea wave.

We need to take one more factor into account. Wind blowing on a water surface creates a group of waves. At the front of the group, the leading wave disappears as it is drained of energy by having to set the undisturbed water in front of it in motion. However, at the rear of the group a new wave is generated from energy in the form of water-particle

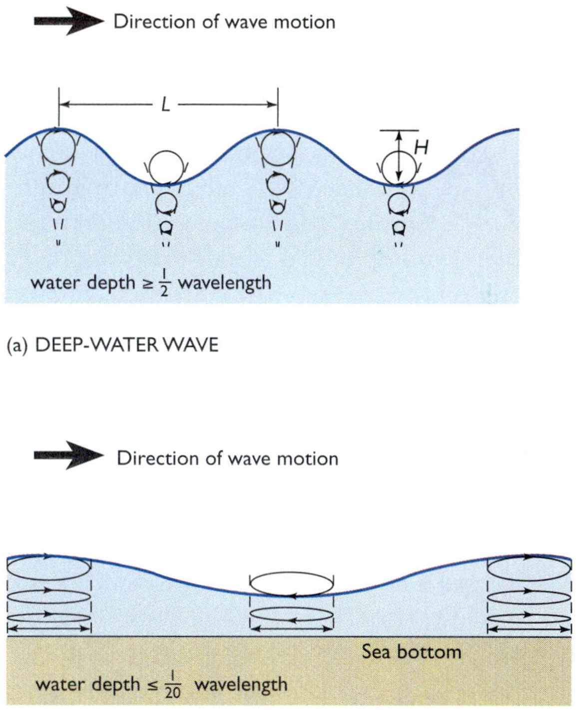

FIGURE **7-4**

The distortion of water-particle orbits in shallow water. (a) For deep-water waves, the orbits described by water particles beneath waves are circles. (b) For shallow-water waves, the orbits of water particles are greatly compressed vertically into elongated ellipses.

orbits left behind by the advancing group of waves. Because of this, the formula for speed applies to the individual waves traveling through the group. The wave group itself, and therefore wave energy, travel across an undisturbed sea surface at a speed that is less than the celerity of the individual waves.

When you study waves, you learn that their speed depends directly on their length and period. The longer the wavelength and period of the waves, the faster they travel. However, this is correct only for deep-water waves, which are not in contact with the sea floor. As we will discover, the speed of intermediate- and shallow-water waves is regulated by water depth, because bottom friction comes into play whenever the sea floor is shallower than the wave base.

7-3

The Life History of Ocean Waves

We can trace the history of waves, beginning with their generation in the fetch area, continuing with their travel across the open ocean, and ending with their collapse as breakers on some distant shore. By examining waves in this manner, you will gain a fuller understanding of their behavior.

GROWTH OF WAVES IN THE FETCH AREA

The fetch area is where the wind is in contact with the sea and is rippling its surface. This is the generating area, where waves are "born." If there is a storm at sea, the fetch coincides with the area of the ocean affected by the storm winds, where wind transfers energy to the water surface and raises waves. As the storm winds increase in strength, the waves grow longer and higher. Remember that their maximum size at any time depends on the size of the fetch, the wind speed, and the wind duration.

People sailing through the storm would be in the fetch area. There they would experience the uncomfortable, jerky motion associated with chaotic waves that impart an irregular, disordered pattern to the sea surface (**Figure 7–5a**). The water motion is turbulent with sea spray, whitecaps (foaming wave crests), and even breaking waves (unstable waves that collapse). These confused seas are really a composite of numerous waves having different lengths, heights, and periods that are continually merging and separating as they move in different directions and at different speeds through the fetch area. The interaction of several waves is called **wave interference**. The irregular shape of a profile of the sea surface in the fetch is due to the interference of progressive waves of different sizes. In fact, it is possible to separate out mathematically all the interfering waves that contribute to the disordered pattern of the profile by a process called *harmonic analysis* (**Figure 7–6a**). When one does this, each separate wave is even and "wavy" in shape. But blended together, they create an irregular sea surface. Let's explore these ideas more thoroughly.

Waves interact with one another in constructive and destructive ways. **Constructive wave interference** occurs whenever and wherever several wave crests or troughs coincide. This results in a composite wave with the crests of the different waves "building up" on one another and their troughs "building down" on one another. This additive process creates a composite wave that is much larger than any of the individual wave components (**Figure 7–6b**). Sometimes it creates **rogue waves**, unusually large breaking waves that are really composed of several large waves that have merged briefly because of constructive wave interference. **Destructive wave interference** occurs when the crest of one wave coincides with the trough of a second wave in such a way that cancellation results (**Figure 7–6c**). This cancellation produces a composite wave that is smaller than the individual component waves. Because speed depends on wavelength, waves in the fetch area are continually merging and separating, producing ever changing wave-interference patterns (**Figure 7–6d**) called **seas**. Waves will continue to grow in size in the fetch area until the seas become either fully developed or fetch- or time-limited for the prevailing wind conditions.

STORM WAVES OUTSIDE THE GENERATING AREA

In contrast to the confused or chaotic seas in the fetch area, ocean waves are much more regular and ordered outside this region (Figure 7–5b). Why would this be? The explanation is simple. The waves raised by the storm move through the fetch at a celerity that depends on their length and period. Short-length, short-period waves move slowly, whereas long-length, long-period waves move rapidly, combining and separating all the time and creating chaotic seas in the fetch area. Once the waves are outside the generating area, no further energy is imparted to them by the storm winds, so their wavelength, celerity, and period remain unchanged. Consequently, waves with longer periods and lengths will outrun waves with shorter periods and lengths. This effect imposes order on the seas, as the waves begin to sort themselves according to celerity and, hence, according to length and period. The process of wave separation, termed **dispersion** (**Figure 7–7a**), produces a regular swell, a wave that is free of the fetch and has a regular up-and-down and back-and-forth motion. If the distance of travel from the fetch to the shoreline is several thousand kilometers, the longer waves will reach the coast several days ahead of the shorter waves, because the former move so much faster than the latter do. This means that a boat sailing beyond the edge of the fetch will sense only ocean swell, because there the waves are dispersed, that is, sorted according to celerity. The leading ocean swell from the storm will be large and long, and, with time, the waves leaving the fetch area will become progressively smaller and shorter—all because of dispersion.

WAVES IN SHALLOW WATER

Once waves enter shallow water, the sea bottom exerts a strong influence on their form and celerity. Three important changes are brought about by shallow-water conditions: transformation of the wave's properties, wave refraction, and collapse as a shore breaker. Each is treated separately here.

FIGURE 7-5

Types of Waves. (a) The chaotic pattern of seas are produced by waves of many sizes interfering with one another. (b) A regular ocean swell produced by dispersion that separates waves on the basis of their celerity. (c) Breakers dissipate their energy at the shoreline by wave collapse.

(a) CHAOTIC WAVES

(b) OCEAN SWELLS

(c) OCEAN BREAKERS

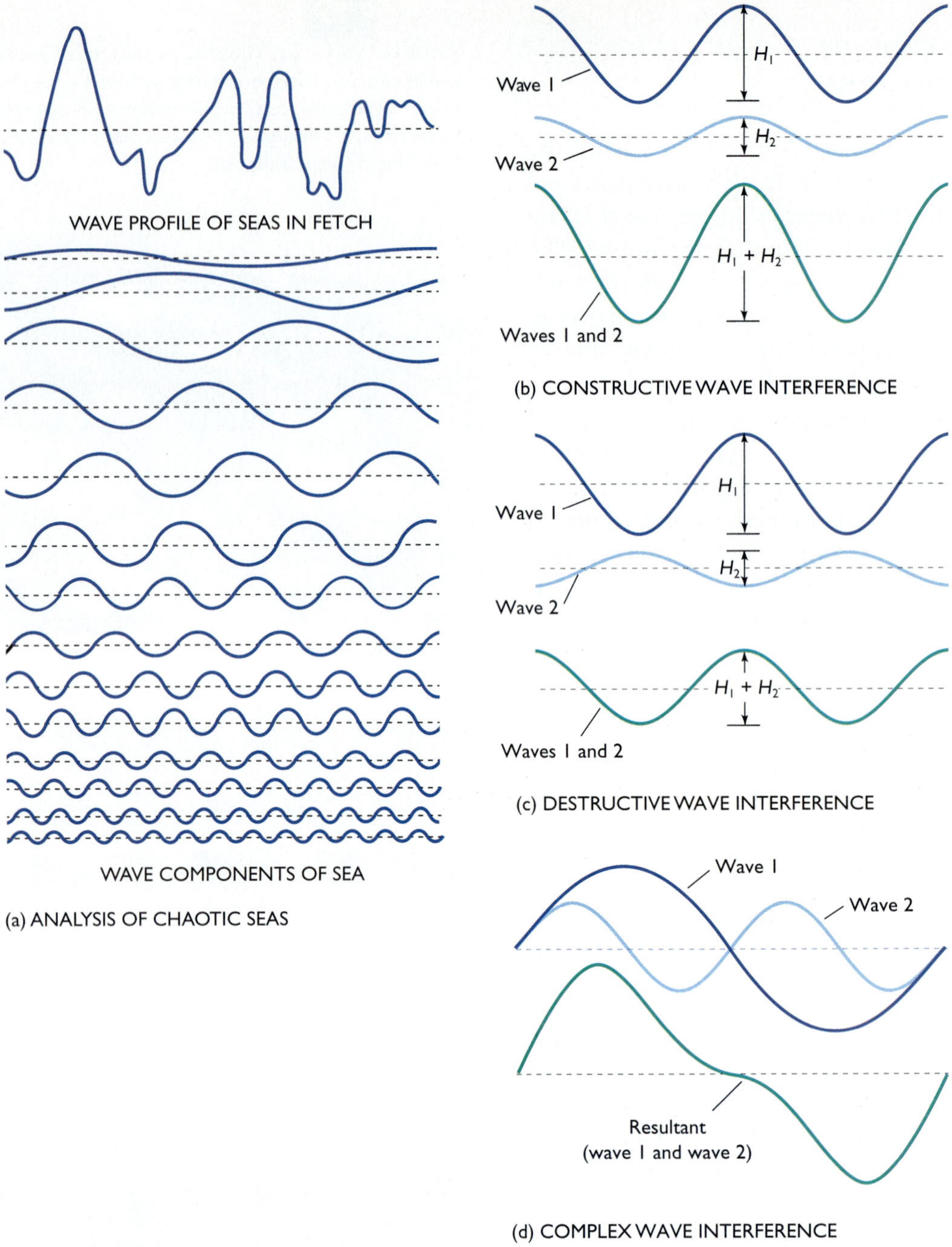

FIGURE 7-6

Wave interference. (a) Seas in the fetch appear chaotic, as indicated by the irregular wave profile at the top of the diagram. However, the irregular seas are created by the interaction—wave interference—of 14 regular waves. The profile of each of these waves is shown beneath the composite profile. Once out of the fetch area, each of these waves will separate into a regular swell, because the longer waves travel faster than the shorter waves. The process of wave separation is called dispersion. [Adapted from G. E. R. Deacon, "The Sea and Its Problems" in *The World Around Us,* G. Sutton, ed. (London: English University Press, 1960), 77–97.] (b) Constructive wave interference deepens the wave troughs and raises the waves crest, producing a larger wave. (c) Destructive wave interference reduces the height of the larger wave. (d) Typically, both constructive and destructive wave interference occur simultaneously, resulting in a complex, irregular wave form.

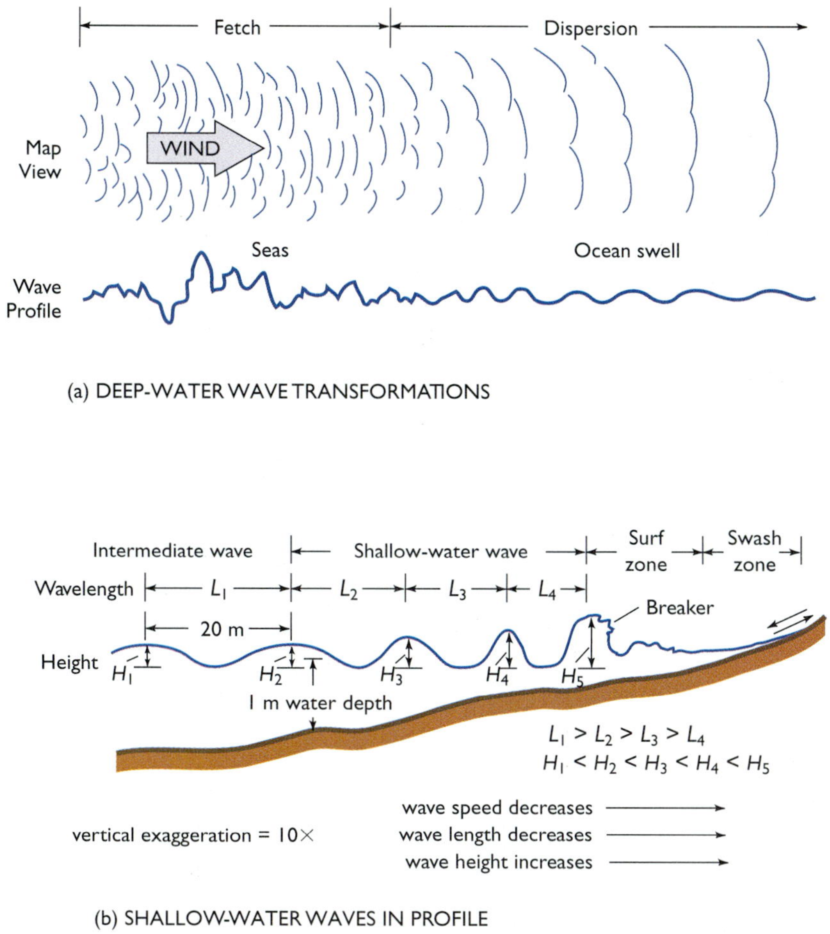

FIGURE 7-7

Wave transformation. (a) Seas are irregular in the fetch area because of constructive and destructive wave interference. Outside the fetch, a regular ocean swell develops as waves sort themselves according to celerity by the process of dispersion. (b) This diagram shows how shallow-water waves are transformed by shoaling of the sea floor. The swash zone is alternately covered and uncovered by water that rushes up the beach and back down the beach as each wave breaks.

SHALLOW-WATER WAVE TRANSFORMATIONS

Far out at sea, even the largest waves generated by strong winds are unaffected by the sea bottom, because their wave base lies far above the deep ocean floor. Near shorelines, however, the sea bottom interferes with the motion of water particles beneath the wave form. These bottom effects become most pronounced where water depths are less than one-twentieth the wavelength, at which point the waves change into shallow-water waves. The influence of the bottom alters the wave form and its celerity. Let's examine these changes.

When waves begin to interact with the bottom, they slow down, and their celerity, which in deep water depended on wavelength and period, now is regulated directly by water depth. The formula for calculating celerity of a shallow-water wave is

$$C = \sqrt{gd} = 3.13\ \sqrt{d}\ \text{m/sec} = 7.01\sqrt{d}\,\text{mph},$$

where C is celerity, g is gravitational acceleration, and d is water depth in meters. The formula says that, regardless of their height or length, the celerity of all shallow-water waves—in teacups, pools, or ocean basins—is governed entirely by water depth. Note that the shallower the water, the slower is the celerity of a shallow-water wave. This means that since they were generated in the fetch, waves travel at their slowest rates at the time that they break on the beach. Also, the waves "bunch up" in shallow water, as their wavelengths decrease, because the leading waves, which are in shallow water, travel more slowly than those trailing behind them in slightly deeper water. As waves approach a shoreline, their height increases and their troughs flatten out, giving rise to an asymmetric (irregular or less ordered) wave profile. This change in shape and growth in height reflect the redistribution of wave energy as water depth decreases. All of these shallow-water transformations are summarized in **Figure 7–7b**.

What doesn't change at all, however, is the period, a fundamental property of a wave. This means that the period of a wave measured with a stopwatch off a beach is identical to its period a the time of its generation in the storm area far out at sea. A 10-second-period wave is always a 10-second-period wave, whether it is in deep or shallow water; and this is true for a 4-second- and 13-second-period wave as well.

WAVE REFRACTION

One of the most important results of a wave's entering shallow water is **refraction**, that is, the bending of the wave crest in response to changes in wave speed. As waves approach a coast, rarely are their crests parallel to the shoreline. Rather, the crestline of the wave lies at some angle to the shore, so the water depth beneath the crest varies. Because the speed of shallow-water waves depends directly on depth, *different parts of the same crest,* which lies at an angle to the shore and to the bottom contours, are traveling shoreward at variable speeds. The inshore part of the crest, which is in the shallow water, moves more slowly landward than the offshore part of the crest, which is in deeper water. This differential speed along the crest causes the wave to reorient itself by refracting, or "bending," so that it becomes more nearly parallel to the shape of the shoreline. If the shoreline is irregular, with embayments and headlands, the wave crests because of refraction will tend to mimic the contours of the coast. Let's flesh out this idea with a concrete example.

Wave refraction is not difficult to understand. For example, consider the refraction pattern of a set of waves advancing toward a jagged, rocky coast (**Figure 7–8a**). We can study the effect of wave refraction in two ways. First, we can examine the wave crestlines themselves. When we do this we see that the offshore wave crests are fairly straight, whereas the inshore wave crests are "bent" around the irregularities of the shore. This "bending" is the result of wave refraction. Imagine a straight crest approaching a headland and a nearby embayment. The part of the crest off the headland is in shallower water than the part of the same crest off the embayment. Because the speed of a shallow-water wave is regulated directly by water depth, the wave crest off the headland will move more slowly (because the water is shallower) than the same crest off the embayment (where the water is deeper). Because of this differential speed the wave bends (see Figure 7–8a).

Second, we can assess refraction by drawing **wave rays** or **wave orthogonals**—imaginary arrows always drawn *perpendicular* to the wave crests that divide an unrefracted wave (the wave in deep water) into equal crestal segments. Because the spacing between orthogonals is constant, and assuming a uniform wave height along the crest, the amount of energy between any two wave orthogonals is similar along the entire crest. This kind of wave analysis is quite simple . Find a wave offshore and divide its crest into even intervals. Then trace a line landward, always bisecting the wave crests encountered at a 90-degree angle. What happens to the wave ray as the waves refract in shallow water? Well, they bend, and some even cross one another (**Figure 7–8b**)! Where wave orthogonals diverge from one another, the wave energy is spread out (defocused), and the breakers are small. Wherever wave orthogonals converge, the wave energy is concentrated (focused), and the waves are large.

Using the idea of focused and defocused wave energy by wave refraction, explain why on any given day large breakers strike headlands while smaller breakers strike embayments (see Figure 7–8a). Then, as a check, predict where the breakers will be the largest along the Cape Cod shore (see Figure 7–8b). If you have trouble doing these exercises, carefully reread the previous two paragraphs.

SHORE BREAKERS

A simple ratio that divides the wave height by the wave length (H/L) gives the **wave steepness**, which helps predict when and where waves will become unstable and break. As waves enter shallow water, their height (H) increases and their wavelength (L) decreases. This means, of course, that the steepness of the wave increases, and you can readily see this as a wave gets bigger and steeper just before it

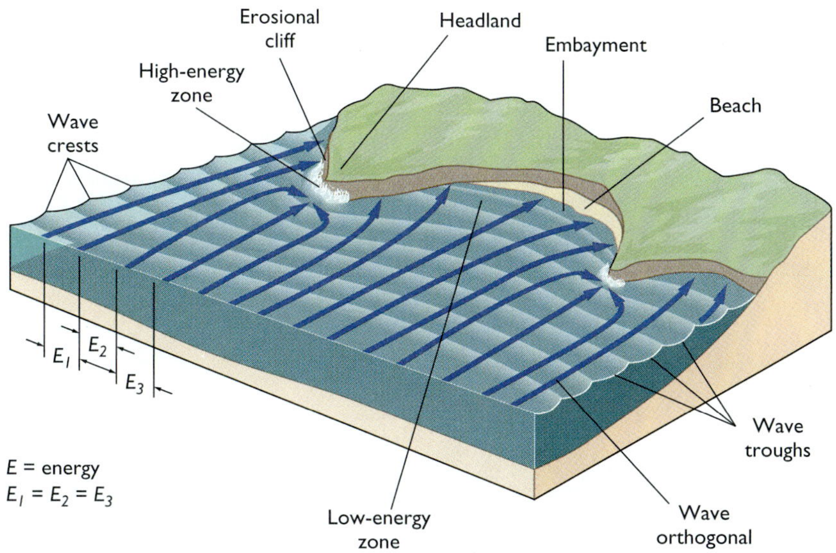

(a) WAVE REFRACTION

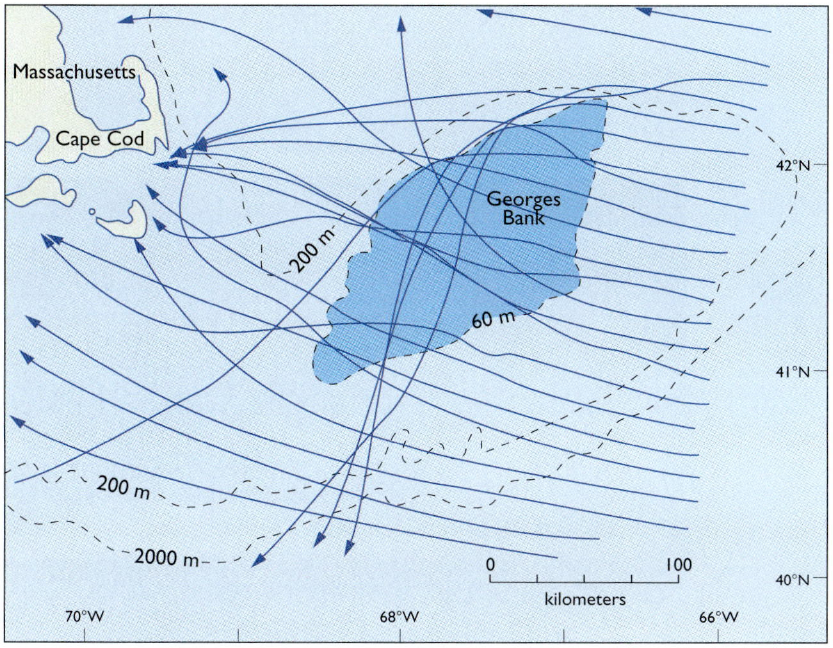

(b) WAVE REFRACTION OFF CAPE COD

FIGURE **7-8**

Wave refraction. (a) This block diagram shows the pattern of wave refraction along an irregular shoreline. Notice that the relatively straight wave crests offshore become bent as they proceed onshore and mimic the general shape of the coastline. The wave orthogonals (wave rays), which are drawn perpendicular to the wave crests, indicate that wave energy is focused on the headlands (high wave energy) and defocused in the embayments (low wave energy). (b) Georges Banks, an important fishing ground, is located several hundred kilometers offshore of Cape Cod, Massachusetts. The refraction of waves over the bank creates a very irregular pattern of wave rays. Note that many wave rays crisscross one another. Some are so sharply bent to the south that they miss Cape Cod entirely. Also, observe how Georges Bank focuses wave energy on the southeastern end of Cape Cod. [Adapted from M. D. Earle and O. S. Madsen, in *Georges Bank,* R. H. Backus and D.W. Bourne, eds. (Cambridge, Mass.:MIT Press, 1987), 78–87.]

collapses as a **breaker** at the shoreline. A critical wave steepness occurs when the wave height is about equal to one-seventh the wavelength (H/L = 1/7, which means that for a 1-meter-high wave, H = 1 meter, and L = 7 meters). Then, the crest is oversteepened and unstable, and bottom friction retards the base of the wave permitting the top of the wave to get ahead. This results in the wave's "breaking."

At least three types of breakers are recognized (**Figure 7–9**):

1. In a **spilling breaker**, the upper part of the crest becomes oversteepened and "spills" down the front side of the advancing wave, continually breaking, and slowly losing its energy across the surf zone.

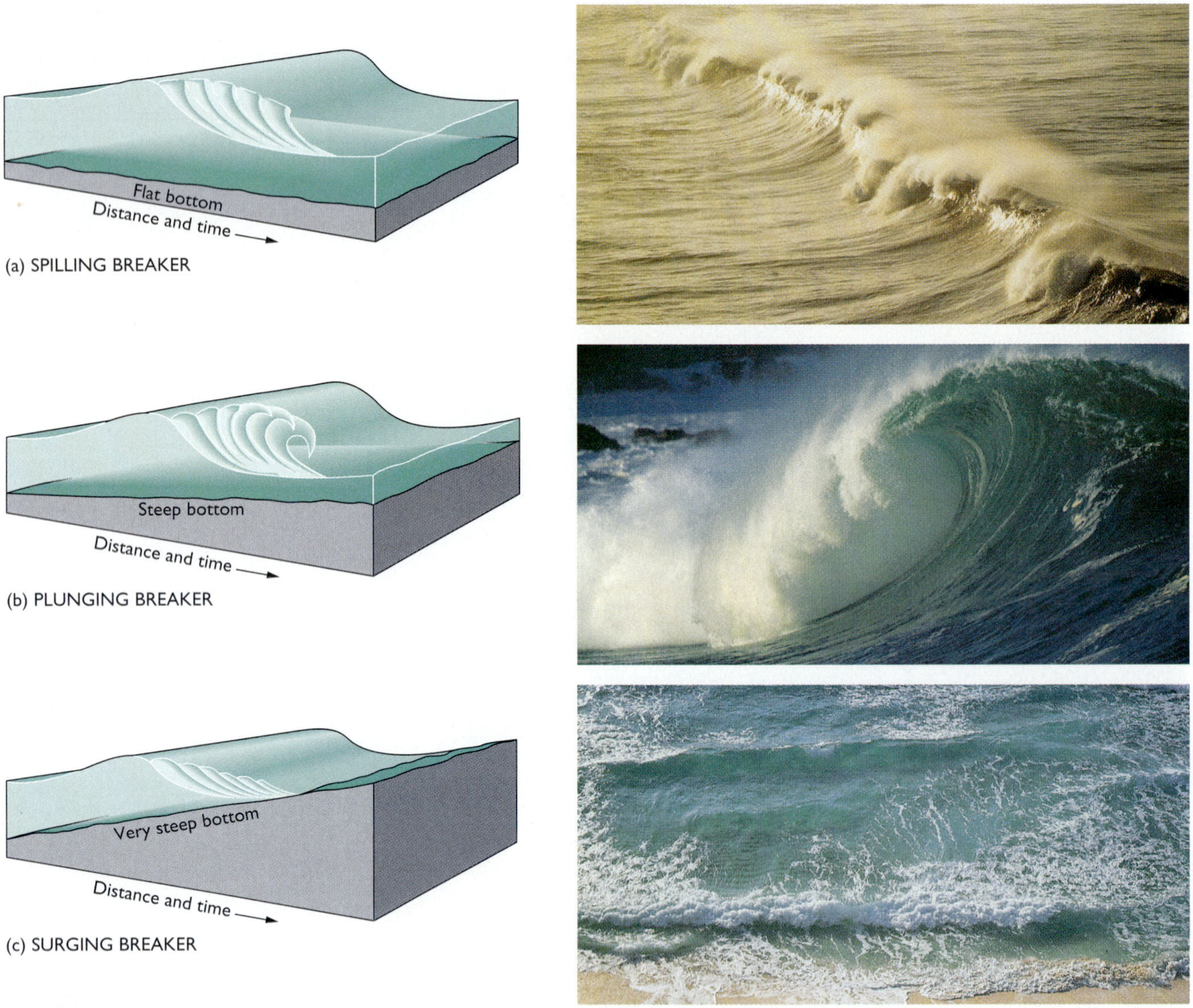

FIGURE 7-9

Types of breakers. (a) The crest of the spilling breaker becomes oversteepened, breaks, and cascades down the front of the wave as it proceeds through the surf zone. Wave energy is released gradually across the entire surf zone. (b) The plunging breaker evolves into the classic "pipeline" shape, as the crest curls over the front of the advancing wave. This wave collapses instantaneously, and much of the wave energy dissipates quickly. (c) The surging breaker does not break, because it never attains a critical wave steepness (H:L—1:7). Rather, the breaker diminishes in size and loses momentum as it rushes up the beach face. Some proportion of the incoming wave energy is reflected back offshore from the beach. [Adapted from C. J. Galvin, *Journal of Geophysical Research* 73 (1968): 3651–59.]

2. In a **plunging breaker**, the entire wave front steepens, curls, and collapses, or "plunges" forward, releasing much of its energy instantaneously.

3. In a **surging breaker**, the flat, low waves do not become oversteepened or actually break; instead they move smoothly up and then down the face of the beach, reflecting much of their energy seaward.

The most common breaker is the spilling type, which tends to form along shores on which a great deal of sand is spread out over a gently sloping sea bottom. Here the surf zone tends to be wide. Steeper bottom gradients tend to produce plunging breakers that collapse instantaneously, creating a relatively narrow surf zone. Surging breakers are associated with the steepest beaches, where the crest does not attain a critical wave steepness, fails to break, and is reflected off the beach and back into the sea. Surging breakers are also found off steep seawalls and sea cliffs, which serve as good reflective surfaces.

SHORELINE UNDER STORM CONDITIONS

Storms affect the shore in a variety of ways. First, the water surface rises in response to the low atmospheric pressure of the storm. Second, strong winds blowing onshore drive water landward and cause it to be stacked against the shore. The combined effect of these two processes results in unusually high water levels that flood the land. These sudden changes of coastal water levels produced by storms are called **storm surges**. They create deep water near the shore and allow waves to break much farther inland than usual. Storm surges are particularly destructive when they happen to coincide with unusually high tides.

The most devastating storm surges are associated with hurricane winds (**Figure 7–10**). For example, in 1900, a storm surge caused extensive flooding of the low-lying coastal areas of Galveston, Texas. The surge, which raised the water level 5 meters (~17 feet) higher than the predicted high tide, killed over 5,000 people. A more recent occurrence of storm surge devastated the northern Bay of Bengal of India in 1970, when over one-half million people died in the floods created by a 6-meter-high (~20 feet) storm surge. In September of 1989, Hurricane Hugo struck Charleston, South Carolina; the storm surge, which amounted to 5 meters (~17 feet) and was superimposed on a high tide, resulted in extensive flooding and damage that amounted to

(a) STORM DAMAGE

(b) STORM SURGE EFFECT

FIGURE **7-10**

Hurricane Damage. (a) Damage caused by the 1900 Galveston hurricane and storm surge. (b) Boats in Gulfport. Mississippi were washed up on shore by the force of Hurricane Camille in 1969.

Tiny Waves and Giant Waves

The most common type of water wave undoubtedly is the capillary wave, a tiny wavelet a few centimeters in wavelength and a few millimeters in height, and with a period of less than half a second. Capillary waves are most apparent when water is flat and calm, although they occur on the backs and fronts of large waves as well. A gentle breeze disturbs the water surface ever so slightly, raising a series of these tiny waves that will decay and disappear as soon as the puff of wind dies out. Their presence is indicated by dark patches of water where the puff of wind strikes the water surface, rippling it and causing sunlight to be reflected irregularly. Because they are so tiny and superficial on the water surface, their restoring force is surface tension rather than gravity. The restoring force for the larger waves, including ripples, swell, breakers, and tides, is gravity.

What exactly is surface tension? A water surface can be visualized as a thin layer of H_2O molecules that are attracted to one another by hydrogen bonds, constituting an electrical force that arises as a consequence of the dipolar nature of the water molecules (see Figure 5-3c in Chapter 5). The surface tension results from the fact that each H_2O molecule at the surface is "pulled" laterally and downward by the hydrogen bonds that it shares with the surrounding molecules (**Figure B7–3a**), stretching the water surface and creating a remarkably strong elastic "skin" that resists deformation.

When wind blows, air molecules drag across this water skin, wrinkle its surface, and produce capillary waves that have properties governed by the elasticity and tension of the water surface rather than gravity. Because of this stretching of the surface film by surface tension, the crests of capillary waves are not peaked but rounded (**Figure B7–3b**). Furthermore, the elastic nature of this skin of water accounts for the almost instantaneous disappearance of capillary waves once the slight breeze dissipates. Also, the speed of capillary waves, unlike that of gravity-induced waves, decreases as the wavelength increases. As the wind intensifies, the capillary waves grow into the more familiar gravity waves (chop and swell) that are the main topic of discussion in this chapter.

Other relatively smaller waves are found in very shallow water. These are called **solitary waves**, because each acts independently of others. In profile, they consist of one-half of the classic sinusoidal wave form, having a distinct crest, but no trough (**Figure B7–4a**). Water moves forward beneath the crest in the direction of wave propagation. Because there is no trough, there is no corresponding water flow in the opposite direction, as there would be in a complete sinusoidal wave. The water to either side of the crest of a solitary wave is virtually at rest. Hence, solitary waves cause water to move in the direction of wave advance. Because they are shallow-water waves, their speed depends merely on water depth and is given simply by $C = \sqrt{g(H + d)}$, where d is the water depth and H is the wave height.

In the surf zone, collapsing breakers produce **waves of translation**, whereby water tumbles forward in a foamy, bubble-filled, turbulent mass, propelled by its own momentum. Any

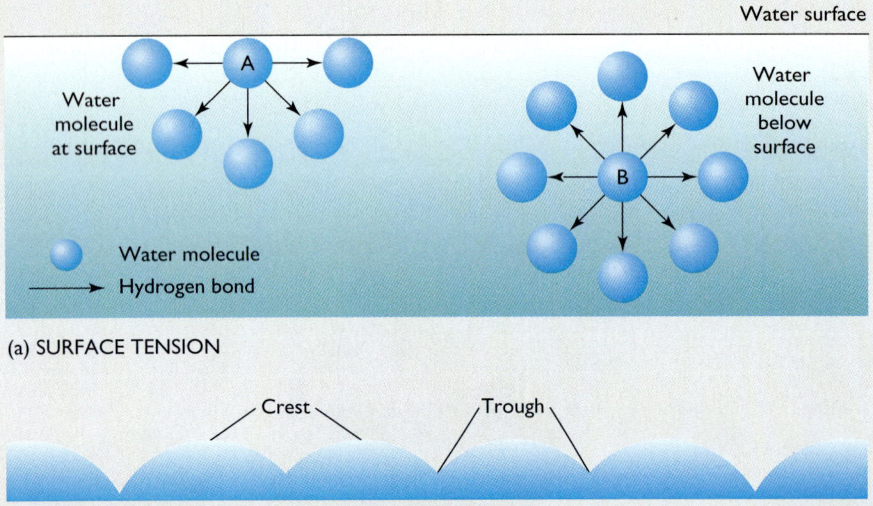

(a) SURFACE TENSION

(b) CAPILLARY WAVE

FIGURE B7-3

Surface tension of water. (a) Water molecule B in the interior of the liquid is attracted on all sides and on the top and bottom by hydrogen bonding to surrounding water molecules. Water molecule A at the surface of the liquid, by contrast, is attracted only to water molecules to either side and below it by hydrogen bonding. This creates surface tension, which produces a strong, elastic "skin" at the water surface. (b) The restoring force for capillary waves is surface tension. The capillary waves with their rounded crests and grooved troughs distort the elastic "skin" of the water surface. Note how different the profile of capillary waves is from the profile of swell.

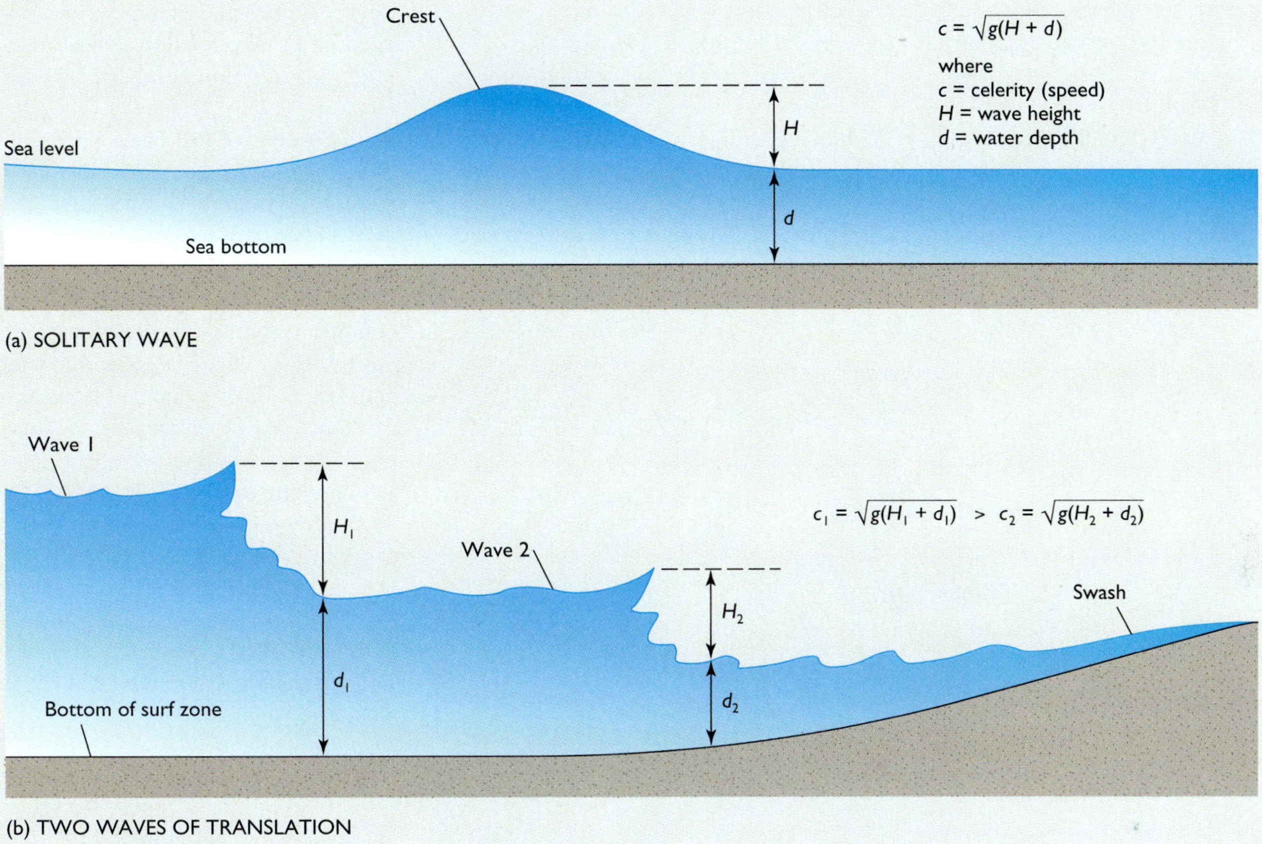

FIGURE **B7-4**

Ocean waves in very shallow water. (a) A solitary wave has a crest and virtually no trough, and resembles half of a sine wave. (b) In waves of translation in the surf zone, momentum carries water toward the shore. Because their celerity depends directly on water depth (d), trailing waves catch up to the waves ahead of them as they ride up on top of them.

floating objects caught by a wave of translation are pushed shoreward by the moving water, until the wave spends its momentum rushing up the beach face and spreading out as swash. In profile, a wave of translation is step-shaped (**Figure B7–4b**), and its speed, like that of the solitary wave, depends on water depth and wave height. Trailing waves of translation can catch and overrun the leading waves because, by traveling on the "backside" of the wave ahead of them, they are traveling through deeper water and therefore move faster than the wave ahead of them.

Shifting to the other end of the wave spectrum, we encounter the *giant waves,* described as such because of their height, not because of their length. Typically, they are encountered in the fetch area where storm winds are raising seas. Here waves of wide-ranging size are moving at different speeds, sometimes merging with others and collectively growing in size by wave construction (see Figure 7–6). Some of these, which seem unbelievably tall and are in fact much bigger than wave theory predicts for the wind conditions, are aptly termed *rogue waves* (**Figure B7–5**). They are especially dangerous if they become oversteepened and collapse, sending a wall of turbulent water with all of its mass and momentum crashing into a helpless vessel.

Surprisingly, even large supertankers are vulnerable to rogue waves. There are documented cases of large ships sustaining structural damage or breaking apart in gales. One foundered as it was raised up to the crest of a towering wave, where its front third was suspended in air, causing the hull to break apart because of the immense weight that was unsupported by water. Smaller vessels can ride up and down giant waves unharmed. Because of their length, supertankers cannot do this; rather, they behave like rigid, floating seawalls, absorbing the pounding of waves. The prolonged slamming of waves against the hull of these supertankers has caused serious structural damage and even failure in a few of these large vessels.

Giant waves are also created by the force of strong currents that flow in opposition to the direction of wave advance. If a series of large waves encounters an opposing current, they undergo refraction because of the surface variation in the speed of the current (**Figure B7–6**). The stronger flow of the current retards wave advancement more than the weaker flow, inducing the wave crests to bend. This refraction focuses energy and causes the waves to grow in height. This very condition creates dangerous seas off South Africa, where large storm waves generated off Antarctica move northward and grow even bigger once they encounter the southerly flowing Agulhas Current.

Giant waves can even occur near shorelines. Large chunks of breakwaters, seawalls, and jetties, weighing many tons, have been washed away at many coastal localities by the onslaught of giant storm waves. A classic illustration of the power of storm waves took place at the northern shoreline of Oregon, just to the south of the Columbia River. Here the base of a lighthouse on Tillamook Rock stands 28 meters (~92 feet) above low water, its top standing about 42 meters (~139 feet) above the sea. Dur-

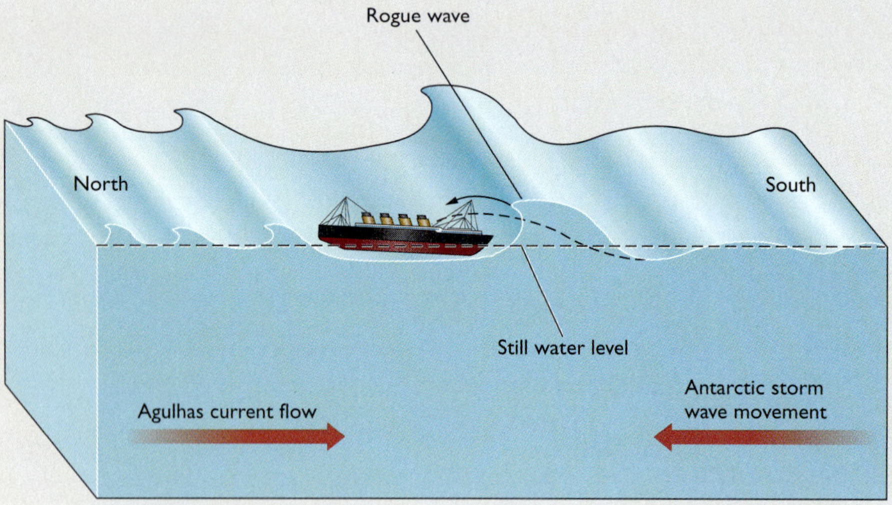

FIGURE **B7-5**

Rogue wave. Constructive wave interference (see Figure 7-6b) can create waves that are much higher than expected for the wind speeds and sea state.

over $3 billion. In October 1992, southern Florida was ravaged by a storm surge created by Hurricane Andrew. Its storm winds blew persistently well over 100 mph, and resulted in widespread coastal flooding, wave erosion, and wind destruction that caused almost $12 billion-worth of damage.

Much of Chapter 11, "The Dynamic Shoreline," reviews the erosive power of storm waves and storm surges. Also, the boxed feature, "Tiny Waves and Giant Waves," describes other interesting wave phenomena in the ocean.

7-4

Standing Waves

Seas, swell, and breakers are progressive waves. Their crests and troughs travel from one point on the sea surface to another. **Standing waves** do not move horizontally, but remain stationary, "trapped," so to speak, as water moves beneath them. These waves oscillate back and forth about ing a winter storm, a boulder weighing 61 kilograms (135 pounds) was tossed by a giant wave *over* the top of the lighthouse and landed on the keeper's house! This is equivalent to picking up and tossing a person upward to a height equivalent to almost half the length of a football field!!

Visit www.jbpub.com/oceanlink for more information.

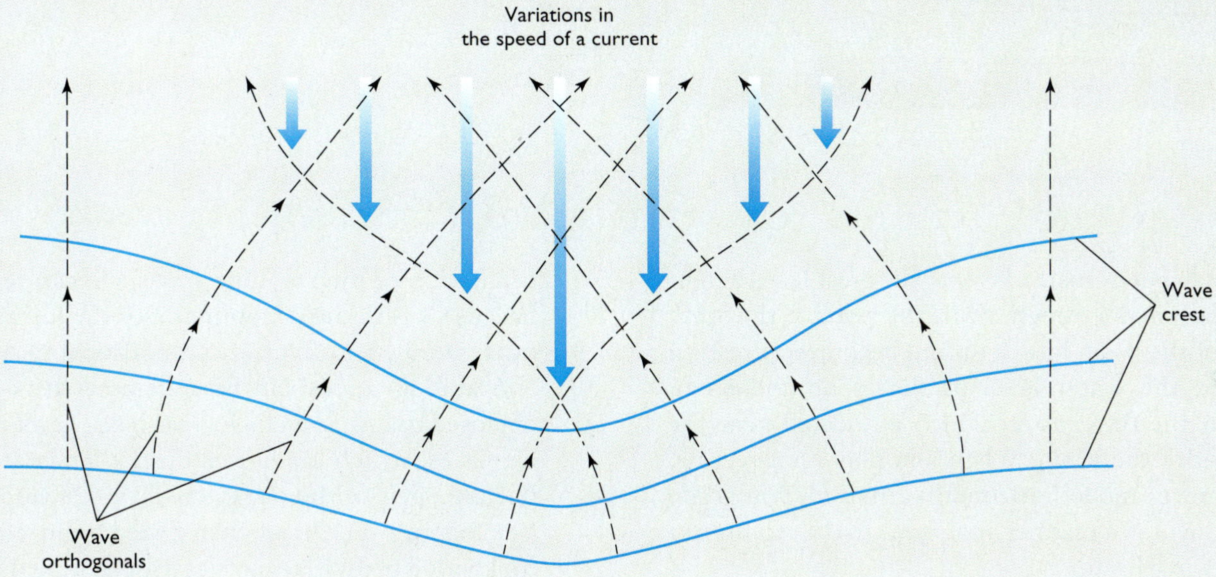

FIGURE B7-6

Steep, high waves created by an opposing current. A wave undergoes refraction when flowing against a current that has a variable surface speed. This causes the buildup of giant waves and confused seas wherever wave orthogonals converge because of wave refraction. [Adapted from Open University Course Team, *Waves, Tides, and Shallow-Water Processes* (New York: Pergamon Press, 1989).]

a fixed point called a **node** (**Figure 7–11**). Think of a seesaw, with one end raised up to a "crest," the other dipped down to a "trough." The node would be the pivot point of the seesaw. Many of you have created standing waves in soup bowls, ice trays, dishpans, and bathtubs as you sloshed the water in these containers back and forth, creating *seiches*. As explained earlier, seiches raise the water level at one end of a basin or container while simultaneously causing the water level to drop at the other side. The maximum vertical displacement of the water level occurs at the sides of the basin; these are called the antinodes (see Figure 7–11). They are the opposite of nodes, where the water level does not change at all. If we go back to our seesaw analogy, the **antinodes** are the two ends where you sit, and the node, as you recall, is the balance point where the seesaw pivots. Note how the height above the ground varies most at the two antinodes of the seesaw and does not vary at all at its node. The same is true for water sloshing back and forth as a seiche in a lake or in a harbor.

It is not possible, as it is with a progressive wave, to follow the crest or trough of a standing wave.

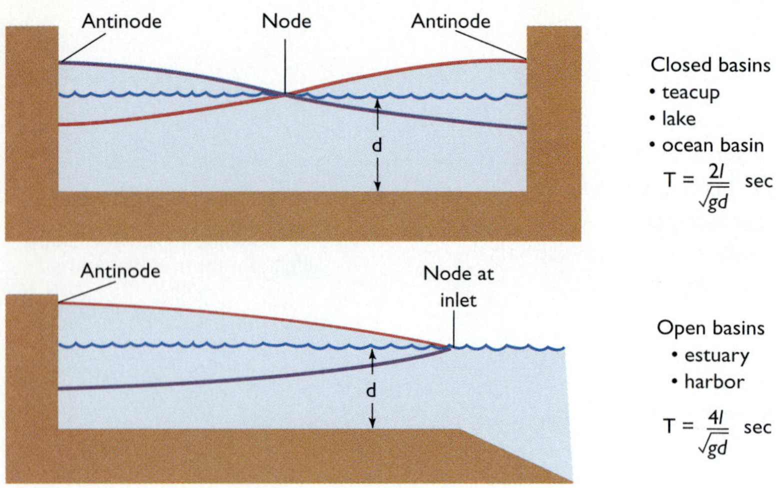

FIGURE **7-11**

Natural period of standing waves. The natural period of standing waves in closed and open basins is a function directly of the basin length and inversely of water depth. The bigger the basin and the shallower the water, the longer is the natural period of wave oscillation.

What happens instead is that the crest loses height and becomes a trough, while the water at the other end of the basin builds up and becomes a crest. In effect, the water level oscillates up and down about the fixed node, which is located near the center of the basin. When you play on a seesaw, you never move horizontally; instead, you go up and down in a manner analogous to the seiching of water in a basin.

The properties of a standing wave depend on the geometry of the basin. The larger the container, the longer its characteristic standing wave will take to oscillate back and forth as the water surface tilts first to one side and then to the opposite side. You can test this yourself by sloshing water in a coffee cup, an ice tray, and a bathtub, and comparing the rates of oscillation. Each will have a natural period of oscillation that is controlled by the length of the basin and its depth of water as follows:

$$T = 2l / \sqrt{gd} \text{ sec},$$

where l is the length of the container, g is gravitational attraction, and d is the depth of water in the container (see Figure 7–11). If we examine this simple formula, we see that as l, the container length, gets larger, the natural period of oscillation increases. So, the back-and-forth sloshing takes longer in a bay than it does in a pond, and longer in a bathtub than it does in an ice tray. Also, the formula shows us that for the same container, the deeper the water, the slower is the natural period of oscillation of the standing wave. We can tell this because d is in the denominator of the formula. In other words, T is inversely proportional to d.

Standing waves can be generated in partially enclosed basins as well, such as bays, harbors, and estuaries, which are closed on the landward side but are open to the ocean on their seaward sides (see Figure 7–11). In an open-ended basin, the natural period of oscillation is easily calculated by the formula

$$T = 4l / \sqrt{gd} \text{ sec},$$

where the variables are the same as in the previous formula. The only difference between the two formulas is that one multiplies the length of the container by 2 for the closed basin and by 4 for the open-ended basin. Also, if you compare the two types of basins, you will notice that the node in an open-sided basin occurs at the mouth of the basin rather than at its center, as it does in the case of a closed basin (see Figure 7–11).

A standing wave in a lake, harbor, or estuary is called a *seiche*. Storm winds blowing persistently in one direction drag and pile up water at the downwind end of a basin, creating a storm surge. When the wind dies after the storm, the water surface in the basin may slosh back and forth. Usually, such oscillations are minor and short-lived. Seiches become real dangers, however, under the condition of **resonance**, which occurs whenever the period of the force (such as the wind or the tides) that stacks the water on one side of the basin equals the natural period of oscillation of the basin as calculated by the formulas given. Let me try to clarify this idea by using a common example.

Some of you may recall setting up resonance in bathtubs as children. By pushing your hands and body against the water with a certain rhythm, you would get the water to "seesaw" back and forth in the tub. If you chose the cadence carefully, the wave moving from one end of the tub to the other would get higher and higher, and slosh over the sides, covering the floor with puddles. What an achievement, or so you thought, until you saw the fire in your parent's eyes! Your sloshing wave was a seiche, and the reason that this standing wave grew in size was because of resonance. You, the forcing element, instinctively pushed the water at the right cadence, the natural period of your tub, and by so doing induced resonance. Pushing the water too fast or too slow would not have created resonance in your tub. Another apt analogy is the timing you use to push a friend on a swing. If your timing is correct, you reinforce the swinging effect, and your friend swings back and forth higher and higher. This is resonance. In the next chapter we will discuss resonance as a cause of unusually large tides in certain basins.

7-5

Other Types of Progressive Waves

To complete our survey of ocean waves, we need to discuss two additional wave types. The first—an **internal wave**—occurs underwater and not at the surface; such waves move along the pycnoclines, which are surfaces that separate water masses of different densities. This is analogous to ocean swell which forms at the contact between two fluids—air and water—where there is a sharp density discontinuity. The second, the **tsunami** , is a series of seismic sea waves that have nothing to do with the tides despite their familiar name of *tidal waves*. Other wave types in the ocean are discussed in the boxed feature, "Tiny Waves and Giant Waves."

INTERNAL WAVES

Energy travels along density discontinuities, or pycnoclines, as wave pulses. As mentioned, the most marked density change occurs at the air-sea contact, where two fluids—air and water—abut each other. Here we find wind-generated waves of all sizes. As you know, the water column is not homogeneous, but stratified, as revealed by prominent thermoclines, haloclines, and pycnoclines that mark boundaries between water masses. Like the sea surface, these underwater surfaces between water masses are ideal locations for the generation of internal waves. Unlike ocean swell, which travels over the water, internal waves move below the water surface. **Figure 7–12** shows a series of internal waves detected off the continental slope of Mozambique, Africa.

Internal waves travel at much slower speeds than do surface waves, because the difference in density between two water masses is much less than it is between air and water. However, their physical size can be much grander than ocean swell. For example, the periods of internal waves are measured in minutes, rather than in seconds, and their wavelengths in hundreds rather than in tens of meters (see Table 7–1). Furthermore, although they characteristically have heights on the order of several meters, the larger internal waves can attain and even exceed the height of 100 meters (~330 feet)—almost three times higher than some of the larger rogue waves created by constructive wave interference at the sea surface. Take a football field and upend it. That is how tall such an internal wave would be!

How are internal waves generated? This is still an active area of research. The period of some internal waves approximates the period of the tides, suggesting a cause-and-effect relationship. Apparently, the tidal movement of water over an uneven bottom can cause flow instability and create waves under water along pycnoclines. Also, friction of one water mass slipping over another water mass—a process somewhat similar to wind friction on the sea surface—may raise internal waves along their boundaries. This process occurs along the halocline of estuaries, the boundary separating fresh river water from salty ocean water,

FIGURE 7-12

Internal waves. Echo returns from water layers containing dense assemblages of small organisms (plankton) reveal a wavelike pattern that is attributed to the passage of internal waves. Sometimes the organisms are lifted to the crest and sometimes dropped to the trough of internal waves that are progressing along pycnoclines. These profiles were measured over the continental slope of Mozambique, Africa. [Adapted from R. Saetre and R. dePaula e Silba, 1979, as seen in A. R. Longhurst and D. Pauly, *Ecology of Tropical Oceans* (New York: Academic Press, 1987).]

which is discussed in Chapter 12. Internal waves may be created by other mechanisms, including slumping of the sea floor, ocean storms, and even motorized ships. Once generated, internal waves can refract, undergo constructive and destructive interference, become unstable, and break, all of this occurring under water!

The U. S. Navy has been investigating the nature, causes, and effects of internal waves ever since they have been suspected of causing the sinking of several submarines. In 1963, for example, the nuclear submarine USS *Thresher* was lost with all hands on board on its shakedown cruise. Prior to the sinking, there had been no indication of debilitating equipment failures or of unusual storm weather. While submerged, submarines attain neutral buoyancy by flooding or discharging seawater from ballast tanks. A common way for submarines to avoid detection by surface vessels while submerged is to cruise silently along density discontinuities, which tend to reflect sonar pulses. Navy scientists speculate that the USS *Thresher* was probably cruising along a pycnocline when it was struck by a large internal wave also traveling along this boundary. Because of its neutral buoyancy, the submarine probably dropped suddenly to greater depths, as the craft literally slid down the back of the towering internal wave. Evidently, the incident occurred too rapidly for the crew to arrest their fall, and it is likely that the submarine plunged down to a depth that exceeded the pressure that its hull could withstand, and the craft imploded.

TSUNAMIS

In earlier times, seismic ocean waves were called *tidal waves*, incorrectly implying that they were related to the tides. To eliminate this confusion, scientists worldwide decided to use the Japanese word *tsunami* to refer to seismic ocean waves.

A stone dropped in a puddle of water generates a series of waves that radiate out from the impact point. Similarly, a sudden shifting of the sea bed caused by explosive volcanic eruptions, earthquakes, or submarine slumping can cause the water over the area affected to drop and generate a tsunami—a series of low waves with long periods and long wavelengths (**Figure 7–13a**). These surface waves travel radially away from their point of origin at remarkable speeds (**Figure 7–13b**). Because of wavelengths greater than 100 kilometers (~62 miles), waves in a tsunami are shallow-water waves over most parts of the ocean, except over deep-sea trenches, where they are intermediate-water waves. As you remember, the celerity (speed) of shallow-water waves depends directly on water depth. In the open ocean, tsunamis travel at a rate

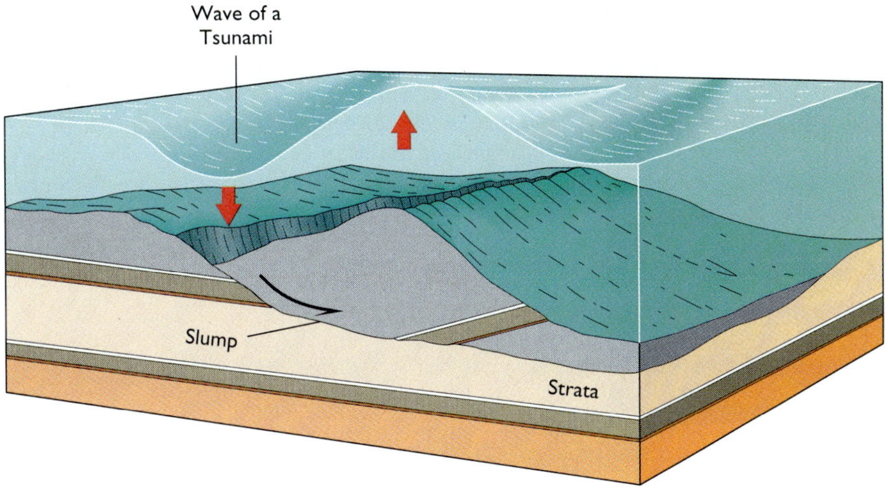

(a) GENERATION OF A TSUNAMI

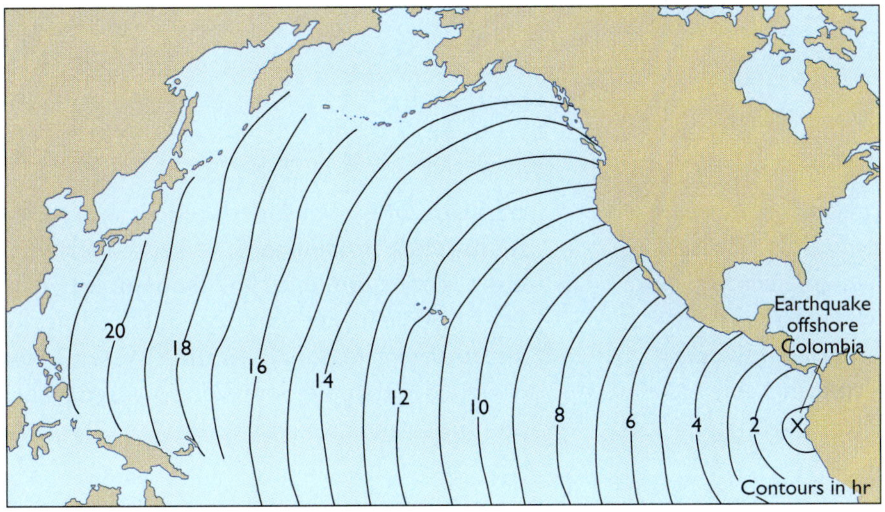

(b) REFRACTION PATTERN OF 1979 TSUNAMI

FIGURE **7-13**

Tsunami. (a) Slumping of a large mass of sediment disturbs the overlying water surface and produces a series of flat, long-period waves, known as a tsunami. (b) This diagram shows the position of the leading wave of a tsunami generated by a 1979 earthquake off-shore Colombia, South America. [Adapted from T. Hatori, "Colombia-Peru Tsunamis Observed Along the Coast of Japan" in *Tsunamis: Their Science and Engineering,* K. Ilda and T. Iwasaki, eds. (Boston: D. Reidel, 1983), 173–83.]

of around 760 kilometers per hour (~483 mph), a speed comparable to that of modern jet aircraft! Of course, in shallow water, they slow down appreciably, because the water depths that control their speed are so slight.

Despite these incredible speeds, tsunamis pose no danger to vessels in the open ocean, because, as we know, energy and not mass is traveling at this speed. In fact, sailors on the deep sea would not even notice the passage of a tsunami beneath their vessel; the regular ocean swell would hide the presence of these flat, low seismic waves, which in the open ocean are rarely higher than a meter or two. However, the waves in a tsunami can grow to a height of much greater than 10 meters (~33 feet) when they reach the shallows of a shoreline. Then they flood the coast, sometimes with catastrophic results, including widespread property damage (**Figure 7–14**) and loss of life.

Japan has a long history of being struck by tsunamis that dates back to A.D. 684; more than 150 tsunamis have struck its islands since then. The Meiji Sanriku tsunami of 1896 killed over 27,000 people. In more recent times, a tsunami generated in the Moro Gulf of the Philippine Islands struck Japan and resulted in 8,000 deaths, 10,000 injuries, and 90,000 people left homeless.

Although little can be done to avert the generation of tsunamis, efforts can be made to minimize their destructive effects. One approach—the International Tsunami Warning System—is a network of stations that measure and record the vibrations

FIGURE 7-14

Tsunami damage in Riangkroko, Indonesia. A sandy beach is all that remains after the waves removed all trace of Riangkroko. An extremely large tsunami height (26m) was measured at this small rural village on Flores Island, and 137 people lost their lives to the earthquake and the tsunamis.

of the crust to detect large earthquakes in the ocean and predict the occurrence of tsunamis. Coastal areas likely to be struck by the seismic sea waves are then alerted, permitting the evacuation of people and ships in harbors. This international warning system, coupled with evacuation training, the relocation of coastal houses and villages, and engineering structures to absorb or reflect the energy of tsunamis, has proved to be quite effective in some cases. For example, a 1957 tsunami killed no one in Hawaii, despite excessive coastal flooding. Recently, a number of regional warning systems have been designed and implemented in Japan, Alaska, Hawaii, and the former Soviet Union.

STUDY GUIDE

KEY CONCEPTS

1. A *wave* is a disturbance that moves along or beneath the sea surface. It is described by its *wave height, wavelength, wave period,* and *celerity.*

2. The size of waves depends on the speed and duration of the wind, and the *fetch*. Large waves are generated by strong winds that blow persistently for an extended period of time across an expansive fetch.

3. Two kinds of motion are associated with sea waves: the movement of the wave form itself and the orbital motions of the water particles beneath the wave. At a depth below the water surface equal to one-half the wavelength, wave motion is no longer detectable. This level is called the *wave base*.

4. Waves traveling in water deeper than one-half their wavelength are called *deep-water waves*.

The speed, or *celerity*, of deep-water waves depends directly on their period or wavelength. Waves in water shallower than one-twentieth of their wavelength are called *shallow-water waves* because they interact with the sea bottom. Their celerity depends solely on water depth.

5. As waves enter shallow water, their period remains constant, their celerity and wavelength decrease, and their wave height increases. Also, the orbital motion of water particles changes from a circular shape beneath deep-water waves to an elliptical shape beneath shallow-water waves.

6. *Standing waves* are stationary wave forms that oscillate up and down about a fixed point called a *node*. *Seiches* are standing waves in estuaries and harbors that have natural periods of oscillation proportional to the basin length and water depth. *Resonance*, the amplification of a standing wave, occurs when the period of the forcing element, such as the tide, approximates the natural period of oscillation of the basin.

7. *Internal waves* are submarine disturbances that travel along pycnoclines in the water column. They have much longer periods and wavelengths, greater wave heights, and slower wave speeds than wind-generated surface waves.

8. A *tsunami* is a series of seismic sea waves that contain a tremendous amount of energy. Tsunami waves have very long periods and wavelengths. They grow in height in shallow water and flood the shoreline, sometimes causing catastrophic destruction and loss of life.

KEY WORDS

antinode
breaker
celerity
constructive wave interference
deep-water wave
destructive wave interference
dispersion
fetch
internal wave
node
progressive wave
refraction
resonance
seiche
shallow-water wave
standing wave
storm surge
swell
tsunami
wave base
wave height
wavelength
wave period

QUESTIONS

REVIEW OF BASIC CONCEPTS

1. Clearly distinguish between a progressive and a stationary wave. To which of these two wave categories do the following belong: tsunami, seiche, swell?

2. What is meant by constructive and destructive wave interference? Which of these accounts for rogue waves?

3. What are chaotic seas? How does wave dispersion create swell from a chaotic sea?

4. Describe and account for the changes that waves undergo as they move from deep water and break on a shoreline.

5. What is a storm surge, and how does it affect a shoreline?

6. Distinguish among types of breakers and the conditions that tend to give rise to them.

7. How do the properties of internal waves—height, wavelength, period, celerity—differ from those of ocean swell?

CRITICAL-THINKING ESSAYS

1. Discuss the differences between the motion of the wave form and the motion of water particles that are affected by the wave.

2. Why under the same wind conditions will waves in an estuary be larger than waves in a pond?

3. Sketch the pattern of wave refraction that a series of straight-crested shallow-water waves will undergo as they pass over a deep, broad hole in the sea bottom. Now repeat the analysis for waves passing over a large shoal (submarine hill) on the bottom.

4. Why is it proper to refer to the speed of an ocean wave as celerity rather than velocity?

5. Is it possible to generate a deep-water wave in a puddle of water that is no deeper than 8 centimeters? Explain. Is it possible to generate a shallow-water wave over a deep-sea trench where the water depths exceed 12 kilometers? Explain.

6. What would be the nature (period, wavelength, height) of the first swell to reach a beach that is located far from the storm center? How would the physical appearance of the swell from this distant storm change with time along this beach?

DISCOVERING WITH NUMBERS

1. Calculate the celerity of a wave in water depths of 5 meters and 1 meter, if it has (a) a 100-meter wavelength and (b) a 150-meter wavelength.

2. Draw a progressive wave. Using arrows, depict the motion of water particles on the crest, trough, and front and back sides. Then contrast the motion of water at the crest and trough. If mass transport is 1 centimeter per wave period, calculate in meters the total mass transport over 10 hours, assuming the wave has a 10-second period.

3. Assume that, just off a beach, a 10-second swell has a wavelength of 50 meters and a wave height of 2 meters. Calculate the wave steepness. Now assume that its wavelength will no longer change, and estimate the height of this swell as it becomes unstable and collapses as a breaker. What type of breaker will it likely be if the slope of the bottom just off the beach is moderately steep?

4. Assume that an estuary is 300 kilometers long and has an average water depth of 10 meters. What is its natural period? If the tides in this area have a period of about 12.5 hours, will they set up resonance in this basin? Why or why not?

5. Examine Figure 7–12. Estimate the wavelength and wave height of the deepest internal waves. Are these shallow-water waves? Why or why not? Calculate the vertical exaggeration of the profile (consult Appendix IV).

SELECTED READINGS

Barnes-Svarney, P. 1988. Tsunami: Following the deadly wave. *Sea Frontiers* 34 (5): 256–63.

Bascom, W. 1980. *Waves and Beaches: The Dynamics of the Ocean Surface.* Garden City, N.J.: Doubleday.

Changery, M. J., and Quale, R. G. 1987. Coastal wave energy. *Sea Frontiers* 33 (4): 260–61.

Dudley, W. C. 1998. *Tsunami.* 2d ed. Honolulu, University of Hawaii Press.

Gonzalez, F. I. 1999. Tsunami! *Scientific American* 280 (5); 56–65.

Giese, G., and Chapman, D. 1993. Coastal seiches. *Oceanus* 36 (1): 38–46.

Kampion, D. and Brewer, A. (eds.). 1997. *The Book of Waves: Form and Beauty on the Ocean.* Robert Rinehart.

Land, C. 1975. Freak killer waves. *Sea Frontiers* 21 (3): 139–41.

LeBlanc, J. 1986. Tales of the wild ocean. *Oceans* 19 (4): 36–41.

McCrede, S. 1994. Tsunamis—The wrath of Poseidon. *Smithsonian* 24 (12): 28–38.

Myles, D. 1986. *The Great Waves.* London: Robert Hale.

Newman, J. N. 1979. Power from ocean waves. *Oceanus* 22 (4): 38–45.

Pararas-Carayannis, G. 1977. The International Tsunami Warning System. *Sea Frontiers* 23 (1): 20–27.

Robinson, J. P. 1976. Superwaves of southeast Africa. *Sea Frontiers* 22 (2): 106–16.

Smail, J. 1982. Internal waves: the wake of sea monsters. *Sea Frontiers* 28 (1): 16–22.

Smail, J. 1986. The topsy-turvy world of capillary waves. *Sea frontiers* 32 (5): 331–37.

Truby, J. D. 1971. Krakatoa—the killer wave. *Sea Frontiers* 17 (3): 130–39.

TOOLS FOR LEARNING

Tools for Learning is an on-line review area located at this book's web site OceanLink (**www.jbpub.com/oceanlink**). The review area provides a variety of activities designed to help you study for your class. You will find chapter outlines, review questions, hints for some of the book's math questions (identified by the math icon), web research tips for selected Critical Thinking Essay questions, key term reviews, and figure labeling exercises.

8 Tides

The tide rises, the tide falls,
The twilight darkens, the curlew calls;
Along the sea-sands damp and brown
The traveler hastens toward the town,
and the tide rises, the tide falls.

Darkness settles on the roofs and walls,
But the sea, the sea in darkness calls;
The little waves, with their soft, white hands,
Efface the footprints in the sands,
And the tide rises, the tide falls.

The morning breaks, the steeds in their stalls
stamp and neigh, as the hostler calls;
The day returns, but nevermore
Returns the traveler to the shore,
And the tide rises, the tide falls.

—Henry Wadsworth Longfellow,
The Tide Rises, the Tide Falls, *1879*

web navigator

critical thinking on the web

math tutor on the web

www.jbpub.com/oceanlink

PREVIEW

TIDES ARE WAVES with very long wavelengths—much longer than ordinary wind waves—that cause sea level to rise and fall with extraordinary regularity. In fact, tides are the most uniformly varying phenomenon in the ocean. The daily rise and fall of the tide influences all life along the seashore, including humans.

In this chapter, we examine the fundamental properties of tides, including the forces that give rise to them. Then we develop elementary models of tidal circulation in broad ocean basins and in narrow coastal basins. The chapter ends with a description of electricity generation using the power of the tides.

One of the most fascinating aspects of the ocean is the tide—the slow, up-and-down movement of sea level that occurs each day. What could possibly cause such a pulse? To this day, I remember clearly my parents telling me as a child that the tides are caused by the Moon and the Sun—and my unwillingness to accept such a bizarre explanation! How could objects as distant as the Moon and Sun possibly cause the ocean to pulse in such a regular manner? It made no sense to me, because the connection seemed so tenuous, so mysterious. Now that I understand that this explanation is in fact rational, the tides still and nevertheless retain a magical quality for me. They serve as a reminder of how planetary bodies are bound into a unified whole and of the extraordinary quality of the human mind that is able to perceive the invisible interconnections.

8-1 Tidal Characteristics

The simplest and most economical way to measure the tides is to mount a graduated pole on a pier and note the position of the water level as a function of time. This technique is not practical, however, when continuous long-term records are needed. Certain simple modifications solve the problem. First, the position of the sea surface is kept track of by the use of a float placed in a well that is connected to the ocean by a small pipe (**Figure 8–1**). Because the well eliminates waves, the water enclosed in the well is flat and moves slowly up and down only in response to the tides. Second, a wire leading from the float is fastened to a pen, which inscribes the tidal curve onto a recording paper mounted on a clock-driven rotating drum. Such tide gauges yield accurate, permanent records of the vertical movements of local tides.

Pressure sensors are used to record tidal oscillations farther offshore, where the erection of a structure to house a tide gauge is impractical. These sensors are set in the water below the lowest level of the tide. From there, they transmit a record of pressure variations that accompany the rise and fall of the tide. The more water there is above the sensor, the higher the pressure recorded. Therefore, the variations of pressure over time reflect the actual fluctuations of the tide.

Plotting the vertical position of sea level over a day reveals clearly that the tide is a wave with a crest (high tide) and a trough (low tide) and with a distinctive period of 12 hours, 25 minutes, or 24 hours, 50 minutes, depending on where you are on the Earth. The wave height of the tide, referred to as the **tidal range**, is the vertical distance between high tide (the crest) and low tide (the trough). The tidal range varies from region to region, but is typically between a fraction of a meter (~3.3 feet) and several meters (**Figure 8–2**); the range of the greatest tides, however, can exceed 15 meters (~50 feet).

In their simplest form, tides are single waves that stretch across the entire ocean, causing its water to move up the front of its crest on one side of the basin and down its back into the trough on the opposite side. Also, they are shallow-water waves even over the deep ocean, because their wavelengths dwarf the seas' water depths. It should be noted, however, that tides do differ from ordinary ocean waves in several important respects. First, tides arise in response to complex, but pre-

FIGURE 8-1

The measurement of tides. The water level in the well responds only to tidal fluctuations. A pen, mechanically attached to a float which rises and falls with the tide, transcribes a tidal record onto a piece of paper mounted on a rotating drum.

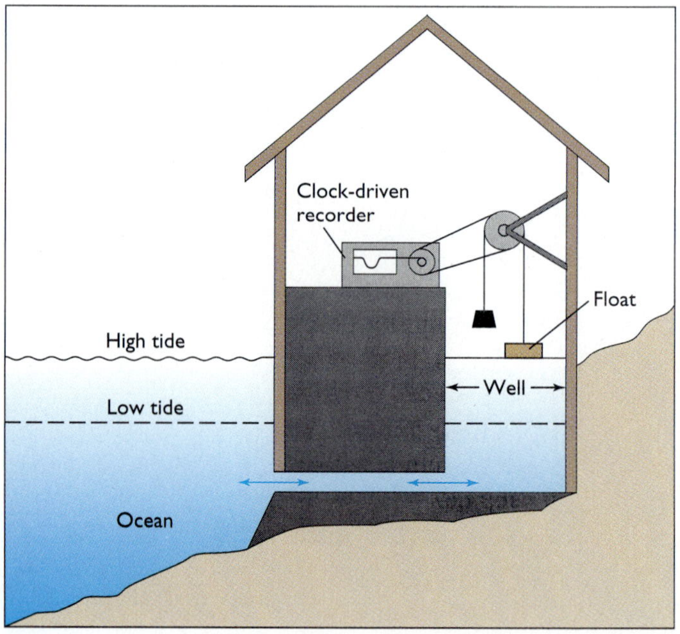

cisely known, lunar and solar interactions with the Earth's oceans. Also, tides carry much more total energy than even the largest rogue waves or storm breakers. Rogue waves are very high, but limited in area; by contrast, tides are relatively low waves, but they stretch across entire oceans. The total amount of energy contained in any wave is a function of its wave height and its wavelength. Hence, tides because of their long crests and wavelengths contain much more energy than even the tallest wind waves, which have relatively short wavelengths. These are the reasons why a separate chapter is given over to the study of ocean tides.

A comparison of tidal records for the world's shorelines reveals that tides can be classified into three categories based on their periods and regularity. Let's consider the U.S. coastline. For example, in Pensacola, Florida (**Figure 8–3a**), tides are **diurnal**. They have a period of about one day, meaning that there is one high tide and one low tide daily. In Boston (**Figure 8–4**) by contrast, tides are **semidiurnal**, because they occur twice instead of once daily, so that there are two high tides and two low tides each day. The up-and-down motion of the tides is uneven along the Pacific coast of the United States, where **mixed tides** (**Figure 8–3b**) vary irregularly twice daily. These high and low tides are of unequal shape each day. In other words, the tidal range (wave height) between successive high tides is variable. To see this clearly, go back to Figure 8–3b and estimate the magnitude of the two tidal ranges for one day in the San Francisco area.

When tidal records for any region are examined over a month, the tidal range varies slightly, but systematically, each day. The phase when the tidal range is at a maximum (that is, the highest high tide and the lowest low tide) is referred to as a **spring tide**. This is the time when the greatest expanse of beach is exposed at low tide and the greatest expanse of beach is flooded at high tide. By contrast, there are the **neap tides** when the tidal range is at a minimum, meaning that the high tide is unusually "low" and the low tide is unusually "high." To see this clearly, examine the tidal records in Figure 8–4 and find the times of spring tides and neap tides for Boston, Key West, and Pensacola. Spring tides and neap tides everywhere occur twice monthly and vary directly with the phases of the Moon, as will be explained shortly. All of these tidal fluctuations are so regular that

FIGURE **8-2**

Tidal range. (a) Low tide at a bay of Campobello Island, Canada. (b) High tide at the same bay.

FIGURE 8-3

The measurement of tides. (a) Tidal variation (in centimeters) measured at Pensacola, Florida, is of a diurnal tide: one high tide and one low tide occur during a 24-hour period. (b) Tidal variation (in meters) measured at San Francisco is of a mixed tide: two high tides and two low tides of unequal tidal range occur during a 24-hour period.

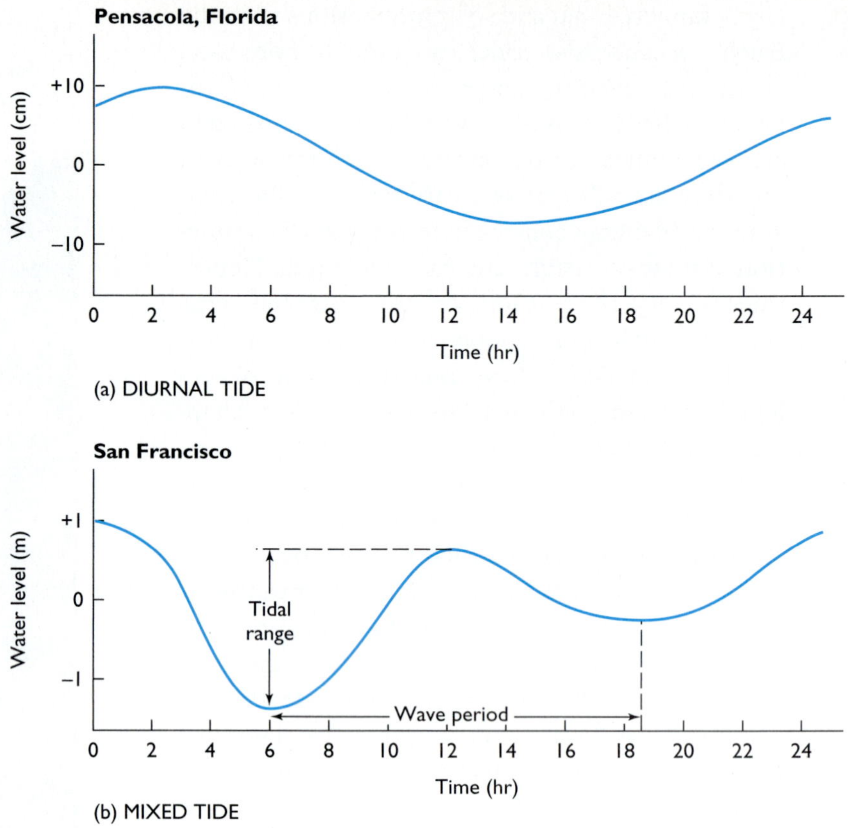

the magnitude and timing of the daily rise and fall of the tide can be predicted with accuracy far into the future for any coastal locality where a tidal record has been measured for about one year.

Once physical oceanographers obtain a tidal record of reasonable duration from a coastal site, they can use mathematical techniques to identify which of the astronomical components related to the Sun and Moon are most responsible for affecting the tides at this locality. What remains unexplained after the analysis is attributed to the local geography of the basin—its depth and shape, which distort the tide from its ideal form. Once completed, these data are recombined by powerful computers to predict the tidal range and the time of high and low water for any day in the future. This information is then tabulated into tide tables and made available to anyone who needs the information.

You are now familiar with the types and the common properties of the ocean's tides. Let's go on to consider how tides arise and why they vary in some of their regional characteristics. Throughout the analysis, we will rely on a conceptual approach to explain the origin of the oceans' tides.

8-2

Origin of the Tides

The tidal pulse of water anywhere on the Earth is caused by two principal factors: gravitational attraction and centrifugal force. All masses are drawn to one another, and exert a gravitational pull. This is, of course, why we don't fall off the Earth. The Earth "pulls" on us because of its mass, and we "pull" on the Earth because of the mass of our body. In fact, the strength of gravitational attraction varies directly with the masses of the interacting bodies. Remember how the astronauts were able to jump about the Moon easily, bounding effortlessly with enormous leaps of Olympian proportion. This was a direct result of the Moon's small mass and its consequent weak gravitational field.

The strength of gravity also varies with the distance separating any two masses. Actually, the magnitude of the tide-raising forces varies *inversely* as the cube of the distance between them, meaning that as you double the distance between the objects, the tide-raising forces decrease proportionately by a factor of 8, that is, 2^3. This means that the distant planets in the solar system exert very little gravitational pull on the Earth because of the enormous distances that separate them from the Earth. As a spaceship moves into space, the Earth's gravitational "pull" on the vehicle and its crew decreases rapidly with distance (that is, it falls off as the square of the distance). Far out in space, gravity caused by the Earth's mass is negligible, and the astronauts and all their gear float about the spacecraft in a very weak gravitational field. There is no longer an "up" and a "down," and ceilings, walls, and floors have little meaning to the inhabitants of the spacecraft.

Because the Moon is much closer to the Earth than the Sun, it has more than twice the gravitational effect, despite the Sun's immense mass. Although it is 10^7 times (10 multiplied by itself seven times, which is 10 million) more massive than the Moon, the Sun is about 390 times farther

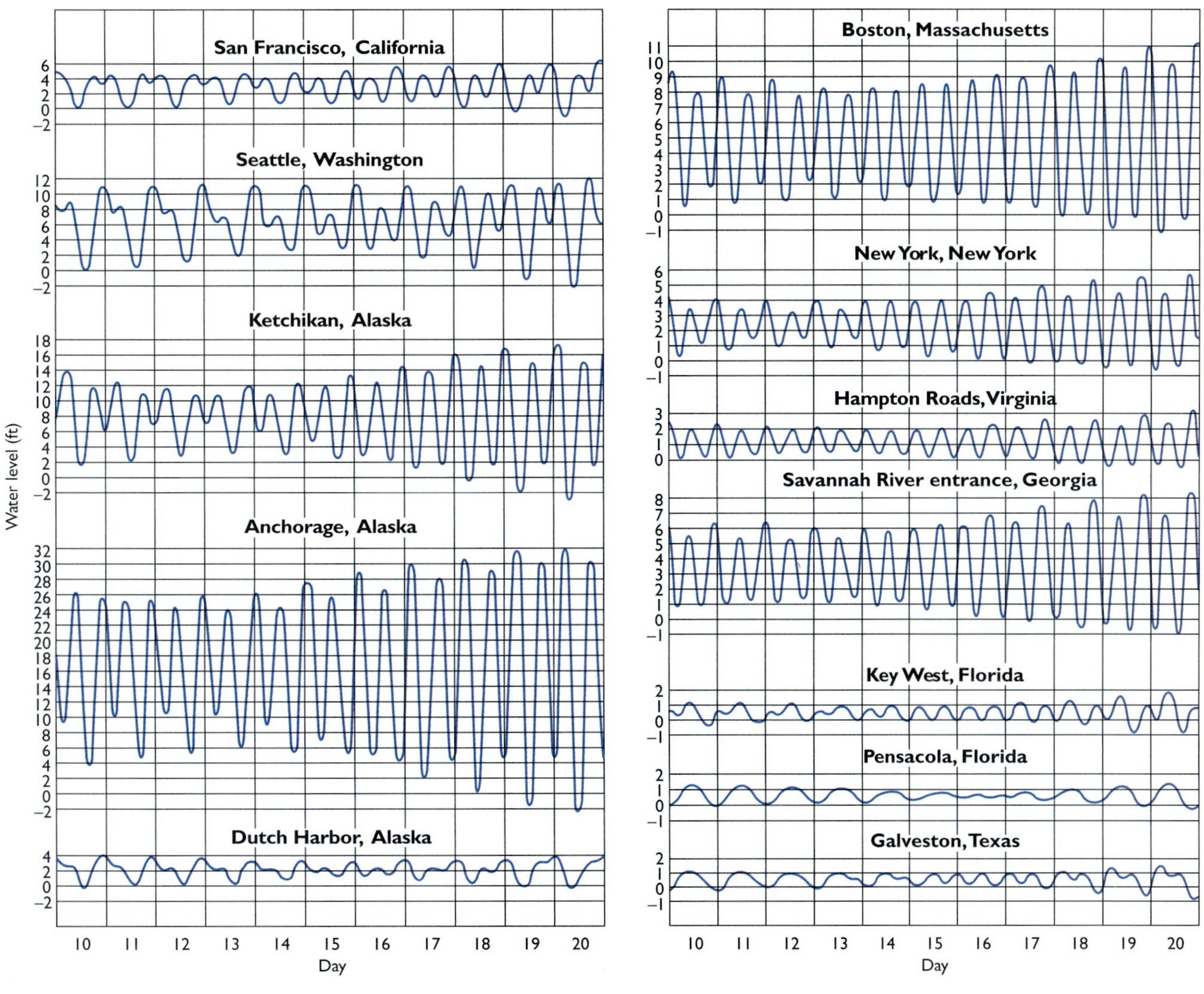

FIGURE **8-4**

Tidal records. These are tidal records for selected sites along the Pacific, Atlantic, and Gulf of Mexico coasts of the United States. Note that the tidal range is specified in feet, rather than in meters, because these are the units used in the National Ocean Survey tide tables. [Adapted from the National Oceanographic and Atmospheric Administration, 1988.]

away from the Earth than the Moon is. Therefore, the Moon, because of its closeness to the Earth, is the main regulator of the planet's tides. It exerts about twice the tide-raising force on the Earth as that of the Sun.

The lunar gravitational force "pulling" on the oceans causes water to be drawn toward the side of the Earth that faces the Moon, creating a **tidal bulge** of water (**Figure 8–5a**). A second tidal bulge exists on the opposite side of the Earth. This one results from the second important tide-raising factor—the centrifugal effect that arises as the Earth and Moon revolve about each other. Moving bodies travel in straight lines. Imagine yourself as a passenger in a car being driven down a straight stretch of road. Suddenly the driver abruptly turns left onto a side road. As the maneuver is executed, you feel your body being "pushed" hard against the door on the right side of the car. This, basically, is centrifugal force. The strong "pull" you feel is actually your body continuing to move in a straight line, but the car has turned into it, giving the impression of a force that is pressing you hard against the door frame. This is exactly the same effect you experience when you ride the "whip" at an amusement park or the roller coaster as it follows the curve of a track at high speed.

Both the Earth and the Moon are rotating simultaneously about a common center of mass. Because all rotating bodies are subject to centrifugal effects, the water of the oceans shifts away from the center of rotation, creating a second tidal bulge on the side of the Earth that faces away from the Moon (**Figure 8–5b**). If the masses of the Earth and the Moon were equal, the rotation point or the center of mass would be located exactly midway between the two bodies. Think of a seesaw with two people, one on each end, with equal masses; the pivot point of the seesaw is dead center. Inequalities of the masses of the Earth and Moon place the center of rotation, that is, the center of mass, beneath the Earth's surface, because the Moon has a much smaller mass than the Earth does. This is equivalent to an adult and a child seesawing together. In order for the adult to be raised off the ground by the small child, the balance point (the center of mass) of the seesaw has to be shifted toward the adult. Otherwise, the seesaw cannot move up and down. In any case, the Earth and Moon rotate mutually about their common center

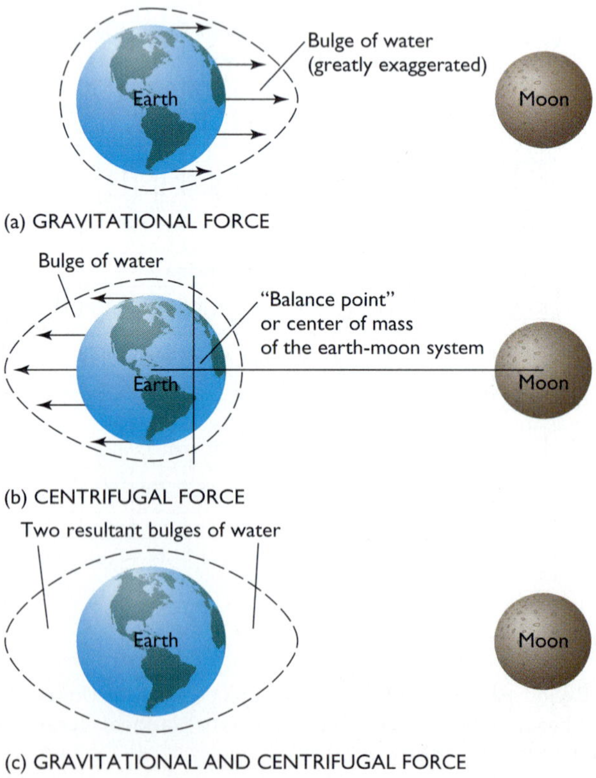

FIGURE **8-5**

Tidal bulges. (a) Because gravitational attraction varies with the distance separating the masses, the gravitational force of the moon is weaker on the far side of the Earth than its near side. Ocean water is drawn toward the moon by gravitational attraction. Note that the mound is greatly exaggerated in the drawing. (b) Water is also displaced to the side of the Earth that faces away from the moon. This mound results from rotation about the center of mass of the Earth-Moon system. (c) Gravitational attraction (as described in a) and centrifugal force (as described in b) produce two tidal bulges of water of about the same size, positioned on opposite sides of the Earth. Both the size of the water bulges and the distance and size of the Moon are greatly exaggerated to bring out the features being discussed.

of mass, which actually is located in the Earth (see Figure 8–5b).

What we've established so far is that in theory, water in the oceans will tend to be arranged into two tidal bulges. One bulge will face the Moon and is a consequence of gravitational attraction. The other bulge, which will be on the side of the Earth that is opposite to the Moon, arises because of the centrifugal effect associated with the rotation of the Earth and the Moon about their common center of mass (**Figure 8–5c**).

We can now proceed to models of the tides. We will begin by considering the equilibrium model of tides, a simplified but quite useful depiction of tidal variations on an ideal Earth. Next, we will study the dynamic model of tides, a more complex and therefore realistic representation of tidal oscillations in the various oceans. As you continue, what you need to keep in mind are the two basic concepts of gravitational attraction and centrifugal force.

EQUILIBRIUM MODEL OF TIDES

Models are simplifications of the real world. The **equilibrium model of tides** makes the following key assumptions, which represent such simplifications:

1. The Earth's surface is considered to be completely covered by seawater to an infinite depth, so that the tides are unaffected by the sea bottom. In other words, in this model, there are no landmasses or effects of the sea floor, both of which greatly distort the tidal bulges and complicate the analysis.
2. The waves associated with the tides are assumed to be progressive waves.
3. The water is assumed to be in equilibrium with the tide-generating forces—gravitational attraction and centrifugal effect—at all times.

Given these three simplifications, it is possible to predict tidal variations on an idealized Earth that is covered everywhere by an infinitely deep ocean. To begin with, we will also consider only the effect of the Moon and ignore the Sun's influence until later. The gravitational attraction of the Moon's mass is "pulling" on the water everywhere in this globe-circling ocean, but it is strongest on the side of the Earth that is closest to the Moon. This effect, of course, is caused by the gravitational attraction between any two masses, which decreases as the square of the distance that separates them. Water on the far side of the Earth is farther away from the Moon than it is on the near side, and its gravitational attraction to the Moon is weaker there than on the near side. The result of this attraction is the creation of a tidal bulge of water on the side of the Earth that is facing the Moon (see Figure 8–5a).

The water on the side of the Earth that faces away from the Moon feels a small "tug" because of the Moon's gravitational field, but at the same time the water here bulges outward as a consequence of the centrifugal effect. Because the centrifugal effect dominates the gravitational attraction here, the water is drawn out into a second tidal bulge (see Figure 8–5b). The net result is that these two interacting forces distort the uniformly deep ocean into two tidal bulges: one on the side near the Moon; the other on the side opposite the Moon (see Figure 8–5c). So far, you might say, we have not covered any new ground, and besides, how do the tidal bulges relate to the daily rhythms of the tides on the real Earth?

The Earth spins daily around its axis, giving rise to the daylight and nighttime halves of a day. Well, as the Earth rotates about its axis, a point on the planet's surface passes into and out of the bulges of water, creating the tides (**Figure 8–6**). As a point on the Earth enters a water bulge, the tide rises; as it leaves a water bulge, the tide falls. Depending on the angle, called *declination*, between the Earth's axis and the Moon's orbit around the Earth, different points on the planet's surface pass through various parts of the tidal bulges—in some places right through their centers, in others through their edges.

Examine Figure 8–6 carefully and note how location on the globe determines the period of the tides in the equilibrium model. Note that diurnal tides occur in the high latitudes, mixed tides in the midlatitudes, and semidiurnal tides in the low latitudes. In reality, the Earth's tides do not vary systematically as they do in this model. But this is a good beginning for our analysis.

As illustrated for Port Adelaide, Australia, and Pakhoi, China (**Figure 8–7a**), the tidal range varies systematically over the course of a month, being maximum twice monthly (the spring tides) and minimum twice monthly (the neap tides). The

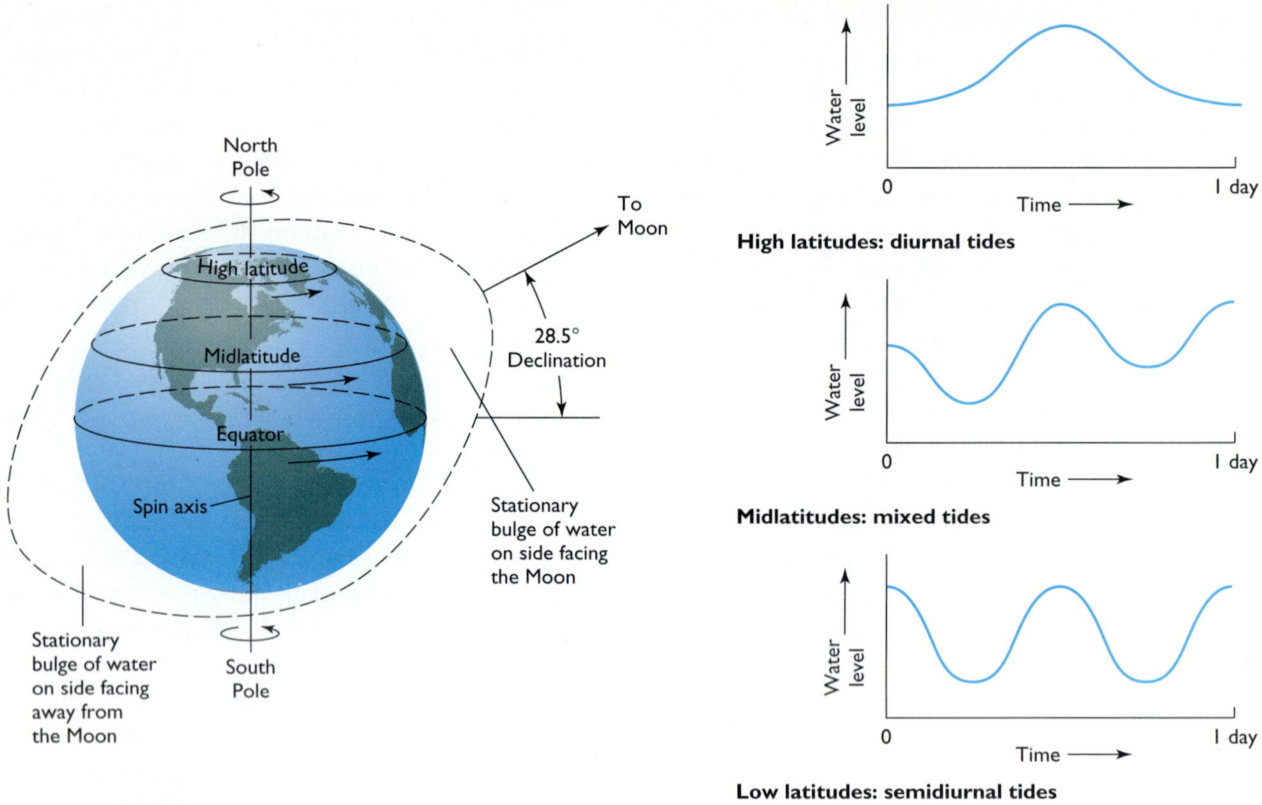

FIGURE 8-6

Equilibrium tides. Fixed points on the Earth's surface rotate into and out of the "stationary" tidal bulges, creating daily tides everywhere on the planet. In this equilibrium model, high latitudes are characterized by diurnal tides, midlatitudes by mixed tides, and low latitudes by semidiurnal tides.

equilibrium model of the tides does a nice job of accounting for neap and spring tides, if we now consider the effect of the Sun. The Sun raises two tidal bulges on the Earth for the very same reasons that the Moon does—one the result of gravitational attraction, the other, the centrifugal effect. The Sun's bulges are not, however, as pronounced as the Moon's are because of its much greater distance from the Earth. So, in effect, there are four bulges of water to keep track of in our equilibrium model. At first glance, this seems very complicated, but, for our purposes, it need not be. Every two weeks, the Earth, Sun, and Moon are aligned in their orbits about one another, producing a new Moon and a full Moon. At such times, the two tidal bulges caused by the Moon are superimposed on the two tidal bulges caused by the Sun. The net result of the combined bulges is the spring tide, with a higher than normal tidal range (**Figure 8–7b**). The smaller tidal ranges associated with the neap tides occur during the quarter-Moon phases. At such times, which occur twice monthly, the Sun and the Moon are oriented at a 90-degree angle to each other, relative to the Earth, so that the two pairs of tidal bulges are separated from each other (see Figure 8–7b). In effect, we are seeing the result of constructive (spring tides) and destructive (neap tides) wave interference (see Figure 7-6 in Chapter 7).

There is an additional factor that needs to be addressed. If the Earth is rotating in and out of these tidal bulges, why is the period of a diurnal tide 24 hours and 50 minutes and, for mixed and semidiurnal tides, 12 hours and 25 minutes? Shouldn't they be 24 hours and 12 hours, respectively? The explanation is actually quite simple. After one Earth day, the Moon has moved forward slightly in its orbit. So an additional 50 minutes is necessary for a spot on the Earth to regain its position relative to the Moon each day. Hence, the times of high tide and low tide shift forward each day, as anyone who reads a tide table in the newspaper knows.

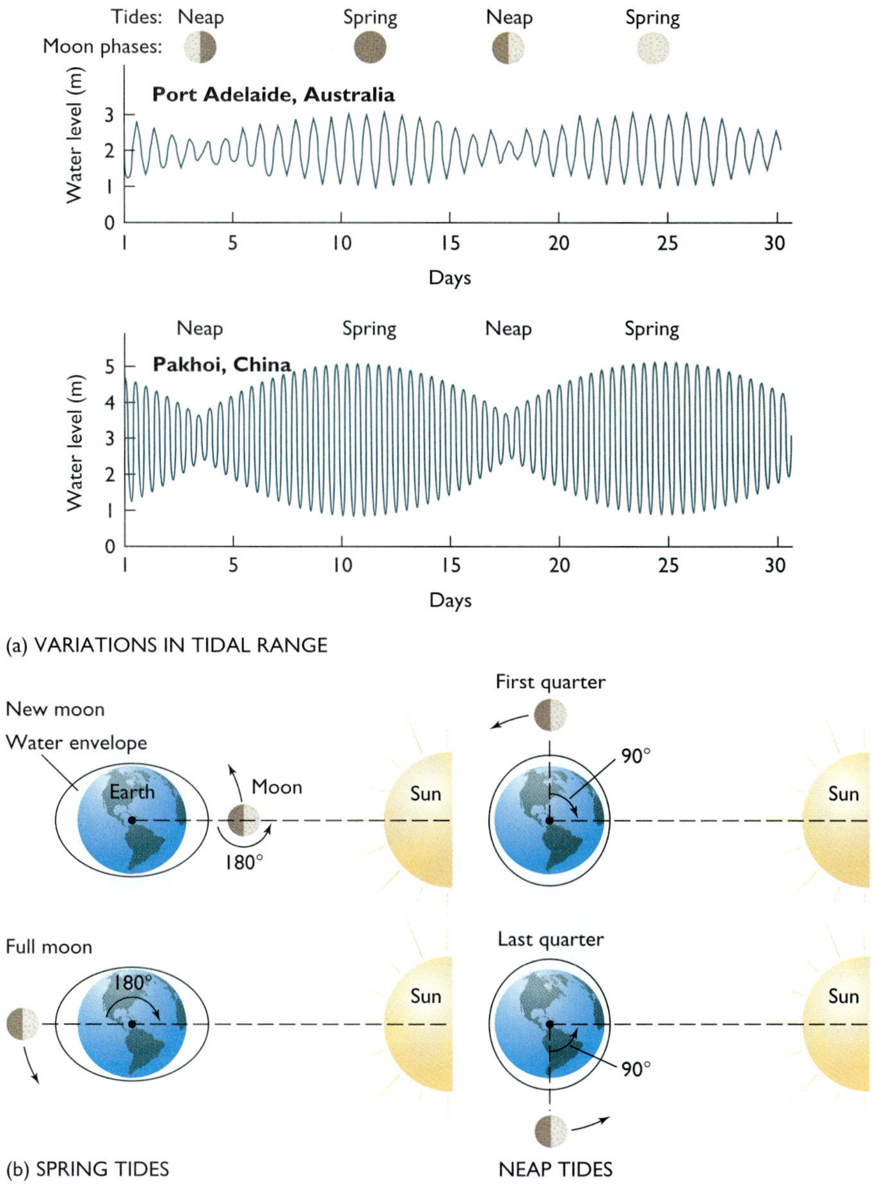

FIGURE **8-7**

Spring and neap tides. (a) Month-long records of tides in Port Adelaide, Australia, and Pakhoi, China, show systematic variations in tidal range. The maximum tidal range (the spring tide) occurs near the new and full phases of the Moon. The minimum tidal range (the neap tide) occurs near the quarter-Moon phases. [Adapted from *American Practical Navigator,* H.M. Publication no. 9 (Washington, D.C.: U.S. Naval Oceanographic Office, 1958).] (b) During the new and full Moons, spring tides occur as the result of the alignment of the Moon, Sun, and Earth. Under these conditions, tide-raising forces of both the Sun and Moon act in concert (constructive wave interference). Tidal ranges are minimal during neap tides, because the 90-degree geometry of the three bodies produces opposing tide-raising forces (destructive wave interference).

The equilibrium model of tides is a good first step toward understanding tidal oscillations. But because of its simplifications, it does not give us a detailed and accurate description of tides in the real world, which occur in oceans that are separated by continents and that are not infinitely deep. To portray tides realistically, we must introduce additional factors; and this leads us to the dynamic theory of tides.

DYNAMIC MODEL OF THE TIDES

A much more accurate depiction of tides is possible if we assume that the oceans are dynamic (active), and not static (still). Thus, in the **dynamic model of tides**, water is assumed to respond actively, rather than passively, to the tide-generating forces, so that we no longer have stationary (static) tidal bulges of water that stay aligned with the Moon as the Earth's surface rotates through them. Also, we consider the effects of continents in this model. In this case, tidal bulges of water are trapped in ocean basins and dragged away from the fixed positions they would occupy in the absence of landmasses. In our equilibrium model, with a continuous ocean, the tidal bulges remained fixed in place (static) relative to the Moon, and the Earth rotated through them (see Figure 8–6). In the dynamic model, the ocean basins drag the tidal bulges with them each day as they rotate with the Earth around its spin axis.

To begin, we will simplify our analysis by imagining a hypothetical Earth having a single ocean basin on its surface that is uniform in depth everywhere and is surrounded on all sides by land (**Figure 8–8a**). As before, we will ignore the effect of the Sun. Also, we will assume that the Moon is positioned directly above the ocean basin and that its water surface is distorted into a broad tidal bulge because of the Moon's gravitational attraction. As this hypothetical ocean moves from west to east with the rotating Earth, the tidal bulge, in an attempt to remain positioned beneath the Moon, is forced against the western margin of the basin (**Figure 8–8b1**). As the basin continues to move to the east, the tidal bulge is "squashed" more and more against the basin's western edge with the water level tilted down to the east toward the center of the basin. The slope of the sea surface, of course, is a pressure gradient, and at some point, the water will flow downslope in response to this pressure gradient. Once set in motion, water will be influenced by the Coriolis effect. Because the hypothetical basin is in the Northern Hemisphere, the flowing water will be deflected to the right and pile up against the southern edge of the basin (**Figure 8–8b2**). This in turn will create a water surface that slopes to the north toward the center of the basin. Eventually, the water current, in response to this new pressure gradient, will reverse and flow northward, be deflected to the right by the Coriolis effect, and be piled up against the basin's eastern border (**Figure 8–8b3**). This complex water motion continues (**Figure 8–8b4**) until the pile of water circles the basin and arrives back at its start-

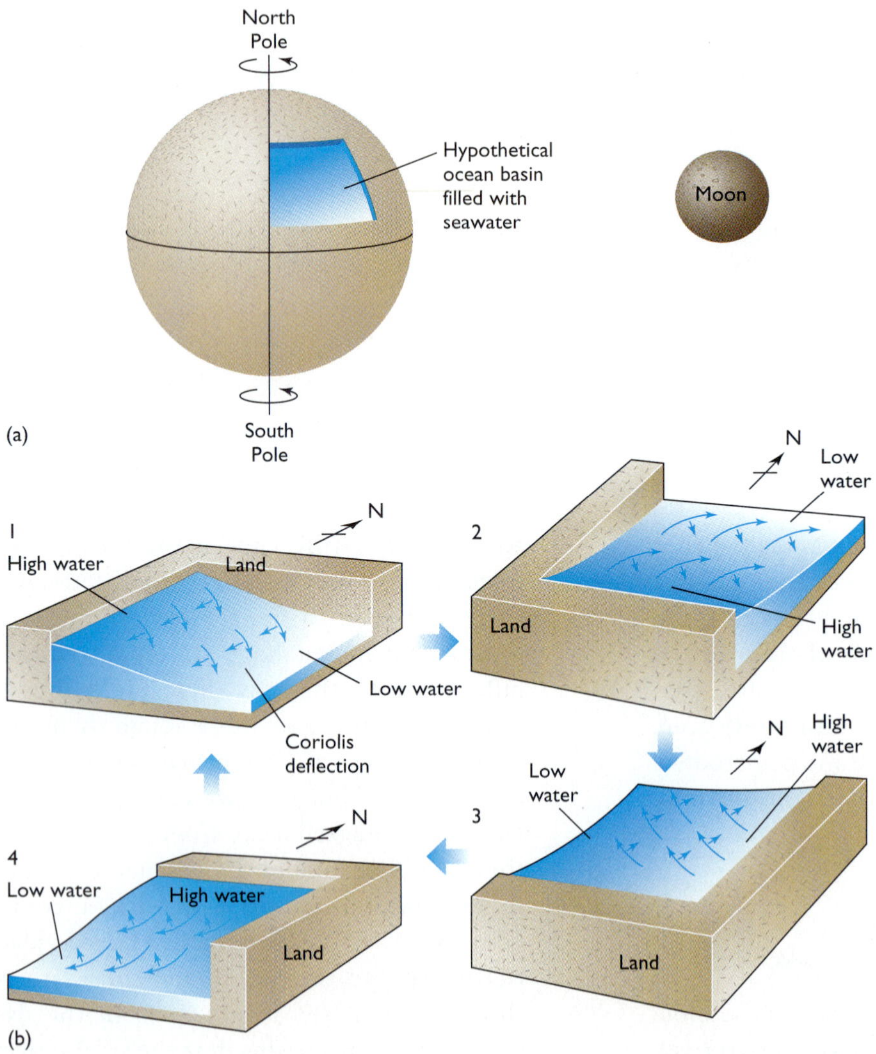

FIGURE 8-8

The dynamic model of the tides. (a) Tidal bulges are not stationary relative to the Moon. Rather, they are trapped in ocean basins and move around the Earth as the planet rotates once daily about its spin axis. (b) As the Earth rotates from west to east, the tidal bulge of water is forced against the western side of the basin, where it piles up and creates a pressure gradient. Water flows downslope in response to the pressure gradient and in the Northern Hemisphere is deflected to the right by the Coriolis effect. The deflection causes water to stack up against the southern edge of the basin, where the pressure gradient causes currents to reverse and flow northward and bend toward the easter side of the basin. The final effect of this motion is the creation of a rotary system in which high and low tides occur on opposite sides of the basin and in the Northern hemisphere rotate counterclockwise.

ing point at the western perimeter of the basin, at which point the circuit begins once again. Study Figure 8–8b until you grasp what has just been described. This concept is crucial in understanding real tides. If the basin had a diurnal tide, the bulge would rotate around the basin once each day. If it had a semidiurnal or mixed tide, the bulge would undergo two complete circuits around the basin each day.

Once a balance is reached, what was a tidal bulge becomes a **rotary wave** similar to the type of wave that you can produce easily by swirling water in a cup. Try it. Note that this rotary wave consists of a crest on one side of the basin (cup) and a trough on the opposite side, both of which progress around the basin (cup). Thus, the rotary wave creates high tides (the crest) and low tides (the trough) each day (**Figure 8–9a**). Viewed in this manner, the rotary wave is a progressive wave, because you can follow the crest or the trough as it "progresses" around the basin. But if you examine the tidal fluctuations along a cross section of the basin (**Figure 8–9b**), you see that the water surface moves up and down about a node in the center of the basin like a seesaw. This, as you know, is the classic motion of a standing wave. The antinodes are located against the landmasses where the maximum high tides and low tides occur, and the node is positioned at

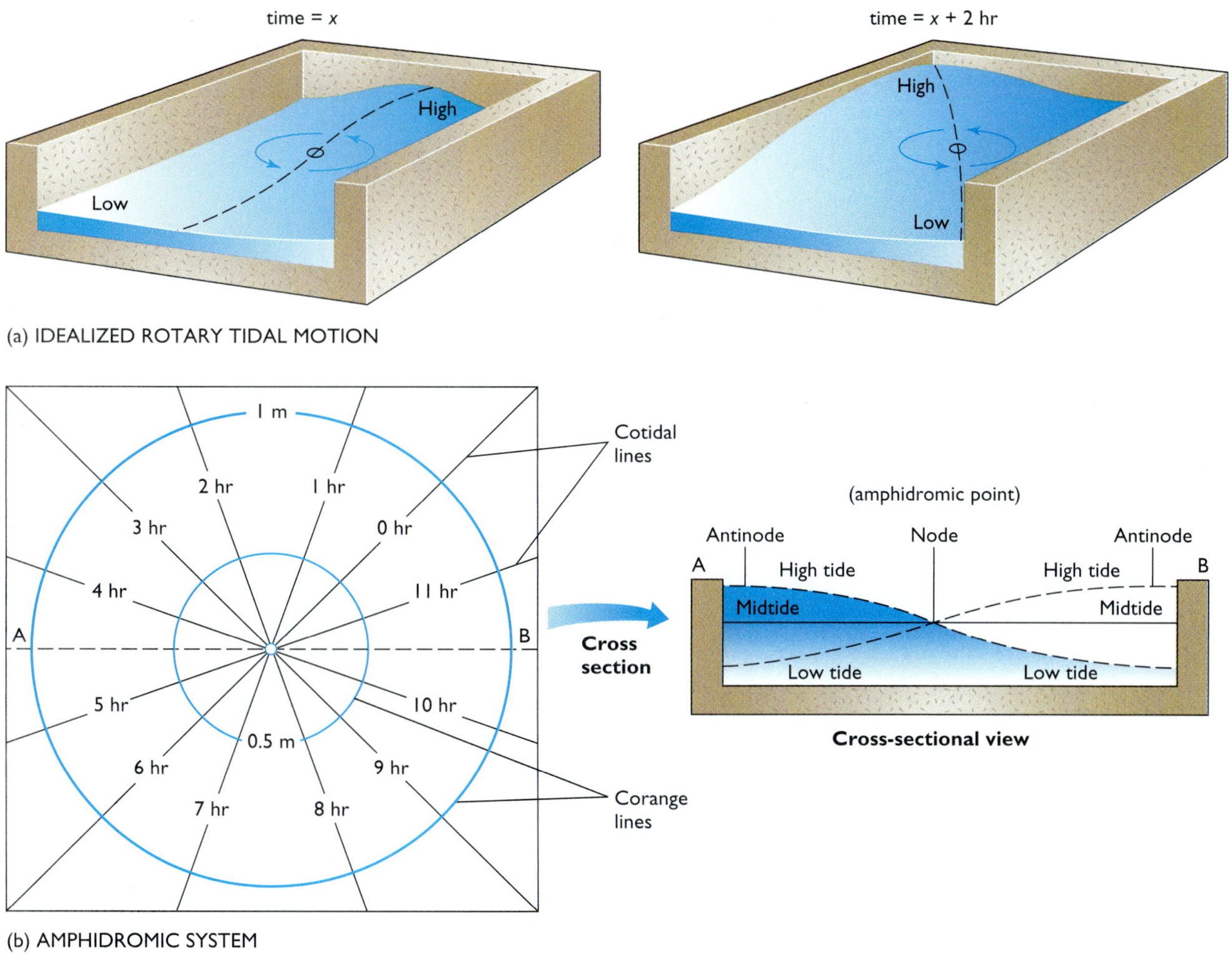

FIGURE **8-9**

Amphidromic systems. (a) In the Northern Hemisphere, the tidal bulge created by the tide-raising forces circulates counterclockwise around the basin—once daily for diurnal tides and twice daily for mixed and semidiurnal tides. The water level does not change at the center of the basin, which is a node referred to as an amphidromic point. (b) An idealized amphidromic system is pictured for one component of a semidiurnal tide in the Northern Hemisphere, showing cotidal and corange lines. A cross section through the center of the basin shows the up-and-down motion of the sea surface, caused by the tides, about a node.

Currents through Tidal Inlets

Along coastlines, tidal currents can be fast and powerful because of the effect of a shallow bottom and an irregular shoreline, which constrict the flow of water and thereby accelerate the currents. This is similar to what happens when you place your thumb across the nozzle of a hose and restrict the sluggish flow; suddenly the water accelerates and a strong "jet" flow results. In estuaries that are semienclosed, tidal currents are forced to flow through narrow tidal inlets between islands or harbor entrances. These constrictions cause the currents in the channels to speed up.

Many tidal currents in inlets are **hydraulic currents**, a special kind of coastal current created by the buildup of water at the tidal inlet (**Figure B8-1**). As the tide floods in the open ocean outside the inlet, water piles up against the entrance way to the estuary because of the narrowness of the inlet. Therefore, the water level outside the inlet will become higher than the water level inside the inlet (**Figure B8–1b**). The result is, of course, a water slope (a pressure gradient), which causes the flood current to flow even more swiftly through the inlet than simple flow constriction would indicate; this is a hydraulic current. When the tides reverse, the ebb currents drain water out of the estuary, but the narrow passageway of the inlet causes water to be stacked just inside the inlet, creating a water slope that is now tilted seaward through the inlet (Figure B8–1b).

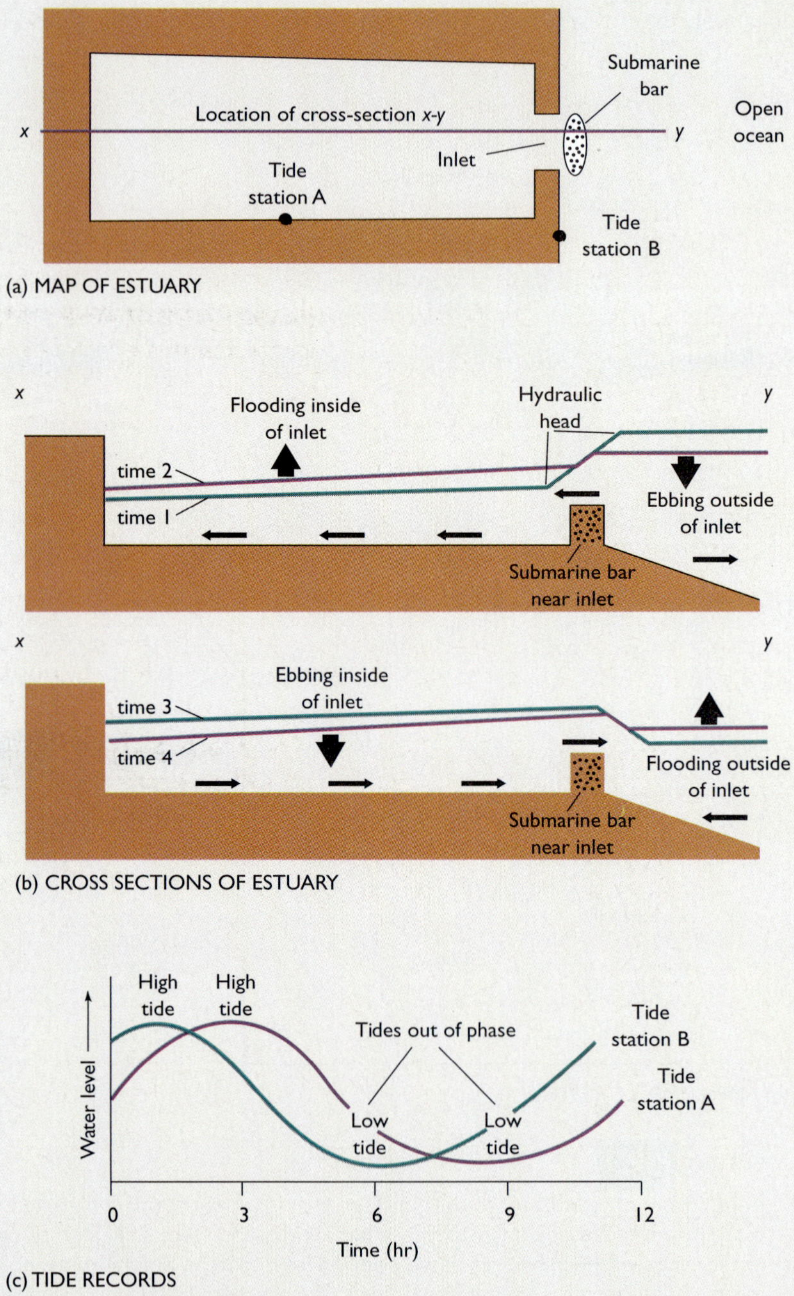

FIGURE **B8-1**

Hydraulic currents. (a) A map of a hypothetical estuary with a narrow tidal inlet is shown. (b) Cross sections show how the inlet constriction creates tidal currents that are driven by a hydraulic head (a pressure gradient). Note that the tides in the estuary and the open ocean are out of phase. (c) A comparison is shown of tidal records obtained at Station A in the estuary and Station B outside the estuary (locations are shown in part a). Note that the times of high tide and low tide at the two stations are different. Tides outside the estuary are regulated by amphidromic systems. By contrast, tides in the inlet and in the estuary are driven by the hydraulic head that forms at the inlet.

As during the flood, ebb currents in the inlet flow strongly in a hydraulic current in response to the pressure gradient.

A spectacular example of hydraulically driven tidal currents are the reversing falls of the Saint John River in New Brunswick in eastern Canada (**Figure B8–2a**), where the tidal range is very large approaching 7 meters (~23 feet). Here on the west side of the Bay of Fundy, a narrow gorge in the river (**Figure B8–3**) obstructs the tidal flow and produces a pile of water that is so steep that water actually cascades down the front slope as a waterfall (**Figure B8–2b**). As the tide reverses its direction with the ebb, so do the falls!

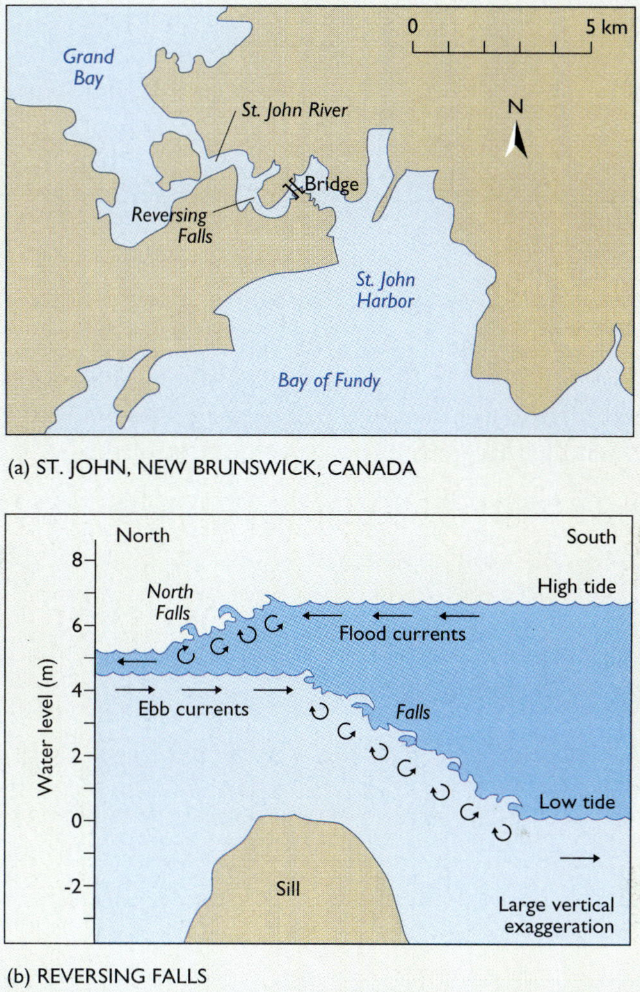

FIGURE **B8-2**

The reversing falls. (a) A map of the north-central coast of the Bay of Fundy near the St. John River of New Brunswick, Canada, is shown. (b) A sill impedes tidal flow, creating a sloped surface of cascading water during both the flood and ebb tides—the reversing falls of St. John. [Adapted from A. C. Redfield, *Introduction to Tides: The Tides of the Waters of New England and New York* (Woods Hole, Mass. : Marine Science International, 1980).]

FIGURE **B8-3**

St. John River. Located in New Brunswick, Canada, the St. John River is noted for its remarkable reversing falls.

the center of the basin. This is precisely why tides are so complicated and difficult to understand. In effect, tides have attributes of both progressive and standing waves. At first, this may be difficult to grasp, but a detailed study of Figure 8–9 and a rereading of this section should help clarify the concept.

The idealized map of Figure 8–9b depicts an **amphidromic system**, which describes a tide that rotates about a fixed node. This circular motion of the tide results in part from the Coriolis deflection and occurs in large embayments, in seas, and in the open ocean. Before considering how an amphidromic system arises, we need to examine its various parts in a bit more detail. The lines on the map that radiate outward from the node of an amphidromic system (like the spokes of a bicycle wheel) are called **cotidal lines**; they connect points at which high tide occurs at the same time of the day. At zero hours, (0 hr), for example, high tide (the crest of the rotary wave) occurs everywhere along the "0 hour" cotidal line. One hour later, the crest has rotated counterclockwise to the "1 hour" cotidal line; two hours later to the "2 hour" line. At 12 hours, the rotary wave has returned to its starting point in the basin, bringing with it the second high tide of the day to that zero-hour point. Now the rotation continues through another 12 hours.

Note that the tidal range is zero (no tides!) at the node and increases systematically to a maximum value at the antinodes located at the edges of the basin. The tidal range anywhere in the basin is indicated by **corange lines**, which link the points on the water surface that have equal tidal ranges. They appear in amphidromic systems as irregular circles that are centered about the node (see Figure 8–9b). Also, note that the rotary wave in amphidromic systems travels in a counterclockwise direction in the Northern Hemisphere because Coriolis deflection is to the right (see Fig 8–8b). In the Southern Hemisphere the Coriolis deflection is to the left, inducing clockwise rotation of the rotary wave in amphidromic systems. In order to check and see if you grasp these ideas, you should do an analysis similar to the one in Figure 8–8b, demonstrating the resultant clockwise circulation of the rotary wave for the Southern Hemisphere.

This theory has been simplified in that it considers only a single tidal component, in this case the main lunar constituent (**Figure 8–10**). Other tidal components, with different periods and heights, exist as well. Any actual tide is the sum of as many as sixty-five distinct tidal compo-

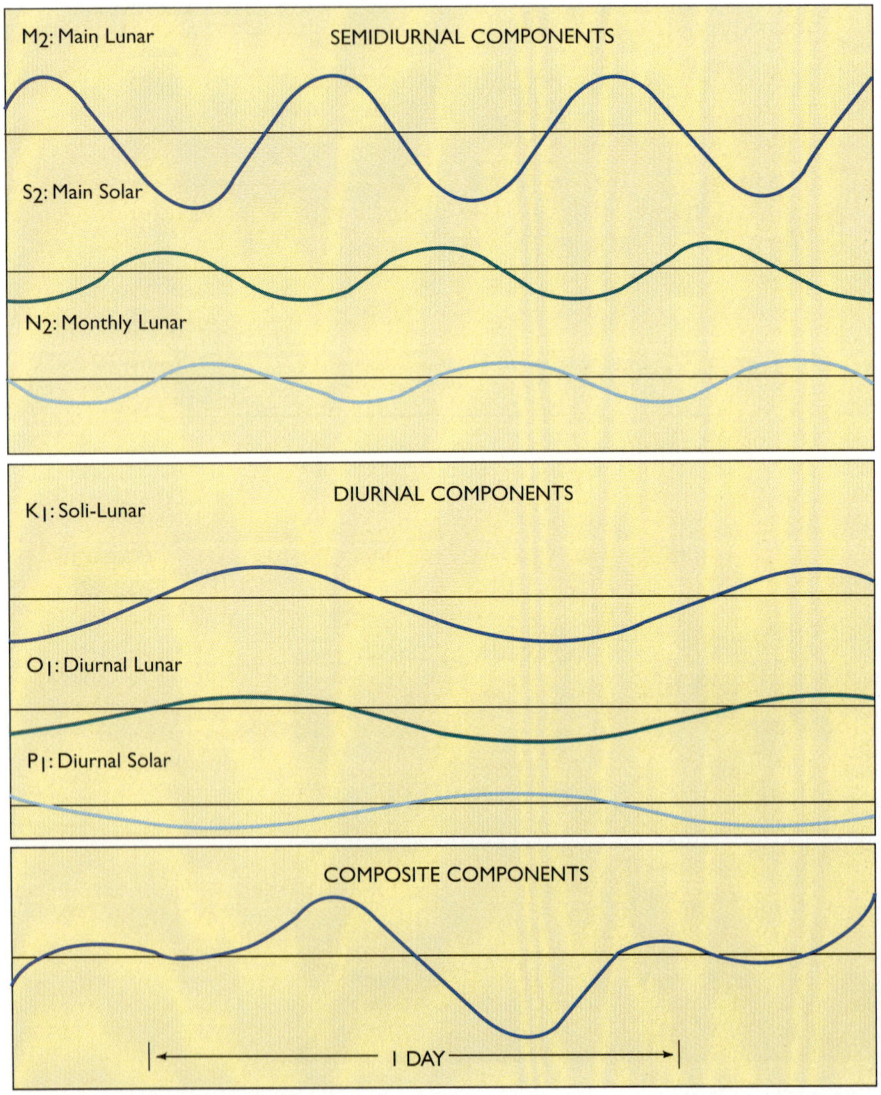

FIGURE 8-10

Tidal predictions. The lower tidal curve is a composite of a number of semidiurnal and diurnal tidal components that results from constructive and destructive wave interference. The main lunar constituent M2 has the greatest effect on the resultant tidal curve because of its large tidal range. Note how irregular the composite tidal curve is as a result of wave interference. [Reproduced with permission of R. E. Johnson, 1988.]

nents. It is not important that you understand what each component is. It's sufficient to know that they interact by constructive and destructive wave interference, so that the final wave form is a composite of all the tidal components. Examine the bottom profile in Figure 8–10; it is the composite profile of the individual tidal components shown above that comes about by wave interference. Note that in this example there are two high tides and two low tides with different tidal ranges in one day, a classic mixed tide.

Figure 8–11 shows the global amphidromic system for the main lunar component. The lines in this map are cotidal lines that converge on nodes. Notice that the rotary waves rotate counterclockwise in the Northern Hemisphere and clockwise in the Southern Hemisphere because of Coriolis deflection. Also, unlike in our idealized amphidromic system of Figure 8–9b, the cotidal lines are not evenly spaced. The reason for these distortions is that tides are shallow-water waves, so their speed of rotation (their celerity) depends directly on water depth, which varies tremendously, even in the deep ocean, because of the presence of ocean-spreading ridges. Therefore, the complex bathymetry in conjunction with the irregular shapes of the various ocean basins causes wave refraction, which distorts the rotary motion of the circulating wave. Additional distortions occur near the equator, where Coriolis deflection changes direction, and along the continental margins and around large islands, where the sea bottom is shallow and the wave is slowed down. The combined effect of all these factors is to produce considerable regional variations in the type of tide and the tidal range along the world's coastline (**Figure 8–12**).

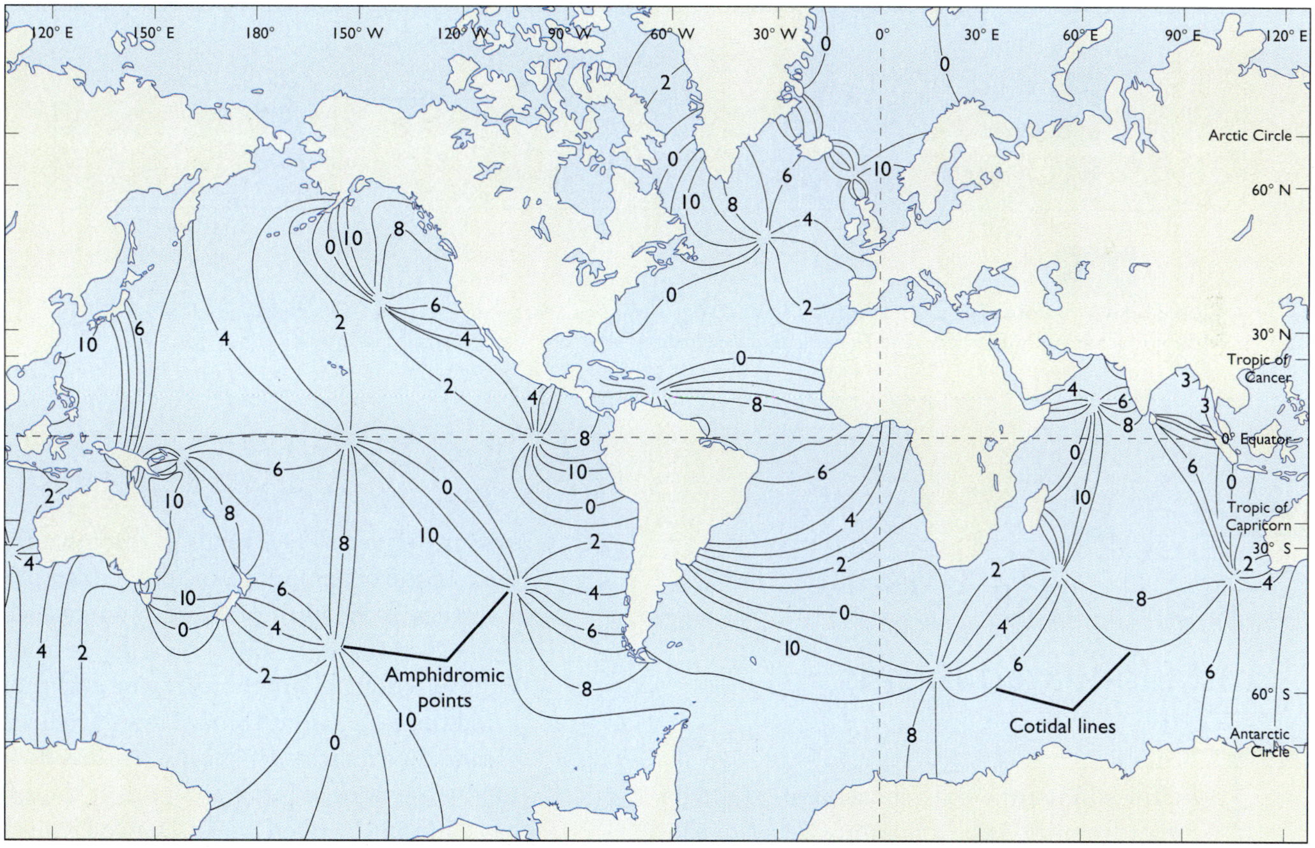

FIGURE **8-11**

Global amphidromic systems for the main lunar component. The amphidromic systems in the world's oceans show clockwise rotation in the Southern Hemisphere and counterclockwise rotation in the Northern Hemisphere because of the Coriolis effect. The bending of the cotidal lines reflects wave refraction. [Adapted from D. E. Cartwright, "Deep Sea Tides," *Science Journal* 5 (1969): 60–67.]

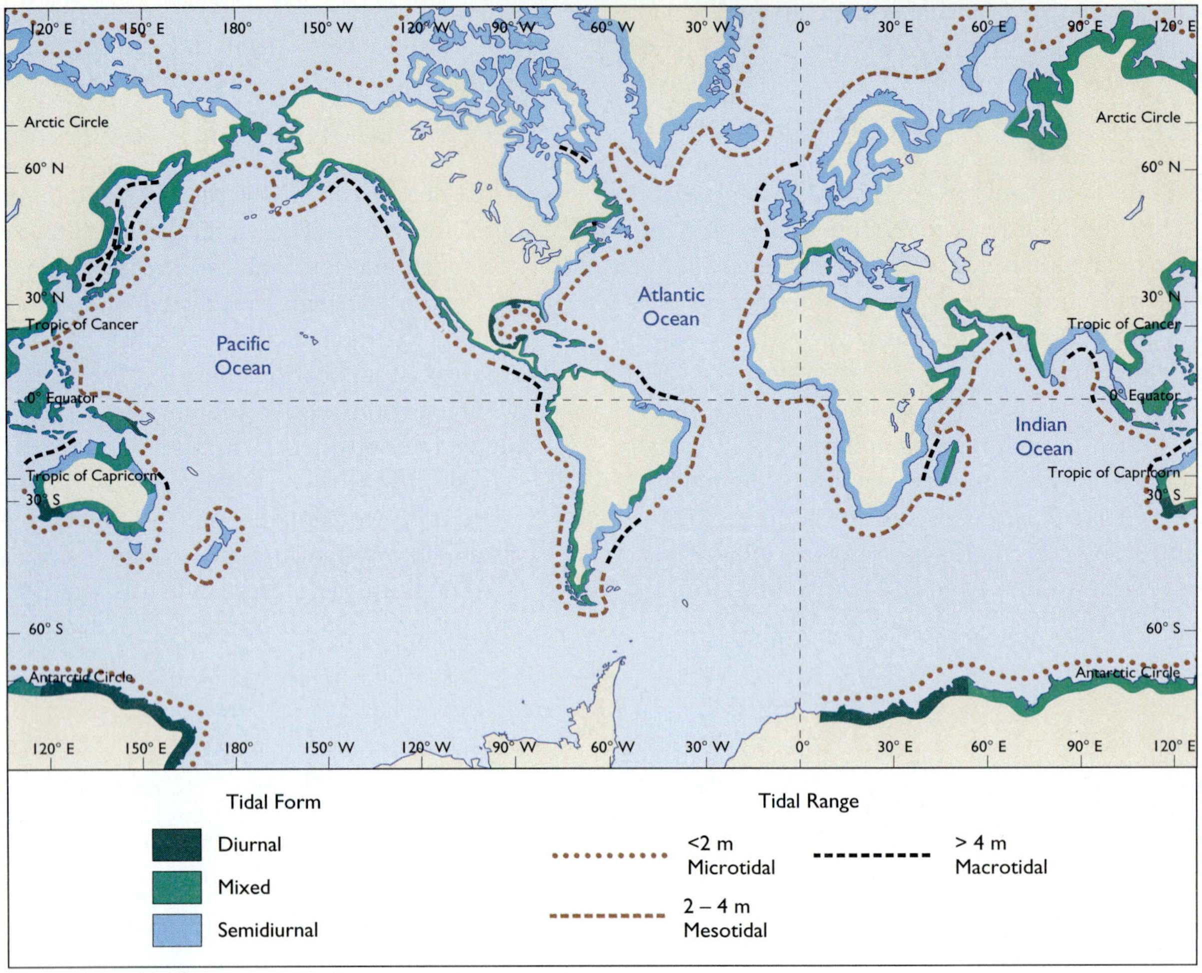

FIGURE **8-12**

Global tidal variations. The types of tide and tidal range vary considerably along the world's coastlines. [Adapted from A. Couper, ed., *The Times Atlas of the Oceans* (New York: Van Nostrand Reinhold, 1983).]

8-3

Tides in Small and Elongated Basins

The tides in narrow coastal basins are quite different from the large amphidromic systems of the much wider and deeper ocean basins. These differences result from the smallness, shallowness, and shape of many estuaries and bays. If a basin is broad and symmetrical, the resultant amphidromic system (**Figure 8–13a**) tends to resemble those that form in the open ocean. An actual example of such a system is the Gulf of Saint Lawrence of eastern Canada (**Figure 8–13b**). Nearby is the Bay of Fundy, which is much narrower and more elongated than the Gulf of Saint Lawrence. In this type of narrow (restrictive) basin (**Figure 8–13c**), the tidal wave form cannot rotate as it does in larger bays and in the open ocean. Instead, the tide simply moves in and out of the estuary, and the tidal currents reverse their flow direction rather than rotating around a distinct node. This behavior is evident in the Bay of Fundy in Canada (**Figure**

SCIENCE BY NUMBERS

Speed of the Tide

The most obvious effects of the tides to a sailor are the rhythmic rise and fall of the ocean surface and the reversing ebb currents and flood currents. The very long wavelengths of the tides make them shallow-water waves even in the deep water of the open ocean. This being the case, their speed is independent of the wave characteristics and varies directly with water depth as follows:

$$C = 3.13\sqrt{d}\ \text{m/sec},$$

where d is water depth in meters. Because the average depth of the ocean is 4 kilometers (4,000 meters), we can easily estimate the average speed of the tide:

$$C = 3.13\sqrt{d}\ \text{m/sec} = 3.13\sqrt{4{,}000}\ \text{m/sec} = (3.13)(63.3)\ \text{m/sec} = 198\ \text{m/sec}.$$

That's equivalent to two lengths of a football field in one second. How fast is that in more familiar units such as miles per hour? Let's do the conversion (Appendix II contains the relevant information):

$$(198\ \cancel{\text{m}}/\cancel{\text{sec}})\,(2.24\ \text{mph}/1\ \cancel{\text{m}}/\cancel{\text{sec}}) = 444\ \text{mph}.$$

What would be the speed of the tide as it crosses the crest of the midocean ridges, which lie about 1 kilometer below the sea surface? Let's do the calculation:

$$C = 3.13\sqrt{d}\ \text{m/sec} = 3.13\sqrt{1{,}000}\ \text{m/sec} = (3.13)(31.5)\ \text{m/sec} = 99\ \text{m/sec}.$$

Converting this value to mph, we get

$$(99\ \cancel{\text{m}}/\cancel{\text{sec}})\,(2.24\ \text{mph}/1\ \cancel{\text{m}}/\cancel{\text{sec}}) = 222\ \text{mph}.$$

Because the speed of a shallow-water wave depends directly on water depth, the wave crest will bend, a process called refraction. Wave retraction will be most intense when the wave crest moves from the deep water at the base of the continental slope onto the shallow continental shelf. Note the effect of refraction on the shape of the cotidal lines along the western and eastern continental margins of North America in Figure 8–11.

8–13d), where the cotidal and corange lines are all nearly parallel.

The Bay of Fundy is noted for its unusually large tidal range, which can be more than 15 meters (~50 feet) at the bay's most interior locations. This extreme tidal range is a consequence of **tidal resonance**—a condition, as you may remember, that arises when the period of a main tidal component almost equals the natural period of the basin. In effect, resonance produces a seiche (as you did in your bathtub, remember!), causing the water to slosh slowly back and forth as a standing wave. The natural period of the Bay of Fundy, which depends on the length of the basin and its average water depth, is approximately 12 hours; this is very close to the period of the main lunar component, which is 12 hours and 25 minutes. the semidiurnal tides moving in and out of the Bay of Fundy set up resonance, and the resultant seiche creates an unusually large tidal range. Furthermore, the bay tapers inward, so that the water is funneled into a progressively narrower area, thereby increasing the tidal range even further. As the Bay of Fundy narrows landward, the tidal range during a spring tide attains 16 meters (~53 feet), the largest tidal range anywhere in the world.

A spectacular tidal event in some rivers is the **tidal bore**, a "wall" of water that surges upriver with the advancing tide. Three factors appear to contribute to the occurrence of tidal bores: (1) a large tidal range, typically greater than 5 meters (~17 feet); (2) a tapering basin geometry; and (3) water depths

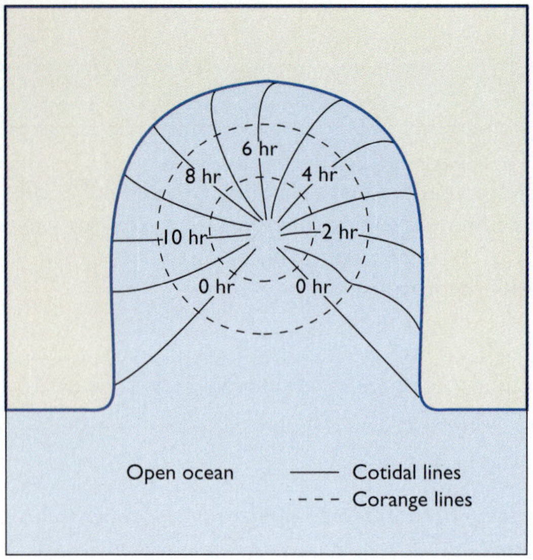

(a) BROAD BASIN

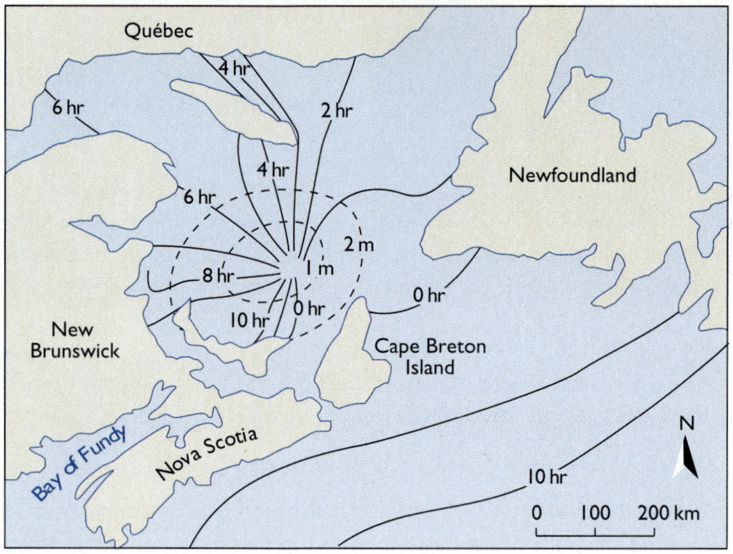

(b) AMPHIDROMIC SYSTEM: GULF OF ST. LAWRENCE

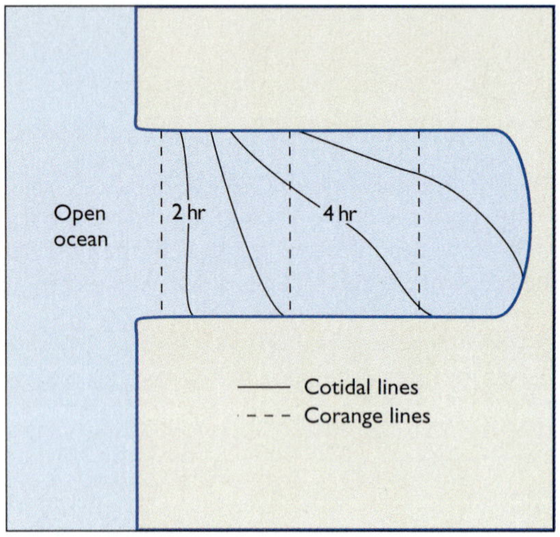

(c) NARROW BASIN

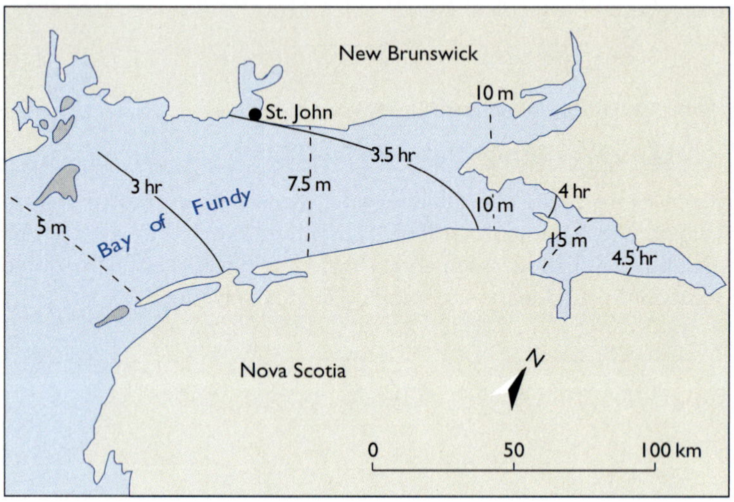

(d) COTIDAL AND CORANGE LINES: BAY OF FUNDY

FIGURE **8-13**

Tides in restricted basins. (a) An idealized amphidromic system is calculated here for a semidiurnal tide in a broad basin. (b) This map depicts the actual amphidromic system for the Gulf of St. Lawrence, a broad basin located between New Brunswick and Newfoundland along the southeastern coast of Canada. (c) In narrow basins, flow is restricted, so true amphidromic systems do not develop. (d) Cotidal and corange lines are generalized for the Bay of Fundy (see part b), a narrow basin. [Adapted from A. C. Redfield, *Introduction to Tides: The Tides of the Waters of New England and New York* (Woods Hole, Mass. : Marine Science International, 1980).]

that decrease systematically upriver. Under such conditions, the water of the incoming tide is constricted by the sides and bottom of the river basin so that the front of the tide steepens and is forced to move much faster than it normally would. Most tidal bores are small, measuring less than 20 centimeters (~8 inches) in height, which is hardly noticeable (**Figure 8–14**). However, in certain rivers that have unusual channel geometries, the wall of water becomes a towering, fast-moving wave front, and all watercraft in its path must be pulled from the river or be swamped. Examples are the 5-meter-high (~17 feet) bore in the Amazon River, which can travel upstream at more than 20 kilometers per hour

FIGURE 8-14

Tidal bore. A tidal bore moving upriver from the upper Bay of Fundy, Nova Scotia, Canada.

(~25 mph), and the spectacular 7- to 8-meter-high (~23 to 26 feet) bore in the Fu-Ch'un River in northern China, which used to move at a speed of more than 25 kilometers per hour (~31 mph)! Dredging of the river bottom and stabilization of the river banks reduced the size of this tidal bore. Now, it reaches no more than 4 to 5 meters (~13 to 17 feet) in height and then only during the highest spring tides.

8-4

Tidal Currents

As sea level moves up and down with the tide, water currents are generated. **Flood currents** transfer water toward the coast, and the tide rises. **Ebb currents** flow away from the coast, and the tide falls. The two create a pronounced reversal pattern to the flow of the tidal currents in the shallow water near shorelines. Tidal flow through the inlets of harbors and estuaries tends to be unusually strong. The specific reasons for this are described in the boxed feature, "Currents through Tidal Inlets." Also, at the ocean's edge, the daily oscillation and currents of the tides

THE OCEAN SCIENCES BIOLOGY

Tidal Rhythms in Organisms

Many coastal organisms are cyclically attuned to the rhythms of the tides. Experiments indicate that many can sense and even precisely measure the passage of time. One of these is the diatom *Hantzschia virgata,* a species of unicellular (one-celled) alga of microscopic size that lives in the water-filled spaces between the grains of beach sand. Being a plant, this diatom species needs sunlight to photosynthesize; otherwise it will die. Consequently, as the tide drops each day, *H. virgata* migrates upward through the water-saturated beach sand to the beach surface, where it stays throughout the ebb tide. These algae impart a distinctive golden brown color to the beach wherever they occur in dense concentrations. When the flood tide comes in, they descend back into the water-filled sand, relying on the sand layer above them to protect them from the surf.

The behavior of fiddler crabs (genus *Uca*) is also synchronized to the daily flooding and ebbing of the tides. During the flood phase of the tide, the fiddler crab is secure in a burrow, safe from predators that come in with the

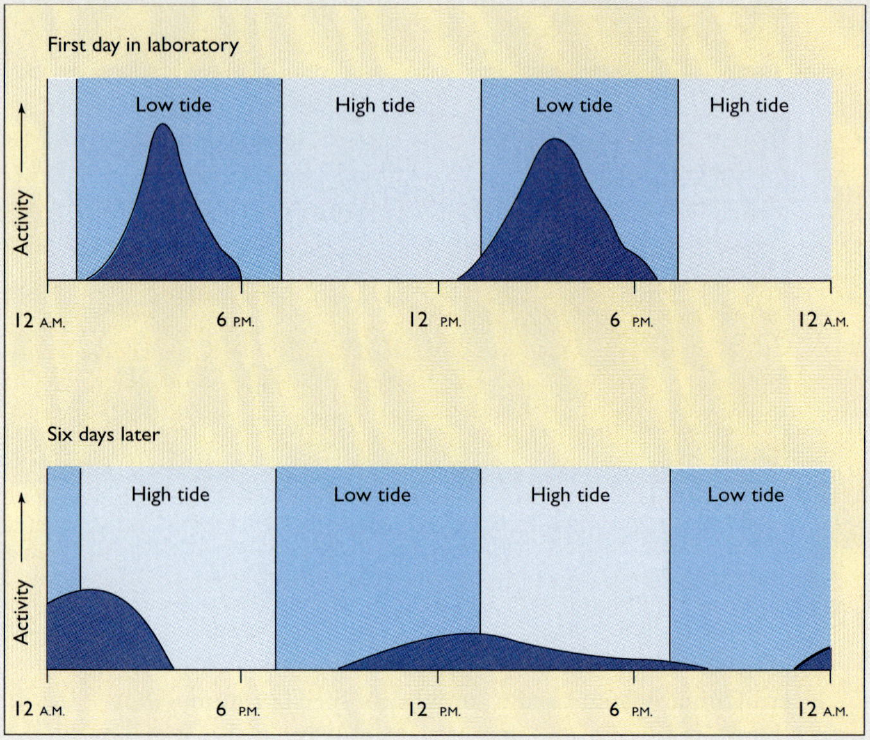

FIGURE B8-4 **Diurnal activity of fiddler crabs.** In their natural tidal habitat, fiddler crabs remain inactive in burrows during high tide and emerge and search for food during low tide. This behavior is maintained in the laboratory despite the absence of tides. Over time, however, the duration of the crabs' active phase lengthens, and the time of peak activity becomes out of phase with the tides. [Adapted from J. D. Palmer, "Biological Clocks of the Tidal Zone," *Scientific American* 232 (1975): 70–79.]

influence the behavior of many organisms (see boxed feature, "Tidal Rhythms of Organisms"). In the open ocean, far from the confinement of land, and in response to the Coriolis effect, tidal currents describe a rotary pattern rather than the back-and-forth flow pattern so typical of coastal waters.

As an illustration of rotary flow, near-surface current velocities over a tidal cycle were measured each hour off southern New England and are plotted as a current "rose" diagram (**Figure 8–15a**). The orientation of the arrows indicates the flow direction of the tidal current at a specific hour of the day; the length of the arrow, the speed of the current (the longer the arrow, the faster the current). Because the measurements are made in the Northern Hemisphere, where the Coriolis deflection is to the right, the tidal currents display a clockwise rotary motion, sweeping from the north advancing tide. When the tide ebbs, fiddler crabs emerge from their burrows and begin to search for small bits of food in the mud and sand. When taken to a laboratory, these crabs continue to engage in rhythmic activity, despite the absence of tides. They are most active at the time of day when low tide occurs in their natural habitat on a tidal flat! However, as time passes, the correlation between their activity and the phase of the tide in their home waters weakens and then disappears (**Figure B8–4**).

There are many odd behavioral traits of shore-dwelling organisms that are synchronized to tidal fluctuations. For example, a green flatworm, *Convoluta,* inhabits sand flats along the coast of western France. The tissue of this worm contains numerous microscopic algae that, as is true for all plants, require regular doses of sunshine to conduct photosynthesis. Therefore, *Convoluta* digs out a burrow during high tide to avoid being harmed by waves and surf, but lies exposed on the sand surface at low tide during the day so that its algae can photosynthesize. The worm benefits from this association by obtaining all of its food from the algae. In turn, the delicate algal cells are supplied with the nutrients (the worm's waste products) they need to photosynthesize while the flatworm's tissue protects them from the harsh tidal environment.

FIGURE **B8-5**

Grunion spawning. Grunion spawn in dense clusters on the beaches of southern California during the maximum spring high tide that occurs in the spring and summer months. The eggs hatch about nine days later, just after they are freed from the sand by flooding tidal currents.

From March through September, small fish known as grunion (*Leuresthes tenuis*) come ashore in large numbers to spawn along many beaches of southern California, just before the maximum spring tide (**Figure B8–5**). The female lays thousands of eggs just below the sand surface, while the male deposits milt to fertilize the eggs, which then become deeply buried by sand that is deposited as the tide falls. Nine days later, the eggs are freed from the sand by the currents of a high tide and hatch after about a 3-minute exposure to seawater. The hatchlings then swim offshore.

Visit www.jbpub.com/oceanlink for more information.

to the east, to the south, to the west, and back again to the north in about 12 hours, the period of the semidiurnal tides in this area. These measurements show how the tidal currents change speed and direction at a fixed point.

Another way to view these data is to place systematically each hourly measurement of the tidal currents head to tail, creating a progressive-vector diagram (**Figure 8–15b**). This shows the flow pattern of a parcel of water as it is moved about by the tides. Notice that the pattern is a broad ellipse, rather than a circle. This reflects the fact that the rotary currents here have different speeds for different phases of the tide; otherwise, when arranged head to tail they would describe a circle. Also, note that the elliptical pattern is not quite closed, indicating the presence of another current unrelated to the tides, most probably a wind-driven current.

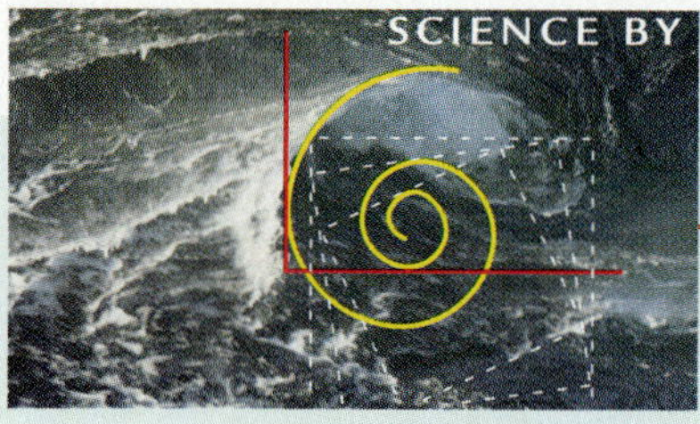

SCIENCE BY NUMBERS

Depth of the Bay of Fundy

Very high tides occur in the Bay of Fundy as a result of resonance. This means that the natural period of the basin approximates the period of the tides (~12 hours and 25 minutes), and this causes seiching in the basin. Because the Bay of Fundy is an open-ended basin, its natural period is

$$T = 4l/\sqrt{gd}\ \text{sec},$$

where l is the basin length in meters, d is water depth in meters, and g is gravitational acceleration which is 9.8 m/sec^2. How deep is the Bay of Fundy on the average? We can easily answer this question by using the formula. We know that its natural period is about 12 hours and 25 minutes (the period of the tide), and its basin length is about 200 kilometers (2×10^5 m), which we determine by inspecting the map in Figure 8-13d. We can now solve the formula for d and insert the proper values for all the variables. Let's begin.

$$T = 4l/\sqrt{gd}.$$

By multiplying both sides of the formula by $\sqrt{gd}$, we get

$$T\sqrt{gd} = 4l\sqrt{gd}/\sqrt{gd},$$

and

$$T\sqrt{gd} = 4l.$$

Next, we divide both sides of the formula by T, and we get

$$T\sqrt{gd}/T = 4l/T,$$

and

$$\sqrt{gd} = 4l/T.$$

FIGURE **8-15**

Tidal currents. (a) Each arrow represents the direction and the speed of hourly currents measured from the Nantucket Lightship, located off the southern coast of Massachusetts. (b) The placement of the head and tail of each arrow in a time sequence shows the circular path of tidal currents during a full tidal cycle. Failure of the ring to close results from the effect of a nontidal current that also moved water during the tidal cycle. [Adapted from A. C. Redfield, *Introduction to Tides: The Tides of the Waters of New England and New York* (Woods Hole, Mass.: Marine Science International, 1980).]

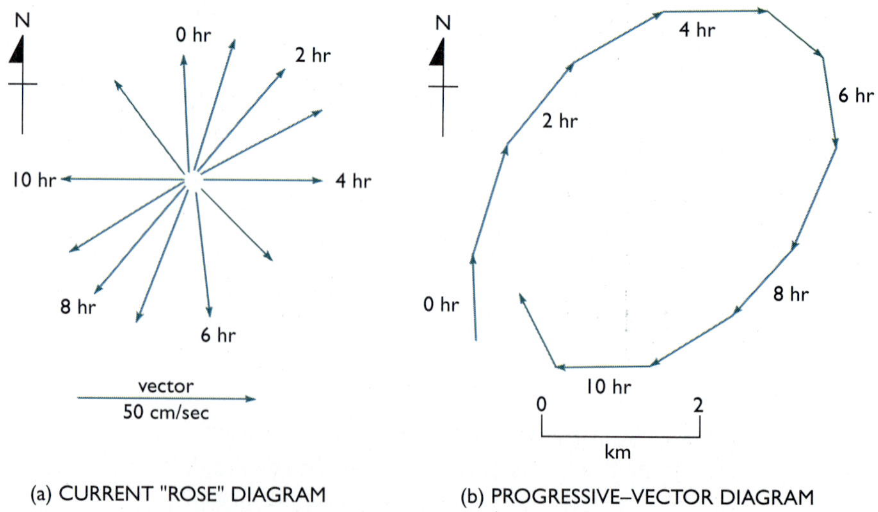

We now square both sides in order to eliminate the square root sign.

$$(\sqrt{gd})^2 = (4l/T)^2,$$

and

$$gd = 16l^2/T^2,$$

and

$$d = 16l^2/gT^2.$$

We now have the formula as we want it for determining the average depth of the Bay of Fundy. The variables are easily determined as follows:

$$g = 9.8 \text{ m/sec}^2$$

$$l = 200 \text{ km} = 2 \times 10^2 \text{ km} = (2 \times 10^2 \cancel{\text{km}})(10^3 \text{ m}/1 \cancel{\text{km}}) = 2 \times 10^5 \text{ m}$$

$$T = 12 \text{ hr } 25 \text{ min; } 12 \text{ hr} = (12 \cancel{\text{hr}})(60 \text{ min}/1)(1 \cancel{\text{hr}}) = 720 \text{ min}$$

and

$$T = 720 \text{ min} + 25 \text{ min} = 745 \text{ min} = (745 \cancel{\text{min}})(60 \text{ sec}/1 \cancel{\text{min}}) = 44{,}700 \text{ sec} = 4.47 \times 10^4 \text{ sec}.$$

The final step is to substitute the values for the variables and solve for d in meters.

$$d = 161^2/gT^2 = (16)(2 \times 10^5 \text{ m})^2/(9.8 \text{ m/sec}^2)(4.47 \times 10^4 \text{ sec})^2$$

$$= (16)(4 \times 10^{10} \text{ m}^2)/(9.8 \text{ m/sec}^2)(1.998 \times 10^9 \text{ sec}^2)$$

$$= (64 \times 10^{10} \text{ m}^3/\cancel{\text{sec}^2})/(1.96 \times 10^{10} \cancel{\text{sec}^2}) = (64/1.96) \text{ m} = 32.7 \text{ m}.$$

We conclude from this analysis that the average depth of the Bay of Fundy is just a bit more than 30 meters.

8-5

Power from the Tides

Under the right conditions, the rise and fall of the water level caused by flooding and ebbing tides can be used to generate electricity. For economic reasons, the coastal location must have a substantial tidal range, ideally more than 5 meters (~17 feet), and the water must flow through a constricted inlet into a large bay or estuary. Where these conditions are met, the basin can be closed off at its inlet with a dam. Then a series of gates are opened during the flood part of the tidal cycle and closed during the ebb, trapping a large volume of water behind the dam at the level of high tide. Later the gates are opened, allowing water to drain from the pool back to the ocean. Large, reversing turbines at the base of the dam generate electricity during parts of both the flood and ebb tides as the seawater flows across their blades. These special turbines are designed with adjustable blades; their pitch or angle can be varied so that they rotate at a uniform rate despite changes in the speed of the tidal currents. This assures that the electric generators operate at a constant speed.

Although worldwide more than 150 coastal sites satisfy the physical requirements for power generation, only a few tidal power plants have been constructed. The oldest and most successful, the La

(a) LA RANCE RIVER PLANT

(b) TURBINES AT LA RANCE

FIGURE 8-16

La Rance River Plant. (a) This power plant generates electricity by harnessing the energy of the tides, which have a maximum tidal range of over 13 meters (approximately 43 feet). (b) Flowing seawater spins these huge turbines in the plant.

FIGURE 8-17

Power from the tides. The generation of power during an entire cycle is shown. Notice that the plant produces power for only about one-half of the tidal period, when the water levels in the ocean side and basin side are correctly positioned for power generation.

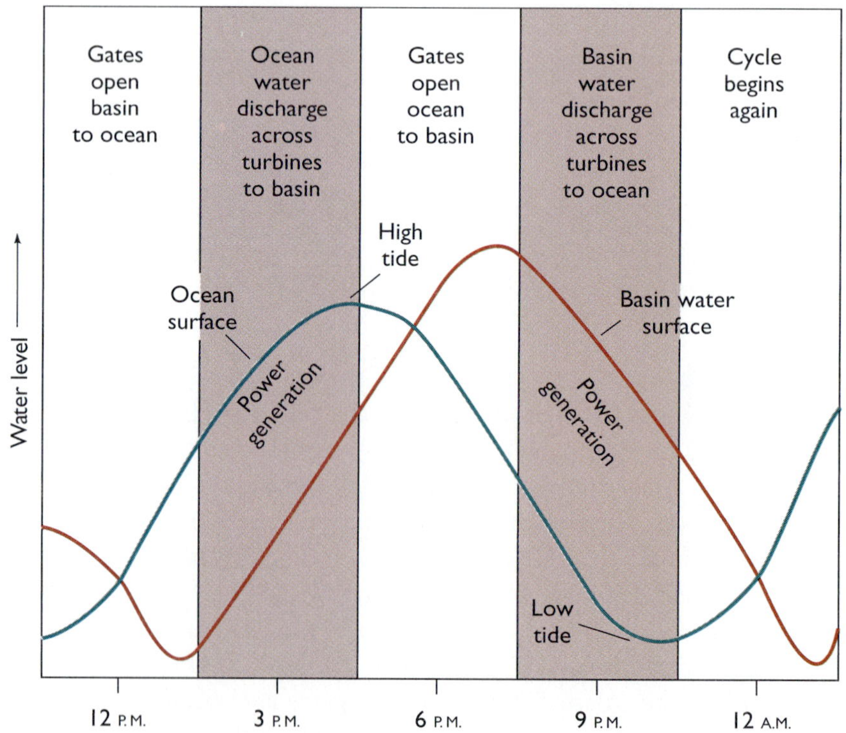

Rance River Plant (**Figure 8-16**), is located just south of Saint-Malo, France, where the tidal range varies between a neap tide of 9 meters (~30 feet) and a spring tide of over 13 meters (~43 feet). There a total of twenty-four reversible turbines, mounted about 10 meters (~33 feet) below the water surface at the lowest tide, generate electrical power for about one-half of the tidal cycle (**Figure 8–17**).

STUDY GUIDE

KEY CONCEPTS

1. *Tides* are very long shallow-water waves generated by the gravitational attraction of the Moon and the Sun on the ocean's water.

2. Discounting the effect of the Sun, the *equilibrium model of the tides* relates the formation of two bulges of water on opposite sides of the Earth to the interplay of gravitational attraction between the Earth and the Moon and of centrifugal force that results from the Earth and the Moon revolving about their common center of mass. The result is two tidal bulges of equal size. The Earth rotates through these bulges, which are fixed relative to the Moon, creating *diurnal, semidiurnal,* and *mixed tides* for different areas, depending on their position on the globe.

3. The *dynamic model of tides* examines tides when water is confined to ocean basins and responds dynamically, rather than passively, to the tide-raising forces. Under these conditions, the tidal wave rotates about a node, where the tidal range is zero. In the Northern Hemisphere, the tides rotate in a counterclockwise direction around the basin because of Coriolis deflection. The edges of the basin represent antinodes where the tidal range is maximum. The *rotary wave* completes one circuit around the basin in 12 hours and 25 minutes (semidiurnal and mixed tides) or 24 hours and 50 minutes (diurnal tides).

4. When the natural period of a basin, which depends on the basin's length and water depth, is very nearly equal to the period of the tides, *tidal resonance* results, producing a *seiche* —a standing wave that may create a tidal range in excess of 10 meters (~33 feet).

5. In the open ocean, tidal currents describe a rotary pattern. In shallow water and near coastlines, tidal currents do not rotate, but reverse back and forth. *Flood currents* direct water landward; *ebb currents* direct water seaward.

6. In a basin with an appropriate geometry and tidal range, electrical power can be generated by having the flood and ebb currents flow over specially designed turbines.

KEY WORDS

amphidromic system
corange line
cotidal line
diurnal tide
dynamic model of tides
ebb current
equilibrium model of tides
flood current
mixed tide
neap tide
semidiurnal tide
spring tide
tidal bore
tidal range
tidal resonance

QUESTIONS

REVIEW OF BASIC CONCEPTS

1. Distinguish clearly among mixed, spring, ebb, high, semidiurnal, flood, low, diurnal, and neap tides.

2. Describe the equilibrium and dynamic models of the tides. What simplifications are made in each of these theoretical models?

3. What is an amphidromic system, and how does it arise?

4. What is a tidal bore, and how is it generated?

5. What is tidal resonance, and how is it related to a seiche?

CRITICAL-THINKING ESSAYS

1. Why do different shorelines have different kinds of tides and different tidal ranges?

2. Using diagrams, discuss the development of an amphidromic system in a large ocean basin of the Southern Hemisphere.

3. What general effects on tides would result if (a) the Moon's mass were doubled, or (b) the distance between the Earth and the Moon were halved?

4. How and why can tides at any one coastal site vary each day, each week, each month?

DISCOVERING WITH NUMBERS

1. Examine Figure 8–3. What are the period and the range of the tides at Pensacola, Florida, and at San Francisco?

2. Examine Figure 8–9b. What is the wavelength of the tide if the width of the basin is 2,800 kilometers? Is the tide a shallow-water or deep-water wave? Explain. In meters per second, what is the celerity of this wave if the average depth in the basin is 4 kilometers?

3. Examine Figure 8–15a. You're sailing a 26-foot sailboat off the southern shore of Massachusetts. Your speed is a steady 4 knots through the water as you work your way to the northeast; the tides are flowing as shown for the 2-hour current on the diagram. What is your actual boat speed over the bottom? Calculate what your boat speed over the bottom would be if you were sailing through the water at a steady 4 knots to the northeast, but with the tides running as shown for the 8-hour current.

4. Now study Figure 8–15b, and assume that your boat started a journey at the spot on the map equivalent to the tail end of the "0-hour vector." Calculate how far you would go and where you would be on this map if you sailed for 11 hours to the northeast at a steady rate of 4 knots.

SELECTED READINGS

Clancy, E. P. 1969, *The Tides: Pulse of the Earth.* Garden City, New York: Doubleday.

Fisher, A. 1989. The model makers. *Oceanus* 32 (2): 16–21.

Garrett, C., and Maas, L. R. M. 1993. Tides and their effects. *Oceanus* 36 (1): 27–37.

Goldreich, P. 1972. Tides and the earth–moon system. *Scientific American* 226 (4): 42–57.

Greenberg, D. A. 1987. Modeling tidal power. *Scientific American* 257 (5): 128–131.

Lynch, D. K. 1982. Tidal bores. *Scientific American* 247 (4): 146–156.

Open University Course Team. 1989. *Waves, Tides and Shallow Water Processes.* New York: Pergamon Press.

Pugh, D. T. 1987. *Tides, Surges and Mean Sea-Level.* New York: Wiley.

Redfield, A. C. 1980. *Introduction to the Tides.* Woods Hole, MA: Marine Science International.

Ryan, P. R. 1979. Harnessing power from the tides: State of the art.*Oceanus* 22 (4): 64–67.

Sobey, E. 1982. What is sea-level? *Sea Frontiers* 28 (3): 136–142.

Zerbe, W. B. 1973. Alexander and the bore. *Sea Frontiers* 19 (4): 203–08.

TOOLS FOR LEARNING

Tools for Learning is an on-line review area located at this book's web site OceanLink (**www.jbpub.com/oceanlink**). The review area provides a variety of activities designed to help you study for your class. You will find chapter outlines, review questions, hints for some of the book's math questions (identified by the math icon), web research tips for selected Critical Thinking Essay questions, key term reviews, and figure labeling exercises.

9

Marine Ecology

The world below the brine,
Forests at the bottom of the sea, the branches
and leaves,
Sea-lettuce, vast lichens, strange flowers and seals,
the thick tangle, openings, and pink turf,
Different colors, pale gray and green, purple, white,
and gold, the play of light through the water,
Dumb swimmers there among the rocks, coral, gluten,
grass, rushes, and the aliment of the swimmers,
Sluggish existences grazing there suspended, or slowly
crawling close to the bottom,
The sperm-whale at the surface blowing air and spray,
or disporting with his flukes,
The leaden-eyed shark, the walrus, the turtle, the
hairy sea-leopard, and the sting-ray,
Passions there, wars, pursuits, tribes, sight in those
ocean depths, breathing that thick breathing air,
as so many do,
The change thence to the sight here, and to the subtle
air breathed by beings like us who walk this sphere,
The change onward from ours to that of beings who
walk other spheres.

—*Walt Whitman,*
The World below the Brine, *1860*

web navigator

critical thinking on the web

math tutor on the web

www.jbpub.com/oceanlink

PREVIEW

LIFE IN THE OCEAN is extremely diverse and unevenly distributed on the bottom and in the water column. Most of this biological richness is not even apparent, because many of the plants and animals are microscopic and can be seen only when magnified. Under a microscope, clear seawater can be swarming with life. The oceans consist of an astonishing variety of habitats populated by a bewildering assortment of organisms.

Because so many different types of vegetation and animals live in the sea, an introductory book on oceanography must rely on a general treatment of groups of organisms rather than focusing on individual species. The main purpose of this chapter is to develop a few basic ecological principles so that we can begin to understand the interconnections between environments and organisms. These notions will then be reinforced by a discussion of some adaptive strategies of both plant and animal dwellers of the sea. With this broad ecological background, we can then proceed to examine the biological productivity of selected marine environments, the central topic of the following chapter.

Ecology is a word that leaped into the limelight during the early 1970s, when many people in Europe and North America became concerned about the quality and, in the case of some individuals, the sanctity of terrestrial and marine environments. In this text the word will be used in a very specific way—to denote the study of the interrelationships between the physical and biological aspects of an environment. It is a complicated field of study, because organisms adapt to, and in turn actively change, their environment; there is a constant back-and-forth interplay between organisms and the physical and chemical setting. However, before we can deal with this level of biological complexity, we had better begin with some basic information about ocean environments. A good place to start is to consider how biologists classify marine habitats.

9-1

Ocean Habitats

An obvious and useful way to subdivide the ocean is to separate the sea bottom, the **benthic province**, from the water column, the **pelagic province**. This is a first-order breakdown of the ocean (**Figure 9–1a**) that broadly partitions marine life into water-dwelling (e.g., fish) and bottom-dwelling (e.g., clams) organisms. Each of these provinces can be further subdivided. Because water depth and illumination (the amount of light) directly and indirectly affect the distribution of organisms, biologists use them to subdivide the benthic and pelagic provinces into smaller zones.

The pelagic province (the water column) includes the **neritic zone**, the shallow water that overlies the continental shelves, and the **oceanic zone**, the deep water in the open sea beyond the shelf break. Light tends to penetrate to the sea bottom of the shallow neritic zone, so there is no need for a further subdivision there. It is an entirely different matter farther offshore in the oceanic zone because of the much greater water depths that lie beyond the edge of the continental shelf. The water column there is subdivided into five distinct horizons. The surface layer, the **epipelagic zone**, extends down from the sea surface to a depth of about 200 meters (~660 feet). It is illuminated throughout its depths, although faintly in its deeper parts. The **mesopelagic zone** underlies the epipelagic zone and extends down to about 1,000 meters (~3,300 feet). Here there is hardly a trace of sunlight. Between 1,000 meters and 2,000 meters (~6,600 feet) lies the **bathypelagic zone**, which is underlain by the **abyssalpelagic zone** down to a water depth of 6,000 meters (~19,800 feet). Water deeper than 6,000 meters, found in deep-sea trenches, is termed the **hadalpelagic zone**. The vertical extent of each of these zones is depicted in Figure 9–1a, and **Table 9–1** summarizes the relative proportion of each of theses zones based on volume of water. Note that the abyssalpelagic zone is by far the dominant environment of the ocean in terms of water volume.

Now, how can the sea floor be subdivided into regions? Well, it's possible to classify the benthic province into a fivefold division based on depth limits that roughly coincide with the zones of the pelagic province. The **sublittoral zone**, which encompasses the floor of the continental shelf, extends from the beach to the shelf break. It is flanked on the landward side by the **intertidal zone**, the part of the shoreline between high and low tides that biologists sometimes refer to as the **littoral zone**. Seaward of the shelf break beginning at a water depth of about 200 meters (~660 feet) is the **bathyal zone**, which extends to a water depth of 2,000 meters (~6,600 feet), and represents the sea bottom that underlies the mesopelagic and bathypelagic zones of the pelagic province (see Figure 9–1a). The **abyssal** and **hadal zones** represent the deepest sea bottom; they are found, respectively, beneath the abyssalpelagic and hadalpelagic zones. The relative percentages by area of each of these bottom habitats are specified in Table 9–1.

It will help to remember that these depth terms correspond grossly to physical locations. For example, the sublittoral generally is the environment of the continental shelf; the bathyal zone represents the continental slope and rise; and the abyssal depths corresponds to the average deep-ocean bottom, exclusive of the deep-sea trenches, which represent the hadal zone.

There are many other ways and schemes for subdividing the ocean. One that is commonly used by oceanographers relies solely on illumination. It provides a threefold breakdown of the water col-

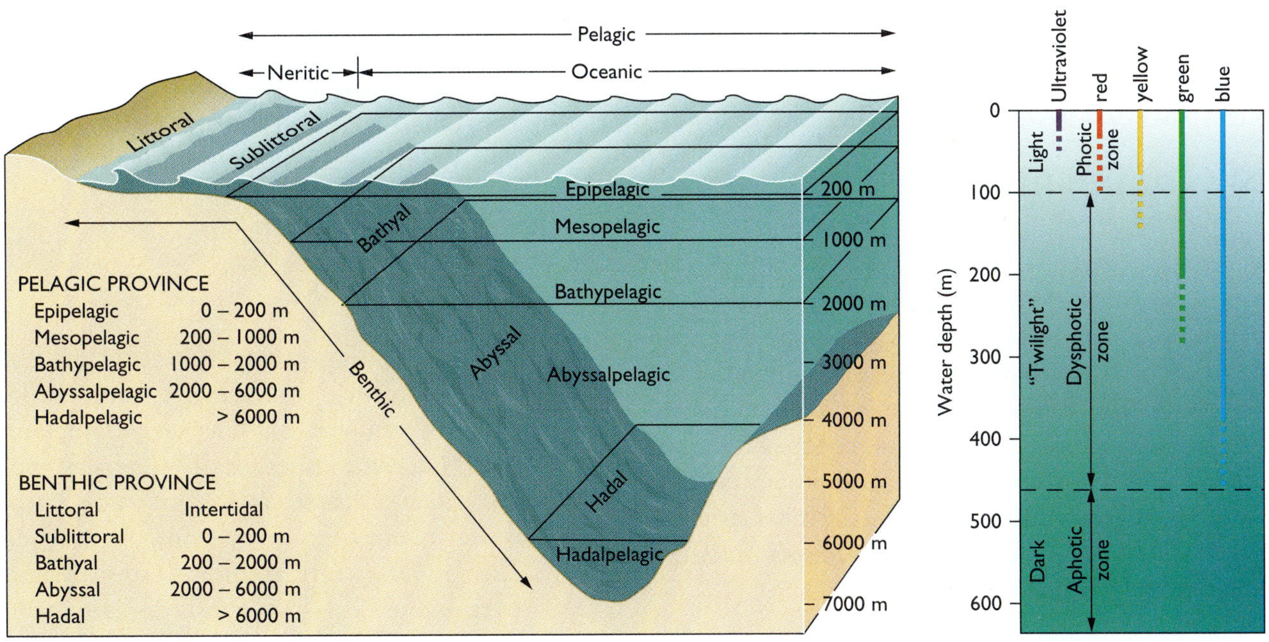

(a) BIOZONES (b) LIGHT ZONES

FIGURE **9-1**

Ocean habitats. (a) Both the water column (the pelagic province) and the sea bottom (the benthic province) are divided into discrete zones, largely as a function of water depth. (b) On the basis of light, the water column is separated into a three-part division: a well-illuminated photic zone, a "twilight" zone, and a totally dark aphotic zone. Also shown is the approximate depth of penetration for different wavelengths of sunlight. The green and blue wavelengths penetrate more deeply into the water before being scattered and absorbed than the red and yellow wavelengths do.

umn (**Figure 9–1b**). **The photic zone** is well lit, so that plant photosynthesis is possible during daylight. The photic zone extends from the water surface down to a depth that ranges between 20 and 100 meters (~65.4 and 327 feet), depending on water clarity. I use 100 meters as the depth limit to this zone because it is representative of very clear water of the oceanic province and is an easy number to remember.

At water depths of more than 100 meters (~330 feet), two additional zones are determined by illumination (see Figure 9–1b). The **dysphotic zone**, sometimes referred to as the twilight zone, has very low levels of illumination, because more than 95 percent of the sunlight has been absorbed by the water above. Photosynthesis under these dim conditions is not possible. Deeper still is the **aphotic zone**—the vast part of the ocean that is in total and perpetual darkness.

Now that we have a common vocabulary about marine environments, we need to consider the classification of organisms. Once we have done this, we can discuss ecological principles.

TABLE **9-1**

Percentages of marine habitats.

PELAGIC ENVIRONMENTS*

Zone	Depth (m)	Volume (%)
Epipelagic	0–200	3
Mesopelagic	200–1000	28
Bathypelagic	1000–2000	15
Abyssalpelagic	2000–6000	54
Hadalpelagic	>6000	<1

BENTHIC ENVIRONMENTS*

Zone	Depth (m)	Area (%)
Sublittoral	0–200	8
Bathyal	200–2000	16
Abyssal	2000–6000	75
Hadal	>6000	1

* Excludes the Arctic Ocean and includes all seas adjacent to other oceans.

9-2

Classification of Organisms

Numerous varieties of organisms have adapted well to life in the ocean. To make it easier to identify these organisms and to simplify the massive amounts of taxonomic description, biologists have arranged the biota into broad groupings that share common traits. Proceeding from the most general to the most specific, and using the human animal as an example, the taxonomic categories are:

Kingdom *(Metazoa)*
Phylum *(Chordata)*
Class *(Mammalia)*
Order *(Primate)*
Family *(Hominidae)*
Genus *(Homo)*
Species *(Homo sapiens)*

A few organisms are indexed in this manner in **Table 9–2** as well. This classification scheme is based on the binomial system developed by the Swedish naturalist Carolus Linnaeus in 1735. Organisms are identified by their species name. This consists of the genus, first letter capitalized, and the trivial name, all in lowercase letters. Both the genus and the trivial name are underlined or italicized. For example, the scientific designation of ourselves is *Homo sapiens*.

It is simply impossible to review all of the varieties of marine life in each of the taxonomic categories. **Table 9–3** lists the important marine phyla and some representative organisms belonging to the five kingdoms. What follows is a brief description of the common traits of each of the five kingdoms as they relate to the marine biota.

KINGDOM MONERA

Monera includes the bacteria and the blue-green algae, simple, microscopic, unicellular (one-celled) organisms that have no cell nucleus (**Figure 9–2**). *Bacteria* are critically important to marine biological processes in several ways. They secrete powerful enzymes that decompose dead plants and animals and convert them into inorganic nutrients. Bacteria also synthesize organic compounds that are dissolved in seawater and can then be taken up by larger organisms. *Blue-green algae* or cyanobacteria are primitive plants whose exact role is still being established, but they are significant in at least three important ways: (1) they conduct a significant amount of photosynthesis, (2) they convert ammonia and nitrogen into nitrite and nitrate, which are important plant nutrients, and (3) they dominate environments that are too stressful for most organisms. Also, blue-green algae were likely responsible for the infusion of oxygen to the primitive atmosphere early in the Earth's geologic history.

KINGDOM PROTISTA

Protista consist of single-celled organisms that possess a true nucleus and include both plants and ani-

TABLE 9-2 Classification and marine biota.

Organism	Diatom	Copepod	Haddock	Sperm Whale	Human
Kingdom	Protista	Metazoa	Metazoa	Metazoa	Metazoa
Phylum	Chrysophyta	Arthropoda	Chordata	Chordata	Chordata
Class	Bacillariophyceae	Crustacea	Pisces	Mammalia	Mammalia
Order	Centrales	Calanoida	Teleostei	Cetacea	Primate
Family	Chaetoceraceae	Amphascandria	Gadidae	Physeteridae	Hominidae
Genus	*Chaetoceros*	*Calanus*	*Melanogrammus*	*Physeter*	*Homo*
Species	*C. decipiens*	*C. finmarchicus*	*M. aeglefinus*	*P. catodon*	*H. sapiens*

TABLE 9-3

Principal marine organisms.

Kingdom	Phylum	Class	Organism
Monera	Cyanophyta		blue-green algae
	Schizophyta		bacteria
Protista	Chrysophyta		diatoms, coccolithophores
	Protozoa		foraminifera, radiolaria, flagellates
	Pyrrophyta		dinoflagellates, zooxanthellae
	Ciliophora		ciliates
Fungi	Mycophyta		fungi, lichens
Metaphytae	Rhodophyta		red algae
	Phaeophyta		brown algae
	Chlorophyta		green algae
	Tracheophyta		salt-marsh grasses, eel grasses, mangroves
Metazoa	Ctenophora		comb jellies
	Cnidaria	Hydrozoa	hydras
		Scyphozoa	jellyfishes
		Anthozoa	corals, sea anemones
	Porifera		sponges
	Bryozoa		moss animals
	Platyhelminthes		flatworms
	Chaetognatha		arrow worms
	Annelida		polychaete worms
	Brachiopoda		lamp shells
	Mollusca	Amphineura	chitons
		Gastropoda	snails, limpets
		Bivalvia	clams, oysters, mussels, scallops
		Scaphopoda	tooth shells
		Cephalopoda	octopuses, squids, nautiluses
	Arthropoda	Merostomata	horseshoe crabs
		Arachnida	marine mites
		Pycnogonida	sea spiders
		Crustacea	copepods, barnacles, krill, shrimp, crabs, lobsters, isopods, amphipods
	Echinodermata	Asteroidea	starfish
		Echinoidea	sea urchins, sand dollars
		Holothuroidea	sea cucumbers
		Ophiuroidea	brittle stars
		Crinoidea	sea lilies
			pterobranchs, acorn worms
	Protochordata	Urochordata	tunicates, salps
	Chordata	Cephalochordata	lancelets
		Pisces (superclass)	cartilaginous, bony, and jawless fishes
		Reptilia	sea turtles, sea snakes
		Aves	sea birds
		Mammalia	seals, sea otters, manatees, whales, porpoises, dolphins, walruses

mals, as well as intermediate forms that possess characteristics of both plants and animals. The more common marine members of this kingdom are the foraminifera, flagellates, dinoflagellates, coccolithophores, and diatoms (**Figure 9–3**). Despite their microscopic size, the protista collectively constitute a very large portion of the living mass of the ocean and are critical for the ecology of the ocean. They take up and release chemicals in seawater, make food and serve as food for larger organisms, and contribute to the pelagic sedimentary deposits of the deep sea (e. g., foram and diatom oozes; see Figure 4–14 in Chapter 4).

(a) BACTERIUM

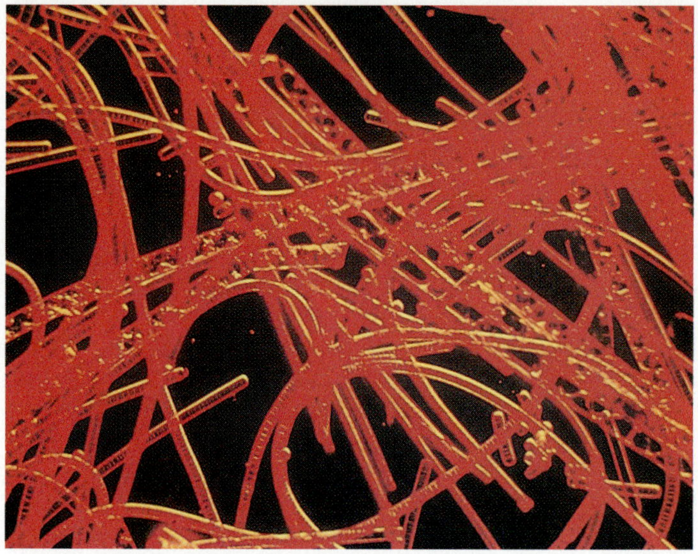

(b) BLUE-GREEN ALGA

FIGURE 9-2

Kingdom Monera. Photomicrographs show (a) a marine bacterium (~5 μm; ~0.0002 in.) and (b) a blue-green alga (50 μm; ~0.002 in.).

FIGURE 9-3

Kingdom Protista. Photomicrographs show (a) foraminifera (~1mm; ~0.04 in.); (b) dinoflagellates (100 μm; ~0.004 in.); (c) diatoms (100 μm; ~0.004 in.). Forams and diatoms are major producers of sediment (oozes) in the ocean.

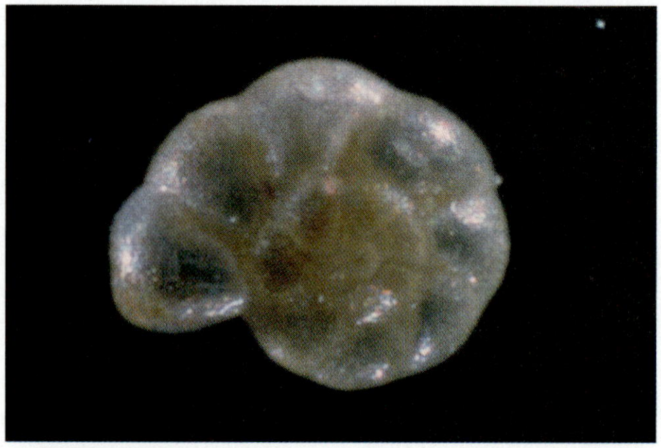

(a) FORAMINIFERA

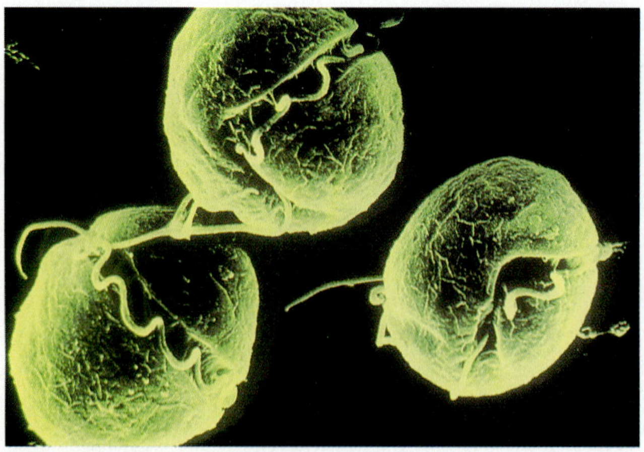

(b) DINOFLAGELLATES

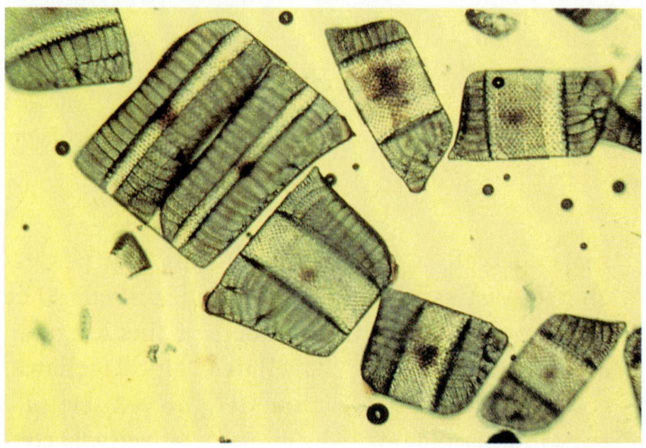

(c) DIATOMS

KINGDOM FUNGI

Although **fungi** are widely dispersed in the ocean, they are not as diverse there as they are on land. They are very common in the intertidal zone, where they help to keep algae from drying up and dying during low tide. As with bacteria, the primary role of fungi is to decompose organic matter. Remarkably few studies have focused on the ecology and general life habits of marine fungi (**Figure 9–4**).

KINGDOM METAPHYTAE

Metaphytae are plants attached to the sea floor and confined to the shallow water of the shoreline and the inner continental shelf. Deeper water simply does not provide sufficient light for these plants to survive. The principal attached plants in the ocean are the primitive red, brown, and green algae, which are immersed in seawater, and the flowering, seed-bearing plants of salt marshes (**Figure 9–5**) and mangrove swamps, which are exposed to the air except at high tide.

KINGDOM METAZOA

Metazoa are marine animals, the most familiar sea-dwelling organisms, because they are easily seen with the naked eye. The numerous and diverse organisms in this taxonomic group include sponges, mollusks (such as clams, oysters, snails, and squids), arthropods (such as barnacles, crabs, and shrimp), and chordates (such as fishes, sea turtles, seals, and whales). Less familiar marine animals are annelids (polychaete worms), ctenophores (comb jellies), brachiopods (lamp shells), and echinoderms (such as sea cucumbers, starfish, sea lilies, sea urchins), among others. Table 9–3 and **Figure 9–6** give information on some of these animals.

FIGURE 9-4
Kingdom Fungi. Photomicrograph shows a marine fungus.

9-3 Classification by Lifestyle

A more informal way to categorize marine biota is according to lifestyle. Thus, **plankton** consist of organisms that drift or swim weakly and are hence powerless to counteract currents. Although they drift slowly with the ocean currents, a surprising number of plankton do engage in vertical migrations, ascending and descending through the water column as they are transported by currents. Plankton are divided into **phytoplankton** (plants) and **zooplankton** (animals). A few of the techniques used to sample the plankton, the majority of which are microscopic in size, are described in the boxed feature, "Sampling the Biota."

The **nekton**, which include the active swimmers such as fishes, squids, reptiles, birds, and mammals (e. g., seals, manatees), contrast markedly with the tiny, passively drifting plankton. Typically, the members of the nekton are large animals that possess specialized muscles for locomotion (movement). Although some migrate long distances, most nekton have a restricted range that is controlled largely by water temperature and salinity.

The **benthos**, organisms that are attached to or move on or beneath the sea bottom, are the third informal group of the marine biota. Like all plants, benthic flora—attached algae (seaweeds), seed-bearing plants, and bottom-dwelling diatoms—require illumination and are therefore restricted to the shallow sea bottom of the photic zone. In

(a) RED ALGA

(b) BROWN ALGA

(c) GREEN ALGA

(d) EELGRASS

(e) SALT-MARSH GRASS

FIGURE 9-5

Kingdom Metaphytae. Photographs show (a) a red alga; (b) a brown alga; (c) a green alga; (d) eelgrass; (e) salt-marsh grasses. The grasses in photographs d and e are not seaweed but true flowering, seed-bearing plants with a root system.

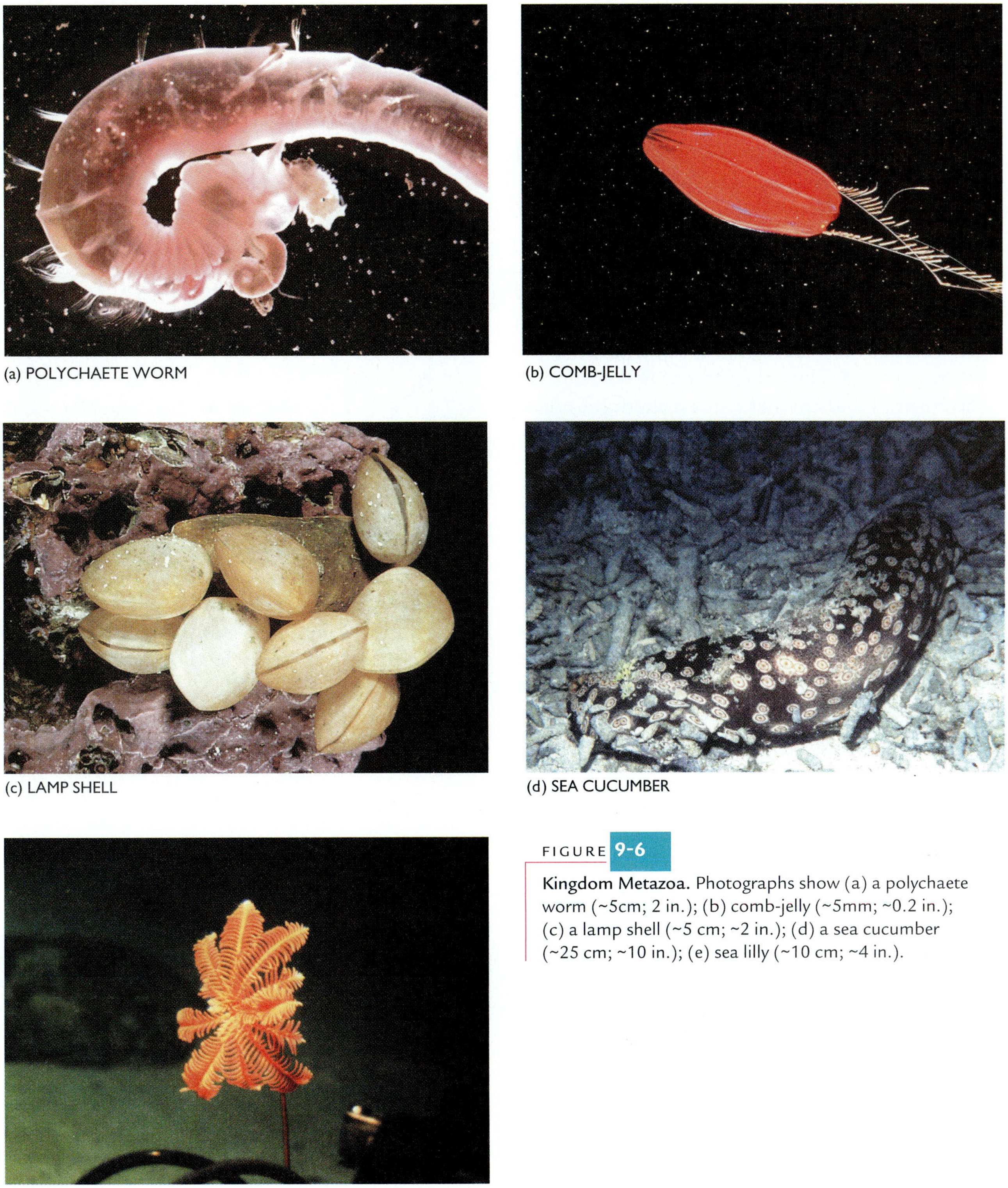

(a) POLYCHAETE WORM

(b) COMB-JELLY

(c) LAMP SHELL

(d) SEA CUCUMBER

(e) SEA LILLY

FIGURE 9-6

Kingdom Metazoa. Photographs show (a) a polychaete worm (~5cm; 2 in.); (b) comb-jelly (~5mm; ~0.2 in.); (c) a lamp shell (~5 cm; ~2 in.); (d) a sea cucumber (~25 cm; ~10 in.); (e) sea lilly (~10 cm; ~4 in.).

contrast, benthic bacteria and animals (most of which are invertebrates) live at all depths on the ocean floor, including the deep-sea trenches of the hadal zone. The benthos are divided further (**Figure 9–7**) into the **epifauna** and **epiflora** (animals and plants, respectively, that live *on* the sea bottom), and the **infauna** (animals that live *in* the sea bottom).

This threefold division of marine life is useful and works quite well generally. However, its applications to some organisms can be troublesome at times. For example, some flatfishes, such as halibut and flounder, lie quietly on the ocean floor a great deal of the time (**Figure 9–8**) but swim away when disturbed. Thus, they engage in the habits of the benthos and the nekton. Furthermore, their lifestyle changes with age. As juveniles (a young, immature growth stage), flatfishes have a symmetrical body shape with an eye on each side of the head. At this young stage of their life, they clearly are nekton, actively swimming in search of food. After maturing into adults, however, they undergo a rather remarkable metamorphosis. They assume a flattened body shape, and differential growth of the skull causes one eye to migrate around the body until it is adjacent to the others. The result is a flattened fish with both eyes located on one side of the body, an ideal form for lying passively on the sea bottom. In fact, the changing of lifestyle with age is quite common among marine organisms. Many benthic invertebrates (barnacles, oysters, and polychaete worms, to name a few)—perhaps as many as 70 percent of the species—produce planktonic larvae (early immature growth forms) for dispersal by currents (**Figure 9–9**).

9-4

Basic Ecology

Now that we have a vocabulary to describe the marine biota and their environments, we can proceed to the study of ecology. The physical and chemical factors that characterize ocean environments are many. Examples include temperature, salinity, water pressure, nutrients, dissolved gases, currents, light, suspended sediment, substrate (bottom surface), river inflow, tides, waves, and many more. It would be quite tedious and somewhat boring merely to list these factors and explain how each affects organisms. A better way is to select a few—temperature, salinity, and water pressure—and discuss in some detail how

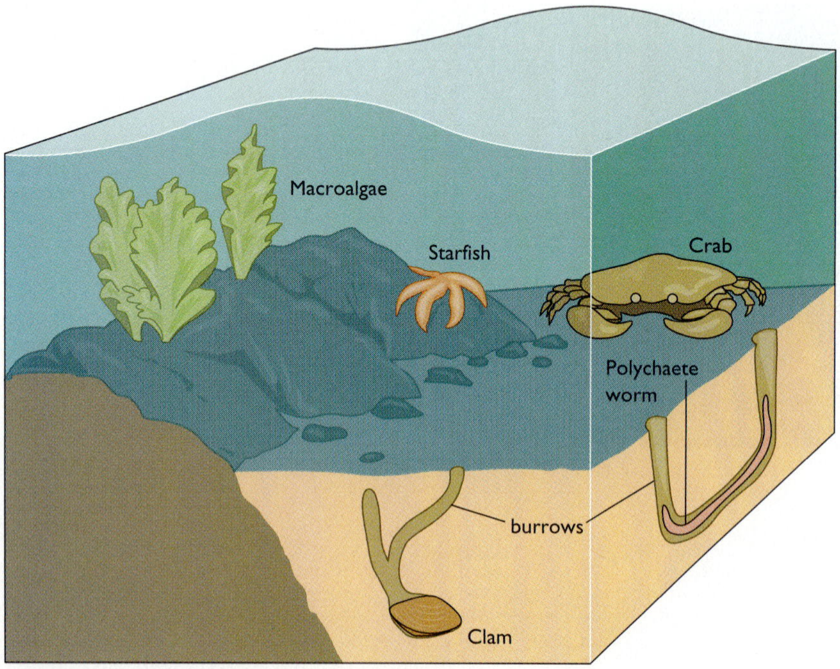

FIGURE 9-7

A benthic community. Common bottom-dwelling (benthic) organisms include the epiflora and epifauna, both of which live on the sea bottom, and the infauna, which live within the sediment or rock substrate.

FIGURE 9-8

Bottom fish. This flounder has almost completed its metamorphosis and is lying on the sea bottom.

FIGURE 9-9

Plankton. (a) Planktonic larval forms of a variety of benthic organisms are magnified here 35 to 55 times. (b) Planktonic anchovy eggs collected during a plankton tow (~1mm; ~0.04 in.).

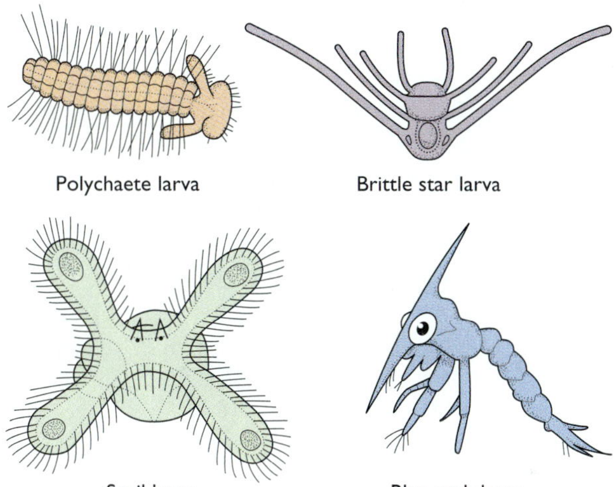

(a) LARVAE

(b) EGGS OF ANCHOVY

they regulate the distribution and behavior of a variety of organisms. Then, later in the chapter, we can examine strategies that members of various **ecosystems** use to adapt to the physical and chemical aspects of environments and to one another as well.

The word *ecosystem* refers to the totality of an environment, which includes all of its living (the biota) and nonliving (physical and chemical) parts. Examples of ecosystems are salt marshes, estuaries, a kelp (seaweed) forest growing on the floor of the inner continental shelf (see the boxed feature, "Ecology of the Giant Kelp Community"), coral reefs, epipelagic or abyssalpelagic waters (if you don't recall these terms, refer to Figure 9–1), an abyssal sea bottom, and many, many more. Ecologists study ecosystems in an effort to uncover the numerous interrelationships and interconnections that bind all their parts into a whole. It is a complicated but, as you shall see, fascinating field of study. Let's begin our examination of marine ecology by considering some of the effects that temperature, salinity, and water pressure have on organisms.

TEMPERATURE

Many biologists consider the temperature of water to be the most important single regulator of the distribution and activity of organisms in the ocean. This belief is supported by the distribution of many marine species, which closely mirrors the shape of

Sampling the Biota

■ PLANKTON

Filtering nets are used to sample plankton at sea. These **plankton nets**, which come in a variety of sizes and designs, are towed behind ships (**Figure B9–1**). The size of the mesh openings in the net determines which phytoplankton will be sampled, because organisms smaller than the mesh size will flow through the net and not be collected. Counting the plant cells collected and monitoring the volume of water that has passed through the net enable biologists to calculate the plankton concentrations for the area being investigated. Another sampling technique draws water from a certain depth with a hose and passes it through a filter system consisting of a series of nets with different mesh sizes. In this way, many different sizes of phytoplankton can be separated, identified, and counted.

■ NEKTON

Nektonic organisms are mobile and therefore more difficult to capture than floating plankton. Typically, a large net of coarse mesh size is towed behind the ship to catch these large swimmers. For example, **trawling gear** (**Figure B9–2a**), consists of a net with attached "otter boards"—large "doors" that flare out when towed and keep the mouth of the trawl net open. The **purse seine** (**Figure B9–2b**) is a large net that is set out around a school of fish and drawn in at the bottom by the "purse string," effectively preventing fish from escaping by diving beneath the net.

■ BENTHOS

Benthic organisms are sampled by two principal methods, using the

FIGURE **B9-1**

Sampling the plankton. This Clark-Bumpus plankton net has a closing device for sampling discrete water depths and a flowmeter for determining the volume of water passing through the net. This information is then used to determine the concentration of plankton in water sampled by the net.

isotherms. (Isotherms, as you recall from Chapter 5, are contour lines of equal temperature.) For example, the amphipod species *Parathemisto gaudichaudi* in the Atlantic Ocean is rarely found in water much warmer than 10°C. When it is found in warm water, it is living "sluggishly" and cannot reproduce successfully. (An *amphipod* is a small crustacean a few centimeters—that is, an inch or so—in length, with a flat body from side to side.)

Temperature exerts a strong control on the rates of chemical reactions. This in turn affects the chemical synthesis of enzymes in the cells of organisms and thus the growth rates, reproductive success, and general activity of organisms. This is particularly true for the majority of marine organisms, because their cell or body temperature is the same as the temperature of the water that surrounds them. Therefore, their activity varies greatly with water

anchor dredge or the **grab sampler,** both illustrated in Chapter 4 in the box feature "Probing the Sea Floor" (Figure B4-1a-c). The more primitive method employs the **anchor dredge** , which is dropped over the side of the ship and dragged along the ocean floor. The dredge collects indiscriminately, scraping up benthic organisms, rock fragments, and sediment, and tossing all this material into a collection bag typically composed of metal to prevent its tearing. The **grab sampler** is used to collect benthos living on or in soft substrates such as mud. The better grab samplers are spring-loaded, allowing them to bite deeply and cleanly into the bottom and to close securely before being lifted off the ocean floor. Unlike the anchor dredge, which mixes the sample haphazardly, grab samplers yield a discrete bite of the sea bottom, enabling biologists to determine accurately the species composition and concentration of the benthos.

FIGURE **B9-2**

Fishing gear. (a) The trawling gear employs a pair of "otter boards," large "doors" that move apart when the assembly is towed, keeping the mouth of the net open. (b) The purse seine can surround a school of fish. Drawing in the "purse string" then prevents the fish from diving beneath the net assembly and escaping.

(a) TRAWLING GEAR

(b) PURSE SEINE (CLOSED)

temperature, as demonstrated by the rates of cirri beats for different barnacle species (**Figure 9–10**). (*Cirri* are small filaments used to "catch" particles in the water.) Note that their activity changes much more rapidly for temperature changes at the warmer than at the colder side of the plot. Generally, rates of biological activity of marine organisms tend to double for every 10 degree rise in water temperature. Consequently, the biota of polar seas grow much more slowly, reproduce less frequently, and live longer than do organisms of the tropical oceans. However, cold water has more nutrients (nitrate, phosphate) than warm water has; consequently, there are more marine plants growing in polar than in equatorial waters.

Tolerance to changes in water temperature varies enormously among marine species. In most cases, eggs, larvae, and juveniles are considerably more

FIGURE 9-10

Activity of barnacles. Many organisms are not able to regulate their body temperatures, which therefore vary with the temperature of the surrounding water. Consequently, physiological activity, such as the rate of cirri beats, also varies strongly with water temperature. For both barnacle species, the rate of cirri beats decreases from a maximum value at about 20°C for *Balanus balanoides* and at about 33°C for *B. perforatus.* [Adapted from R. V. Tait and R. S. DeSanto, *Elements of Marine Ecology* (New York: Springer-Verlag, 1972).]

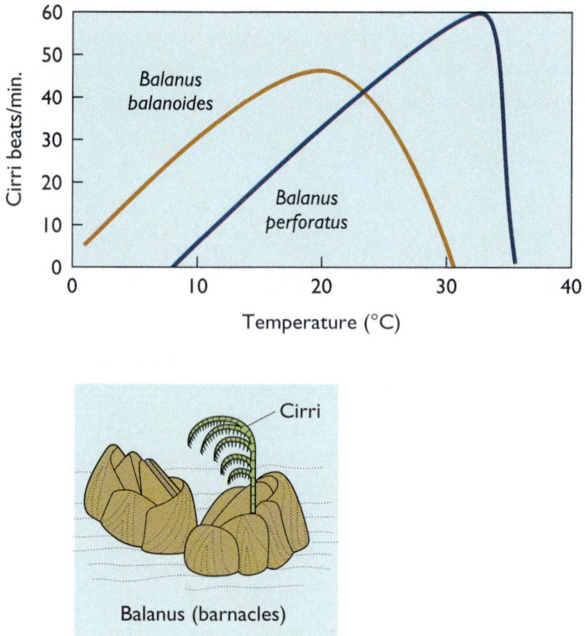

sensitive to temperature fluctuations than are mature adults of the same species. The eggs of the vast majority of invertebrates and fishes hatch in water rather than in the body. Therefore, the temperature of the water that surrounds the eggs strongly influences their embryonic development and survival rate. Even after hatching, larvae and juveniles typically have little tolerance for temperature changes. For example, the eggs and larvae of the sardine species *Sardinops sagax* found offshore California cannot survive in water colder than 13°C. When the temperature drops below this value, as it often does in that region, mass mortality of eggs and larvae drastically reduces the future abundance of the adult sardine in the area. Studies of the Pacific cod, *Gadus macrocephalus,* have pinpointed the narrow temperature range of 3°C to 5°C as best for embryo development. Within this temperature range, salinity and the amount of dissolved oxygen can vary significantly with little effect on the embryo, a fact that underscores the importance of temperature in their survival.

Although ocean temperatures range between 2°C and 40°C, more than 90 percent of the ocean's water is colder than 5°C. Nevertheless, both plants and animals thrive in this water provided they receive adequate nutrients, or food. In fact, there is generally a much greater quantity of living matter in the polar than in the tropical seas because the supply of nutrients in the warmer waters of the ocean is limited, a phenomenon that is the main topic of the next chapter.

Water temperature may also have an indirect effect on organisms, although this is much harder to study than the direct connections. A case in point is the annual variation of shellfish yields in Casco Bay, Maine. Long-term records of fishing indicate that there is a strong inverse relationship between the average water temperature for the year and the annual catch of clams. In other words, when the annual average temperature of water in the bay is low, clam harvests for that year are high, and vice versa (**Figure 9–11**). At first glance, the inverse relationship suggests a direct cause and effect—high temperatures reduce the survival rate of clams in such a way that there are fewer clams to catch during warm than cold years. Research, however, has revealed a different story. High water temperatures are associated with the appearance in Casco Bay of above-normal numbers of green crabs, which prey heavily on the soft-shelled clams, leaving fewer shellfish for humans to catch. Apparently, then, the green crabs are directly controlled by water temperature, whereas the soft-shelled clams are indirectly affected by water temperature through a relationship between prey (clam) and predator (crab). This illustrates nicely the strong ecological interconnections between the environment (water temperature) and various members of a benthic community (crabs, clams, and humans).

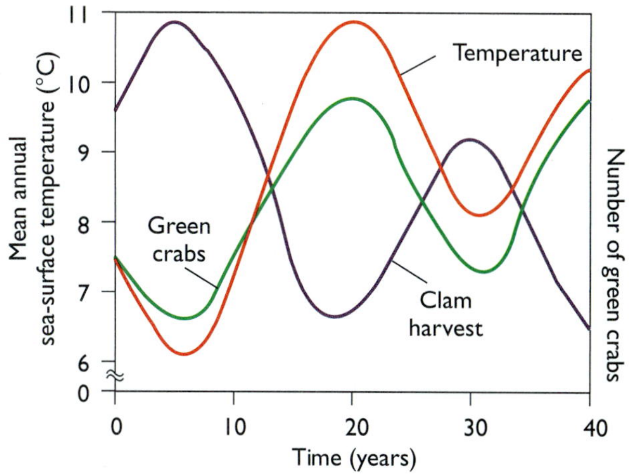

FIGURE **9-11**

Shellfish harvest in Maine. The clam yield in Casco Bay, Maine, is inversely related to the mean annual sea-surface temperature (SST), suggesting that warm seawater adversely affects the soft-shell clam population. Research has revealed that the correlation between clam yield and SST is an indirect one, because a higher SST results in the proliferation of the green crabs in the area. The green crabs are voracious predators that compete with humans for the soft-shell clams.

SALINITY

The salinity of the ocean is one of its most widely recognized properties. Like temperature, salinity affects marine organisms. For example, many species of plankton extract dissolved chemicals in seawater to construct hard, protective parts, such as shells made from silica (SiO_2) or calcium carbonate ($CaCO_3$). Also, marine species show a wide range of abilities to withstand salinity changes, so that their distribution is determined in part by salinity tolerances.

Because of shifts in the volume of river inflow, nearshore animals that are intolerant of salinity changes must migrate in and out of the region as salinity varies during the course of the year. In contrast, many benthic animals are firmly attached to the bottom or have limited means of locomotion, and so they cannot move easily to more favorable water. These organisms, such as barnacles, clams, oysters, have high tolerances for salinity changes. They can remain closed, locked securely in their shells, and refrain from feeding when salinity conditions are unfavorable. A prolonged period of adverse conditions will, however, cause a massive kill of shellfish, highlighting the ecological significance of salinity even for these durable animals.

In many ocean regions, a distinct, permanent halocline stratifies the water column. The surface layer is usually well mixed by waves and currents, producing water with a uniform salinity that changes regularly with the passage of the seasons. The layer beneath the halocline is less well mixed and it is unaffected by seasonal variations of weather. Hence, the surface-water layer, the epipelagic zone, has a higher proportion of organisms that can tolerate salinity changes than does the layer beneath the halocline, the mesopelagic zone, where salinity is stable over time.

The chemistry of the body fluids of marine invertebrates and vertebrates is very similar in relative proportion to the salt composition of seawater. This does not mean that the salinity of the two is the same. The relative amounts of the various salts are alike, but not their absolute amounts. The absolute salinity of the body fluids of animals may be identical to, lower than, or higher than the salinity of ocean water, even though the relative amounts of salts are similar in all cases. In other words, the percentage of sodium in very salty water is identical to the percentage of sodium in less salty water or less salty body fluid. If there is a party of

100 or 10 people, both can have the same percentage of females, even though there are more females in attendance at the larger party of 100 people than at the smaller gathering of 10.

Differences between the concentration of salts in the fluid of a cell and the surrounding seawater create chemical gradients, with the salt content of the cellular fluids higher or lower than the salt content of the seawater outside the cell. This setup results in **diffusion**, a physical process whereby molecules, such as salts, move from areas of high concentration to areas of low concentration (**Figure 9–12a**). This is exactly what happens when you're standing in the far side of a room and someone with a lot of perfume or aftershave lotion enters. Even if that person remains standing at the door, in a short time you smell perfume or lotion molecules that have diffused through the air in the room, moving down the chemical gradient from the high concentrations of perfume molecules on the person to the low levels in the far side of the room. The process of diffusion is at least partly responsible for the transfer of dissolved nutrients in seawater (where they are relatively abundant) into a plant cell (where they are less concentrated, but needed for photosynthesis). The nutrient molecules diffuse from seawater and pass through a semipermeable membrane (a membrane that allows only certain molecules to pass through it) into the cell. Diffusion also is the means by which organisms dispose of unwanted waste. When the level of waste products due to metabolism is greater in the cell than outside the cell, chemical wastes simply diffuse through the cell membrane and out into the surrounding fluid, which in effect is a natural "plumbing" system for the transfer of chemicals.

Diffusion is responsible for the passage of water molecules through cell membranes whenever there is an unequal concentration of water molecules inside and outside the cell. The diffusion of water molecules through a semipermeable membrane is called **osmosis**. At first, it may seem impos-

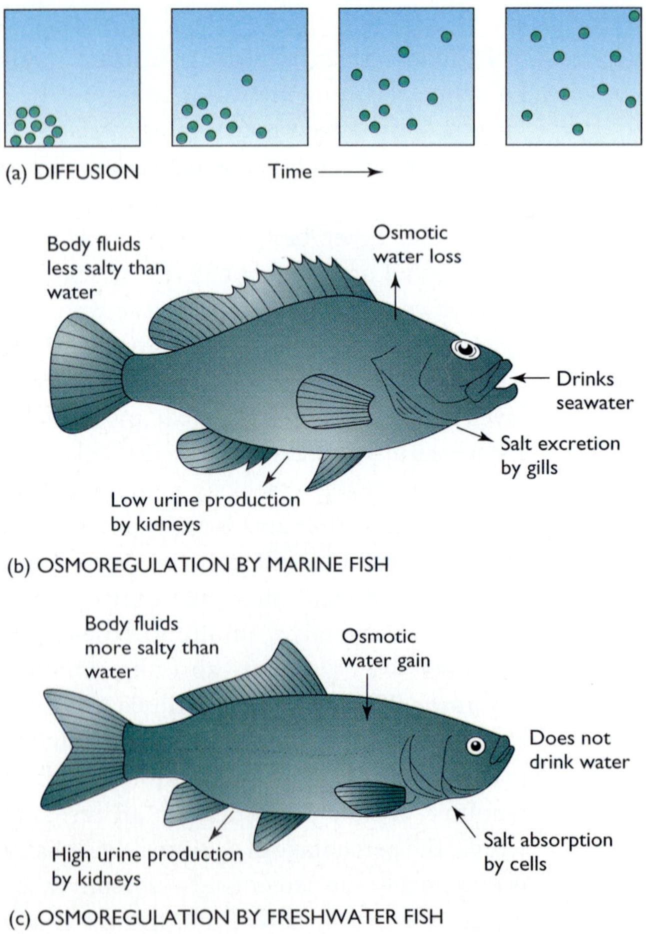

FIGURE 9-12

Osmoregulation. (a) Random molecular motion results in the movement of substances from points of high concentration to points of low concentration, a chemical process called diffusion. (b) Marine fishes counteract the osmotic diffusion of water out of their cells by drinking seawater, urinating infrequently, and excreting salt ions through chloride cells in the gills. (c) Freshwater fishes osmoregulate by not drinking water, urinating frequently, and absorbing salt ions through chloride cells in the gills.

sible to have different concentrations of water across a membrane. But think of it this way: if the body fluid of a fish is less salty than the surrounding seawater, this means that a volume of body fluid must have more molecules of water than an equal volume of seawater, because the latter has more salt molecules in that volume than does the cell fluid. In other words, there is less "room" in that volume of seawater for water molecules. Osmosis will cause a net transfer of water molecules from the animal's cells, where there is a high water concentration, into the saltwater, where there is a lower level of water. In effect, this will quickly lead to dehydration as the fish loses a lot of water to the ocean. The reverse flow, that is, the net transfer of water from the ocean into an animal's cells by osmosis, results from a chemical gradient that is reversed, meaning that the animal's body fluids have a higher salinity than the surrounding water. This will lead to swelling of the cells as they continue to be filled with water by osmosis.

Osmosis across cell membranes can have dire consequences for animals and plants unless they can counter the effect by a process termed **osmoregulation** (meaning, literally, regulation of the process of osmosis). Most marine fish do have cellular fluids with salinities that are lower than the salinity of seawater. If they couldn't counteract (osmoregulate) the continued osmotic water loss to the external environment, they would become dehydrated and perish in short order. They do this by drinking large quantities of seawater, urinating infrequently, and excreting excess salt ions through specialized chloride cells located in the gills (**Figure 9–12b**). In contrast, freshwater fishes have bodily fluid compositions that are saltier than the surrounding water. This condition results in the osmotic transfer of water from the outside of the cell into its interior, causing swelling and the eventual bursting of the cell walls and death of the animal. Freshwater fishes osmoregulate by not drinking water (they get plenty of water by osmosis through the cell walls), expelling large quantities of dilute urine, and absorbing salts through the gills (**Figure 9–12c**).

HYDROSTATIC PRESSURE

We are hardly ever aware of the pressure exerted by the enormous height of the column of air above us, because we are adapted to living in this gaseous environment. In water, however, it is an entirely different matter, as anyone who dives down just a few meters (a few yards) below the sea surface will learn. The pressure exerted by the water on the eardrums, even at these shallow depths, is enormous and can cause a great deal of pain.

The pressure created by the height of a stationary column of water is called **hydrostatic pressure**. It is a function of the density of seawater (which, as you know, depends directly on temperature, salinity, and turbidity) and the total height of the water column. A simple calculation shows that the pressure associated with a 10-meter-high (~33 feet) column of water equals the pressure exerted by the full height of the atmosphere above the Earth's surface—a pressure value that is referred to as one atmosphere (1 atm). If, then, 10 meters of water is equivalent to 1 atm, 1 meter (~3.3 feet) of water produces 0.1 atm of pressure, 100 meters (~330 feet) 10 atm of pressure, and 10 kilometers (~33,000 feet) 1,000 atm of pressure (**Figure 9–13a**). An aircraft with a cabin pressure of 1 atm must be structurally sound at an altitude of 10 kilometers (~30,000 feet), so that it will not explode (burst outward). Conversely, a submarine with a cabin pressure of 1 atm must be structurally sound so that it will not implode (burst inward).

The huge hydrostatic pressure combined with the absence of any light and the presence of very cold water at the bottom of the deep sea would appear to make life an impossibility there. Most biologists of the nineteenth century believed just that—the deep ocean must be lifeless because of the "terrible" living conditions that exist there. We now know that the great depths of the ocean, even the deep-sea trenches, are populated by a richly diversified community of animals. So in the context of this discussion, the question now becomes, how do animals at abyssal depths cope with such high hydrostatic pressures?

The explanation lies in the fact that gases, which are highly compressible, are absent in the bodies of most of the inhabitants of the deep sea. They are composed mainly of water, a fluid that resists compression under pressure. If you could fill a submarine with seawater instead of with air, its superstructure would not have to be strong because the pressure inside and outside the craft would be identical. Hence, bottom-dwelling organisms (**Figure**

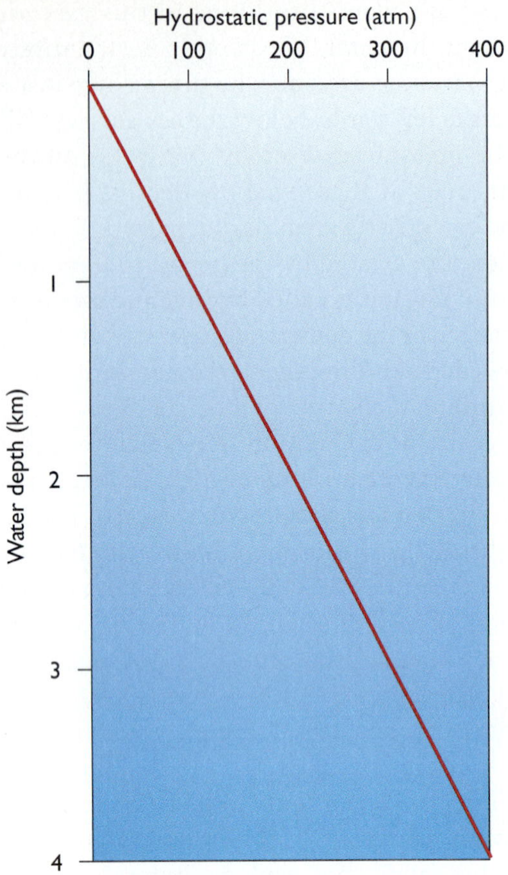

(a) WATER PRESSURE

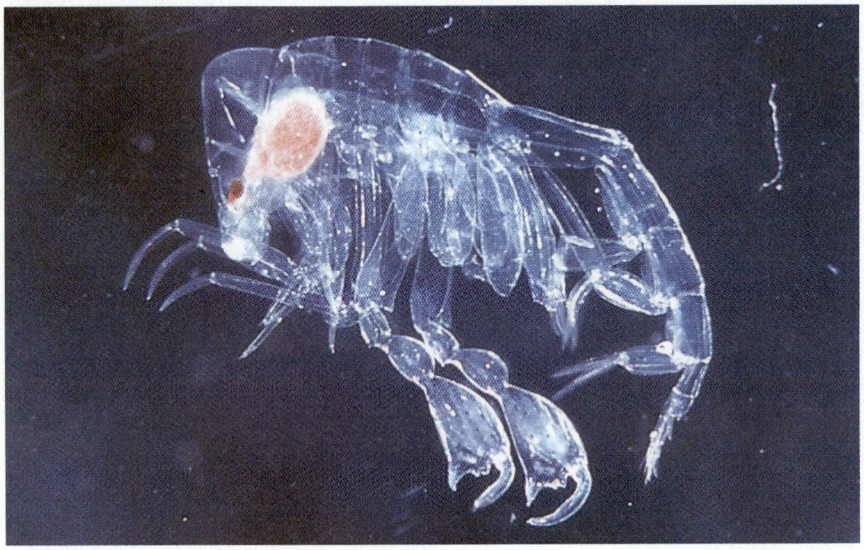

(b) AMPHIPOD

FIGURE 9-13

Hydrostatic pressure. (a) On the average, hydrostatic pressure increases with water depth at a rate of about 1 atm per 10 meters of the water column. The specific hydrostatic pressure at any depth varies slightly from site to site due to differences of water density, which depends on water temperature and salinity. (b) This photograph shows an abyssal amphipod (~0.5 cm; ~0.02 in.). Such organisms are unaffected by the high hydrostatic pressures found at great depths, because their body fluids and cavities contain no compressible gases.

9–13b) are not influenced by or aware of the excessive hydrostatic pressure that exists at the bottom of the sea. In fact, many of these animals can be sampled from the deep ocean and transferred to saltwater aquaria that have the appropriate temperature and salinity levels, and survive the experience of now living under what, for them, is extremely low hydrostatic pressure. This is not the case for many mesopelagic fishes—that is, those in the region between 200 and 1,000 meters (~660 to 3,300 feet) below the sea surface—which have gas-filled swim bladders to help them remain neutrally buoyant. When these fishes are caught and brought rapidly to the sea surface, the consequent drop of hydrostatic pressure may cause, among other physiological stresses, expansion of the gas bladder, which will force internal organs out of their mouth and cause death.

9-5

Selected Adaptive Strategies

In the previous section, we studied some physical and chemical aspects of ocean environments and their general effects on organisms. With this background, we are now prepared to study more sophisticated interactions between the living and nonliving parts of ecosystems. This can best be done by considering some adaptive strategies of selected organisms that are members of the plankton (free-floaters), the nekton (active swimmers), and the benthos (bottom dwellers). Read this section with an ecological mindset, that is, by keeping in mind how organisms are affected by and in turn affect the environment.

More than 90 percent of the ocean's plant species are **algae**, anatomically simple plants that do not have true roots, stems, leaves, flowers, or seeds. The vast majority of these are *phytoplankton* (microscopic, unicellular plants that drift with the currents). This free-floating, virtually invisible plant life of the oceans contrasts markedly with the large, rooted vegetation of the land. Why is this so? What characteristics of the marine environment produced a community of marine plants dominated by drifting, minute algae rather than large, rooted plants?

Let's begin to answer this question by considering the general life requirements of plants. The most vital activity of all plants is **photosynthesis**, the process whereby food is synthesized chemically from inorganic matter. In order to photosynthesize, plants require sunlight as an energy source, and matter—nutrients, water, and carbon dioxide. In the ocean, plants are bathed in water that contains bountiful quantities of dissolved carbon dioxide. Thus, the availability of sunshine and nutrients, not water and carbon dioxide, limit the growth of marine plants. On land, in contrast, it is the availability of water and nutrients that determine the lushness of the vegetation. This is the reason, of course, why people fertilize and water their lawns.

The striking contrast between the size of terrestrial and marine plants is a consequence of evolutionary adaptation to two very different kinds of fluid—air and water. Sunlight penetrates to the very bottom of the atmosphere, warming the Earth's surface. This factor, combined with air's low density, has resulted in the evolution of plants that are attached to the land surface and raise their leaves, which contain **chlorophyll** (the green coloring agent in plants that is essential for photosynthesis) skyward by means of rigid, woody stalks. The classic example is a tree. Terrestrial plants obtain water and essential nutrients, which are scarce in air, by roots imbedded in the soil (**Figure 9–14**).

But in the ocean there are significant differences. For example, sunlight is readily absorbed by water, so light decays rapidly with water depth. In

FIGURE 9-14

Terrestrial and marine plants. Plants on land are relatively large in size and transfer nutrients and water from the soil to their photosynthesizing leaves via an efficient vascular system. By contrast, marine plants are typically microscopic, unicellular organisms that absorb essential nutrients dissolved in seawater through their cell walls.

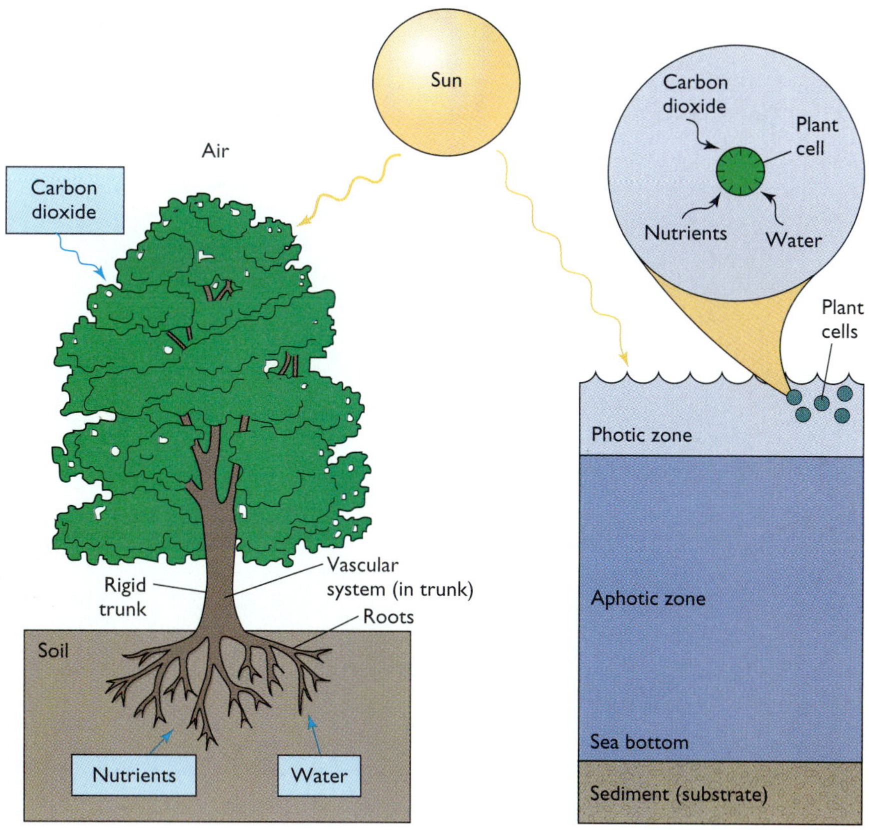

the clearest water in bright sunshine, light levels sufficient for conducting photosynthesis may extend downward to a depth of about 100 meters (~330 feet) (see Figure 9–1b). Plants, therefore, if they are to exist in the ocean where the bottom is deeper than 100 meters, must be able to remain suspended in the photic zone, the well-lit upper reaches of the water column. Fortunately, nutrients are dissolved in water, so specialized roots to obtain nourishment and water are not needed in the ocean as they are on land. Marine plants merely absorb water and nutrients through their cell walls (see Figure 9–14). Even multicelled algae do this. What appear to be roots on seaweed are **holdfasts**, rootlike masses used only to anchor the plant to the sea bottom.

Marine plants, in order to photosynthesize, must remain afloat near the water surface where sunlight is available. The problem is that plant cells, being slightly more dense than seawater, sink rather than float. One way that a plant can reduce its rate of settling is to maximize its surface area, so that friction retards sinking. As an object grows in size, its surface area expands as the square, its volume as the cube. This means that a bigger object has less surface area *relative to its volume* than does a smaller object of the same shape. Restating this idea, a smaller object will have a greater surface area and more frictional drag *per unit volume* than will a larger object (**Figure 9–15**). Consequently, plants of microscopic proportions are well suited to a life suspended in seawater for two reasons: (1) slow settling rates because of their small mass, and (2) high frictional drag, because their surface area is immense *relative* to their volume. Other means that marine plants use to enhance flotation include pores and spines on shells to increase surface area and, hence, drag, and the formation of chains of attached plant cells, a colonial habit of some species whereby individual cells are linked to one another. These factors, in addition to the turbulent motion of moving water, keep plankton suspended in the water of the photic zone—the area that receives sunlight—where they can photosynthesize. Phytoplankton in nutrient-poor regions are necessarily small to survive. Being small, they have a large surface area relative to their volume, enabling them to absorb sufficient nutrient even when it occurs in low concentrations.

Relative size	(small circle)	(medium circle)	(large circle)
Diameter (unit^1)	1	2	4
Area (unit^2)	$1^2 = 1$	$2^2 = 4$	$4^2 = 16$
Volume (unit^3)	$1^3 = 1$	$2^3 = 8$	$4^3 = 64$
$\frac{\text{Area}}{\text{Volume}}$	$\frac{1}{1} = 1$	$\frac{4}{8} = \frac{1}{2}$	$\frac{16}{64} = \frac{1}{4}$
Conclusions	1. Most area per unit volume 2. Most drag per unit volume 3. Slowest rate of settling	← ← →	Least area per unit volume Least drag per unit volume Fastest rate of settling

FIGURE **9-15**

Area-to-volume ratios. Marine plants must remain suspended in the photic zone in order to photosynthesize. Because settling rates depend directly on body size and frictional drag, minute phytoplankton are more buoyant than large plants are. A small cell settles slowly and has a large surface area per unit volume which maximizes friction with the water.

Because of evolutionary pressure over geologic time, the sea has become necessarily populated by phytoplankton. Seaweeds that are attached to the sea bottom and thereby restricted to the nearshore and the shallow water of the continental shelves are dwarfed in numbers by the abundant, but virtually invisible phytoplankton. In fact, most of you have never seen one of the more common plant members of the ocean—the unicellular diatoms!

What exactly are diatoms, and how do they live in the ocean? The single cell of each diatom is encased in a rigid shell, or **frustule** (**Figure 9–16**) composed of silica (SiO_2). These rigid frustules resemble "pillboxes," consisting of a "lid," the **epitheca**, that covers a "box," the **hypotheca** (see **Figure 9–16a**). The frustule is perforated with numerous tiny holes through which the interior cell can exchange nutrients and gases and discharge waste products into the water. Diatoms reproduce largely by cell division, whereby the epitheca and hypotheca separate, so that each contains half of the soft, organic cell contents. A new shell that fits into the original shell half is secreted by both daughter cells. The result is that one daughter (half of the parent cell) remains the same size as the parent cell, whereas the second is smaller than the parent (**Figure 9–17**). This reduction in size of many of the daughter cells continues through the succeeding generations. Hence, the average cell size of a growing population of diatoms decreases over time. At some critical minimum size for an individual plant cell (about one-third the size of the original parent), cell division ceases and body size is restored by the formation of a naked cell that reproduces sexually and then encapsulates itself with a new frustule (see Figure 9–17). Under optimum conditions with adequate light and nutrients, diatoms divide once every 12 to 24 hours, so that their numbers swell in short order, producing a dense **plankton bloom** (**Figure 9–18**). Diatoms flourish in cold, nutrient-rich water, particularly in the polar and subpolar latitudes and the inshore

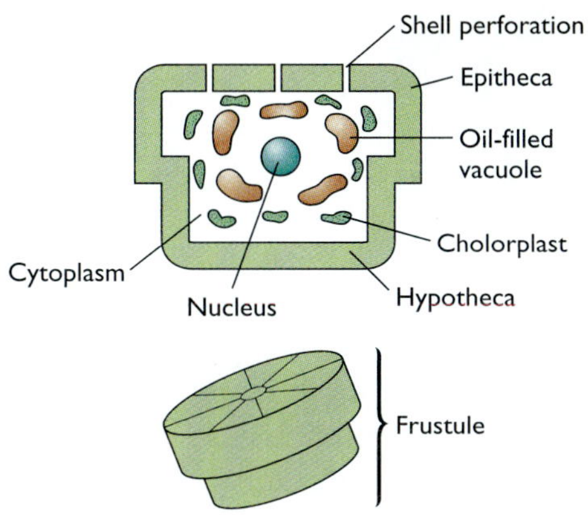

(a) DIATOM FRUSTULE

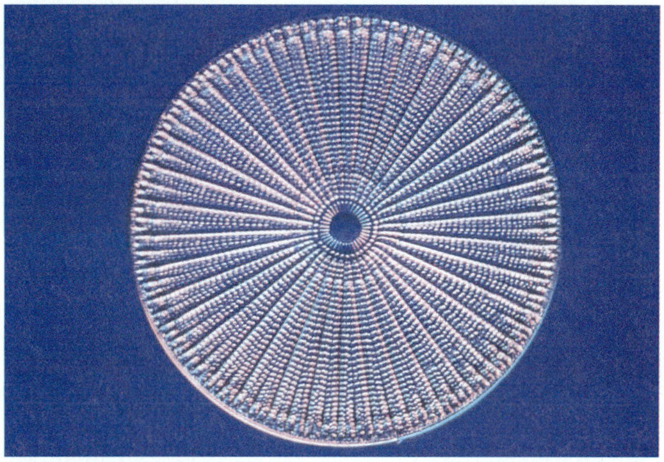

(b) CENTRIC DIATOM

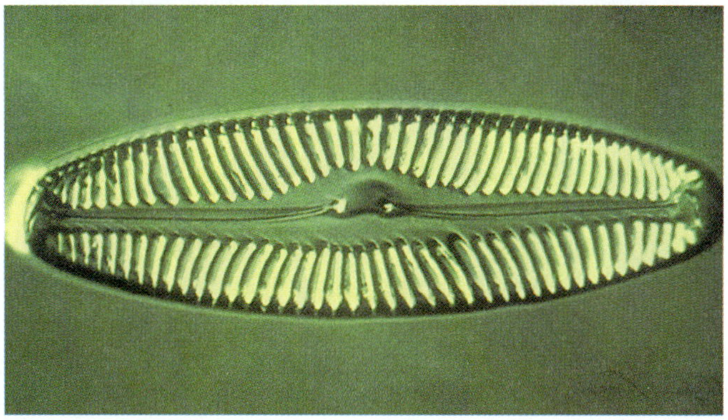

(c) PENNATE DIATOM

FIGURE **9-16**

Diatoms. (a) The silica frustule that surrounds the diatom cell consists of an epitheca (a "lid") and a hypotheca (a "box"). (b) Photomicrograph of a centric diatom (~80 μm; ~0.003 in.), which usually are phytoplankton. (c) Photomicrograph of a pennate diatom (~100 μm; ~0.004 in.), which is typically a benthic plant (lies on the sea bottom).

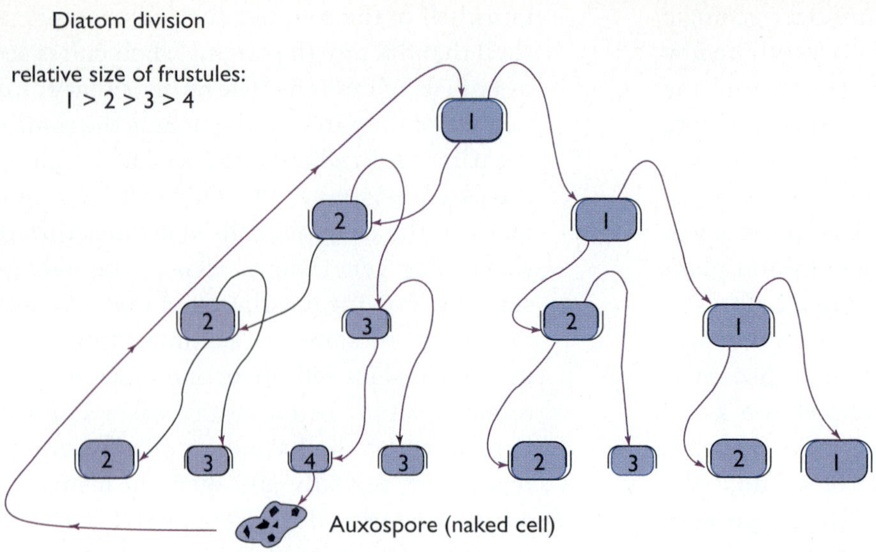

FIGURE 9-17

Life cycles of diatoms. Cell division and the secretion of a new half of a shell (a hypotheca) by each daughter cell leads to a progressive decrease in the average body size of the diatom population. At some critical minimum size, a naked cell (an auxospore) enlarges itself to a normal diatom size before secreting an entirely new silica frustule.

FIGURE 9-18

Plankton bloom in the Gulf of Alaska. The greenish color indicates that plankton abound in the water.

and shelf waters of the midlatitudes.

Dinoflagellates are another important group of marine plants that belong to the phylum Protista (see Figure 9–3b). These single-celled organisms use two whiplike flagella to propel themselves short distances through the water. Their **theca**, or cell wall, is composed typically of cellulose, an inert carbohydrate that is a common cellular component of plants and wood. Some types of dinoflagellates are "naked"; that is, their cell lacks a protective theca. A few species of yellow-green dinoflagellates actually live inside the cell walls of animals, such as coral; we will explore this important ecological relationship in a subsequent chapter. Although dinoflagellates are found in all areas of the ocean, they abound in warm, tropical waters. When dissolved silica concentrations are low in the water, diatom growth is limited. Under these conditions dinoflagellates may outnumber diatoms until such time as the dissolved silica levels in the water increase and the diatoms can resume growth.

The animal counterpart of the phytoplankton are the zooplankton, such as copepods, tiny crustaceans that feed voraciously on diatoms and zooplankton. Specialized appendages generate localized currents that direct food to the copepod's mouth (**Figure 9–19a**). The quantity of water that can be filtered in this way increases with body size and, for large, adult copepods, can be as high as 1 liter per day (1 l/day ~ 1 qt/day)—a very large volume of water compared with the small size of the animal. Growth of the copepod is rapid, provided that water tem-

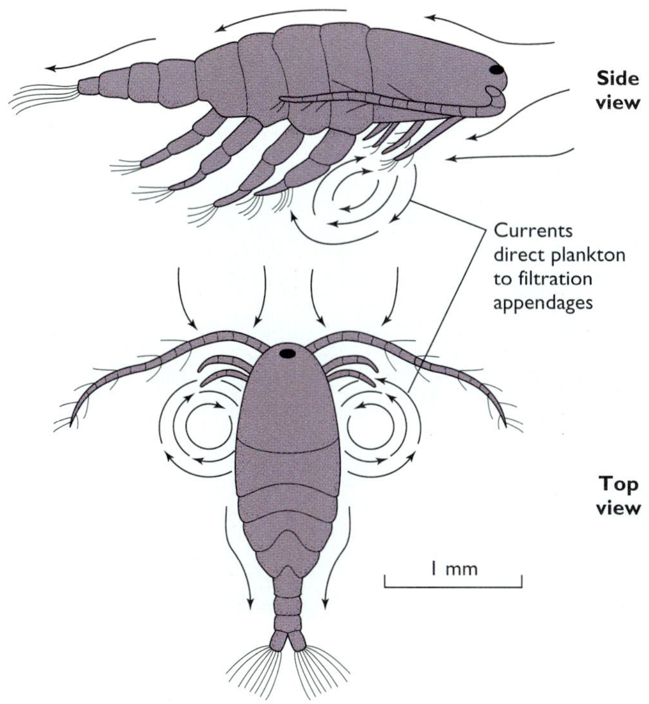

(a) FILTER FEEDING IN COPEPODS

(b) FORAMINIFERA

FIGURE 9-19

Zooplankton. (a) A copepod increases its filtering efficiency by moving its cephalic appendages to generate a current that directs phytoplankton to the mouth region. (b) The shells of forams are perforated in order for the animal to catch food.

peratures are adequate and plant cells abundant. As a copepod grows to an adult, it passes through a dozen intermediate stages. Each stage is separated by a **molt**, during which the outer skeleton (exoskeleton) is shed for a newer, larger one. Under ideal conditions, a newly hatched copepod will reach the final adult stage in fewer than 30 days. Mature copepods sink to greater water depths during the winter months, become dormant, and return to the surface waters to breed in the late spring when diatoms, are undergoing plankton blooms.

Foraminifera (*forams* for short), which belong to the phylum Protista, are single-celled species of zooplankton that secrete a chambered shell of calcium carbonate (**Figure 9–19b**). (It should be noted that there are probably more species of benthic forams than there are of planktonic forams.) The planktonic forams, which attain body lengths of 2 millimeters (~0.08 inch) or a little less, are found in the top 100 meters (~ 330 feet) of the oceans. Their shells are pockmarked with numerous pores through which they extend the protoplasm of their cell and capture small organisms, such as diatoms, or other bits of organic matter. This food is then engulfed by the cytoplasm of the cell and chemically broken down by powerful enzymes. In a sense, forams are essentially "armored" amoebas, and the protoplasm "extensions" are the same as the amoeba's pseudopods.

FUNCTIONAL MORPHOLOGY OF FISHES (NEKTON)

Fishes swim through a three-dimensional world of water. Predator fish must be adept at moving their body through the water swiftly and efficiently in order to capture prey (**Figure 9–20**). Prey fish need the same mobility in order to avoid organisms that are hunting them. Given the number of species and abundance of fishes in the ocean, particularly along the edges of continents where food abounds, it is

FIGURE 9-20

Large predators. Jacks (a) and sharks (b) have evolved hydrodynamically efficient body designs for capturing prey.

(a) JACKS

(b) SHARK

obvious that fishes have had remarkable success at exploiting the swimming lifestyle. This is no simple evolutionary feat. Regardless of the swimming style, it is difficult to move a body cleanly through water, because of the frictional and turbulent drag of the fluid itself. To grasp the magnitude of this problem, attempt to run a short distance in water up to your shoulders. You will quickly acquire an appreciation of how difficult it is to propel a body, even a powerfully muscular one, through water. Reducing frictional resistance is accomplished by lying prone in the water and swimming the distance in a manner similar to a fish. A person can swim 50 meters (~170 feet) much, much faster than run that same distance through water. Let's examine the reasons for this.

An organism must overcome three distinct types of drag to become an efficient swimmer (**Figure 9–21**). First, a fish must reduce **surface drag**, the friction between the moving body and the surrounding water, by assuming a shape that minimizes the surface area exposed to the fluid. The principle is a simple one—the less surface area, the less surface drag. A simple calculation shows that a sphere has the least surface area for a given volume of any other geometric shape. To get at this idea intuitively, imagine a cube and a sphere with the *same* volume. Which has the greater surface area?

Obviously, fish do not resemble grapefruits! So, surface drag cannot be the sole or even the most important determinant of swimming ability. Fish bodies, if they are to move efficiently through water, need to deal with **form drag** as well as surface drag. Form drag is a function of the volume of water that must be pushed aside by a moving body, which means that it is proportional to the cross-sectional area of the fish's body. The larger the cross-sectional area, the greater is the amount of water that has to be shoved aside. For example, a human moving upright through water would reduce form drag by moving sideways rather than frontward. For a fish, a long, narrow, pencil-shaped body is a good design to minimize form drag (see Figure 9–21b). However, despite a few notable exceptions, such as the needlefish, fishes do not even come close to looking like pencils. Most fishes have bodies that resemble lean cylinders with tapered ends. This cylindrical form, which can be viewed as a compromise between the sphere and the pencil, nicely minimizes the combined effects of surface and form drag.

The last factor that reduces swimming efficiency is **turbulent drag**, a retarding force that strongly robs speed from a moving object. The turbulent flow of water around a body is reduced by having a blunt leading end and a tapered back end—the classic hydrodynamically efficient torpedo shape (see Figure 9–21). Moving objects that are shaped like a torpedo displace water with minimal disturbance,

FIGURE 9-21

Fish bodies. (a) Swimming efficiency in fishes is achieved by minimizing the drag created by friction, turbulence, and body form. The "torpedo" shape, hydrodynamically efficient for cruising at high speeds, is the final result of taking these three types of drag into account. (b) A redfin needlefish has a good body design for minimizing form drag.

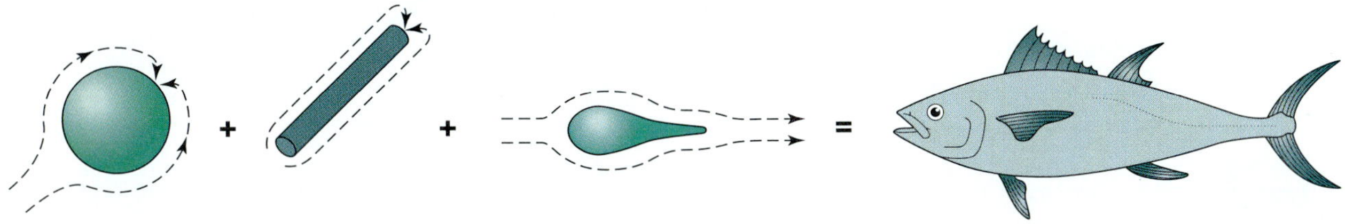

(a) FLUID DRAG ON FISH

(b) REDFIN NEEDLEFISH

so that the energy is used for propulsion and not dissipated uselessly in the generation of turbulence. The body shape of the tuna, which minimizes surface, form, and turbulent drag, is superbly adapted for high-speed swimming.

The speed of a fish depends directly on the shape of the **caudal fin** (tail fin). A useful ratio for assessing the style of swimming of fish is the **aspect ratio** (AR). It is defined simply as

$$AR = \frac{(\text{height of the caudal fin})^2}{\text{area of caudal fin}}.$$

This simple expression is a comparison of the height squared to the area of the caudal fin. Imagine a perfectly square-shaped tail fin, 10 centimeters (~3.9 inches) to a side. What is its aspect ratio? Well, you simply divide the area of the fin (area = 10 cm × 10 cm = 100 cm^2) into the tail height squared (height^2 = (10 cm)^2 = 100 cm^2). The aspect ratio for a square-shaped fin is 1. Compare this to a narrow, tall caudal fin with a height of 10 centimeters (as in the previous example), but a width of only 1 centimeter. If you do the computation, you should get an aspect ratio of 10. Do you agree? This short analysis indicates that for the *same* height, a larger tail area will have a smaller aspect ratio than will a smaller tail area. This mathematical stuff is all fine and good, but what can you learn about fish mobility from it? Well, let's analyze the problem concretely.

A tail with a broad surface area (that is, a low aspect ratio—think of the square shape) provides powerful thrust for quick acceleration or maneuvering, but the large frictional drag of the broad tail interferes with high-speed movement. Consequently, fish that cruise fast, such as tuna and swordfish, have tail fins with high aspect ratios (AR ~7+)—a design that is inefficient for fast acceleration or for darting maneuvers, but effective for rapid, sustained movement. In other words, it's hard for fish with high aspect ratios to get up to speed (there is only a small surface area of the caudal fin to push against the water), but once they are underway there is minimal surface friction because of the small tail area. **Figure 9–22** shows five types of caudal fins distinguished by their aspect ratios, along with representative examples of fishes.

Both theory and experiments indicate that the swimming speed of fishes depends mainly on the body length, the beat frequency (the number of back-and-forth sweeps of the tail fin per unit time), and the aspect ratio of the caudal fin. The fastest fish have long bodies, high beat frequencies, and high aspect ratios. Swordfish and tuna are prime examples. They specialize in fast cruising (sustained rapid swimming) to search for and to capture prey. Other fish use different techniques for appre-

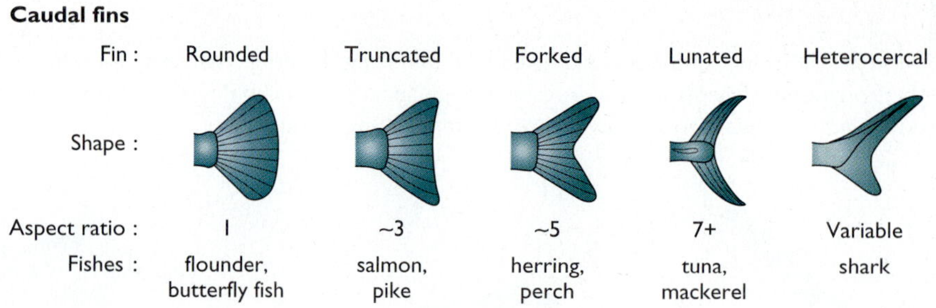

FIGURE **9-22**

Caudal fins. Fast-swimming fish possess caudal fins with high aspect ratios. The caudal fins of fish that are highly maneuverable, however, have low aspect ratio—the tail forces the animal downward, which compensates for the lift it gets from a pair of pectoral fins located on either side just behind the head region. [Adapted from H. V. Thurman, *Essentials of Oceanography* (Columbus, Ohio: Merrill Publishing, 1987).]

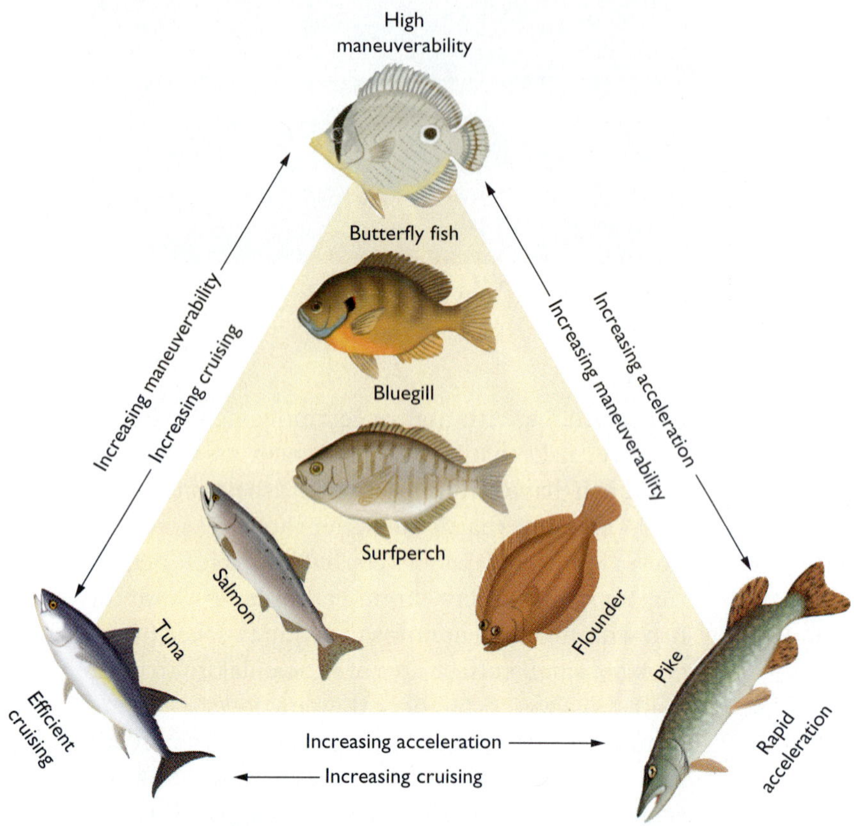

FIGURE **9-23**

A fish-classification scheme. Based on mode of locomotion, fishes can be arranged into a triangular diagram, each corner of which represents a distinct specialty (cruising, acceleration, and maneuvering). Most fishes are generalists; they combine these three locomotive modes but do not perform any of them as effectively as a specialist does. These fishes are not drawn to scale, and the bluegill and pike are freshwater species. [Adapted from P. W. Webb, "Form and Function in Fish Swimming," *Scientific American* 251, no. 1 (1984): 74–82.]

hending prey or escaping from predators. Two common strategies are making precise, crisp turning maneuvers, and rapid but brief darting maneuvers. These types of fish turn or accelerate quickly by using the pushing force of their caudal fins, which have small aspect ratios. But once they have turned or are sprinting ahead, the large surface area associated with the low aspect ratios of their tail fins slows them down because of frictional drag. In other words, they cannot cruise effectively.

Using a simple triangular diagram, we can represent the cruising, maneuvering, and accelerating types of fishes at the corners of a triangle (**Figure 9–23**). Note how the body shapes of these three "specialists" differ. The hydrodynamically efficient torpedo design characterizes the tuna, an efficient high-speed cruiser, and the pike, a quick lunger that relies on explosive acceleration. In contrast, the slow-moving butterfly fish possesses a caudal fin with a low aspect ratio and long fins at the bottom (ventral) and top (dorsal) of the body that systematically undulate to power the fish through precise maneuvers. Most fishes, however, do not resemble any one of these three body types. Rather, they are "generalists" that combine many of the design features of the specialists (see Figure 9–23), so that they can execute all three styles of movement to various degrees. The generalists cannot perform any specialized swimming technique as well as the specialists do, but the generalists' range of motions is much greater than that of the specialists.

Studies reveal a good correlation between predation success and type of swimming. On the one

Killer Whales

The species name of the killer whale is *Orcinus orca.* It is easily distinguished from other whales by its distinctive color pattern, a black top broken irregularly by a white underside and a distinctive white spot near each eye (**Figure B9–3**). Males are quite large, the bigger ones reaching a length of over 30 feet and weighing up to 9 tons. They frequent all the oceans, and are particularly abundant along coastlines with cold water, such as the inner continental shelf and nearshore waters of northern Washington State and British Columbia, Canada. These animals have strong social instincts, associating in pods comprising between three and thirty individuals.

Although referred to informally as a whale, *Ocinus orca* actually is a large dolphin possessing between ten and twelve conical teeth in each jaw. These teeth curve into the mouth and can effectively grasp large prey, mainly fish and squid. The killer whale also hunts penguins, dolphins, porpoises, seals, sea turtles, and even large whales, including the very largest animal anywhere on the Earth, the blue whale. Its hunting prowess results in part from its ability to move rapidly through water (note its streamlined shape) and from its habit of hunting in packs.

A curious animal, it is often attracted to small vessels. However, despite numerous opportunities, there has never been a confirmed incident of a human being killed by a killer whale inthe wild. Many scientists who study killer whales characterize them as intelligent, even gentle creatures.

Visit www.jbpub.com/oceanlink for more information.

FIGURE **B9-3**

Killer whale. Orca, the killer whale, has distinct color markings.

hand, for example, only 10 to 15 percent of all attacks mounted on prey by tuna are successful. This is quite understandable. Although this cruising specialist covers a lot of territory and encounters a large number of prey, it has very limited maneuvering and accelerating abilities—critical traits for catching fish once sighted. Despite its low success rate, however, the tuna does catch enough prey, because it cruises and searches across large areas and makes many sightings of potential prey. On the other hand, the pike, with a remarkable 70 to 80 percent success rate, is particularly adept at striking prey, Its attacks are few in number, because the pike is not a cruiser as the tuna is, but waits quietly for its prey to approach within striking distance. Thus, the pike has many fewer prey contacts per day than does the tuna. Generalists, such as the bass and trout, which engage in a variety of swimming skills to catch prey, have a success rate of between 40 and 50 percent—a value that is intermediate between the success rate of the pike and the tuna.

Whales are not fish, but warm-blooded mammals. Yet, like fish, they must be adept at swimming in order to survive in the ocean. Some characteristics and behavior of killer whales are discussed in the boxed feature, "Killer Whales."

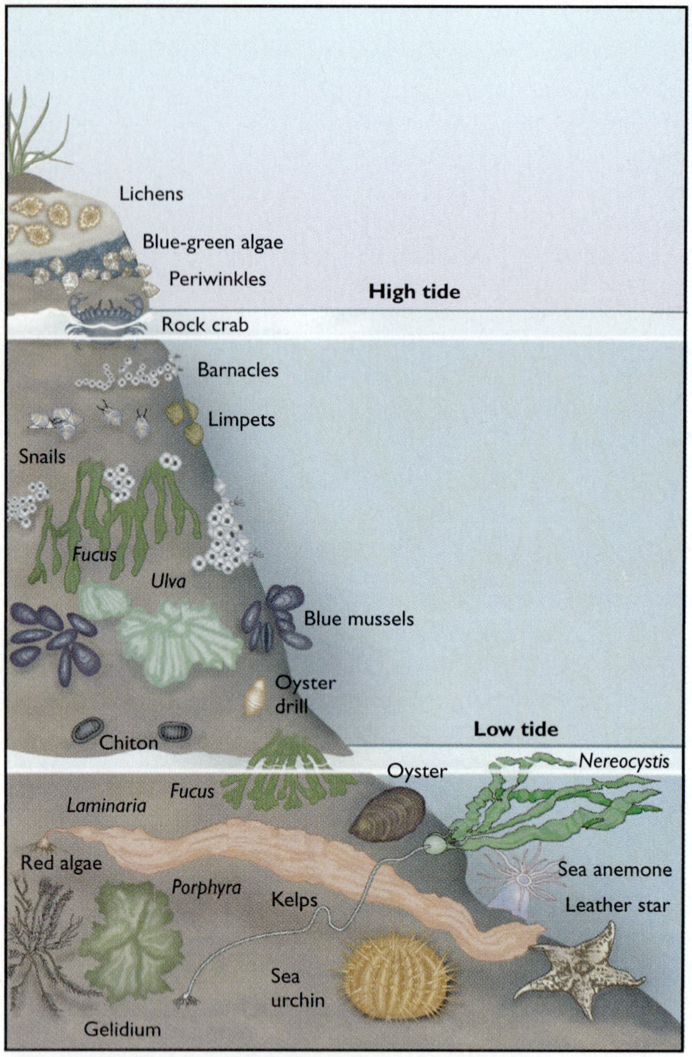

(a) BANDED BENTHIC COMMUNITIES

BENTHIC COMMUNITIES

A trip to a rocky shoreline at low tide often reveals that the plants and animals that live on the bottom are arranged in distinctive bands. The arrangement of these parallel bands, termed **vertical zonation**, represents distinct communities of benthic organisms. The topmost band, which occurs at the spring high-tide level, is dominated by blue-green algae (cyanobacteria) and snails (periwinkles). The algae are encrusted (grow as a hard coating) to the ground and impart a characteristic black smudge to the rock, which most people misinterpret as an oil stain. This upper zone gives way downward to a band of barnacles that farther down gets replaced by mussels and brown seaweed (*Laminaria*). This generalized pattern is representative of vertical zonation along much of the midlatitudinal rocky coasts of the Atlantic and Pacific Oceans of North America (**Figure 9–24a**).

What exactly causes the vertical zonation of the bottom biota in these areas? Remember that the intertidal zone is not truly marine, nor truly terrestrial, because it is regularly covered and uncovered by seawater as the tide rises and falls. As such, it represents an environment that is very difficult

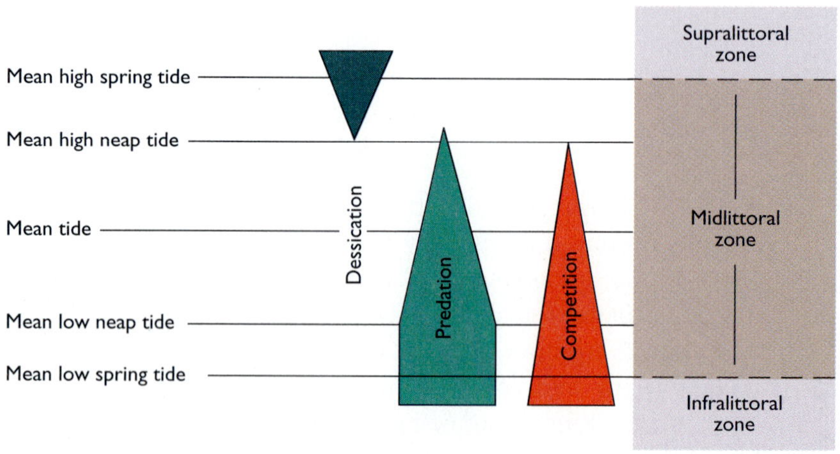

(b) CONTROLS ON VERTICAL ZONATION

FIGURE 9-24

Vertical zonation. (a) Along the rocky shorelines of the midlatitudes, the benthic biota may display a banded distribution termed vertical zonation. Each of the bands possesses a unique assemblage of organisms. (b) The upper boundary of each zone is related to the degree of tolerance to exposure to air. The lower boundaries reflect competition and predation pressure.

for both animals and plants to adapt to. It appears that the upper limit of the topmost band reflects the organisms' ability to avoid desiccation (drying out and dehydration) when exposed to the air for most of each day. Also, the biota living there are subjected to freezing temperatures during the winter, a season when the water is much warmer than the air. Additionally, the community in this top band is bathed in freshwater whenever it rains, and being marine organisms, they prefer saltwater. Here the rocks are wetted by seawater only during spring high tides or during storms when large breakers crash against the shoreline. The two dominant organisms—the blue-green algae and the periwinkles—of this zone use different means to cope with the desiccation. The blue-green algae consist of entangled filaments covered by a gelatinous secretion that protects them from drying out while exposed to air. During prolonged exposure to air and sunshine, the periwinkle can seal itself tightly to the rock, and this airtight seal reduces the loss of vital body fluids.

Although seaweed belonging to the genus *Fucus* can be present, the middle band (see Figure 9–24a) tends to be covered principally by barnacles. This zone's upper limit is determined by the ability of the particular resident species of barnacle to tolerate desiccation. The more it can tolerate drying out, the higher up on the rocks it is. Interestingly, the barnacles inhabiting the upper region are smaller than those in the lower fringe of this zone. Their small size is presumably the effect of significantly reduced submersion time in seawater—and, hence, less opportunity to filter enough food to growing to a normal adult size. The lower limit of the barnacle band is determined by intense competition for living space with other organisms, particularly mussels. Furthermore, predation pressure, especially by carnivorous snails and starfish, is intense here, and this helps determine the position of the lower boundary (see **Figure 9–24b**).

The lowest band of the intertidal zone is exposed to air only during spring low tides, which occur twice monthly. Living among the dominant laminaria seaweed of the zone is an extremely rich bottom community. Sea urchins are particularly abundant. The species composition of the seaweeds changes in this band as a result of (1) variations in the tolerance of different plants to decreasing light levels at different water depths and (2) the selective grazing pressure of the many benthic herbivores. In some areas, the growth of red and brown algae is so luxuriant that they provide a dense shelter for animals such as starfish, crabs, and sea slugs

Sandy or muddy bottoms are populated by an entirely different community of benthic organisms than that of the rocky shoreline. Rocks are stable, but difficult to cling to; sand and mud are less stable than rock—they can easily be eroded—and are relatively easy materials to burrow into. Except for protected bays and estuaries, seaweeds are usually absent on sandy substrates (bottoms) because of the lack of a firm bottom for attachment.

By far the most dominant organisms in soft substrates of sand and mud are *infauna,* animals that burrow into sediment. Sediment is a solid medium that differs radically from the fluid environment of fishes and plankton. Infauna live surrounded by solid grains of sediment and, if mobile, must overcome the resistance of dense, often compacted sand and mud. These animals not only must displace sediment particles, but also protect their soft tissue from the abrasive edges of sharp grains, construct homes (burrows) that are structurally sound and do not collapse, and obtain adequate food while imbedded in sediment. There are clear advantages to their lifestyle of living buried in sediment. The infaunal residents of the intertidal zone are protected from predators, from damage caused by breaking waves and current scour, and from desiccation at low tide (because they remain buried in watery sediment).

The razor clam (*Ensis directa*), a sediment-dwelling inhabitant of sand flats, has developed a remarkable ability to propel itself by burrowing downward into the substrate (**Figure 9–25a**), even faster than most shellfishers can dig! How does the razor clam accomplish this remarkable feat? To a large degree, its mobility, like that of fish, is related to functional morphology, that is, a body style and shape adapted to its particular mode of movement. The shell of the adult razor clam can exceed a length of 20 centimeter (~8 inches). However, despite its large size, this shell is well crafted for rapid movement through sand. It is narrow and slightly tapered, and the surface is very smooth. Hence, form and frictional drag are minimized, and the shell offers little resistance to rapid burrowing.

Ecology of the Giant Kelp Community

Macrocystis (**Figure B9–4a**), a brown alga, grows abundantly along the Pacific coast of North America, where the immensely long kelp, some up to 40 meters (~132 feet) in length, form a unique forestlike growth (**Figure B9–4b** and **B9–5**) along the inner continental shelf. The kelp beds support a rich community of animals.

Studies have documented an inverse relationship between the abundance of sea urchins (**Figure B9–4c**) and the density of kelp growth (**Figure B9–6**). This means that as the number of sea urchins goes up, the density of the kelp falls off. The relationship is easily explained. Sea urchins are herbivores and graze voraciously on kelp, detaching them from the bottom and setting them free to wash up onto the beach. Large populations of sea urchins can destroy kelp beds quickly, not only deforesting the ocean floor, but also eliminating important habitats for the many fishes that dwell among the kelp fronds. The sea otter (**Figure 9–4d**), a skillful mammalian predator, feeds mainly on benthic invertebrates, including sea urchins. Wherever the sea otter abounds, as it does off Amchitka Island, Alaska, the number of sea urchins is low, and kelp forests and their associated fish communities flourish. In contrast, the scarcity of sea otters on the nearby island of Shemya

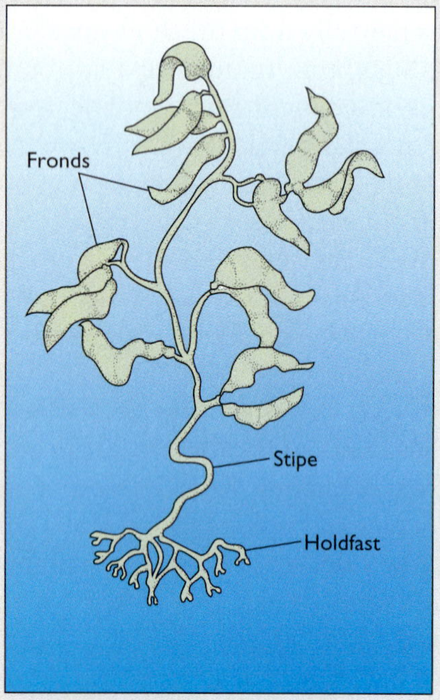

(a) *MACROCYSTIS*

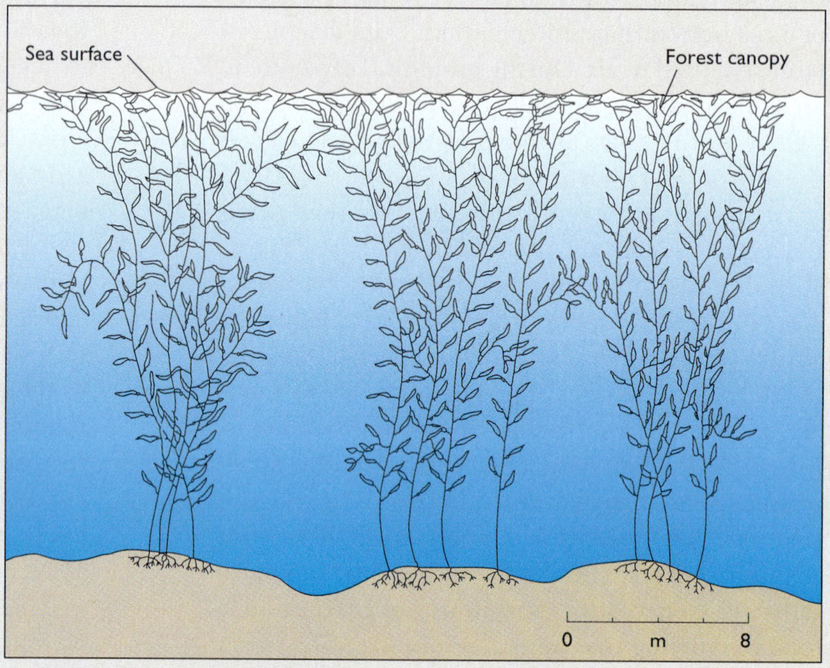

(b) KELP *(MACROCYSTIS)* FOREST

(c) SEA URCHINS

(d) SEA OTTER

FIGURE **B9-4**

Kelp forests. (a) The giant kelp *Macrocystis* flourishes on the Pacific continental shelves of North and South America wherever a hard substrate permits the plant to attach its holdfasts securely to the sea bottom. (b) *Macrocystis* can attain lengths of 40 meters, producing dense kelp forests. These thick canopies of intertwined fronds at the sea surface provide shelter for large numbers of invertebrates and fishes. (c) Sea urchins are shown grazing on giant kelp. (d) A sea otter is shown preying on a sea urchin.

FIGURE B9-5

Underwater forest. A view of a dense growth of kelp off shore of California.

allows large numbers of sea urchins to exist, and they graze down the kelp beds (see Figure B9-4c).

Table B9-1 summarizes a simplified ecological model of these interactions. Basically, high predation pressure by sea otters keeps the number of sea urchins low, enabling kelp forests to thicken and expand and attracting an abundant population of fishes. Because of the scarcity of sea urchins, sea otters must spend a great deal of time preying on the elusive fishes that dwell among the kelp fronds. Conversely, nearshore areas populated by few sea otters are characterized by large numbers of sea urchins, and their intense grazing obliterates kelp beds and their fish communities in short order. The few sea otters in these deforested areas forage for only a small part of the day because it is easier for them to catch sea urchins than to capture the elusive kelp-dwelling fishes.

For almost a century, otter populations along the coast of Alaska have been rebounding from overhunting. Kelp forests in the region have also recovered and expanded because the otters feed on sea urchins, limiting their grazing effect. Recently, this century-long trend of natural kelp reforestation has reversed unexpectedly. In large areas of the western Alaskan seaboard the number of otters has plummeted and the density of sea urchins increased. The kelp forests, as a result, are being destroyed; in some areas less than a twelfth of the previous kelp cover survives. Biologists speculate that killer whales, which normally eat fish, are now hunting the otters, possibly because the fish have decreased in numbers or shifted out of the area. Although the reasons are difficult to verify, biologists suspect that the decline in the fish stocks is caused by a combination of ecological factors, most notably overfishing by humans and elevated water temperatures. This new trophic linkage has caused a sudden shift in the dynamics of the food web in the shelf waters of western Alaska: Sea urchins, no longer regulated by otter predation, are devastating the kelp forest ecosystem.

Visit www.jbpub.com/oceanlink for more information.

TABLE B9-1

Sea otter/sea urchin/kelp ecosystem.

Characteristic	Shemya Island	Amchitka Island
Sea otters	scarce	abundant
Sea urchins	large and abundant	small and scarce
Kelp	thin or absent	abundant
Prey of sea otters	sessile (urchins); available all day	motile (fishes); most easily caught at twilight
Feeding activity of sea otters	small amount of time spent feeding; no diurnal cycle	large amount of time spent feeding; diurnal cycle

Source: Adapted from A. J. Estes, R. J. Jameson, and E. B. Rhode, "Activity and Prey Election in the Sea Otter: Influence of Population on Community Structure." *The American Naturalist* 120, no. 2 (1982): 247–57.

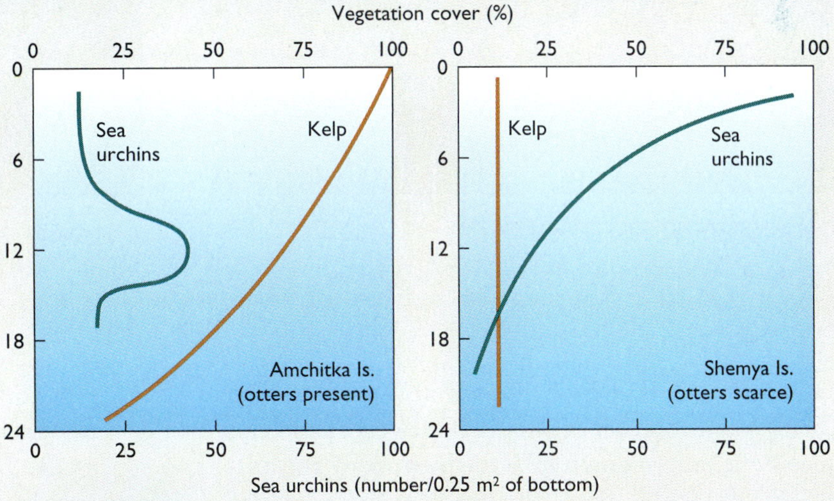

FIGURE B9-6

The ecology of kelp forests. In the Amchitka Islands of Alaska, sea otters control the size of the sea-urchin population, which in turn regulates the extent and growth of the kelp beds there. The specific ecological interactions among the dominant members of this ecosystem are summarized in Table B9-1. [Adapted from J. A. Estes and J. F. Palmisan, "Otters: Their Role in Structuring Nearshore Communities," *Science* 185 (1974): 1058–60.]

FIGURE 9-25

Infaunal organisms. (a) The razor clam and its burrowing technique. (b) The U-shaped dwelling of some species of polychaete worms is illustrated. (c) This photograph shows the feeding posture of sand dollars.

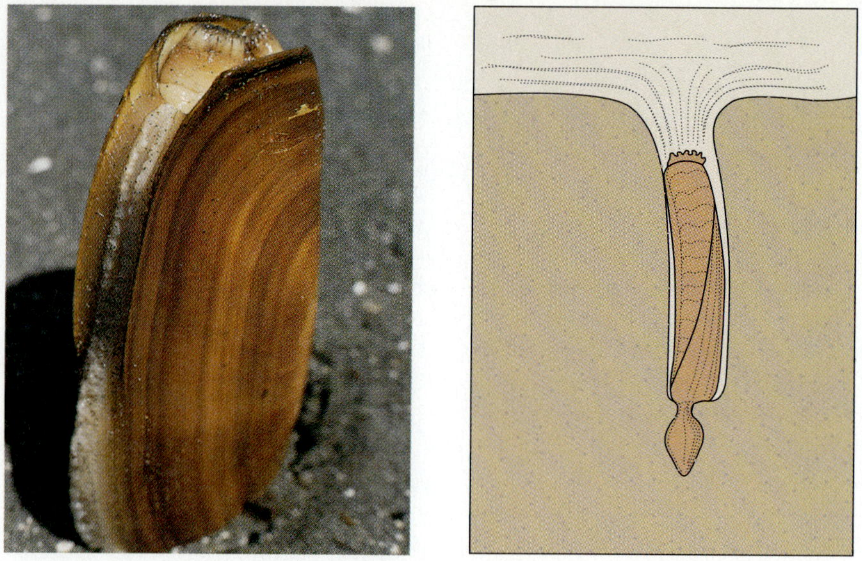

(a) BURROWING RAZOR CLAM

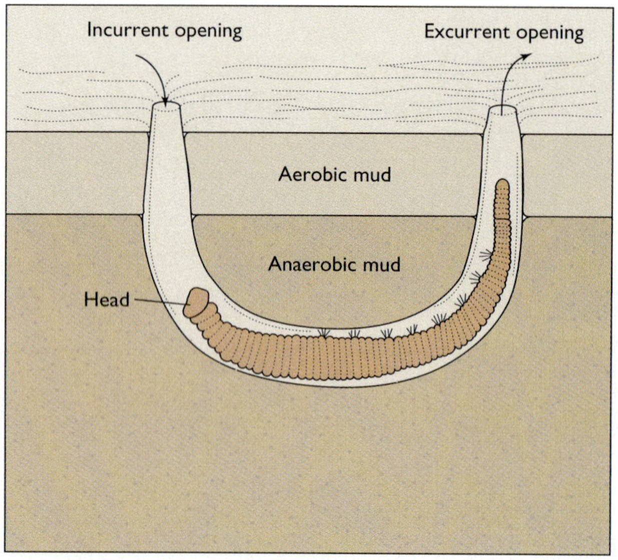

(b) BURROWING POLYCHAETE WORM

(c) FEEDING SAND DOLLARS

Also, the shell, which houses and protects the clam's soft internal organs, is lightweight and can easily be pulled down into the burrow by the animal's muscular foot, the swollen base of the clam's body, which serves as a digging implement. The razor clam attains rapid propulsion by physically extending and forcing its muscular foot into the sand. This activity causes the tip of the foot to become engorged with blood and to swell to several times normal size, pushing sediment aside. The swollen foot serves as a secure anchor, enabling the foot to pull the shell down into the new excavation. The rapid repetition of this digging sequence enables the razor clam to move down through half a foot of sand in a matter of a few seconds.

Some *bivalves* (two-valved critters) use a combination of burrowing activity and water currents to propel themselves. The small Atlantic coquina clam (*Donax variabilis*) of the surf zone lives in a shallow burrow not much deeper than one shell length. Wave turbulence regularly washes these clams out of their burrows and, while they are exposed, transports them with the sand grains. Once the wave has passed, the clam energetically digs a new burrow, which it occupies until another wave surge exposes it again. In this way, whole populations of *Donax* migrate slowly down the beach in the direction of longshore drift.

The polychaete worm (**Figure 9–25b**), another important member of the infauna, has a long, narrow, tubular body and a tapered head with a streamlined shape to reduce form drag. This worm forces its head into the sediment, using a combination of muscular contractions and *fluidization* of the substrate (suspending the sediment by a stream of water). The contracting muscles cause the head region to expand, forming an anchor from which the worm pulls the remainder of its body into the burrow. These worms also pump water through their U-shaped burrows, keeping them aerated with dissolved oxygen, bringing bits of food through the incurrent end, and washing out wastes through the excurrent opening (see Figure 9–25b).

Sand dollars such as *Mellita*, with their coin-shaped bodies, are **deposit feeders**, meaning they obtain food by processing mud and selectively removing organics that are mixed into it. They feed by two principal means. First, they "plow" slowly through sediment, digging with the flattish "spines" on the underside of their shells. The spines push away sediment particles, so that the animal sinks gradually downward while moving slowly forward. Bits of food in the sediment stick to the mucous coating on its body and are moved toward its mouth by the wavelike motion of its beating cilia. Sand dollars also feed by standing on edge, partially extended out of the sediment, and allowing small food particles wafted by currents to brush against and stick to their mucous-coated cilia (**Figure 9–25c**). Sections of Chapters 12 and 13 cover a variety of interactions of benthic organisms and substrates.

9-1

Future Discoveries

A great deal of the ocean's biological diversity has yet to be documented properly. New studies of the ocean's remarkable biological communities are wide ranging. They include efforts to describe what is present, the ecological roles played by both small and large organisms, and the extent of species, genetic, and ecosystem diversity. A few examples of some of this exciting research includes examining the diversity of marine bacteria, midwater fishes, and marine mammals. Marine microbes are ubiquitous and abundant, and their presence is essential for the survival of all other life forms. For example, some bacteria called extremophiles live beneath hundreds of meters of marine sediment, virtual underground bacterial assemblages. How do they live and what role do they play in the chemistry of the water trapped in the sediment? One of the least explored sea habitats is the midwater region with its assortment of odd-looking fishes. What is and what regulates the diversity of midwater communities, and what broader ecological role does this fauna play as a link between plankton at the sea surface and the tremendous diversity of invertebrates that live on the deep-sea bottom? How have human activities affected mammals such as manatees, sea otters, polar bears, whales, and dolphins, and is it possible to mitigate such effects in the future so that they are not endangered? Should the ban on the commercial hunting of species of whales that have rebounded from low population numbers be lifted, and, if so, how can this harvesting be managed sustainably?

Because each marine species possesses unique genetic information, many current biological studies rely increasingly on DNA molecular techniques that provide accurate information about the genetic code of any species. These powerful, new laboratory techniques provide rapid, relatively inexpensive means of identifying genes and their specific function in organisms, and therefore have wide application in research problems. They are invaluable, for example, for estimating microbial diversity, for establishing the population distribution and migratory routes of marine organisms, and for identifying the nature and source of human pathogens in contaminated water.

STUDY GUIDE

KEY CONCEPTS

1. The oceans are divided into two parts: the *pelagic province* (all the ocean water) and the *benthic province* (all the ocean bottom). Finer subdivisions of each province are explained in Table 9–1.

2. A useful way of broadly classifying the marine biota is by lifestyle. *Plankton,* which include the *phytoplankton* (plants) and the *zooplankton* (animals), float passively and drift with the ocean currents. By contrast, *nekton* are active swimmers, moving easily through the water. The *benthos* live in contact with the sea bottom, either securely attached to it or moving freely on or within the substrate.

3. *Ecology* deals with the interrelationships among organisms and their physical and chemical surroundings. Water temperature directly affects organisms in a variety of ways: nutrient and food uptake, growth rate, survival, and reproductive success. Salinity also exerts controls on the activity and distribution of organisms. If the salinity of an organism's body fluid is different from that of the surrounding water, it must osmoregulate in order to maintain a proper water balance in its cells. Hydrostatic pressure increases by about 1 atm for every 10 meters (~33 feet) of water depth. Many deep-sea organisms are immune to high hydrostatic pressures because their bodies are free of gases, which are easily compressible.

4. Phytoplankton are minute plants that, as a result of their high surface area to volume, possess low settling rates. This allows them to remain suspended in the *photic zone,* where they photosynthesize. Diatoms and dinoflagellates are the most important members of the phytoplankton. Under favorable light and nutrient conditions, rapid cell division of plant cells creates dense *plankton blooms.* Copepods, important members of the zooplankton, graze on diatoms.

5. The style of swimming adopted by different species of fish depends on body form and the shape and beat frequency of the *caudal fin.* Some fishes specialize in cruising, others in maneuvering, and still others in accelerating. Fishes that are generalists combine all three styles of movement but do not perform any one motion as well as the specialists do.

6. The benthos of rocky intertidal shorelines commonly display *vertical zonation,* parallel bands of bottom communities. The upper limit to the bands is regulated by desiccation, the lower limits by competition for space and predation pressure. Burrows protect the infauna from damage or death from predation or the bottom scour of swift currents or breakers.

KEY WORDS

abyssal zone
aphotic zone
aspect ratio
bathyal zone
benthic province
benthos
caudal fin
diffusion
ecology
epifauna
form drag
hydrostatic pressure
infauna
nekton
neritic zone
oceanic zone
osmoregulate
osmosis
pelagic province
photic zone
phytoplankton
plankton bloom
surface drag
turbulent drag
vertical zonation
zooplankton

QUESTIONS

REVIEW OF BASIC CONCEPTS

1. Compare the lifestyles of planktonic, nektonic, and benthic organisms. Cite several examples of each group.
2. What is a plankton bloom? What is its effect on the average size of a diatom population with the succeeding generations?
3. Explain why the benthos of an intertidal rocky coast commonly display vertical zonation.
4. Contrast how a fish and a razor clam attain rapid mobility.
5. Describe a direct and an indirect effect that temperature can have on marine organisms.
6. Why must most plants in the ocean be microscopic in size?
7. What principal role do bacteria play in the ocean?

CRITICAL-THINKING ESSAYS

1. How does the physical environment differ over a year for the neritic as distinguished from the abyssal sea bottom? Be specific and explain the differences.
2. Contrast the advantages and disadvantages associated with an epifaunal and an infaunal mode of existence.
3. Examine the drawing of the North Pacific pink salmon in the boxed feature entitled "Migrants" in the next chapter. Make some deductions about its swimming style, and defend your inferences.
4. Design a fish that would be reasonably adept at maneuvering through holes and crevices of a coral reef while also relying on a darting motion to catch prey. Explain your model.
5. Why are roots unnecessary for seaweeds but essential for trees?

DISCOVERING WITH NUMBERS

1. Given that the hydrostatic pressure increases at a rate of 1 atm per 10 meters of water depth, calculate the hydrostatic pressure for a sea bottom that lies at a depth of 1.5 kilometers. Is the hydrostatic pressure for a water depth of 3 kilometers simply twice the hydrostatic pressure for 1.5 kilometers? Why or why not?
2. Examine Figure 9–15. If you increase the diameter of an organism to 16 units, what will be the ratio of its area to volume?

3. Assume that a copepod that is 0.5 centimeters long processes 1 liter of water each day in order to feed. How much would a man 2 meters tall (average height) have to process for the volume to be proportional with that handled each day by the copepod?

4. Calculate the aspect ratio of a caudal fin that is 12 centimeters high and 36 centimeters square in area. For the same height, determine the area of the caudal fin that is necessary for a fish to be a fast cruiser.

5. Assume that phytoplankton are undergoing cell division once each day, meaning that the population is doubling its numbers every day. What will be the concentration of diatoms after a five-day-long bloom, if their initial concentration is 10 diatom cells per liter of seawater?

SELECTED READINGS

Anderson, D. M. 1994, Red tides. *Scientific American* 271 (2): 62–68.

Barnes, R. S. K., and Hughes, R. N. 1982. *An Introduction to Marine Ecology*. Boston: Blackwell .

Broad, W. 1997. *The Universe Below: Discovering the Secrets of the Deep Sea.* New York: Simon and Schuster.

Burgess, W., and Shaw, E. 1979, Development and ecology of fish schooling *Oceanus* 22 (2): 11–17.

Childress, J. J., Felbeck, H., and Somero, G. N. 1987. Symbiosis in the deep sea. *Scientific American* 256 (5): 114–20.

Dring, M. J. 1982. *The Biology of Marine Plants.* London: Edward Arnold.

Eastman, J. T., and DeVries, A. L. 1986. Antarctic fishes. *Scientific American* 255 (5): 106–14.

Feder, H. A. 1972. Escape responses in marine invertebrates. *Scientific American* 227 (1): 92–100.

Horn, M. H., and Gibson, R. N. 1988. Intertidal fishes. *Scientific American* 258 (1): 64–70.

Koehl, M. A. R. 1982. The interaction of moving water and sessile animals. *Scientific American* 274 (6): 124–35.

Lalli, C. M., and Parsons, T. R. 1993. *Biological Oceanography: An Introduction.* Oxford: Butterworth-Heinemann.

Lerman, M. 1986. *Marine Biology: Environment, Diversity, and Ecology.* Menlo Park, Calif.: Benjamin-Cummings.

Levinton, J. S. 1982. *Marine Ecology.* New York: Prentice-Hall.

McKeown, B. A. 1984. *Fish Migration.* Portland, Ore.: Timber Press.

Milne, D. H. 1995. *Marine Life and the Sea.* Belmont, Calif.: Wadsworth.

Nelson, C. H., and Johnson, K. R. 1987. Whales and walruses as tillers of the sea floor. *Scientific American* 256 (2): 112–17.

Niiler, E. 2001. The trouble with turtles. *Scientific American* 285(2): 80–85.

Nybakken, J. W. 1996. *Marine Biology: An Ecological Approach.* New York: Addison-Welsey.

O'Shea, T. J. 1994. Manatees. *Scientific American* 271 (1): 66–72.

Oceanus. 1978. The deep sea. Special issue 21 (1).

Oceanus. 1978. Marine mammals. Special issue 21 (2).

Oceanus. 1980. Senses of the sea. Special issue 23 (3).

Oceanus. 1981–82. Sharks. Special issue 24 (4).

Oceanus. 1992. Biological oceanography. Special issue 35 (3).

Oceanus. 1995. Marine biodiversity I. Special issue 38 (2).

Oceanus. 1996. Marine biodiversity II. Special issue 39 (1).

Partridge, B. L. 1982. The structure and function of fish schools. *Scientific American* 246 (6): 114–23.

Roper, C. F. E., and Bass, K. J. 1982. The giant squid. *Scientific American* 246 (4): 96–105.

Sumich, J. L. 1992. *An Introduction to the Biology of Marine Life.* Dubuque, Iowa: Brown.

Steele, J. H. 1980. Patterns in plankton. *Oceanus* 23 (2): 2–8.

Teal, J., and Teal, M. 1975. *The Sargasso Sea.* Boston: Wm. C. Brown.

Thorne-Miller, B. 1998. *The Living Ocean: Understanding and Protecting Marine Biodiversity.* Washington, D.C.: Island Press.

Webb, P. W. 1984. Form and function in fish swimming. *Scientific American* 251 (1): 72–82

Wong, K. 2002. The mammals that conquered the seas. *Scientific American* 286 (5): 70–79.

TOOLS FOR LEARNING

Tools for Learning is an on-line review area located at this book's web site OceanLink (**www.jbpub.com/oceanlink**). The review area provides a variety of activities designed to help you study for your class. You will find chapter outlines, review questions, hints for some of the book's math questions (identified by the math icon), web research tips for selected Critical Thinking Essay questions, key term reviews, and figure labeling exercises

10 Biological Productivity in the Ocean

All that is told of the sea has a fabulous sound to an inhabitant of the land, and all its products have a certain fabulous quality, as if they belonged to another planet, from seaweed to a sailor's yarn, or a fish story! In this element the animal and vegetable kingdoms meet and are strangely mingled.
—*Henry David Thoreau,* Cape Cod, 1865

web navigator

critical thinking on the web

math tutor on the web

www.jbpub.com/oceanlink

PREVIEW

ECOLOGY INVOLVES the study of natural systems, including the interconnections that exist among all of their countless living and nonliving parts. Life exists because of the cycling of matter and the exchange of energy. Ecosystems depend on the activity of plants. In the ocean, the common plants are the ever present unicellular phytoplankton. Through the biochemical process of photosynthesis, these microscopic plants begin the nutritional cycle by using the energy of sunlight to synthesize (manufacture) food from simple inorganic substances dissolved in seawater. Because plant photosynthesis is such a critical link between the living and nonliving worlds, we focus in this chapter on this crucial biological process. Without plants, the oceans would not be populated with animals.

The theme of this chapter is food: specifically, how it is manufactured by tiny plant cells and passed on to large fish and human beings. We will be employing an ecological approach to our survey of marine productivity. Rather than dealing with an individual species, we will treat the ocean and its many inhabitants as a marine system and strive to understand how its parts work together. This will provide a solid basis for understanding the biology of any specific marine species of plant or animal.

Marine organisms are not static, isolated units that merely occupy space in the water column and on the sea bed. Ecology teaches us that every living thing is affected by and, in turn, affects the nonliving and living elements of its surroundings. All species, including humans, are imbedded firmly in this complex system of chemical and physical/biological interactions. Each organism must compete for living space and food with individuals of its own species, as well as with members of other species, sometimes directly, more often indirectly. Animals prey and are preyed upon. These interconnections, which operate on a microscopic and a macroscopic (visible to the eye) level, and over different spans of time, are the "stuff" of ecology.

10-1

Food Webs and Trophic Dynamics

The term **ecosystem** represents the totality of an environment, encompassing all of its chemical, physical, and geological parts, as well as all of the plants and animals that live there. Although marine ecosystems differ widely in terms of size, complexity, and diversity of their organisms, they share common characteristics. It is these that we will focus on in this chapter.

Regardless of type and location, ecosystems operate from day to day by exchanging energy and matter. Typically, plants obtain energy from the sun and transform simple inorganic chemicals into food, and by so doing they grow and multiply. Plants are grazed by herbivores, which in turn are consumed by predators. So organic matter and chemical energy are passed through the living part of the ecosystem. Yet it doesn't end there! Other types of organisms, principally bacteria and fungi, decompose and convert dead organic matter into simpler inorganic compounds—nutrients—that can be taken up by plants again for photosynthesis. And thus the cycle of life is completed.

Let's construct a simple conceptual model of an ecosystem that shows the pathways of energy and matter (**Figure 10–1**). What powers ecosystems is solar energy. Without sunshine there would be no life on Earth. Solar radiation is harnessed by plants, which contain **chlorophyll,** chemical pigments that enable them to photosynthesize. The process of **photosynthesis** converts solar radiation into chemical energy, stored in the molecules of food. Animals eat food in order to exploit its stored chemical energy.

Plants are the primary producers and as such are called **autotrophs.** They are organisms capable of self-nourishment by synthesizing food from inor-

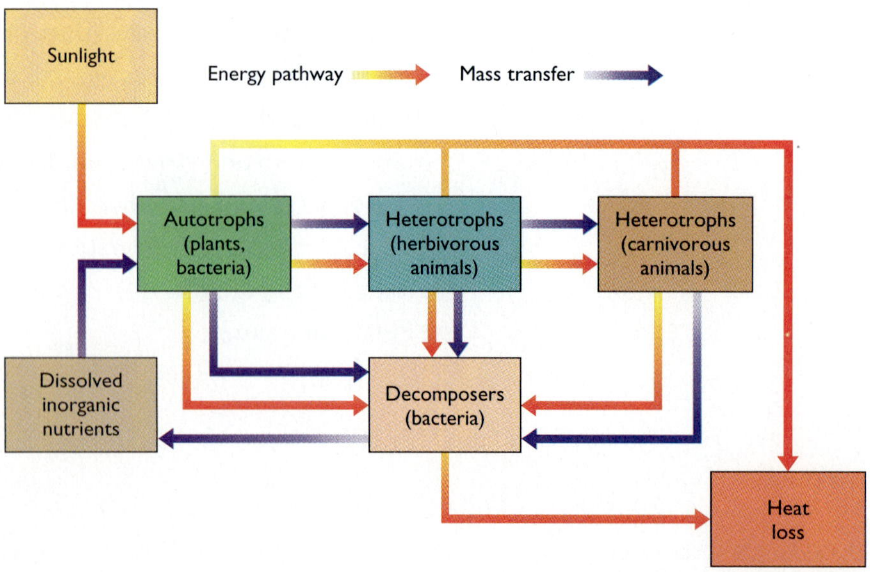

FIGURE 10-1

Ecosystem model. Matter is recycled through the ecosystem as nutrients are incorporated into food by plant photosynthesis, and dead organic matter is transformed back into nutrients by bacterial decomposition. By contrast, energy cannot be recycled. Rather, it is dissipated as heat into the environment.

ganic nutrients. All other organisms are **heterotrophs.** They require prefabricated food because they are unable to obtain energy directly from the Sun. Basically, autotrophs (plants) are producers, whereas heterotrophs (animals) are consumers.

As food is consumed, both energy and matter (chemicals) pass through the various levels of the ecosystem. From the standpoint of matter, ecosystems are self-contained, meaning that chemical substances are recycled. Nutrients are consumed by plants, plants are eaten by animals, and these animals by other animals. Bacteria decompose dead plant and animal matter, transforming them into nutrients. This brings us back to the beginning of the process, which starts all over again.

The path taken by energy through ecosystems is a different matter entirely. Energy is obtained from an external source, the Sun, and it must be replenished all the time. It cannot be reused or recycled in the way that chemicals are. Solar energy is transformed by photosynthesis into chemical energy (food). Once this food is consumed, organisms metabolically convert the bound chemical energy into physiologic processes (movement, digestion, reproduction, growth) and into heat, which is dissipated into the atmosphere and eventually into space. This means that, unlike matter which is recycled, energy flows into an ecosystem, drives its biological processes, and flows out of that ecosystem. In our conceptual model (see Figure 10–1), you can visualize matter as an arrow that is bent back on itself, creating a loop (a cycle), and energy as a straight arrow that enters the ecosystem as sunlight and leaves the ecosystem as heat, passing right through it as it plays its life-supporting role.

ENERGY AND TROPHIC DYNAMICS

The word trophic refers to nutrition, the process of nourishing or being nourished. Nutrition is a vital aspect of all ecosystems, and it serves as a basis for studying the interrelationships among organisms. Such an approach is called **trophic dynamics;** it is the study of the nutritional interconnections among the parts of an ecosystem. Organisms can be arranged into a series of **trophic levels,** beginning with plants at the first trophic level and ending with large animals at the fifth or sixth trophic level.

As autotrophs, plants provide matter and energy to the other members of the ecosystem. They occupy the first trophic level and are considered to be the primary producers. Above the primary producers are the consumers. These include **herbivores,** which graze the plants, at the second trophic level and the small and large **carnivores,** which eat animal flesh, at the third and higher trophic levels.

We can visualize trophic levels as being arranged into a **food chain** (see **Figure 10–2a**), a simple, linear, stepwise succession of the members of an ecosystem based on nutritional relationships. A diatom photosynthesizes and grows; it is eaten by a copepod, which grows; the copepod is consumed by a small fish, which grows. The concept of the food chain is useful for understanding how energy and matter are passed through the trophic levels of any ecosystem, whether it is a small tidal pool along the rocky shores of Oregon or thousands of square kilometers of sea bottom beneath the North Atlantic Ocean. Food chains also remind us that all of the energy that an animal expends during its lifetime is ultimately dependent on the photosynthesis of plants. In the ocean, the largest fishes and mammals (whales), despite their incredible bulk and strength, cannot survive without the photosynthetic activity of the autotrophs—the microscopic, unicellular phytoplankton!

Long food chains have five or six trophic levels. Their top predators have much more distant connections to the plants than do the top predators of short food chains . In reality, food chains are rarely as simple or as linear as shown in Figure 10–2a, because many animals have varied diets and feed on two or more trophic levels. For example, humans are **omnivores,** organisms that eat both plants and animals, and therefore feed on virtually all trophic levels. Given the variety of dietary preferences of many animals, it is appropriate to depict trophic connections as a **food web**—a network of interlaced and interdependent food chains (**Figure 10–2b**).

Another means of portraying the complexities of a community of organisms is to arrange species based on their nutritional habits. This leads to the concept of an *energy pyramid* (**Figure 10–2c**). Note that the size of any compartment in the energy pyramid indicates the relative amount of energy contained in that trophic level. Over a long time period, the total energy at any level of the pyramid is rarely greater than the energy contained in any

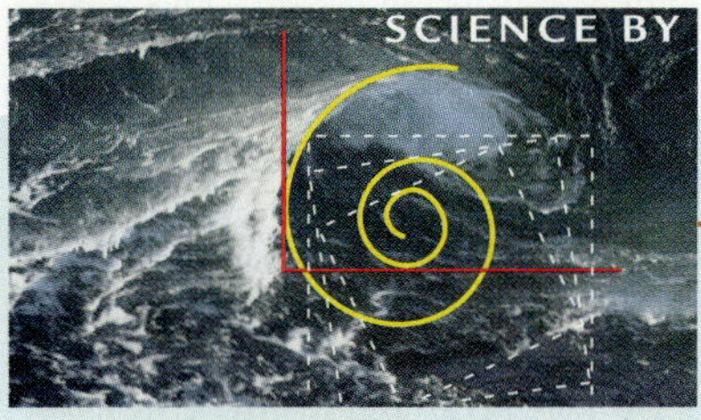

SCIENCE BY NUMBERS

Doubling Rates

Exponential growth results when populations double in size at regular intervals. Every time diatom cells divide, you get two daughter cells for each parent. This doubles the population in one generation. When the daughter cells in turn divide, each produces two of its own daughter cells. This has doubled the population again. In short order, the number of diatom cells in the surface water of the ocean has increased dramatically, producing a diatom bloom. However, this verbal description fails to adequately portray the enormous potential of exponential growth. A specific doubling-rate problem might do better.

Imagine a very large square piece of paper of normal thickness (~0.01 millimeter). Assume you take this piece of paper and fold it in half. What happens? Well you've doubled the thickness (0.02 millimeter). Fold the paper in a half a second time. Now you've doubled the thickness once again, so you have a pile of paper in your hands that is 0.04 millimeter thick (four thicknesses). If you fold the pile in a half a third time, you create eight thicknesses of paper that is 0.08 millimeter thick. Folding it in half for a fourth time creates sixteen thicknesses of paper that is 0.16 millimeter thick.

This doubling-rate problem can be expressed simply as a power to the base 2. In other words,

No doubling:	$2^0 = 1$	This is our one sheet of unfolded paper, indicated by the 0 exponent.
1st doubling:	$2^1 = 2$	This is the first folding in half; we've gone from one to two thicknesses.
2nd doubling:	$2^2 = 4$	This is the second folding, indicated by the exponent 2; this produces four thicknesses from the previous two thicknesses.
3rd doubling:	$2^3 = 8$	This is the third folding indicated by the exponent 3; this produces eight thicknesses from the previous four thicknesses.
4th doubling	$2^4 = 16$	This is the fourth folding, indicated by the exponent 4; this produces sixteen thicknesses from the previous eight thicknesses.

In other words, we can express the doubling rate simply as 2^n, where n is the number of times we fold the paper.

FIGURE 10-2

Trophic relationships. (a) This simple food chain consists of four trophic levels arranged in a linear array. The tuna prey on anchovies, the anchovies feed on copepods, and the copepods graze the diatoms. (b) Actual feeding habits are not simple food chains, but food webs, which are networks of interconnected and interdependent food chains. (c) The energy pyramid depicts trophic relationships within a community. Energy passes upward as organisms in the higher compartments feed on organisms in compartments below them. Note that most of the energy of an ecosystem is located at the base of the pyramid, where the primary producers, the plants, photosynthesize. Only about 10 percent of the energy is transferred from level to level.

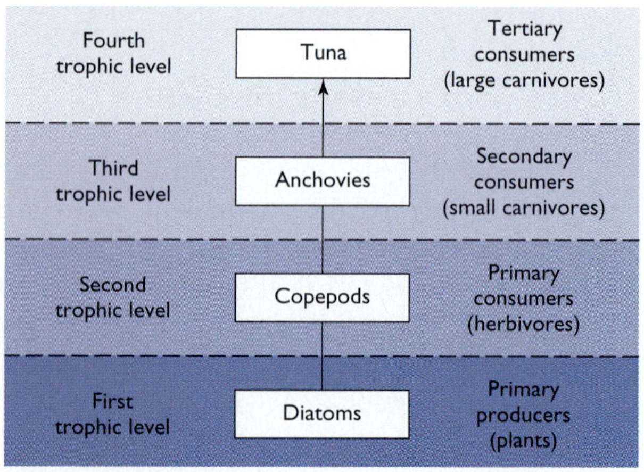

(a) SIMPLE FOOD CHAIN

My question to you is this: How thick would the last thickness of paper be if we folded the paper 50 times (in other words, 2^{50})? Would it be 1 meter thick? Would it be as tall as a person (about 2 meters)? Could it possibly be as high as a one-story house (about 4 to 5 meters)? Would it be 100 meters tall? This certainly seems to be getting beyond what seems probable. But maybe not. Let's do the simple calculation. I will use my calculator as an aid.

First, I understand that the thickness of the original sheet of paper is ~0.01 millimeter. Second, I realize that each time I fold the paper I'm doubling the thickness, and that this can be expressed simply as 2^n. As I'm folding it 50 times, the expression is simply the original thickness times 2^{50}, or 0.01×2^{50} mm. This is where I use my calculator. I discover that

$$2^{50} = 10^{15}.$$

Therefore, I conclude that the eventual height of the column of paper will be

$$0.01 \times 10^{15} \text{ mm}.$$

I don't really have a feel for this number, so let me convert it to meters:

$$(0.01 \times 10^{15} \cancel{\text{mm}})(1 \cancel{\text{cm}}/10 \cancel{\text{mm}})(1 \text{ m}/10^2 \cancel{\text{cm}}) = [(0.01 \times 10^{15})/(10)(10^2)]\text{m} = 0.01 \times 10^{(15-3)} \text{ m}$$

$$= 0.01 \times 10^{12} \text{ m} = 10^{10} \text{ m}.$$

What does this number mean? It seems quite large. Let's convert it into kilometers:

$$10^{10} \text{ m} = (10^{10} \cancel{\text{m}})(1 \text{ km}/10^3 \cancel{\text{m}}) = 10^{10-3} \text{ km} = 10^7 \text{ km}.$$

This is 10,000,000 kilometers! One sheet of paper 0.01 millimeter thick folded on itself merely 50 times creates a pile that is 10,000,000 kilometers high. That is 6,210,000 miles! The Moon's distance from the Earth averages about 348,000 kilometers, and our pile of paper is two orders of magnitude higher.

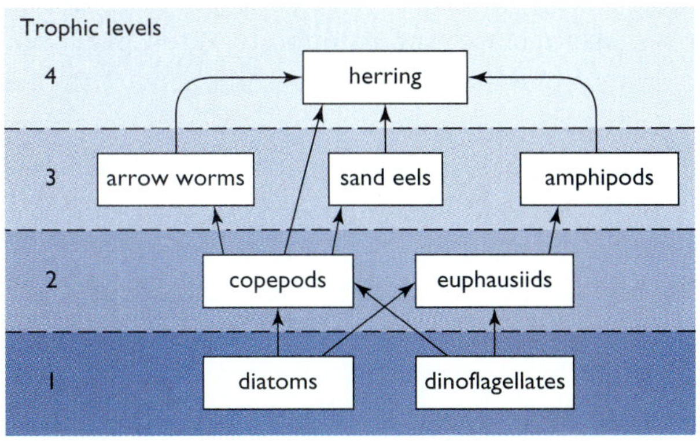

(b) FOOD WEB

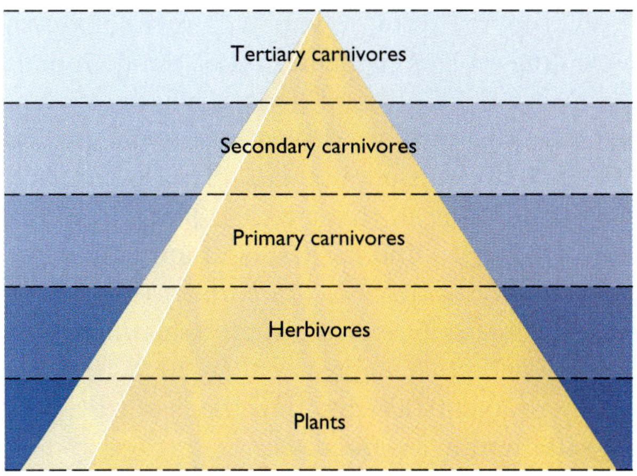

(c) ENERGY PYRAMID

part below it. This shows how large animals located high in the food chain depend on smaller animals and plants lower in the food chain.

PLANTS AS PRIMARY PRODUCERS

As noted previously, plants require sunlight, nutrients, water, and carbon dioxide for photosynthesis. Limited supplies of the first two—sunlight and nutrients—are the principal factors that limit primary production in the ocean. The remaining two requirements—water and carbon dioxide—are abundant in the ocean and do not limit plant growth.

Photosynthesis involves a series of chemical reactions that can be summarized as follows.

$$\underset{\text{(GAS)}}{6CO_2} + \underset{\text{(WATER)}}{6H_2O} + \text{solar energy} \rightarrow \underset{\text{(SUGAR)}}{C_6H_{12}O_6} + \underset{\text{(GAS)}}{6O_2}$$

This shows that photosynthesis results in the synthesis of food (in this case $C_6H_{12}O_6$, the simple sugar glucose) from water and carbon dioxide, and in the release of oxygen as a by-product of the chemical reaction. Note that *inorganic* compounds—water and carbon dioxide—have been converted into an *organic* molecule—sugar—with its store of chemical energy. In addition to carbohydrates such as glucose, plants manufacture other organic compounds, including proteins, fats (lipids), and nucleic acids, such as DNA and RNA.

Plankton blooms occur whenever and wherever light and nutrients abound (see Figure 9–18). During a bloom, rapid cell division causes diatom populations to increase dramatically and rapidly, discoloring the water to yellow, brown, or green, depending on the pigmentation of the dominant species. Under such conditions, a dense growth of plant cells is evident that at times can exceed 1,000,000 diatoms per liter (10^6 diatoms per liter) of seawater. (Recall that a liter is just a bit more than a quart!) Dinoflagellates, another diverse group of phytoplankton, can surpass this rate of cell division under optimum growth conditions, creating cell clusters greater than 20×10^6 cells per liter (20,000,000 dinoflagellates per liter) of seawater within several days!

ANIMALS AS CONSUMERS

It is very difficult to make generalizations about the trophic relationships of marine animals, because their ecological habits are so diverse. We can say, however, that all animals are heterotrophs; they satisfy their nutritional needs by eating preexisting food. This complex process is described chemically by the following formula,

$$\underset{\text{(SUGAR)}}{C_6H_{12}O_6} + \underset{\text{(GAS)}}{6O_2} \rightarrow \underset{\text{(GAS)}}{6CO_2} + \underset{\text{(WATER)}}{6H_2O} + \text{energy.}$$

Look familiar? This reaction, called **respiration,** is the reverse of plant photosynthesis. In this case, animals eat food (the sugar glucose) and combine it with oxygen (glucose is oxidized) to yield carbon dioxide and water. Most importantly, chemical energy is released by the oxidation reaction, and this energy is what animals use to fuel their metabolic processes (maintain themselves) in order to grow, move about, and reproduce. Note that plants also respire in order to obtain energy to live their lives.

Marine heterotrophs range in size from the unicellular forms of zooplankton to the immense bodies of whales. There is a positive correlation between body size of animals and their trophic position in a food web. Typically, herbivorous zooplankton such as copepods and forams are small, because their food, the phytoplankton, is small. (Compare this with cows on land; they are large herbivores, because the grasses they feed on are large in size.) Proceeding to higher trophic levels in the food web, the body size and the weight of animals increase proportionately. A few exceptions to this trend are notable. The blue whale, the largest animal ever to evolve since life first appeared on Earth, feeds mainly on krill, small herbivorous crustaceans and, despite its immense bulk, is a secondary consumer.

Animals in the ocean feed in one of five basic ways. Herbivores are **grazers** that consume plants directly. Many species of copepods are grazers that feed on diatoms, dinoflagellates and other phytoplankters. Some grazers, however, are much larger, such as snails (**Figure 10–3a**) and sea urchins that

(a) A HERBIVOROUS SNAIL

(b) SEA STAR

(c) BARRACUDA

(d) CRAB

(e) BARNACLES

(f) MOON SNAIL

FIGURE 10-3

Feeding strategies. (a) An herbivorous snail grazes on large kelp. (b) A sea star pulls apart the valves of a clam and devours its soft tissue. (c) A large barracuda hunts members of a school of small fish. (d) A ghost crab feeding on sea urchin. (e) Feather duster worms obtain food by filtering bits of organic detritus out of the water. (f) Moon snail feeding on algae from seaweed.

THE OCEAN SCIENCES BIOLOGY

Large Sharks

Marine *vertebrates* (reptiles, fish, birds, and mammals), animals that belong to the phylum Chordata (see Table 9–3 in Chapter 9), tend to be large. Fish are the most widely distributed vertebrates of the ocean, found in surface and deep water, in very warm and very cold water. Fish are divided into two major groups, depending on whether they have bone or cartilage for a skeleton. The more familiar seafood offerings are bony fish, such as tuna, cod, halibut, swordfish. Sharks have cartilaginous skeletons and rank among the largest animals in the sea.

There are almost 300 species of shark, each adapted in some particular way to its environment. Many sharks are skillful hunters and possess teeth and jaws designed for biting and tearing the flesh of their prey. Their rows of serrated teeth are very sharp, able to slash or bite off large hunks of meat from the body of prey. These fish also possess remarkable eyesight and well-developed senses of smell, hearing, and touch. For example, experiments have shown that some sharks are able to detect certain chemicals that are as dilute as one part per billion. They also possess organs arranged along the sides of their bodies that can sense tiny vibrations in the water and thus track potential prey, particularly if the animal has been seriously injured. Of course, sharks also are scavengers, devouring prey that has been killed or wounded by others.

All sharks act instinctually and, like large carnivores on land—grizzly bears, large cats, wolves—they can be dangerous to humans and should always be treated with respect and caution if encountered in the water. Shark attacks in U.S. waters are uncommon. Most have occurred in northern California along rocky stretches of shoreline that support large populations of seals and sea lions. Presumably surfers, when lying prone on their surfboards, resemble seals. About fifteen shark attacks have been reported for this area since the early 1970s.

Because of popularization in books and films, the great white sharks (**Figure B10–1a**) are pictured as ruthless devourers of human flesh, the foremost man-eaters of the deeps. They are gigantic fish, reaching lengths of over 7 meters (~23 feet) and weighing almost 1,400 kilograms (~1.5 tons). Like any large carnivore, great whites hunt instinctively for large prey and they have attacked and devoured people. Great whites, however, are dwarfed by the whale shark (**Figure B10–1b**), which grows to a length longer than 17 meters (~56 feet) and to a weight over 40,000 kilograms (88,400 pounds)! However, unlike the great whites, which are dangerous to humans, whale sharks are docile and harmless. They feed on plankton, using gill rakers in their enormous mouths to filter it out from the water.

Visit www.jbpub.com/oceanlink for more information.

feed on large seaweed (see the boxed feature, "Ecology of the Giant Kelp Community" in the preceding chapter). Carnivores are **predators**, which typically hunt animals that are smaller than they are (**Figures 10–3b** and **10–3c** and the boxed feature, "Large Sharks"). **Scavengers**, among which are many benthic invertebrates, such as crabs and snails, search for dead organic matter (**Figure 10–3d**). **Filter feeders** obtain food by filtering suspended particles of living and dead organic matter out of the water that flows by them. Common filter feeders include barnacles (**Figure 10–3e**), mussels, oysters, and even some infauna, such as polychaete worms and clams, that live in burrows and create water currents to sweep bits of food toward their mouths. **Deposit feeders** obtain food either nonselectively, by ingesting (swallowing) sediment and extracting food particles that are mixed into the mud (**Figure 10–3f**), or selectively, by searching for specific bits of food contained in the sediment.

The size of the population of any organism is usually a direct function of its food supply. The number and growth rate of animals increase when food is abundant, and decrease when food is limited. Typically there is a time lag between the occurrence of peak food abundance and the maximum size of the animal population that is exploiting this food resource. Usually the relationship between predator and prey fluctuates in a regular cycle (**Figure 10–4**). As predators feed on a large population of animals, the number of predators increases as a consequence

FIGURE **B10-1**

Sharks. (a) The great white shark is an efficient and effective hunter. It is a quick, powerful fish with a jaw and teeth designed for biting and tearing flesh. (b) The whale shark is almost twice as large as the great white shark, but it is a docile, harmless plankton feeder.

(a) GREAT WHITE SHARK

(b) WHALE SHARK

of the ample supply of food. In time, the number of prey is reduced by the increasing predation pressure, and the reduction in prey causes mass starvation among the predator population. Their numbers increase only after the prey species resurges once again, initiating a new cycle.

BACTERIA AS DECOMPOSERS

Bacteria—a diverse group of microscopic, unicellular organisms—were once believed to be rare in the ocean. They are not. They thrive in the sea on the surfaces of just about everything, including dead organic matter, living organisms, and sediment grains in the water column or on the sea bottom.

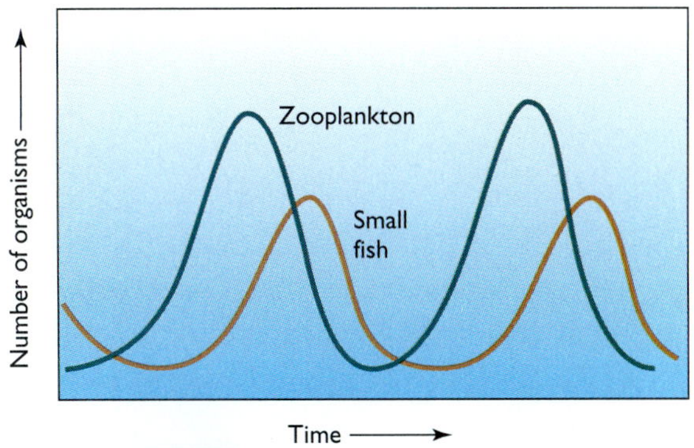

FIGURE **10-4**

Prey-predator relationships. The size of a population of small fish, the predators in this model, fluctuates in direct response to the abundance of its prey, zooplankton. When food is abundant, the population increases. The increase in predation pressure then reduces the zooplankton population, and as the food supply decreases, so does the population size of the fish. Hence, predation pressure also declines, and allows the zooplankton to rebound and their numbers to increase.

Some species are free floating and are found suspended in the water column, commonly in concentrations of 10^6 to 10^8 microbes per liter of seawater. Many microbes, including bacteria, serve as food for certain species of zooplankton, including copepods, ciliates, and flagellates. They consume bacteria in two principal ways. The most common method is by feeding directly on the bacterial cells themselves. The second method is by ingesting mud particles and their microbial populations.

Bacteria perform a crucial trophic function in all ecosystems: they decompose dead tissue, thereby releasing essential inorganic nutrients into the water for recycling by plants (**Figure 10–5a**). Without decomposition, nutrients would remain chemically bound in the tissue of dead plants and animals and would be lost to the ecosystem as they became buried in sediment. The outcome of this loss would be a tragic one indeed. As nutrients became exhausted, the autotrophs would not be able to photosynthesize, and the entire ecosystem would collapse in short order.

Bacteria secrete a wide assortment of powerful enzymes that react chemically with organic matter. The most easily degradable chemical compounds are lipids (fats), sugars (carbohydrates), and protein. The most difficult to degrade are (1) cellulose, a carbohydrate common in the cell walls of

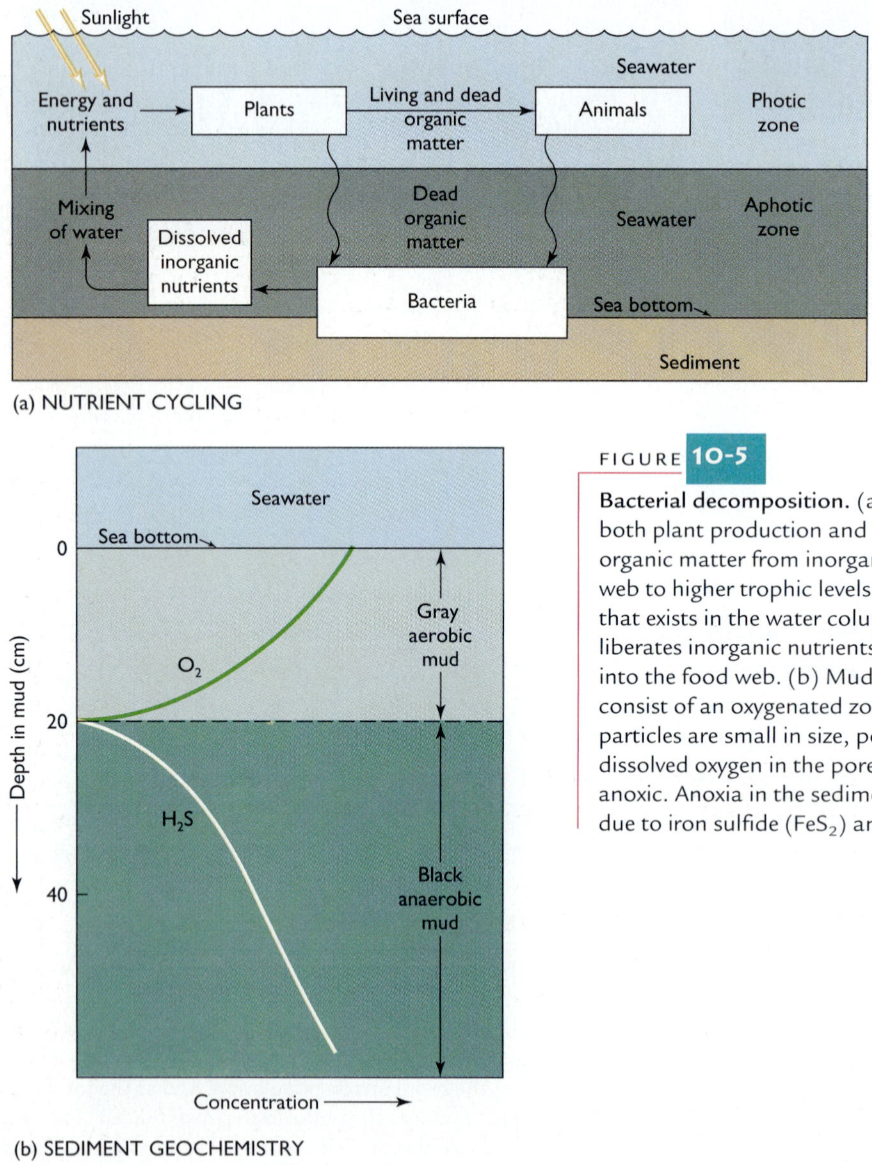

FIGURE **10-5**

Bacterial decomposition. (a) The food cycle in an ecosystem relies on both plant production and bacterial decomposition. Plants manufacture organic matter from inorganic nutrients. This matter is passed up the food web to higher trophic levels. Bacteria decompose dead organic matter that exists in the water column, on the sea bed, and in the sediment. This liberates inorganic nutrients, which are then recycled by the plants back into the food web. (b) Mud deposits tend to be zoned chemically and consist of an oxygenated zone that overlies an anoxic zone. Because mud particles are small in size, pore water is trapped between the grains. The dissolved oxygen in the pore water is used up by bacteria, making the mud anoxic. Anoxia in the sediment is indicated by the black color of the mud, due to iron sulfide (FeS_2) and the presence of hydrogen sulfide (H_2S).

plants and wood, (2) chitin, a celluloselike substance that forms the exoskeleton (an outer skeleton) of insects, crustaceans, and other arthropods, (3) cartilage, and (4) bone. The by-products of bacterial decay are left behind in the environment as food for other animals, and released as dissolved nutrients and gases for plant uptake.

Very few species of bacteria are capable of degrading dead matter entirely by themselves. Rather, various species work in concert or in succession in order to break down the wide assortment of organic molecules in dead matter. An example of an important chemical reaction conducted by bacteria is

$$NH_3 + 2O_2 \rightarrow H^+ + NO_3^- + H_2O.$$

In this simple reaction, ammonia (NH_3—a nitrogen-bearing compound), which is either secreted directly into the water by zooplankton (usually in urine) or produced by bacteria themselves as they decompose organic matter, is oxidized into nitrate (NO_3^-). Nitrate is a vital nutrient for plant photosynthesis. Plants require nitrogen for growth, but they cannot use the nitrogen that is bound to the ammonia molecule. Rather, they require nitrate, which is provided largely by bacteria. Hence, the role of bacteria is as vital as that of the primary producing plants in the existence of an ecosystem over time. In fact, bacterial decomposition is the starting point (because this activity provides the essential nutrients for plant photosynthesis) and the end point (decay of organic matter) of the food cycle, providing the critical link between nonliving and living matter.

All of the chemical reactions that we have discussed involve **aerobic bacteria,** which require oxygen in order to respire and decompose dead matter. However, decay does occur in the absence of free oxygen, a condition termed **anoxia.** This activity is conducted in oxygen-depleted environments by **anaerobic bacteria.** In the absence of free oxygen, anaerobes separate oxygen from various compounds, such as the sulfate ion (SO_4^{2-}), in order to respire and decompose organic matter. The reaction for the sulfate ion is

$$SO_4^{2-} \rightarrow 2O_2 + S^{2-}.$$

Note that the sulfate reaction above liberates free oxygen (which the bacterium uses to respire) and sulfide (S^{2-}). At the pH of normal seawater, the sulfide ion combines readily with two hydrogen ions to produce hydrogen sulfide gas by the reaction

$$2H^+ + S^{2-} \rightarrow H_2S.$$

Anaerobic conditions can develop in the water that is trapped in the pores of sediment (minute spaces between the sediment grains), particularly mud (**Figure 10–5b**). This can be confirmed easily by digging into the substrate of a coastal salt marsh and noting the black coloration (due to the high organic content and the presence of iron sulfide) of the mud and the pungent rotten-egg odor of hydrogen sulfide, both of which are clear signs of anoxia.

Most bacteria are heterotrophs and obtain energy from decomposing organic matter. Some—the **cyanobacteria,** or blue-green algae—are photosynthetic and use sunlight directly to manufacture food from dissolved nutrients. Other bacteria are **chemosynthetic** instead; in total darkness, they use the chemical energy released by the oxidation of inorganic compounds to produce food. This very biochemical process—discussed in the boxed feature, "Volcanic Vent Communities"—occurs in the blackness of the deep sea at hydrothermal vents scattered along segments of spreading ocean ridges. The vent communities at such sites owe their existence to the chemosynthetic activity of countless bacteria that oxidize sulfides that seep out of volcanic vents. The base of this food web is occupied not by plants, but by microbes, because there is no sunlight in that deep water!

FOOD CHAINS AND ENERGY TRANSFER

Some chemical energy is passed upward in a stepwise fashion to the higher trophic levels of food webs. This energy is used by animals to grow and to reproduce. The result is an increase in the biomass of a region, which is the quantity of living matter expressed either as grams per unit volume of water (g/l) or as grams per unit area (g/m^2) of the sea surface or sea bed. A generalized food chain in the ocean is

$$\text{phytoplankton} \rightarrow \text{zooplankton} \rightarrow \text{nekton}.$$

Some striking characteristics are evident at each step of the food chain. For example, as we proceed up the chain, the organisms become larger

Volcanic Vent Communities

The Galapagos Ridge is an arm of the spreading ocean-ridge system of the equatorial Pacific Ocean (**Figure B10–2a**). Here seawater is being heated as it circulates underground in the fractured basaltic rock. The hot fluids escape from the bottom through hydrothermal (literally, hot water) vents (**Figure B10–2b**). Geochemists studying the chemistry of the escaping hydrothermal fluids used the submersible Alvin to make their observations and to get their samples. To their astonishment, the geologists discovered small clusters of dense benthic communities surrounding some of the active vents. These were literally oases of life in a desertlike environment where few organisms are able to live because of the lack of food. Moreover, some of the organisms, including tube worms, mussels, and clams, were gigantic in size. How could these giants be living in this dark environment without phytoplankton or much other food?

Later cruises to the area by biologists confirmed that these benthic communities were unique, composed in some cases of new species of invertebrates. Since these 1979 expeditions, vent communities have been discovered elsewhere on the midocean ridges, including the East Pacific Rise, the Juan de Fuca Ridge, ridge segments in the Gulf of California, and the Mid-Atlantic Ridge (see Figure B10–2a).

Geologists infer that hot magma (molten rock) in a shallow chamber,

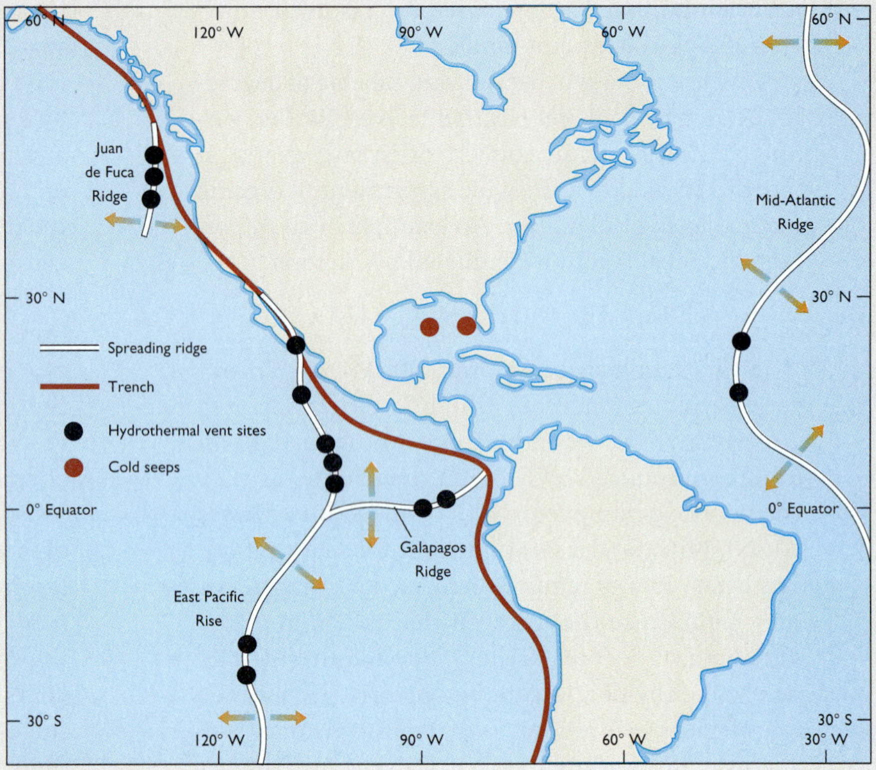

(a) DISTRIBUTION OF VENTS

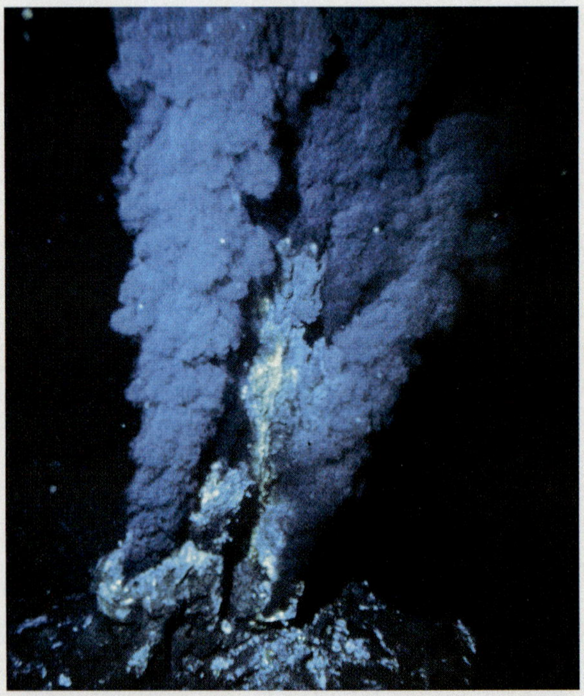

(b) BLACK SMOKER

FIGURE **B10-2**

Volcanic vent sites. (a) Vent sites have been located in both the Pacific and Atlantic Oceans. With the exception of cold seeps in the Gulf of Mexico, all the vents shown discharge fluid that is hotter than seawater. (b) The sulfides spewing out of this vent discolor the water and give the "black smoker" its name.

perhaps 1 to 3 kilometers (~0.6 to 1.9 miles) below the sea floor, heats seawater that fills the fractures and cracks in the basalt crust. Heating raises the temperature of the subterranean water, causing it to convect upward to the sea bottom, where it escapes as a submarine spring (**Figure B10–3a**). The exiting hydrothermal fluids have high concentrations of metals that they dissolved from the basalt. As the water seeps out of the bottom and is cooled, various metal sulfide compounds are precipitated close to the vent. A buildup of metallic sulfides around the exit hole results in the construction of a tall chimney or spire, with a diameter of several meters (several yards) and a height of about 20 meters (~66 feet) (**Figure B10–3b**). Many such geologic features have been extensively studied on land in similar volcanic settings.

What is truly surprising are the associated vent communities, consisting mainly of filter feeders. How do these numerous types of heterotrophs feed themselves? There is no evidence of plants to support a food web. The clustering of the communities around active hydrothermal vents suggests a cause-and-effect relationship. Water extruded from vents has been found to contain large numbers of anaerobic bacteria, ranging from 10^5 to 10^9 cells per milliliter of water (a milliliter equals abut 10^{-3} quarts) (**Figure B10–4a**). These particular species of bacteria were discovered to be autotrophs, producing food from inorganic nutrients by **chemosynthesis**. Unlike true plants, which obtain energy for photosynthesis from sunlight, chemosynthetic

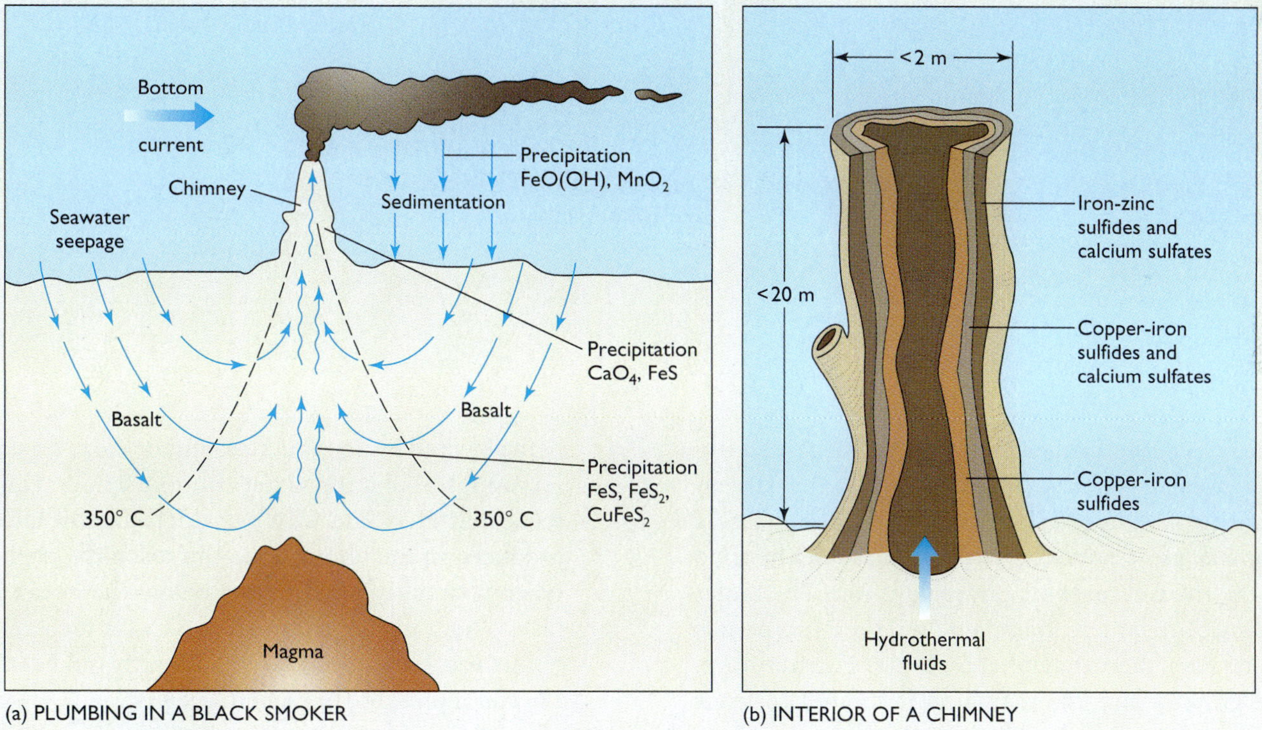

FIGURE **B10-3**

Volcanic vents. (a) Heat conducted from a shallow magma chamber causes the convective flow of seawater through fractured and faulted basalt. From the basalts the heated seawater leaches metals that are discharged with the hot fluid from a chimney. The fluids cool and chemical reactions precipitate sulfide and sulfate minerals. [Adapted from H. W. Jannasch and M. J. Mottle, *Science* 229 (1985): 717–25.] (b) Mineral precipitation causes chimneys to grow upward, sometimes to heights of more than 20 meters. In cross section, chimneys are layered with distinctive bands of metal sulfides and sulfates. [Adapted from R. M. Haymon and K. C. Macdonald, *American Scientist* 73, no. 5 (1985): 441–49.]

organisms oxidize the gas H_2S (hydrogen sulfide) to form sulfur compounds. In the process, chemical energy is released and used by the bacteria to synthesize food. Therefore, the base of the food webs in deep-sea vent communities is occupied by chemosynthetic bacteria rather than by photosynthesizing plants! These sulfur-oxidizing bacteria thrive in the vent itself. Dense aggregates of bacteria are torn from the interior rock face of the chimney by the flowing hydrothermal water and are swept in suspension to the vent outlet, where filter-feeding invertebrates (**Figure B10–4b**) catch and consume them. The invertebrates arrived at the vents as larvae transported by bottom currents. When circulation of the hydrothermal water stops in a vent, bacteria no longer receive a supply of H_2S, and so they die—undermining the food web and causing the death of the entire vent community.

Visit www.jbpub.com/oceanlink for more information.

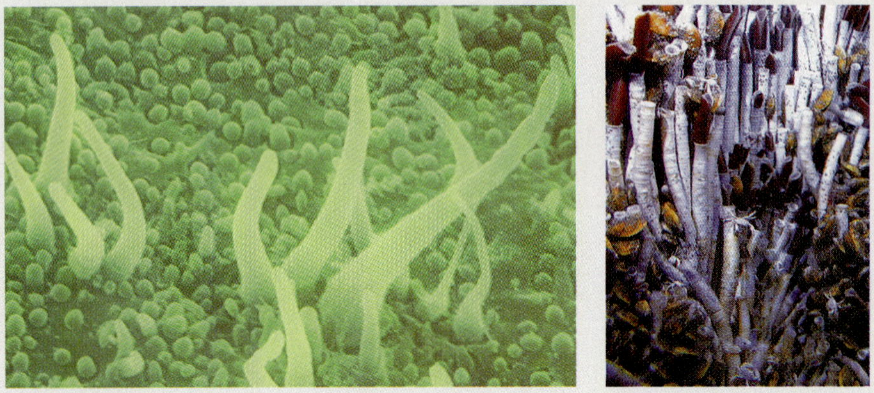

(a) BACTERIA

1 mm

(b) TUBE WORMS

FIGURE B10-4

Volcanic vent communities. (a) This scanning electron micrograph shows a bacterial mat living on the shell of a mussel. (b) Tube worms grow near a vent on the Galapagos Ridge in the eastern Pacific Ocean.

and fewer in number, and the biomass usually decreases markedly. In other words, the largest number of individual organisms and the greatest biomass are associated with the base of the food chain, the microscopic phytoplankton. The larger animals are far less abundant, and their biomass is much less than that of the plants. Furthermore, the growth rate of organisms is related to their position in the chain. The smaller organisms low in the food chain are able to double their biomass at much faster rates than are the larger animals that occupy higher trophic levels.

The classic trophic relationship is the **grazing food chain,** whereby herbivores graze plants. Another trophic interaction is the **detritus food chain,** whereby nonliving organic matter (detritus), such as dead cells and tissues and fecal matter, forms the base of the food chain. For example, after being ejected by zooplankton and nekton, fecal pellets sink through the water column. The settling of fecal matter is one way by which organic matter produced in the photic zone can reach the deep-sea bottom thousands of meters below the ocean surface. There it is consumed by deposit feeders and enters the detritus food web of the abyssal benthos. We can represent this schematically.

Grazing Food Chain

Phytoplankton → Zooplankton → Nekton

(Phytoplankton, Zooplankton and Nekton each → Detritus)

Detritus → *Deposit Feeder* → *Nekton*

Detritus Food Chain

Bear in mind though that the detritus food chain ultimately is connected to plant production, because the organic detritus is derived either from plants or from animals that fed on plants directly (herbivores) or indirectly (carnivores).

The transfer of energy from one trophic level to the next in a food web is not an efficient process. This explains the drastic decrease of biomass from the base to the top of the chain. As we know, plant photosynthesis stores chemical energy in food. When plants are grazed by herbivores, most of the energy released by digesting food is expended as *kinetic energy* (energy of motion) or is used in the manufacturing of nonnutritional tissue, such as shell, bone, chitin, scales, and the like. Only the small remaining fraction of the assimilated energy from the food is used to increase mass, either by growth of the individual organism or by reproduction (**Figure 10–6**). The same relationship exists at each step of the food chain.

Studies indicate that the efficiency of energy transfer at each level of a food chain ranges between 3 and 23 percent, and averages about 10 percent. Using a 10 percent (10% = 0.10) transfer efficiency, it becomes apparent that

100,000,000 g diatoms →
10,000,000 g copepods →
1,000,000 g small fish →
100,000 g large fish → 10,000 g human

A 10,000-gram human (10 kilograms or ~22 pounds) is the size of a small child. Obviously, the longer the food chain, the greater is the amount of plant biomass needed to supply the nutritional needs of the large organisms at the top of the food chain. In the preceding example, 100,000,000 grams of plants are required to sustain a 10,000-gram human. If we caught copepods instead of large fish, then in theory 100,000,000 grams of diatoms would support 1,000,000 grams (~2,200 pounds) of humans.

Given this, you might wonder why large fishes don't feed on zooplankton or why humans don't hunt small fish or harvest zooplankton. The answer is quite simple: they would expend more energy in catching the smaller prey than they would gain by consuming it. The result of this energy deficiency would shortly lead to death by starvation. Think about the amount of energy it

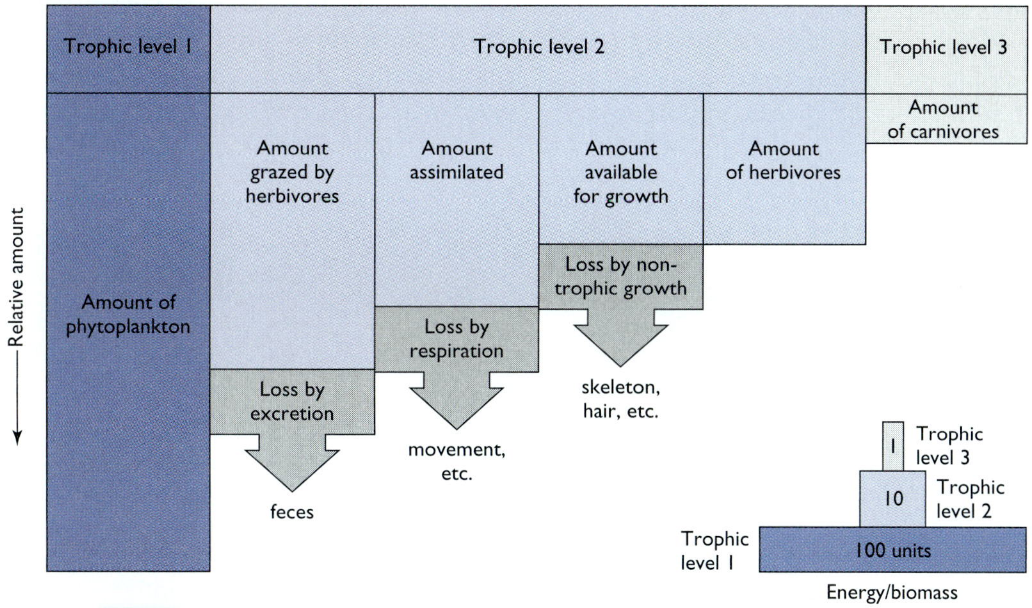

FIGURE **10–6**

Energy transfer between trophic levels. This diagram accounts for the energy/biomass distribution shown as an inset in the lower righthand corner of this figure. The biomass of plants is much greater than the biomass of herbivores; herbivores, in turn, are more abundant than carnivores. The reason for the decrease in biomass and number of organisms at each step of the food chain is the inefficient transfer of energy between trophic levels. Of the total amount of plant cells grazed by a herbivore, only a small fraction (about 10 percent) is used for growth. Consequently, the herbivore biomass is smaller than the plant biomass by a factor of about 10. The same inefficient transfer of energy occurs between higher trophic levels as well. [Adapted from W. D. Russell-Hunter, *Aquatic Productivity* (New York: Macmillan, 1979).]

would take you to gather 1,000 grams of microscopic plant cells as opposed to catching one fish that weighed 1,000 grams.

10-2

General Marine Productivity

Given this basic understanding of food chains and food webs, and the nutritional linkages between the different levels, we can now discuss the factors that limit primary production in different parts of the ocean. Also, we will review global variations in animal production and an assessment of fish production and its implications for the world's fisheries.

PRIMARY PRODUCTION IN THE OCEAN

As you know, plant photosynthesis converts the inorganic carbon of carbon dioxide into the organic carbon of sugar. Primary productivity, then, can be defined as the total quantity of carbon fixed by plants, expressed as grams of carbon fixed per square meter of sea surface per unit of time ($gC/m^2/yr$). If plants abound in the water of an area, so will animals. Conversely, the absence of plants means a scarcity of animals.

We begin our appraisal of ocean productivity by evaluating the factors that impose a limit on plant growth. There are primary and secondary limiting factors.

Primary Factors	Secondary Factors
Solar radiation	Upwelling and turbulence
Nutrient concentrations	Grazing intensity
	Water turbidity

Although these factors are discussed separately, many of them are interrelated, meaning that change in one brings about change in another.

SOLAR RADIATION

Numerous studies indicate that phytoplankton can photosynthetically transform no more than 0.1 to 0.2 percent (0.001 to 0.002) of the available solar energy into chemical energy—an incredibly small amount! As light penetrates the water surface, passing from air into water, its intensity falls off rapidly with depth as a result of absorption and scattering (**Figure 10–7a**). **Net primary productivity,** the amount of carbon fixed by photosynthesis that exceeds the respiration demands of the plant and goes into growth, occurs in the water column down to the **compensation depth,** below which there is no net productivity. Roughly speaking, the compensation depth, which should not be confused with the *calcium carbonate compensation depth* (CCD), occurs where light levels are reduced

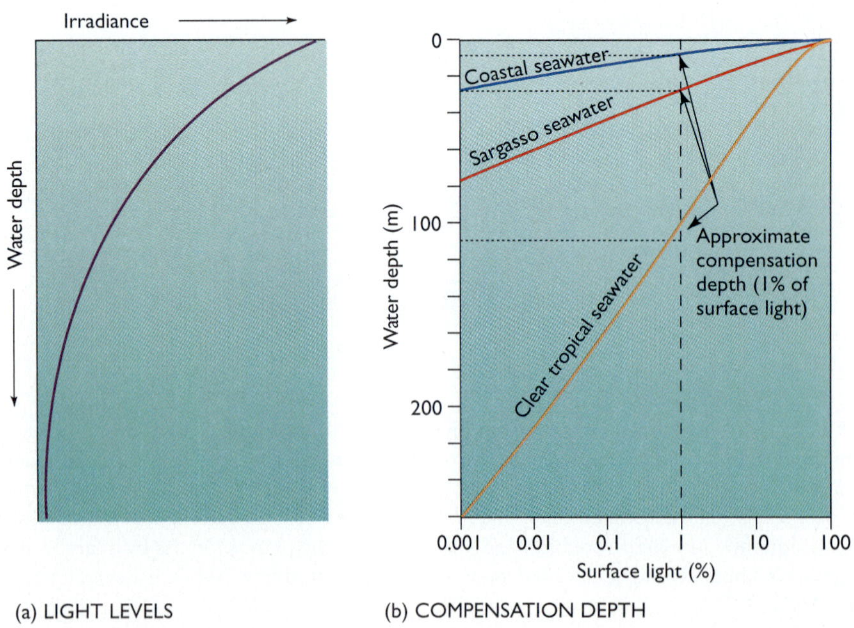

FIGURE **10-7**

Light absorption by seawater. (a) As a result of absorption and scattering, light levels decrease sharply with depth. (b) The compensation depth is the point at which the photosynthetic production by plants exceeds their respiratory needs. It occurs at a water depth where light intensity is reduced to about 1 percent of its surface value. In turbid coastal water, the compensation depth is typically shallower than 20 meters. In clear water it can be as deep as 110 meters. [Adapted from G. L. Clarke and E. J. Denton, "The Sea," *Physical Oceanography*, vol. 1, M. M. Hill, ed. (New York: John Wiley, 1962), 456–68.]

to about 1 percent of their surface value. This means that the compensation depth will be deeper in the clear, transparent water of the open ocean—about 110 meters (~363 feet) than in the turbid, mud-laden water of estuaries—less than 15 meters (< ~50 feet) (**Figure 10–7b**).

Much of the incoming solar radiation is used to heat water. Water temperature influences the photosynthetic activity of plants. Generally, net primary production will be negligible in water with temperatures of less than 0°C and more than 40°C. Within those limits, the rate of chemical reactions in the plant cell increases as water temperature increases. Because the quantity of solar radiation that reaches the sea surface decreases with increasing latitude, in theory primary activity *should* diminish from the equator to the poles. In fact, just the opposite is true, an effect of other factors that regulate primary productivity, notably the concentration of critical nutrients.

NUTRIENT CONCENTRATIONS

Plants require a steady supply of essential nutrients to sustain food production. **Macronutrients,** which are compounds that contain phosphorus, nitrogen, and silicon, are required in large doses, and it is usually these elements that limit the growth of plants on a regional scale. **Micronutrients** are chemical substances that are indispensable to plant life but are needed only in very small doses. Micronutrients are quite varied in type and include iron, copper, manganese, zinc, boron, and cobalt, among others. Inadequate concentrations of micronutrients usually limit biological production on a local scale.

Numerous studies on the uptake of nutrients indicate that phytoplankton extract carbon, nitrogen, and phosphorus from seawater in the proportion of about 116:16:1. This means that a plant while photosynthesizing requires 116 times more carbon than phosphorus, and 16 times more nitrogen than phosphorus. At first glance, it would seem that carbon, which is required in such large doses, would be the limiting macronutrient. Yet it is not, because bicarbonate (HCO_3^-), the source of the carbon, abounds in seawater in concentrations that are more than adequate to sustain even the largest plankton blooms. Nitrogen and phosphorus, even though they are needed in much smaller quantities than carbon, are the nutrients that typically limit primary productivity of ecosystems. Also, silicon can be a limiting macronutrient for diatoms and silicoflagellates (flagellates with silica shells), as they require large amounts of dissolved silicon for the construction of their frustules (shells).

Nitrogen in seawater occurs in the form of molecular nitrogen (N_2) and compounds, such as nitrate (NO_3^-), nitrite (NO_2^{2-}), and ammonia (NH_3). Most algae are unable to use molecular nitrogen and ammonia, so they rely on nitrate and nitrite for their supply of nitrogen. Dissolved inorganic phosphate (PO_4^{2-}) is the principal source of phosphorus for phytoplankton. During plankton blooms, uptake of these essential macronutrients reduces their concentrations in seawater to such low levels that supplies are no longer adequate to meet the metabolic requirements of the growing and multiplying plant cells. This results in the death of the plants, which then sink, removing these assimilated nutrients (they are now part of the organic makeup of the plant) from the photic zone. As more and more plants die and sink, the water of the photic zone loses its supply of essential nutrients. Therefore, the capacity for sustained primary plant production in most regions of the ocean depends on the renewal rate of macronutrients in the photic zone.

Recall that nutrients are regenerated by the bacterial decomposition of dead cells and tissue and of fecal matter. Nitrogen renewal by bacterial decay takes longer to complete than does phosphorus regeneration. The former is a three-step procedure involving three distinct bacterial species. The latter, in contrast, is a one-step reaction involving a single species of bacteria. Hence, nitrogen, which is required in much greater quantities than phosphorus (remember the N:P ratio of 16:1) and which takes much longer to regenerate by the bacterial decay of dead matter, is usually the limiting nutrient in marine habitats.

Because organisms sink when dead, decomposition and renewal of nutrients occur below the photic zone. This implies that a means of transporting the dissolved nutrients back to the sunlit photic zone for use by phytoplankton must exist in order for the sunlit water to remain fertile and support primary productivity. This leads us to a discussion of upwelling and turbulence of water.

UPWELLING AND TURBULENCE

The most productive areas of the ocean are characterized by **upwelling**—the slow, persistent rising (a few meters or yards per day) of nutrient-rich water toward the ocean surface. This vertical flow of water is produced in two principal ways. First, there is equatorial upwelling in the low latitudes. Ekman transport (see Figures 6–6 and 6–7 in Chapter 6) causes water to move away from the equator, because Coriolis deflection is to the right in the Northern Hemisphere and to the left in the Southern Hemisphere (**Figure 10–8a**). Surface water diverging from the equator is replenished by water rich in dissolved nutrients that upwells from below. Second, there is coastal upwelling. Winds blowing persistently parallel to the edge of a landmass can generate Ekman transport that moves water offshore away from the land. This seaward flow of surface water is compensated by upwelling water from the ocean depths (**Figure 10–8b**). Both of these upwelling mechanisms assure that nutrients regenerated in the ocean depths by bacterial decay are continually injected into the photic zone for plant use. As a result, upwelling areas are some of the most naturally fertile habitats on the Earth.

The upwelling of nutrients is comparable to homeowners in a neighborhood continually fertilizing their lawns during the spring and summer. The result of this effort would be the sustained, rapid growth of grass everywhere in the neighborhood. If the grass clippings are removed from the area and disposed in the local landfill, then the

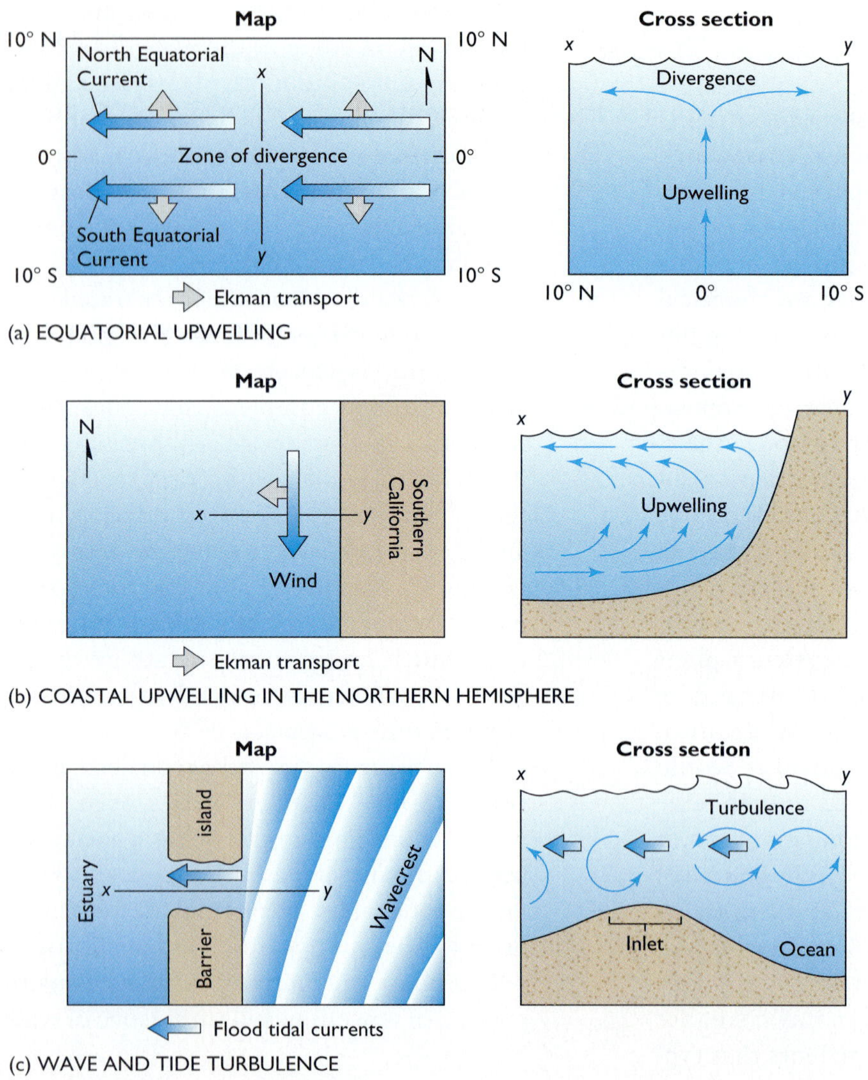

FIGURE 10-8

Upwelling and mixing. (a) Because of the Coriolis effect, Ekman transport is directed poleward on either side of the equator. This current divergence causes the upwelling of nutrient-rich water, which sustains large populations of phytoplankton. (b) Along the eastern sides of ocean basins in both hemispheres, prevailing winds blowing parallel to the shoreline generate Ekman transport away from the coastline. In turn, this induces upwelling and nutrient regeneration of the surface water. (c) Nutrients can be resuspended from the bottom by flow turbulence induced by waves and strong tidal currents. Tidal currents flowing through inlets or over shoals are particularly turbulent, because their speed is increased by these flow constrictions.

lawns must be regularly fertilized with chemicals in order to sustain the rapid growth of the lawn. (As an aside, it makes much more sense and is much less expensive to leave the grass clippings on the lawn as mulch, so that bacteria can decompose the dead grass and return the nutrients to the soil, where they can be reused by the grasses.)

In nearshore and shelf waters, storm waves and strong tidal currents cause a great deal of water **turbulence**—the irregular, chaotic flow of fluids—which mixes the water column (**Figure 10–8c**), so that nutrients from the sea bottom are brought to the surface and made available to plants. Turbulent flow is most pronounced when the sea floor is irregular and the currents are strong. Such conditions exist around submarine bars and shoals, particularly near tidal inlets, where channel constrictions accelerate the flow and thus the turbulence of the water. In such regions, bottom water, with its dissolved nutrient load, is mixed upward into the photic zone.

GRAZING

Many herbivorous zooplankton, like the copepods, time their egg laying to coincide with plankton blooms. As the size of the zooplankton population expands in response to the abundant plant food, the grazing intensity of the voracious copepods can arrest or even reduce the growth rate of the plants, causing the bloom to end. In turn, the reduced amount of plant food causes the large population of copepods to crash, and the plants, provided that nutrients are available, can rebound because the grazing pressure has been reduced.

WATER TURBIDITY

Rivers tend to supply muddy (turbid) water to the adjoining shore and shelf. This **water turbidity** reduces the amount of sunlight that can penetrate the water column and this depresses primary productivity, even if dissolved nutrients are abundant (see Figure 10–7b). As an example, the inshore waters of Georgia (**Figure 10–9**) are discolored brown by large suspended loads of mud supplied in the spring by swollen rivers. Under these conditions, underwater visibility is severely limited, often to less than a foot! Primary production by the phytoplankton at such times is negligible, even in the nutrient-rich surface water, because of the lack of sunlight in the water.

VARIATIONS IN PRODUCTIVITY

The ocean, with its average depth of 4 kilometers (~13,200 feet), represents an unimaginably large volume of water. Only the topmost, illuminated area—the photic zone—is capable of supporting plant production and a dense population of animals. This productive zone is not uniform, however. Marked variations in the physical and chemical makeup of the water affect the size and diversity of the native biological communities. The purpose of this section is to describe and explain variations in

FIGURE 10–9

Turbid water. Rivers flowing into the coastal waters of Georgia tend to have high loads of suspended mud. The turbidity of this water greatly reduces light penetration into the water column, which limits the productivity of these waters despite their very high content of nutrients.

Satellite Oceanography

When a biologist at sea takes a water sample or makes a plankton tow, the measurements are made at a tiny point in a vast ocean. The question then becomes, how representative are these numbers for the study area, which may cover thousands of square miles? Also, a ship takes time to sail between the various sampling stations in a research area. Can one genuinely compare estimates of plankton productivity taken at stations that are days and even weeks apart? Yet what other recourse is there? Biologists have had to assume that the measurements they make at a few sites are representative of the entire region—an unlikely possibility given the characteristic patchy distribution of marine plants and animals in both space and time. Quite clearly, to understand marine productivity, biologists must be able to conduct simultaneous measurements over a broad area of the ocean. The application of remote-sensing technology from space is now providing such data. Let's consider one such system, known as the Coastal Zone Color Scanner (CZCS), that is used to assess variations in the primary productivity of the oceans.

The CZCS is an elaborate, television-like instrument mounted on satellites that detects and collects different bands of visible light—blue, green, yellow, and red (**Figure B10–5**)—emitted from the ocean's surface water layer. The CZCS converts this radiant energy—the visible light—into an electronic signal that is transmitted to ground-receiving stations and displayed visually as a colored map.

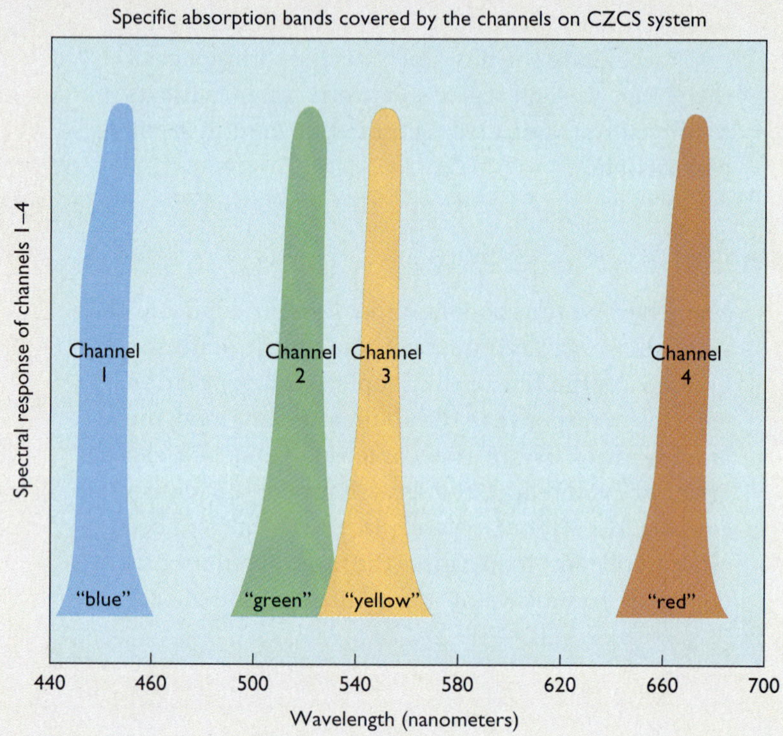

FIGURE **B10-5**

Coastal Zone Color Scanner (CZCS). The CZCS system detects radiant energy from discrete wavelength bands of visible light, emitted from the topmost 2 meters of the ocean. Because these colors largely reflect the concentration of chlorophyll a in the ocean, their relative intensities are an indirect measure of primary productivity.

biological productivity across the globe and over time. Regional patterns of primary productivity are easily obtained from satellite imagery (see the boxed feature, "Satellite Oceanography").

The amount of solar radiation varies with latitude. It also varies at any one location with the seasons. In the tropical climates of the low latitudes, the sun is high overhead and the day is long, providing plants with prolonged doses of high-intensity light. Despite these light levels, however, the phytoplankton biomass of tropical and subtropical seas is so inconsequential that the area turns out to be one of the least productive regions of the ocean (**Figure 10–10a**). In fact, the absence of plant cells explains the clarity and blueness of tropical water. There is little plant matter suspended in the water to scatter and reflect light and produce green and yellow color tones. Tropical oceans are infertile

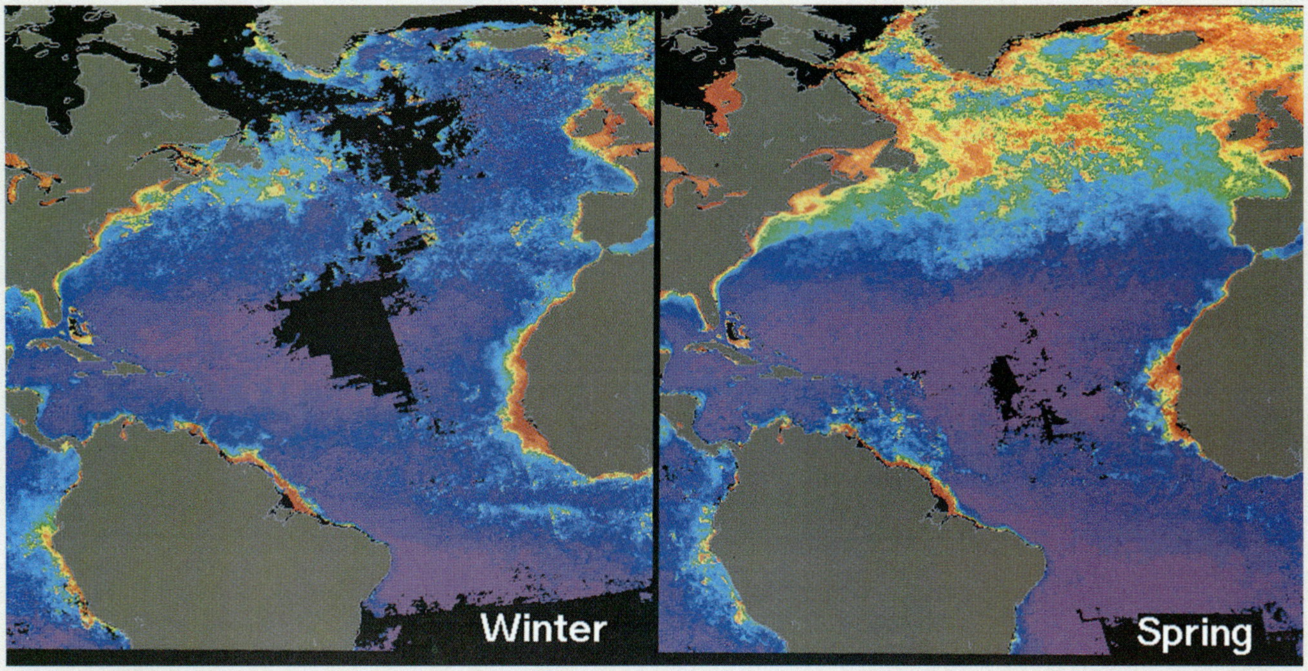

FIGURE B10-6

Satellite images of primary productivity. These two CZCS photographs, taken at different seasons, reveal variations in plant production in time and space. The red hues indicate surface water with high levels of chlorophyll a (high productivity), and the blue hues indicate low levels (low productivity). No data were measured in black areas of ocean.

From a spacecraft altitude of about 1,100 kilometers (~ 660 miles), a swath of ocean area of more than 1,500 kilometers (~930 miles) can be scanned by the CZCS system. After making appropriate corrections, variations in color reflect mainly differences in the concentration of **chlorophyll a**, the principal photosynthetic pigment in plants. This is an indirect measure of the abundance of phytoplankton and, hence, primary productivity. Large concentrations of phytoplankton are indicated on the images by red, yellow, and green colors and sparse concentrations by blue and purple colors. An example of such images is presented in **Figure B10–6**, in which primary productivity for the entire North Atlantic Ocean is compared during the winter and spring seasons.

Visit www.jbpub.com/oceanlink for more information.

because of the thermal structure of the water column, which results from the intense tropical sunlight heating the surface layer. This creates a sharp, permanent thermocline (**Figure 10–10d**) between about 100 and 1,000 meters (~33 to 3,300 feet) below the sea surface. The stable water column inhibits upwelling, and prevents the exchange of warm, nutrient-poor surface water with cold, nutrient-rich deep water below the thermocline. Consequently, nutrient levels in the surface water of tropical oceans are always low.

In summary, tropical oceans are biologically impoverished because of the scarcity of nutrients. There are however, three exceptions to this generalization. Primary production is quite high in (1) coastal upwelling areas along the western edges of landmasses, (2) the narrow but long zones of equatorial upwelling in the Pacific and Atlantic

FIGURE **10–10**

Variations in primary productivity. (a) In the tropics, primary production is low throughout the year because a sharp, permanent thermocline prevents the water column from overturning. This results in poor nutrient renewal in the photic zone, and very low biological productivity. (b) In the North Atlantic Ocean, strong plankton blooms in the spring and weaker plankton blooms in the fall are controlled by seasonal variations in solar radiation and nutrient infusion into the water column. (c) In polar seas, plant production occurs throughout the summer due to high nutrient levels and long daylight periods. (d) In polar seas, the water column is isothermal. In temperate latitudes, the upper water column is stratified during the summer and near isothermal during the winter. In tropical seas, it is permanently stratified due to the presence of a sharp thermocline. [Adapted from A. K. Heinrich, *Journal of Cons. Int. Explor. Mer.* 27 (1962): 15–24.]

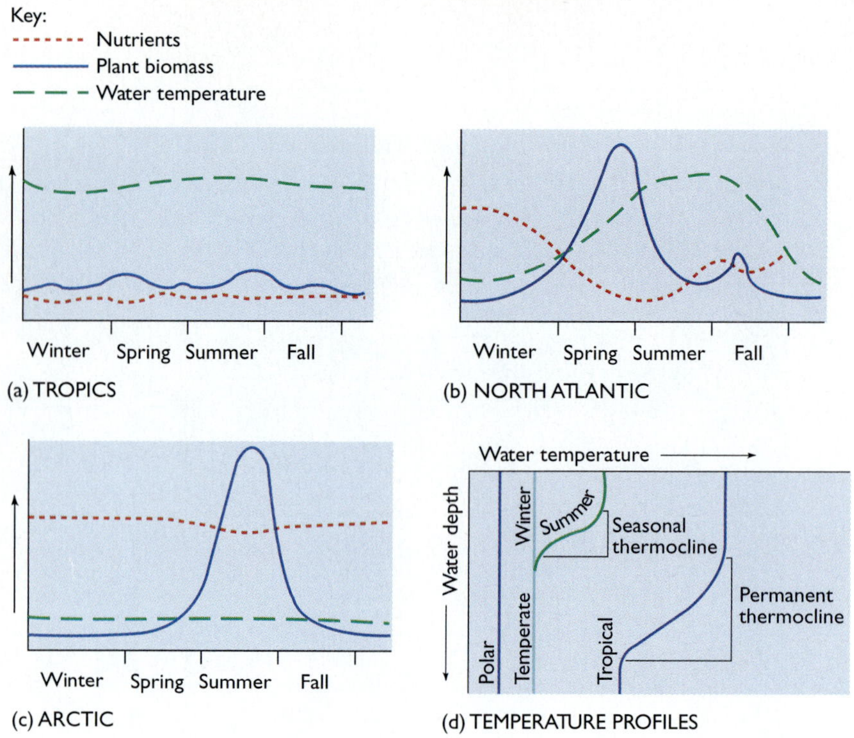

Oceans, and (3) coral reefs. Despite the unusually high fertility of water in these three settings, they are not, contrary to popular belief, representative of the tropical seas, which typically support very low concentrations of life for the reasons stated above.

The results of recent long-term studies of primary production in the subtropical gyre of the North Pacific are causing marine biologists to reconsider their characterization of tropical oceans as biological deserts. It turns out that the annual primary production in this area of the Pacific Ocean is about one-half of what is measured in the fertile coastal waters! Furthermore, the rate of primary production is very uneven, varying by as much as five to six times from month to month. Researchers have established that phosphorus and not nitrogen is the limiting macronutrient in these subtropical waters. Following these new findings, it appears that our understanding of tropical and subtropical marine production may have to be revised as more studies are conducted.

The temperate regions of the middle latitudes have distinct seasons and, not surprisingly, show strong seasonal variations in primary production (**Figure 10–10b**). In the North Atlantic, for example, plant productivity is very low in the winter, despite a well-mixed water column and high nutrient concentrations in the photic zone. This is because sunlight is weak and therefore limited at that season. By spring, however, a diatom bloom of enormous proportion occurs in response to the increasing levels of sunshine. This momentous event is termed, for obvious reasons, the **spring diatom bloom.** By midsummer, the bloom has stopped because of (1) intense grazing pressure by zooplankton in response to the ample plant food, (2) the formation of a thermocline between the depths of 10 to 20 meters (~33 to 66 feet) as surface water is warmed by the summer sun, and (3) the uptake of nutrients from the photic zone by the rapidly multiplying plant cells.

Sometime in the early fall, plant production may increase to a modest level as the thermocline weakens and nutrient-rich water from below is mixed upward into the photic zone at a time when light levels are still sufficient for photosynthesis. By winter, production is minimal once again as a result of the weak sunlight. It is at this time that storms mix water from below and recharge the photic zone with vital nutrients that will eventually sustain a spring diatom bloom.

Polar seas, despite their cold temperatures and bleak, storm-ridden winters when darkness may be continuous for months at a time (that is, there is no daylight!), are very productive during the summer (**Figure 10–10c**). Because the water column of the polar seas is well mixed, it tends to be isothermal, meaning that it shows little variation of water temperature with depth (see Figure 10–10d). As a result, the overturning of surface and deep water occurs continuously. This keeps nutrient levels in the photic zone high—even during the summer, when diatom blooms create a high demand for dissolved chemical compounds. Because biological productivity in polar seas is limited not by nutrients, but by solar energy, the result is a short, but highly productive summer season.

10-3

Global Patterns of Productivity

Photosynthesis converts inorganic carbon (the carbon in carbon dioxide—CO_2) into organic carbon (the carbon bound in sugar—$C_6H_{12}O_6$). Therefore, the primary productivity of any region is measured by the amount of inorganic carbon removed from the water. This quantity is expressed as grams of carbon per square meter of sea surface area per year, or gC/m^2/yr. The uptake of carbon by marine plants through photosynthesis varies considerably, ranging from a low of ~25 gC/m^2/yr to a high of ~1,250 gC/m^2/yr. This variation is similar to the range of values measured in terrestrial environments (**Table 10–1**). The highest rates of marine plant growth occur in shallow estuaries and are comparable to the productivity of prime farmland. At the other extreme, the low productivity of the open ocean corresponds to that of deserts.

SPATIAL VARIATIONS

The broad distribution pattern of primary productivity in each ocean roughly resembles a bull's-eye, with primary production decreasing toward its center (**Figure 10–11a**). The centers of oceans contain sparse plant populations and can be regarded as tracts of ocean deserts. Here the water is an extremely clear azure blue and contains few large animals, with the exception of some large fish that migrate through the region on their way to feeding and breeding grounds (see the boxed feature, "Migrants"). Although solar radiation is abundant in these desert regions of the oceans, nutrients are not, which explains their low biological productivity (<100 gC/m^2/yr). These sites are the centers of the large ocean circulation gyres, where Ekman transport causes water to converge and to downwell. In downwelling, the opposite of upwelling, water sinks and moves away from the sea surface, taking nutrients with it, away from the photic zone (**Figure 10–12a**). Also the water in the center of the gyres is highly stratified, because the intense sunlight creates a deep and permanent thermocline. These water layers are stable and do not allow overturning of the water column. Consequently, the deep, nutrient-rich water never reaches the photic zone (**Figure 10–12b**).

Moderate rates of plant productivity (50 to 200 gC/m^2/yr) characterize the continental shelves and coastal regions, which tend to receive dissolved nutrients from rivers, and also tend to be well

TABLE 10-1 Gross primary productivity.

Quantity (gC/m2/yr)	Ocean Area	Terrestrial Area
<50	Open ocean	Deserts
50–150	Continental shelves	Forests; grasslands; croplands
150–500	Upwelling areas; deep estuaries	Pastures; rain forests; moist croplands; lakes
500–1250	Shallow estuaries; coral reefs	Swamplands; intensively developed agricultural areas

FIGURE **10-11**

Global variations in primary and secondary production. (a) Phytoplankton abound in the surface water of continental shelves and upwelling areas and are scarce in the centers of oceans. (b) Because zooplankton are dependent on plants for food, their distribution mimics the pattern of primary plant production. (c) Even abundances of benthic animals of the ocean parallel the pattern of primary plant production. [Adapted from A. Couper, ed., *The Times Atlas of the Oceans* (New York: Van Nostrand Reinhold, 1983).]

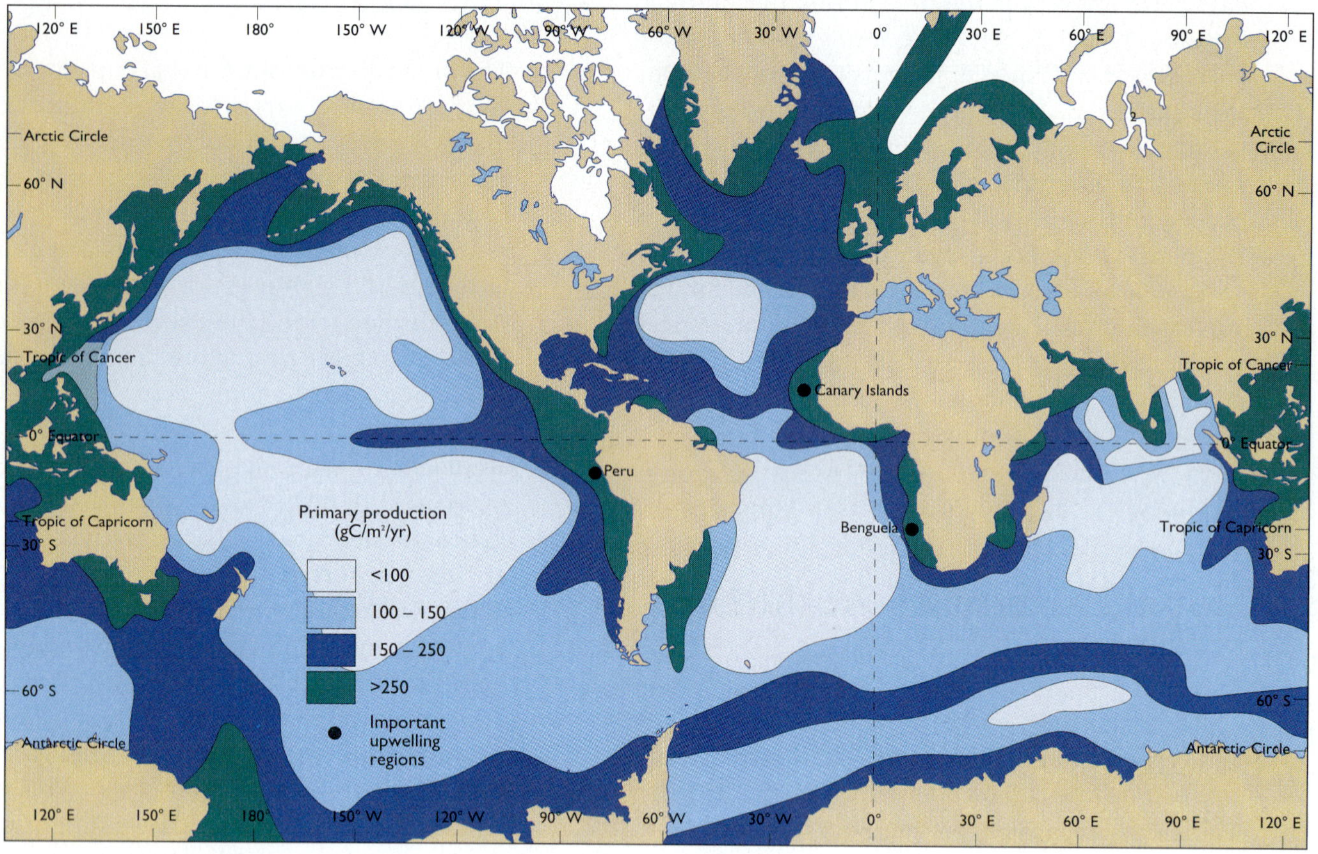

(a) PRIMARY PRODUCTIVITY

mixed by wave activity and tides because of the shallow sea bottom. The polar seas, where cold waters promote vertical mixing and nutrient replenishment from below, and the equatorial upwelling belt, where the surface water is kept fertile by the slow rising of relatively nutrient-rich water from below, are also moderately productive.

The most fertile areas of the ocean, the coastal upwelling regions, tend to be located along the western edges of landmasses in the low to middle latitudes. Here, primary production is greater than 250 $gC/m^2/yr$ and locally can even exceed 2,000 $gC/m^2/yr$! The most notable coastal upwelling sites include the waters adjacent to Peru, the Canary Islands, and Benguela in southwestern Africa (see Figure 10–11a). At all of these sites the upwelling of water and high biological production suggest a cause-and-effect relationship, which we will explore in the last section of this chapter.

The biomass present at each step of food webs depends ultimately on the growth of the primary producers, the autotrophs. Wherever plants abound, animals abound. This means that zooplankton are most plentiful along coastlines (particularly estuaries), over continental shelves, and in polar seas and upwelling areas. A comparison of the maps in Figures 10–11a and **10–11b** shows how the zooplankton biomass mimics closely the biomass

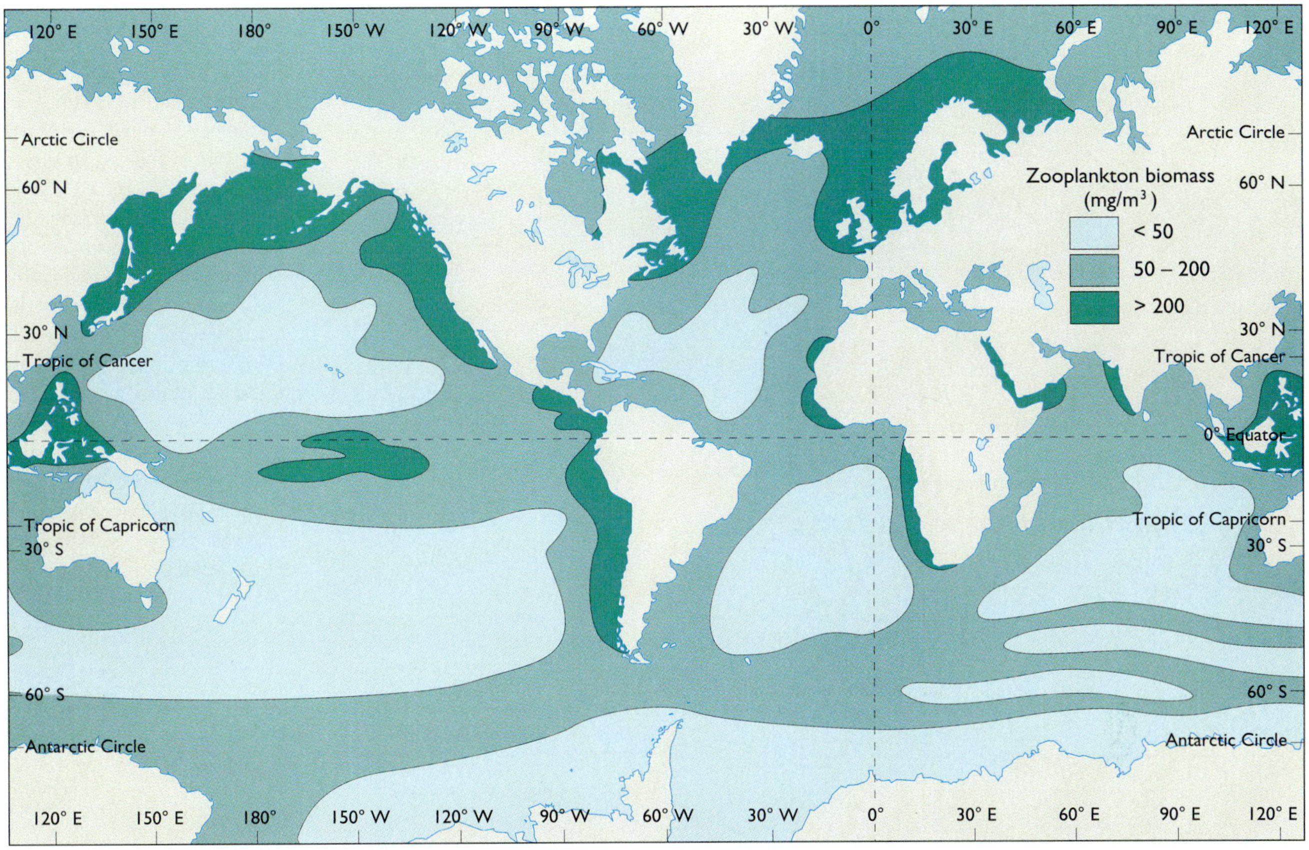

(b) ZOOPLANKTON BIOMASS

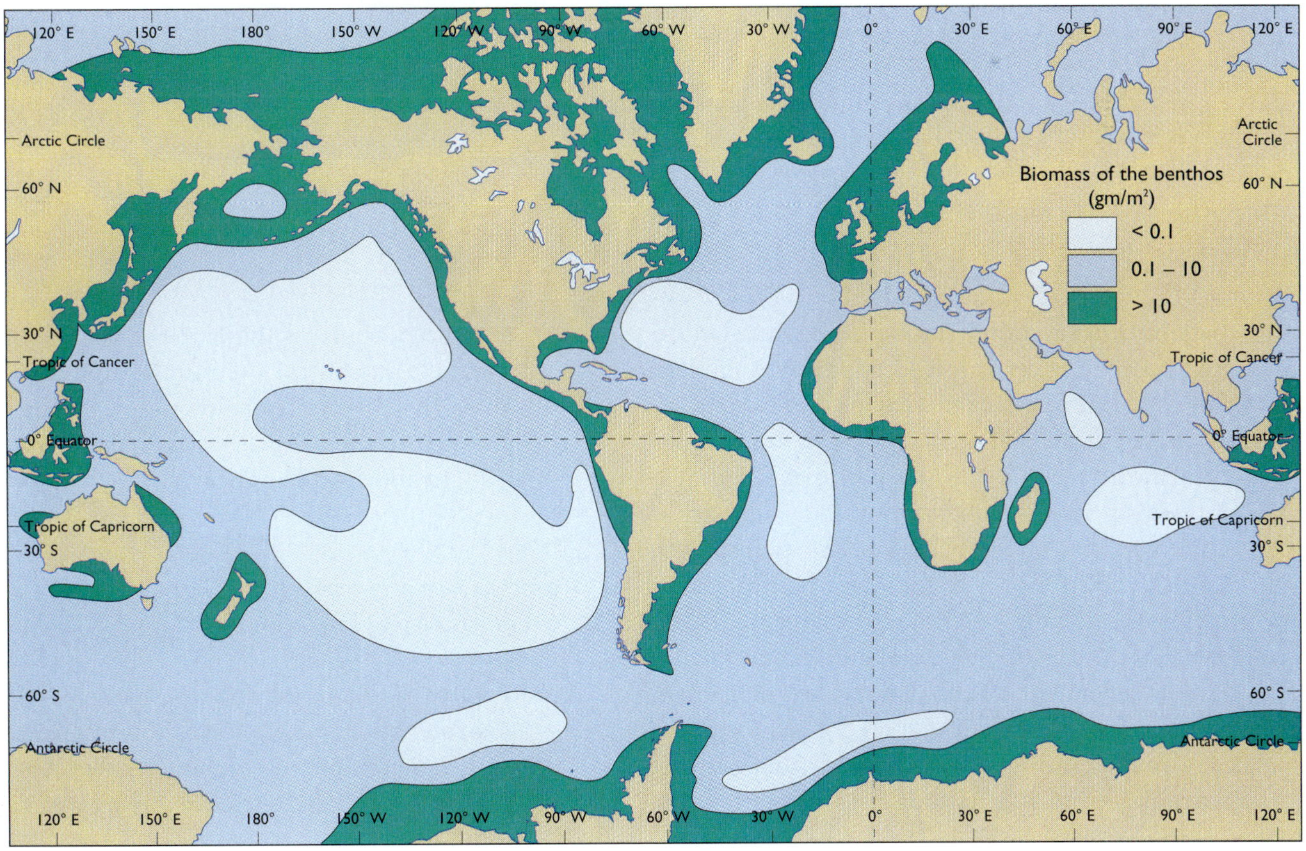

(c) BENTHIC BIOMASS

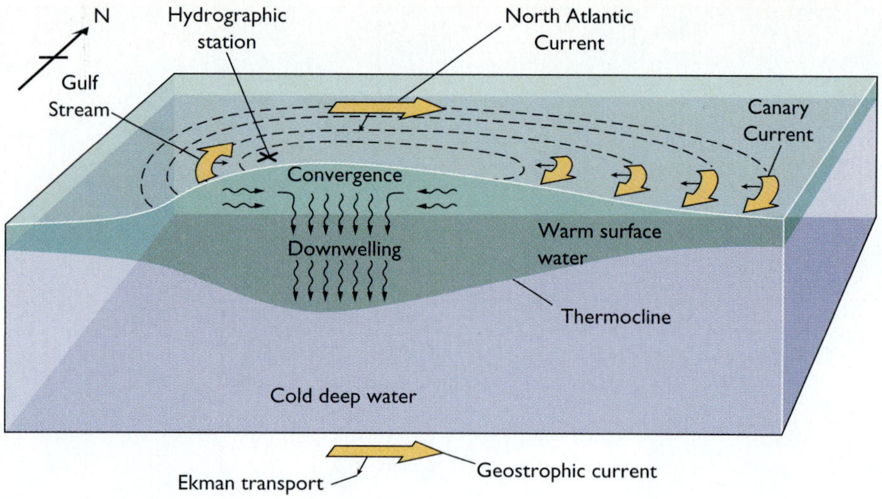

(a) CUTAWAY VIEW OF THE SARGASSO SEA

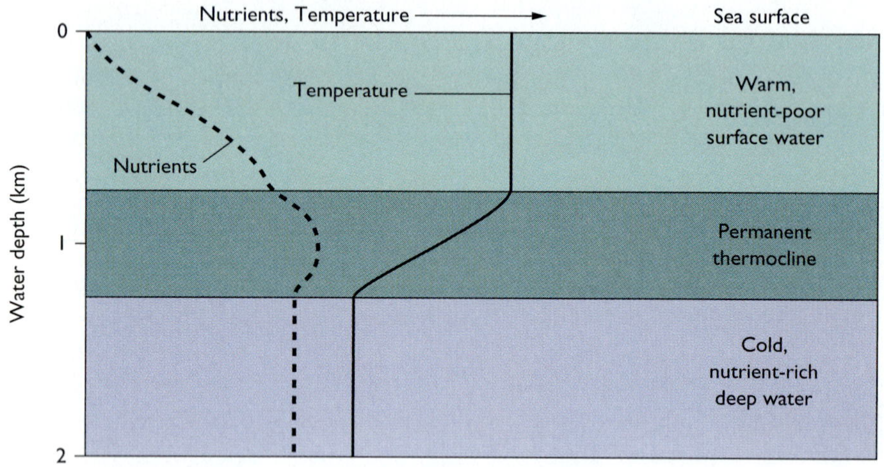

(b) VERTICAL PROFILES

FIGURE **10-12**

The Sargasso Sea. (a) The prevailing currents along with Coriolis deflection direct water toward the center of the gyre in the North Atlantic Ocean, an area known as the Sargasso Sea (see Figure 5–5a). This causes the surface water to downwell, producing a thick lens of warm water that overlies cold deep water. (b) The stratified nature of the water column in the center of the Sargasso Sea does not allow nutrient replenishment of the photic zone, as shown by the vertical profiles. Therefore, primary productivity in the center of gyres is very low, being comparable to deserts on the land.

distribution of the phytoplankton. Zooplankton are scarce in the centers of oceans and are abundant along the edges of the landmasses, a pattern identical to that of the plants.

Even the benthic animals living on the sea bed of the deep sea display distribution trends that are similar to that of the plankton. For example, the biomass distribution of the large benthic fauna, expressed as grams per square meter of sea bed (g/m^2), reinforces the idea that all animals are ultimately dependent on the plants—even organisms on the deep-sea bottom that are far removed from the thin, illuminated surface water that supports phytoplankton. It's hard at first to believe that microscopic plants living in the photic zone of the open ocean control the numbers of large sea animals that live miles below them on the deep-sea floor. Yet they do, as a comparison of Figures 10–11a and **10–11c** indicates.

ESTIMATES OF PLANT AND FISH PRODUCTION

Food webs describe how the biomass of animals depends on primary production by plants. Given this, it's possible to estimate the biomass of large predators such as fish indirectly by measuring the rate of primary production by plants. In effect, we are tracing what happens to the carbon as it is passed to the higher trophic levels of a food chain. As an example, let's do one of these calculations for estimating the yearly global production of fish. Many generalizations and assumptions enter these fish-production estimates, but they can serve as a general guide for assessing natural fish resources in a region and even for the globe.

The most cited study of global fish productivity, published by John H. Ryther in 1969 (**Table 10–2**), divides the world's oceans into three provinces—the oceanic, coastal, and upwelling regions—according to the rate of carbon uptake by phytoplankton as a result of photosynthesis. If these averages for primary production in these three provinces are representative, then the calculation of fish production is a cinch, as illustrated in **Figure 10–13**. For any region, one takes the rate of primary production in g $C/m^2/yr$ and multiplies it by the area; this yields the total primary production per year. Then one estimates the *transfer efficiency* of carbon for each step in the food chain—this is a measure of how much of the carbon that is passed on to the next trophic level of the food chain is used for growth. For reasons discussed earlier (see Figure 10–6), transfer efficiencies are low, on the order of 10 to 20 percent (0.1 to 0.2).

In the food chain of Figure 10–13, three trophic levels separate the anchovies from the diatoms. The transfer efficiency between the diatoms (plants) and the copepods (herbivores), and between the copepods and the anchovies (carnivores) is 10 percent (0.1). The total anchovy production is estimated

TABLE **10-2**

Global fish production.

Area	Primary Production ($gC/m^2/yr$)	Ocean Area (km^2)	Ocean Area (%)	Total Primary Production (tons C/yr)	Total Primary Production (%)	Average Number of Trophic Steps	Material Transfer Efficiency per Trophic Level (%)	Fish Production (tons/yr)	Fish Production (%)
Oceanic	50	325 × 106	90.0	16.3 × 109	81.5	5	10	1.6 × 106	<1
Coastal	100	36 × 106	9.9	3.6 × 109	18.0	3	15	120 × 106	50
Upwelling	300	0.36 × 106	0.1	0.1 × 109	0.5	1.5	20	120 × 106	50

Source: Adapted from J. H. Ryther, *Science* 166 (1969): 72–76.

FIGURE **10–13**

Fish production. Estimates of fish populations in a region are easy to calculate, provided reliable data are available for (1) rates of primary production per unit area, (2) the number of steps in the food chain, and (3) the material transfer efficiency between each trophic level. An example of such a calculation is shown at the bottom of the figure.

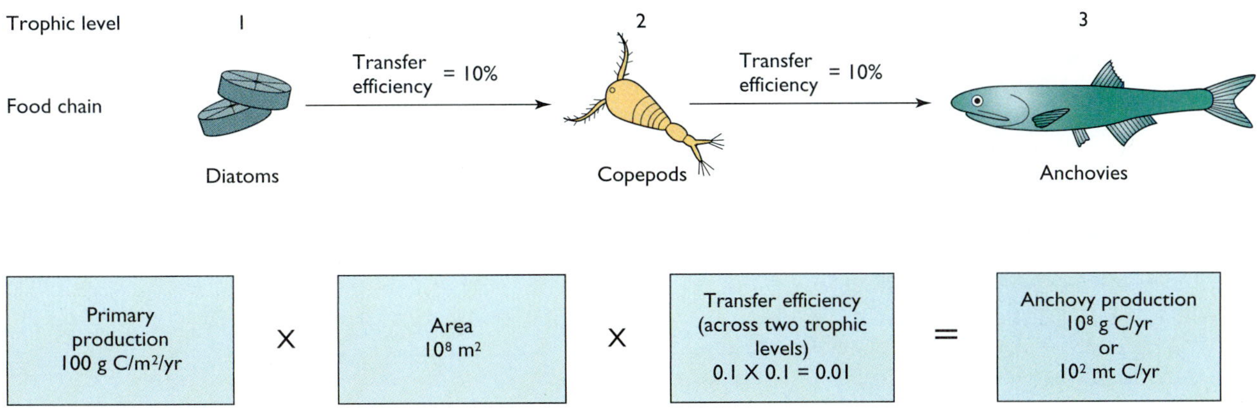

Migrants

Some marine animals, mainly large fishes, are migrants, moving across the vast distances that separate their feeding and breeding grounds. These migrants generally travel well-defined, established routes during regular times of the year. Eels, for example, live in freshwater streams but breed in saltwater. Sexually mature North Atlantic eels (**Figure B10–7a** and **7b**) abandon their freshwater home ponds and lakes and travel seaward down rivers of North America (the species *Anguilla rostrata*) and Europe (the species *A. anguilla*). They embark on these two separate long-distance journeys to their spawning grounds in the Sargasso Sea (**Figure B10–7c**), located southwest of Bermuda some 400 to 700 meters (~1,320 to 2,210 feet) below the ocean's surface . Once they breed and spawn in this area, the adults die.

The free-floating eggs hatch, producing transparent, leaf-shaped larvae (see Figure B10–7a) that drift with the surface currents of the Sargasso Sea for between one and three years. Once within reach of coastal waters, the larvae metamorphose into *elvers,* which resemble small adults 5 to 10 centimeters (~2 to 4 inches) long. The elvers migrate upstream to lakes and ponds, where they remain for about a decade

FIGURE **B10-7**

Eel migration. (a) The eel *Anguilla* inhabits freshwater rivers but breeds and spawns in saltwater. Eel larvae *(leptocephali)* hatch from pelagic eggs in the Sargasso Sea. They have large eyes, transparent bodies, and large surface areas to help them float as they drift with the surface currents. (b) Photograph of the American eel. (c) North American *(Anguilla rostrata)* and European *(A. anguilla)* eels breed in the Sargasso Sea. The pelagic eel larvae drift to their respective shores and then migrate upstream to lakes and ponds. [Adapted from C. E. Bond, *Biology of Fishes* (Philadelphia: Saunders, 1979).]

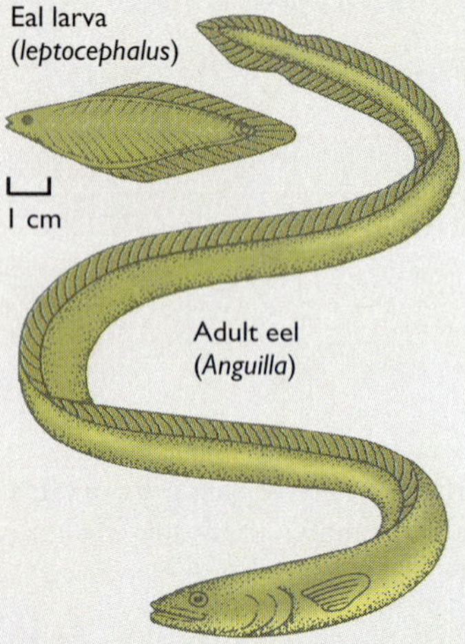

(a) *ANGUILLA*

(b) AMERICAN EEL

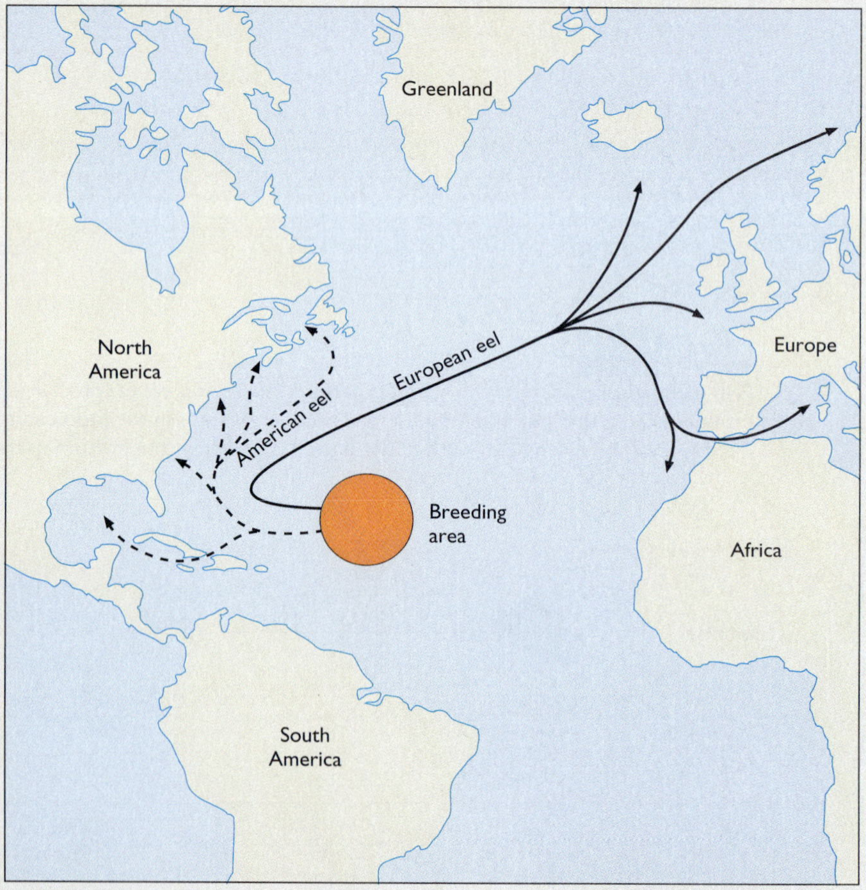

(c) MIGRATION ROUTES

before returning to the ocean to spawn and die.

In contrast to the eels, salmon (**Figures B10–8a** and **b** and **B10–9a**) live in saltwater but spawn in freshwater streams. Two genera are *Salmo,* the Atlantic salmon, and *Oncorhynchus,* the Pacific salmon. Both are wide-ranging in the northern temperate latitudes of their respective oceans (**Figures B10–8c** and **B10–9b**). After ascending rivers, adult salmon lay comparatively few eggs (less than 5,000) in gravel that is bathed by fast-moving water. The eggs hatch after incubating for between one and four months. Salmon depart from their home streams as juveniles, venturing to feeding grounds in the ocean, where they remain three to four years (see Figures B10–8c and B10–9b). There the salmon feed on small fish, grow rapidly, and mature into adults. When they are about 10 kilograms (~22 pounds) in weight and about 1 meter (~3.3 feet) in length, they migrate thousands of kilometers back to their natal (i. e., birth) streams, where the spawning cycle is repeated once again. After reproduction, Pacific salmon die in the stream beds; Atlantic salmon survive to repeat their migratory and spawning cycles several more times.

Visit www.jbpub.com/oceanlink for more information.

(a) NORTH PACIFIC PINK SALMON

(b) NORTH PACIFIC PINK SALMON

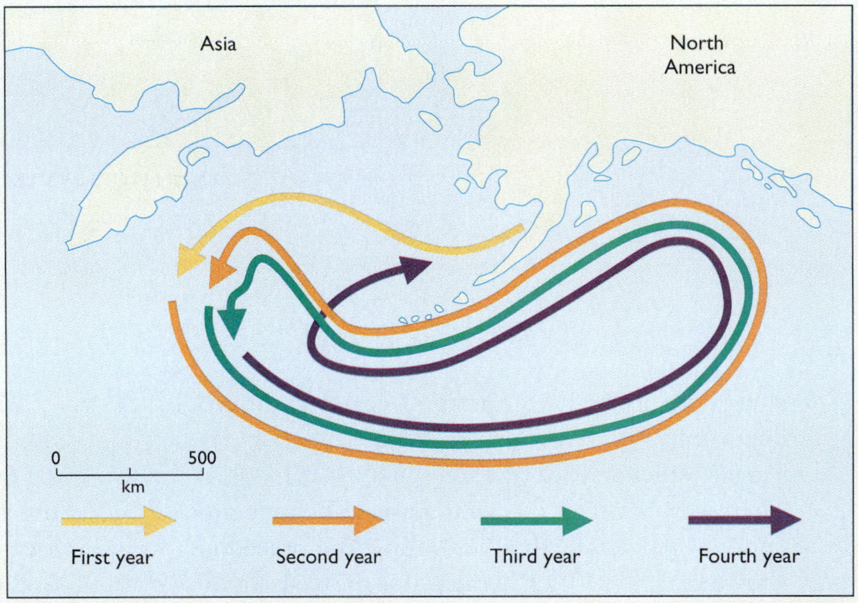

(c) PACIFIC SALMON MIGRATION ROUTES

FIGURE **B10-8**

Salmon migration. (a) The North Pacific pink salmon lives in saltwater and breeds in freshwater streams. (b) North Pacific Pink salmon. (c) The principal migratory routes of the North Pacific salmon require these fish to swim thousands of kilometers between their feeding and spawning grounds. [Adapted from P. A. Larkin, in *Fish Population Dynamics,* J. A. Gulland, ed. (London: John Wiley, 1977); and D. H. Cushing, *Climate and Fisheries* (New York: Academic Press, 1982).]

FIGURE B10-9

(a) Atlantic salmon. (b) The Atlantic salmon spend their adult lives in the open ocean and return to their natal streams to spawn. [Adapted from A. Netboy, *The Salmon: Their Fight for Survival* (London: Andre Deutsch, 1974).]

(a) ATLANTIC SALMON

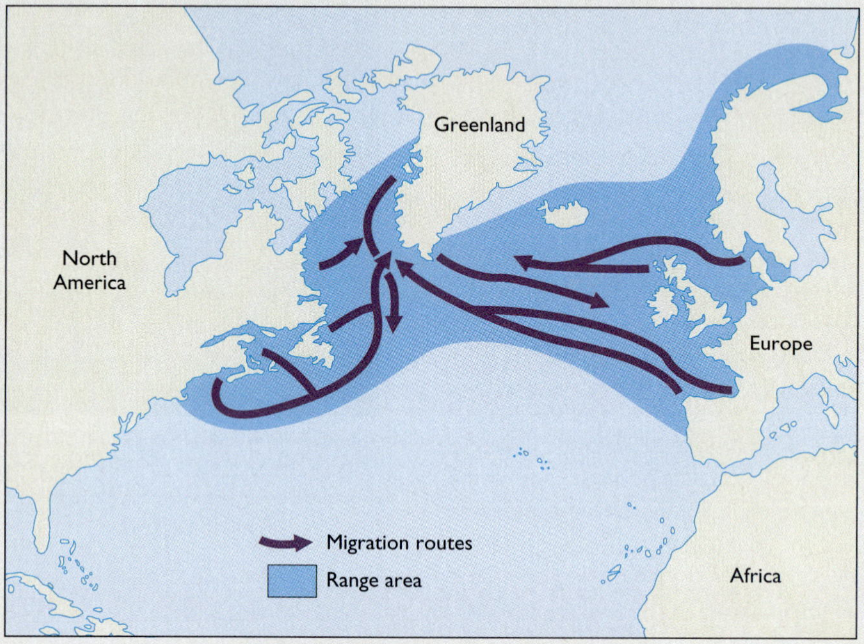

(b) ATLANTIC SALMON MIGRATION ROUTES

simply by multiplying the total yearly plant production by the two transfer efficiencies (i.e., the transfer efficiency for two steps is $0.1 \times 0.1 = 0.01$). There you have it—the total annual anchovy production! Be certain that you follow the procedure shown in Figure 10–13, because it is the basis for the values tabulated in Table 10–2. Let's return to Ryther's estimates of global fish production.

When you study Table 10–2, it is apparent that the overwhelming amount of primary production (81.5 percent) occurs in the oceanic province and the least amount (0.5 percent) in the upwelling areas. This seems to contradict our characterization of the open ocean as a desert with sparse life and the upwelling areas as regions of remarkably high fertility. However, the contradiction is not a real one, because we must keep the area of each province in mind. Although the waters of the open ocean contain little living matter, they are extensive in area. Their combined plant biomass is immense, far exceeding the plant biomass of the more productive coastal and upwelling provinces, which together amount to about 10 percent of the total ocean's area.

Perhaps an analogy will help you better understand this important concept. Imagine a large desert, thousands of square miles in area, with sparse vegetation, and then think about an irrigated farm tract, several acres in size, with dense vegetation. Which of the two has the highest *rate* of primary productivity? Which of the two will have the greatest *total amount* of plant production for a year? Well, the answer to the first question is easy—clearly the rate of primary productivity is in orders of magnitude greater for the farmland than for the desert. Yet the total annual production can be much, much greater for the desert than for the farmed acreage. Why is this so? Consider that a tiny bit of vegetation across thousands of square miles collectively can far exceed the total growth of the vegetation on a small farm plot. If still confused, think of it this way. If you scatter a dollar's worth of change in each acre of a two-acre farm, and 10 cents in each acre of a thousand-acre desert, which will yield the greatest total amount of money—the farm or the desert? A simple calculation indicates that the farm will yield \$2 (2 acres × \$1), the desert \$100 (1,000 acres × 10 cents).

Food chains are longer and the carbon transfer efficiency is smaller in the oceanic province than they are in either the coastal and upwelling provinces (see Table 10–2). Because of these marked differences, fish production is divided almost equally between the coastal and upwelling provinces, and is less than 0.1 percent for the open ocean. In other words, the fish biomass of the open ocean is of no significance. What is truly impressive, though, is that the upwelling regions, which represent about 0.1 percent of the ocean surface, produce 50 percent of all marine fishes (see Table 10–2)! In summary, 90 percent of the ocean is represented by the oceanic province, which consists of veritable deserts, with sparse plant life and few fish. Essentially 100 percent of the fish biomass is found in the waters of the continental shelf and the upwelling provinces.

Ryther's calculations suggest that about 240 million tons of fish are produced annually in the world's oceans. This valuable resource, however, is not exclusively for human use. It is shared with other large carnivores (larger fishes, mammals, and birds). Also, a minimum number of fish from each stock must be maintained from year to year in order that recovery from predation pressure is possible. Given these factors, Ryther suggests that no more than about 110 million tons of fish should be harvested by humans each year. Fishing records indicate that the global fish harvest stabilized at about 70 million tons during the 1970s and rose to about 98 million tons during the 1990s (**Figure 10–14a**)—a figure that is quite close to Ryther's inferred 110–million-ton limit.

In light of these calculations, can humans significantly increase their annual fish catch? The answer to this query depends on the assumptions that are made. Current fishing is limited to stocks that are large enough and sufficiently concentrated to make harvesting economically practical. Ironically, the very traits that make these fishes easy prey for humans —large stocks concentrated in schools, usually near coasts—render them vulnerable to overfishing and to pollution stresses (**Figure 10–14b**). New fisheries are likely to develop in the coastal oceans of the Southern Hemisphere, such as the continental shelf waters of South America, Africa, Australia, and Southeast Asia. Some optimistic fishery biologists believe that, with judicious management of traditional and newly discovered stocks, humans could raise the annual global fish catch to well over 110 million tons, perhaps to as much as 1,000 million tons! Others, like Ryther, argue that food production from the sea is limited, with little prospect of increasing global harvest in any significant way. Besides, they point out that many traditional fisheries, such as haddock, sole, cod, clams, and scallops in New England waters, have collapsed or are collapsing because of overfishing. Although only time will tell, the realistic answer probably lies somewhere between these two extreme positions. However, given our ignorance about actual fish production, the lower, conservative value should be used to establish policy and to set fish quotas.

FIGURE **10–14**

Fish harvests. (a) Between 1950 and 1970 the world catch of fish climbed steadily from about 20 million tons to about 70 million tons. It has since climbed to about 98 million tons. [Adapted from W. E. Odum, in *Fundamentals of Aquatic Ecosystems,* R. S. K. Barnes and K. H. Mann, eds. (Boston: Blackwell Scientific Publications, 1980), 143–61.] (b) Excessive fishing pressure without proper management can seriously affect a fish stock. Haddock recovered quickly during the war years, when fishing was not possible in the North Sea. However, the stocks were quickly reduced as soon as unrestrained fishing was resumed. This resulted in the collapse of the fishery around 1950. [Adapted from W. D. Russell-Hunter, *Aquatic Productivity* (New York: Macmillan, 1970).]

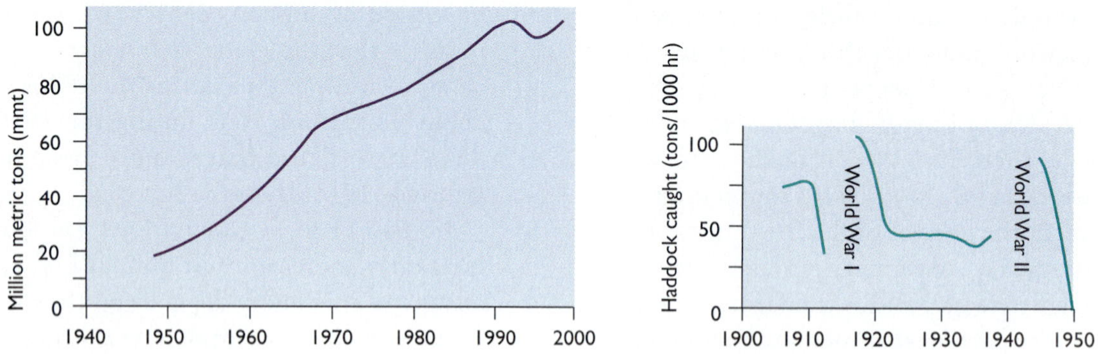

(a) GLOBAL FISH HARVEST

(b) HADDOCK CATCH IN NORTH SEA

10-4

Biological Productivity of Upwelling Water

Largely because nutrient levels are persistently low in the photic zone, the open ocean is generally unproductive. The two principal exceptions to this state are the shallow waters of the continental shelves and the upwelling regions. In the former, nutrients are supplied by rivers and by wave mixing of surface and bottom water, particularly during storms. Upwelling regions owe their unusually high fertility to the slow but persistent upward flow of deep water, which continually charges the photic zone with nutrients. Not surprisingly, some of the world's largest fisheries are located in upwelling areas.

One of the best studied upwelling regions of the world is offshore Peru in western South America. Here, the northward-flowing Peru Current (**Figure 10–15a**) transports a tongue of cold water from far to the south, making the *sea-surface temperature* (SST) unusually cold for Peru's tropical latitudes. SST is lowered even more by Ekman transport. The prevailing southerly and southeasterly winds blowing parallel to Peru's shore cause surface water to flow offshore in Ekman transport, because Coriolis deflection is to the left in the Southern Hemisphere. As a consequence, cold, nutrient-rich water upwells from below within a relatively narrow coastal zone some 10 to 20 kilometers (~6.2 to 12.4 miles) wide (**Figure 10–15b**). Water wells up slowly, but continually, at a rate of about 8 to 9 meters (26 to 30 feet) per day. This upwelling process continually fertilizes the photic zone and assures a bountiful crop of phytoplankton.

The plants, in turn, support incredibly large populations of anchoveta, a small fish that is harvested, dried, ground into fishmeal, and then sold on the world market as feed for livestock and poultry. At one time, the local Peruvian waters were so fertile that the fish harvest of 1971 amounted to about 12 million tons, almost 20 percent of the total fish catch of that year for the

FIGURE **10–15**

Coastal upwelling. (a) Coastal upwelling (see figure 10-8b), in conjunction with cold water from the Peru Current, lowers the sea-surface temperature (SST) offshore Peru and Chile. (b) The prevailing winds blow out of the south, parallel to the Peruvian coastline. Due to Coriolis deflection to the left, Ekman transport produces an offshore drift of surface water that induces the upwelling of cold, nutrient-rich water from below. This upwelling effect is clearly indicated by the isotherms, which are bent upward in response to the ascent of cold subsurface water. (c) Subtracting the SSTs during an El Niño from normal SST yields an SST-anomaly map. A tongue of unusually warm water, which characterizes El Niño, was evident during the spring of 1983. [Adapted from M. A. Cane, "Oceanographic Events during El Niño," *Science* 222 (1983): 1189–95.]

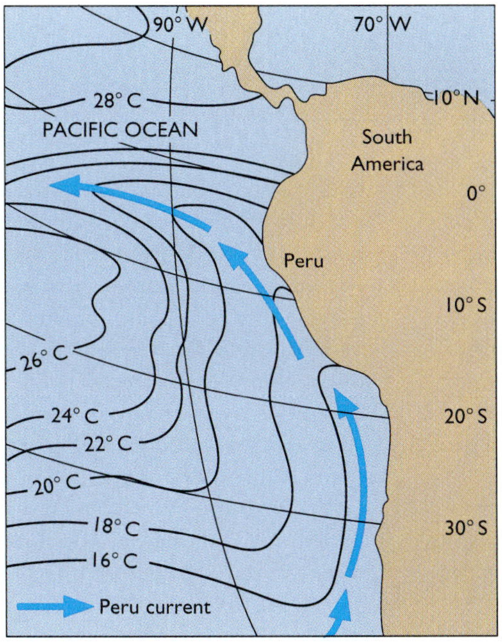

(a) MEAN SST OF THE EASTERN PACIFIC

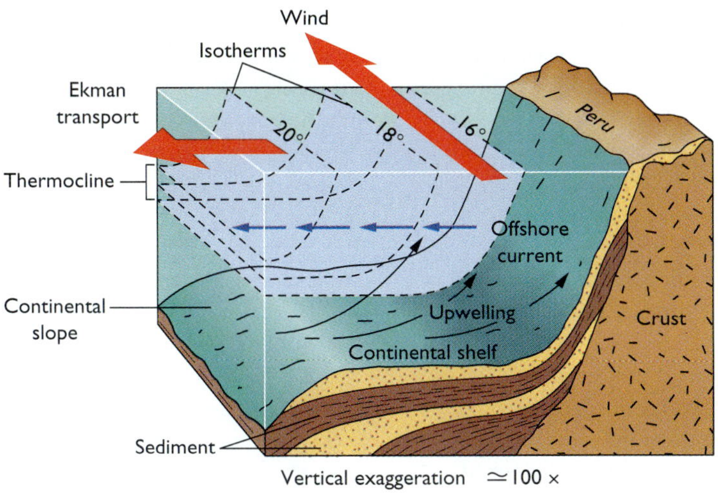

(b) COASTAL UPWELLING

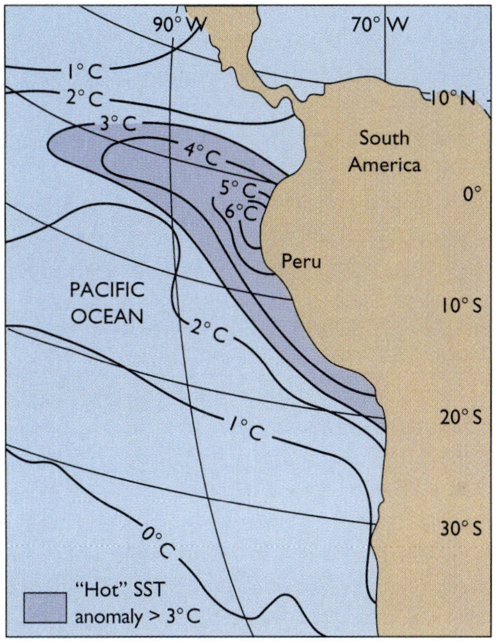

(c) SST ANOMALY DURING EL NIÑO

entire world! Periodically (every three to seven years), however, **El Niño**—meaning "the (Christ) Child," because it commonly appears at Christmas time—occurs. Its appearance is indicated by a change in global weather patterns, which results in the intrusion from the west of warm, nutrient-depleted water across the Peruvian continental shelf (**Figure 10–15c**). This brings about an abrupt decline in the anchoveta catch, with disastrous economic consequences for Peru.

When El Niño sets in, the chemistry and biology of the surface water change drastically. The sea-surface temperature becomes abnormally warm, and nutrient levels, diatom populations, and fish abundances decline drastically (**Figure 10–16**). Also, a tropical marine fauna not native to the cold waters of the area appears as well. These unusual conditions last for 6 to 18 months offshore Peru. Surprisingly, during an El Niño event, upwelling currents continue, bringing up warm, nutrient-poor water

THE OCEAN SCIENCES PHYSICS

What Causes El Niño?

Energy and matter are exchanged regularly between the sea and the land. Much of this interchange occurs by atmospheric processes, as wind and air masses sweep over land and water, transporting gases, solid and liquid particles, and heat across the globe. An example of one such global linkage is the peculiar nature of weather worldwide that accompanies the occurrence of a strong El Niño offshore Peru. At first, oceanographers tended to treat El Niño as a regional event, not realizing that its cause is located far away in the equatorial Pacific and that its ramifications are global in extent. Let's explore our current understanding of this climate.

During a strong El Niño event, warm, nutrient-poor tropical water displaces the normal cold, nutrient-rich water of the Peruvian continental shelf (see Fig. 10–15). This leads to economic catastrophe as the anchovy fisheries in the area collapses. One of the strongest El Niño events on record occurred in 1982–83 and was accompanied by major weather disturbances worldwide. To name a few, they included record rainfall in Ecuador and northwestern Peru (a desert); drought in the agricultural sectors of Bolivia and southern Peru; severe drought in eastern Australia, southern India, Indonesia, and southern Africa; an unusual number of typhoons (hurricanes) in the southern Pacific Ocean; and severe coastal storms along California, drought in the north central states, and mild winters in the eastern parts of the United States.

Scientists discovered that changes in the flow pattern of the southeast trade winds trigger El Niño. Normally, the southeast trade winds flow vigorously from a high-pressure zone in the eastern Pacific to a low-pressure zone in the western Pacific (see Figure 6–3a). These strong and persistent winds drag warm water westward and pile it against Southeast Asia. Under these conditions, the sea surface slopes up to the west and cold, nutrient-rich water occurs in the eastern equatorial Pacific and upwells along Peru (**Figure B10–10a**). At times, such as during 1982–83, the atmospheric pressure gradient that controls the trade winds breaks down, and air pressure rises in the eastern Pacific and drops in the western Pacific. This causes the trades to slacken and sometimes to even reverse direction, so that equatorial winds blow from west to east rather than from east to west. Then, warm surface water is dragged eastward by these anomalous winds. This raises sea level in the eastern Pacific, lowers it in the western Pacific, and allows warm, nutrient-poor water to flow against South America, replacing the cold, nutrient-rich water off Peru (**Figure B10–10b**). These peculiar oceanographic conditions give rise to El Niño. Eventually, usually over one to two years, normal atmospheric pressure gradients and trade winds reestablish themselves in the tropical Pacific, and cold, nutrient-rich water moves in from the south offshore Peru. This back-and-forth reversal of air pressure in the equatorial Pacific is called the Southern Oscillation, and its correlation to El Niño occurrences, the El Niño Southern Oscillation or ENSO for short.

The ENSO model has potential predictive value. By monitoring air pressure in the tropical Pacific it may be possible to forecast the weakening or reversal of the trade winds and, hence, the onset of an El Niño event offshore Peru. It is still unclear what specific conditions give rise to ENSO. Some scientists believe that ENSO is triggered by seasonal changes; others by winter monsoons. Because climate and marine interactions are so complex, dynamic, and intertwined, it has not yet been possible to distinguish clearly between the driving forces and their effects. Despite such complications, most scientists agree that the ENSO model, once refined, has tremendous potential for making long-range weather forecasts for widely separated parts of the Earth.

Visit www.jbpub.com/oceanlink for more information.

FIGURE **B10-10**

El Niño Southern Oscillation. (a) When the trade winds blow strongly, cold, nutrient-rich water occurs at the ocean surface in the eastern equatorial Pacific and offshore Peru. Upwelling supports high biological productivity in Peruvian waters under these oceanographic conditions. (b) On occasion, normal air-pressure patterns break down. This causes the trade winds to weaken and even reverse their direction, dragging warm, nutrient-poor water to the east as far south as Peru and initiating an El Niño event.

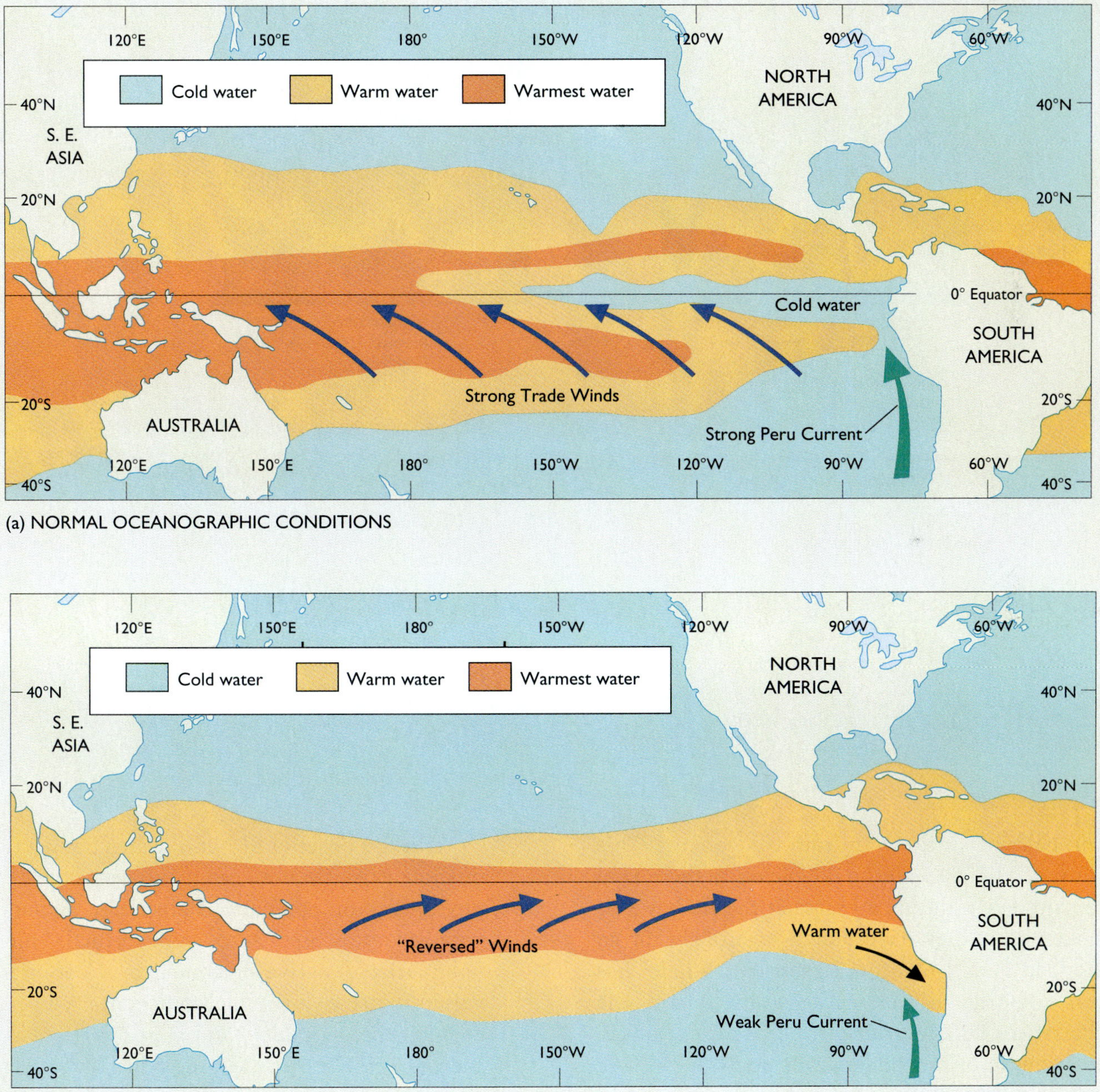

(b) EL NIÑO CONDITIONS

FIGURE **10–16**

El Niño. (a) Normal conditions are contrasted here with those that prevail during El Niño in Peruvian waters. (b) El Niño affects the composition of the Peruvian coastal fauna: sardines and hake—the native fishes of the area—disappear and are replaced by shrimp. (c) The yield of the Peruvian anchoveta fisheries increased dramatically during the 1960s, with only minor reductions in the fish catch associated with El Niño events. By 1972, however, the fisheries collapsed, presumably due to the combination of El Niño and overfishing. The fisheries still have not recovered from this disaster. [Parts b and c adapted from R. T. Barber and F. P. Chavez, "Biological Consequences of El Niño," *Science* 222 (1983): 1203–10.]

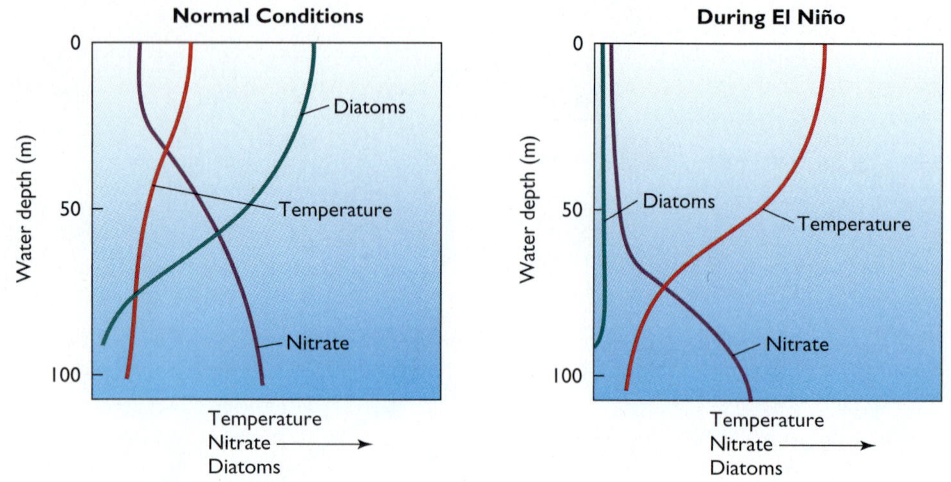

(a) WATER CONDITIONS

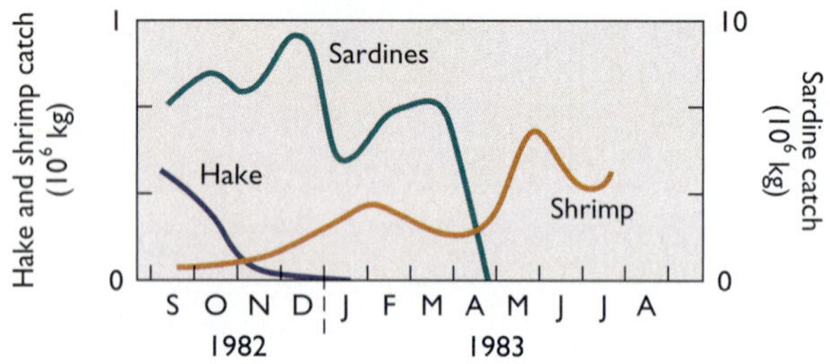

(b) IMPACT OF EL NIÑO ON PERUVIAN BIOTA

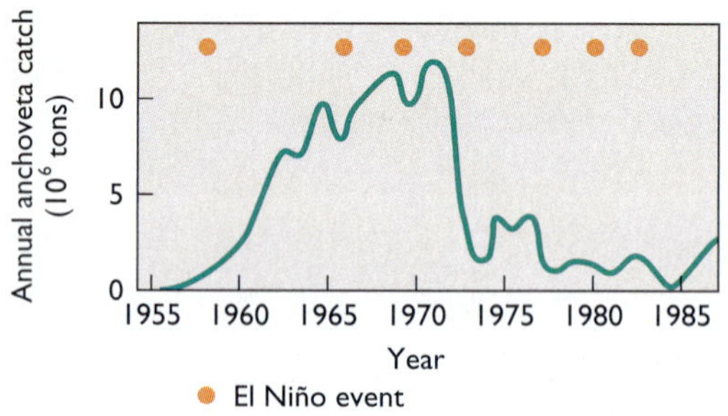

(c) ANNUAL CATCH OF PERUVIAN ANCHOVETA

from below that flows in from the equator. This inflow of tropical water to the region offshore Peru seems to occur during a year that the trade winds slacken after they have blown vigorously for an extended period of time. Recently, a model has been constructed that successfully relates the occurrence of El Niño offshore Peru to variations in the strength of the trade winds and the effect of such fluctuations of wind speed on the movement into the area of tropical water masses.

With the onset of El Niño, water temperatures offshore Peru rise noticeably, nutrient levels drop, phytoplankton populations plummet, and anchoveta become scarce. Also, some animals, such as hake (a large, active bottom fish of commercial value) migrate down the continental slope into colder, more suitable water (**Figure 10–16b**). Furthermore, the growth and reproductive success of other fishes, many seabirds, and some mammals are adversely affected by the lack of food. Among the large bird populations, for example, the blue-faced booby and the great frigate abandon their nestlings, resulting in mass starvation of the young. If the El Niño event is prolonged, even adult birds die from starvation. Seal pups also perish in large numbers, probably because their parents are unable to find food, as their five-day (compared with their normal one- or two-day) foraging trips to the sea suggest. Shrimps, which are normally absent offshore Peru, are transported to the area with the tropical water masses (see Figure 10–16b).

The most serious economic consequence of El Niño in the past has been its devastating effect on the Peruvian anchoveta fisheries. The immense fish stocks of anchoveta that supported a thriving and enormous fishery during the 1960s were directly linked to the sustained high plant productivity of the region—a biological response to the upwelling of nutrient-rich water by Ekman transport (see Figure 10–15b). The cells of many species of diatoms and dinoflagellates in the upwelling area are large or are joined together into colonies. Because of this, adult anchoveta graze the phytoplankton, producing a very efficient one-step food chain.

Commercial fishing of the Peruvian anchoveta began in earnest in the late 1950s. By 1964 the fish yield was approaching 10 million tons. By 1971 it had exceeded 12 million tons (**Figure 10–16c**), becoming the world's largest fishery. Twice during the 1960s (in 1965 and 1969), the fish catch dropped noticeably, despite intense fishing effort. El Niño was present during these years. The fish population recovered quickly from both lean fishing years. A few enlightened biologists feared that the incredible predation pressure on the anchoveta from humans, large fishes, and seabirds, combined with the injurious effects of an intense El Niño episode, could reduce the anchoveta stock to such critically low numbers that recovery might be difficult, if not impossible. Their worst fears were realized in 1972 and 1973 when the fish catch dropped drastically from more than 10 million tons to less than 2.5 million tons during an El Niño occurrence. To date, the anchoveta still have not recovered to their pre-1972 levels, despite vigorous efforts on the part of the Peruvian government to regulate fishing in their territorial waters.

10-5

Future Discoveries

Primary productivity at the base of food webs provides the foundational support of any ecosystem. Therefore, biologists are conducting research to understand what factors promote and limit biological production in a wide variety of marine habitats. What specifically determines spatial and temporal variations of primary production in coral reefs, in coastal zones including estuaries and lagoons, in the kelp forests of continental shelves, in polar seas, in the open ocean, in hydrothermal vents? How do climatic events like El Niño and water contamination by humans affect marine primary productivity? Can primary production be increased by artificial means so that a greater fish harvest is possible? These are critical questions that need answers if we are ever going to manage the ocean's biological resources responsibly and over the long term.

Biological oceanographers are still being surprised by new discoveries. A case in point is the potential significance of picoplankton, very tiny plankton that can occur in concentrations between millions and billions cells per ounce of seawater. Picoplankton are so small that microscopic diatoms and dinoflagellates seem to be giant plant cells by comparison. Picoplankton are dominated by species of cyanobacteria, which are photosynthetic organisms. The obvious question is what contribution do picoplankton make to the overall global productivity of the oceans? A few preliminary studies have indicated that picoplankton may be responsible for almost 70 percent of the productivity of the open ocean. What does this imply for marine productivity estimates?

Recent work suggests that enormous "rivers" of dust, microbes, and chemicals flow in the atmosphere, the best documented being large wind storms in the Sahara and Sahel regions of northern Africa (**Figure 10–17**) that carry fine particulate matter across the Atlantic Ocean into the Caribbean and Amazonian regions of the Americas. Nutrients carried from Africa by the wind settle into the dense canopy of the Amazon rainforest where they promote plant production, and into the reef ecosystems of the Caribbean Sea where they induce the growth of seaweeds that smother the coral. Also, since the early 1970s, some of the reefs of the Caribbean Sea and Florida have become affected by fungal infections and other diseases believed to have been caused by pathogens in the dust blown across the Atlantic Ocean from Africa. Such discoveries indicate how interconnected are the land, the sea, and the atmosphere, and how the decline of marine ecosystems, like the dust-dispersed diseases of the Caribbean's coral reefs, may be caused by processes and pathogens located thousands of kilometers away from the affected area.

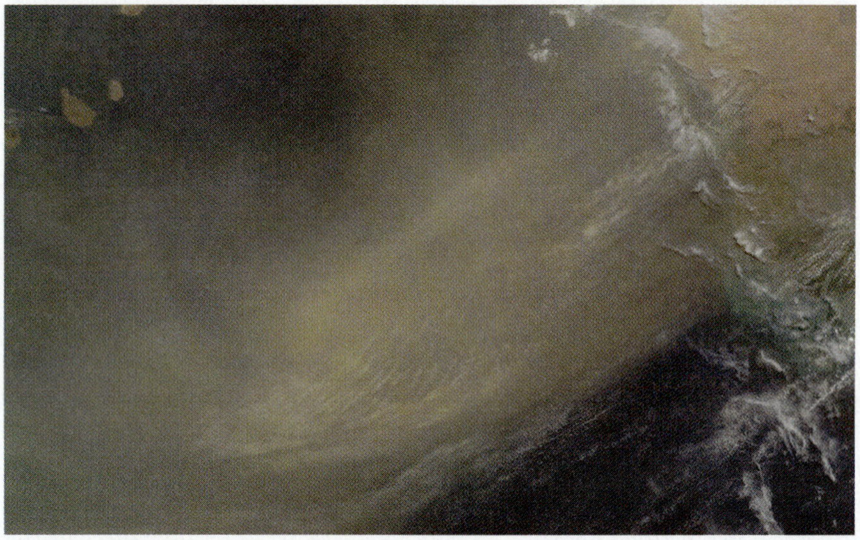

FIGURE 10–17

Dust storms. Large plumes of airborne dust extend over the Atlantic Ocean from northwestern Africa.

STUDY GUIDE

KEY CONCEPTS

1. An ecosystem consists of all the living and nonliving components of an environment. By the process of photosynthesis, *autotrophs*—principally plants—manufacture food out of inorganic nutrients, using sunlight as a source of energy. *Heterotrophs*—animals—are consumers that feed on organisms. Microbes, such as bacteria, decompose organic matter into simpler inorganic chemicals that plants reconstitute into food by photosynthesis. Hence, matter is reused within an ecosystem. Energy is not recycled and must be continually replenished by the Sun.

2. A *food chain* represents a simple, linear trophic (nutritional) arrangement; its base is occupied by the primary producers, the plants. Energy and matter are passed up the food chain to herbivores and carnivores. A *food web* consists of a system of interconnected food chains and describes a more complex and realistic energy-flow network through an ecosystem.

3. In the ocean, *primary production* is limited by the availability of sunshine and critical nutrients, particularly nitrates, phosphates, and silica. Nutrients regenerated by the bacterial decay of dead organic matter that has sunk below the photic zone are resupplied to plants by *upwelling* and wave turbulence, both of which transfer nutrient-rich bottom water upward to the photic zone.

4. Tropical waters are infertile, despite the abundant and intense sunlight, because a strong, permanent thermocline prevents the upwelling of nutrient-rich bottom water. Primary production in the temperate latitudes varies with the season. Winter production is low because of light limitations. A *spring diatom bloom* is triggered by increasing

levels of solar radiation. It is succeeded by a mid-to-late summer low in primary productivity as nutrients have been exhausted and the water column has become thermally stratified. Short-lived productivity may occur in the early fall. Plant production in the polar latitudes is intense during the summer, when solar radiation is adequate, but is nonexistent during the dark winter months.

5. The centers of oceans are biological deserts, because of distance from river-supplied nutrients and the presence of downwelling water in the center of circulation gyres, which removes nutrients from the photic zone. Shelf and coastal waters are moderately productive because of the supply of nutrients by rivers and by resuspension off the sea bed by waves. The most fertile areas are coastal upwelling regions, such as offshore Peru. Here, nutrient infusion by upwelling water supports high and persistent biological production.

6. Calculations of primary production and the study of food transfer demonstrate that coastal and upwelling regions produce essentially all of the world's fish. The upwelling regions are particularly fertile, yielding 50 percent of the world's fish production, although they represent merely 0.1 percent of the ocean's surface area.

7. Offshore Peru, wind-induced Ekman transport moves water away from the coastline and allows cold, nutrient-rich water to upwell to the surface. This process of fertilization sustains high diatom production, which in turn supports high anchoveta stocks. During El Niño events, sea-surface temperature (SST) becomes abnormally warm and nutrient levels drop sharply, causing phytoplankton and fish to die.

KEY WORDS

autotroph
chemosynthesis
detritus food chain
downwelling
ecosystem
food chain
food web
grazing food chain
heterotroph
photosynthesis
plankton bloom
respiration
spring diatom bloom
trophic dynamics
upwelling

QUESTIONS

REVIEW OF BASIC CONCEPTS

1. Describe how energy and matter flow through an ecosystem. Which of the two is recycled and which is not?

2. Why is nitrate rather than phosphate likely to limit plant production in the ocean?

3. Why is energy transfer between trophic levels such an inefficient process?

4. Contrast a detritus food chain with a grazing food chain. How are they linked?

5. What critical role(s) do bacteria play in the food cycle of the sea?

6. How and why does primary production vary with distance from land?

7. What is El Niño? What are the chemical and biological consequences of a strong El Niño occurrence offshore Peru?

CRITICAL-THINKING ESSAYS

1. Discuss how photosynthesis, respiration, nutrients, and sunlight are linked in ecosystems.

2. On a graph, plot the annual variation of solar radiation, water temperature, phosphate levels, plant biomass, and fish biomass that would characterize

   a. A midlatitude inner continental shelf
   b. An Arctic shelf
   c. A tropical shelf
   d. The center of a large ocean

3. The Peruvian anchoveta must have survived many El Niño events in the recent geologic past. Speculate on the reasons that their numbers have not recovered from the effect of recent El Niño occurrences, notably those since 1972.

4. What are the chances of discovering large fish stocks in the center of the Indian Ocean, an area that has not been previously fished? Provide a solid argument for your assessment.

5. Why are seaweeds large in size and phytoplankton microscopic in size?

6. If you were in a submersible, how would you quickly establish whether a newly discovered hydrothermal vent was active? Explain your reasoning.

7. Examine Figure B10–6 in the boxed feature, "Satellite Oceanography." Account for the time variations in plant productivity for the

   a. Center of the North Atlantic circulation gyre (the Sargasso Sea)
   b. North Atlantic waters off the east coast of Canada
   c. Continental shelf waters off western Africa, near the Canary Islands (hint: see Figure 10–11a)

DISCOVERING WITH NUMBERS

1. Assume a three-step food chain (diatoms → copepods → small fish). If the transfer efficiency is 15 percent (0.15) between the plants and copepods and 10 percent (0.10) between the copepods and small fish, determine the quantity of fish that can be supported by 1,000 grams of diatoms.

2. If the diatoms in Question 1 are fixing carbon by photosynthesis at the rate of 50 $gC/m^2/day$, estimate the daily primary production over an area of 100,000 square meters (10^5 m^2). Now estimate the amount of carbon fixed by the diatoms over the total area for a year.

3. Given that a fishery harvests the small fish in the three-step food chain described in Questions 1 and 2, estimate the average daily (gC/day) fish production across the $10^5 m^2$ area. If you need help, study Figure 10–13.

4. Assume that only 40 percent of the annual fish production can be harvested in the $10^5 m^2$ area described in Question 3 if this fishery is to be sustainable from year to year. At what rate in gC/yr can fish be harvested from the region?

5. Assume that during one year the fishery in Question 4 is affected by an event like El Niño, which reduces diatom production to 10 $gC/m^2/day$ for that year. How many gC of fish can be caught for that year in order to maintain a sustainable harvest?

SELECTED READINGS

Anderson, D. M. 1994. Red tides. *Scientific American* 271 (2): 62–68.

Arntz, W. E. 1984. El Niño and Peru: Positive effects. *Oceanus* 27 (2): 36–39.

Barnes, R. S. K., and Mann, K. H. 1980. *Fundamentals of Aquatic Ecosystems.* Oxford: Blackwell .

Caron, D. A. 1992. An introduction to biological oceanography. *Oceanus* 35 (3): 10–17.

Caviedes, C.N. 2001. *El Niño in History*. Miami: University Press of Florida.

Chisholm, S. W. 1992. What limits phytoplankton growth? *Oceanus* 35 (3): 36–46.

Edmond, J. M., and Damm, K. V. 1983. Hot springs on the ocean floor. *Scientific American* 284 (4): 78–93.

Esaias, W. E. 1981. Remote sensing in biological oceanography. *Oceanus* 24 (3): 32–38.

Govindjee and Coleman, W. J. 1990. How plants make oxygen. *Scientific American* 262 (2): 50–58.

Harrison, D. E., and Cane, M. A. 1984. Changes in the Pacific during the 1982–1983 (El Niño) event. *Oceanus* 27 (2): 21–28.

Jumars, P. A. 1993. *Concepts in Biological Oceanography: An Interdisciplinary Primer.* New York: Oxford University Press.

Leetman, A. 1989. The interplay of El Niño and La Nina. *Oceanus* 32 (2): 30–34.

Lerman, M. 1986. *Marine Biology: Environment, Diversity, and Ecology.* Menlo Park, Calif.: Benjamin-Cummings.

Levinton, J. S. 1982. *Marine Ecology.* Englewood Cliffs, N.J.: Prentice-Hall.

Oceanus. 1984. Deep-sea hot springs and cold seeps. Special issue 27 (3).

Oceanus. 1992. Biological oceanography. Special issue 35 (3).

Pomeroy, L. R. 1992. The microbial food web. *Oceanus* 35 (3): 28–35.

Ramage, C. S. 1986. El Niño. *Scientific American* 254 (6): 76–83.

Rasmusson, E. M. 1985. El Niño and variations in climate. *American Scientist* 73 (2): 168–77.

Schweid, R. 2002. *Consider the Eel.* Chapel Hill, N.C.: The University of North Carolina Press.

Stoecker, D. K. 1987. Photosynthesis found in some single-cell marine animals. *Oceanus* 30 (3): 49–53.

Sumich, J. L. 1992. *An Introduction to the Biology of Marine Life.* Dubuque, Iowa: Wm. C. Brown.

Youvan, D. C., and Mairs, B. C. 1987. Molecular mechanism of photosynthesis. *Scientific American* 256 (6): 42–48.

TOOLS FOR LEARNING

Tools for Learning is an on-line review area located at this book's web site OceanLink (**www.jbpub.com/oceanlink**). The review area provides a variety of activities designed to help you study for your class. You will find chapter outlines, review questions, hints for some of the book's math questions (identified by the math icon), web research tips for selected Critical Thinking Essay questions, key term reviews, and figure labeling exercises.

11

The Dynamic Shoreline

I have seen the hungry ocean gain
Advantage on the kingdom of the shore,
And the firm soil win of the wat'ry main
Increasing store with loss, and loss with store.
—*William Shakespeare,* Sonnet LXIV

PREVIEW

A VISIT TO THE SEASHORE can be enriched by an awareness of the natural forces that control and mold coastal landscapes. The purpose of this chapter is to describe the dynamic balances that exist among these driving forces. Emphasis is placed on the physical and geologic processes of the coastal zone that operate on different time scales and their effects in shaping the beaches, coastal dunes, barrier islands, and cliffed coasts. The chapter ends with an overview of engineering techniques used to stabilize and protect shorelines.

The ocean has a long-standing history of being viewed as an adversary, an opponent to be subdued and tamed. It is this incessant struggle between humans and nature that Shakespeare captured. What I hope that you will learn and remember from this chapter is that along a beach "there is nothing permanent except change," an observation made by the early Greek philosopher Heraclitus (c. 540–470 B.C.). We must continually bear this wisdom in mind as we attempt to understand and control the shoreline.

web navigator

critical thinking on the web

math tutor on the web

www.jbpub.com/oceanlink

Shorelines represent the juncture where land meets the sea. Here we have a great variety of unique habitats that are shaped by the interaction of breaking waves, the rise and fall of the tides, the discharge of river sediment, longshore and nearshore currents, biological processes, the slow change of sea level, the sinking and rising of the land, and the activity of humans. Variations in these factors across space and over time create a rich mosaic of restless coastal landforms made of sand and mud that are always shifting about—sometimes eroding here, sometimes building up there.

11-1

Coastal Water Movement

Breaking waves provide most of the energy that changes the shape and texture of beach deposits. Waves also are responsible for generating currents that move sand in the surf zone. In this section, we will briefly review what happens to waves as they enter shallow water and interact with the sea bottom. Then we will examine the water circulation and sand-dispersal systems that originate because of breakers crashing on the shoreline. You will grasp the ideas in this chapter better if you understand the preceding text about ocean waves (Chapter 7), tides (Chapter 8), and the erosion and deposition of sediment (Chapter 4, the section entitled Factors that Control Sedimentation).

SHOALING WAVES AND REFRACTION

In shallow water, waves interact with the sea bottom. This significantly affects their motion and shape. As water depths decrease, shoaling waves slow down, wavelength diminishes, and wave height increases (see Figure 7–7b in Chapter 7). Most waves approaching the shoreline from far offshore are not parallel to the shore and undergo **refraction**. This means the wave crests bend (refract) and become more nearly aligned with the contours of the shallow sea bottom (see Figure 7–8a). In regions where the nearshore bottom is uneven, refraction may focus wave energy on specific parts of the shore. In very shallow water, waves become oversteepened and unstable and then collapse as **breakers**. The type of breaker—spilling, plunging, or surging—depends to a great extent on the slope of the sea bottom (see Figure 7–9).

CIRCULATION IN THE SURF ZONE

When waves strike a beach and release energy, they generate currents in the surf zone that flow parallel to the shoreline; these are called **longshore currents** (**Figure11–1a** and **b**). The greater the **angle of wave approach**, the stronger the longshore current is for the same wave height. Because refraction causes waves to bend and conform to the shape of the shoreline, the angle at which waves strike the shoreline is usually slight, rarely exceeding 10 degrees. The flow direction of longshore currents will shift back and forth as the direction of wave approach changes from day to day. As you face the ocean, sometimes the longshore current is moving to your right, on other days to your left, depending on the direction that waves strike the beach. However, in most regions there is a dominant direction of longshore current.

When wave crests are parallel to the shore, and wave rays or orthogonals intersect the shore at right angles, a longshore current, in theory, should not exist. However, under these conditions, longshore currents can develop because of **wave setup**, a process that creates piles of water in the surf zone. Big waves cause a greater volume of water to move shoreward by mass transport than do small waves. If you look carefully at the crest of a single breaker, you'll notice that its height along the beach varies considerably. Because waves rarely are of uniform height along their crestal length, different amounts of water drift against the shoreline. This creates piles of water at spots along the beach where breaking waves are the largest. Consequently, the water surface of the surf zone is often uneven. High water elevations (mounds) occur where breakers are large, and low water elevations (depressions) occur where breakers are small. The slopes along the uneven water surface represent pressure gradients. Water reacts to the pressure gradients and flows off the sides of the piles, generating two diverging (moving apart) longshore currents in the surf zone (**Figure 11–1c**).

Wave setup can also be created by wave refraction. In this case, wave rays are focused on a stretch of beach, causing the breakers there to be high, resulting in wave setup. The resulting mound of water produces diverging longshore currents in the surf zone.

If a stretch of shoreline has variable wave setup because of uneven breaker heights, a nearshore circulation system will develop. It will consist of a series of diverging and converging longshore currents (see Figure 11–1c). Longshore currents move apart from the piles of water created by wave setup. Between these piles, typically where the breakers are smallest, longshore currents will converge. Here the water will be forced to flow seaward as a narrow, swift **rip current** that drains the excess water out of the surf zone (**Figure 11–1d**).

Try to imagine all of this as a gigantic natural plumbing system. Excess water brought into the surf zone by waves is moved parallel to the shore by longshore currents, and this water is eventually flushed offshore by rip currents. At their seaward ends, the rip has a bulbous head where the flow of the current weakens and water spreads and mixes with the surrounding offshore water. Hence, incoming waves not only create shore-parallel currents (longshore currents), but also shore-normal currents (rip currents).

11-2

Beaches

The sediment of beaches and nearshore zones is continually being moved by longshore currents and rip currents. As breakers stir up water, sand grains, even large ones, are lifted off the bottom by wave turbulence and get pushed along by the current before settling back down to the sea bottom. This up-and-down motion is repeated incessantly with each breaking wave, transporting sand grains along the shore for remarkably long distances. Even gravel can be rolled along the bottom by breaking waves. Typically, sand particles, once they enter a beach system, are transported parallel to the shore in the direction of the prevailing longshore current. When sand grains are swept into rip currents, they are transported beyond the surf zone into the offshore. Strong rip currents, if they persist, can cause severe erosion of beaches.

Before beginning a survey of coastal processes, however,we need to subdivide the shoreline environment. The term *beach* is used loosely to refer to the part of the land that touches the sea. The beach is an active zone of sediment transportation that lies between the erosional area above the water level and the depositional area below the water level. It is subdivided into a number of environments. The **nearshore zone** (**Figure 11–2**) extends from the **breaker zone** (where waves begin to break) across the **surf zone** (where most of the wave energy is expended) to the **swash zone** (where the beach is covered and uncovered by water with each wave surge). The nearshore zone shifts back and forth as the tide floods and ebbs, covering and uncovering the beach. The **berm**, a prominent wave-deposited feature of most beaches, is an accumulation of sand having a flat top surface and a relatively steep seaward slope (**Figure 11–3**). The **offshore zone** is the open water that lies seaward of the nearshore zone, and the **backshore zone** is the land that adjoins the nearshore zone.

BEACH PROFILES

An important technique for studying sand deposition and erosion is the measurement of **beach profiles**, which delineate the shape of the beach surface along a survey line or transect. By comparing profiles of the same transect taken over time, it is possible to determine gains and losses of sand. When the beach has gained sand between surveys, the profile appears "swollen." Erosion of the beach between surveys will show the profile to be "shrunken." Under steady-state conditions, of course, the profiles will be identical.

The interpretation of many beach profiles from all over the world has shown that beaches typically undergo regular seasonal variations that cause them sometimes to grow in size and at other times to shrink. From such studies, we know that fair weather and the arrival of low, flat swell for several days result in the accumulation of sand and the growth of the beach. Under these weather and wave conditions, the resulting beach profile is concave

FIGURE 11-1

Nearshore currents. (a) Where waves approach a shoreline at an angle, a shore-parallel current called a longshore current is generated in the surf zone. The speed of the longshore current is highest in the midsurf region, and increases as the wave height and angle of wave approach increase. (b) Aerial view of oblique wave approach. (c) Waves that are parallel to the beach can generate longshore currents, provided wave height varies along the waves' crest. In such cases, high wave crests produce more mass transport toward the shore than do low wave crests. This creates wave setup and a mound of water on the parts of the beach affected by the higher waves. Longshore currents diverge from zones of maximum wave setup and converge at points of minimum wave setup, where rip currents discharge the excess water out of the surf zone. (d) Aerial view of rip currents off the California shore.

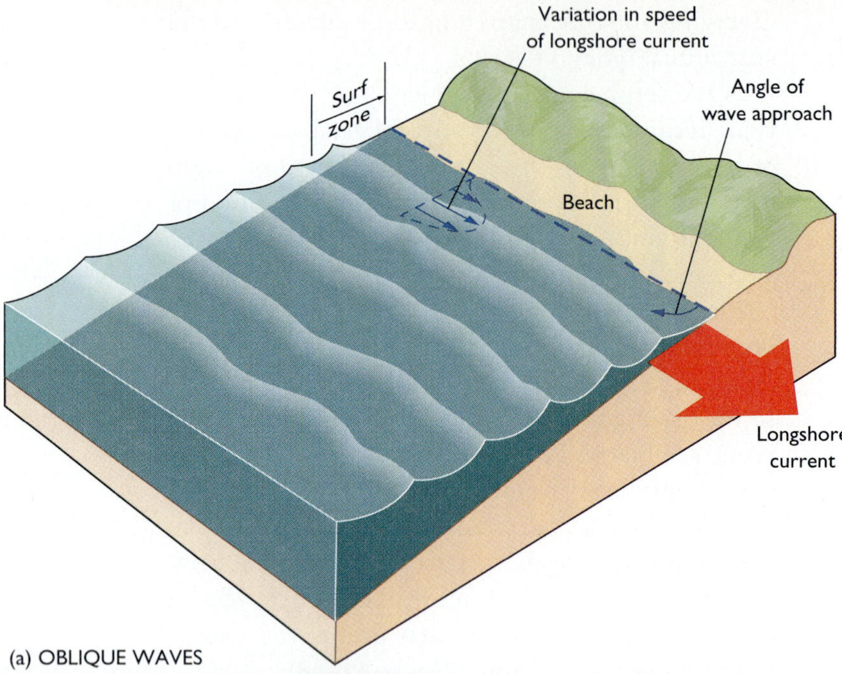

(a) OBLIQUE WAVES

(b) OBLIQUE WAVE APPROACH

upward with a broad **berm** and a steep **beach face** of the intertidal zone (**Figure 11–3a**). In other words, the beach is wide and steep. This configuration, referred to as a **swell profile**, is best developed during the calm, low-energy waves of summer, and it is the beach shape that is most familiar to tourists (**Figure 11–3b**). However, under storm conditions, which tend to typify the winter season, high, steep waves packed with energy attack the beach. The resulting erosion cuts away at the berm, reducing its width, and flattens the intertidal beach face, creating a **storm profile** (see Figure 11–3a). Eroded sand is moved seaward by rips and other shore-normal currents, where waves and currents shape it into one or more **longshore bars**.

Not only does the beach get eroded during a storm, but also the average grain size of its sand increases. This makes sense, because the high-energy waves and the rip currents selectively transport finer sand offshore, leaving the coarser sand on the beach. After the storm, the reappearance of the flat, fair-weather swell begins to "bulldoze" the fine

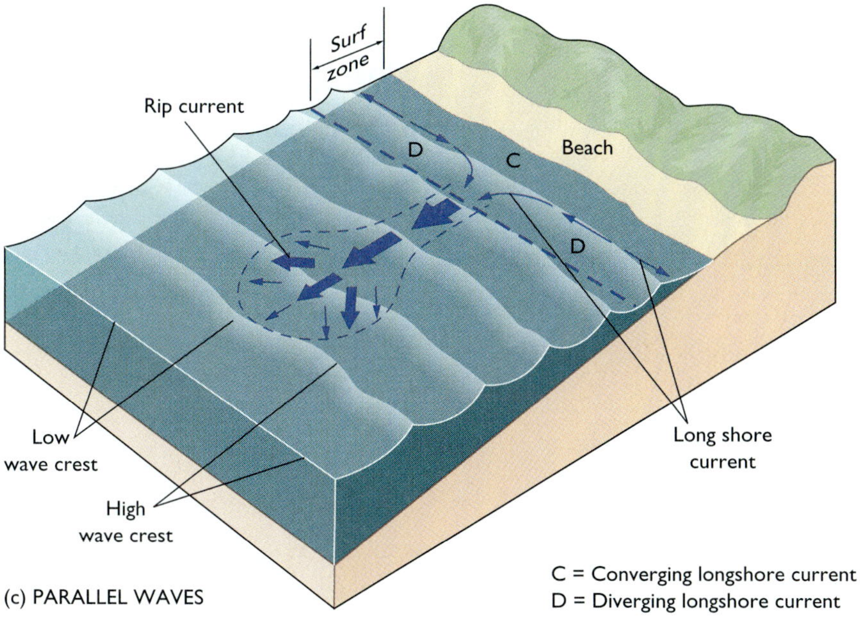

(c) PARALLEL WAVES

(d) RIP CURRENTS

FIGURE 11-2

The coastal zone. The nearshore zone is subdivided into breaker, surf, and swash zones. Waves begin to collapse at the breaker zone, generating the turbulent, foamy flow of the surf zone. The swash zone is the portion of the beach that is alternately covered by the swash (the uprush of water) and uncovered by the backwash (the downrush of water) of the surging water. Seaward of the nearshore is the offshore, and landward is the backshore.

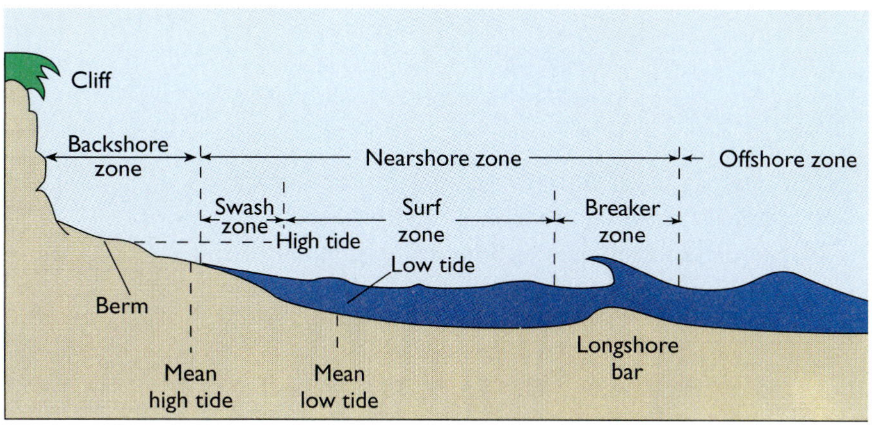

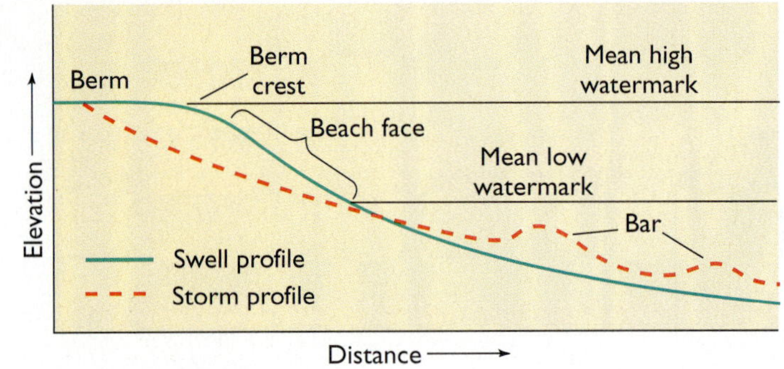

(a) SWELL PROFILE

(b) STEEP SUMMER BEACH

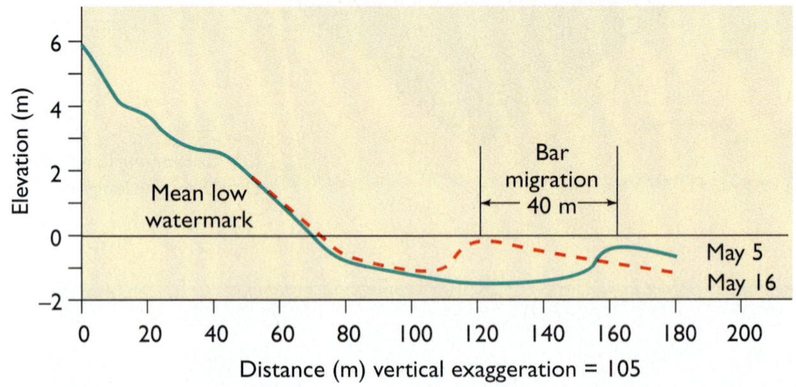

(c) STORM PROFILE WITH ONSHORE BAR MIGRATIONS

FIGURE **11-3**

Beach profiles. (a) A swell profile (broad berm, steep beach face, concave profile) is compared with a storm profile (narrow beach, gentle beach face, longshore bars). (b) A beach displaying the steep beach face that is characteristic of the swell profile. (c) Following a storm, the eroded beach tends to undergo accretion as longshore bars migrate onshore and become welded to the beach face, causing the berm to widen over time.

sand of the longshore bars back onto the beach. In fact, regular beach profiling after a storm shows that the bar migrates landward (**Figure 11–3c**). Sand from the bar is eventually added to the berm of the beach under these weather conditions.

In summary, the analysis of beach profiles suggests that sand in most beach systems is simply exchanged between the nearshore and offshore zone in response to weather and wave conditions. Storms cause beach sand to move offshore; fair weather causes it to move onshore. Accordingly, the width of the beach shrinks and expands with the change of the weather and the seasons.

SAND BUDGETS

Coastal engineers and geologists keep track of changes in the sediment volume of a beach by the use of **sand budgets**, which are estimates of the principal sand sources (credits) and sand losses (debits) for a stretch of shoreline (**Figure 11–4a**). This kind of analysis is comparable to keeping track of your money in a saving account at the bank. Sand sources (major credit items) include river input, sea-cliff erosion, and longshore and onshore sand transport onto the beach. These sources add sand to the beach and are equivalent to putting money into your bank account. The removal of sand from a stretch of beach (debits in the form of losses) occurs by longshore and offshore transport and by wind erosion. These activities subtract sand from the beach, which is equivalent to withdrawing money from your savings account. If the input of sand equals the output, you have a **steady-state condition**—that is, a balance—and the beach is stable for a time. More input of sand than output results in deposition and widening of the beach. More output than input causes the beach to erode and become narrow.

A sand budget for a hypothetical beach is presented in **Figure 11–4b**. Comparing the beach's total input to its output of sand shows that there is a net volume loss of sediment to the system, indicating that this stretch of shoreline is eroding over time. Gains and losses of sand can be markedly affected by human activities, and the sand budget of a beach can vary over different time scales (e.g., a one-year sand budget compared with a ten-year or a century-long sand budget) as climate and storm activity vary (see the boxed feature, "The Oregon Coastline").

11-3 Coastal Dunes

Broad beaches are flanked by sand dunes. These features are formed by wind blowing sand off the dry part of the beach, a process that represents a debit in the sand budget of a beach system. Well-developed dunes typically display a smooth, sinusoidal profile. The primary dune ridge is located at the landward edge of the beach and may be flanked by a secondary and, in rarer cases, a tertiary dune ridge, each located farther away from the shoreline and less influence by coastal processes (**Figure 11–5**). The general morphology of coastal dune systems consists of sand ridges that trend parallel to the shoreline (**Figure 11–6**), each separate from the other by distinct swales or valleys. In some cases, coastal dunes can extend landward for up to 10 kilometers (~6.2 miles). Small, irregularly distributed, multiformed foredunes commonly develop on the upper part of the beach in front of the primary dunes (see Figure 11–5), where litter, cobbles, and patches of vegetation tend to trap wind-blown sand. Under the right conditions (no storms, dry weather, strong onshore winds, and ample supply of dry sand), foredunes can grow rapidly in size and even migrate landward into the primary dune field.

INPUTS	+	**OUTPUTS**	=	**BALANCE**
Longshore transport into beach		Longshore transport out of beach		Accretion
River supply		Offshore transport		Erosion
Cliff erosion		Wind transportation into dunes		Steady state
Onshore transport				

(a) PRINCIPAL COMPONENTS OF A SAND BUDGET

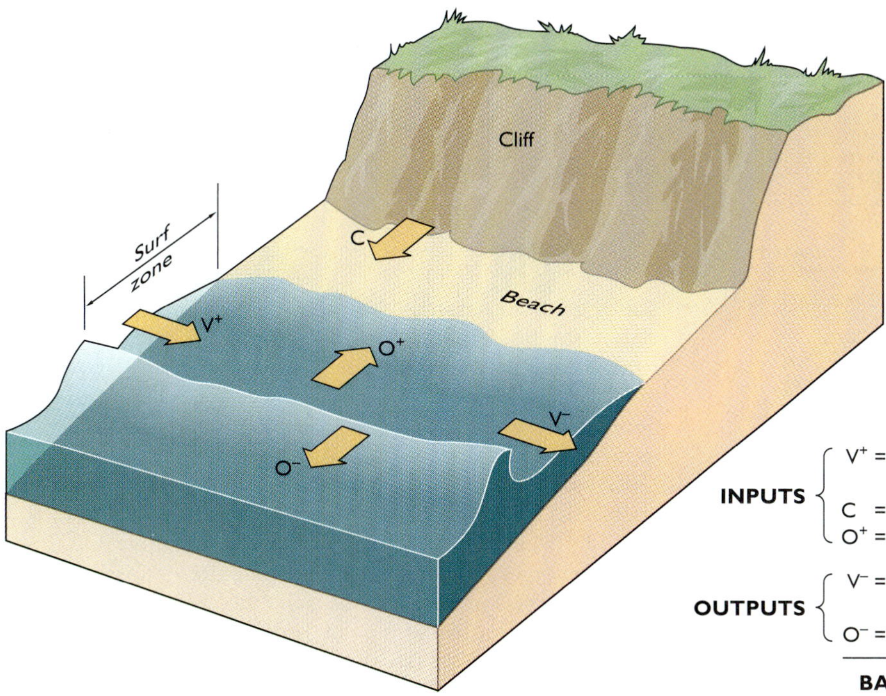

INPUTS
- V^+ = longshore transport into beach: $+50{,}000\ m^3/yr$
- C = cliff erosion: $+5{,}000\ m^3/yr$
- O^+ = onshore transport: $+5{,}000\ m^3/yr$

OUTPUTS
- V^- = longshore transport out of beach: $-55{,}000\ m^3/yr$
- O^- = offshore transport: $-15{,}000\ m^3/yr$

BALANCE: $-10{,}000\ m^3/yr$ (net erosion)

(b) HYPOTHETICAL SAND BUDGET

FIGURE 11-4

Sand budgets. (a) The state of a beach can be assessed by considering the credits (inputs) and debits (outputs) of sand to a stretch of shoreline. Beach accretion (buildup) results when sand inputs exceed outputs. Erosion results when sand outputs are greater than sand inputs. A balance between inputs and outputs produces no change in the volume of beach sand, a condition called steady state. [Adapted from P. D. Komar, *Beach Processes and Sedimentation* (Englewood Cliffs, N.J.:Prentice-Hall, 1976).] (b) In this example of a hypothetical sand budget, sand inputs are less than sand outputs. The result is acute erosion of the beach at an estimated rate of 10,000 m^3/yr.

Cape Cod, Massachusetts

Cape Cod resembles a flexed arm of sand thrust out into the Atlantic Ocean. It owes much of its origin to glaciers, which were active in the area as recently as 14,000 years ago. Since that time waves and currents have extensively reshaped the sedimentary deposits left by those glaciers into sandy beaches, barrier islands, and saltwater marshes. The coastline of Cape Cod is the longest natural coastline of New England, and it has hardly been modified by shore-protective structures.

About 15,300 years ago, all of New England was covered by a huge sheet of ice that flowed south from Canada. As this ice mass crept across the continental shelf, one of its ice lobes—the Cape Cod Bay Lobe—deposited sediment along its edge and formed a ridge, called a **terminal moraine** because it marks the farthest (terminal) position of the glacier. This terminal moraine can now be traced across Martha's Vineyard and Nantucket Island, the two principal islands of the Cape (**Figure B11–1a**). As the climate became warmer, the ice lobe melted and retreated northward, depositing a series of **recessional**

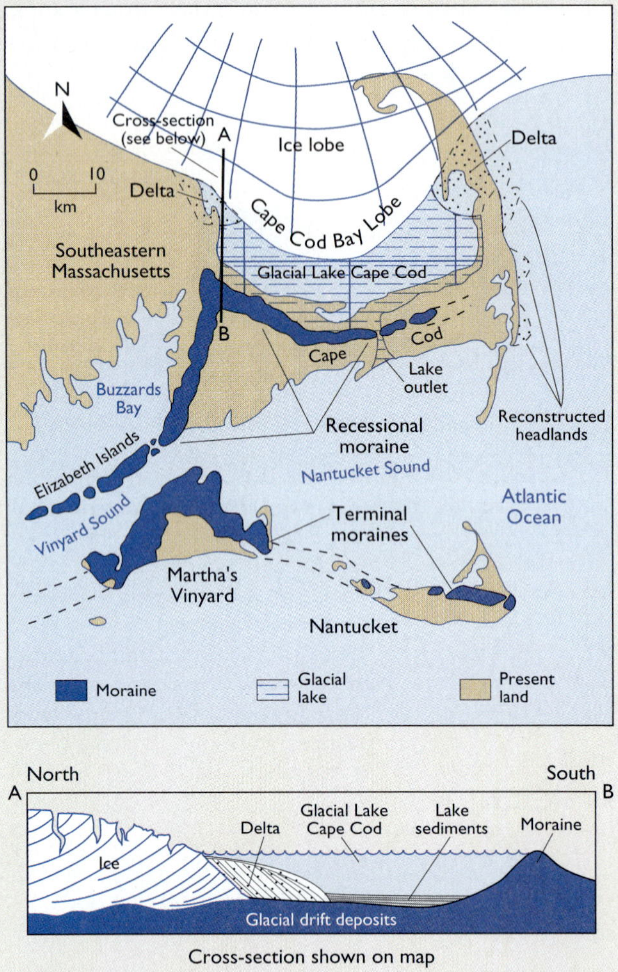

(a) DEVELOPMENT OF CAPE COD

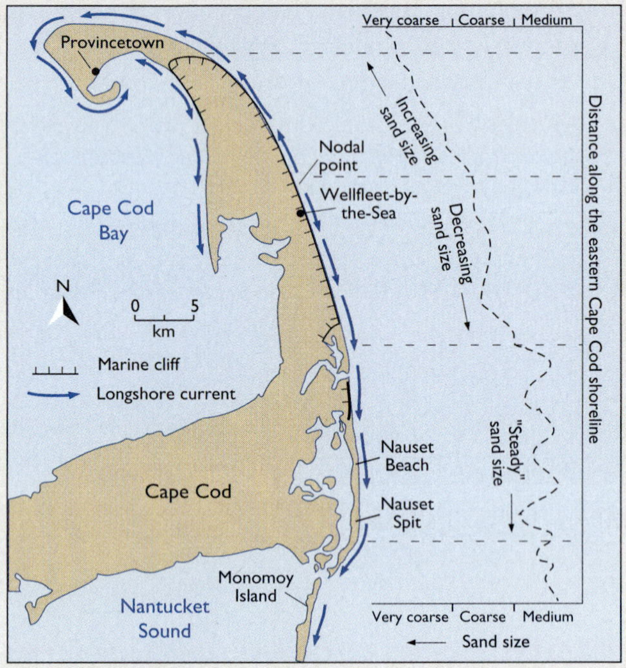

(b) LONGSHORE DRIFT

FIGURE **B11-1**

Cape Cod, Massachusetts. (a) The presence in the past of a glacial ice lobe in southeastern Massachusetts is indicated by terminal and recessional moraines, ridges of glacial till that mark the position of the ice front. As the ice sheet retreated northward, meltwater trapped by the recessional moraine formed Glacial Lake Cape Cod. [Adapted from G. L. Larson, *Late Wisconsin Glaciation of New England,* G. L. Larson and B. D. Sonte, eds. (Dubuque, Iowa: Kendall/Hunt Publishing, 1982), 101–14.] (b) Because of wave refraction, the longshore drift of sand diverges at a nodal point. The grain size increases with distance to the north of the nodal point, but decreases with distance to the south of this point. [Adapted from J. J. Fisher, in *Environmental Geologic Guide to Cape Cod National Seashore,* S. P. Leatherman, ed. (Tulsa, Okla.: SEPM, 1979).]

moraines that stretch along the Elizabeth Islands and the northern coastline of Cape Cod Bay (see Figure B11–1a). During the retreat of the ice sheet, meltwater from the glaciers was trapped between the end of the ice lobe and the recessional moraine, forming the huge Lake Cape Cod (Figure B11–1a). Beds of mud and silt and delta sands were deposited in this glacial lake and covered the glacial sediments.

With the melting of glaciers worldwide, sea level rose. Erosion of the glacial deposits by waves produced steep cliffs called **bluffs**, many of which are still retreating at alarming rates, some by as much as 1 meter (~3.3 feet) per year. The glacial sediment eroded from the bluffs is poorly sorted and is moved about by waves and currents according to grain size. Gravel and boulders tend to accumulate at the base of the bluffs. Clay and silt particles are suspended in the water and swept offshore or into sheltered bays. Sand is kept in the nearshore zone and moved parallel to the shoreline by longshore currents, supplying many of the beaches of the area. To the north of Wellfleet-by-the-Sea, sand drifts northward and gets coarser towards Provincetown; to the south of Wellfleet-by-the-Sea, it drifts southward and gets finer-grained with distance of transport (**Figure B11–1b**). The reason for the opposing grain-size trends is unclear.

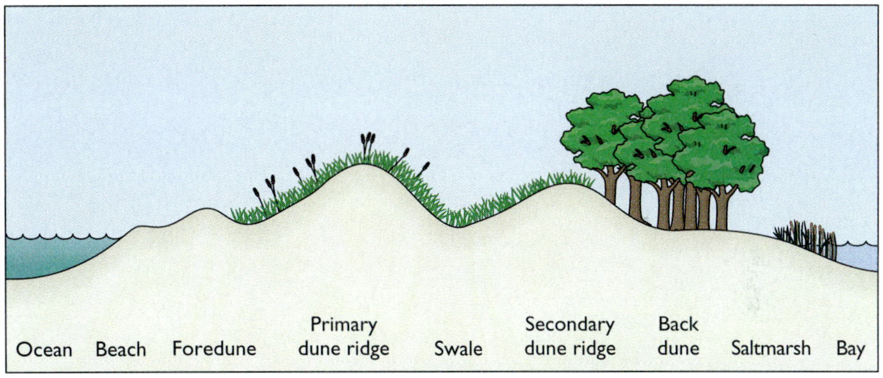

FIGURE **11-5**

Topographic profile of dune ridges across a barrier island. Simple plants, such as grasses and shrubs, dominate the primary dune ridge, and complex tall shrubs and trees dominate the secondary dune ridge and back dune area. The primary dune ridge consists of younger dunes with little soil, while the older, secondary dune ridge and back dune is composed of mature soil.

FIGURE **11-6**

Coastal dunes. Aerial view of primary and secondary dune fields that trend parallel to the beach. Note the recent and extensive damage by storm waves.

The Oregon Coastline

The coast of Oregon consists of sandy beaches that are separated from one another by prominent headlands or by long stretches of cliff rock (**Figure B11–2**). In large part, the majestic shore of Oregon owes its distinctive landscape directly to the collision of lithospheric plates. The sea floor is being subducted beneath Oregon, Washington, and northern California as oceanic crust is created along the ocean-spreading Juan de Fuca Ridge (see Figure B3–1). Evidence for this ongoing collision between massive lithosperic plates includes (1) the volcanoes of the Cascade Range (e.g., Mount Hood, Mount Ranier, Mount Saint Helens), (2) the presence of deformed marine sedimentary rocks that are plastered to the coast, and (3) a series of elevated wave-cut terraces (**Figure B11–3**) that represent sudden uplift of the crust in response to subduction. At present, the southern half of the Oregon shore is being pushed upward tectonically at a rate faster than the worldwide rise of sea level. This has resulted in the land emerging from the sea at a rate of about 2 to 3 millimeters (~0.08 to 0.12 inches) a year. In contrast, the northern half of the Oregon coast is stable, neither rising nor sinking relative to the ocean, because tectonic uplift here is about equal to the rate of global sea-level rise. According to Paul Komar and his colleagues at Oregon State University, the strong earthquakes that, fortunately, occur infrequently in the area can cause sections of the coast to sink abruptly, allowing the ocean to flood and erode the affected land. Six such catastrophic earthquakes have happened in the past 4,000 years, the most recent one occurring about 300 years ago. Obviously, more such earthquakes will occur in the future, although it is impossible to predict exactly when these seismic events will happen or what specific shoreline areas they will affect.

Paul Komar regards the plate tectonic setting as exerting the dominant regional control of the landscape pattern of coastal Oregon. Local, second-order effects that are superimposed on and therefore modify this broad regional pattern include (1) the morphology of pocket beaches between rock headlands, (2) the eroding capacity of large rip currents, and (3) the composition and geologic structure of the sedimentary rocks of the coastal cliffs. These secondary factors tend to exert strong control over the rate at which specific cliffs erode and retreat. Cliffs recede fastest along stretches of the Oregon shore where the beaches are steep, low-lying, and narrow, where large rip currents create embayments that allow wave runup to lap against and erode the cliff base, and where the cliff rock consists of relatively soft mudstones that are tilted seaward, allowing the cliffs to collapse periodically in massive, catastrophic landslides. Cliffs retreat more slowly along segments of

FIGURE **B11-2**

Pocket beach. Aerial photograph of Oregon shoreline showing a pocket beach nestled between large headlands.

FIGURE **B11-3**

Signs of tectonism. Note the series of elevated wave-cut terraces, common along the Oregon coast.

the shore where the fronting beach is topographically broad and high and slopes gradually seaward, where large rip currents are uncommon, and where the cliff rocks are composed of durable sandstones that lie flat and are, therefore, less prone to sliding.

Recently Paul Komar attributed the short-lived but unusually rapid coastal erosion along Oregon to the exceptionally severe 1982–83 El Niño. Apparently El Niño events create high sea levels off Oregon, generate storms that create monster coastal breakers as high as 7 meters (~23 feet), and generate an atypical northward-directed longshore drift of sand along local beaches. These factors when combined permit the overtopping of beaches by waves that can then directly attack and undermine the cliff base. The result is rapid and widespread beach and cliff erosion for a year or so until the El Niño event ends. The more recent 1997–98 El Niño has been devastating for the Oregon shore as well, as indicated by frequent coastal flooding, storm surges of unusual magnitudes, and widespread collapse of cliff faces. **Figure B11–4** consists of two aerial photographs of a segment of the northern shore of Oregon; one was taken on November 1997 and the other April 1998. During the winter, unusually large breakers generated by El Niño storms eroded the entire dune field fronting the large house at the extreme right of the photographs.

FIGURE **B11-4**

Flooding. Severe coastal flooding of the Oregon shoreline occurred during the 1997–98 El Niño.

(a) NORTHERN SHORE OF OREGON IN NOVEMBER 1997

(b) NORTHERN SHORE OF OREGON IN APRIL 1998

Vegetation has a profound influence on the development of dunes. Dune plants, such as marram grass (*Ammophila arenaria*) in Washington, Oregon, and California and American beach grass (*A. brevigulata*) along the Atlantic coast stretching from the Carolinas to Massachusetts, are salt tolerant, have deep root systems in order to tap groundwater, and have the ability to withstand being buried in sand. Grasses and bushes tend to cause windblown sand to be deposited in their wind shadows, and by so doing, they promote the growth of dunes. Blowouts, which are breaks or deep depressions in the dune ridge (**Figure 11–7**), commonly occur where the dune's vegetation cover has been destroyed by natural processes, by vehicles, such as dune buggies, or by foot traffic, allowing wind to scour out the loose sand. Blowouts can deepen and widen rapidly under the right weather conditions, leading to deflation and eventual destruction of the dune itself.

Coastal dunes are best developed along shores that have ample quantities of sand, moderately strong (15 to 30 knots) and persistent onshore winds, wide and gradually sloping beaches, and high tidal ranges, all factors that assure a regular and ample supply of sand to the backshore. Windblown sand grains move by rolling, skipping, bouncing, and jumping along the ground, a process called **saltation**. The term does not refer to salt, but is derived from the Latin "to leap." When saltating grains hit the ground, they can bounce back into the air on impact or their momentum can cause other sand particles to skip upward. When the wind blows vigorously and persistently over dry fine sand, the air near the ground is filled with saltating sand grains that are bouncing along rapidly downwind. Saltating sand particles are driven up the gentle windward side of a dune and are deposited on the dune's upper backslope, where the wind currents slacken. With accumulation these sand deposits become unstable, and gravity causes the collected sand to slide and slump down the steep leeward slope of the dune (**Figure 11–8**). This process allows dunes to migrate slowly downwind, as sand is transported by the wind up the gentle windward slope and cascades down the steep leeward face of the dune.

Under storm conditions, dunes can be eroded by waves. This eroded dune sand can then be dispersed through the swash zone, where is it incorporated into submarine storm bars or is fluxed offshore by strong rip currents or parallel to the shore

FIGURE **11-7**

Dune Blowout. A dune ridge blowout at Pea Island, South Carolina. The dune in the forefront shows considerable storm damage.

FIGURE **11-8**

Coastal dune. Wind causes sand grains to saltate up the dune's gentle windward slope. Sand that accumulates beyond the crest in the dune's wind shadow slumps down the steep leeward slope. This process causes the dune to migrate slowly downwind.

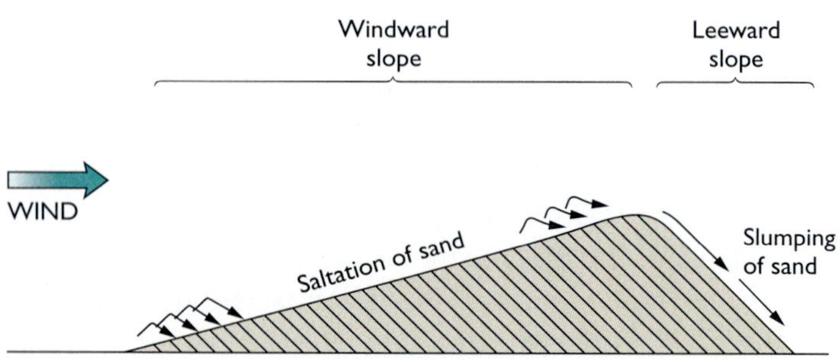

by longshore currents. With sustained wave attack, a steep scarp is carved into the dunes and gets taller as it is cut back into the dunes (**Figure 11–9**). This erosional feature itself mitigates storm erosion in two ways. First, the near-vertical face of the scarp reflects wave energy, sending it back offshore. Second, the surging water associated with storm breakers is deflected straight up by the near-vertical scarp wall, where it stalls and falls back down under the influence of gravity and flows harmlessly offshore. Furthermore, coastal dunes are natural "sea walls," so to speak, barriers that effectively prevent widespread flooding of the interior of barrier islands during severe storms or unusually high tides.

Human activity, if not carefully regulated, tends to degrade dunes in short order. Foot traffic can destroy grass that stabilizes the dune and make it susceptible to wind scour. Dune buggies are particularly harmful, because they quickly destroy the vegetation cover that anchors the sand; this can lead to large blowouts that become self-propagating as they become enlarged by wind that is funneled through the breach in the dune ridge. Also, storm water can surge through the blowouts, transporting sand from the beach and forming washover fans in the backshore area of the barrier island. The various ecosystems of barrier islands are best protected by limiting recreational activities that destroy dune grasses, by preventing building and general development along any of the dune ridges, by not allowing the mining of sand, and by curtailing foot passage to gain access to the neighboring beaches (**Figure 11–10**). The bayside area of the barrier island is less affected by human activity and this is where land development for human settlement is appropriate.

FIGURE 11-9

Dune erosion. A steep scarp has been cut into the primary dune ridge by storm waves.

FIGURE 11-10

Erosion control. Wooden foot ramps are used to direct foot traffic through dunes in order to minimize damage to grasses and shrubs.

11-4

Barrier Islands

Where coastal sand supplies are abundant and the sea floor slopes gently, the mainland is commonly bordered by a series of **barrier islands**. These islands are large deposits of sand that are separated from the mainland by the water of estuaries, bays, and lagoons. Prime examples are the barrier islands that stretch along the Atlantic, Pacific, and Gulf coasts of the United States. The water in the estuaries and bays is regularly exchanged with the water of the open shelf by flow through **tidal inlets**, which are water gaps between the barrier islands.

BARRIER-ISLAND LANDSCAPE

Barrier islands consist of a series of discrete environments (**Figure 11–11a**). Crossing the barrier from the ocean side, one typically finds the nearshore zone, beach, dune field, back-island flat, salt marsh, and water of the bay, lagoon, or estuary. Particularly broad barriers, such as the Sea Islands of Georgia, also display gently rolling to flat, forested interiors. Let's examine each of these.

The nearshore zone and beaches of the seaward side of the barrier island are reworked continually by breaking waves. These high-energy environments

FIGURE **11-11**

Barrier islands. (a) This is an aerial photograph of a broad, mature barrier island—Hatteras Island, North Carolina. (b) Several distinct land types comprise a barrier-island system. Note that the barrier island has migrated landward in response to rising sea level, as indicated by the fact that ancient salt-marsh deposits (peat) underlie the sand of the beach, dunes, and overwash fans. [Adapted from P. J. Godfrey, Barrier islands of the East Coast, *Oceanus* 19, no. 5 (1976): 27–40.]

(a) HATTERAS ISLAND, NORTH CAROLINA

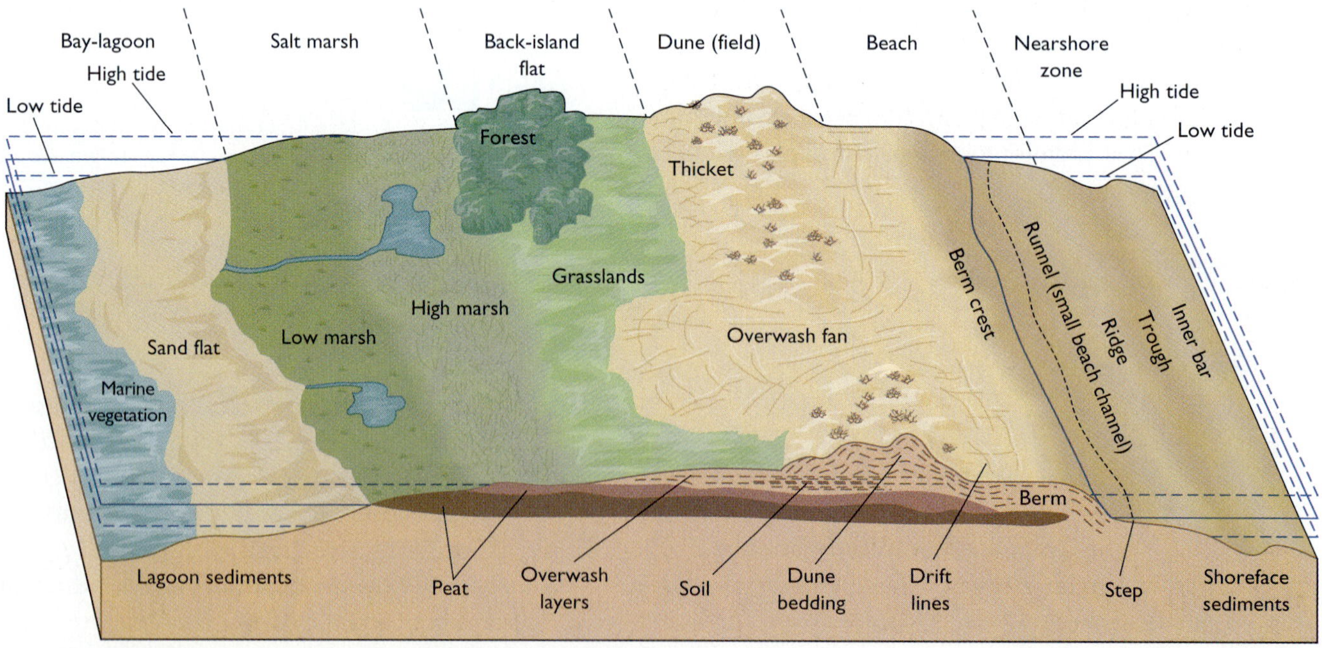

(b) MORPHOLOGY OF A BARRIER-ISLAND SYSTEM

are sites of longshore, and offshore-and-onshore transport of sand by wave-induced circulation, as discussed in the previous sections. The berm grades landward into a dune field, which is a system of sand dunes created by sea breezes that blow sand off the beach into the backshore. The dunes may be distributed irregularly or be arranged in parallel or almost parallel rows along the shoreline. If the sand is not stabilized by grasses, dunes will migrate downwind as sand is blown up the gentle windward side of the dune, over its crest, and down the steep downwind slope. Back-island flats generally consist of sand deposits shaped like pancakes or lobes that may or may not be covered by terrestrial or salt-marsh grasses. The sand lobes of the back-island flats are created during storm events by **washover**, or **overwash**, processes, whereby large, powerful storm waves overtop the low-lying parts of the barrier island and transfer beach and dune sand into the backshore zone. This overtopping by water moves sand landward and creates **washover** (or **overwash**) **fans** (see **Figure 11–11b**). Salt marshes fringe the bayward side of the barrier and represent low-energy environments where mainly mud accumulates and where salt-marsh grasses get established.

ORIGIN OF BARRIER ISLANDS

Comparisons of the geologic development of barrier islands in many places of the world indicate that they are created in a variety of ways. Many seem to have originated long ago, about 5,000 years before the present, when elongated coastal sand ridges were submerged during the Holocene rise of sea level (**Figure 11–12a**). As sea level rose, the barrier islands migrated landward and the shoreline shifted position (see the boxed feature, "Cape Cod, Massachusetts"). Other barrier islands seem to have developed from the longshore extension of **sand spits** along an irregular coastline (**Figure 11–12b**). A sand spit is a narrow "tongue" of sand extending from the shoreline; it is created by longshore drift. As the sand spit grows and becomes elongated, it may be breached during coastal storms. This breaching cuts tidal inlets and, as a consequence, creates sand barriers that are no longer attached to the mainland. Still other types of barrier islands may have resulted from the vertical growth of a longshore sand bar, so that its crest emerges out of the water well above high tide and becomes an island (**Figure 11–12c**).

Whatever their origin, barrier islands contain enormous quantities of sand. Once formed, the barriers migrated landward by overwash processes, which transferred sand from the front to the back side of the barrier island as sea level rose (**Figure 11–12d**). This is indicated clearly by the organic-rich black muds and peat containing root fragments of salt-marsh grasses that underlie (see Figure 11–11b) or are exposed on the seaward side of many barrier islands. Carbon-14 dating indicates that these mud deposits are several thousands of years old and must have formed behind barrier islands when they were located tens of kilometers seaward of the present-day shoreline. These salt marshes were engulfed, their grasses killed, and their muds and peat deposits covered by the sand of the migrating barrier islands.

STORM EFFECTS

The strong onshore winds that accompany storms drag and stack water against the ocean side of the barrier island, creating a **storm surge,** an extraordinarily high water level that floods the shore. Low-lying and flat coastal areas are particularly susceptible to damage by such flooding, especially if the storm surge coincides with a high spring tide when the total rise of sea level may be between 2 and 10 meters (~7 to 36 feet) above normal. Furthermore, storm surges create water that is much deeper than normal in the nearshore zone, enabling large waves, which would normally break offshore, to break directly on the beach and at times even to breach the island. As discussed previously, this overtopping process results in the transfer of sand from the seaward side to the back side of the barrier, producing a system of washover fans in the back-island flat (**Figure 11–13a**). The washover process, whereby sand is moved from the front to the backside of the barrier, is the principal mechanism by which a barrier-island system migrates landward in response to the rising level of the sea (see the boxed feature, "Padre Island, Texas").

Another possible storm effect is the excavation of new tidal inlets through a barrier (**Figure 11–13b**). Surprisingly, this type of barrier breaching is

FIGURE 11-12

The origin of barrier islands. (a) A coastal sand ridge develops when sea level is reasonably stable. Subsequently, a rise in sea level floods the backside of the coastal sand ridge, forming a lagoon and creating a barrier island. (b) The rapid longshore drift of sand promotes spit growth. Following a storm, a new tidal inlet separates the end of the elongated spit from its main trunk, creating a barrier island. (c) A longshore bar is built upward to sea level. During low tide, the exposed bar crest is reshaped by wind and during high tide by nearshore currents, creating a barrier island. (d) Sea level has risen continually since the last glaciation, a worldwide event known as the Holocene sea-level rise. Coastal geologists believe that many barrier islands originated about 5,000 years ago when the rate of sea-level rise decreased sharply in most areas of the world.

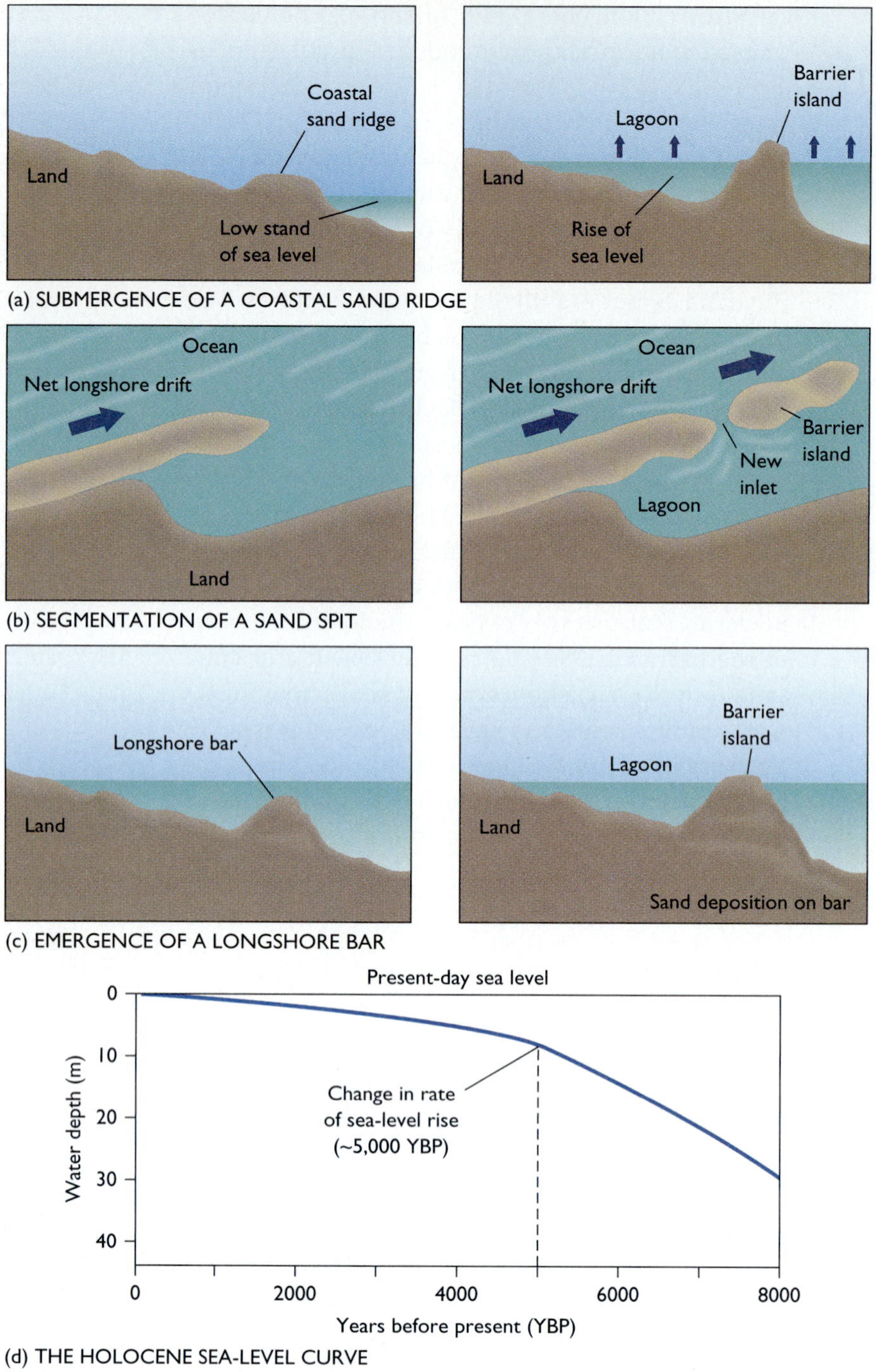

more likely to originate from the bay or lagoon than from the ocean side. The cutting of an inlet and the resultant segmentation of the barrier results from a combination of factors. First, the bay becomes swollen with water because of (1) the heavy rainfall that accompanies storms, (2) the increased run-off from both the mainland and the barrier island, and (3) a large tidal-inlet inflow of seawater driven by the storm surge pushing water onshore. Second, the strong onshore winds that lash the area create a storm surge not only against the seaward side of the barrier, but also in the bay on the mainland side, so that the water surface of the bay during the storm is tilted toward the barrier island. Once the storm winds end, water in the bay sloshes back against the barrier as a seiche, occasionally overtopping the island at its most vulnerable, low-lying points and cutting channels through it. Third, the process of inlet formation may be enhanced by winds that veer from an on-shore to an offshore direction as the

(a) WASHOVER FANS

(b) NEW TIDAL INLET

(c) STORM DAMAGE

(d) STORM DAMAGE

FIGURE 11-13

Storm modifications of barrier islands. (a) This is an aerial view of a barrier island with recently formed washover fans that carried sand into the salt marsh. Arrows indicate large washover fans. (b) This tidal inlet formed when part of the barrier island was breached during a recent storm. (c) Storm waves are eroding the beach and shore-front property. (d) Wave attack has undercut this coastal road.

storm center passes the site, blowing bay water through a newly created inlet and thereby helping to deepen and widen it.

In most cases, the breached inlet is short lived, because the longshore drift of sand quickly closes the passage as sand drifts across the inlet mouth. Some breached inlets may deepen and widen to become well-established features of the coastline. This is most likely to occur along narrow barriers that are long and continuous and hence have few preexisting tidal inlets for the exchange of bay and ocean water. Even if an inlet is not formed, storm waves can cause extensive damage (**Figure 11–13c** and **d**).

11-5

Cliffed Coasts

Perhaps the most spectacular and emotionally energizing sight along a stretch of coastline is the cliff under storm attack. From a cliff top, you can experience the raw, primordial energy unleashed by storm waves as they batter the base of hillside outcrops. Robert Frost captured, as few have, the natural power of storm waves attacking a cliff in his poem "Once by the Pacific."

Padre Island, Texas

Padre Island (**Figure B11–5a**), which extends along the Texas Gulf Coast for more than 180 kilometers (~112 miles), is the longest barrier island in the United States. Most of the island remains undeveloped, in part because of congressional legislation that created the Padre Island National Seashore in 1962. Despite its impressive length, Padre Island is vulnerable to wave erosion because most of its land is no higher than 6 meters (~20 feet) above sea level. The long, thin, low-lying island of sand is susceptible to washover, particularly during hurricanes.

From the evidence of core samples, geologists believe that Padre Island first emerged from the sea several thousand years ago when sea level was ~5 meters (~17 feet) lower than it is today. The geologic history of the island is interpreted from a geologic cross section (**Figure B11–5b**). As the seas invaded the land after the last glaciation, Padre Island responded in two ways. First, the barrier grew upward with the addition of sand. Second, storm waves overtopped the low-lying barrier, and strong winds blew sand onshore. Both of these processes transported a great deal of sand into the lagoon on the backside of the island, and created a thick wedge-shaped deposit of washover and windblown sand. At first the lagoon was open to the ocean, because of the presence of numerous tidal inlets. However, it eventually was blocked off from the ocean as strong longshore currents closed and filled the tidal inlets, transforming the chain of barrier islands into a single, very long barrier island. Today, Laguna Madre is very salty, the result of limited mixing by the tides (few tidal inlets), little freshwater inflow, and high evaporation rates.

Viewed from above (**Figure B11–5c**), Padre Island has a curved shape. From October through January, winds out of the northeast generate strong longshore currents that move sand predominantly southward. In contrast, southeasterly winds blow during February and throughout the summer, producing a prevailing northward drift of sand. This seasonal shift in the direction of longshore drift is reflected in the mineralogy of the sand (see Figure B11-5c). North of 27°N latitude, beach minerals can be traced to rivers located to the north of Padre Island, such as the Brazos and Colorado Rivers. The beach sands along the southern half of the barrier island are composed of minerals supplied by the Rio Grande, the river that separates Mexico from Texas.

Hurricane season runs from June through October. These storms cause tremendous damage to this thin, low-lying barrier. During a hurricane attack, much of Padre Island is flooded, and washovers are common. At times, scouring of sand from the island carves out tidal inlets. However, most of these quickly close up during fair weather because of the longshore drift of sand.

Visit **www.jbpub.com/oceanlink for more information.**

FIGURE **B11-5**

Padre Island, Texas. (a) This is an aerial view of the developed part of Padre Island, located on the Texas Gulf Coast between Mexico and Corpus Christi Bay. Padre Island is the longest uninterrupted barrier island in the United States. (b) This cross section reveals that Padre Island has grown upward since it first formed. Storm washovers and strong onshore winds have transported large amounts of sand landward, filling in Laguna Madre. (c) During a year, longshore currents converge near the center of Padre Island in response to seasonal changes in the direction of the wind. [Adapted from B. R. Weise and W. A. White, *Padre Island National Seashore,* Guidebook 17 (Austin, Tex.: University of Texas, Bureau of Economic Geology, 1980).]

The shattered water made a misty din.
Great waves looked over others coming in.
And thought of doing something to the shore
That water never did to land before.
The clouds were low and hairy in the skies,
Like locks blown forward in the gleam of eyes.
You could not tell, and yet it looked as if
The shore was lucky in being backed by a cliff,
The cliff being backed by a continent;
It looked as if a night of dark intent
Was coming, and not only a night, an age.
Someone had better be prepared for rage.
There would be more than ocean water broken
Before God's last Put out the Light *was spoken.*

Unlike the case of a beach undergoing storm erosion, a single storm rarely produces a noticeable change in the appearance of a cliff composed of

(a) PADRE ISLAND

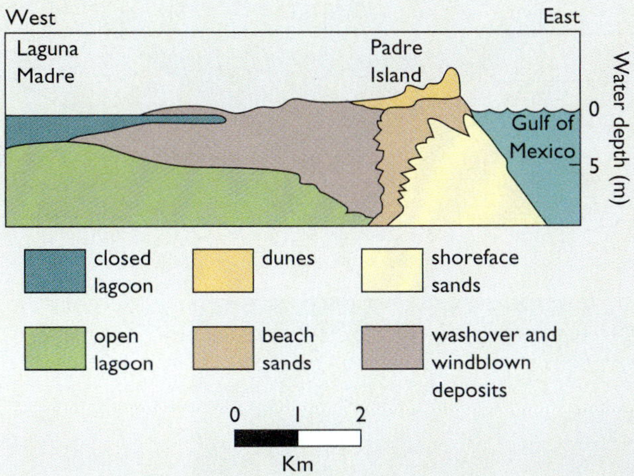

(b) GEOLOGIC CROSS SECTION

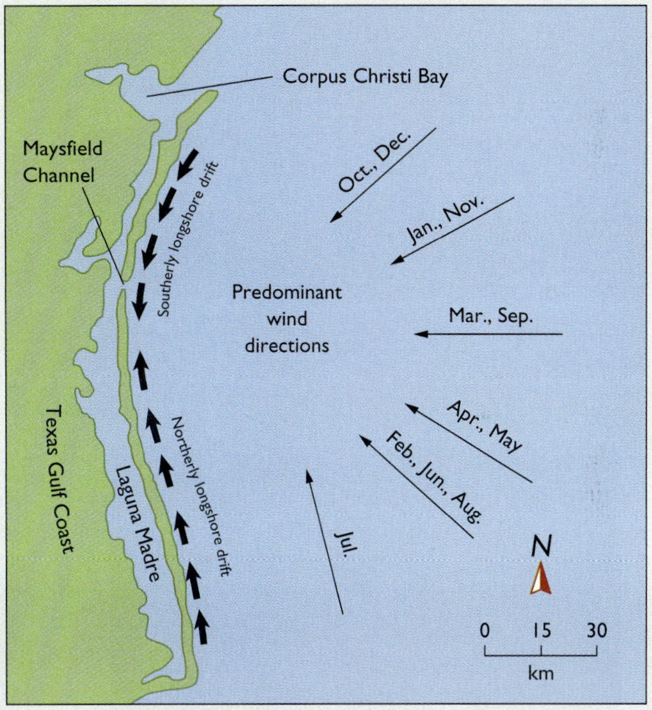

(c) LONGSHORE SAND DRIFT

rock or consolidated sediment. After all, this is the reason that cliffs exist where they do—because of the hardness and strength of the rock that composes them. When changes do occur, however, it is frequently spectacular and catastrophic, with gigantic slabs of rock and debris sliding off the cliff and falling into the sea (see the boxed feature, "San Diego County, California").

Cliffs are most vulnerable to the sea at their bases. Here waves exert powerful forces that compress the air trapped in the cavities of the rock. As the wave recedes, these cavities expand explosively, stressing and eventually shattering the cliff material. Also, the impact of sand, gravel, and cobbles tossed against the base of the cliff by breakers wears down the rock. In addition, chemical attack by seawater can dissolve rock, particularly limestone. Eventually, these destructive processes undercut the cliff base to form a deep notch, and the oversteepened rock rips away from the cliff face and collapses into the sea (**Figure 11–14**). For a time, the fallen rock debris protects the base of the hillslope from further wave attack. Once the material is worn down and carried away by longshore transport, though, the cliff face becomes exposed again to wave attack, and the process of notching of the cliff base is renewed. In this way, collapse and retreat of the cliff are episodic, with little change for a long time, then drastic change suddenly.

Large quantities of cliff material are moved downslope by mass movement, including the slumping, sliding, and flowing of rock, debris, and mud (**Figure 11–14a** and **b**). The type and frequency of mass movement on a cliff are influenced by the composition and durability of the cliff material, as well as by such geologic structures as joints,

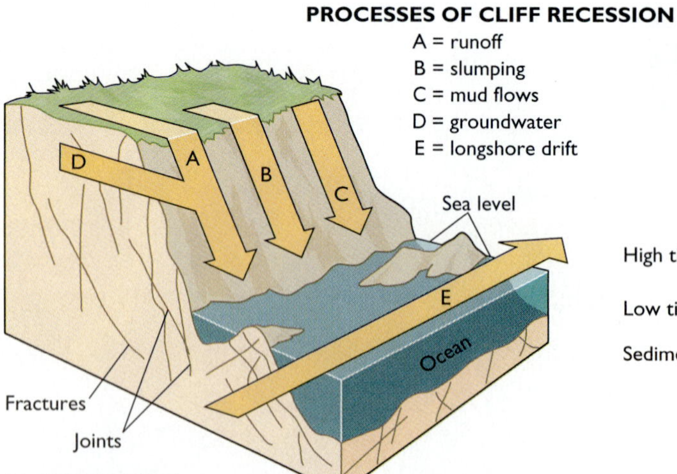

(a) CLIFF EROSION

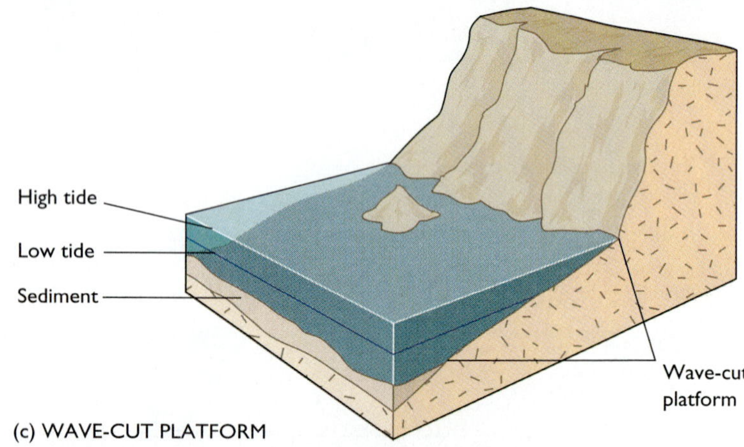

(c) WAVE-CUT PLATFORM

(b) CLIFF FAILURE

(d) TECTONIC UPLIFT OF WAVE-CUT PLATFORM

FIGURE **11-14**

Cliff erosion. (a) Surface runoff, slumping, mud flows, and groundwater seepage all erode cliffs, as waves destabilize the cliff face by undercutting its base. [Adapted from V. J. May, in *The Coastline,* R. S. K. Barnes, ed. (New York: John Wiley, 1977).] (b) Recent collapse of cliffs along Puget Sound after storm. (c) If sea level is stable for a time, waves will erode a wave-cut platform as the cliff face retreats landward. A new platform is being cut at sea level by breakers. (d) Crustal uplift has raised a wave-cut platform by about 20 meters (~66 feet) at Newport, Oregon.

San Diego County, California

Commercial and private development of the coastal bluffs north of San Diego is extensive and directly affects the rates of sea-cliff recession. Between 1973 and 1978 large masses of rock fell to the beach (**Figure B11–6**). Coastal engineers and geologists believe that many of these cliff failures are the direct result of human activity in the following ways:

1. *Rise of groundwater levels.* Because of agricultural irrigation, the watering of lawns, and leakage from swimming pools, septic tanks, and cesspools, the level of water in the ground has risen and remains at a high level even during the dry summer months when normally it would be low. This surplus groundwater has increased the weight of the cliff rock, lubricated cracks and fractures in the rock, and increased the chemical dissolution of certain rocks, particularly limestone. All of these factors weaken the cliff and eventually trigger its failure.
2. *Coastal defense measures.* Construction of groins, seawalls, and concrete embankments along the cliffed coastline to retard erosion may have had the opposite effect and increased the rate of shoreline retreat. Artificial structures affect the shore in a variety of undesirable ways. In San Diego County, for example, **riprap** (a pile of boulders placed at the base of a cliff or beach) and seawalls have changed the bottom profile of the shore, causing beaches to become narrower and steeper. As a result, storm waves now break closer inshore and are able to attack and erode the base of the cliff more directly.
3. *Canyon-head erosion.* Drainage culverts that control the flow of rainwater have replaced natural runoff. The concentrated discharge of water from these drains has not only reactivated gullying of old river canyons, but also created new gullies that cut deeply into the coastal bluffs.
4. *Bluff-top grading.* The removal of soil and vegetation for construction purposes has led to the immediate retreat and, in some cases, to the collapse of bluffs as surface runoff erodes the unprotected cliff tops. A case in point is the decision by condominium owners at Solana Beach to have engineers install a seawall at the base of their bluff and a concrete wall along the upper edge of the cliff face—projects that were to be completed at a cost in excess of $1.5 million. In May 1979 the almost-completed upper wall rotated outward and, along with part of the cliff, collapsed onto the seawall below. This certainly is an expensive way to place riprap at the bottom of a cliff!

FIGURE **B11-6**

Cliff collapse. These photographs show before (a) and after (b) cliffs along a stretch of cliff in San Diego County collapsed. Note the wave-cut platform at the base of the cliff in part a.

(a) BEFORE CLIFF COLLAPSE (b) AFTER CLIFF COLLAPSE

fractures, and faults that weaken the rock. Hard, crystalline rock, such as granite, resists wave attack much better than unconsolidated sediment like sand, clay, and glacial till (**Table 11–1**). The amount of rainfall can also be a factor; rainwater seeps into and saturates the ground, weakening and lubricating the rock and increasing the potential for failure.

As the cliff is worn back, wave erosion produces a **wave-cut platform** in front of the cliff (**Figure 11–14c**). The platform is an eroded surface that has been cut by the power of breaking waves. Typically, such platforms slope gently seaward and have variable widths that rarely exceed 1 kilometer (~0.6 mile). Striking examples of wave-cut platforms are scattered along the Pacific shore of the United States, where they are evident not only at sea level, but also well above the ocean surface due to tectonic uplift of the land (**Figure 11–14d**).

11-6 Deltas

Where rivers supply large quantities of sediment to the shore, the estuaries and bays become completely filled, and rivers then dump sediment directly into the ocean. If the rate of sediment supply exceeds the rate of sediment removal by waves and tidal currents, a buildup of sediment occurs at the river mouth. This deposit, which commonly has a triangular shape when seen from high above, is called a **delta**, because it resembles the Greek capital letter, *delta* Δ.

Large deltas are divided broadly into three zones (**Figure 11–15a**). The **delta plain** is the flat lowland that lies at or above sea level; it is drained by a network of distributary channels. The areas between the distributary channels are vegetated or flooded with ponded water. The delta plain grades seaward into the **delta front**, which consists of the shoreline and the broad submerged "front" of the delta. Typically, the delta front slopes gently seaward and is gullied with submarine channels. Farther offshore, the bottom flattens out into **prodelta** (meaning "in front of") deposits of the inner continental shelf.

The deposits in each of these delta areas are distinctive in composition and arrangement (Figure 11–15a). For example, delta plains are underlain by sequences of flat-lying beds of sand and mud deposited in the distributary channels and interchannel (between channels) areas of the river; they are referred to collectively as **topset beds**. The delta front consists of thick sands and silts that dip seaward; these are called **foreset beds**. The foreset beds, in turn, grade into the **bottomset beds** of the prodelta, which are flat-lying beds composed of fine river silts and clays that have settled out of suspension farther offshore. As sediment accumulates, the delta grows seaward by the deposition of foreset beds, and the topset beds of the delta plain overlap and bury the foreset beds of the delta front.

Actually, not all deltas display the classic delta form (**Figure 11–15b**). Only at river mouths where waves and tides are ineffective in rearranging the river deposits does this characteristic shape develop. Such delta systems, exemplified by the Mississippi Delta, are called **river-dominated deltas**. In coastal areas, where wave energy is high relative to the supply of river sediment, deltas become **wave-dominated**. Wave erosion and strong longshore currents disperse the sediment away from the river mouth, producing a relatively straight coast with only a slight bulge of the shoreline. In some regions, a large tidal range overshadows river and wave effects, creating a **tide-dominated delta**. The strong ebb and flood currents of the tide rearrange the river-supplied sediment into long, linear sub-

TABLE 11-1 Sea-cliff recession rates.

Cliff composition	General rates of erosion
Granite and crystalline rocks	$<10^{-3}$ m/yr
Limestone	10^{-3}–10^{-2} m/yr
Shale and sandstone	10^{-2}–10^{-0} m/yr
Unconsolidated sediment	10^{-0}–10^{-1} m/yr
Location and material	**Measured rates of erosion**
New England: crystalline rocks	0 m/yr
Cape Cod, Massachusetts: glacial deposits	0.3 m/yr
New Jersey: gravel, sand, and clay	1.8 m/yr
Louisiana: sand and clay	8–38 m/yr
Southern California: alluvium	0.3 m/yr

Source: Compiled by D. F. Ritter, *Process Geomorphology* (Dubuque, Iowa: William C. Brown, 1986).

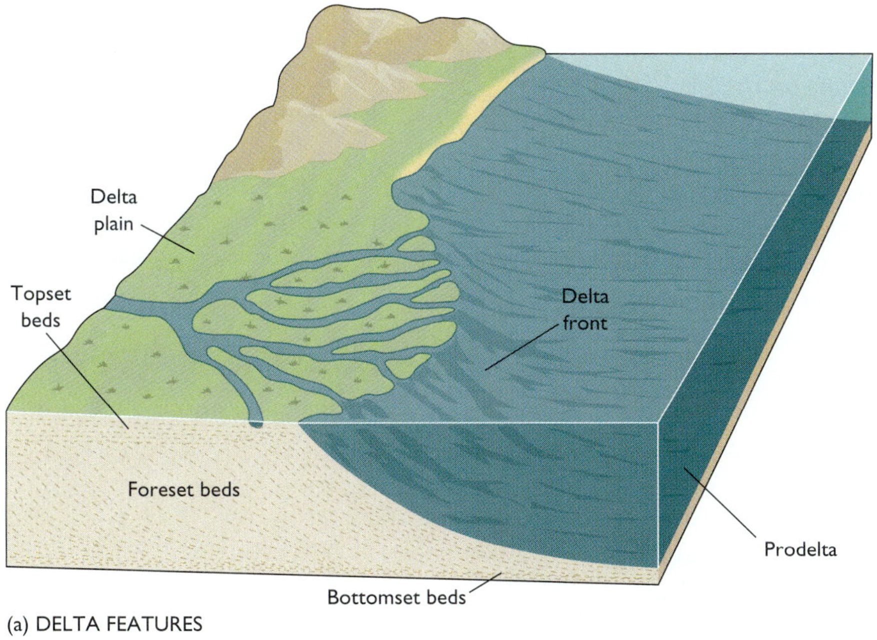

FIGURE 11-15

River deltas. (a) Where river-supplied sediments are abundant, a delta grows seaward. Typically, the delta consists of a delta plain, a delta front, and a prodelta. (b) Depending on the relative effects of river, waves, and tides, deltas assume a variety of shapes. [Adapted from Open University Course Team, *Waves, Tides, and Shallow-Water Processes* (New York, Pergamon Press, 1989).]

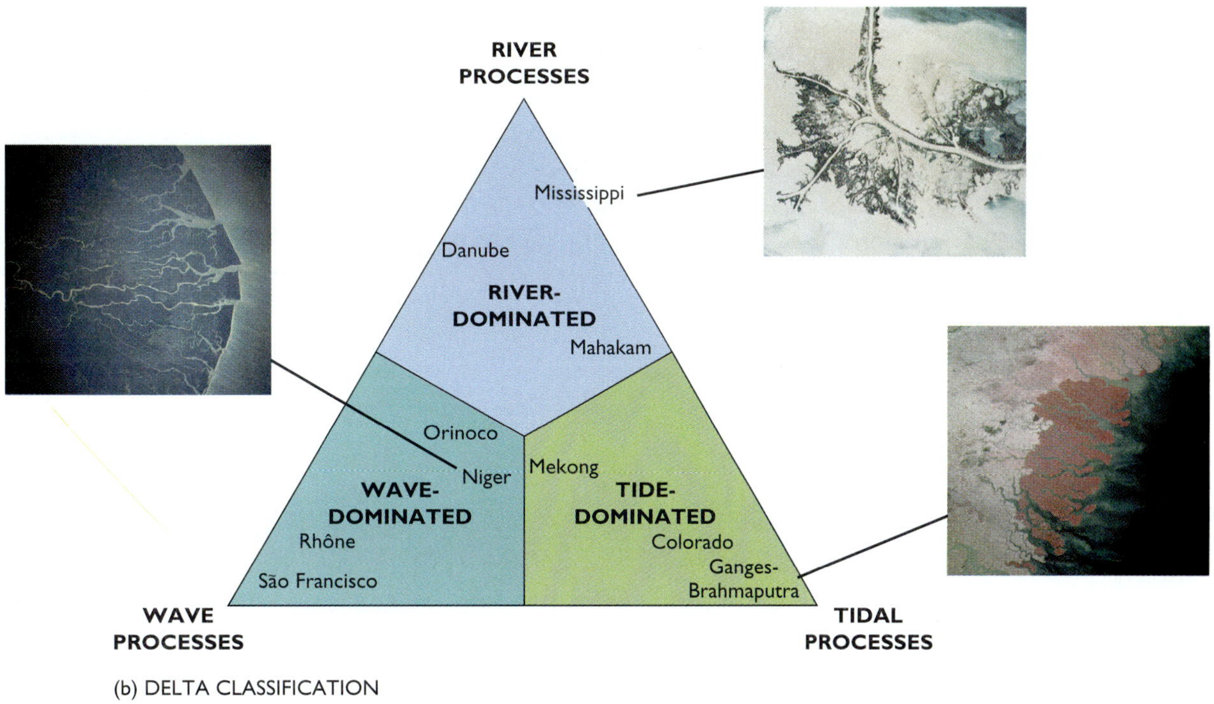

marine ridges and islands that tend to fan out from the river mouth.

Figure 11–15b is a triangular plot with the corners representing deltas that are clearly dominated by river, wave, and tidal effects. Note that the area of the triangle is divided into three regions—river-dominated, wave-dominated, and tide-dominated —which intersect at a single point in the center of the triangle. A delta that plots near this point, such as the Orinoco Delta of Venezuela, is affected equally by rivers, waves and tides. The Mahakam Delta of Borneo, which is plotted on the right side of the triangle, midway between the corners representing river and tidal dominance, displays a fan-like arrangement of linear islands that collectively form a classic delta shape. This shape reflects an equal impact by river and tidal processes and almost no influence by waves.

Deltas represent accumulations of muds and sand that are largely supplied from the parent river.

A reduction in the supply of sediment typically has deleterious effects for the delta region. For example, navigation and flood control measures on the Mississippi River—such as the construction of dams, the dredging of river channels, and the building of levees—and the increased subsidence of the Mississippi Delta itself—a result of the compaction of muds and of pumping out groundwater—have caused widespread erosion and flooding in southern Louisiana. As a result, sea level there is rising at a rate of 1 centimeter (~0.39 inches) per year, about ten times (one order of magnitude!) the world average. Some segments of the Louisiana shore are retreating faster than 20 meters (~65.7 feet) each year. In 1992, Hurricane Andrew devastated the southern Louisiana coast by (1) severely eroding and overwashing the many low-lying barrier islands, (2) smothering in sediment more than 80 percent of the oyster reefs that flourished in the estuaries behind the barrier islands, and (3) eroding almost 70 kilometers (~43.4 miles) of coastal dunes that protect the backshores of barrier islands. Some coastal geologists fear that the barrier islands and coastal wetlands of Louisiana may disappear entirely by 2020 if measures are not immediately taken to mitigate the severe erosion.

TABLE 11-2 Shoreline erosion rates.

Region	Average Rate of Shoreline Change (m/y)	Maximum Accretion (m/y)	Maximum Erosion (m/yr)
Atlantic Coast	-0.8	25.5	-24.6
Maine	-0.4	1.9	-0.5
New Hampshire	-0.5	-0.5	-0.5
Massachusetts	-0.9	4.5	-4.5
Rhode Island	-0.5	-0.3	-0.7
New York	-0.5	-0.3	-0.7
New Jersey	0.1	18.8	-2.2
Delaware	0.1	5.0	-2.3
Maryland	-1.5	1.3	-8.8
Virginia	-4.2	0.9	-24.6
North Carolina	-0.6	9.4	-6.0
South Carolina	-2.0	5.9	-17.7
Georgia	0.7	5.0	-4.0
Florida	-0.1	5.0	-2.9
Gulf of Mexico	-1.8	8.8	-15.3
Florida	-0.4	8.8	-4.5
Alabama	-1.1	0.8	-3.1
Mississippi	-0.6	0.6	-6.4
Louisiana	-4.2	3.4	-15.3
Texas	-1.2	0.8	-5.0
Pacific Coast	0.0	10.1	-5.0
California	-0.1	10.1	-4.2
Oregon	-0.1	5.0	-5.0
Washington	0.5	5.0	-3.9
Alaska	-2.4	2.9	-6.0

Note: Negative values denote erosion; positive values, accretion.
Source: Adapted from S. K. May, R. Dolan, and B. P. Hayden, *EOS* 64 (34) (1983): 551–53.

11-7

Impact of People on the Coastline

The resilience of the shoreline stems from its dynamic capacity to respond to changes in wave energy and variations in current flow. Daily, seasonal, annual, and even longer cycles of sand erosion and deposition affect beaches. They are never stable for long. Shorelines worldwide are migrating landward as a result of the Holocene rise of sea level (**Table 11–2**). The coast is exciting and scenic, and it provides recreational opportunities as well (for swimming, boating, etc.). Consequently, it is a desirable area for habitation by people and hence a prime target for real-estate development. The livability and value of buildings, such as beach homes, condominiums, hotels, and marinas, depend, however, on coastal stability—a requirement that is at odds with the dynamic state of coastal systems. The English philosopher Francis Bacon (1561–1626) observed that "Nature is often hidden, sometimes overcome, seldom extinguished . . . [and] to be commanded, must be obeyed." This fundamental principle is only now beginning to be understood by more people than merely a handful of coastal geologists.

Humans endeavor to impose their will on the coastline principally in one of two ways: (1) by interfering with the longshore transport of sand, and (2) by redirecting wave energy striking a stretch of shore. Intervention in the drift of sand involves the erection of jetties and groins perpendicular to the shoreline (**Figures 11–16a** and **11–16b**). **Jetties** are built to prevent or, at the very least, to diminish sediment deposition at the mouth of harbors, estuaries, and tidal inlets due to the longshore drift of sand. Channelizing the flow between jetties

FIGURE 11-16

Coastal engineering structures. Coastlines can be modified by trapping the longshore drift of sand by groins (a) or jetties (b), and by redirecting the energy of incoming waves with breakwaters (c) or seawalls (d).

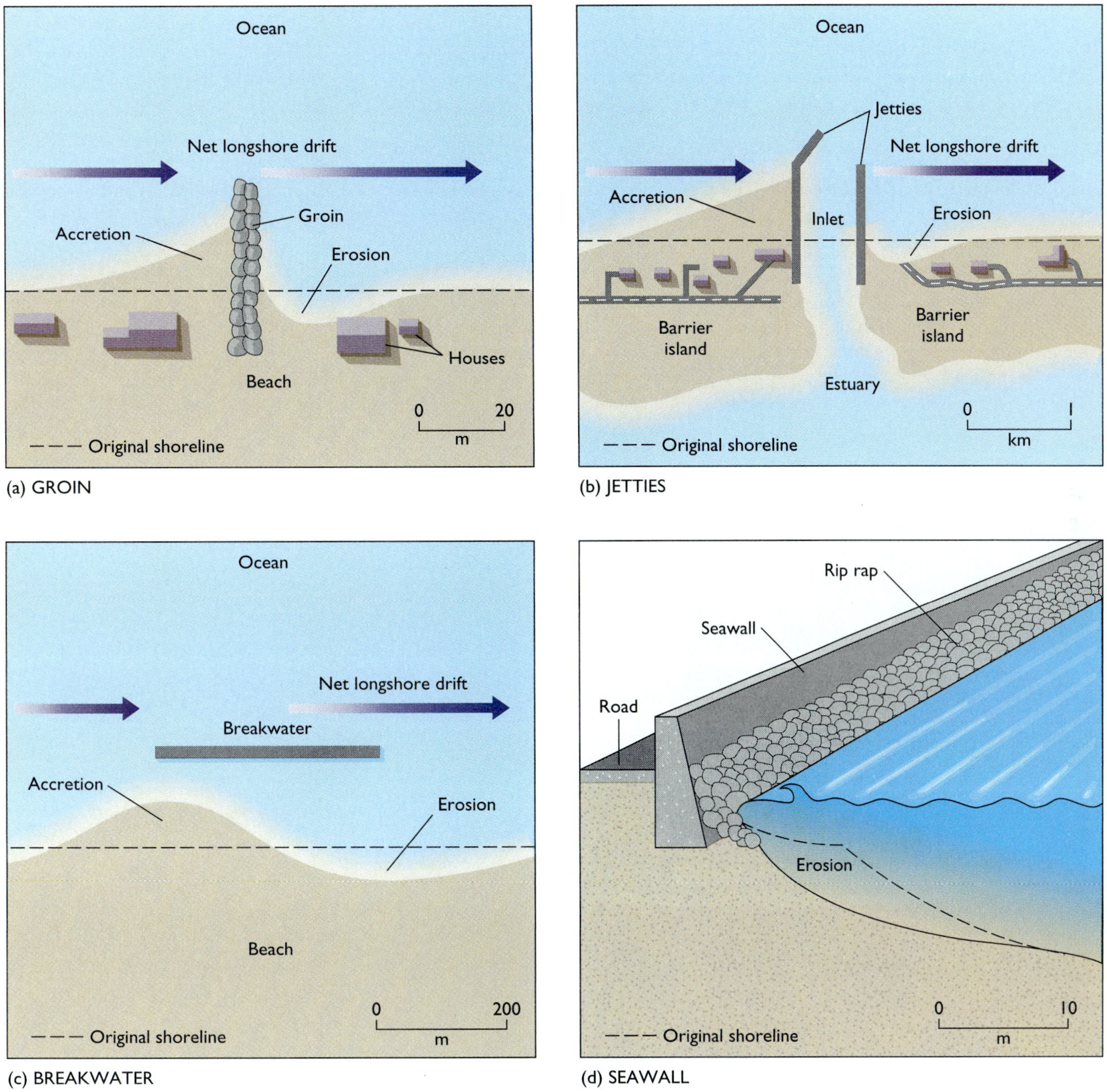

invariably increases current velocities in the inlet. This accelerated flow prevents the deposition of river sediment and may cause scour and deepening of the channel bottom between the jetties. If they are effective, however, jetties prevent the dispersal of sand across the mouth to the downdrift shore, with the unfortunate result of causing a debit in its sand budget and, hence, net erosion over time (see the boxed feature, "Ocean City and Assateague Island, Maryland"). Also, as sand piles up against the updrift side of the jetty and exceeds its retaining capacity, sediment (if it is not dredged) can "leak" around the end of the jetty and spill into the harbor or estuary mouth. Then the inlet must be dredged in order to keep it safe for shipping.

Groins, much smaller and cheaper than jetties, are positioned on eroding beaches to trap sand artificially and to promote accretion (buildup). They

Ocean City and Assateague Island, Maryland

A tragic example of the damaging consequences of uncontrolled urbanization of barrier islands is Ocean City, Maryland (**Figure B11–7a**). Here, land developers, so-called city planners, and engineers have either failed or been unwilling to recognize that the various parts of a barrier-island system are interdependent. They continue to formulate and implement policies on the mistaken belief that coastal islands are isolated, self-contained systems that can be stabilized and controlled if enough money is spent.

Ocean City is located on Fenwick Island, a barrier island nourished by the southerly drift of sand eroded from

FIGURE **B11-7**

Ocean City, Maryland. (a)This is an aerial view of Ocean City looking south. The eroding Assateague Island can be seen in the extreme upper right corner of the photo. (b) Jetties which were constructed shortly after a hurricane opened Ocean City Inlet in 1933 interfered with the southerly longshore drift of sand. The beaches on the southern end of Fenwick Island accreted as sand was trapped on the updrift side of the jetties. The northern end of Assateague Island became sediment-starved and rapidly eroded. [Adapted from S. P. Leatherman, *Overwash Processes* (Stoudsburg, PA: Hutchinson, 1981); and S. P. Leatherman, *Shore and Beach* 52 (3) (1984): 3–10.]

(a) OCEAN CITY

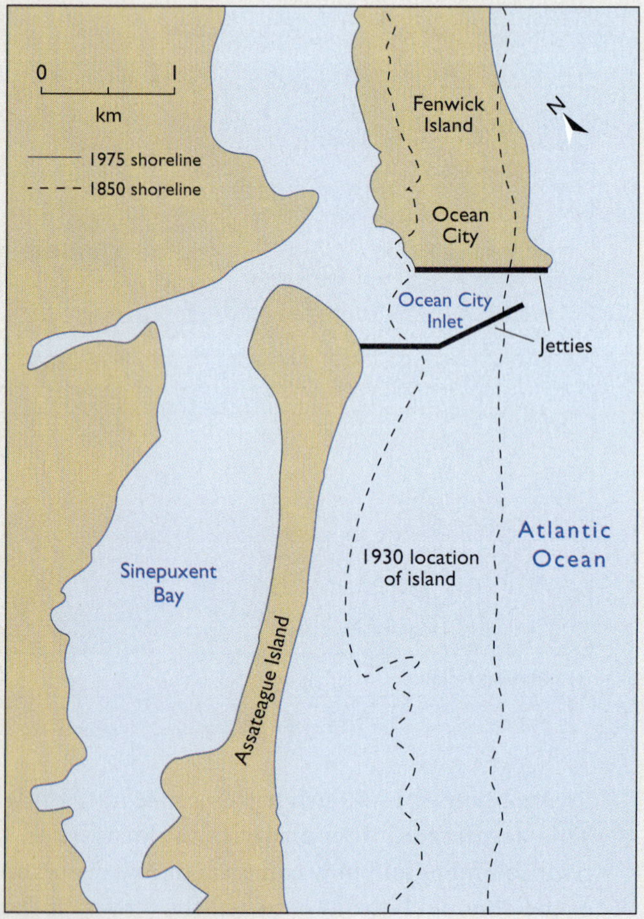

(b) HISTORICAL CHANGES

the Delaware shoreline to the north. In 1934 and 1935, jetties were installed to stabilize the inlet (**Figure B11–7b**) —a decision that profoundly altered the Maryland shoreline. Sand was deposited on the updrift side of the northern jetty. This rapid beach growth provided a false sense of security to the inhabitants of Ocean City, initiating a period of prosperity and rapid growth. The population currently swells to over 250,000 people during the peak summer season in Ocean City.

Historical records and maps indicate that Fenwick Island is migrating landward in response to the worldwide rise of sea level. Sand is being transferred from the front to the back of the barrier by washover processes, and the island is migrating at a rate of about 0.6 meters (~2.0 feet) per year. This rate is typical of much of the Maryland shoreline. Because the southern end is broad and anchored to the northern jetty, Ocean City is being stranded as the remainder of the island migrates landward. City residents, in an effort to stem the loss of sand and the consequent encroachment of the sea, have installed a system of fifty groins, each costing about $400,000. In addition, a sand-nourishment effort for Ocean City has been recently completed at a cost of between $75 million and $100 million.

The Army Corps of Engineers warns that sand nourishment of Fenwick Island will likely be unaffordable thirty or forty years from now. The long-term outcome of such "management" policies seems obvious— destruction of the shoreline and eventual bankruptcy of the city. Ocean City will become a human-made fortress, armored with seawalls, groins, and jetties, but no beaches. The continuously rising ocean level dooms the city, which will, without any doubt, be devastated by storm surge. It's simply a matter of time.

As is usually the case, the damage that has resulted from "stabilizing" Fenwick Island is not limited to Ocean City. Across the inlet lies Assateague Island, a largely undeveloped state parkland. The northern end of this island has been eroding at an alarming rate of up to about 12 meters (~40 feet) per year ever since the mouth of the nearby inlet has been "protected" by jetties (see Figure B11–7a and b). This rate of retreat is almost *a hundred times* faster than the average rate of coastal recession for the Maryland shore. In fact, Assateague Island's northern end has become detached from the southern jetty because of its rapid retreat into the bay. If the current rate of island migration continues, the northern end of Assateague Island will drift onto the mainland sometime during the early 21st century. According to S. P. Leatherman, a coastal geologist, this is the most severe human-induced erosion anywhere in the United States.

Visit www.jbpub.com/oceanlink for more information.

essentially are dams designed to impound sand rather than water. However, they suffer from the same shortcoming as jetties in that they starve the downdrift stretch of beach of sand and aggravate erosion there. In response to this effect, more groins are built, until the shore is littered with these structures. In many cases, the effort proves fruitless, because the efficient entrapment of sand by the updrift groins denies a natural supply of sand to the downdrift structures. Groins also can be dangerous to swimmers, because deep holes in the bottom can be scoured out near their seaward ends and they tend to deflect longshore currents offshore, turning them into rip currents.

A second strategy to control shorelines is the redirection of wave energy away from sites that are eroding by the construction of breakwaters and seawalls (**Figures 11–16c** and **11–16d**). **Breakwaters**, which are barriers built in front of harbors or shorelines, are expensive structures with high maintenance costs. Built offshore and parallel to a beach or harbor mouth, they are designed to absorb the pounding of breakers or reflect waves back to sea. However, the quiet water behind the breakwater impounds sand drifting alongshore and denies this sediment to the downdrift portion of the shoreline.

Seawalls are built on the shore, on the assumption that they protect beaches, roads, and bluffs from erosion by storm waves. They affect the shoreline in a bewildering variety of ways. First, by reflecting and redirecting wave energy, seawalls

SCIENCE BY NUMBERS

Sand-Budget Study

A beach is located downdrift from a tidal inlet between two barrier islands. A great deal of sand bypasses the inlet and supplies the downdrift beach with an average of 85,000 m^3/yr. Studies indicate that this longshore drift is in fact the only source of sand to the beach. Sand losses result from longshore transport out of the beach (72,000 m^3/yr) and offshore transport (9,000 m^3/yr).

A sand-budge analysis indicates the following:

Net effect = (total sand inputs) - (total sand outputs).
Net effect = (longshore drift in) - (longshore drift out + offshore transport).
Net effect = (85,000 m^3/yr) - (72,000 m^3/yr + 9,000 m^3/yr) = 4,000 m^3/yr.

We conclude that the beach is accreting (widening) over time.

Because of the boat traffic through the tidal inlet, coastal engineers plan to build jetties in order to stabilize the inlet. They realize that this will eliminate the only supply of sand to the beach located downdrift of the inlet. Their plan is to nourish the beach each year with an amount of sand that will keep it in a steady-state condition. This means that they want to supply a volume of sand each year such that the beach will neither erode nor accrete over time. Its width will remain fixed, because the input of sand will balance the output of sand each year. The engineers estimate that sand will cost about \$5.00/m^3. What will be the annual cost for nourishing the beach so that it remains in a steady-state condition?

The problem is straightforward. To keep the beach at a steady state, we need to supply a volume of sand that equals the volume of sand loss each year. If we do this, then the beach will remain in steady state, because the input will equal the output of sand. The critical assumption, of course, is that the sand inputs and outputs estimated from the recent past will continue into the future. The output of sand is

72,000 m^3/yr + 9,000 m^3/yr = 81,000 m^3/yr.

This is the amount of sand that must be supplied each year in order to achieve steady-state conditions. The annual cost will be

(81,000 m^3/yr) (\$5.00/m^3) = \$405,000/yr.

This is a reasonable annual cost, because the tourist trade brings in an excess of \$5,000,000 each year to the local economy.

increase the turbulence of the water at the front of the barrier. This increases the erosion of sand, and causes the beach at the foot of the wall to become narrower and steeper, and the nearshore zone to deepen. Consequently, large storm waves, which normally break offshore, now reach the beach and seawall—an effect that increases, rather than retards, storm erosion of the shore the structure was designed to protect! Ultimately, the seawall is weakened, undermined, and destroyed by the incessant pounding of waves. Maintenance costs for these structures are very high and often prohibitive. Second, the seawall and the steep bottom of the nearshore cause wave reflection, which under certain conditions can generate strong rip currents that erode the beach even more. Third, as rock or concrete slabs break off the seawall, they become submerged in the nearshore zone, making swimming and diving perilous to the point where they may be forbidden by town ordinance. Fourth, a seawall armors the shore, removing a potential source of valuable sand to the local beaches. A shoreline that is eroding often supplies sand to the downdrift beaches. Obviously, reducing or eliminating sediment input to a beach disrupts its sand budget, shifting it to the deficit side, which aggravates rather than abates erosion.

A useful although expensive way to reduce beach erosion is to introduce sand fill artificially by importing and placing large quantities of sand on the beach and in the nearshore zone. This engineering technique, called **beach nourishment**, relies on sand quarried on land or, where such sources of sediment are unavailable, on sand dredged from offshore deposits. A well-designed nourishment program widens the beach and naturally protects shorefront property from the destructive effects of large storm waves. However, the placement of sand fill on a beach is *always a temporary solution at best.* Longshore and offshore transport result in the systematic loss of this valuable sand over time and a consequent return of the beach to its original eroded state. Rarely does the growth of the beach by sand nourishment last more than five years.

Since implementation of the 1962 Water Resources Act, many beaches along the east coast of the United States have been artificially nourished. Such projects require large expenditures of funds provided by federal, state, and local agencies. **Table 11–3** lists the volume of sand and the cost for a few of these projects. Both Florida and New Jersey have recently completed extensive beach-nourishment programs in order to "reverse" beach erosion.

Building homes and cottages along sea cliffs and beaches is a very risky proposition, not only because of the dynamic nature of these environments, but also because of the worldwide rise of sea level. The encroachment of the sea onto the land is likely to continue well into the foreseeable future. Many climatologists believe that the increasing levels of carbon dioxide and other gases injected into the atmosphere by the combustion of fossil fuels will raise global temperatures (the *greenhouse effect*) and accelerate the melting of glaciers. This, along with the thermal expansion of seawater as sea surface temperatures climb up, will increase the rate of sea level rise, causing more land to be flooded by seawater.

To educate land investors about such pitfalls, a series of excellent books has been published by Duke University Press under the editorship of O. H. Pilkey and W. J. Neal. Each book details the specific coastal oceanography of a state or region of the United States and offers guidelines for the erection of shoreline structures. The books already published are listed at the end of the chapter. They are excellent, concise resources that are invaluable for learning about specific, regional problems related to shore construction.

11-8

Future Discoveries

Coastal oceanographers are working hard at understanding the many intricate and interactive processes that make the shore such a dynamic environment. The combination of waves, tides, currents, storms, multiple sources of sediment, rising sea level, and coastal development by humans create a complex interactive system that responds to perturbations on many different scales of space and time. Coastal engineers are designing structures such as groins and jetties in ways that minimize their impact on sediment dispersal to

TABLE 11-3

Selected beach-replenishment projects on the U.S. East Coast.

Beach	State	Year	Volume of Sand (yd^3)	Cost at Time of Construction
Great South Beach	N.Y.	1962	993,500	$ 844,100
Jones Beach	N.Y.	1927–61	>40,000,000	?
Sea Gurt	N.J.	1966	425,211	552,774
Long Beach Island	N.J.	1979	1,000,000	4,600,000
Avalon	N.J.	1987	1,300,000	24,000,000
Ocean City	Md.	1963	1,050,000	1,517,600
Atlantic Beach	N.C.	1986	3,600,000	4,750,000
Myrtle Beach	S.C.	1986–87	850,000	4,500,000
Tybee Island	Ga.	1976	2,300,000	3,600,000
Cape Canaveral Beach	Fl.	1975	2,715,000	1,050,000
Pompano Beach	Fl.	1970	1,076,000	1,873,437
Hollywood-Hallandale	Fl.	1979	1,980,000	7,743,376
Miami Beach	Fl.	1979–82	12,000,000	55,000,000
Key Biscayne	Fl.	1987	360,000	2,600,000

Source: Adapted from O. H. Pilkey, Jr., and T. D. Clayton, Summary of beach replenishment experiences on U.S. East Coast barrier islands. *Journal of Coastal Research* 5 (1) (1988): 147–59.

coastal sectors located downcurrent from these stabilizing structures. Programs for the long term monitoring of engineered coastal sites are in place, which should yield important information about the long term impact of shore stabilization projects. Efforts are also underway to develop better techniques of beach nourishment so that the artificially introduced sand remains longer in beach systems. Recently, field studies involving coastal ecologists and geologists have begun to document the long-term environmental effects of hurricanes on beaches, salt marshes, mangroves, and coral reefs. Finally, many managers and planners are educating the public about the fragility of coastal environments, striving to implement codes and zoning laws that protect the diversity and resiliency of coastal ecosystems so that future generations of people can enjoy their beauty.

STUDY GUIDE

KEY CONCEPTS

1. Circulation in the nearshore is powered by the energy of incoming waves, and consists of the *shore-parallel flow of longshore currents and the shore-normal flow of rip currents*. The taller the waves and the greater the angle of wave approach, the stronger are the currents of the nearshore. Longshore currents transport sand parallel to the beach, whereas rip currents transfer sand from the beach face to the offshore.

2. Whether a beach erodes or widens depends on the net balance between sand loss and gain over time, expressed as a *sand budget*. Sand losses along a stretch of beach occur by longshore and offshore transport and by wind erosion. Sand sources to a beach include river influx, cliff erosion, longshore drift, and onshore transport of sediment.

3. Low, fair-weather swell produces a broad *berm* of fine sand and a steep *beach face*. By contrast, storm waves create a narrow, eroded, coarse-grained beach with a flattened beach face and *longshore bars* in the nearshore zone.

4. Coastal dunes are created by onshore winds blowing dry fine sand off the beach. The shore-parallel dunes protect the backshore of the barrier island from storm-wave attack.

5. *Barrier islands* are slowly moving landward in response to the current worldwide rise of sea level. Rising sea level allows more frequent washover of barrier islands by storm surges. This washover process transfers sand from the front side to the back side of barrier islands, causing them to shift landward.

6. Storm waves attack cliffs at their base, undercutting and eventually causing the cliff face to collapse. The rate of cliff retreat reflects the durability of the cliff material, which depends on its composition and degree of fracturing.

7. Where rivers supply large quantities of sediment, estuaries are filled and a *delta* is formed. The shape of the delta depends on whether the system is river-, wave-, or tide-dominated.

8. Efforts to stabilize shorelines usually do not work in the long run, because of the naturally dynamic, ever changing character of coastal environments. Structures designed to redirect wave energy are costly and include *breakwaters* and *seawalls*. These are expensive to maintain and usually aggravate rather than retard coastal erosion. *Groins*, which are placed on beaches, and *jetties*, which are built at river mouths or tidal inlets, are designed to trap sand being transported by longshore currents. However, the entrapment of sand on the updrift side of these structures starves the downdrift stretch of the shoreline and causes severe erosion there. *Beach nourishment*, a process of artificially providing sand to a beach, is a viable although costly means to lessen erosion over the short term.

KEY WORDS

beach nourishment
beach profile
breaker zone
breakwater
delta
dune
groin
jetty
longshore bar
longshore current
nearshore zone
offshore zone
overwash
refraction
rip current
saltation
sand budget
sand spit
seawall
storm profile
storm surge
surf zone
swell profile
washover fan
wave-cut platform

QUESTIONS

REVIEW OF BASIC CONCEPTS

1. Contrast the following properties of deep-water and shallow-water waves: celerity (speed), wavelength, period, and height. (Consulting Chapter 7 will be useful.)

2. What are the connections between angle of wave approach and the direction and strength of longshore currents? Be as specific as possible.

3. Compare the mode of origin and the properties of longshore and rip currents.

4. Compare and account for the different appearance of a swell and a storm profile of a beach. How does a storm-ravaged beach recover from its eroded state?

5. What are the principal sources and sinks of beach sand?

6. What is a sand budget, and how is it useful for studying beaches over time?

7. How are barrier islands formed?

8. Why and how do barrier islands migrate landward over time?

9. Why does sea-cliff erosion tend to be catastrophic in nature? By what processes do waves cause the collapse of cliff faces?

10. What main factors lead to the buildup of coastal dunes?

CRITICAL-THINKING ESSAYS

1. Using a diagram and orthogonals (wave rays), show how refraction over shoals causes wave rays to converge, but refraction over canyons or deeps causes wave rays to diverge. (Consult Chapter 7.)

2. How does storm surge affect both the seaward and bay sides of a barrier island?

3. What would you recommend be done to promote the growth of coastal dunes.

4. What shape will a delta display if it is affected equally by wave and river processes? By tidal and river processes? By wave and tidal processes?

5. Explain the potentially devastating effects of installing seawalls and breakwaters to protect a stretch of shoreline.

6. Why are groins and jetties rarely beneficial to the coastal system they are designed to protect? Be specific in your analysis.

7. Discuss the advantages and disadvantages of sand nourishment as a means of promoting beach accretion.

■ DISCOVERING WITH NUMBERS

1. See Figure 11–3c. Estimate the time in days it will likely take the submarine bar to migrate to the berm crest.
2. Assume that a beach system receives 13,500 m^3 of sand annually from cliff erosion and 8, 200 m^3 of sand annually from river input, and that it loses each year 18,500 m^3 of sand to longshore and offshore transport. Is this beach eroding, accreting, or in a steady-state condition? What will happen to this beach if a seawall is built against the cliff to reduce erosion?
3. Examine Figure 11–12d. Calculate the rate of sea level rise in centimeters per year and meters per 1,000 years before and since 5,000 years B. P. (before present). Convert these answers to inches per year and feet per 1,000 years (see Appendix II).
4. Study Table 11–3. Assume that Atlantic Beach in North Carolina loses on average 7,200,000 cubic yards of sand every two years. If you were to maintain this beach in a steady-state condition by sand nourishment until the year 2050, how much would this cost?.

SELECTED READINGS

Bascom, W. 1980. *Waves and Beaches: The Dynamics of the Ocean Surface*. Garden City, N. J.: Doubleday.

Davis, R. A. 1994. *The Evolving Coast.* New York: Scientific American Library.

Davis, R. A. 1996. *Coasts.* Saddle River, N. J.: Prentice-Hall.

Dean C. 1999. *Against the Tide: The Battle for America's Beaches.* New York: Columbia University Press.

Dean, R. G. 1988. Managing sand and preserving shorelines. *Oceanus* 31 (3): 49–55.

Dolan, R., and Lins, H. 1987. Beaches and barrier islands. *Scientific American* 257 (1): 68–77.

Fischetti, M. 2001. Drowning New Orleans. *Scientific American* 285 (4): 76–85.

Fox, W. T. 1983. *At the Sea's Edge: An Introduction to Coastal Oceanography for the Amateur Naturalist*. Englewood Cliffs, N. J.: Prentice-Hall.

Fulton, K. 1981. Coastal retreat. *Sea Frontiers* 27 (2): 82–88.

Godfrey, P. J. 1976. Barrier Islands of the East Coast. *Oceanus* 19 (5): 27–40.

Jackson, T. C. 1981. *Coast Alert: Scientists Speak Out*. San Francisco: Friends of the Earth.

Kaufman, W., and Pilkey, O. H. 1979. *The Beaches Are Moving: Drowning of America's Shoreline.* Garden City, N. J.: Doubleday.

Kemper, S. 1992. If you can fish from your condo, you're too close. *Smithsonian* 23 (7): 72–86.

Lowenstein, F. 1985. Beaches or bedrooms—The choice as sea level rises. *Oceanus* 28 (3): 20–29.

MacLeish, W. H., ed. 1980–1981. The coast. *Oceanus* 23 (4):1–79.

Milliman, J. 1989. Sea levels: Past, present, and future. *Oceanus* 32 (2): 40–43.

Oceanus. 1993. Coastal science and policy, I. Special issue 36.

Oceanus. 1993. Coastal science and policy, II. Special issue 36 (2).

Pilkey, O. H. 1990. Barrier islands. *Sea Frontiers* 36 (6): 30–36.

Viles, H., and Spencer, T. 1995. *Coastal Problems.* London: Edward Arnold.

Walker, J. 1982. Walking on the beach, watching the waves and thinking on how they shape the beach. *Scientific American* 247 (2): 144–48.

Wanless, H. R. 1989. The inundation of our coastline. *Sea Frontiers* 35 (5): 264–71.

Westgate, J. W. 1983. Beachfront roulette. *Sea Frontiers* 29 (2): 104–09.

SPECIAL READINGS ON SHORELINE CONSTRUCTION

Canis, W. F., et al. 1985. *Living with the Alabama-Mississippi Coast*. Durham, N. C.: Duke University Press.

Carter, C. H., et al. 1987. *Living with the Lake Erie Shore*. Durham, N. C.: Duke University Press.

Clayton, T. D., et al. 1992. *Living with the Georgia Shore*. Durham, N. C.: Duke University Press.

Doyle, L. J., et al. 1984. *Living with the West Florida Shore*. Durham, N. C.: Duke University Press.

Griggs, G., and Savoy, L. 1985. *Living with the California Coast*. Durham, N. C.: Duke University Press.

Kelley, J. T., et al. 1988. *Living with the Coast of Maine*. Durham, N. C.: Duke University Press.

Kelley, S., et al. 1985. *Living with the Louisiana Shore*. Durham, N. C.: Duke University Press.

Lennon, G., ed. 1996. *Living with the South Carolina Coast*. Durham, N. C.: Duke University Press.

McMormick, L. R., et al. 1984. *Living with Long Island's South Shore*. Durham, N. C.: Duke University Press.

Mason, K. 1998. *Living with the Coast of Alaska*. Durham, N.C.: Duke University Press.

Morton, R. A., et al. 1983. *Living with the Texas Shore*. Durham, N. C.: Duke University Press.

Neal, W. J., et al. 1984. *Living with the South Carolina Shore*. Durham, N. C.: Duke University Press.

Nordstrom, K. F., et al. 1986. *Living with the New Jersey Shore*. Durham, N. C.: Duke University Press.

Pilkey, O. H., et al. 1982. *From Curritock to Calabash: Living with North Carolina Barrier Islands*. Durham, N. C.: Duke University Press.

Pilkey, O. H., et al. 1985. *Living with the East Florida Shore*. Durham, N. C.: Duke University Press.

Terich, T. A. 1987. *Living with the Coast of Puget Sound and the Georgia Strait*. Durham, N. C.: Duke University Press.

Ward, G., et al. 1989. *Living with the Chesapeake Bay and Virginia's Ocean Shores*. Durham, N. C.: Duke University Press.

TOOLS FOR LEARNING

Tools for Learning is an on-line review area located at this book's web site OceanLink (**www.jbpub.com/oceanlink**). The review area provides a variety of activities designed to help you study for your class. You will find chapter outlines, review questions, hints for some of the book's math questions (identified by the math icon), web research tips for selected Critical Thinking Essay questions, key term reviews, and figure labeling exercises.

12 Coastal Habitats

Sand-strewn caverns, cool and deep,
Where the winds are all asleep;
Where the spent lights quiver and gleam;
Where the sea-beasts rang'd all around
Feed in the ooze of their pasture-ground.
—*Matthew Arnold,*
The Forsaken Merman, *1849*

web navigator

critical thinking on the web

math tutor on the web

www.jbpub.com/oceanlink

PREVIEW

SHORELINES ARE COMPLEX environments because they represent a transition boundary between what is truly land and what is truly ocean. The *shore* represents the juncture between land and water, the virtual edge of the sea, depending on your perspective. The term *coast* denotes a broader zone, one that includes not only the shore, but also many other habitats and ecosystems that are affected by processes associated with this unique setting between land and water. So, the shore may be a gravel beach abutting a rocky cliff, but the coast of that region may include bays, coves, salt marshes, tidal inlets, estuaries, and mudflats. In the previous chapter we focused mainly on beaches and barrier islands, stressing the physical processes—waves, currents, tides—that shape these sand deposits. Here, we will survey the general characteristics, processes, and biota, of five important coastal settings: (1) estuaries—crossroads between the freshwater of rivers and the saltwater of oceans; (2) lagoons—isolated bodies of shallow coastal water; (3) salt marshes—plant-covered, intertidal fringes of coastline; (4) mangrove swamps—dense tree growths of tropical and subtropical shorelines; and (5) coral reefs—calcareous ridges produced by coral, a group of benthic coelenterates.

Shorelines of the world are populated by a rich variety of organisms that are well adapted to one another and to the demands of their physical surroundings. Because of their shallowness, coastal waters respond strongly to the effects of waves and tides and to the daily and seasonal variations of weather. Also, shore habitats being close to land receive a regular infusion of river water that locally causes pronounced changes in the salinity and water temperature of the environment. Despite these demanding conditions, certain coastal habitats, including salt marshes, mangrove forests, estuaries, and coral reefs, rank among the most biologically productive ecosystems anywhere on the Earth.

12-1

Estuaries

An important coastal environment is the **estuary** located between the mainland and barrier islands. It is here that the rivers begin to sense the rhythm of the tides. The word "estuary" is derived from the Latin word *aestuarium,* which means "tidal." Oceanographers define an estuary as a semienclosed body of water where river water mixes with ocean water. This is in contrast to **lagoons**, which are likewise semienclosed coastal water bodies, but which receive essentially no inflow of river water, so they are as salty or saltier than ocean water. Most lagoons have reasonably simple water circulation, unlike the complex water flow associated with estuaries.

A glance at a world map shows how vital the influence of estuaries has been for the location of large and small cities throughout the world. Estuaries provide good natural harbors and waterways for shipping food, raw materials, and manufactured items—an activity invaluable for economic and commercial enterprises. Furthermore, because vegetated wetlands that surround estuaries, such as salt marshes, were once considered useless land, they were used as convenient sites to dispose of industrial and domestic wastes. Now, cleaned up more or less and reclaimed by landfill, they have become valuable land for urban expansion at the edge of the sea.

Not too long ago, city dwellers were unconcerned with their impact on the water quality of estuaries and wetlands. After all, it was argued, coastal environments are flushed daily by the tides, which should keep them clean and pure. Tragically, this is not the case. Estuaries, even very large ones, have a limited capacity to absorb pollutants. If that capacity is exceeded, damage can be widespread. It has recently been discovered that estuaries actually can trap and store pollutants. In this section, we will examine the unique circulation processes that allows this to happen as sediment and waste accumulate.

ORIGIN OF ESTUARIES

Many estuaries are **drowned river valleys** (**Figure 12–1a**). With deglaciation and the recent Holocene rise of sea level, seawater flooded river valleys as the seas advanced onto the land. But estuaries form by other means as well. In high latitudes, for example, glaciers have carved the soil and rock of the land, gouging out steep-sided valleys. The mouth of many of these valleys contains mounds of glacial till called **moraines**. As sea level rose and flooded over the glacial moraines, their basins were filled with seawater, creating **fjords**, long, narrow arms of the sea that are bordered by steep cliffs (**Figure 12–1b**). Circulation in some fjords can be sluggish near the bottom because of the morainal barrier. This sill traps bottom water, which eventually becomes anoxic as the oxygen is used up by organisms. Estuaries are also created by longshore currents that form a sand spit or sand bar across an embayment, producing a **bar-built estuary** (**Figure 12–1c**). Last, there are **tectonic estuaries**, which form along coastlines that lie on active plate boundaries, where faulting and folding can create coastal basins that fill up with seawater (**Figure 12–1d**).

Estuaries are short-lived basins in a geologic sense, because they tend to be quickly filled in with river and marine sediment. Their circulation not only traps sediment, but also results in the importation of sand and mud from offshore areas. The filling-in process commonly begins at the head of the estuary with the accumulation of sediment in river deltas, which, with time, grow toward the estuary mouth. The fact that many estuaries are not yet filled has much to do with sea level that has been rising and flooding the land since the last glaciation.

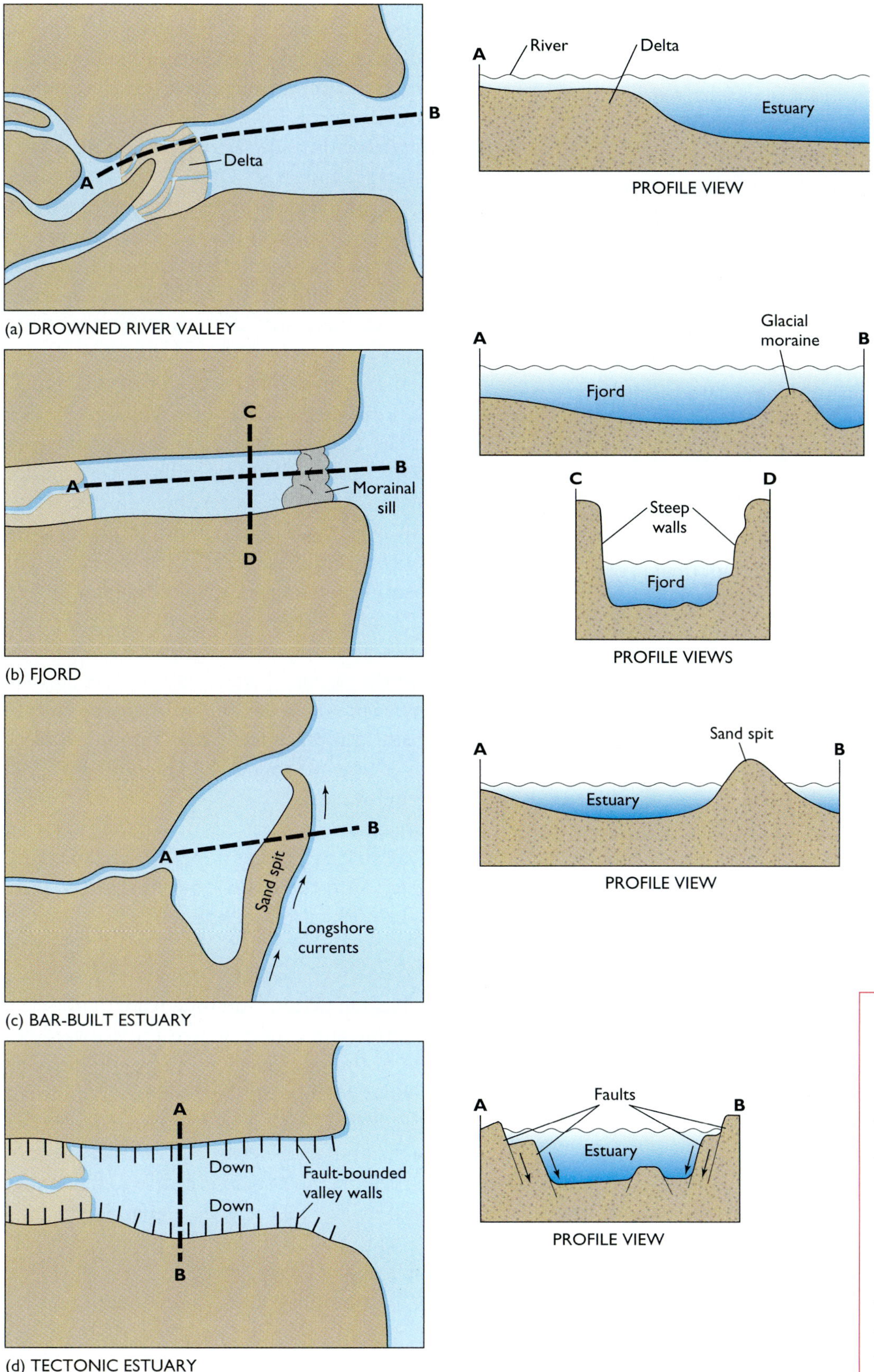

FIGURE **12-1**

Origin of estuaries. All types of estuaries have resulted from the rise of sea level during the Holocene epoch. (a) Perhaps the most common origin is the drowning of the mouth of a river valley. (b) In the high latitudes, glaciers have carved deep, narrow, steep-walled valleys and many have glacial moraines (ridges of sediment scraped off the land by moving glaciers) at their mouths. Flooding of these valleys with seawater creates fjords. (c) Bar-built estuaries evolve by spit extension across an embayment. (d) Tectonic estuaries commonly result from block faulting.

CIRCULATION AND SEDIMENTATION IN ESTUARIES

Estuaries, like beaches, are dynamic systems. They are greatly influenced by two important factors. The first is the inflow of river water, which creates a salinity gradient extending from the point of river entry, where salinity is low, to the ocean inlet at the estuary mouth, where salinity is the same as it is in the ocean. A supply of freshwater from rivers stratifies the water column with a sharp halocline. This feature separates the less saline surface water from the more saline bottom water. The second factor is tidal flow, which provides energy that mixes freshwater and saltwater. If tidal flow is strong in an estuary, then the stratification of the water column is weakened by the turbulent mixing of the water.

The relative effects of river input and tidal mixing determine the character of the circulation in an estuary. When the river inflow dominates the tides, the estuary is a river-dominated system, and its water column is highly stratified; in this case it is called a **salt-wedge estuary**. When tidal mixing dominates the river input of freshwater, the estuary is a tide-dominated system, and the water column is unstratified; in this case it is called a **well-mixed estuary**. Between these two extremes is the **partially mixed estuary**, which displays weak stratification of the water column. Because the relative effects of river and tidal flow vary with time, many estuaries assume different circulation characteristics as conditions change. For example, a partially mixed estuary may become well mixed for a time, and a salt-wedge estuary may become partially mixed for a time. Let's examine the general circulation characteristics of each of these three types of estuaries.

SALT-WEDGE ESTUARIES

Coastal basins that receive a high river discharge and are affected by relatively weak tidal currents (usually this means a small tidal range) have strongly stratified water columns. The surface body of low-density freshwater floats above a bottom layer of high-density saltwater (**Figure 12–2a**). Weak tidal currents in salt-wedge estuaries cannot mix the two water masses, because they are separated by a sharp and stable halocline. The result is a distinctive salt-wedge profile with a tongue (a wedge) of seawater lying beneath a layer of freshwater. Currents in the surface layer are vigorous and turbulent because of the high river discharge, while flow in the bottom salty layer is slight because of the weak tidal currents.

The strong river flow in the surface layer "rubs" or shears across the halocline. Shearing creates friction and generates internal waves on the halocline, a process similar to wind shear generating waves on the sea surface. Once created, these internal waves grow in size, and can become oversteepened and break. This results in the transfer of a small amount of saltwater from the bottom layer to the surface layer as each internal wave breaks (**Figure 12–3a**). In turn, this process induces a net inflow of bottom water through the estuary mouth in order to replace the saltwater that has mixed upward into the top freshwater layer. This compensating landward-directed bottom current is weak, because the breaking of internal waves is an inefficient means of moving saltwater upward into the surface layer. Thus, only small volumes of ocean water are drawn in along the estuary bottom to replace the small amount of water that moves upward across the halocline by breaking internal waves.

The weak bottom currents of salt-wedge estuaries are not able to transport significant quantities of sediment into the basin from offshore sources. Rather, the bulk of the sedimentary fill is supplied by river transport from inland sources. The river sand tends to accumulate in shoals (a series of irregularly shaped sand and mud bars) that form at the tip of the salt wedge, where the seaward bottom flow of the river is canceled by the landward bottom flow of the salt layer. Fine-grained suspended sediment, such as clay and silt, move into the estuary and then settle out of suspension as currents weaken. As as result, coarse to fine sand is concentrated in the upper estuary and mud is deposited in the lower estuary. Most of this sand and mud is derived from the river.

PARTIALLY MIXED ESTUARIES

An increase in the strength of the tidal currents relative to river discharge promotes mixing of the water column. Therefore, water stratification and

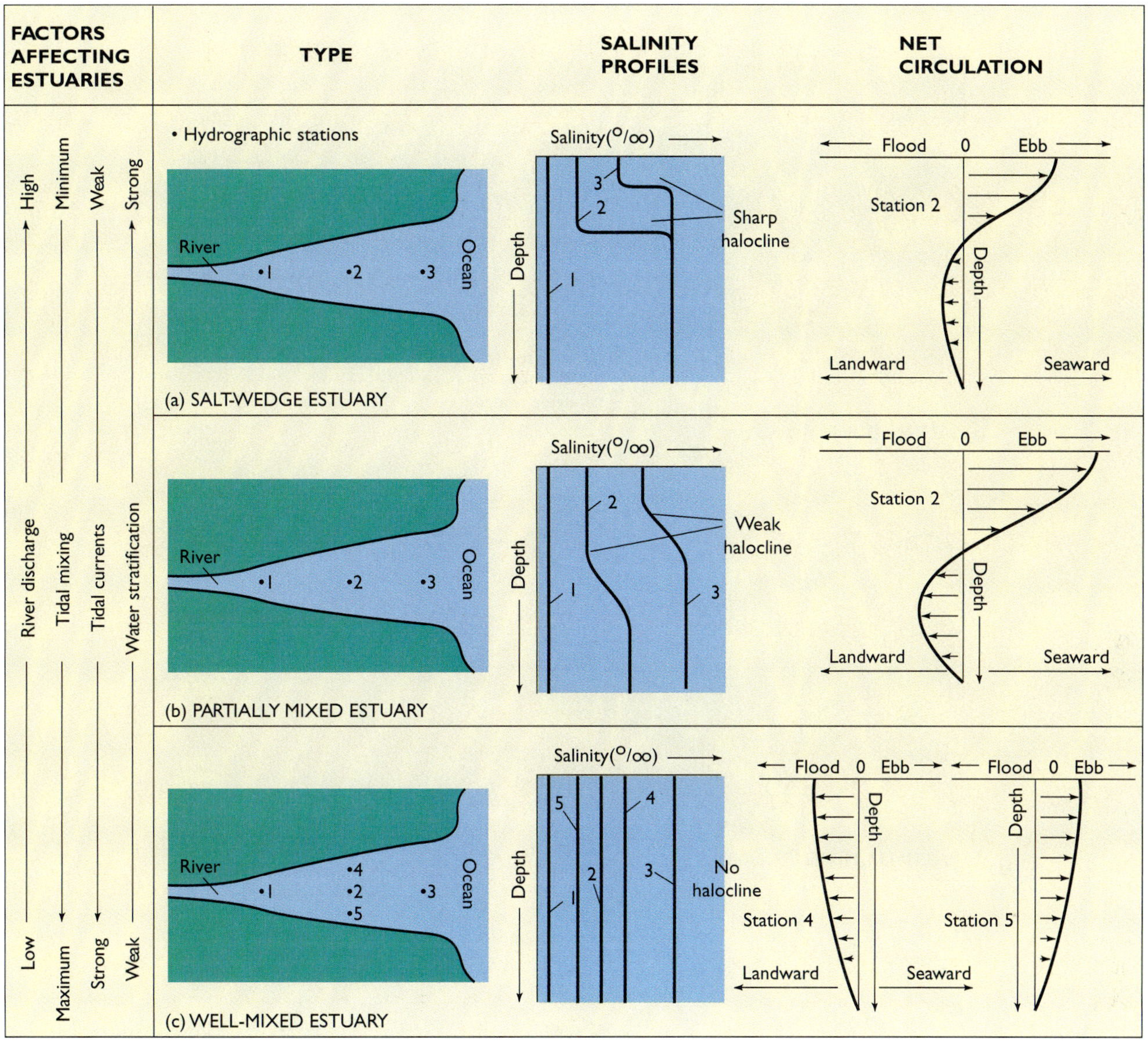

FIGURE 12-2

Types of estuaries. Estuaries are classified on the basis of circulation, which depends on the degree of freshwater and seawater mixing. Water mixing is controlled by the relative influence of tidal currents and river discharge as they produce a net circulation. (a) Salt-wedge estuaries are river-dominated and, as such, are highly stratified, as shown by the sharp halocline. Surface currents flow out of the estuary and bottom currents into the estuary. (b) Partially mixed estuaries have a weakly developed halocline, because tidal flow mixes surface and bottom water. This results in strong incoming bottom flow and outgoing surface flow. (c) Well-mixed estuaries are tide-dominated. Consequently, they are thoroughly mixed and have no halocline. Water flows into the estuary on one side and out of the estuary on the other side.

the halocline are weakened to the point that salinity varies by no more than a few parts per thousand (‰) from the surface to the bottom of a partially mixed estuary (**Figure 12–2b**). As the strong tides flood into and ebb out of the basin, friction with the bottom and sides of the estuary generates widespread turbulence, which mixes water much more effectively than the breaking internal waves. Although details are complicated, the net result is the transfer of a substantial volume of saltwater up into the layer of surface water. This, of course, necessitates a compensatory landward-directed bottom flow (**Figure 12–3b**), similar to, but much stronger than, the one in the salt-wedge estuary. The seaward flow above and the landward flow below the halocline are much stronger in the par-

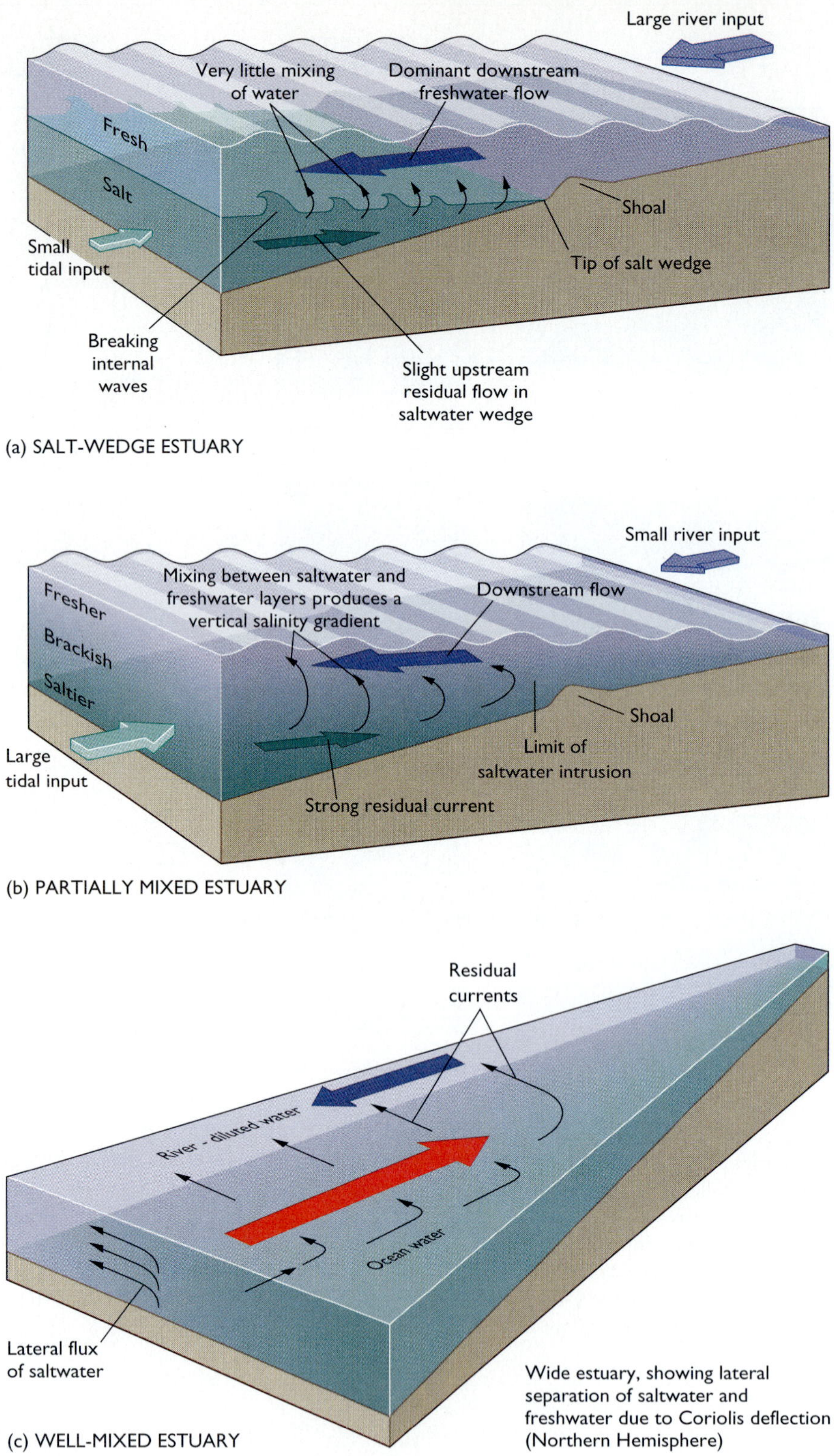

FIGURE 12-3

Water motions in estuaries. (a) Salt-wedge estuaries are characterized by a net landward-directed bottom current and a net seaward-directed surface current. Breaking internal waves along the halocline produce an upward flow of water. (b) Partially mixed estuaries have strong currents, with net landward-flowing bottom currents and net seaward-flowing surface currents. (c) Well-mixed estuaries have net currents that are landward-directed at all depths on one side of the estuary and seaward-directed flow at all depths on the other side. [Adapted from J. Pethick, *An Introduction to Coastal Geomorphology* (Baltimore, Md.: Edward Arnold, 1984).]

tially mixed than in the salt-wedge estuary. The reason for that is the more effective interchange and mixing of bottom with surface water in the partially mixed than in the salt-wedge estuary. (See the boxed feature, "Chesapeake Bay," for a discussion of the circulation character of what is a classic partially mixed estuary.)

Because of the strong bottom flow that draws water from offshore, a partially mixed estuary receives a substantial amount of sand and mud from the inner continental shelf. There is a strong tendency for the lower reaches of a partially mixed estuary to be floored predominantly by sediment derived from the offshore region. That bottom material gets progressively diluted by river-supplied sediment toward the head of the estuary.

A great deal of suspended mud gets concentrated near the halocline where currents are weak, creating a **turbidity maximum** in the water column. Many of these mud particles are ingested by filter-feeding organisms, which abound here and, when excreted as fecal pellets, are deposited in extensive mud shoals. Also, clumping of fine suspended particles occurs by **flocculation**, whereby clay particles in seawater "stick" together and form aggregates that then settle to the bottom more easily than the individual and much smaller mud particles do. More flocculation of mud occurs in salty than in fresh water. These processes create widespread mud shoals in this part of the estuary. The zone of mud deposition shifts up and down the estuary with the spring and neap tides and as the river discharge varies seasonally.

WELL-MIXED ESTUARIES

In estuaries dominated by tides, the well-mixed estuaries, the halocline is obliterated because the surface and bottom waters are completely mixed by the strong turbulence associated with vigorous tidal flow (see **Figure 12–2c**). In wide, well-mixed estuaries, Coriolis deflection tends to divert freshwater against one side and saltwater against the opposite bank. Both are deflected to the right, *looking downcurrent* in the Northern Hemisphere, which forces them to the opposite sides of the estuary. This produces a gradual increase in salinity from one side of the estuary to the other. Strong tidal mixing causes saltwater to be mixed laterally across the width of the estuary. In salt-wedge and partially mixed estuaries, the mixing is vertical; in a well-mixed estuary, the mixing is lateral. The excess saltwater, which flows laterally across the basin, does two things. It draws in saltwater from offshore at all depths on the saltier side of the estuary, and it causes a discharge of this water out of the estuary at all depths on the less salty side of the estuary (**Figure 12–3c**).

The strong tidal currents associated with a well-mixed estuary import great quantities of both fine- and coarse-grained sediment from offshore. The contribution of river-supplied sediment to this estuary type is slight, because river discharge is relatively insignificant.

THE BIOLOGY OF ESTUARIES

The water chemistry and current patterns in estuaries are greatly influenced by the relative effects of river discharge and tidal mixing (see Figure 12–2). The widely fluctuating environmental conditions in estuaries make them difficult places for plants and animals to inhabit, so species diversity is low (**Figure 12–4**). However, despite these harsh living conditions, estuaries have a large *carrying capacity*, meaning that they are very fertile and can support large populations of organisms.

There are several reasons for their remarkable fertility. First, rivers supply large quantities of dissolved nutrients. Also, tides and waves keep the water well ventilated with oxygen and churn up sediment, continually resuspending nutrients and organic matter from the shallow sea bottom. These processes keep high levels of nutrients in the water, thus sustaining phytoplankton production. As a result of the estuary's demanding living conditions, the relatively few animal species that have adapted to the setting flourish there because of an abundant food supply, a reduction in competition for food and space, and relatively limited predation.

Phytoplankton species in estuaries have growth, reproduction, and survival requirements similar to those of other marine-plant communities. They need both macronutrients and micronutrients, adequate sunlight, and suitable water quality (temperature, salinity, oxygen, pH, water clarity). Because these factors vary considerably from season to season and from spot to spot, plankton blooms are irregular. In areas where muddy water severely limits the penetration of sunlight into the water

Chesapeake Bay

Chesapeake Bay (**Figure B12–1a** and **b**) is less than 300 kilometers (~186 miles) long, no wider than 65 kilometers (~40 miles), and averages about 20 meters (~72 feet) in depth. Studies of its currents (**Figure B12–1c)** and salinity (**Figure B12–1d**) show it to be a classic partially mixed estuary, with strong net upstream flow along the bottom and strong net downstream flow at its surface. The Susquehanna River supplies most of the freshwater to the estuary, with significant contributions by the Potomac River and the James River. Salinity in the estuary may vary vertically by as much as 10 ‰ during the spring snowmelt.

During the summer, when the water column in the estuary is strongly stratified, some of the bottom water may contain very low levels of dissolved oxygen, a condition called **hypoxia**, or may contain no dissolved oxygen at all, a condition called **anoxia** (**Figure B12–2a**). Usually, the onset of anoxia

(a) CHESAPEAKE BAY

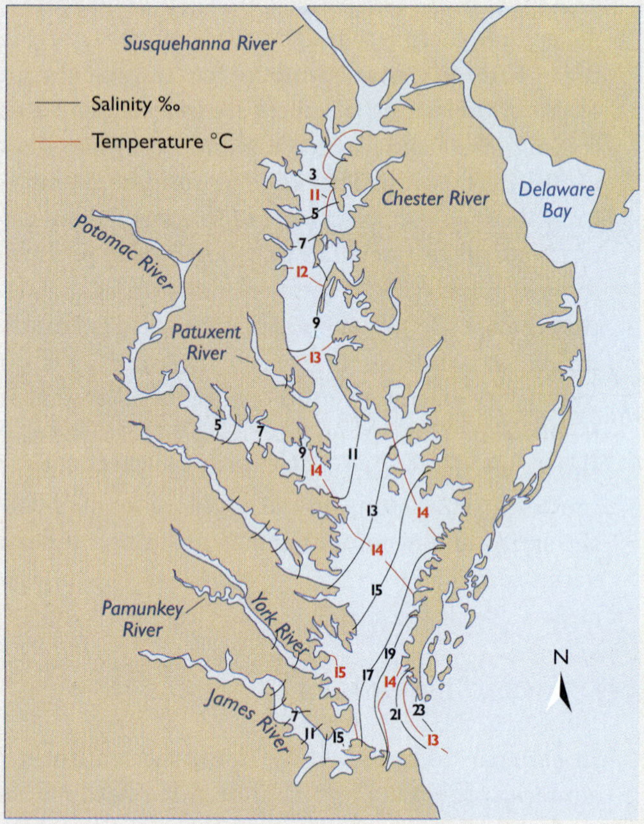

(b) SALINITY AND TEMPERATURE OF CHESAPEAKE BAY

FIGURE **B12-1**

Chesapeake Bay. (a) Satellite image of Chesapeake Bay. (b) Chesapeake Bay is a drowned river valley where freshwater and saltwater mix. (c) Net circulation consisting of landward-flowing bottom currents and seaward-directed surface currents identify the bay as a partially mixed estuary. (d) A longitudinal profile down the center of Chesapeake Bay shows the presence of several haloclines that result from the discharge of freshwater by several large rivers that enter along the bay's western margin. [Adapted from D. W. Pritchard, in *The Estuarine Environment: Estuaries and Estuarine Sedimentation,* J. R. Schubel, ed. (Washington, D.C.: American Geological Institute, 1971).]

in the bottom water coincides with an extensive microscopic plant growth (**Figure B12–2b**). Also, during the late spring and summer, large quantities of dead organic matter settle to the bottom and accumulate in the mud. Microbes decompose this organic matter and use up all the available dissolved oxygen. This occurs at a time when the water column is highly stratified and reoxygenation by mixing of water is not possible. What is most alarming, however, is that the volume of anoxic water has increased steadily over the past three decades (**Figure B12–2c**). This trend cannot be attributed to more intense water stratification, because river discharge, although varying from year to year, did not increase during that time interval. Rather, the increase of anoxic bottom water with time has apparently resulted from the increased dumping of nutrients into the bay. This fertilization of the water has stimulated larger and larger microscopic plant growth, and these plant cells, once dead, sink to the bottom and decompose, requiring more and more oxygen each year.

The ecological implications of hypoxia and anoxia for Chesapeake Bay are serious. The anoxic water causes massive kills of shellfish (oysters and blue crabs) and causes finfish (shad, striped bass, alewife, and white perch) to go elsewhere for feeding and breeding. All studies to date indicate clearly that the water and the sediment of the bay have been fouled by the dumping of raw sewage and nutrients in its water. These cause the massive growth of the algae that die and then decompose on the bottom, using up all the oxygen. Also, the recent destruction of oyster and rockfish populations appears to be linked to the discharge of strong chemical toxins into the bay's water. In 1984 Virginia, Pennsylvania, Maryland, and the District of Columbia joined forces to establish the Chesapeake Bay Program in an effort to revitalize the bay. Numerous grassroots organizations have joined the effort as well as the bay attempts its "comeback." Some progress has been made, particularly with the chemical quality of the water. In 1998 the Environmental Protection Agency reported that improved management and regulation of wastewater and agricultural practices and bans on certain types of phosphorus-rich detergents have caused nutrient levels in Chesapeake Bay waters to decrease, resulting in the partial recovery of the bay's bottom vegetation.

Visit www.jbpub.com/oceanlink for more information.

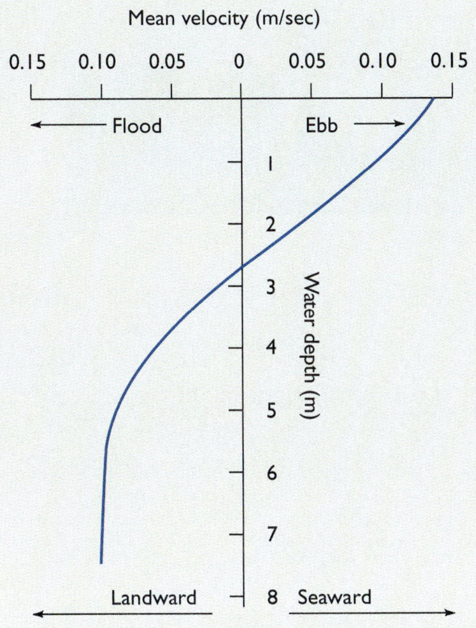

(c) NET CIRCULATION IN CHESAPEAKE BAY

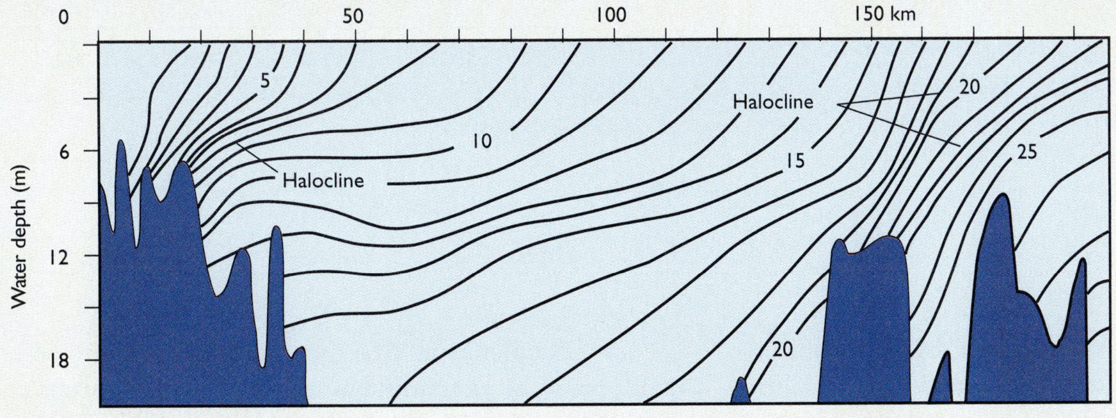

(d) SALINITY (°/oo): SPRING AVERAGE

FIGURE **B12-2**

Anoxia in Chesapeake Bay. (a) The buildup of anoxic water is seen by comparing the 1980 to the 1950 profile. (b) Anoxia of bottom water is associated with high plant productivity in the late spring and summer. These plants die and decompose, using up the dissolved oxygen. The strong heating of the surface water at this time prevents overturning of water and aeration of the bottom. (c) Since 1950 the volume of anoxic water in Chesapeake Bay that builds up during the summer has increased at an alarming rate. This is attributed to the increase in nutrients discharged into the bay's water. [Adapted from C. B. Officer et al., *Science* 223 (1984): 22–27.]

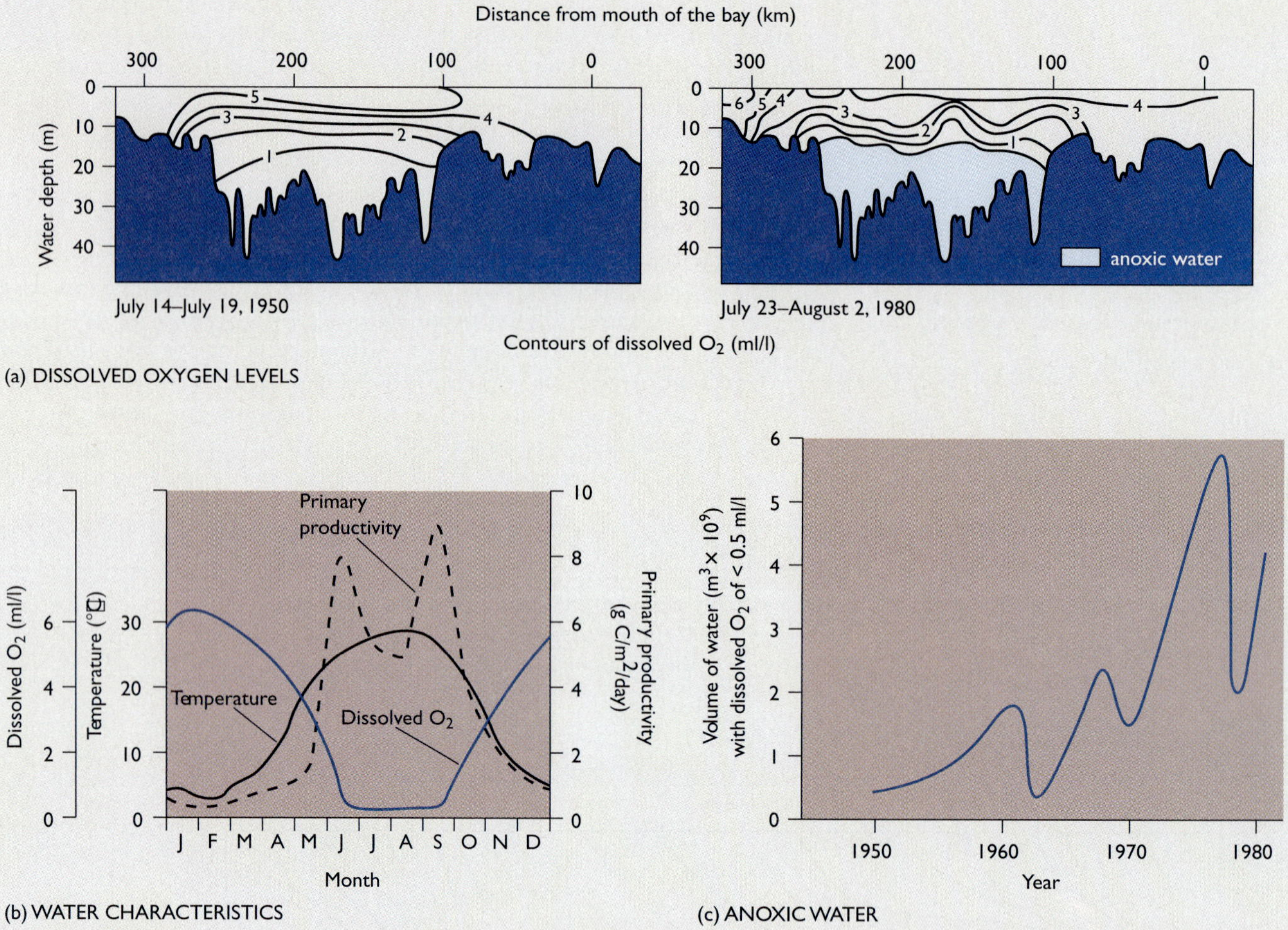

column, phytoplankton are not the dominant primary producers. Rather plant *detritus* (debris) exported in suspension out of the nearby salt marshes by ebb tidal currents supports the base of food webs in many estuaries. Here the detritus food chain may be more significant than the grazing food chain.

Zooplankton, like the phytoplankton, are hampered by the turbid water common in estuaries. Their distribution tends to be regulated by salinity gradients, with freshwater species in the headwaters, brackish water species in the middle reaches, and marine species near the mouth of the estuary. Temperature also plays a role in their distribution

FIGURE 12-4

Species diversity. The number of species is low in estuaries compared to that in rivers and oceans because the stressful environmental conditions of estuaries make adaptation difficult for organisms . [Adapted from A. Remane, in *Biology of Brackish Water*, A. Remane and C. Schlieper, eds. (Stuttgart:Schweizerbartsche Verlagsbuchhandlung, 1971).]

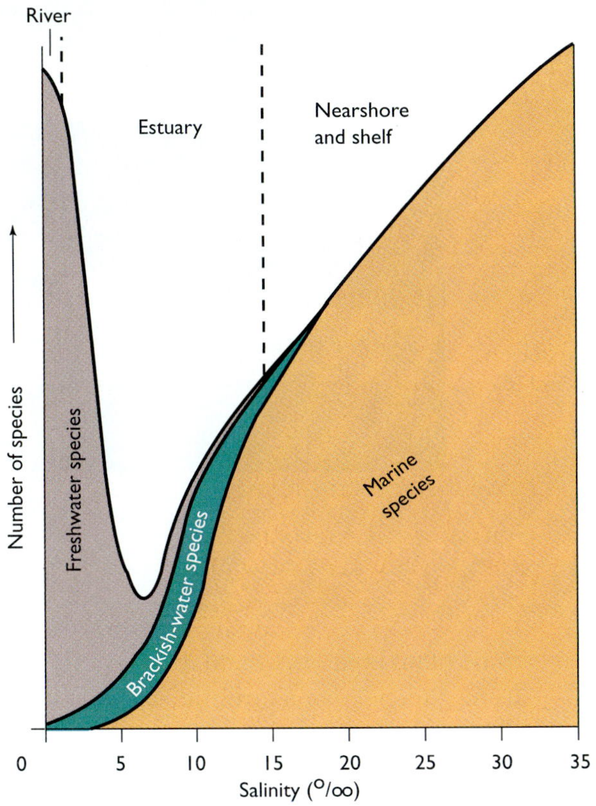

and behavior over time. For example, the females of some copepod species lay resting eggs that lie dormant on the estuary floor during the cold months of the winter and spring.

The biomass of the estuarine benthos is typically large because of the abundance of food resources. Where the bottom is rocky, benthic communities are dominated by *sessile* (permanently attached) filter feeders such as bivalves (mussels, oysters), sponges, barnacles, and bryozoans (small invertebrates that resemble coral) (**Figure 12–5**). Mud and sand substrates support an epifauna (sea urchins, starfish, snails) and an infauna (polychaete worms, clams) of suspension and deposit feeders.

Oysters are bivalved mollusks that inhabit the shallow and unpolluted bottoms of estuaries, bays, and tidal creeks. Most Pacific and Atlantic coast species thrive in water with a salinity of about 26 to 28‰ and a water temperature between 0° and 25°C. Also, being benthic filer feeders, oysters need an ample supply of plankton and a hard substrate for attachment. Native oysters have disappeared from the coastal waters of Oregon. Currently a Japanese species, the Pacific oyster (*Crassostrea gigas*), is being farmed (aquaculture) in Oregon's estuaries, including Coos Bay, Tillamook Bay, and Yaquina Bay among others. Because the cold Oregon coastal water does not permit natural spawning, the Pacific oyster are spawned in specially designed tanks in which the quality, temperature, and salinity of the water are regulated. After the spat settle on and attach themselves to old oyster shells, they are transferred to the estuaries where they mature to harvestable size in two or three years.

As is true for the plankton and benthos, estuarine fishes are typically abundant, but represent very few species, largely because of the stressful nature of their environment. Fish species with high physiological tolerances to environmental changes favor estuaries for their reproductive, spawning, and feeding activities. Many estuaries are considered "nursery" grounds because the vast majority of fishes in estuaries are juveniles. These young, immature fish feed voraciously and grow rapidly under the relatively low predation pressures there.

12-2

Lagoons

Unlike estuaries, which are greatly affected by river inflow, lagoons are isolated or semi-isolated bodies of shallow coastal water that do not receive an appreciable input of freshwater. In other words, dilution of salt water in a lagoon is negligible. As a consequence, lagoons tend to have much simpler water-circulation patterns and salinity distributions than do estuaries. A lagoon that is cut off from the ocean is basically a body of standing water. If connected to the sea, however, tidal water flows in and out of the lagoon

(a) OYSTERS

(b) SPONGES

(c) STARFISH

(d) POLYCHAETE WORM

FIGURE 12-5

Benthic invertebrates. Various organisms are adapted to living on the bottom of estuaries. These include (a) oysters, (b) sponges, (c) starfish, and (d) polychaete worms.

through its inlet, the amount and intensity of the flow depending largely on the range of the local tide. The larger the tidal range, the greater the exchange of water between the lagoon and the open ocean. Also, it is a common misperception in part propagated by the Hollywood media that lagoons are strictly tropical water bodies, usually associated with coral reefs. This is not the case, because lagoons occur anywhere on the Earth where there are irregularities in the shoreline or where sand spits and barrier islands isolate seawater from the nearshore (**Figure 12–6**). Therefore, lagoons are a common coastal feature of the polar and temperate latitudes as well as of the tropical climes.

Lagoons are best characterized as calm, shallow bodies of water with salinities that range between brackish and hypersaline, depending on the local hydrology and climate. They may be vegetated by sea grasses, seaweeds, mangrove, or salt-marsh plants or have edges and bottoms that are barren. It is not unusual, particularly in the tropics, for the water column to be isothermal, meaning that temperature does not vary with water depth. They tend to be blanketed with sand and mud that is eroded from the nearby shoreline or that is imported through tidal inlets from offshore.

An interesting phenomenon of subtropical lagoons is their distinctive circulation pattern that is driven by the high evaporation rates characteristic of these latitudes. Here intense evaporation combined with little rainfall results in the formation of lagoonal water that becomes saltier than the open water of the shelf. This dense surface water sinks to the bottom of the lagoon and flows out through the inlet as a seaward-directed bottom current. This, in turn induces shelf water to flow into the lagoon on the surface. This current-flow pattern—inflow at the surface and outflow along the bottom—is opposite to that of salt-wedge and partially mixed estuaries, and is referred to by oceanographers as **inverse flow.** You may recognize that this is the exact flow pattern of the Mediterranean Sea (see Figure 6–17), the classic example of inverse flow. Laguna Madre, a large, shallow, and hypersaline lagoon located behind Padre Island, Texas, displays this inverse flow: high salinity

FIGURE 12-6

Map showing lagoons (no freshwater inflow) and an estuary (significant freshwater inflow). Water is exchanged with the open ocean through tidal inlets that are located between sand spits and barrier islands. The central lagoon on the map is isolated from the ocean.

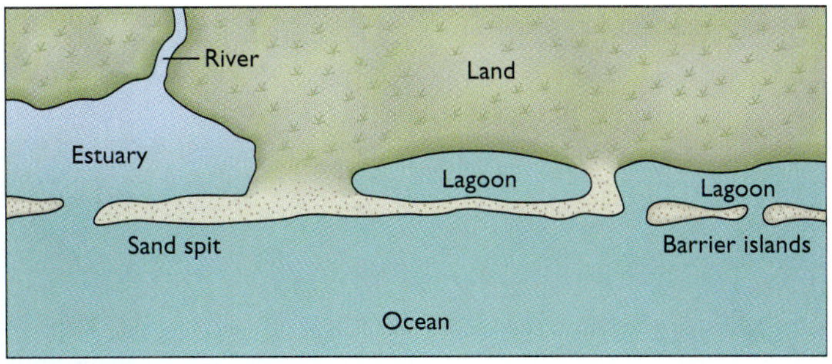

lagoonal water flows seaward along the bottom and less salty and therefore less dense Gulf of Mexico water enters the lagoon at the surface.

Numerous lagoons dot the shore of Baja California. Filled with warm, shallow, protected water, they are ideal for the breeding and birthing activities of gray whales. The eastern Pacific populations of gray whales spend summers feeding in the Bering, Chukchi, and Beaufort Seas and then make the longest known migration of any mammal, traveling each year over 17,000 kilometers (~10,540 miles) to their calving grounds in Baja California. These remarkable creatures are very social and today support whale watching, a vast and locally important ecotourist activity (**Figure 12–7**).

Some lagoons are pristine, fecund environments filled with uncontaminated, clear blue water. This is particularly true of the lagoons that lie at the center of Pacific atolls (**Figure 12–8**). Which range in size from more than 1,000 kilometers^2 (~310 miles^2) to less than 1 kilometer^2 (~0.3 miles^2). Because many lagoons are, however, located at the edge of continents and large populated islands and have relatively sluggish circulation, their water is susceptible to pollution. Some of the most fouled lagoons, which are tucked behind a system of barrier islands, occur along West Africa. Originally these lagoon waters were biologically productive and supported important local and commercial fisheries. Today residents refer to them as "biological graveyards." Little marine life lives in them, because they have been used for decades as convenient dumping sites for municipal and industrial waste. For example, before 1970 fisherman harvested over 5,000 metric tons (~5,513 English tons) of fish each year from Aby Lagoon in the Côte d'Ivoire. In 1981, the catch had plummeted to 500 metric tons (~551 English tons); today nothing is caught in Aby Lagoon and its waters breed diseases such as typhoid and cholera.

12-3

Salt Marshes

Although they are limited to the innermost strip of the nearshore zone, higher plants have colonized marine environments. A case in point are the tall grasses of salt marshes, which are vegetated intertidal flats. Unlike the simple seaweeds, salt-marsh grasses possess all the characteristics of terrestrial plants, including a root mass and a vascular system for the circulation of internal fluids. Marsh grasses are typically rooted in mud, which accumulates wherever sheltered, low-energy conditions prevail along a shoreline. Therefore, they tend to abound behind barrier islands and along the edges of estuaries (**Figure 12–9a**). Because of their intertidal location, salt-

FIGURE 12-7

Gray whales. A whale-watch cruise to observe gray whales in the Gulf of California.

FIGURE 12-8

Atoll. Aerial photo of an atoll and its central lagoon in the South Pacific Ocean.

marsh meadows are continually covered and uncovered by seawater each day as tidal currents flood and ebb. Saltwater drainage is controlled largely by a meandering network of tidal channels (**Figure 12–9b**), through which is flushed sediment, plant detritus, dissolved nutrients, plankton, and many small fish into and out of the marsh. Despite the harsh physical conditions—daily immersion in seawater and exposure to air—salt marshes are one of the most productive natural environments on the Earth.

Based on a distinctive zonation pattern of grasses, salt marshes are subdivided into two principal zones: the low and high marshes (**Figure 12–9c**). The **low salt marsh** extends roughly from mean low tide to the neap high tide. Along the Atlantic seaboard and the Gulf Coast of the United States, the low marsh is dominated by the smooth cordgrass *Spartina alterniflora* (**Figure 12–10a**), a knee- to waist-high grass that spreads rapidly. The cordgrass produces new grass stalks from **rhizomes**, buried, rootlike stems that send out roots and grass shoots as they grow out horizontally. The **high salt marsh**, which extends from neap high tide to the level of the highest spring tide, is flooded only during unusually high tides and during storm surges. Therefore, unlike the low marsh, the high marsh leans more toward a terrestrial than a true marine environment. Its floral community is much more diversified than that of the low marsh, consisting of salt hay grass *Spartina patens*, the saltwort grass *Salicornia* sp. (species), and the spike grass *Distichlis spicata* (**Figure 12–10b–d**).

Most marsh ecologists agree that frequency of flooding plays a major role in the distribution of plants in salt marshes. Low marshes are covered by seawater and exposed to air repeatedly each day. *Spartina alterniflora* flourishes in this stressful setting because there is no competitive pressure from plants that are less tolerant of such harsh environmental conditions. By contrast, much of the high marsh is infrequently flooded by seawater. This permits *S. patens*, *Salicornia* species, and *Distichlis spicata* to compete with and thus exclude *Spartina alterniflora* from the high marsh. The manner by which low marshes evolve into high marshes is described in the boxed feature, "Salt-Marsh Evolution."

Food, protection, and frequency of tidal flooding appear to be the dominant ecological controls of the distribution, density, and activity of invertebrates in salt marshes. The fiddler crab, *Uca pugnax*, for example (**Figure 12–11a**), excavates a dwelling burrow with a single opening (**Figure 12–11b**), which it occupies during tidal flooding in order to escape predators. During ebbing tides, the fiddler crab feeds by searching the mud for food scraps. Other animals of the marsh include snails, mussels, and many birds and large terrestrial mammals like raccoons and people!

The most productive region of saltwater wetlands is the low marsh. The annual primary production of *S. alterniflora* ranges between 800 and 2,600 grams of organic matter per square meter (g/m^2), and seems limited in most areas studied by the availability of nitrate. Almost none of this plant growth is grazed by animals. Rather, the plants die during the autumn and are partially decomposed by bacteria and fungi. The resulting plant detritus is consumed by **detritivores**, animals that ingest detritus,

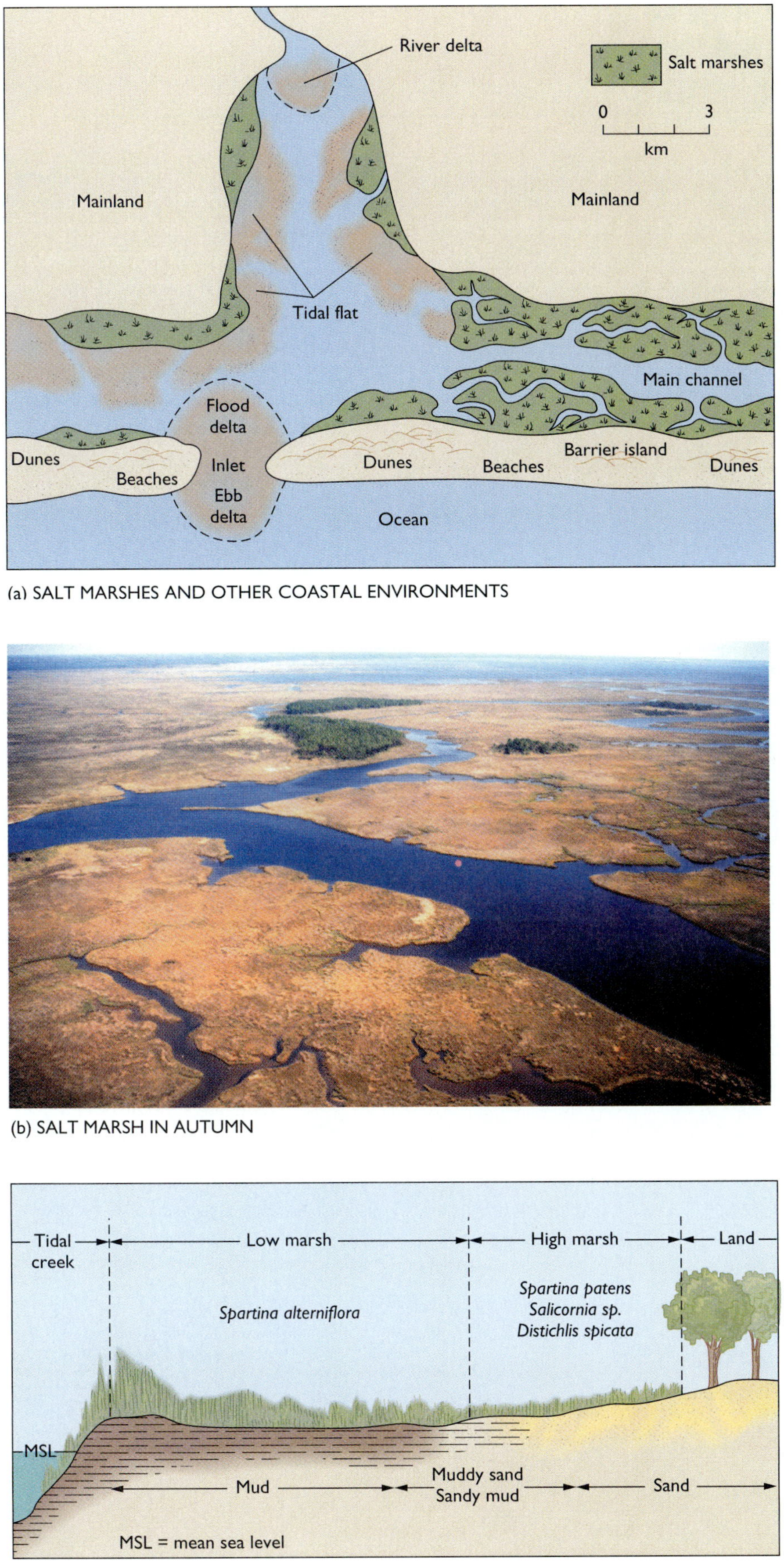

(a) SALT MARSHES AND OTHER COASTAL ENVIRONMENTS

(b) SALT MARSH IN AUTUMN

(c) SALT-MARSH PROFILE

FIGURE **12-9**

Salt marshes. (a) Salt marshes grow in muds and sands that are sheltered by barrier islands. (b) This view of a mature salt marsh in New England shows a well-developed network of sinuous tidal channels. These channels allow flood and ebb currents to transport saltwater, nutrients, plankton, plant detritus, and sediment into and out of the marsh. They are also important feeding areas for many young fish. (c) Based on topography and characteristic plant assemblages, salt marshes are classified as low or high marshes. [Adapted from J. M. Edwards and R. W. Frey, *Senckenbergiana Maritima* 9 (1977): 215–59.]

Salt-Marsh Evolution

Salt marshes evolve over time. Youthful marshes (**Figure B12–3a**) consist predominantly of the low-marsh cordgrass *Spartina alterniflora*. Flood tides bring nutrients into the low marsh, and the grasses grow thickly and luxuriantly. Much of the drainage of seawater is accomplished through a system of meandering tidal channels. The dense growth of the cordgrass dampens the effect of waves and tidal currents, which increases the depositional rate of mud. Also, the roots and rhizomes of the plants bind the mud and reduce erosion of the marsh surface. The net effect is rapid vertical growth of the marsh surface with time, until it is built above the neap high-tide level. At this stage, the plants of the high marsh invade, outcompete, and replace the low-marsh cordgrass. When the amount of high marsh and low marsh is about equal, the wetland is in a mature stage of development (**Figure B12–3b**). Eventually, the continued deposition of mud converts most of the low marsh to high marsh, a stage called *old marsh* (**Figure B12–3c**). In old marshes, water drainage through tidal channels is minimal, the marsh surface is remarkably level, and tidal effects are inconsequential because of the high elevation of the high marsh. Then streams and rivers deposit sand and mud on the high marsh, and con-

FIGURE **B12-3**

Salt-marsh evolution. (a) The youthful stage of salt-marsh evolution is dominated by low marsh. As the marsh surface is built upward by sediment deposition, a high-marsh community of plants develops. (b) The stage of mature development is reached when about one-half of the salt marsh consists of high marsh. (c) Additional deposition of sediment causes more vertical growth of the marsh surface, until the entire wetland is high marsh, a stage called old marsh. [Adapted from P. B. Basan and R. W. Frey, in *Trace Fossils 2. Geological Journal,* Special Issue 9, T. P. Crimes and J. C. Harper, eds. (Liverpool: Seel House Press, 1977), 41–70.]

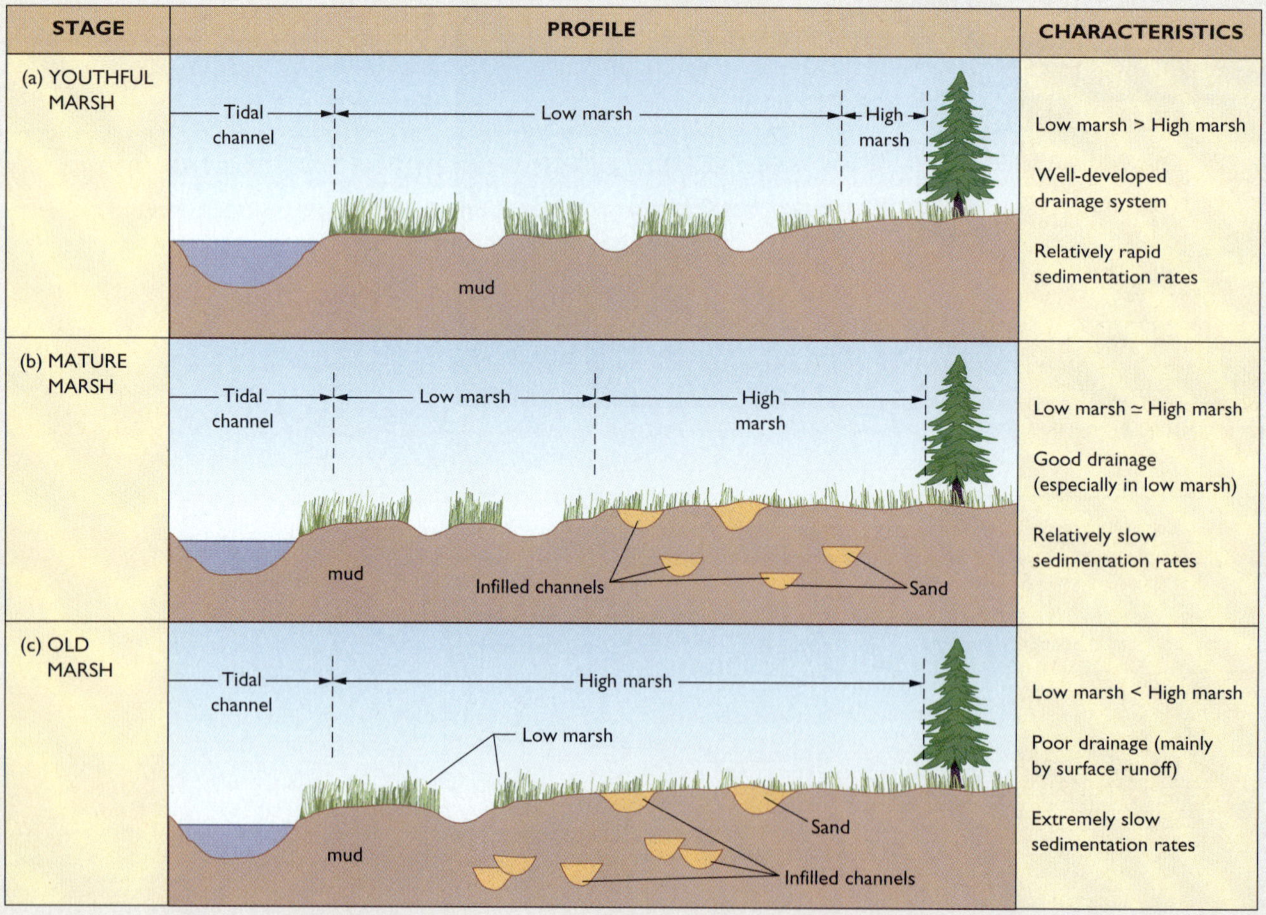

vert it into dry land that is no longer directly affected by the ocean.

This very sequence of marsh development has occurred at Sandy Neck, Barnstable, Massachusetts, on Cape Cod (**Figure B12–4**). About 3,300 years ago, the easterly longshore drift of sand produced a large sand spit that grew in length with time, protecting an ever-enlarging body of water and creating an estuary. Initially, youthful marshes were sparsely developed and were limited to the protected western embayment (see **Figure B12–4a**). With continued elongation of the sand spit and vertical growth of the marsh, the system evolved into a mature marsh (see **Figure B12–4b**) and then an old marsh (see **Figure B12–4c**). Sedimentary cores taken from the marsh show a complete chronology of the development—high-marsh peat at the top, low-marsh peat in the middle, and nearshore bay sand at the bottom (**Figure B12–4d**). Ancient marsh peats buried on land or on the continental shelf are important deposits for documenting the former positions of sea level.

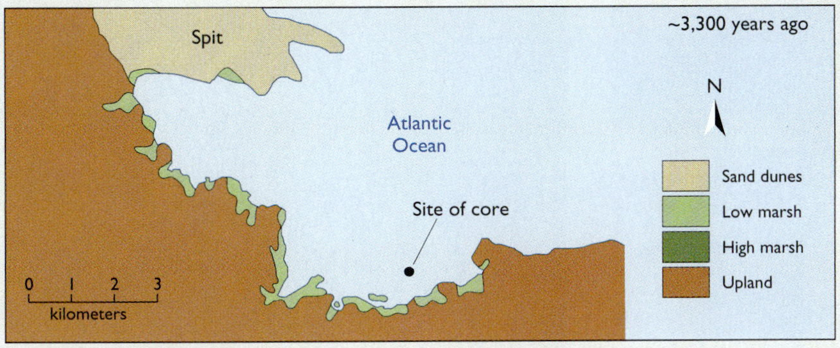

(a) ABOUT 3,300 YEARS AGO

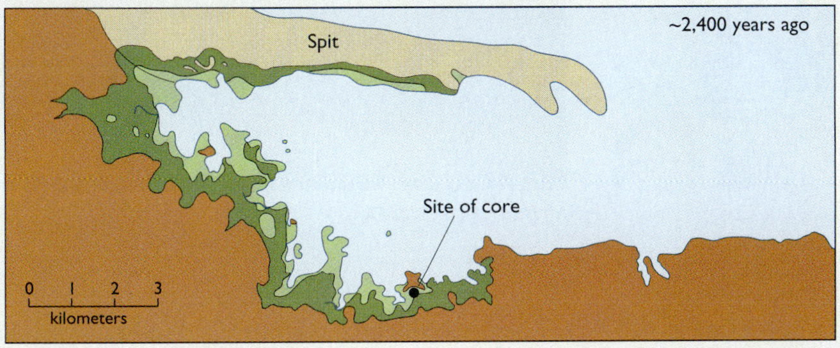

(b) ABOUT 2,400 YEARS AGO

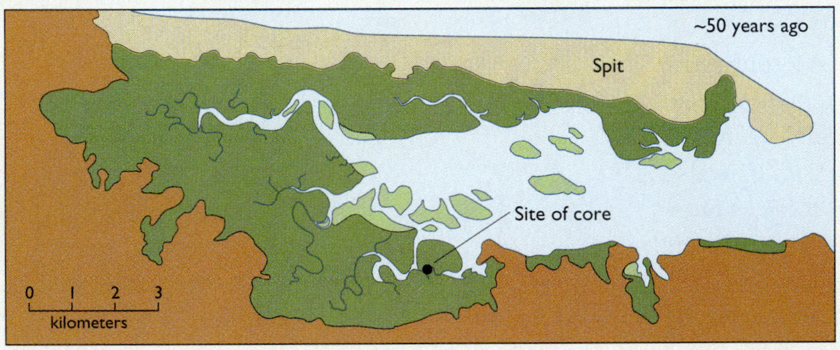

(c) ABOUT 50 YEARS AGO

FIGURE **B12-4**

Barnstable salt marsh, Massachusetts. (a) The longshore drift of sand created a spit that protected a cove from direct wave attack. Low marshes became established at many spots in the bay. (b) By 400 B.C., the spit had elongated considerably, and high marsh had extended into the bay covering older low-marsh deposits. At this time, the marsh was in the mature stage of development. (c) Spit growth has continued to the present. A great deal of mud deposition and the vertical growth of the wetlands have resulted in dominantly high-marsh communities, the stage known as old marsh. (d) A sediment core (located on maps a, b, and c) clearly shows the stages of marsh development. Note that the marked change in the sand content about 1.25 meters from the top of the core marks the time when the high-energy sand environment of the open bay changed to a low-energy mud environment as the spit closed off the bay. [Adapted from A. C. Redfield, *Ecological Monographs* 42 (1972): 201–37.]

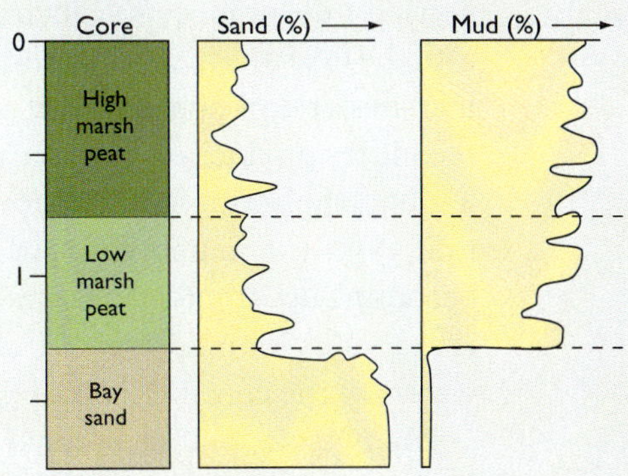

(d) SEDIMENT CORE

(a) *SPARTINA ALTERNIFLORA*

(b) *SPARTINA PATENS*

FIGURE 12-10

Salt-marsh grasses. Common plant varieties found in the salt marshes of temperate latitudes include (a) *Spartina alterniflora,* the dominant species of the low marsh, and in the high marsh, (b) *Spartina patens,* (c) *Salicornia species,* and (d) *Distichlis spicata.*

(c) *SALICORNIA SP.*

(d) *DISTICHLIS SPICATA*

such as filter-feeding shellfish and zooplankton. The unused components of the plant accumulate in the sediment and form peat (a richly organic soil of partially decomposed vegetation), or are exported into the nearby estuary by ebbing tidal currents. Salt marshes, it now appears, play an indirect role in the secondary production of estuaries by providing feeding and sheltering grounds for juvenile organisms that enter the marsh with the flood tide. It is also becoming apparent that salt marshes serve as nursery grounds for many species of fish that live in the waters of the continental shelf as adults.

Many salt marshes have been damaged by human activity. Now that their ecological value is recognized, efforts are being made in many places to restore coastal wetlands. A case in point is the salt-marsh reclamation project in San Francisco Bay (see the boxed feature, "San Francisco Bay").

12-4 Mangrove Swamps

The dominant intertidal plants of the tropics and subtropics are several species of **mangrove** (**Figure 12–12a**). These distinctive plant communities, rainforests by the sea, occupy more than 20 million hectares (~49 million acres) of coastal land. Mangrove swamps support an abundant and diverse fauna, including mollusks, crabs, fish, birds, lizards, turtles, and manatees.

FIGURE **12-11**

Fiddler crabs. (a) Males of the fiddler crabs, *Uca pugnax;* possess a single large claw (the "fiddle"), whereas females do not. (b) As the tide floods, fiddler crabs descend into their burrows, which protect them from marine predators that enter the low marsh with the advancing tide.

(a) FIDDLER CRAB

(b) BURROWS OF FIDDLER CRAB

The mangrove is a large woody, treelike plant with a thick, partially exposed network of intertwined roots that grow down from the branches into the water and sediment (**Figure 12–12b**). Immersed in saltwater, some mangrove species exclude salt from entering their tissues, whereas other species take in salt but excrete it from leaves, branches, and roots. The distribution, density, and species composition of mangrove forests depend on water and air temperatures during the winter, exposure to wave attack and tidal currents, the range of the tide, the type of substrate, and the chemistry of seawater. Hurricanes can be particularly damaging. Mangroves are hardly affected by normal storms, but suffer widespread structural damage from powerful hurricane winds, which can completely defoliate them and break their limbs and trunks.

The detritus food chain is the principal pathway for the flow of energy and matter between the trophic levels of food webs of mangrove forests. Leaf fall occurs throughout the year, but at a rate faster during the summer than during other times, because rain during short but intense downpours batters the canopy, dislodging leaves from the branches. About 60 percent of this dead-leaf biomass is consumed by detritivores; a mere 1 to 2 percent is lost as peat. The remaining 38 to 39 percent is flushed out by tidal currents into nearby bays, lagoons, and estuaries. Insects, particularly ants, are very important primary consumers that feed directly on living mangrove leaves. Also, crabs are particularly varied in mangrove forests, probably because the intricate root network of the trees provides them with a large variety of ecological niches. Some crabs feed directly on the live leaves and wood of the mangroves; others graze the leaf litter. The particular species of crab found at any locality reflects largely the salinity of the water at that site.

Large sea mammals like manatees (**Figure 12-13a**) also inhabit the tidal creeks of mangrove forests. They are enormous creatures, the larger males weighing more than one-and-a-half tons. However, despite their huge size, they are gentle, solitary animals that feed primarily on aquatic plants. In the 1970s, the Marine Mammal Protection Act was enacted. This proved fortunate for manatees, whose numbers were dwindling rapidly in U.S. waters for a variety of reasons. First, widespread coastal development and pollution of water in coves and bays have killed large tracts of sea grass, the prime food staple for manatees. Second, many manatees have inadvertently ingested rubber, plastic, and metal waste products, in some cases fatally. Finally, between forty and fifty manatees are killed annually by collisions with motorboats and by deep wounds inflicted by boat propellers (**Figure 12-13b**). Efforts to protect manatees have worked for some populations. Since 1975, a group of 60 individuals living in the clear waters of Crystal River, Florida has grown to almost 300.

San Francisco Bay

San Francisco Bay, located at the confluence of the Sacramento and San Joaquin Rivers (**Figure B12–5**) and bounded by upraised and faulted crustal blocks on its westward and eastward sides, owes its origin to tectonic processes associated with strike-slip motion along the San Andreas Fault Zone, a transform plate boundary that runs through southern California (see Figure B3-1 in Chapter 3). In addition, during the Pleistocene when sea level was low, the Sacramento River eroded a system of channels in the bay although there was no San Francisco Bay as such during glacial times, the shoreline being located tens of kilometers to the west on what is now the continental shelf. At present, San Francisco Bay is best characterized as a partially mixed estuary, but a very complex one.

There is a 150-year-long history of human settlement in the San Francisco Bay area, and this prolonged activity has over time severely affected this coastal ecosystem, including its natural biota, sedimentation patterns, hydrology, and water chemistry. Nevertheless there is reason for optimism, because numerous federal and state agencies, as well as grass-roots organizations, are working hard and cooperatively to develop a scientific understanding of this complex ecosystem in order to mitigate and even reverse the degradation of the bay. The goal is to distinguish the natural from the human-induced changes, so that effective management and regulatory strategies can be implemented to protect the natural integrity and beauty of the Bay Area for future generations.

What specific chemical, biological, and geological changes has San Francisco Bay undergone since 1850? Principally, the damage has been the result of: loss of wetlands, the diversion of water, overenrichment with nutrients, deterioration of water quality, and the introduction of exotic species.

WETLAND LOSS

Since 1950 over 95 percent of the estuary's wetlands have been lost (**Figure B12–6**) to development, levee construction, pollution, erosion, and rising sea level. Sediment infilling of a bay and its wetlands is a natural process, but it can be greatly accelerated by human activity, such as the gold mining by the Forty-niners during the nineteenth century, when huge quantities of sediment were washed into the bay. Because wetlands are critical habitats for many fish and wildlife, destruction there has had grave consequences for the area's natural biota. For example, the decline of fish diversity in the estuary has been so extreme that today commercial fisheries are limited to harvesting anchovies and herring. The U.S. Geological Survey, the U.S. Army Corps of Engineers, and the San Francisco Bay Conservation and Development Commission are striving to establish new wetlands in parts of the bay that have been diked to create farmland and salt evaporation ponds. More such reclamation projects have been proposed by conservation groups.

FIGURE **B12-5**
San Francisco Bay. Satellite image of the San Francisco Bay Area.

WATER DIVERSION

Today, about 40 percent of the freshwater discharge of the Sacramento-San Joaquin watershed is diverted for local consumption. An additional 24 percent is pumped out of the upper bay and exported by aqueducts to southern California. Most of this diverted water is used to irrigate farmland and the demand for additional water is expected to increase because of the profitability of agricultural production. The reduced inflow of freshwater has had grave ecological consequences for the estuary, preventing some fish species from spawning, reducing the phytoplankton biomass, and generally disrupting food webs. The reduced freshwater discharge into the bay is also likely to have modified circulation and sedimentation patterns and the capability of the estuarine water to dilute and flush out contaminants. Several agencies are extensively researching and monitoring the bay to document the effect on the estuary's biological system and water circulation of this reduced inflow of freshwater.

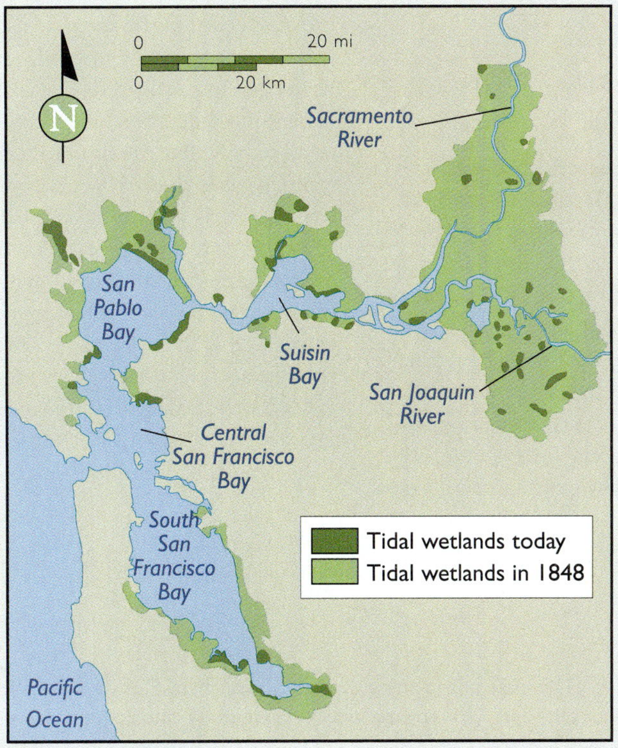

FIGURE **B12-6**

Bayfill. Wetlands lost during the past century and a half in the San Francisco Bay Area.

NUTRIENT ENRICHMENT

The urbanization of the area and agricultural practices have substantially increased the levels of nutrients such as ammonia, nitrate and phosphate in the waters of the bay. For example, more than 800 million gallons of municipal wastewater containing 60 tons of nitrogen are discharged annually into San Francisco Bay. This influx of nutrients has not, however, led to eutrophication, because of the richness of the benthos (clams, mussels, crustaceans), which filter out huge quantities of algae and thereby control the phytoplankton biomass of the water. State-of-the-art wastewater treatment has had immediate results in reducing the influx of nutrients to portions of San Francisco Bay, and more such measures and water treatment technology are scheduled to be in place in the near future.

WATER QUALITY

Bay waters receive variable loads of a great variety of toxic contaminants from agricultural, industrial, and urban activities. Locally, particularly near wastewater discharge pipes, the concentrations of toxins in bottom sediment commonly is very high. Many of these substances, such as pesticides, PCBs, petroleum hydrocarbons, and trace metals, have long-range consequences for the viability of estuarine organisms, particularly of benthic communities, mammals, such as seals, and birds that feed on marine invertebrates and fish. The high concentrations of cadmium, silver, and lead that have been measured at certain localities in bay muds are dangerous, because these metals are known to interfere with the growth and reproduction of fish, mammals, and birds. Furthermore, shellfish and finfish biomagnify such toxins, so there is concern for the health of humans who consume them. Modern waste treatment procedures and technology can substantially reduce the inputs of contaminants to the bay. Models are being developed and monitoring in the field is currently being used to investigate the way in which contaminants in muds are recycled through ecosystems.

THE INTRODUCTION OF EXOTICS

The introduction of exotic species has had devastating effects on the natural populations of San Francisco Bay by displacing or outcompeting them for space and resources. More than 130 species of plants, invertebrates, and fish that are not indigenous to the region have been released into the bay waters since the mid-nineteenth century, some inadvertently (foreign vessels, for example, expel their bilge water with its alien organisms into the bay), some on purpose. Today, just about all of the common large invertebrates that dwell in shallow water are exotic species; of the forty-two species of fish that inhabit the upper bay, twenty have been introduced by humans. Recent investigations have demonstrated convincingly that the damage caused by exotic species to the bay's natural ecosystem can be as great or even greater than the effect of chemical contaminants. Many studies are underway to investigate the specific ecological changes that are associated with the invasion of exotic species, and the results of this research will form the basis for a management program to mitigate what some refer to as "biological contamination."

Nonetheless, the future of San Francisco Bay has never looked better, as private and public individuals and agencies work hard and cooperatively to identify the area's environmental problems. Clever management strategies based on solid scientific research are being designed and implemented with the long term in mind and with a proper balance between the needs of people and the needs of the bay's fauna and flora. Many hope that the general lessons learned here can be used to avert similar problems elsewhere.

Visit www.jbpub.com/oceanlink for more information.

FIGURE 12-12

Florida mangroves. (a) This view of a mangrove forest in southern Florida illustrates the dense tree growth and lush canopy that characterize these environments. (b) Intricate networks of intertwined prop (above-ground) roots of mangrove trees provide habitats for a large variety of invertebrates and fish.

(a) DENSE MANGROVE FOREST

(b) ROOTS OF MANGROVE TREES

Mangroves are a vital natural resource. They help reduce the erosional impact of hurricanes, serve as breeding and feeding grounds for juvenile fish, trap silt that could smother offshore coral reefs, and cleanse nearshore water by the uptake of nutrients and pollutants. Also, they provide local people with food and medicine, as well as wood for heating, cooking, and house building. Unfortunately, many governments and developers consider mangrove swamps to be virtual wastelands with little economic value for human beings except when cut back for their timber, for the development of shrimp aquaculture, or for the establishment of tourist resorts. This human-induced devastation is increasing at an alarming rate, and groups are organizing to protect the remaining stands of mangrove.

12-5

Coral Reefs

A **coral reef** is an organically constructed, wave-resistant rock structure created by carbonate-secreting animals and plants. The biodiversity and trophic dynamics of a reef ecosystem are so distinctive and spectacular that they deserve serious consideration. Accordingly, this section is devoted to an extended discussion of the biology, ecology, and geology of coral reefs.

BIOLOGY OF CORALS

The vast bulk of a coral reef consists of the build-up of loose to well-cemented organic debris—fragments of shells and skeletons—composed of calcium carbonate ($CaCO_3$). In rock form, it is referred to as **limestone**. The living part of the reef is mainly a veneer growing on the surface of massive limestone deposits that record the existence of ancient reef communities. In effect, the thick limestone base is covered by a thin, living "skin." When coral and other carbonate-secreting organisms in this "skin" die, their hard parts are added to the reef structure, helping the reef mass grow in size over geologic time.

Corals (**Figure 12–14a**) are animals with a rather simple anatomical design that belong to the phylum Cnidaria (see Table 9–3 in Chapter 9). The coral animal itself consists of a **polyp**, the body of the living organism, that is housed for protection in a rigid calcium-carbonate **exoskeleton** (an external as distinguished from an internal skeleton) called a **corallite**. The bottom of the corallite is divided into compartments by vertical partitions known as **septa** (**Figure 12–14b**). The polyp resembles a cup or sac, with an opening at the top that is a combined mouth and anus and that leads to the gut. The opening is surrounded by tentacles that may have stinging, mucus-secreting cells for catching prey. Polyps live either as solitary individuals or in a colony. The colonial forms are what create the massive skeletal frameworks of coral reefs.

Careful microscopic examination of coral tissue reveals that numerous unicellular plants, modified dinoflagellates called **zooxanthellae**, are imbedded in the outer layer of the coral's flesh (**Figure 12–14c**). These plant cells are abundant (up to 30,000 plant cells per cubic centimeter of animal tissue) and can represent up to 75 percent of the tissue weight of the coral polyp! The coral animal and the zooxanthellae benefit mutually from this association, a biological interaction called **mutualism**. Being enclosed in coral tissue provides the plants with a stable environment, protects them from predation, and furnishes them with a reliable supply of nutrients (the metabolic waste products of

FIGURE 12-13

Manatees. (a) The manatee is a large marine mammal that grazes on sea grass in coastal waters. (b) During the past few decades, manatees have frequently been injured by motorboat propellers; many of them die from these wounds.

(a) MANATEE

(b) A PROPELLER-CUT MANATEE

THE OCEAN SCIENCES BIOLOGY

Global Decline of Coral Reefs

Coral reefs are resilient ecosystems that support an incredible diversity of marine life. Their fish populations feed many people, they are important for tourism, and their hard limestone structure acts as a natural storm barrier that protects the nearby shoreline from wave erosion. Recently, scientists have been alarmed by a disturbing trend. Perhaps as much as 30% of the world's reefs have been seriously damaged or killed outright. Locally, reef decline is an even more serious problem, as for example in Jamaica and Costa Rica where 90% of the coral reefs are now dead. What factors are responsible for this global devastation? The consensus among scientists is that the recent decline of coral is attributable directly to human activity, as summarized below.

BLEACHING EVENTS

Sea surface temperatures of tropical water have increased during the last few decades, possibly due to global warming. Coral are intolerant of elevated temperatures and respond by ejecting the phytoplankton, the zooxanthellae, which live in their tissue and are responsible for the polyps variegated colors. This, a bleaching event (**Figure B12-7**), causes the coral to go "white." Without its symbiotic zooxanthellae, the coral dies shortly thereafter. Coral bleaching is occurring everywhere.

OVERFISHING

A variety of nonsustainable techniques are used to kill or capture fish that live in reefs. These include trawling the reefs, blasting fish with dynamite, and poisoning them with sodium cyanide. Most of these methods kill organisms indiscriminately and often seriously damage the reef edifice or disturb ecological relationships that affect the reef's health. Moreover, such fishing methods quickly decimate the local fish populations, forcing fishermen to exploit other tracts of the reef to find fish.

AGRICULTURAL PRACTICES

In the tropics, deforestation of the land and the use of fertilizers, herbicides, and pesticides allow soil and chemicals to be swept inadvertently by rivers to the coastal waters, where they can have a serious impact in a variety of ways on the health of coral reefs. For example, river-supplied silt smothers the coral when deposited in the reef. The nutrient input from farmland causes algal blooms and widespread eutrophication of coastal waters, which kills coral. Recently, scientists have detected the buildup of chemical pesticides, such as DDT, Aldrin, and PCBs, in the tissue of coral and fish. These dangerous chemicals interfere with growth and reproduction of organisms. According to the World Health Organization, the consumption of pesticide-tainted fish and invertebrates have killed more than 20,000 people and caused acute poisoning of over three million people.

INDUSTRIAL EFFLUENTS

There are a variety of industrial substances that have an impact on coral reefs. Heavy metals, toxic manufacturing compounds, and leachates from landfills are examples of harmful chemicals that are regularly injected into coastal water by river input. These can cause extensive die-offs of coral. The discharge of untreated sewage into coastal water is another grave problem that leads to eutrophication and the death of marine life, including coral. Waste heat from power plants elevates the seawater temperature locally, which causes bleaching and eventual death of the coral.

COASTAL DEVELOPMENT AND TOURISM

The serenity, climate, and beauty of tropical coastlines attract many people. Tourism is growing exponentially, much of it occurring near pristine tropical reefs. The impact of so many people living or visiting reefs can quickly overwhelm and destroy coral ecosystems, if measures are not taken to protect them. Prudent stewardship is absolutely necessary to assure that reef health is not affected adversely by these activities.

Given that all of the above are human-induced disturbances, it is obvious that people must minimize their unintended impacts, if reef decline is to be stopped. As many suggest, it is imperative that people educate themselves about this crisis. Some organizations, such as the International Coral Reef Initiative and the Coral Reef Alliance, are endeavoring to monitor the problem and develop conservation programs that will protect reefs worldwide.

Visit www.jbpub.com/oceanlink for more information.

FIGURE B12-7

Coral bleaching. This underwater photograph shows the "whiteness" of a dead coral, the typical appearances of bleached coral.

FIGURE 12-14

Coral. (a) Coral polyps are shown in the act of feeding. (b) The coral polyp has a saclike body that sits in a rigid carbonate cup, the corallite. The mouth, located at the top of the animal, also serves as an anus and is surrounded by tentacles that contain stinging cells. (c) The tissue of this brain coral contains many algae (zooxanthellae), which photosynthesize and recycle the metabolic wastes of the coral.

(a) CORAL POLYPS

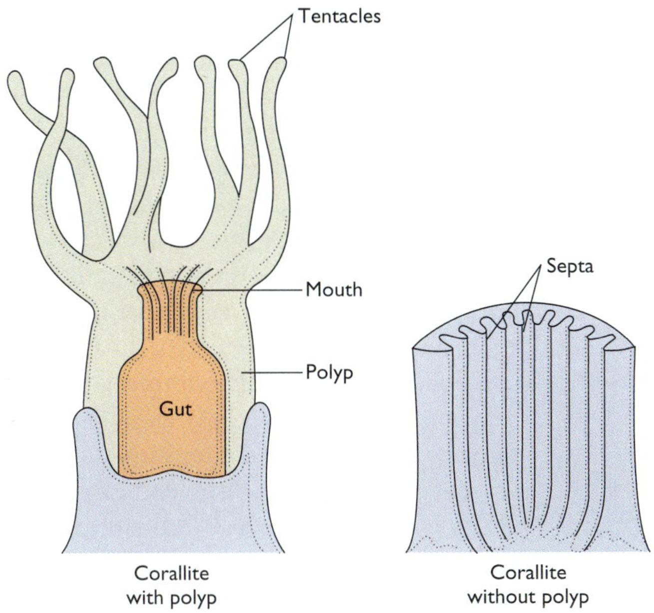

(b) CORAL STRUCTURE

(c) ZOOXANTHELLAE

the coral) essential for photosynthesis. In turn, the algae saturate the coral's tissue with oxygen and food produced by photosynthesis and remove the animal's potentially dangerous metabolic waste products. Zooxanthellae also influence the process of carbonate secretion by the coral, as can be inferred by the reduction of skeletal growth when the algae are expelled from the tissue of the animal. Last, the algae can be digested by the coral when food is scarce.

Coral reefs are a genuine paradox from a trophic perspective, because they support a rich and abundant community of life-forms in a tropical setting that is a "desert" environment. Even more curious is the fact that the clear blue water surrounding tropical reefs is depleted of phytoplankton, which form the base of most marine food webs. The level of dissolved nutrients in reef waters is low as well, which explains the lack of phytoplankton. What supports these conglomerations of life—these fertile marine oases—that thrive in the deserts of tropical seas? Where are the primary producers, the autotrophs?

Well, you have probably guessed the answers to these questions. The explanation, in part, lies within the coral itself, the zooxanthellae that live in the tissue of the animal. These unicellular algae use sunlight, which easily penetrates the transparent tissue of the coral polyp, to produce food photosynthetically using the metabolic waste products of the coral as nutrients. As much as 60 percent of this manufactured food passes through the cell walls of the zooxanthellae and nourishes the coral polyp directly. This system is a self-contained survival capsule in which food is produced and consumed internally and essential nutrients are continually recycled between plant and animal cells. The coral does, of course, supplement its diet by preying on small zooplankton that drift into its stinging tentacles. Biologists believe that large coral reefs could not exist without this mutualistic relationship between coral and zooxanthellae.

Benthic algae also abound in reefs, but they are tiny and hard to see. However, they grow in dense patches. Many that inhabit high-energy zones of the reef are stony, encrusting forms. Their porous carbonate crust protects them from pounding breakers and strong currents. These calcareous algae can be as important as the zooxanthellae in primary production. Herbivores, both invertebrates and fish, graze algae extensively. Also, they are the main contributors of carbonate grains that accumulate as mud deposits in lagoons. Many biologists now refer to reefs as "coralgal" reefs in recognition of the ecological importance of the benthic plants.

When coral flourishes, dense colonies merge to form an intricate network of habitats that provide diverse, specialized niches for other organisms. A listing of the types, colors, sizes, shapes, specialized structures, habitats, and behaviors of all the faunal species that inhabit a large, thriving coral community would truly be encyclopedic. Despite this biological complexity, it is possible to construct a general, but useful food pyramid of a coral reef. This pyramid consists of four tiers (**Figure 12–15**)—one level for the primary producers, and three levels for the animal consumers.

ECOLOGY OF CORAL REEFS

In the broadest sense, coral can be divided into two major groups: **hermatypic** coral, which possess zooxanthellae, and **ahermatypic** coral, which do not. Hermatypes build the colonial reefs found in tropical seas. Ahermatypes, by contrast, tend to be solitary in habit. Unlike the hermatypes, the ahermatypes are not restricted to shallow or warm water. In fact, they can be found in the deep, dark cold waters of the polar seas.

Hermatypic corals flourish in tropical waters, where ecological conditions favor their growth. Water temperature, salinity, water clarity, and nutrient levels are the primary regulators of coral-reef development. Temperature limits reef growth by affecting the zooxanthellae in the coral polyp. Cold kills these plants. Therefore, although reefs can grow at water temperatures as low as 18°C, the growth of zooxanthellae—and thus of reefs—is optimal in water that is warmer than 20°C. This is why coral reefs tend to be most extensively developed along the eastern edges of continents (**Figure 12–16**), which are bathed by the warm equatorial currents of the ocean circulation gyres. Also, because temperature decreases with water depth, carbonate secretion by coral is limited to the warm, sunlit waters of the upper 30 meters (~99 feet) of the water column.

Coral do best in water of normal salinity. In fact, they cannot survive in freshwater or even in brackish water. An additional environmental control on reef development is water turbidity, producing muddy waters. Suspended sediment particles interfere with the ability of the zooxanthellae to photosynthesize by decreasing the amount of light that can be transmitted through the water. Also, sediment may be deposited on the coral polyp itself, burying it and killing the animal. These two factors, combined with the coral's intolerance of freshwater, prevent reef structures from building up near the mouth of large tropical rivers. The Amazon River, which flows into the tropical western Atlantic and where there are no coral reefs, is a case in point.

Phytoplankton flourish in tropical seas that con-

FIGURE **12-15**

Trophic pyramid of a coral reef. Members of a coral reef can be arranged in a simple four-level pyramid, based on trophic relationships. In coral reefs, unlike most marine ecosystems, phytoplankton play a relatively minor role in primary production. [Adapted from M. Lerman, *Marine Biology: Environments, Diversity, and Ecology* (Reading, Mass.: Benjamin-Cummings, 1986).]

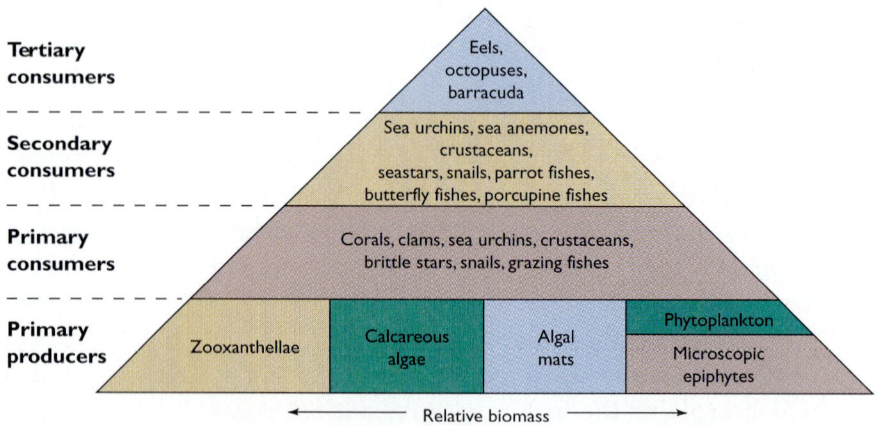

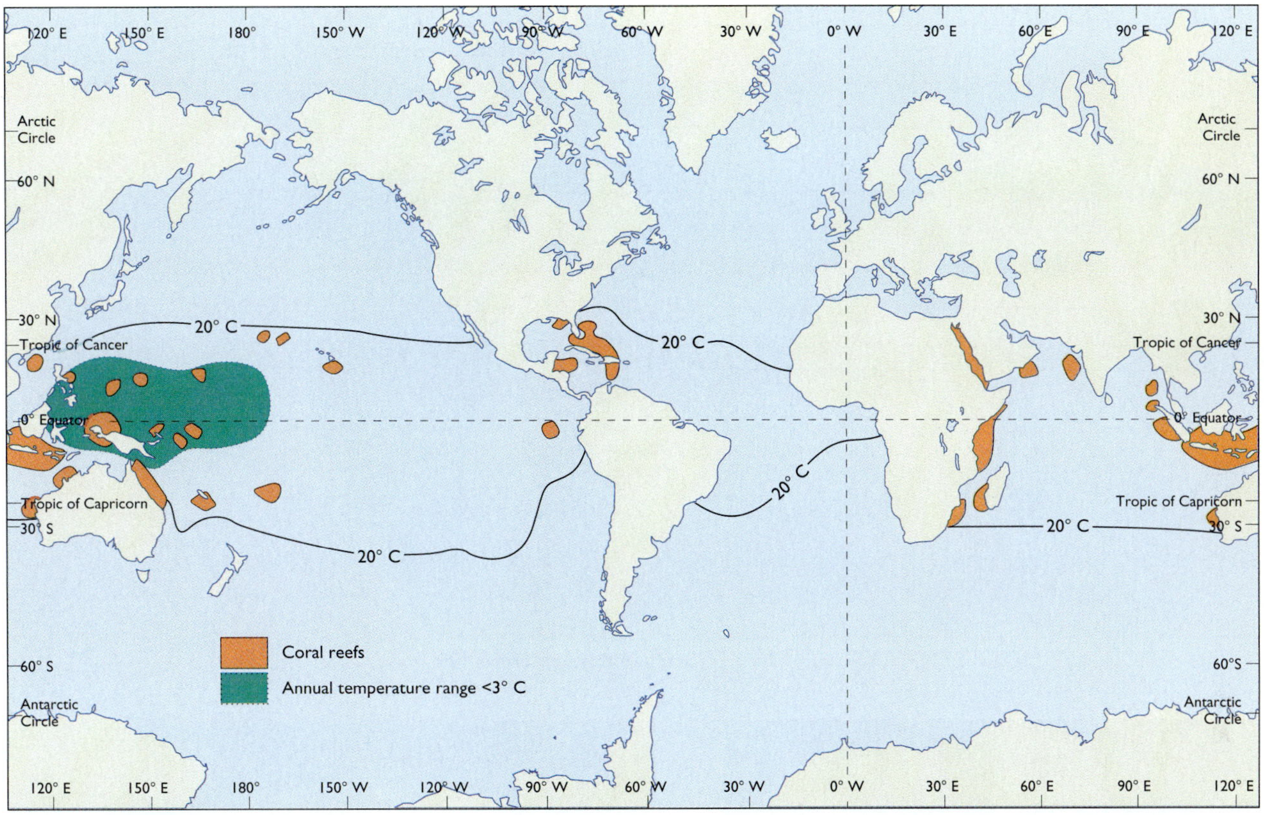

FIGURE 12-16

Distribution of coral reefs. The 20° C isotherm delimits the poleward extent of coral reefs, with few exceptions. The variety of corals is greatest in Indo-Pacific waters, where annual water temperature variations are less than 3° C. [Adapted from M. Lerman, *Marine Biology: Environments, Diversity, and Ecology* (Reading, Mass.: Benjamin-Cummings, 1986).]

tain high concentrations of nutrients. These large phytoplankton populations support dense benthic communities of filter feeders, which competitively exclude coral. Thus, hermatypic coral grow best in warm, clear tropical water that contains low levels of dissolved nutrients.

Many reefs share general topographic characteristics that reflect the degree of exposure to breaking waves. On the windward side, large pounding waves are absorbed by the **algal ridge** (**Figure 12–17a**), which is continuously awash with surf surge (**Figure 12–17b**). Below the algal ridge is the seaward-sloping **buttress zone**, which consists of alternating coral-capped ridges, channels, and furrows. This irregular topography tends to disrupt swell, because an uneven wall face has much more surface area than does a flat wall and, therefore, receives less wave energy per unit area under the same wave conditions. The **reef face** extends downward from the buttress zone (see Figure 12–17a). Here living corals are absent, because they rarely grow in water depths below 40 meters (~132 feet) because there is so little light at those levels.

The **reef terrace** is a broad reef flat just landward of the algal ridge; its floor lies at mean low water, so that much of its surface is exposed at low tide. Encrusting algae flourish here. If particularly broad, the reef terrace may have islands, some vegetated with palm trees. The backside of the reef terrace grades into a shallow—less than 50 meters (~165 feet)—wave-protected lagoon (see Figure 12–17a) with a floor blanketed with carbonate (lime) mud and sand. Numerous small organic knolls (hills) called **patch reefs** (**Figure 12–17c**) grow on the floor of the lagoon. Invertebrates and fish abound in the waters of the lagoon because of the abundance of food and niches.

The growth forms of the hermatypes reflect energy conditions of the various habitats of the reef, which range from the turbulent, powerful

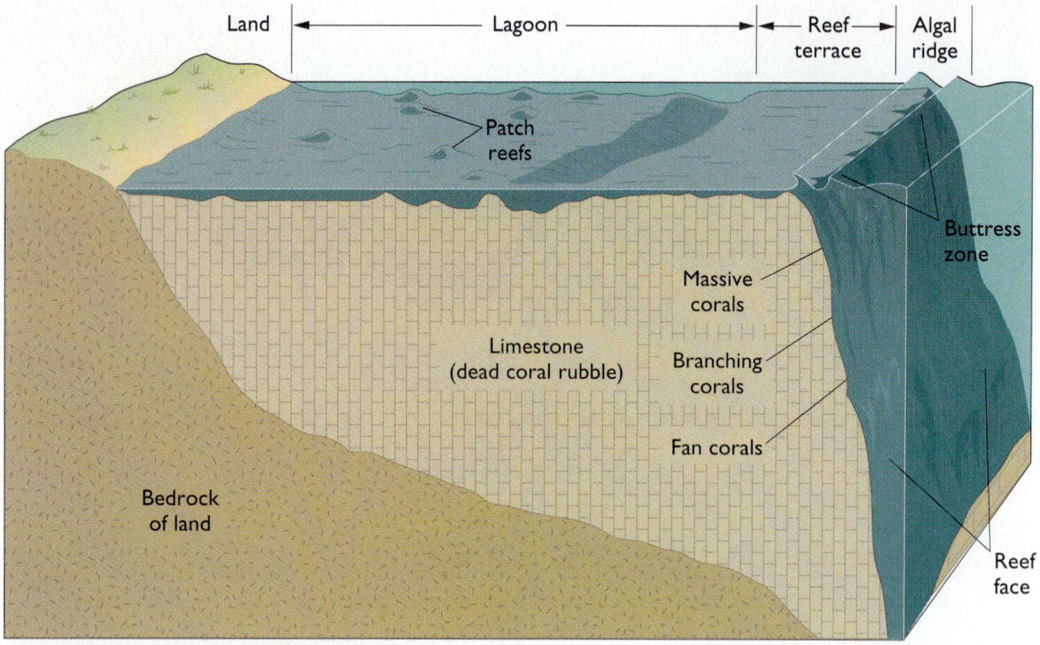

(a) MORPHOLOGY OF A BARRIER REEF Vertical exaggeration ≈ 12×

(b) REEF EDGE

(c) PATCH REEFS IN LAGOON

FIGURE **12-17**

The morphology of coral reefs. (a) The coral reef is divided into distinct topographic elements. The algal ridge and buttress zone are subjected to high wave energy; the lagoon, to low wave energy. (b) The algal ridge is pounded regularly by ocean breakers. Note the furrows in the buttress zone just seaward of the algal ridge. (c) Lagoons commonly contain numerous mounds of coral called patch reefs.

ocean swell that batters the algal ridge to the placid water of the lagoon behind the reef terrace. Encrusting corals (**Figure 12–18a**), which produce thin sheets, or "crusts," of calcium carbonate, are a common growth form of the algal ridge, because they can withstand the constant wave pounding that occurs at the seaward edge of the reef mass. The deep buttress zone is still churned up by waves, and here only the most durable growth forms of coral can exist, such as a few massive branching corals (**Figure 12–18b**) and brain coral (**Figure 12–18c**). The deeper and quieter waters of the reef face are populated by a wide variety of coral forms, including delicately branched (**Figure 12–18d**) and platy (waferlike) corals. The protective waters of the lagoon support a great number

FIGURE 12-18

Coral growth forms. Different species (and sometimes members of the same species) assume a variety of growth forms, depending on the environmental conditions: (a) encrusting coral; (b) flat branching coral; (c) brain coral growing near an orange elephant-ear sponge; (d) delicately branching coral; and (e) robust branching coral.

of coral forms and a large, diverse community of invertebrates and colorful fishes, such as groupers, angel fishes, soldier fishes, lion fishes, and parrot fishes (**Figure 12–19**).

Human impact is adversely affecting the ecology of many of the world's reef systems. Deforestation of rainforests for lumber and farming, for example, results in severe soil erosion; this mud gets transported by rivers to the shore where its deposition can smother large tracts of coral reef. Additionally, overfishing, tourism, and industrial and munincipal effluents are having an impact on coral reefs worldwide as never before. Finally, slight rises in water temperature are forcing coral polyps to expel their symbiotic algae, the zooxanthellae, causing the coral to turn white and eventually die. The number and extent of such "bleaching" events have increased drastically during the last decade. Some reef ecologists believe that over 70% of the world's coral reefs will disappear, if measures are not taken immediately to protect them.

GEOLOGY OF CORAL REEFS

From his study of Pacific coral reefs, Charles Darwin (1809–1882) recognized three distinctive kinds of reefs: the fringing reef, the barrier reef, and the atoll (**Figure 12–20a**). **Fringing reefs** form limestone shorelines around islands and the tropical areas of continents. **Barrier reefs** grow farther off-

(a) GROUPER

(b) ANGEL FISH

(c) SOLDIER FISH

(d) LION FISH

(e) PARROT FISH

FIGURE 12-19

Fish of coral reefs. The color, habits, and body shapes of coral-reef fish vary enormously, reflecting the large number of ecological niches that this ecosystem provides. Common reef fish include (a) groupers, (b) angel fish, (c) soldier fish, (d) lion fish, and (e) parrot fish.

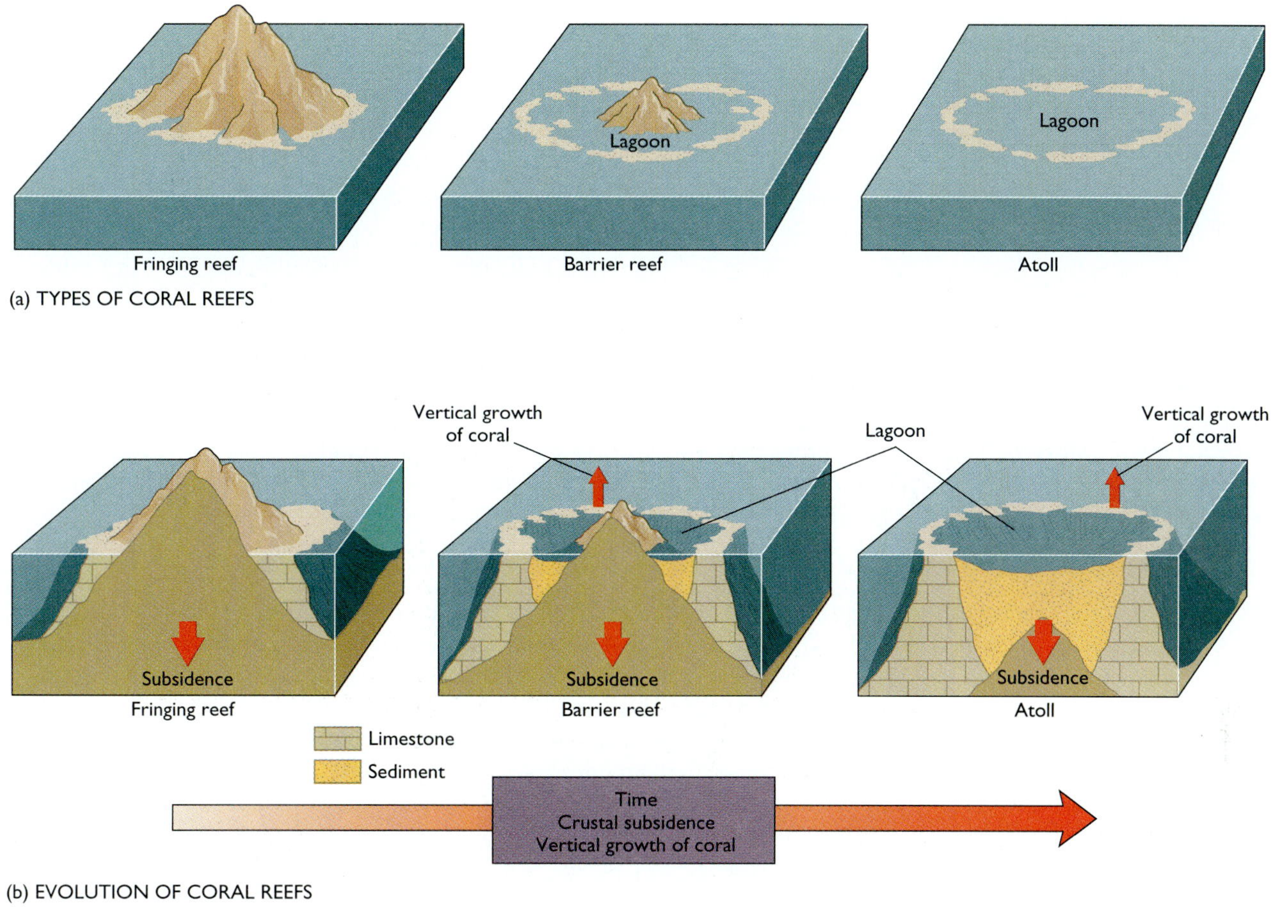

FIGURE 12-20

Evolution of coral reefs. (a) There are three different types of coral reefs; the fringing reef, the barrier reef, and the atoll. Each represents a stage in the morphologic development of a reef. (b) As Charles Darwin proposed, volcanic islands were first populated by organisms living in fringing reefs that grew around their perimeters. Eventually, as the land sank or as sea level rose, the fringing reef grew upward and became detached from the land, creating a lagoon. Finally, as water continued to deepen, the island crest disappeared beneath the sea, and the reef became an atoll that encircled a central lagoon.

shore and are separated from the mainland by a lagoon. **Atolls** are found in the open ocean, far from land and consist of ring-, oval-, or horseshoe-shaped reef structures that enclose a lagoon. Darwin recognized that each type of reef represents a stage in the evolutionary development of a reef mass through geologic time.

In essence, Darwin was visualizing the effect that a sinking sea floor or a rising sea level would have on the natural development of a coral reef (**Figure 12–20b**). Because hermatypic coral can grow only in sunlit waters, colonization by coral occurs in shallow water around the edge of an emerging volcanic island. The gradual buildup of the reef mass around the island's periphery leads to the formation of a fringing reef. As the island subsides and slowly drags the reef edifice down into the water, or as sea level rises, the fringing reef must grow vertically if it is to survive. This transforms it into a barrier reef that is now separated from the island by a broad lagoon. Continued subsidence or the rise of sea level drowns the volcanic island and, provided that the upward growth of the barrier reef has kept pace with the deepening water, creates an atoll with a central lagoon (see Figure 12–20b).

SCIENCE BY NUMBERS

Residence Time

The concept of residence time, introduced in Chapter 5, is basically as follows: if an element or compound is in steady state, then its inputs equal its outputs, and its concentration remains constant over time. Consequently, it is possible to calculate the average time (the residence time) that the element or compound remains in the system from the time of its introduction to the time of its exit.

Think about it this way. Let's assume that there is a popular diner on a well-traveled highway that is open twenty-four hours a day. It has only ten seats, and there are always people waiting to come in. The owner says that about ten people enter each hour. Given this, how long does the average person spend eating a meal in the diner? The answer would be the residence time of the diners. How would we calculate this?

Let's assume steady state. This means that the number of people entering equals the number of people leaving the diner, so that the number of diners is constant (that is 10 diners at the 10 seats). Residence time (*RT*) of the diners is the number of diners (10) divided by their entry rate (10 diners/hr) or

$$RT = 10\ \cancel{\text{diners}}/10\ \cancel{\text{diners}}/\text{hr} = 1\ \text{hr}.$$

This states that the residence time for the diners is 1 hour, meaning that it takes, on the average, an hour for customers to finish their meals. Some, of course, eat faster and leave sooner; others eat slower and leave later. But the average eating time (residence time) is 1 hour.

Let's now calculate the residence time of carbon in marine plants. All we need to do this is to assume steady state for carbon in marine plants and divide the total biomass of carbon by the rate of uptake of carbon due to plant photosynthesis. In other words, *RT* of carbon in marine plants = carbon biomass in plants/primary productivity.

In this case, the carbon biomass is equivalent to the number of diners in the previous example, and primary productivity (carbon uptake by plants due to photosynthesis) is equivalent to the rate at which diners enter the restaurant. If you understand this, then the calculation is a simple matter. It is

$$RT = \text{carbon biomass in plants/primary productivity}$$
$$= 2 \times 10^{15}\ \cancel{\text{g of carbon}}/2.5 \times 10^{16}\ \cancel{\text{g of carbon}}/\text{yr}$$
$$= 2 \times 10^{(15-16)}/2.5\ \text{yr} = 2 \times 10^{-1}/2.5\ \text{yr} = 0.8 \times 10^{-1}\ \text{yr}$$
$$\text{RT} = 0.8 \times 10^{-1}\ \text{yr} = 0.08\ \text{yr} = (0.08\ \cancel{\text{yr}})(365\ \text{days}/\cancel{\text{yr}}) = 29\ \text{days}.$$

This means that, on the average, a carbon atom remains bound to a marine plant for about 1 month before it is released back into the water.

STUDY GUIDE

KEY CONCEPTS

1. Coastal settings, such as estuaries, salt marshes and mangrove swamps, are stressful for organisms because of their location between the land and the ocean. Thus, these environments tend to have relatively low species diversity. But their biological productivity is very high, because the plants and animals that can tolerate the variable conditions of these habitats abound because of the high nutrient levels, plentiful food, and reduced predation pressure there.

2. *Estuaries* are coastal water bodies where freshwater and saltwater intermingle. Their circulation depends on the relative effect of river discharge and tidal mixing. The dominance of river discharge promotes water stratification and

a sharp halocline, creating a *salt-wedge estuary*. The dominance of tides mixes the salt water and freshwater and eliminates the halocline, creating a *well-mixed estuary*. Where the river influx and tides are about equally important, mixing of the water column is partial, creating a *partially mixed estuary*. Each of these estuary types has a distinct circulation pattern (see Figure 12–3).

3. *Lagoons* are isolated or semi-isolated coastal water bodies that, unlike estuaries, do not receive a significant inflow of river water. Their salinity can vary between brackish and hypersaline, depending on the local hydrology and climate; most tend to have relatively simple circulation patterns. In the subtropics where evaporation is intense, inverse flow can develop, whereby ocean water flows into the lagoon at the surface, and dense, salty water flows out of the lagoon at depth.

4. *Salt marshes* are vegetated intertidal flats; the grasses are typically rooted in mud and represent one of the most biologically productive environments on Earth. The salt marsh is divided into a low salt marsh and a high salt marsh, each having a distinctive community of plants and animals. Plant detritus can accumulate to form peat or can be exported to the adjoining estuary where it promotes secondary production.

5. *Mangrove swamps* proliferate in some tropical and subtropical regions. The density, distribution, and species composition of mangrove swamps are a function of air and water temperatures, wave and current activity, tidal range, water chemistry, substrate type, and hurricane frequency. The transfer of energy is accomplished by detritus food chains; common detritivores in mangrove swamps are crabs and various insects.

6. Coral reefs consist of a limestone mass of cemented skeletal debris that is capped with a thin veneer of living organisms. Coral tissue contains unicellular dinoflagellates called *zooxanthellae*. Because the zooxanthellae require sunlight for photosynthesis, reef-building coral (*hermatypes*) flourish in clear, shallow water. Also, water temperatures more than 20°C are optimal for reef growth, as are normal salinity and clear water. Despite low nutrient levels in the water, coral reefs teem with life because of the *mutualistic* relationship between coral and zooxanthellae—a trophic interaction that allows for rapid, efficient recycling of limited resources between producers and consumers. Also, calcareous algae are important primary producers of coral reefs.

7. The three principal types of coral reefs—the fringing reef, the barrier reef, and the atoll—represent distinct stages in the evolutionary history of a coral reef. Initially, coral colonize the edge of a new volcanic island, creating a *fringing reef*. The subsequent sinking of the island or rising of sea level, if accompanied by the upward growth of the reef, creates a lagoon between the island and the reef mass, the classic *barrier reef*. If water deepens, the island is flooded completely, leaving an *atoll*, an irregular ring or oval reef enclosing a central lagoon.

KEY WORDS

ahermatypic coral
anoxia
atoll
barrier reef
detritivore
flocculation
fringing reef
hermatypic coral
high salt marsh
low salt marsh
mangrove
mutualism
partially mixed estuary
salt-wedge estuary
well-mixed estuary
zooxanthellae

QUESTIONS

REVIEW OF BASIC CONCEPTS

1. Contrast the composition of plants in low and high salt marshes. What ecological factors account for the differences between these two floral communities?

2. Why do estuaries have low species diversity?

3. What main factors control water circulation in estuaries?

4. How is a lagoon different from an estuary?

5. What are mangrove swamps, where do they occur, and what physical factors control their character?

6. What is a detritus food chain and what role does it play in salt marshes and mangrove forests?

7. What are zooxanthellae, and how do they interact with coral?

8. Identify the critical environmental factors that limit the growth of coral reefs.

9. Using diagrams, clearly distinguish among fringing reefs, barrier reefs, and atolls.

CRITICAL-THINKING ESSAYS

1. What general role do estuaries and salt marshes play in fish productivity?

2. Speculate about what might happen if you artificially fertilized (added essential nutrients) the water of a barrier reef.

3. Reread the section on the functional morphology of fishes in Chapter 9 (see Figures 9–20, 9–21, 9–22 in Chapter 9), and then deduce the likely swimming characteristics of the trigger fish and the butterfly fish in Figure 12–19.

4. Sketch a topographic profile of a barrier reef, and identify the algal ridge, the lagoon, the buttress zone, and the reef terrace. Specify in which of these environments you would likely find the following coral types, and give reasons.

   a. Patch reefs
   b. Encrusting coral
   c. Fragile branching coral
   d. Brain coral

5. How would the benthos and nekton of an estuary be affected if land reclamation for real-estate development eliminated all of the area's salt marshes?

6. What factors promote and limit high biological production in the following?

   a. Estuaries
   b. Coral reefs
   c. Salt marshes.

DISCOVERING WITH NUMBERS

1. Consult Figure B12–4. Calculate the elongation rate of the sand spit in kilometers and in meters per year. Has the rate varied with time?

2. Consult Figure B12–4d. Assume that at the core site, deposition began about 3,000 years ago in the low-marsh peat and about 2,400 years ago in the high-marsh peat. What is the depositional rate of each peat expressed as centimeters per year? Why is one rate so much higher than the other? Why is there an inverse relationship between the amount of sand and the amount of mud in the core?

3. Examine Figure B12–2c. Estimate the average yearly increase in the volume of anoxic water in Chesapeake Bay between the early 1950s and early 1980s. What simplifications did you make in order to calculate this average annual increase?

SELECTED READINGS

Anderson, D. M. 1994. Red tides. *Scientific American* 271 (2): 62–68.

Boaden, P. J. S., and R. Seeds. 1985. *An Introduction to Coastal Ecology*. London: Blackie.

Bolcourt, W. 1993. Estuaries, where the river meets the sea. *Oceanus* 36 (2): 29–37.

Horn, M. H., and R. N. Gibson. 1988. Intertidal fishes. *Scientific American* 258 (1): 64–70.

Horton, T., and W. E. Eichbaum. 1991. *Turning the Tide: Saving the Chesapeake Bay.* Washington, D.C.: Island Press.

Koehl, M. A. R. 1982. The interaction of moving water and sessile animals. *Scientific American* 274 (6): 124–35.

Kusler, J. A., W. J. Mitsch, and J. S. Larson. 1994. Wetlands. *Scientific American* 270 (1): 64–70.

Levinton, J. S. 1982. *Marine Ecology.* Englewood Cliffs, N.J.: Prentice-Hall.

Lohman, K. J. 1992. How sea turtles navigate. *Scientific American* 266 (1): 68–77.

Milne, D. H. 1995. *Marine Life and the Sea.* Belmont, Calif.: Wadsworth.

Nybakken, J. W. 1988. *Marine Biology: An Ecological Approach.* New York: Harper & Row.

Sale, P. F., G. F. Forrester, and P. S. Levin. 1994. The fishes of coral reefs. *Research and Exploration* (National Geographic Society) 10 (2): 224–35.

Schreiber, E. A., and R. W. Schreiber. 1989. Insights into seabird ecology from a global "natural experiment". *Research and Exploration* (National Geographic Society) 5 (1): 64–81.

Schubel, J. R. 1981. *The Living Chesapeake.* Baltimore, Md.: Johns Hopkins University Press.

Starr, C., and R. Taggart. 1992. *Biology: The Unity and Diversity of Life.* Belmont, Calif.: Wadsworth.

Thorne-Miller, B. 1998. *The Living Ocean: Understanding and Protecting Marine Biodiversity.* Washington, D.C.: Island Press.

Thorson, G. 1971. *Life in the Sea.* New York: World University Library.

Woodard, C. 2002. *Ocean's End: Travels through Endangered Seas*. New York: Basic Books

Woodroffe, C., E. McBean, and E. Wallensky. 1990. Darwin's coral atoll: Geomorphology and recent development of the Cocos (Keeling) Islands, Indian Ocean. *Research and Exploration* (National Geographic Society) 6 (3): 262–76.

TOOLS FOR LEARNING

Tools for Learning is an on-line review area located at this book's web site OceanLink (**www.jbpub.com/oceanlink**). The review area provides a variety of activities designed to help you study for your class. You will find chapter outlines, review questions, hints for some of the book's math questions (identified by the math icon), web research tips for selected Critical Thinking Essay questions, key term reviews, and figure labeling exercises.

13

Ocean Habitats and Their Biota

Oh, 'twas on the broad Atlantic,
'Mid the equinoctal gales,
That a young fellow fell overboard
Among the sharks and whales.
And down he went like a streak of light,
So quickly down went he,
Until he came to a mer-mai-id,
At the bottom of the deep blue sea.
—Oh! 'Twas in the Broad Atlantic,
Anon.

web navigator

critical thinking on the web

math tutor on the web

www.jbpub.com/oceanlink

PREVIEW

IN THIS CHAPTER we present an overview of marine life and some of its habitats that extend across the continental shelf and into the open sea. We begin the survey by examining some general distribution patterns and behaviors of the plankton and fish that inhabit the water of the shelf. Then we investigate the physical nature of the sea floor of the shelf and, in particular, the way in which bottom energy controls the grain size of the substrate and thus the makeup of benthic communities. Next we investigate the water column of the open ocean, focusing on its plankton populations and its odd-looking midwater fishes. Last, we explore the remarkable species diversity of the deep-sea floor and examine the effect that high hydrostatic pressure has on rates of bacterial decomposition.

The ocean, in comparison with many terrestrial environments, is relatively kind and benign to life. Except near the shoreline, the temperature, salinity, and chemistry of seawater are remarkably steady over time. Also, unlike air, water buoys up organisms and prevents their cells and tissues from drying out. The stability and chemical uniformity of the water increases with distance from the continents and with depth below the surface. But, regardless of depth or remoteness, all parts of the ocean are inhabited by a community of organisms that are well adapted to the prevailing environmental conditions. Let's examine some of these offshore communities.

13-1

Biology of the Continental Shelf

The fertile waters of the continental shelves, the *neritic zone* (see Figure 9–1 in Chapter 9), support a rich community of organisms. Plankton include countless species of microscopic plants and animals, as well as the eggs and larvae of many benthic and nekton species. In contrast to the diverse planktonic organisms, the nekton of the neritic zone are dominated overwhelmingly by fish. The shelf bottom, the *sublittoral zone* (see Figure 9–1 in Chapter 9), is occupied by diverse groups of benthic organisms that vary with changes in the bathymetry and sediment cover of the sea bed. In what follows, we will examine plant and animal communities that dwell in this narrow, but biologically productive, strip of shallow ocean.

NERITIC ZONE

The predominant phytoplankton in shelf waters are species of diatoms and dinoflagellates. Seasonal changes in water temperature, salinity, and nutrient input produce a regular succession of phytoplankton species in temperate (**Figure 13–1**) and polar seas. This means that the species composition of plants varies over time as water conditions change, because each species of plant is adapted to a specific range of water temperature and salinity as well as to specific types and levels of nutrients. Zooplankton, which are more diverse in neritic water than are phytoplankton, are dominated by arthropods, particularly species of copepods.

Plankton, whether they be plants or animals, are rarely distributed evenly in space. Rather they tend to be aggregated into relatively dense clusters. This uneven distribution is termed **patchiness**. Some of the patches are related to Langmuir circulation (see Figure 6–8 in Chapter 6). Floating plankton are swept toward the edges of the circulation cells, where currents converge. This concentrates the buoyant organisms into narrow, elongated patches.

FIGURE 13-1

Seasonal succession of phytoplankton. As the temperature and chemistry of the water vary with the season, the species composition of the phytoplankton changes accordingly. This plot displays seasonal phytoplankton succession in the Irish Sea. [Adapted from R. S. K. Barnes and R. N. Hughes, *An Introduction to Marine Ecology* (Boston, Mass. Blackwell Scientific Publications, 1982); and J. Johnstone, A. Scott, and H. C. Chadwick, *The Marine Plankton* (London: Hodder and Stoughton, 1924).]

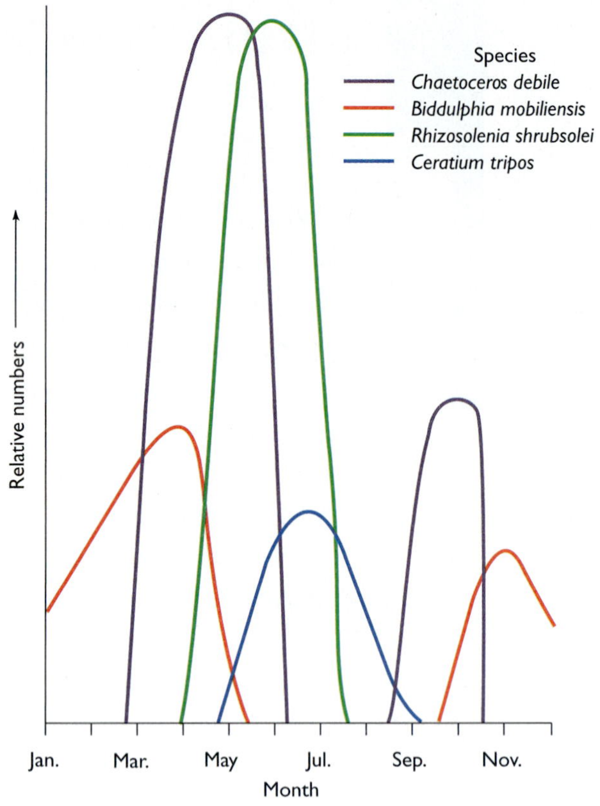

The origin of other types of patchy distributions is not known with certainty. However, if they were not clumped, the concentration of microscopic organisms would be too dilute to serve as an adequate food resource for animals that are higher in the food web. Herbivorous copepods, for example, would expend more energy than they would gain if they had to filter large volumes of seawater in order to capture only a few diatom cells. They would all starve in short order.

Unlike the plankton, which are at the mercy of currents, the nekton are active swimmers. They are able to wander from spot to spot, avoiding unfavorable water and searching for appropriate feeding and breeding grounds. But, like the plankton, many species of fish cluster into groups called **schools**, a habit that gives them, too, a patchy distribution (**Figure 13–2a**). In these compact groupings of fish, each member is the same size and maintains a remarkably fixed three-dimensional spacing in conjunction with its neighbors (**Figure 13–2b**). The entire school swims as a unit, executing precise maneuvers faultlessly as if it were one large organism. No true leaders exist. In fact, with the completion of a turn, the fish that were swimming at the sides of the school find themselves at the head of the cluster. The size and number of fish in a school vary enormously with the species: a school can range from a few individuals to millions of fish that stretch out for tens of kilometers.

What advantages does schooling offer a species of fish? For some species, a likely benefit is protection. Many predators locate their prey visually. A school of small fish may appear to be one large organism from a distance and consequently be avoided by a potential predator. Also, the chance that an individual fish will be discovered by a predator is reduced substantially when fish aggregate in tight clusters. Widely and uniformly distributed, they lack that protection. Predators that attack schools are less efficient hunters than are those that attack solitary fish. Under attack, the individual members of a school dart every which way, making it difficult for the predator to focus on and capture one single fish because of the chaotic motion. As an experiment, attempt to catch a bucketful of rubber balls that are unexpectedly thrown at you; most people fail to catch even a single ball!

Among the more active hunters of schooling fish are the seabirds, many of which dive to impressive depths and swim using their wings in search of food. The boxed feature, "Penguins," discusses the living and hunting habits of Adélie and Emperor penguins.

FIGURE 13-2

Fish schools. (a) This is a wide-angle photograph of a school of glassy sweepers in the Red Sea. (b) Individuals within a school maintain a regular three-dimensional spacing, as shown here by schooling grunts in the Caribbean Sea.

(a) SCHOOL OF FISH

(b) SPACING IN A SCHOOL

SUBLITTORAL ZONE

Now that we have surveyed the biota of the water of the continental shelves, we will turn our attention to the fauna and flora that live on and within its bottom, the sublittoral zone. In the boxed feature, "Ecology of the Giant Kelp Community" in Chapter 9, we examined an important benthic community of the Pacific continental shelf of North America. In this chapter, in the boxed feature, "Bottom Feeding by Whales," we review some recent findings about the bottom-feeding activity of California gray whales. Because the water is shallow over continental shelves, physical factors at the sea floor regulate to a large degree the number, type, and distribution of benthic organisms. So let's begin.

The bottom energy on continental shelves is a function of the wave energy and tidal currents.

Penguins

Most seabirds, including puffins, albatrosses, petrels, shearwaters, and penguins (**Figure B13–1**), eat fish as part of their diet. Some of them, such as albatrosses and penguins, prey on squid at night. Among the seabirds, penguins have lost their ability to fly, and have become specialized in swimming and diving. They have evolved large, heavy bodies, big bones, thick fatty deposits for insulation, greasy feathers that repel water, and streamlined bodies and stubby wings for "flying" underwater. Emperor penguins, for example, have been clocked at speeds of over 15 kilometers per hour (~9.5 mph), and have remained submerged for up to 15 and 20 minutes at a time, diving as deeply as 250 meters (~825 feet) below the water surface while hunting squid.

Penguins, which range from the size of a large seagull to a height of 1.3 meters (~4.3 feet), are native to the Southern Hemisphere only. Of the eighteen species of penguins, the Adélie and Emperor penguins are the only two that have rookeries along the shoreline of Antarctica. The Adélie (**Figure B13–2a**), which is as tall as 60 centimeters (~24 inches), displays the classic tuxedo color pattern seen so often in cartoon characters. They gather in large rookeries to breed during the spring and summer on land and spend the winter at sea. The Emperor, the largest penguin (**Figure B13–2b**), incubates its eggs on land during the Antarctic winter! It survives the incredibly cold temperatures and high winds by huddling in tight groups in order to conserve body heat. Periodically, the "toasty" members in the middle of a group work their way to the outside.

FIGURE **B13-1**

Seabirds. Seabirds are important members of marine food webs. Many of them prey heavily on fish, squid, and shrimp. Examples of some important seabirds include (a) puffins, (b) petrels, and (c) albatrosses.

(a) PUFFIN (b) PETREL (c) ALBATROSS

Bottom energy varies inversely with water depth (see Figure 4–2a in Chapter 4). Given this, the sea bed of the shelf can be subdivided into energy bands: high-energy bottom conditions prevail nearshore, in shallow water, and low-energy conditions prevail offshore, in deep water (**Figure 13–3**).

Bottom energy affects the benthos in at least two ways. First, bottom turbulence continually moves sediment about. This creates an unstable substrate and makes it difficult for an epifaunal community (one living on the sea bed) of any size to become established. In some regions, strong currents strip the shelf floor entirely of its sediment cover, exposing bare rock. There, kelp, mussels, oysters, coral, and anemones anchor or cement themselves securely to the rock, in order not to be swept away by the strong bottom currents. Second, bottom energy directly controls the grain size of sediment that is deposited. Mud, a low-energy sediment deposit, consists of small clay and silt particles and accumulates in wave-protected shore environments or far offshore in quiet water that is below the wave base. Deposits of sand and gravel, high-energy deposits, are common along the inner and

This allows the colder members at the fringe to move toward the center of the group and warm up. Most penguins hunt in water not far from shore. Fishery biologists estimate that penguins take almost 35 million tons of food from the ocean each year.

The Little Blue or Fairy Penguin (**Figure B13-2c**) inhabits the shores of southern Australia and New Zealand. They are noted for the bluish color of their upper feathers, and for their small size (40 cm; ~16 in) and slight weight (1 kg; 2.2 lbs). Little Blue Penguins usually mate for life and tend to reoccupy the same burrow on land each year, having one or two broods per year of two eggs each. Because they are preyed upon at sea by sharks, seals, and killer whales and on land by large seabirds, foxes, dogs, feral cats, rats, snakes, and lizards, local populations can be endangered and need protection to survive.

Visit www.jbpub.com/oceanlink for more information.

FIGURE **B13-2**

Penguins. Two species of penguin that inhabit the waters and shores of the Antarctic continent are (a) the Adélie penguin and (b) the Emperor penguin. The Little Blue penguin (c) inhabits the shores of southern Australia and New Zealand.

(a) ADELIE PENGUIN (b) EMPEROR PENGUINS (c) LITTLE BLUE PENGUIN

middle parts of the shelf, where the shallow bottom is regularly swept clean of mud by wave activity and tidal action (**Figure 13–3a**), leaving clean sand and gravel to accumulate.

The type of bottom sediment exerts a strong control on the makeup of benthic communities. For example, the feeding habits of the benthos clearly correlate with the grain size of the sediment cover (**Figure 13–3b**). Gravel and coarse-sand bottoms are populated overwhelmingly (more than 70 percent) by filter feeders. The strong currents in the area assure that the filter feeders receive a continuous supply of suspended organic detritus and plankton. Besides, the grain size of such deposits is too coarse to be ingested by deposit feeders, and small bits of organic detritus, the principal food of deposit feeders, are too fine to be deposited in such high-energy environments. So the filter feeders rely on detritus suspended in the swift-moving water.

Substrates of fine sand and coarse silt support a mixed faunal assemblage, composed predominantly of detritus and deposit feeders, some infaunal (living in the substrate) filter feeders, and few epifaunal (living on the substrate) filter feeders.

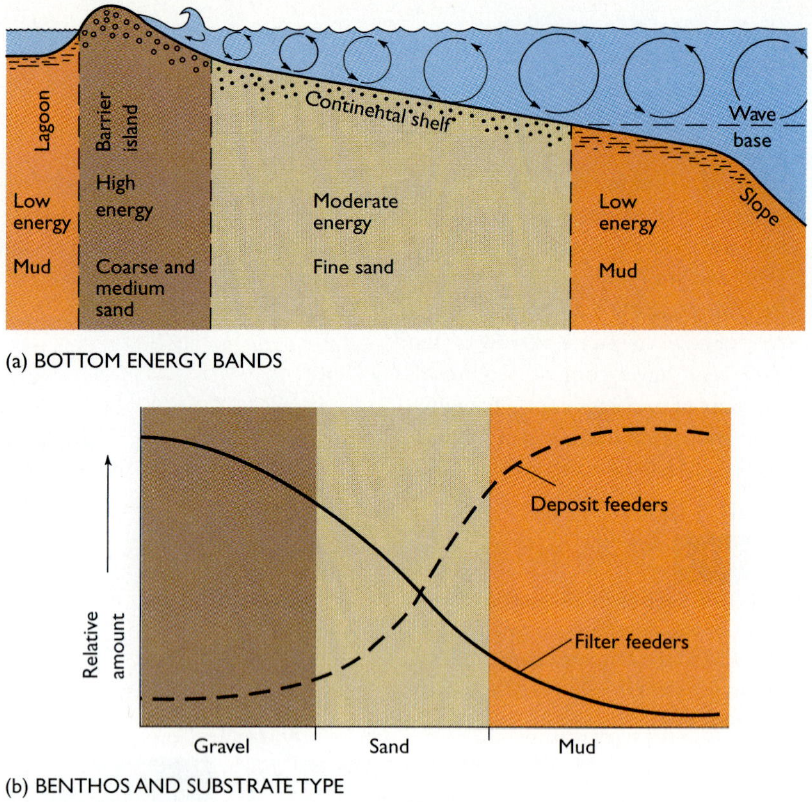

FIGURE **13-3**

The sediment and benthos of the shelf. (a) To a first approximation, the grain size of shelf deposits is regulated by wave-induced bottom energy. Both the energy level and grain size decrease with water depth and distance offshore. (b) Grain size of the sediment influences the feeding characteristics of bottom communities. Deposit feeders tend to predominate in mud; filter feeders, in sand and gravel.

Muddy substrates are almost exclusively inhabited by deposit and detritus feeders, because such substrates tend to have a high content of organic matter. Small particles of both organic detritus and mud are deposited here under low-energy bottom conditions. Furthermore, the weak bottom currents provide insufficient quantities of suspended food for the filter-feeding fauna, and the fine mud and silt particles tend to clog their filters.

Two major benthic communities are defined according to the character of the substrate: hard-bottom and soft-bottom communities. Hard-bottom communities, which are most common in the high-energy intertidal zone, are associated with exposed bedrock and gravel deposits. Here seaweeds tend to grow extensively and, combined with uneven rock surfaces, provide numerous niches, which encourage the growth of a rich bottom fauna. Soft-bottom communities are composed of unconsolidated sand and mud. The biological makeup of these communities varies with the grain size of the sediment. Usually, organisms of a single group dominate a particular type of substrate. Most typically, clean sand is populated by a variety of filter-feeding bi-valves, with fewer species of polychaete worms and snails. The proportion of bivalves decreases in muddier sediment, relative to the population of snails and worms. The *Macoma* community, which dwells in mud, and the *Venus* community, which dwells in sand, are two common soft-bottom communities (**Figure 13–4**) found in many places on the Atlantic continental shelf. In both cases, the nature of the substrate—mud or sand—controls the composition of the species.

13-2

Biology of the Open Ocean and the Deep Sea

The open ocean constitutes the largest division of the Earth's oceans; yet, despite its enormous size, life in the open ocean or on its deep-sea bottom is sparse. This is because the open ocean is located far from river mouths, which supply the

FIGURE **13-4**

Soft-bottom communities. Two common soft-bottom communities are named after the dominant organism in each: (a) The Macoma community of muddy sediment bottoms, and (b) the Venus community of sandy sediment bottoms. [Adapted from G. Thorson, *Life in the Sea* (New York: McGraw-Hill, 1971).]

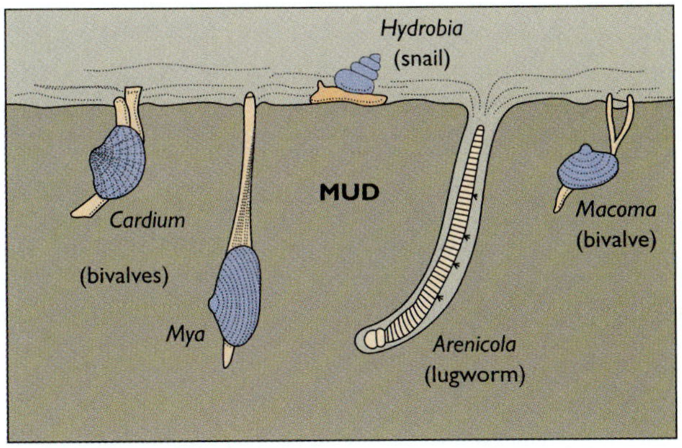

(a) MACOMA COMMUNITY

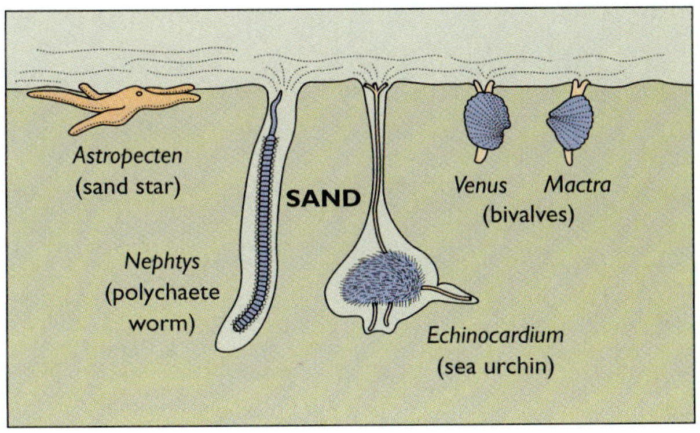

(b) VENUS COMMUNITY

essential nutrients marine plants need for growth. Without many plants, there cannot be many animals. Nonetheless, the open ocean does possess a remarkably diverse biota. In other words, there are not many animals in the region, but the few that exist there are very diverse in species composition. As in the previous section, we will first examine the water column and then the sea floor of the deep sea.

THE OCEANIC REALM

Located far from a supply of river-borne nutrients, the surface water of the oceanic province is poor in nutrients. This condition is aggravated by the tendency of water to downwell in the center of circulation gyres, resulting in low biological productivity. The open ocean is a biological desert, with few plants (**Figure 13–5**). Because of the great depth of the sea bottom here, the sparse plant life is entirely planktonic. There are no seaweeds—with one important exception, the floating sargassum gulfweed discussed in the box "Sargassum Gulfweed." The principal phytoplankton are diatoms, dinoflagellates, and coccolithophores. Diatoms are the predominant plants in the shallow waters of the coast and the shelf. However, their proportion relative to other phytoplankton drops off sharply with distance from land, and dinoflagellates and coccolithophores become equally important (**Figure 13–5b**).

After the plants, the most abundant organisms in the oceanic realm are the herbivorous zooplankton, particularly foraminifera and radiolaria. In turn, forams and radiolaria are preyed upon by larger zooplankton, most notably the copepods. A variety of carnivores, such as arrow worms, sea butterflies, and amphipods, occupy the higher trophic levels. Top predators, including mackerel, squid, jellyfish, tuna, porpoises, sharks, and humans, complete the long food web of the open ocean (**Figure 13–6**). Squids are discussed in the boxed feature of that name.

Deeper, beneath the illuminated photic zone, lie the dysphotic and aphotic zones (see Figure 9–1b in Chapter 9). The dysphotic zone has barely perceptible levels of light—not enough for plant photosynthesis. Seasonal effects of heating and cooling, which play dominant roles in the photic zone, are damped and even absent in these parts of the water column. The water conditions of the dysphotic zone, as well as of the deeper and darker aphotic zone, tend to be very stable and unchanging over time.

Well-represented dysphotic animals include crustaceans, such as prawns, shrimps, copepods, amphipods, and ostracods (small bivalve crustaceans). Squids and fishes are also common. Many fishes and invertebrates have evolved specialized adaptations for living in this dimly lit region. One such adaptation is the large size and light-sensing ability of the eyes, which are able to detect the faintest light. Many of these deep-sea organisms, such as lantern fish, have light-generating organs called **photophores** that produce **bioluminescence** in a way that is somewhat similar to the light of

THE OCEAN SCIENCES BIOLOGY

Bottom Feeding by Whales

While geologists were conducting a geologic hazard survey of a portion of the northeastern Bering Sea (**Figure B13–3a**), they noted that much of the sea floor was inexplicably pockmarked with numerous holes, many arranged in distinct patterns. Subsequent work has linked these pits with the activity of California gray whales, which commonly migrate to this area during the spring and summer to feed on amphipods. Sighting mud plumes near whales, scientists suggested that the whales were feeding on the sea bottom and disturbing the sediment. Underwater observations confirmed this deduction. Gray whales have been seen rolling onto their sides and sucking a large volume of sand and mud into their mouths (**Figure B13–3b**), expelling the sediment through their baleen (the horny plates that hang from their upper jaw), and filtering out the amphipods, creating plumes of turbid water. The linear pits on the sea bottom are gouged out by whales that swim while feeding, whereas the radial pattern of pits is formed by whales feeding systematically at one site.

The ecological and geological consequences of this feeding activity are

FIGURE B13-3

Whale feeding. (a) Bottom pits created by feeding whales in the northern Bering Sea occur in a sea bed composed of clean fine sand that supports a dense population of amphipods. [Adapted from C. H. Nelson and K. R. Johnson, Whales and Walruses as Tillers of the Sea Floor, *Scientific American* 256 (2) (1987): 112–17.] (b) This drawing shows a California gray whale feeding on the bottom of the Bering Sea.

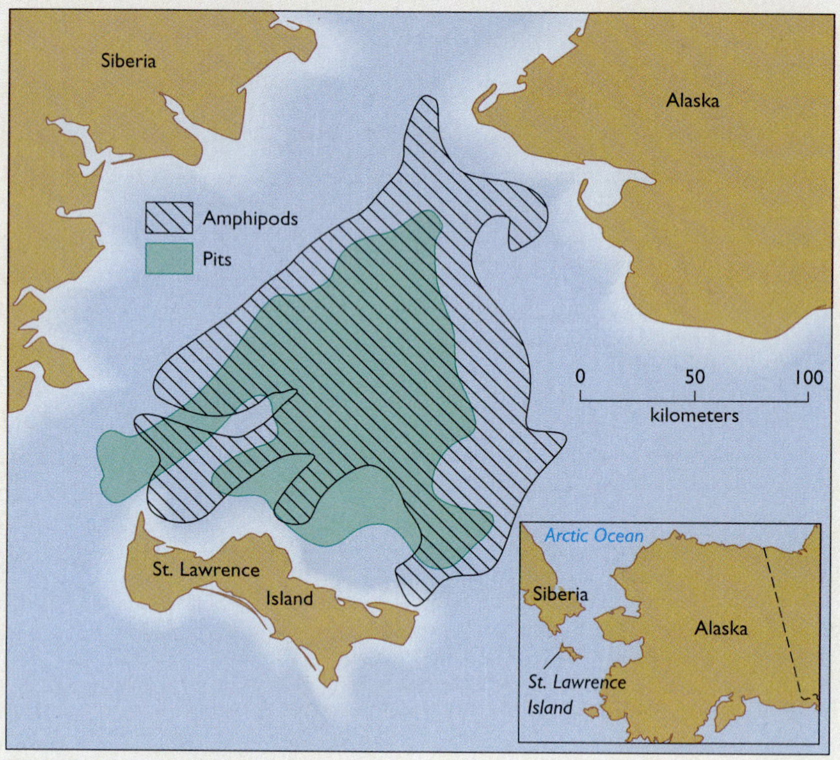

(a) LOCATION OF WHALE FEEDING PITS

fireflies. Cold, bioluminescent light is created as a by-product of biochemical reactions. Some fish have "bacterial photophores," in which light is produced by the metabolic activities of symbiotic bacteria that live in dense concentrations within the photophores (10^{10} bacteria or more per milliliter of glandular fluid, in some cases!). Many fish have distinctive patterns of photophores on different parts of their bodies that help them identify members of their own species in these dimly lit depths. Others carry blinking photophores at the end of lures to entice prey close enough to be captured.

Food in the dysphotic region is scant, because plant production occurs far above in the sunlit waters of the photic zone. Many of the small zooplankters rely on fallout of organic detritus from above. Others, such as euphausiids and lantern fish, engage in **diurnal vertical migration**: they swim upward to the photic zone at night to feed, and descend to depths of between 700 and 900 meters

(b) CALIFORNIA GRAY WHALE FEEDING

most impressive. The volume of sediment placed in suspension in this area by whales feeding in this manner is on the order of 10^8 m^3/yr (~10^8 yd^3/yr). This quantity is almost three times greater than the amount of suspended sediment delivered to the Bering Sea each year by the Yukon River! Furthermore, this feeding activity winnows out mud and creates a clean sand deposit that is the favored habitat of amphipods. Also, nutrients buried in the sediment are released into the water column by the feeding whales. This promotes plankton productivity and increases the principal food source of the bottom-dwelling amphipods, which in turn assures the California gray whales of an ample and dependable food resource. In a very real sense, the whales are "farming" the sea bed. This is an excellent example of a complex ecosystem that achieves a long-term ecological balance by interactions between predator and prey.

(~2,310 to 2,970 feet) during the day. Euphausiids preferentially feed on diatoms while in surface waters, and lantern fish prey on crustaceans, including copepods. Providing a key link between the communities of the photic and dysphotic zones, these nocturnal migrants in turn serve as food for larger predators that dwell deeper than 200 meters (~660 feet) during the daylight hours.

The aphotic zone, which begins at a water depth between 500 and 1,000 meters (~1,650 to 3,300 feet), is constantly dark and cold. Over two thousand species of animals live here. Many of them are colored red or black. They include copepods, ostracods, jellyfishes, prawns, mysids, amphipods, a variety of swimming worms, and an assortment of odd-looking fishes referred to generally as **midwater fishes** (fishes living at "middepth" in the sea). This last group, which displays peculiar traits and structures adapted for living in water that is totally dark and that lacks food, warrants close inspection.

FIGURE 13-5

Phytoplankton decrease with distance offshore. (a) The abundance of phytoplankton in the Sargasso Sea (located in the center of the North Atlantic Ocean) and in continental shelf waters reflects changes in nutrient levels with distance offshore. [Adapted from I. Valiela, *Marine Ecological Processes* (New York: Springer-Verlag, 1984).] (b) The relative proportion of diatoms in the Caribbean Sea decreases markedly with distance offshore. [Adapted from E. M. Hurlburt, *Limnology and Oceanography* 7 (1962): 307–15.]

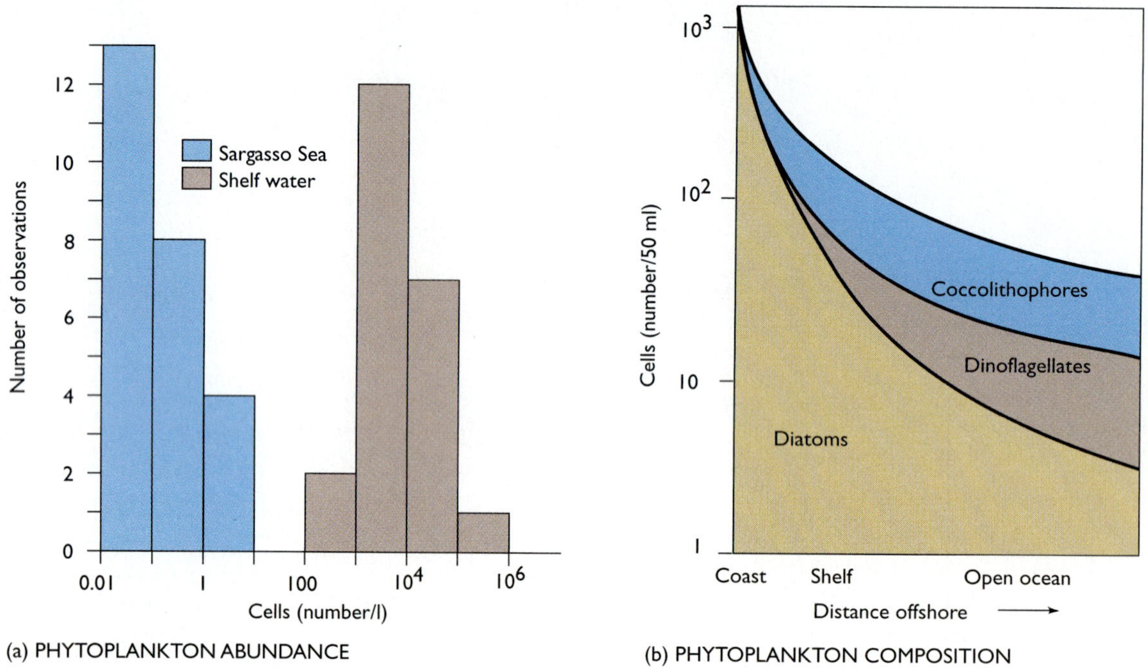

Because their principal prey is one another and because they are not abundant, meals for midwater fishes are probably few and far between. They have evolved effective strategies for capturing prey (**Figure 13–7**). Typically, midwater fishes are small, some only 2 to 10 centimeters (~1 to 4 inches) in length. Many have proportionately enormous mouths lined with sharp, pointed teeth for biting prey firmly, a jaw that can be unhinged, and a body that can be expanded to accommodate a large meal. Some also possess bioluminescent organs to entice prey. We will discuss two such fishes here.

The female deep-sea angler (see Figure 13–7) is noted for her fishing prowess. She dangles a "fishing rod" with an attached luminous lure in front of her mouth, which has an extendible jaw lined with needle-sharp teeth. Instead of swimming energetically after prey, the animal conserves energy by "fishing" quietly in the darkness, using her blinking photophore at the end of her fishing line as an attractant. The dwarfed, parasitic male is fused to her underbelly and serves merely as a sperm bank. Fertilized eggs of the angler fish float up to the photic zone when released, where the hatched larvae feed on copepods and grow to adolescence before descending back to the darkness of the aphotic zone.

Although less diversified in species composition than the angler fish (which comprises about 75 percent of the deep-sea fish species), the most abundant and therefore representative group of fish on the Earth belongs to the genus *Cyclothone*. These fish are small and slight in build. They populate the depths at between 1,000 and 7,000 meters (~3,300 to 23,100 feet). Like many midwater fishes, they possess an impressive tooth-lined jaw, which can be unfolded and rotated open to an angle of almost 180 degrees! Many other types of midwater fishes exist, a few of which are shown in Figure 13–7.

FIGURE **13-6**

Large pelagic predators. The open ocean possesses a variety of large carnivores, including the (a) squid, (b) tuna, (c) jellyfish, and (d) shark.

(a) SQUID (b) TUNA (c) JELLYFISH (d) SHARK

THE DEEP-SEA BOTTOM

The deeper the water, the less biomass there is in the ocean. Water depth, rather than distance from land, is the critical factor that regulates the quantity of biomass, because a deep-sea bottom is always sparsely populated, no matter how close to land it is. As explained earlier, bottom life in the deep ocean is sparse because food is very limited in this region. Photosynthesis requires light; the aphotic zone is totally dark, so there is obviously no plant production on the deep-sea floor. Animal communities there depend on food supplied from the faraway photic zone, by means of (1) slow fallout of both fine and coarse organic detritus from surface waters (plankton and fecal matter), (2) the settling of large animal carcasses (for example, dead fishes and whales) from above, and (3) perhaps the transport of organic matter by turbidity currents to the abyssal plains (**Figure 13–8**).

In addition to their low numbers, several other characteristics typify the biota of the deep-sea bottom. Available information indicates that most fauna of the deep ocean reproduce year round, presumably because there are no seasons. These animals also have smaller broods, grow more slowly, and live considerably longer than do the fauna of the shallow shelf bottoms. These traits are consistent with the effect of the uniform, near-freezing temperatures on reducing the rate of metabolic activity.

Perhaps the least expected discovery has been the species diversity of deep-sea bottom communities. Although the exact degree of diversity is debated, biologists agree that it is surprisingly high, given the low biomass and small populations of the region.

Sargassum Gulfweed

Not all species of phytoplankton are microscopic in size. A few, such as *Sargassum* sp. (species), are macroplankton (large plankton). This unique plant, a brown alga, gives its name to the Sargasso Sea, where floating rafts of sargassum gulfweed (**Figure B13–4a**) drift slowly around the circulation gyre of the North Atlantic Ocean. Biologists estimate that more than 7 million tons of sargassum gulfweed float in the Sargasso Sea.

The sargassum found in the open ocean originates as a benthic seaweed growing in the shallow waters of the Caribbean Sea. There the storm waves wrench plants from the sea bottom and set them adrift. The pelagic sargassum consists of eight species. Two of these—*Sargassum sluitans* and *S. natans*—make up about 90 percent of the gulfweed's biomass. The plants are buoyed up by hollow, gas-filled floats that keep their large fronds suspended near the sea surface where there exists sunlight for photosynthesis. The floats look like grapes, which explains the name *sargassum,* derived from the Portuguese word for "grapes."

While floating with the North Atlantic currents, sargassum reproduces asexually by fragmentation: branches break off the main stem and continue to grow as separate plants. As the alga ages, fewer floats are produced, and the old gulfweed eventually sinks out of the photic zone and dies. Some of the plants have apparently been floating for several hundred years. Large, floating rafts of sargassum are also found in parts of all of the other oceans as well.

The gulfweed does not provide food for herbivores in the Sargasso Sea. Instead, the mass of intricate, intertwined fronds supports an unusually rich and diverse faunal community that depends on it for existence. Over fifty species of encrusted, attached, and swimming organisms have specifically adapted to the floating lifestyle of sargassum. This floating habitat supports a diverse biota including sea slugs, small crabs, juvenile sea turtles, and a variety of fish and seabirds. A few animals, such as the sargassum fish *Histrio histrio* (**Figure B13–4b**), mimic the color and growth form of the gulfweed to such an astonishing degree that they are perfectly camouflaged.

During the summer months, currents and tides can deposit large accumulations of the sargassum weed on beaches, such as along the Gulf Coast of Florida and Texas. There the piles of seaweed help reduce erosion and, as they decompose in the hot sun, provide critical nutrients to the beach and dune ecosystems. Unfortunately, as the seaweed litter rots, it has an unpleasant odor, and this interferes with the tourists' enjoyment of the beaches. Most communities opt to rake the beaches clean in the early morning hours before the arrival of visitors. Other places, such as Galveston, Texas, are dealing with the sargassum litter problem by educating the public about its importance to the ecological well being of the local beaches and dunes; they hope that this knowledge will allow the seaweed and beach visitors to coexist, thereby supporting both the local economy and ecology.

FIGURE **B13-4**

The sargassum gulfweed. (a) The sargassum gulfweed, a large floating seaweed abundant in the North Atlantic Ocean supports a rich ecosystem including juvenile sea turtles. (b) A complex and rich community of organisms is adapted to living among the fronds of the floating sargassum gulfweed. Many of these animals are well camouflaged, as shown by the weedlike appearance of the sargassum fish *Histrio histrio.*

(a) SARGASSUM GULFWEED

(b) SARGASSUM FISH

FIGURE **13-7**

Midwater fishes. Shown here is a sample of the many bizarre-looking fish that are superbly adapted to living in the dark waters of the aphotic zone. Despite looking like monsters, these fish are typically small in size, which reflects the lack of food at these great depths.

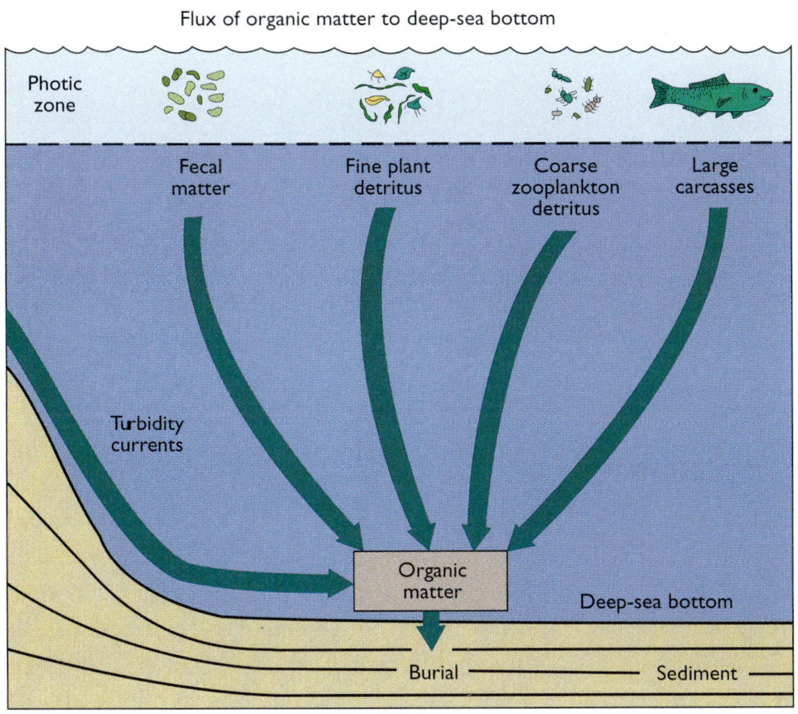

FIGURE **13-8**

Food sources of the abyssal benthos. Abyssal organisms depend on the fallout of food from high above in the photic zone. The types of food most likely to settle to the bottom of the deep sea are fine and coarse organic detritus, small and large animal carcasses, and fecal pellets. These abyssal plains receive an influx of organic matter from turbidity-current transport as well.

Squids

Large carnivorous nekton in the open ocean are dominated by fish, toothed whales, dolphins, marine birds, and squids. Squids belong to the most advanced group of mollusks, the class Cephelopoda. Other cephelopods include the cuttlefish, nautilus, and octopus (**Figure B13–5**). They are characterized by having a well-developed brain, acute eyesight, and *chemosensors,* which enable them to detect, as if smelling, tiny amounts of chemicals in the water. The nautilus is rare, and the octopus and cuttlefish are principally benthic organisms. The octopus does swim, but prefers to scamper along the bottom using its tentacles. The cuttlefish is nocturnal, spending the day buried in mud and the night hunting for prey in the water.

Squids are very intelligent animals and powerful, fast swimmers. Most are no larger than 50 to 60 centimeters (~20 to 24 inches) and weigh a few kilograms (~4.4 pounds). When young they grow rapidly, the rate of growth increasing directly with water temperature. They reproduce sexually; the male transfers sperm to the female using a specialized tentacle. The young offspring are planktonic, floating with the currents and eating large quantities of zooplankton. As adults, squids are nocturnal hunters, preying on small fish, shrimp, crabs, and other squid. They, in turn, are hunted by larger fish

FIGURE **B13-5**

Cephelopods. The most advanced group of mollusks belong to the class Cephelopoda. Some important members of this group are (a) cuttlefish, (b) nautilus, (c) octopus.

(a) CUTTLEFISH (b) NAUTILUS (c) OCTOPUS

They surmise that the stability of the physical and chemical aspects of that environment over long periods of time accounts for this unexpected diversity. In other words, the community of the deep-sea floor is likely to be structured, not by physical variability and perturbations, but by complex biological interactions among community members.

Consider the heavy predation pressure in the deep sea. When predation on a group of animals is low, as often occurs on land, the size of the population can expand, provided other factors do not limit growth. As a result, the favored species are likely to outcompete other animal species for resources, driving them over time to extinction and thus reducing the community's diversity—a concept known as **competitive exclusion**. Perhaps a more familiar example will clarify this concept. Many of you, I am sure, have heard the concern that scientists and environmentalists have voiced about the recent reduction in global biodiversity. This irreversible tragedy is caused by the activity and success of a single species—*Homo sapiens*! This animal species has no natural predators and it is outcompeting other species, both plants and animals, for living space and resources. The result is mass extinction of many species and hence low biological diversity. This is competitive exclusion on a global scale!

Apparently, domination by a species does not occur on the floor of the deep sea where echinoids (brittle stars), cephalopods (squids), and fishes prey heavily on benthic deposit feeders, as it does frequently on the sea bed of the continental shelf. The heavy predation pressure at the deep-sea floor keeps benthic populations in check and serves to (tuna, marlin, shark), seals, penguins, and humans.

The largest species of squid belongs to the genus *Architeuthis.* It has an overall length of 18 meters (~60 feet) and can weigh as much as 4 tons (**Figure B13–6**). Most of this length is, however, in the large body tentacles, which may be as long as 10 meters (~33 feet). They live mainly between the depths of 300 and 600 meters (~990 to 1,980 feet) and are elusive animals to catch. Biologists have studied them indirectly by examining the stomach contents of sperm whales, which prey on them heavily. Also, a few have washed up on beaches. Most adult sperm whales have numerous sucker-shaped scars on their jaws and lips, injuries acquired while hunting the squid.

Seamen have spun many tales about giant squid attacking and sinking ships. Although giant squids do exist, the stories are exaggerated. There is no confirmed sinking of a ship by squid attack. However, in 1941, when the Germans sunk the *Brittania,* a British troopship, in the middle of the North Atlantic, some of the survivors floating in the water were later attacked by giant squids.

Visit www.jbpub.com/oceanlink for more information.

FIGURE **B13-6**

Giant squid. The largest squid belongs to the genus *Architeuthis.* It is an enormous pelagic creature that is a favorite prey of sperm whales.

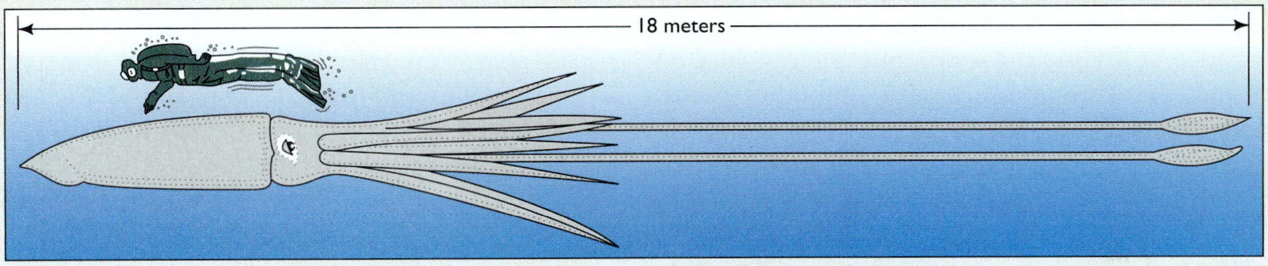

increase biological diversity by suppressing competitive exclusion. According to many biologists, however, there are undoubtedly other causes for the unusually high diversity among deep-sea communities, and the final, definitive explanation has yet to be advanced.

Abyssal depths in all oceans share at least four common traits. They (1) are perpetually dark, (2) are very cold, (3) have high hydrostatic pressures, and (4) have sparse food supplies. Despite these conditions, life exists everywhere at these depths. Microbes, especially bacteria, are particularly abundant. The key ecological function of bacteria, of course, is to break down dead organic matter biochemically into simpler inorganic nutrients, releasing them for recycling through the ecosystem. But does this occur on the abyssal sea bottom, where low temperatures, high hydrostatic pressures, and sparse organic detritus persist? This very question occurred to marine microbiologists after they witnessed the results of an accidental "experiment."

During 1968 the research submersible *Alvin* suddenly sank to the sea bottom while being launched, fortunately with no loss of human life. Because of deteriorating weather and the approach of winter in the North Atlantic, *Alvin* remained on the deep ocean floor for eleven months in more than 1,500 meters (~4,950 feet) of water, some 150 kilometers (~90 miles) off the coast of southern Massachusetts. After *Alvin* was raised off the bottom the following year, an inspection of the interior uncovered a lunchbox containing a bologna sandwich, bouillon broth in a thermos, and an apple. The food, although sopping wet, appeared remarkably fresh. All these food items spoiled within several weeks of their recovery, even though they were immediately refrigerated at 3°C—the temperature of the bottom water where the vessel had sunk. This unexpected result suggested that high hydrostatic pressure and not low temperatures affected the metabolic rate of bacteria and, therefore, the decomposition of the lunch.

A series of experiments confirmed that the rates of bacterial respiration and carbon uptake from organic compounds by bacteria are strongly retarded by high hydrostatic pressures (**Figure 13–9**). Furthermore, in corroborative field experiments, various types of gelatinous materials that were placed on the deep-sea bottom for between twelve and fifteen months and were colonized by bacteria suffered little decomposition in comparison to similar samples placed on the shallow floor of the continental shelf. From this we infer that organic detritus, once it reaches the deep-sea floor, remains in place for long periods of time before it is degraded by microbes. Large scavengers, then, have a better chance of finding the sparse but nutritional food scraps that sink from high above. In other words, even though the fallout of food from above to the abyssal sea bottom is scant, once there, it lasts for a long time before it decays. Bacteria have turned out to be an important link in the detritus food chain of the abyssal zone. These microbes, which dwell in the *pore water* (the water filling the spaces between the grains of sediment) and on small mud particles, are in turn ingested by deposit feeders, and the nutriment so derived is gradually passed up the food web to the top predators.

13-3

Future Discoveries

We know less about the nature and distribution of life on the deep ocean floor than in most shallow-water habitats because of the difficulty and expense in conducting research in this extreme environment. Biologists now understand that the deep-sea floor is inhabitated by a rich variety of benthic organisms. How varied are these bottom communities and what accounts for their biodiversity? What about the hydrothermal vents where chemosynthesis is the only means of primary productivity? How are hydro-

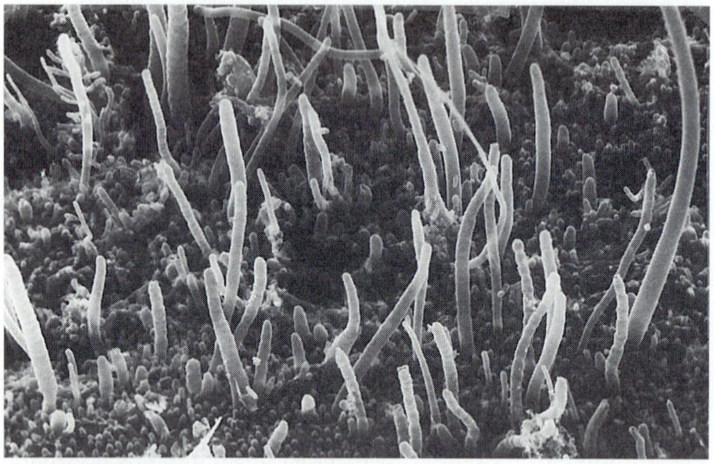

(a) PHOTOMICROGRAPH OF BACTERIA

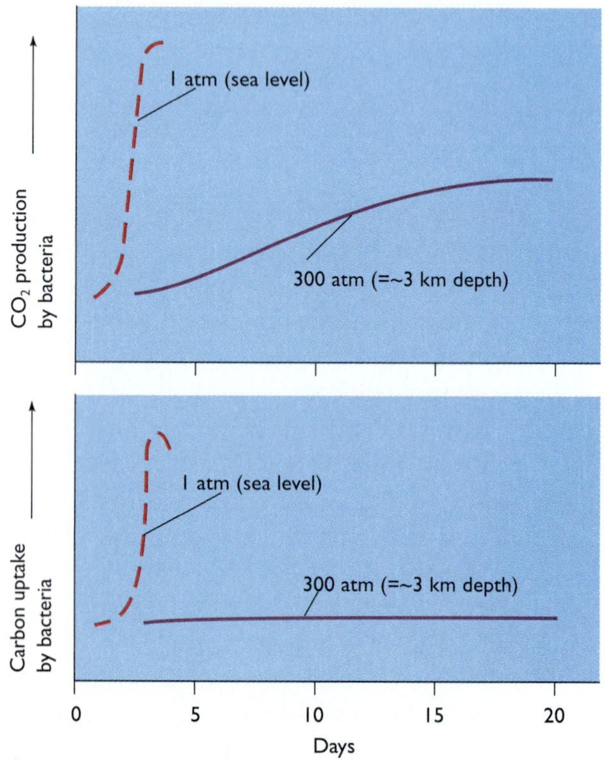

(b) EFFECT OF PRESSURE

FIGURE 13-9

Bacteria and hydrostatic pressure. (a) This photomicrograph shows bacteria that have colonized the surface of a mussel living on the deep-sea bottom. (b) Experiments indicate that the rates of carbon-dioxide production and carbon uptake by bacteria are reduced dramatically by high hydrostatic pressure. Because of this effect, organic detritus is decomposed at much slower rates on the deep-sea bottom than on the shallow continental shelves. [Adapted from H. W. Jannasch, *Oceanus* 21 (1) (1978): 50–57.]

thermal vents colonized and why are chemosynthetic bacteria not inhibited by the prevailing high pressures as are all the other bacteria of the deep sea? Is the glowing light generated by hot fluids and rock at hydrothermal vents sufficient to support photosynthesis on the deep-sea floor, as a few biologists suspect?

Closer to shore on the continental shelves, the marine communities are rich and abundant. However, biologists still need to learn a great deal about specific interactions among the different trophic levels of the food webs that exist in the waters of the continental shelf. What controls biological productivity in these waters and how does it vary over time? What accounts for plankton blooms, the patchiness of plankton, and the schooling of fish? How exactly do coastal birds and mammals fit in to these marine communities and how is the ecology of the region modified by the grazing and preying activity of such large organisms? What role do infrequent storms play in the dynamic ecological stability of these environments, and how have pollution and human activity affected the biota and ecology of these rich habitats? Some marine biologists suspect that the biomass of the sargassum gulfweed has decreased significantly during the last few decades. Has it and, if it has, what are the causes for the decline and what are the ecological implications for the Sargasso Sea? The questions abound and will require research well into the foreseeable future.

STUDY GUIDE

KEY CONCEPTS

1. Plankton teem in shelf waters, but tend to be distributed unevenly in *patches.* Because plants flourish here, so do fish, many of which aggregate into schools. The shelf benthos is controlled by the type of substrate. *Filter feeders* dominate gravel and coarse-sand bottoms, and *deposit feeders* dominate mud bottoms.

2. In the open ocean, most life is confined to the sunlit waters of the photic zone. Animals living below the photic zone are typically small and either prey on one another or depend on the settlement of organic detritus from above for food. Many invertebrates and fishes that live in the dark, deep aphotic zone possess *photophores* that *bioluminesce* to attract potential prey or to identify themselves to one another. Communities on the deep-sea bottom rely on the fallout of organic detritus, carcasses, and fecal matter from the water above.

3. Despite the low biomass of the deep-sea bottom, there is a surprisingly high level of species diversity. The deep-ocean community may be structured by the suppression of *competitive exclusion.*

4. Bacteria are strongly affected by the high hydrostatic pressure of deep water, which inhibits their respiration rates and therefore the rate of decomposition of dead organic matter on the deep-ocean floor.

KEY WORDS

bioluminescence
competitive exclusion
detritivore
diurnal vertical migration
midwater fish
patchiness
photophores
school

QUESTIONS

REVIEW OF BASIC CONCEPTS

1. How does wave activity indirectly control the feeding behavior of the benthos on the floor of the continental shelf?
2. Describe diurnal vertical migration and explain its ecological significance.
3. Discuss the apparent reasons for the low biomass and high species diversity of the deep-sea benthos.
4. How exactly are rates of decomposition of organic matter affected by the high hydrostatic pressures associated with the deep-sea bottom?
5. Contrast the general feeding styles of benthos that occupy a mud substrate and a gravel substrate.
6. Describe how phytoplankton abundances and composition change with distance offshore and explain these changes.
7. What exactly is competitive exclusion and might this concept explain the unusual biodiversity of deep-sea bottom communities?

CRITICAL-THINKING ESSAYS

1. Biologists maintain that all food webs ultimately depend on primary production. How can this be true for communities on the deep-sea bottom where, because of perpetual darkness, there is no primary production?
2. Based on your understanding of fish morphology, which was covered in Chapter 9, design a "new" midwater fish that would be well adapted to life in the aphotic zone. Explain the reasons for your specific design features.
3. Schooling is a common behavior of fish that dwell in water of the photic zone. Speculate why midwater fish do not school.
4. Assume that California gray whales became extinct. What would be the geological and ecological consequences of this extinction for the Bering Sea? Argue logically for your deductions.
5. Using library and Web resources, contrast the predation styles of squid, shark, penguin, and porpoises.

SELECTED READINGS

Childress, J J., H. Felbeck, and G. Somero. 1987. Symbiosis in the deep sea. Scientific American 256 (6): 115–20.

Ellis, R. 1996. Deep Atlantic: Life, Death, and Exploration in the Abyss. New York: Knopf.

Ellis, R., and J. E. McCasker. 1991. Great White Shark. New York: HarperCollins.

Gosline, J. M., and M. E. DeMont. 1985. Jet propelled swimming in squids. Scientific American 252 (1): 96–103.Groves, P. 1998. Leafy sea dragons. Scientific American 279(6): 84–89.

Heyning, J. E. 1995. Whales, Dolphins, and Porpoises: Masters of the Ocean Realm. Seattle, Wash.: University of Washington Press.

Koehl, M. A. R. 1982. The interaction of moving water and sessile animals. Scientific American 274 (6): 124–35.

Kanwisher, J. W., and S. H. Ridgway. 1983. The physiological ecology of whales and porpoises. Scientific American 248 (6): 110–21.

Klaus, A. D., J. S. Oliver, and R. G. Kvitek. 1990. The effect of gray whale, walrus, and ice gouging disturbances on benthic communities in the Bering Sea and Chukchi Sea, Alaska. Research and Exploration (National Geographic Society) 6 (4): 470–84.

Levinton, J. S. 1982. Marine Ecology. Englewood Cliffs, N.J.: Prentice-Hall.

Lohman, K. J. 1992. How sea turtles navigate. Scientific American 266 (1): 68–77.

Mead, J. G. and J. P. Gold. 2002. *Whales and Dolphins in Question.* Washington, D.C.: Smithsonian Institution Press.

Milne, D. H. 1995. Marine Life and the Sea. Belmont, Calif.: Wadsworth.

Nybakken, J. W. 1996. Marine Biology: An Ecological Approach. New York: Addison-Welsey.

Oceanus. 1978. Marine mammals. Special issue 21 (2).

Oceanus. 1995. Marine biodiversity I. Special issue 38(2).

Oceanus. 1996. Marine biodiversity II. Special issue 39(1).

Riedman, M. 1990. The Pinnipeds: Seals, Sea Lions, and Walruses. Berkeley, Calif.: University of California Press.

Robinson, B. H. 1995. Light in the ocean's midwaters. Scientific American 273 (1): 60–65.

Schreiber, E. A., and R. W. Schreiber. 1989. Insights into seabird ecology from a global "natural experiment." Research and Exploration (National Geographic Society) 5 (1): 64–81.

Starr, C., and R. Taggart. 1992. Biology: The Unity and Diversity of Life. Belmont, Calif.: Wadsworth.

Thorson, G. 1971. Life in the Sea. New York: World University Library.

Whitehead, H. 1985. Why whales leap. Scientific American 252 (3): 84–93.

Würsig, B. 1988. The behavior of baleen whales. Scientific American 258 (4): 102–107.

Zapol, W. M. 1987. Diving adaptations of the Weddell seal. Scientific American 256 (6): 100–107.

TOOLS FOR LEARNING

Tools for Learning is an on-line review area located at this book's web site OceanLink (**www.jbpub.com/oceanlink**). The review area provides a variety of activities designed to help you study for your class. You will find chapter outlines, review questions, hints for some of the book's math questions (identified by the math icon), web research tips for selected Critical Thinking Essay questions, key term reviews, and figure labeling exercises.

14 The Ocean's Resources

Full fathom five thy father lies;
Of his bones are coral made:
Those are pearls that were his eyes;
Nothing of him that doth fade
But doth suffer a sea-change
Into something rich and strange.
Sea-nymphs hourly ring his knell:
Ding-dong.
Hark! now I hear them—Ding-dong bell.
—*William Shakespeare,* The Tempest, *1611*

web navigator

critical thinking on the web

math tutor on the web

www.jbpub.com/oceanlink

PREVIEW

THE OCEANS ARE VAST and dominate the Earth. Their water and the seabed contain an abundant assortment of resources that are or are becoming vital for human economic ventures. In the main part of this chapter, we provide an overview of some vital marine resources—petroleum, gas hydrates, sand and gravel, manganese nodules, cobalt-rich crusts, phosphate deposits, and fish (caught and raised)—and review their general distribution, abundance, and management schemes. A key issue pertaining to marine resources is the legal ownership of natural materials that are located far offshore and not under a nation's jurisdiction. Some valuable materials found on the deep-sea floor of the open ocean belong, in the view of some people, to all of humankind. If so, what right does any private company or nation state have to claim a deep-sea resource as its own? These legal problems are addressed in the first section of the chapter, where the basis for the Law of the Sea and Exclusive Economic Zones is explicated.

14-1

Law of the Sea

The oceans are interconnected and occupy vast regions of the Earth. The high seas are located far from land in the open ocean and conflict about the use and ownership of these waters intensified with the rise of modern nation states and their navies, the rapid growth of world commerce that depended on the transport by ship of raw and manufactured goods, and the discovery of a variety of exploitable marine resources, both living (fish) and nonliving (oil, manganese nodules). Who can claim exclusive ownership of a fishing ground located far offshore out of sight of land? Who can claim the right of ownership of rich deposits of metal-rich nodules that lie scattered on the sea bed of the deep sea in what has traditionally been viewed as international waters? What right does any nation have to deny access to a vital seaway located near its territorial borders? These are difficult questions to deal with and have frequently created tension among rival states that in some cases has led tragically to warfare.

Following World War II, President Truman in 1954 claimed American jurisdiction over all the recently discovered biological and physical resources (oil and natural gas) of the "subsoil and seabed of the continental shelf" contiguous to the United States. This proclamation defined the continental shelf as the submarine land that surrounds a continent and that extends seaward out to a water depth of no more than 100 fathoms (~183 meters). Shortly thereafter, other nations followed suit and made similar unilateral proclamations about the ownership of the marine resources of their contiguous continental shelves.

Because of rapidly developing marine drilling technology, the United Nations created a committee within the General Assembly to develop equitable policies for regulating the use of continental shelf seas. The 1958 and 1960 Geneva Conventions on the **Law of the Sea** resulted in a treaty for regulating the exploitation of resources of the continental shelf. In essence the document states that water, seabed, and subseabed resources of any sector of the continental shelf are under the direct control of the country that owns the nearest land. But what about resources far offshore, beyond the continental slope? Ocean exploration throughout the 1960s revealed the mineral richness of the deep sea (mostly manganese nodules) located farther offshore beyond the continental shelf. Questions arose about the legal rights of nations to exploit these resources that some believed are part of the "common heritage of mankind," including, they argued, the people of landlocked countries.

LAW OF THE SEA TREATY

Beginning in 1973, the United Nations organized a third Law of the Sea Conference in New York City to address such pressing legal issues. Extensive deliberations culminated, in 1982, in a Draft Convention of the Law of the Sea that eventually gained the support of most of the member states of the United Nations. Many of the provisions of the treaty have since been ratified and become international law. Japan, Great Britain, and Germany, among others, did not sign the Law of the Sea Treaty. In 1994, the United States signed the treaty, although it has yet to be ratified by the U. S. Senate.

The Law of the Sea treaty guarantees the following:

Territorial seas defined as extending seaward for 12 nautical miles (~22.5 km) are under the direct jurisdiction of the adjoining coastal nation.

An **Exclusive Economic Zone** (EEZ) extends offshore for 200 nautical miles (~370 kilometers) or even farther if the edge of the continental shelf lies beyond 200 nautical miles. The EEZ legally empowers the coastal state to regulate fishing, mineral resources, pollution, and scientific research in these waters.

Vessels have the right of free and "innocent" passage in waters outside the territorial seas and through international straits that lie within a territorial sea.

Private exploitation of mineral resources in the high seas that lie beyond the EEZ in theory requires approval by the International Seabed Authority (ISA), a chartered agency of the United Nations. Mining technology would be shared with the ISA, which is authorized to distribute revenue from resource exploitation of the seabed of the high seas with developing nations.

EXCLUSIVE ECONOMIC ZONES

The Exclusive Economic Zone relegates about 40 percent of the world's oceans to the control of coastal states. The remaining 60 percent of the oceans is deemed the high seas and belongs to everyone in the world. The map of **Figure 14–1** depicts the position of the world's EEZs. Note how islands, even small ones, have large EEZ zones relative to their land areas. The United States has the largest cumulative EEZ, comprising approximately 10 million kilometers^2 (~6.2 million miles^2), an ocean area that exceeds its land area by 30 percent, because Hawaii, Alaska, Puerto Rico, and the U. S. Virgin Islands, as well as its protectorates such as Guam and American Samoa are included (**Figure 14–2**). This places productive fisheries and vast mineral deposits under the direct control of the United States.

14-2 Mineral Resources

OIL AND NATURAL GAS

Petroleum, both liquid oil and natural gas, is composed of hydrocarbons, molecular compounds that

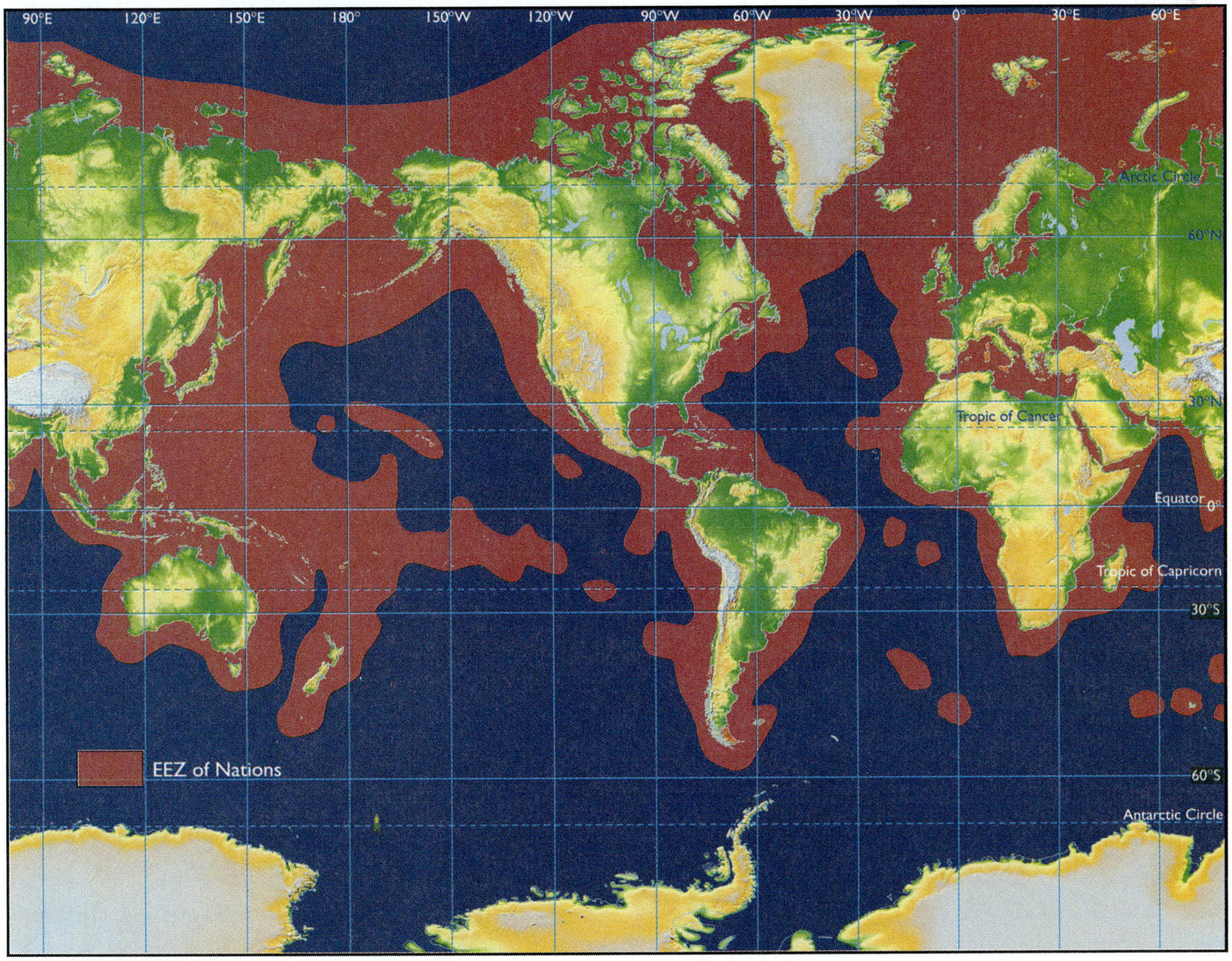

FIGURE **14-1**

Exclusive Economic Zones (EEZ) based on the Law of the Sea Treaty. More than 40 percent of the world's ocean area is controlled by national governments.

FIGURE 14-2

EEZ. Exclusive Economic Zones of the United States.

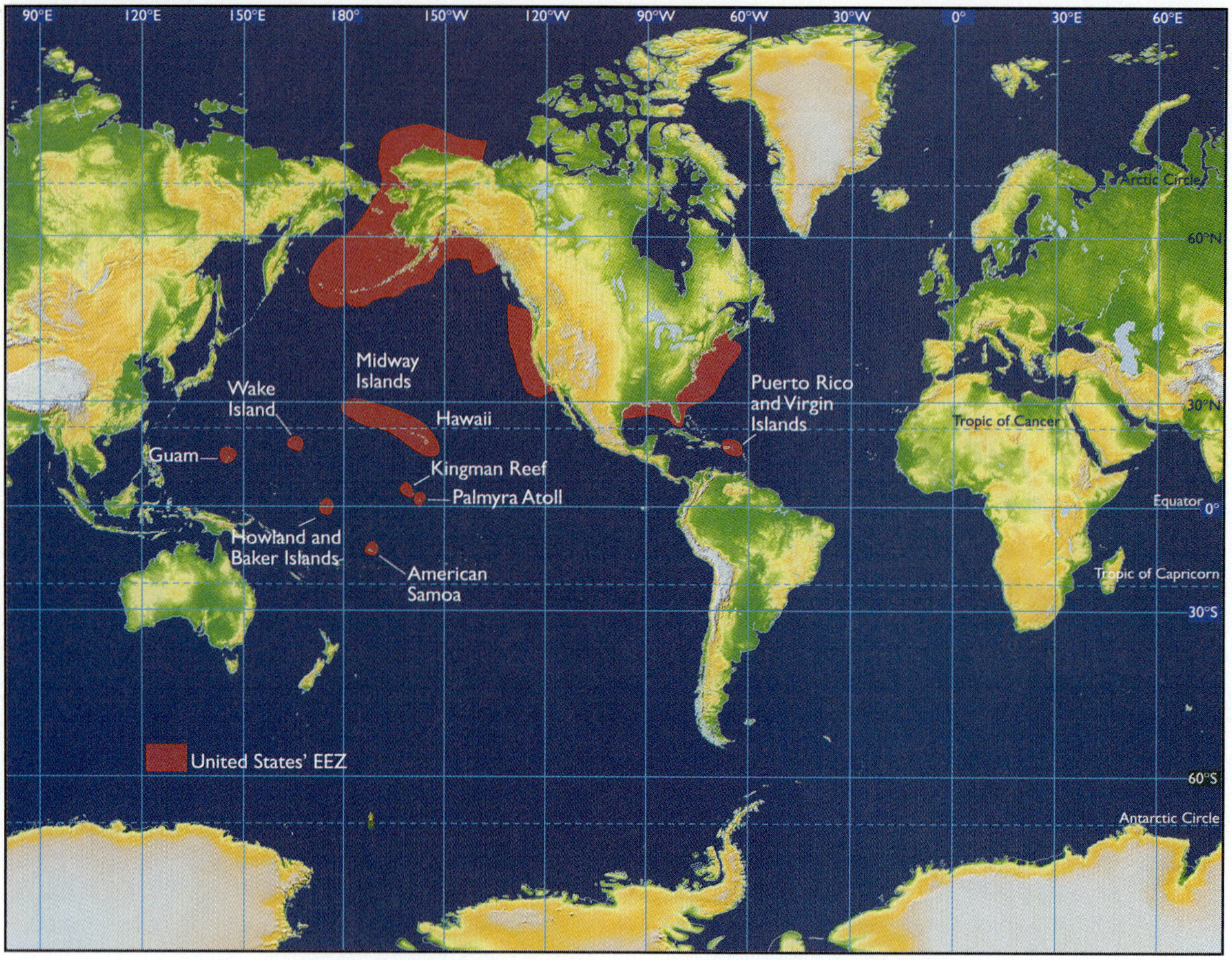

contain principally the elements hydrogen and carbon. Most oil and natural gas is derived from marine sedimentary rocks, notably shales, that contain abundant remains of plankton. Apparently, source rocks that yielded petroleum formed in the geologic past in quiet and biologically productive basins, as mud and the tiny, abundant remains of plankton accumulated on a relatively shallow sea bottom where anoxic bottom water helped preserve the organic material. Subsequent deep burial of these organic-rich deposits subjected them to high temperatures that eventually transformed these organic compounds into hydrocarbons that filled the tiny pore spaces between the sediment particles. After tens of millions of years, the petroleum was extruded from the source beds and, being less dense than the surrounding rock, migrated upward until the oil encountered impermeable layers and became trapped in suitable reservoirs such as a porous and permeable sandstones (**Figure 14–3**). If burial temperatures get very high or if the chemical 'cooking' process is prolonged, oil is converted to natural gas (methane). This means that old or highly baked rocks tend to contain methane gas rather than liquid oil.

Using seismic-reflection techniques (see Figures B2–1 and B2–3 in the box, "Probing the Sea Floor" in Chapter 2), geologists can identify suitable geologic settings and specific sedimentary rocks and traps that may contain economic deposits of oil and gas. Because many petroleum-bearing rocks were formed in shallow marine settings, such as the continental shelf and slope, geologists search for and explore similar sites around the world. The United States, for example, is the most extensively drilled country in the world. According to Joseph P. Riva,

Jr., an expert on the world and U. S. petroleum industry ". . . geologic studies indicate that almost all of the remaining undiscovered large oil accumulations in the United States are likely to be site-specific to a relatively few areas in Alaska and on the outer continental shelves. The Arctic National Wildlife Refuge on the North Slope of Alaska is especially prospective, containing the largest undrilled onshore geologic structures known in the United States. Most of these areas remain off limits to exploration, a questionable policy that has engendered much debate given that there is no oil-based substitute for 80 percent of the oil consumed domestically." Riva's statement defines the current conflict of interests that has arisen between the need to extract natural resources for human welfare and the need to maintain the ecological integrity and resiliency of terrestrial and marine ecosystems. This vital issue is examined in Chapter 15.

In 1996, Shell Oil Company uncovered a large oil field, called Mars, in deep water (850 meters or ~2,790 feet) in the Gulf of Mexico. With its estimated 700 million barrels of oil, the Mars field is the largest oil find in the United States since the discovery of Prudhoe Bay in Alaska three decades earlier. Given that U.S. oil consumption is in the order of 6.6 billion barrels a year, the 700 million barrels of oil in the Mars field actually make only a modest contribution to America's energy needs. The geologic setting of petroleum in the Gulf of Mexico is discussed in the geology box, "Offshore Oil and Gas in the Gulf of Mexico."

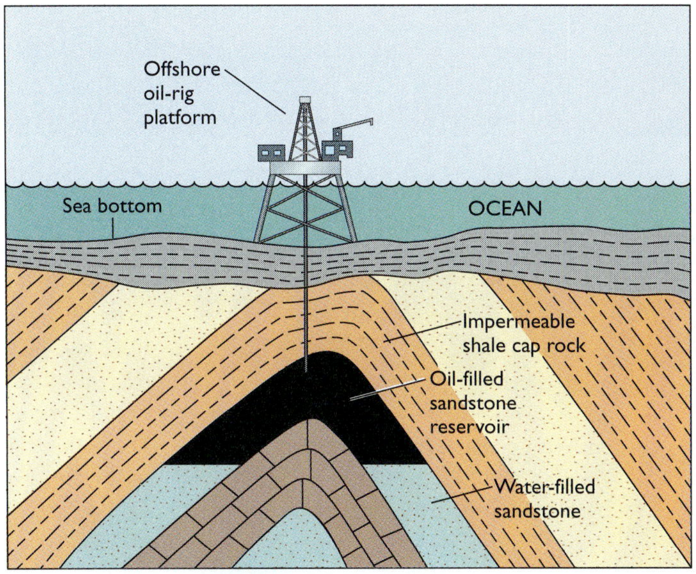

FIGURE **14-3**

Mineral resource. Drilling an oil trap on the continental shelf.

GAS HYDRATES

Extensive deposits of gas hydrates have been discovered in the sediment of polar regions and in deposits of continental slopes between the water depths of 300 and 500 meters (~985 to 1,642 feet), where cold water is in contact with the sea bottom. Gas hydrates are unusual hydrocarbon deposits in that they consist of solid, frozen water molecules that form small cagelike structures in the sediment, each 'cage' trapping a single molecule of the natural gas methane—CH_4. The quantity of natural gas in these gas hydrate deposits is tremendous. The U. S. Geological Survey estimates, for example, that gas hydrates of the continental margin of North Carolina alone contain about 350 times the energy consumed in a single year in the United States. Unfortunately, the technology to extract methane from gas hydrates on the continental shelf has not yet been developed, nor does it appear likely to become economical until well into the twenty-first century when more conventional reserves of petroleum become scarce. Presently, Japan, which ranks as the world's second-largest consumer of hydrocarbon energy, is the leader in developing offshore technology and techniques for extracting methane from gas hydrate deposits in the Sea of Japan.

SAND AND GRAVEL

Sand and gravel deposits are defined as natural aggregations of unconsolidated sediment with grains larger than 0.0625 millimeters (~0.00246 inches) in diameter. Being high-energy deposits, sand and gravel are commonly found blanketing the seabed of the continental shelves of the world where strong wave and current activity prevails. During the ice ages, sea level dropped and exposed the shelf surface to river erosion and deposition; so many deposits of sand and gravel on the shelf are relict in nature, having been deposited long ago under conditions that were very different from those of the present. Although sand and gravel seem to be a mundane

THE OCEAN SCIENCES GEOLOGY

Offshore Oil and Gas in the Gulf of Mexico

The Gulf of Mexico is an immense basin with a surface area of about 564,000 kilometers^2 (~172,584 miles^2) and water depths greater than 3,800 meters (~12,382 feet) below sea level. In plan view (overhead), it resembles a large sports amphitheater with two major entryways, one located between Yucatan and western Cuba, the other between southern Florida and northern Cuba (**Figure B14–1**). Although details are still being worked out by geologists, it appears as if the Gulf of Mexico formed by sea-floor spreading some 200 million or so years ago. Thick salt layers as well as vast quantities of terrigenous sediment derived from the erosion of the surrounding landmasses have accumulated in the basin since its inception. The dominant supplier of sediment to the Gulf has been and continues to be the "mighty" Mississippi River, which contributes about 400 million metric tons of sand and mud annually to the basin.

The buildup of this terrigenous material over geologic time has created a broad continental shelf between Texas and Louisiana. Here the continental slope is not smooth but rises and dips irregularly, being pockmarked by a complex array of hills, mounds, holes and valleys. The rough bathymetry of the continental slope reflects the movement of ancient, deeply buried salt beds that have forced their way upward through the overlying layers of mud and sand and in the process buckled and fractured the seafloor. These salt-created structures have trapped oil and gas in porous reservoirs. These are the geologic features that are targeted for exploration and production by oil companies.

Although oil and gas have been produced in the Gulf of Mexico for fifty years, the outlook for the region has changed dramatically since the mid 1990s because of newly developed

FIGURE **B14-1**

Gulfwaters. The Gulf of Mexico is an enormous sedimentary basin.

resource, they actually are priceless in some regions for constructing commercial and residential buildings, and for improving highways and roads. Sand and gravel mined from the continental shelf may be a particularly important source of building material along shores where there is rapid growth and extensive construction. This coarse sediment can also be used to replenish sand on beaches undergoing coastal erosion and to create salt-marsh wetlands. It is estimated that currently the cost to transport sand and gravel by water is about one-third less than it is by land. At present, only about one percent of the sand and gravel needed each year for construction projects is derived from offshore sources.

Many important shelf deposits of sand and gravel occur along the west coast of the United States stretching from California to Washington and along the east coast from the Carolinas to New England (**Figure 14–4**). Mining the sediment deposits that blanket the continental shelf can pose serious environmental problems, particularly for the benthic and pelagic communities that rely directly or indirectly on stable substrates. Removing sand and gravel from the seabed disrupts the fauna and flora, and can create plumes of mud that interfere with phytoplankton photosynthesis, clog the filtering organs of organisms, and transfer large quantities of nutrients from the bottom to surface waters, chang-

exploration technology and a growing worldwide demand for oil. A special and expensive deepwater technology is required for drilling and production in water depths greater than 400 meters (~1,312 feet). By 1997, there were twelve permanent platforms and twenty-three undersea wells in water depths greater than 300 meters (~985 feet) in the Gulf of Mexico. Currently, in Project Genesis, a floating structure is being erected about 240 kilometers (~150 miles) south of New Orleans for drilling and production in water depths of 790 meters (~2600 feet) (**Figure B14–2**). All production from this massive structure will flow by pipeline to a facility on shore, where handling and processing can be done more safely. Recent estimates place the production potential of the Gulf of Mexico at about 1.82 billion barrels of oil and at about 5.81 trillion cubic feet of natural gas.

FIGURE **B14-2**

Offshore drilling platform. The Mars offshore drilling platform, part of Project Genesis, is the largest in the world. Here, it is being towed through the Corpus Christi ship channel in Port Arkansas, Texas.

ing the ecological character of the system. These disturbances can affect biological production and perhaps even alter the species composition of a shelf ecosystem. So the mining of marine deposits of sand and gravel is done carefully and in a restricted fashion in order to minimize harmful results to coastal ecosystems. If damage is done, the site is restored after the mining enterprise is finished.

MANGANESE NODULES

As discussed in Chapter 4, manganese nodules abound in certain localities of the ocean floor (see Figure 4–15 in Chapter 4), mostly in water depths greater than 2,000 meters (~1.24 miles). Nodules (**Figure 14–5**) on the average contain roughly between 20 and 30 percent manganese dioxide. Although called manganese nodules, they contain a variety of other metals, including between 10 and 20 percent iron oxide, as much as 1.5 percent nickel, and less than 1 percent cobalt , copper, zinc, and lead. Estimates indicate that there are billions of kilograms (1 kilogram = 2.2 pounds) of recoverable metals lying on the sea floor, particularly in the rich and vast deposits of the subtropical Pacific Ocean (see Figure 4–15 in Chapter 4). Interestingly, if such nodular deposits occurred on land, they would be

FIGURE 14-4

Sand and gravel. Sand and gravel deposits of the United States continental shelf. [Adapted from D.A. Segar, *Introduction to Ocean Sciences* (New York: Wadsworth, 1998).]

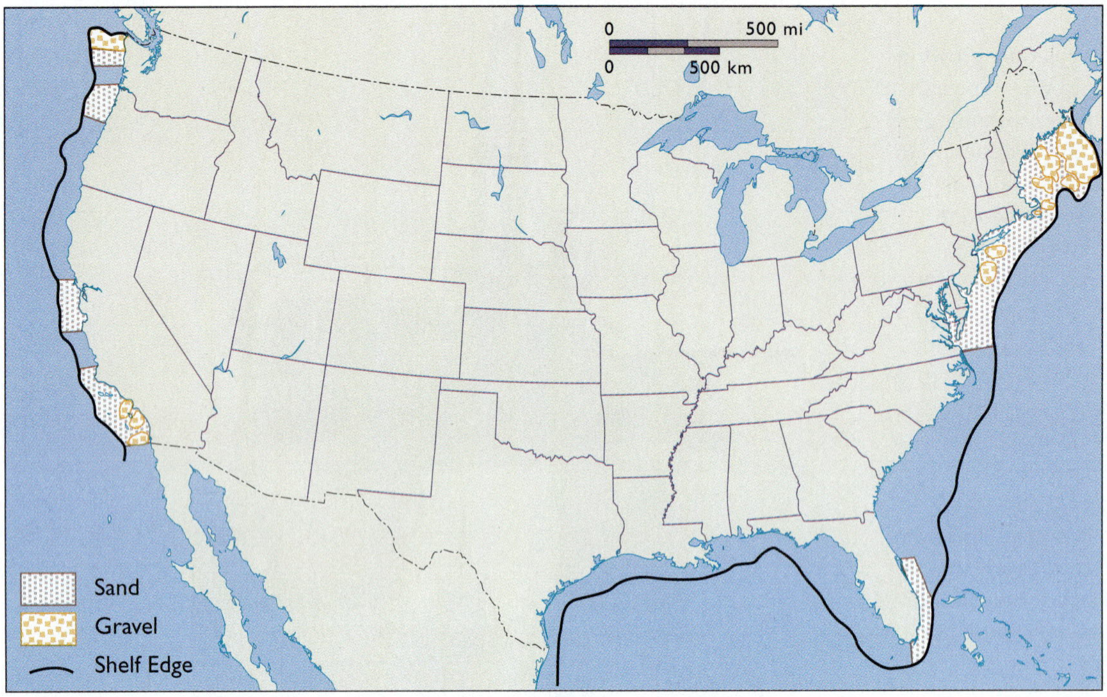

FIGURE 14-5

Manganese nodules. Photograph of a dense concentration of nodules blanketing the sea bottom of the South Pacific Ocean.

valuable metal ores. Being in the open ocean at great water depths, they are quite expensive to mine. Paradoxically, if substantial quantities of deep-sea manganese nodules were recovered from the seabed, the world market for the specific metals would become quickly saturated, causing economic chaos as the value of these metals plummeted.

There are other legal difficulties with mining manganese nodules from the deep-sea floor of the open ocean. Exactly who owns these rich metal deposits, given that most of them lie beyond the Exclusive Economic Zone (EEZ) of nations? The Law of the Sea Treaty has not been formally signed by all nations. Presently, Japan and Korea are in the forefront of developing equipment to recover nodules from the deep seabed.

COBALT-RICH OCEANIC CRUSTS

It has recently been discovered that the sides of many seamounts and even some islands in the Pacific Ocean between the depths of 1 and 2.5 kilometers (~0.62 to 1.55 miles) are composed of rocks that are enriched in cobalt. In the United States cobalt is a strategic mineral, meaning that it is vital for military and industrial enterprises, in particular for producing jet engines, and there is no substitute for it. It cannot be produced domestically in the United States in sufficient quantities to meet internal demands. In 1978 a civil war in Zaire, the single major global supplier of cobalt, disrupted production, and prices for cobalt skyrocketed, increasing from $6 to $40 a pound in a single month. Preliminary investigations by American oceanographers indicate that large numbers of seamounts and islands with flanks composed of cobalt-rich crust are in the Exclusive Economic Zones of the Hawaiian Islands. Each seamount could potentially yield as much as 3 or 4 million metric tons of ore, allowing the United States to control and meet its domestic demand for cobalt.

PHOSPHATE DEPOSITS

All living organisms require phosphorous in order to grow. In humans, for example, bones and teeth are composed of calcium phosphate. Also, phosphate is used extensively in agriculture as a fertilizer to grow food crops and is a compound required in the chemical industry. It appears that phosphate rock forms as an offshore accumulation on submarine terraces where coastal upwelling supports high biological productivity in the surface waters. There, organic phosphorous settles continually on the seabed below. Eventually this organic material is buried and transformed into phosphoritic nodules that appear to grow at a rate of between 1 and 10 millimeters (~0.04 to 0.39 inches) per thousand years—an extremely slow rate of production.

The most extensive offshore deposits of phosphate rock occur in three regions: off the coasts of Morocco, southern California, and the Carolinas. Smaller deposits of phosphate are scattered about, a notable one being in the world's smallest independent nation, Nauru, an island of slightly more than 5,000 acres located in the middle of the Pacific Ocean. The entire economy of Nauru is dependent on this one resource, which is about to be mined out entirely. Because of poor political leadership and a limited economic vision, the future of Nauru as a modern state is uncertain.

The world currently uses about 150 million metric tons of phosphate rock each year, and the demand worldwide is growing at about 4 percent annually. Global reserves of phosphate deposits are estimated to be about 34 billion metric tons. At the current usage and allowing for increasing demand, the reserves will be depleted by about the year 2050. Given that phosphorus is a critical element for plants and that there is no artificial substitute for it, the exhaustion of this vital resource will create a monumental crisis in the industrial production of food crops for humans.

14-3 Living Resources

FISHERIES

Marine finfish are diverse, but can be divided broadly into pelagic fish and groundfish. The former inhabit the water column, whereas the latter

FIGURE 14-6
Various commercially important fishes.

THE OCEAN SCIENCES BIOLOGY

Fish Farming and "Super" Fish

Wild stocks of salmon have been in serious decline along both the eastern and western coasts of Canada and the United States. This has forced government agencies to close many of the traditional river fisheries of wild salmon. The loss of this important natural resource has spurred the development of salmon farms in shallow coastal waters. Norway pioneered this enterprise and today they are by far the largest producers of farmed salmon in the world, accounting for more than 300,000 metric tons annually. By contrast the Canadian farm production of Atlantic salmon in British Columbia (~30,000 metric tons/yr) and in the Maritime Provinces of Canada (~20,000 metric tons/yr) is an order of magnitude less than Norway's yield. Most salmon sold in U.S. and Canadian markets have been farm raised.

The farming technique is a straightforward one. Large numbers of young Atlantic salmon, up to 70,000 individuals on a medium-sized farm, are placed in floating cages that are anchored to the shallow sea bottom in reasonably protected water where they feed and grow in size (**Figure B14-3**). If necessary, the fish are vaccinated with antibiotics to prevent the breakout of bacterial diseases and other infestations. Some salmon farmers rely on the regular use of anesthetics, hormones, and disinfectants to manage their stock of caged fish. Physical factors likewise can pose problems. Severe storms, for example, damage the cages, allowing the farm-bred fish to escape and intermix with wild salmon. In February of 2001, 100,000 farm salmon escaped from their pens along coastal Maine. Seals and seabirds are drawn to the farms and they occasionally damage the floating pens, allowing the salmon to get free. Based on sampling programs at monitoring sites on a few Canadian rivers, farm salmon, both adult and immature fish, have been detected among the wild populations in fourteen rivers of Nova Scotia and New Brunswick.

Biologists and environmentalists fear that ecological and genetic interactions between wild and farm salmon may have dire consequences for the long-term recovery of natural stocks of salmon, which are already in serious decline. For example, predation on wild salmon increases, because farm sites, which are commonly located near river mouths, attract predators that then prey heavily on the wild salmon migrating into or out of the rivers. Also, culturing fish clearly changes their genetic makeup, which affects their fitness and survival abilities. Interbreeding between cultured and wild salmon will clearly reduce the genetic variability of local populations. In addition, farm salmon tend to be larger than natural salmon of the same age, which presumably gives them an advantage in competing for space, food, and mates. In Norway, successful spawning by salmon escapees has produced both farm and hybrid fish. Biologist speculate that farmed and hybrid juveniles have lower survival success in rivers but may have some competitive advantages in certain habitats or during certain stages of their growth or life cycles. Available laboratory and field data, though quite limited, suggest that there is a high probability that wild salmon will be disadvantaged by being outcompeted for food, space, and spawning sites by farm-raised and hybrid salmon.

Recently, an American-based company has produced a strain of genetically engineered salmon, designated by some as "super salmon." Researchers have added one gene to hatchery

dwell near or on the sea bottom. Although a large variety of edible fish exist in the ocean, only a handful accounts for the bulk of the global commercial harvest. **Figure 14–6** portrays some examples of commercially important fish. Although the focus of this section is on fishes, mammals have also been hunted extensively.

As you know well, most of the world's oceans are sparsely populated with organisms because their water has low concentrations of dissolved nutrients and, hence, little primary productivity (see Chapter 10). Consequently, commercially important fish stocks tend to aggregate in two areas (**Figure 14–7**) of the oceans: in the water of the continental shelves where food abounds because of the proximity of the continents and their rich river supply of nutrients, and in a few upwelling regions of the world, such as off western Africa, western United States, and western South America along Peru and northern Chile.

Ocean fishing is very big business because fish

salmon that augments growth hormone production and a second gene that allows them to tolerate cold water. Consequently, rather than growing only during the summer months, genetically engineered salmon grow year round and become three to four times larger than wild fish of the same age. Incredibly, they grow 300 to 400 percent faster than natural salmon in the first fourteen months of their lives. Critics allege that farming these super salmon in floating pens will inevitably result in some of them escaping into the wild. Once free, they will interbreed with wild stock and reduce the genetic biodiversity of native salmon populations. Some fear that they will simply displace what few wild salmon are left. Even the salmon farmers are opposed to the introduction of genetically modified salmon for fear that the public will protest by removing salmon from their diet altogether and cause the economic collapse of their burgeoning industry. So far, nowhere in the world have super salmon been approved for aquacultural production.

Visit www.jbpub.com/oceanlink for more information.

FIGURE B14-3

Fish farm. This photograph shows the general layout of a typical farm sited in the shallow nearshore wasters of Canada.

cluster in the relatively shallow water that surrounds the continents. There they feed and breed and, for the most commercially important species, aggregate tightly into schools composed of thousands and even millions of individual fish. They are relatively easy to detect by sonar, scouting vessels, airplanes, and even satellites. Once the fish have been spotted, vast fishing fleets deploy a variety of nets to catch them, including stern trawl nets, purse nets (both are pictured in the biology box, "Sampling the Biota" in Chapter 9), and drift nets. The use of the last-named is controversial, because drift nets, which are constructed of light, tough plastic filament, can be up to 80 kilometers long and indiscriminately catch not only the prey sought but also other animals, including sea turtles, seals, dolphins, porpoises, small whales, sharks, and even birds. In fact, they efficiently remove all sea creatures that cannot pass through the mesh of the net and, if the waters

FIGURE 14-7

Ocean fisheries. Location of world's major commercial fisheries and the proportion of the total fish catch for 1991.

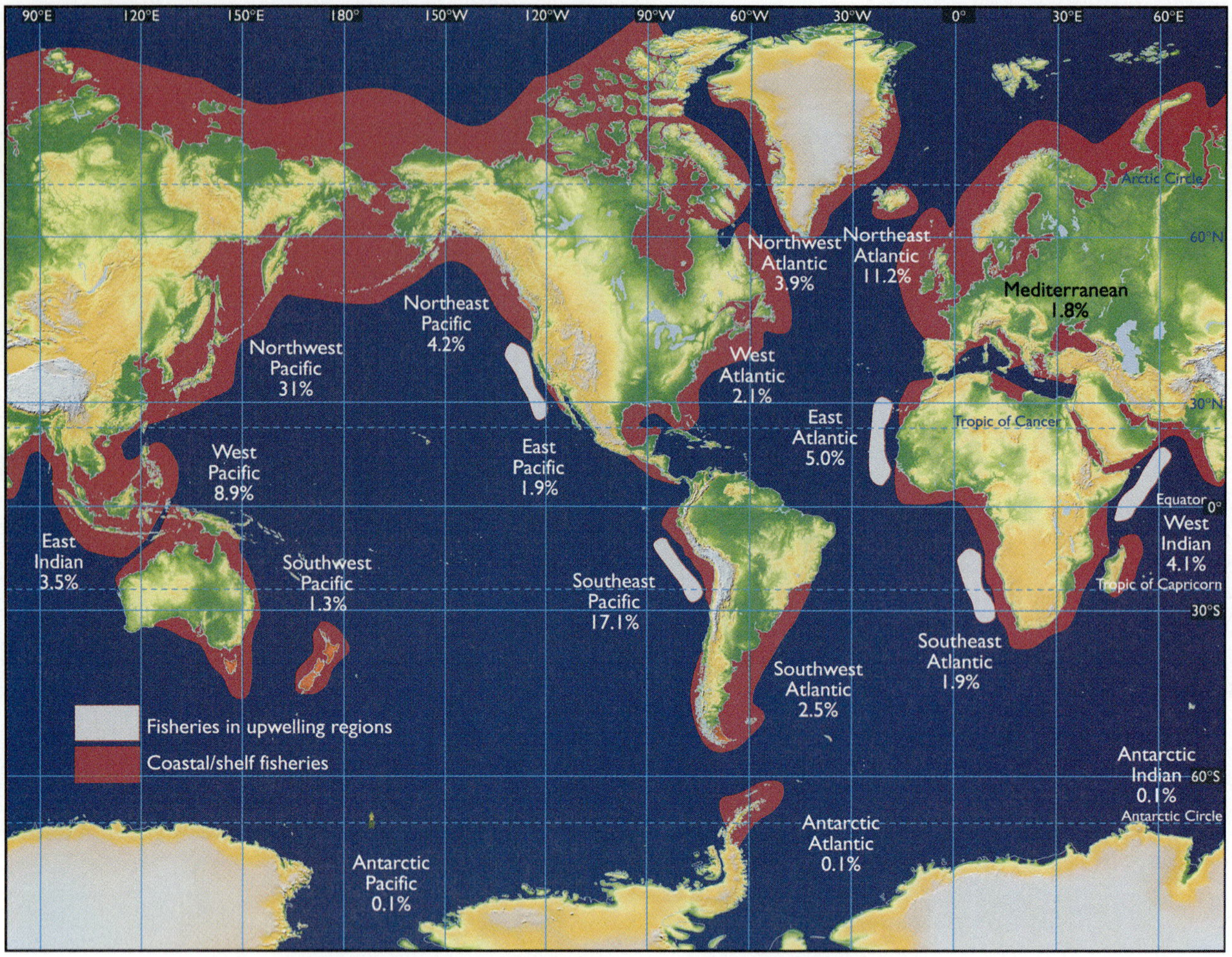

are fished regularly, can cause overfishing in short order. For example, in 1992, Japan reported that its offshore fishing fleet captured over a million squid, the targeted prey, in its drift nets and in the process killed 140,000 salmon, 270,000 sea birds such as petrels, and 26,000 marine mammals. In response to this unintended mass killing of so many animals, some already endangered, the United Nations, at the 1989 Convention for the Prohibition of Long Driftnets, and with the full support of the United States, urged all nations to ban the use of drift nets longer than 2.5 kilometers. In 1992, most fishing nations agreed to comply with the regulation even in the open waters of the high seas. Enforcement of the moratorium is, however, near impossible on the high seas and the drift net ban relies heavily on voluntary compliance by the world's major fishing fleets.

In 1950, the global fish catch was just under 20 million tons. That harvest grew by about 6 percent annually between 1950 and 1970, and by 2 percent annually between 1970 and 1990 (see Figure 15–16). During the early 1990s, the total marine fish catch seemed to have reached a plateau at just over 80 million tons per year, in spite of better technology, larger fishing fleets, and greater fishing intensity. In 1996, the global fish catch climbed to about 93 million tons; this new record tonnage resulted largely from the fishing success of Indonesia and Mexico. Despite world-record fish catches, however, some fishery biologists fear that a finite limit to harvestable fish has been or is about to be reached. This means that it is unlikely that the malnourished and starving humans of the world whose numbers are growing exponentially will be fed by the ocean's bounty, mandating more than ever that growth of the human population be curtailed.

Table 14–1 summarizes the most current data for the gross tonnage of the world's fishing fleet. Except for Europe, the fish harvest for most regions of the world has increased dramatically since 1970. Obviously, continued growth of this magnitude is no longer feasible, given that we have apparently maximized the annual global fish harvest at somewhere between 80 and 95 million tons. Reinforcing this belief is the fact that the world's fisheries are operating under a huge economic deficit, estimated to have been \$54 billion in 1989 and about \$15 billion in 1995. By all accounts, the expense incurred by ocean fleets in capturing fish far exceeds the selling price of fish on the open market. In the late 1980s, economists estimated that for every \$1 earned fishing, \$1.77 had to be spent in catching and marketing the fish. In other words there are no real profits to be made in large-scale commercial fishing and this activity is possible only with the support of major government subsidies and large tax incentives.

MARICULTURE

Fishing is basically a simple hunting process; fish are tracked, captured in nets or by hooks, and killed for food. Mariculture (marine agriculture) is equivalent to farming the seas. Under mariculture, finfish, shellfish, and algae are raised under favorable conditions that promote their growth until they are large enough to be harvested as food for people. As in any farming endeavor, the intent is to maximize the growth of the animal or plant in the shortest possible time. Examples of marine organisms raised by mariculturists include oysters, mussels, shrimp, salmon, and plaice. There is no doubt that interest in mariculture is increasing each year, as it becomes clear that the fish catch of the global ocean fisheries is unlikely to increase for the reasons discussed in the previous section. So fish farming will undoubtedly expand in the future. In fact, it is estimated that one out of every four fish eaten today was raised on either a freshwater or a saltwater fish farm. Almost 40 percent of the salmon eaten today worldwide spent part of their lives on a fish farm (see the biology box "Fish Farming and 'Super' Fish"), and 40 percent of the shellfish (oysters, clams, mussels) consumed presently were farmed in coastal waters.

For mariculture to succeed economically, the species to be raised must be marketable. The chosen species to be farmed must be fairly inexpensive to grow, must be efficient trophically, and must reach harvestable size in short order, usually in no more than a year or two. Also, the species must be resistant to disease and parasites and be protected from predation. A real advantage of mariculture is that the cultivation methods can be chosen to conform with local resources, supply local markets, and promote local business opportunities.

Mussel and oyster mariculture has been very successful as an economic enterprise. These

TABLE **14-1** Size and growth of the world's major fishing fleets.

Region	Gross Tonnage (thousands of tons) 1970	1992	Growth (percentage) 1970–1992
Asia	4,802	11,013	129
Former Soviet Union	3,997	7,766	94
Europe	3,097	3,018	-3
North America	1,077	2,560	138
South America	362	817	126
Africa	244	699	187
Oceania	37	122	230
World	13,616	25,994	91

Source: Adapted from L. R. Brown, C. Flavin, and H. French, eds., *State of the World 1998* (New York: Norton, 1998).

Antarctic Krill

Krill (**Figure B14–4**) are shrimp-like crustaceans or euphausiids that abound in the ocean. There are about eighty-five species of krill distributed worldwide, ranging from the tropics to the poles, each species adapted to specific water temperatures. Krill represent one of the most abundant animal groups that populate the planet. Antarctic krill, known scientifically as *Euphasia superba,* are distributed widely in the circumpolar waters of Antarctica, being particularly bountiful in the East Wind Drift, the Scotia Sea, the Weddell Drift, and off the Antarctic Peninsula (**Figure B14–5**). Adults of *E. superba* measure 6 centimeters (~2.4 inches) or more in length and weigh over a gram (~0.035 ounce). Antarctic krill spawn during late spring and summer (January and February), each female releasing up to ten thousand eggs in the upper 100 meters (~328 feet) of the water column. The eggs are denser than seawater and sink for about ten days before hatching as nauplii at water depths more than 1 kilometer (~0.62 miles) below the sea surface. The nauplii eventually rise to the surface and develop into adults that feed voraciously on krill eggs, larvae, a variety of zooplankton, and large quantities of the diatoms that abound in these fertile waters. It is as adults that krill aggregate into large swarms which can extend over 100 kilometers^2 (~30 miles^2) in very dense concentrations of up to 2 or more kilograms per cubic meter (~67 ounces per quart). Typically, the swarms of krill engage in diurnal vertical migration, remaining at depth during daylight and rising to the surface to feed during the night.

E. superba are considered by biologists to be a **keystone species,** meaning that, being herbivores, they are the single most critical link between phytoplankton and larger animals. They are preyed upon by an incredible variety of carnivores such as fish, squid, seabirds, and baleen whales. Many types of whales could not exist without the presence of dense krill swarms. Otherwise, they would expend more energy than they would acquire from the food they filtered out of the water and they would die in short order. In fact, they could not have evolved in the absence of swarming organisms such as krill to provide their principal food source.

Fishing fleets from the former U.S.S.R. began fishing Antarctic krill in 1961–62 when they harvested about 4 metric tons. Since then, the krill catch in Antarctic waters has increased sharply, reaching 500,000 metric tons by the early 1980s (**Figure B14–6**), the bulk of the fishing (96 percent) being done by Japan, Ukraine, and Russia using large stern trawlers and conical nets. In 1984, the krill catch dropped to about 128,000 metric tons, but rebounded to between 300,000 and 400,000 metric tons between 1986 and 1991. The harvest fell again in 1994 to a mere 81,000 metric tons and is rising slowly once again. It is unclear why the drops occurred.

The commercial harvesting of krill presents several problems. The catch must be processed within three hours of its capture because powerful digestive enzymes in the krill tend to foul the meat soon after death. Furthermore, the shell of krill contains fluoride and it must be removed in order to make the meat suitable for consumption by humans as paste, frozen tails, and "sticks." Krill are also sold as feed for animals and as bait for sportsfishermen.

Fear of a free-for-all fishing mentality devoid of any management procedures led in 1981 to the Convention on the Conservation of Antarctic Marine Living Resources (CCAMLR). The treaty is designed to safeguard the Antarctic ecosystem from the effect of krill fishing and to aid whales and other commercial fish recover from overexploitation.

FIGURE **B14-4**

Krill. An Antarctic krill, *Euphasia superba.*

Empowered by CCAMLR, the treaty commission set a catch limit in 1991 of 1.5 million tons for krill in the South Atlantic where most of the harvest of *E. superba* occurs, and in 1992 a limit of 390,000 tons for the southern Indian Ocean. These harvest limits were much higher than the actual current catch and reflect the enormous potential of this fishery.

According to some fishery biologists, there is cause to be alarmed, because the biology and ecology of krill are not understood at all. Estimates of krill stock in Antarctic waters vary tremendously , ranging between a million and a billion metric tons (3 orders of magnitude!). Hence, the setting of catch limits and attempts to regulate a fishery when there is no basic, never mind a deep, scientific understanding of the working of the complex ecosystem involved seems to be a sure way to cause unwanted effects that could have been avoided with more basic scientific knowledge. A real concern is the fact that Antarctic krill are a keystone species, and if they were to be harvested in a nonsustainable manner, the fishery itself would collapse, and the considerable numbers of squid, whales, fish, and birds that depend on krill as their principal food source

FIGURE **B14-5**

Krill fishery. Location of large krill populations near Antarctica.

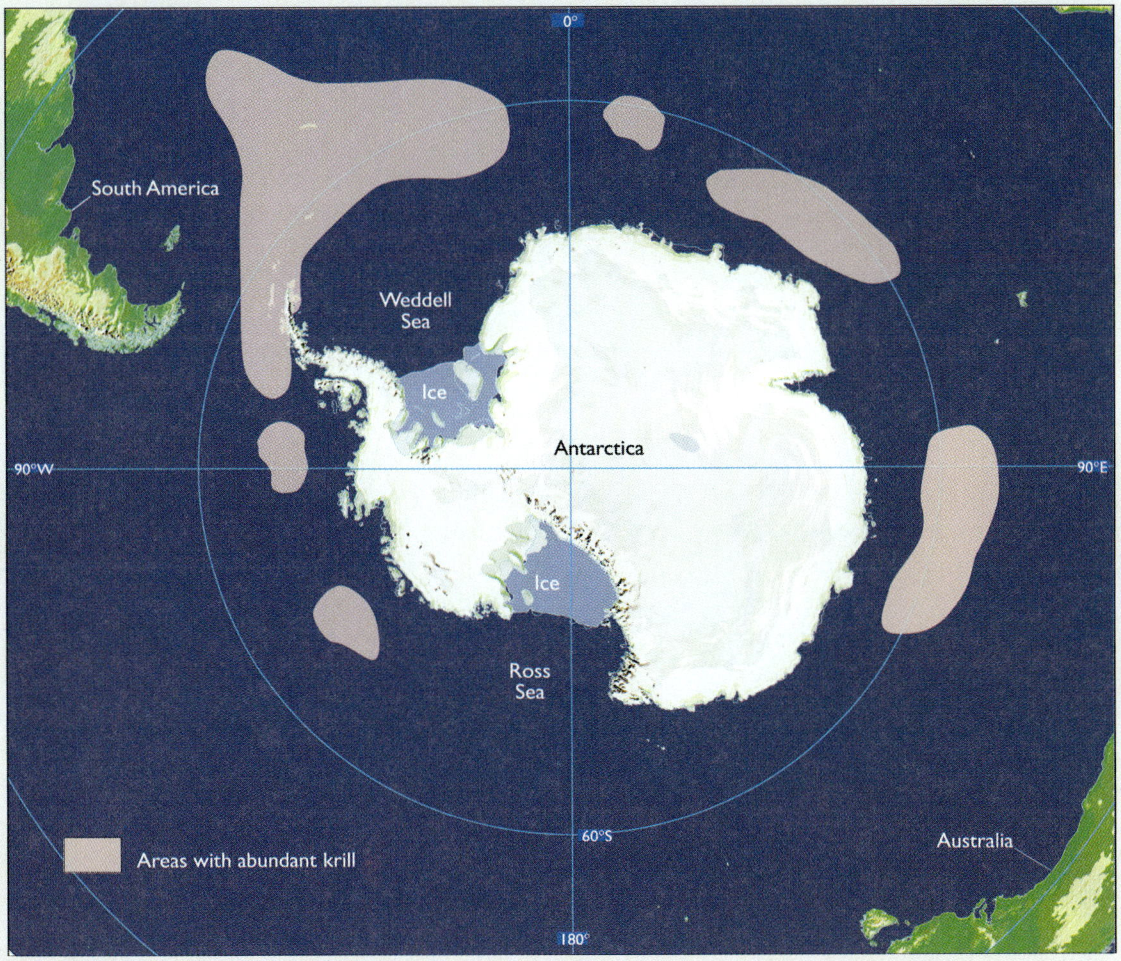

would be harmed. For example, the best estimates put the annual krill consumption by baleen whales at 33 million tons, by seals at 100 million tons, and by seven species of penguins and nineteen species of petrels, albatrosses, and other birds at 39 million tons.

Recently, marine scientists have reported a continuous decline in the population of krill that inhabit the waters adjacent to the Antarctic Peninsula as air temperatures in the area rise and the development of winter ice decreases . According to Roger Hewitt, a marine biologist with NOAA's Antarctic Marine Living Resources program, the Antarctic ecosystem is extremely sensitive to climate change; depressed krill populations and disruption of food chains may be the result of a gradually warming climate. This means that wise, sustainable management of Antarctic krill must factor in, not only predation pressure, but also climatic and other, as yet unrecognized, environmental perturbations.

Visit www.jbpub.com/oceanlink for more information.

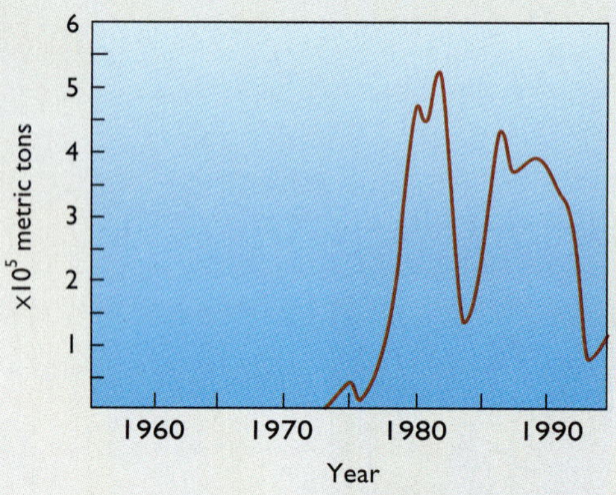

FIGURE B14-6

Krill harvesting. Total harvest of the Antarctic krill, *Euphasia Superba.*

bivalves typically are placed in trays (**Figure 14–8**) or directly on the bottom of bays, inlets, and estuaries, or suspended above the seabed, where they feed by filtering phytoplankton out of the water column. They require a regular supply of seawater having ecologically appropriate temperature and salinity characteristics, dissolved oxygen, and few pollutants. Once mature, they are easily harvested and shipped to markets. In the United States, shrimp and lobster have dominated maricultural production, the farms located principally in Texas, South Carolina, and Hawaii. Shrimp are particularly expensive organisms to farm, because they are raised in brackish-water ponds and must be fed artificially in order to promote their rapid growth.

Mariculture is a developing or thriving industry in much of the world where traditional fisheries have collapsed because of rampart overexploitation of natural fish stocks. For example, in 1993, India, Bangladesh, and Sri Lanka harvested 1.3 million, 230,000, and 100,000 metric tons, respectively, of shrimp and fish raised in brackish-water ponds. The drawback to future maricultural production in these regions is that development is done at the expense of mangrove swamps and coastal wetlands, habitats that are crucial for natural stocks of fish and shrimp. A similar destruction of mangroves for shrimp mariculture has occurred in Panama, Honduras, Ecuador, Indonesia, Thailand, and Malaysia, among others.

A key question pertains to the total amount of food that mariculturists can provide. Currently, fish production by mariculture amounts to roughly 20 percent of the total global harvest provided by the conventional fisheries. Marine biologists who have studied fish farming in coastal oceans do not believe that mariculture will generate a substantial amount of food to help feed the growing human population of the world, many of whom are malnourished. They suggest that the high cost for farming fish produces an expensive product that only the affluent will be able to use as part of their diet. One fishery that appears to have real potential to help feed people is that of the Antarctic krill, as discussed in the biology box, "Antarctic Krill."

FIGURE **14-8**

The practice of mariculture. Oysters being grown in coastal waters.

STUDY GUIDE

KEY CONCEPTS

1. The *Law of the Sea treaty* allows nations to claim jurisdiction over their territorial seas, defined as the contiguous sea bed and its waters that extend offshore for 12 nautical miles (~22.5 kilometers). Also, nations can regulate fishing, mineral exploitation, and scientific research in their *Exclusive Economic Zones,* which typically extend seaward for 200 nautical miles (~370 kilometers).

2. *Oil* and *natural gas* are valuable hydrocarbon resources that are associated with some sedimentary rocks that underlie continental margins. If the proper source and reservoir rocks are present, oil and natural gas can be trapped and concentrated by geologic structures that make them economically exploitable.

3. *Gas hydrates* are unusual hydrocarbon deposits that consist of methane molecules trapped in cagelike structures composed of ice. Although the technology to exploit them has not yet been developed, gas hydrates remain a vast, untapped resource for future use.

4. *Sand and gravel* deposits of the continental shelf are valuable resources for the construction of commercial and residential buildings, beach replenishment, and the building and improvement of roads and highways for coastal communities.

5. Dense concentrations of *manganese nodules* are common in water depths greater than 2 kilometers (~1.24 miles). They represent a major potential source of some important metals, such as iron, cobalt, nickel, copper, zinc, and lead.

6. Many seamounts and some islands are composed of rocks that are enriched in *cobalt*, an element that is quite limited as a mineral deposit on land and is vital for producing jet engines.

7. *Phosphate* is crucial for the chemical and agricultural industries. Major marine deposits of phosphate rock occur off Morocco, southern California, and the Carolinas.

8. The world's fisheries are located in the biologically productive waters of the continental shelves and of coastal upwelling regions. The annual *global fish harvest* has apparently leveled off at about 90 million tons; some biologists believe that this tonnage represents a harvestable limit that will not be exceeded appreciably despite greater fishing effort and the use of more efficient gear.

9. Efforts to raise shellfish (oysters, mussels) and finfish (salmon) by *mariculture* have proven successful. Fish farming is a thriving business that will become increasingly more important as the catch of the world's offshore commercial fisheries declines.

KEY WORDS

cobalt-rich crust
Exclusive Economic Zone
gas hydrates
keystone species
krill
Law of the Sea
manganese nodules
mariculture
petroleum
phosphate deposits

QUESTIONS

REVIEW OF BASIC CONCEPTS

1. What is the difference between a territorial sea and an Exclusive Economic Zone?

2. How are oil and natural gas formed?

3. What is the difference between a gas hydrate deposit and a natural gas deposit?

4. What problems are associated with mining sand and gravel from the sea bed?

5. What economically important metals are contained in manganese nodules?

6. What is the difference between pelagic fish and groundfish?

7. What are drift nets and why do regulatory agencies wish to have them banned outright?

8. What is mariculture and why will fish protein produced in this way not help feed the world's poor?

CRITICAL THINKING ESSAYS

1. Who should benefit from the exploitation of resources that lie seaward of Exclusive Economic Zones? Argue compellingly for your choice.

2. Argue for and against the proposal that landlocked countries should benefit economically from the exploitation of marine resources?

3. Assume that there is a long-standing groundfish fishery on a continental shelf that borders a coast where responsible and necessary house construction is underway. Although the annual groundfish harvest has fluctuated markedly for the past decade, the catch on the average has declined systematically despite more intense fishing and a larger fishing fleet. The sand of the shelf bottom in the fishery area is desperately needed for the

development plans of the local communities. What should or can be done? Develop arguments for your resolution of this conflict.

4. Using your knowledge of biological oceanography, speculate on the possibility that the extensive harvesting of krill in the Southern Ocean could potentially damage Antarctica's marine ecosystem.

5. Assume that you have decided to invest a substantial amount of money in a mariculture project. What sort of organism would you consider and what sort would you not? Specify the reasons for each case.

6. Minke whales have recovered well during the global ban on whaling. Should they now be hunted responsibly and sustainably? Why or why not?

SELECTED READINGS

Berrill, M. 1997. *The Plundered Seas: Can the World's Fish Be Saved?* San Francisco: Sierra Club Books.

Broadus, J. M. 1987. Seabed materials. *Science* 235 (4791): 853–60.

Champ, M. A., Dillon, W. P., and Howell, D.G. 1984–85. Non-living EEZ resources: Mineral, oil, and gas. *Oceanus* 27 (4): 28–34.

Clark, W.C. 1989. Managing planet Earth. *Scientific American* 261 (3): 47–54.

Cronan, D.S. 1992. *Marine Minerals in Exclusive Economic Zones*. London: Chapman and Hall.

Kurlansky, M. 1997. *Cod: A Biography of the Fish that Changed the World*. New York: Walker..

Heath, G.R. 1982. Manganese nodules: Unanswered questions. *Oceanus* 25 (3): 37–41.

Oceanus. 1993. Coastal science and policy I. Special issue 36 (1).

McGinn, A.P. 1998. Blue revolution: The promises and pitfalls of fish farming. *World Watch* 11 (2): 10–19.

McGinn, A.P. 1998. Promoting sustainable fisheries. *State of the World 1998*. New York: Norton.

Sanger, C, 1986. *Ordering the Oceans: The Making of the Law of the Sea*. London: Zed Books.

Thorne-Miller, B. 1999. *The Living Ocean: Understanding and Protecting Marine Biodiversity*. Washington, D. C.: Island Press.

Van Dyke, J.N., Zaelke, D., and Hewison, G., eds. 1993. *Freedom for the Seas in the 21st Century: Ocean Governance and Environmental Harmony*. Washington, D. C.: Island Press.

TOOLS FOR LEARNING

Tools for Learning is an on-line review area located at this book's web site OceanLink (**www.jbpub.com/oceanlink**). The review area provides a variety of activities designed to help you study for your class. You will find chapter outlines, review questions, hints for some of the book's math questions (identified by the math icon), web research tips for selected Critical Thinking Essay questions, key term reviews, and figure labeling exercises.

15 The Human Presence in the Ocean

Roll on, thou deep and dark blue ocean—roll!
Ten thousand fleets sweep over thee in vain;
Man marks the earth with ruin—his control
Stops with the shore.
—*Lord Byron,* Childe Harold's Pilgrimage, *1812*

web navigator

critical thinking on the web

math tutor on the web

www.jbpub.com/oceanlink

PREVIEW

THE OCEANS ARE IMMENSE bodies of water that support a large variety of ecosystems. Quite clearly, biological communities as well as the physical and chemical aspects of these marine systems have changed continually and naturally over time. Humans have used the ocean and its resources in many creative and useful ways. Only recently have scientists come to realize that not all human uses of the ocean are benign and that particular practices are damaging the less resilient parts of some ecosystems. Oceanographers have become aware that some marine resources such as fish are being exploited at a rate that is not sustainable over the long term. These problems are complex and many of them will not be solved by simple measures. However, the first step in dealing with these grave issues is to familiarize oneself with the nature and extent of the effect of humans on marine ecosystems. That is the purpose of this chapter overview.

What follows in this last chapter is a broad overview of the variety of pollutants that are stressing and disrupting marine ecosystems. Although it is depressing and distressing, we must face it squarely and responsibly. So we will discuss these pollutants, how they are being dispersed throughout the ocean's water, and some of their effects on the ecological well-being of these precious environments. For those interested in thinking more broadly about the environmental "problem," we provide at the end of this chapter a list of books that address environmental issues from a philosophical and moral perspective.

15-1

Pollution: What Is It?

We read, hear, or talk about pollution often, but when we actually witness its damaging effects in a favorite beach or cove, the issue becomes much more compelling. The observed effects of pollution sicken us. Also, many pollutants are not visible. We sense that pollution is bad; it is unpleasant; it must be minimized; but most of us would have difficulty in defining the term specifically. What exactly is pollution?

A useful definition of pollution was adopted by the United Nations Joint Group of Experts on the Scientific Aspects of Marine Pollution in 1982:

> *[Pollution is the] introduction by man, directly or indirectly, of substances or energy into the marine environment (including estuaries) resulting in deleterious effects such as harm to living resources, hazards to human health, hindrance of marine activities, including fishing, impairing quality for use of sea-water and reduction of amenities.*

Note that the definition assigns pollution specifically to the activity of humans.

Table 15–1 summarizes some common chemical compounds that occur in the ocean as pollutants but may also be present in the environment because of natural causes. An example is the natural seepage of oil from the sea bed. Sometimes, because of the lack of adequate baseline data, it is very difficult to distinguish between human-induced pollution and the natural occurrence of such substances. We tend to study an area after it shows clear signs of pollution, and how can we realistically assess the effect of pollution if we do not know the characteristics of that environment before it was polluted? This is a particularly serious problem with those chemical substances listed in Table 15–1 that have natural as well as human sources.

Pollutants tend to be concentrated in three parts of the ocean. First, they accumulate on the sea bottom by being chemically attached to particles such as silt and clay or by settling directly as solids onto the sea floor. Burrowing benthic organisms then mix these materials into the sediment. Second, pollutants can be concentrated along pycnoclines that separate water masses of different densities. This commonly occurs in estuaries where a wedge of saltwater is in contact with fresh or brackish water (see Figure 12–3 in Chapter 12). Third, dissolved and solid wastes can collect at the interface between air and sea, a very thin—0.1 to 10 millimeter (0.0039 to 0. 39 inch)—surface microlayer known as the **neuston layer** (see the boxed feature, "The Sea-Surface Microlayer," in Chapter 5). Here, chemicals and solids, including pollutants, tend to aggregate and can affect plankton of all kinds, including the embryonic forms of invertebrates and fishes that use the surface water as a temporary habitat. Once pollutants are in the environment, they are broken down or degraded by a variety of oceanographic and biological processes. These are discussed where appropriate in the following sections.

TABLE 15-1 Some pollutants in the ocean.

Pollutant	Natural Sources	Human Sources
Hydrocarbons	Seeps, rivers, volcanoes, atmosphere, bacteria	Transportation, production, aerosols
Heavy metals	Volcanoes, rivers, sediments, weathering of rocks	Industrial and municipal effluents
Nutrients	Rivers, upwelling, atmosphere, bacterial decomposition	Municipal effluents, agricultural fertilizers
Synthetic chemicals	None	Manufacturing, transportation agricultural fertilizers and pesticides

Source: Adapted from R. A. Geyer, ed., *Marine Environmental Pollution, 1 Hydrocarbons* (New York, Elsevier Scientific, 1980).

15-2 Hydrocarbons in the Sea

Geologists refer to **oil** found in sedimentary deposits as *crude oil* or *petroleum*. Petroleum is a complex mixture of hydrocarbons: molecules of hydrogen and carbon with minor amounts of nitrogen and metals that are bound together in various proportions and in different molecular structures. The exact composition of petroleum depends on its geologic history, and varies considerably from geologic location to location. The chemistry of crude oil is likewise very complex. Whether the oil is "light" or dense depends largely on the size of the hydrocarbon molecules. Because of the tremendous range in the length and shape of hydrocarbon molecules, the various components have different boiling temperatures. This property enables refineries to separate crude oil into "cuts" by systematically drawing off various hydrocarbon components as the crude is heated. This process, called **distillation**, yields kerosene, diesel fuel, gasoline, and many other types of useful petroleum products.

Only a surprisingly small fraction (~3.3 percent) of the oil spilled at sea is the result of tanker accidents (**Table 15–2**), the topic of so much media coverage (see, for example, the boxed feature, "The *Torrey Canyon* Disaster). Almost a third of the oil found in the ocean (~31.1 percent) is supplied by rivers that carry untreated domestic and industrial hydrocarbon wastes to the sea. The second most significant source of oil contaminants (~21.8 percent; see table 15–2) are tankers that transport petroleum from where it is pumped out of the ground to where it is refined and consumed (**Figure 15–1**). For example, **Figure 15–2** shows the principal shipping lanes of petroleum across the world's oceans. Spillage of crude along these sea routes is a function of "standard operational discharge" related to pumping out the bilges and to the ballasting of ships, as well as to incompletely burned fuels from the large engines that power oil tankers. Coastal refineries inadvertently "leak" a lot of oil to the sea (~13.1 percent; see Table 15–2) as they handle large volumes of crude oil, refining it into gasoline, kerosene, heating oil, and diesel fuel for

TABLE 15-2 Relative amounts of petroleum in the oceans.

River runoff	31.1%
Tanker operations	21.8%
Coastal facilities (e.g., refineries)	13.1%
Atmospheric fallout	9.8%
Natural seepage	9.8%
Other transportation activities	9.8%
Tanker accidents	3.3%
Offshore petroleum production	1.3%

Source: Adapted from R. A. Geyer, ed. *Marine Environmental Pollution, 1 Hydrocarbons* (New York, Elsevier Scientific, 1980).

(a) TANKER DELIVERING OIL TO A REFINERY

(b) OIL REFINERY

FIGURE 15-1

Petroleum resources. Once pumped out of the ground, crude oil must be transported to the refineries where it is separated into different products.

The *Torrey Canyon* Disaster

Early on 18 March 1967, a Saturday morning, a series of piloting errors by officers on duty were about to set a 297-meter-long (~970 feet) oil tanker registered in Liberia on a rock reef. The tanker was carrying over 100,000 tons of crude oil and was steaming at full speed (15.75 knots). At 08:50 that morning, the *Torrey Canyon* struck Seven Stones Reef, a natural rock ledge located about 7 nautical miles northeast of the Scilly Isles that lie along the Cornish coast of Great Britain. The oil tanker was impaled on the rocks (**Figure B15–1**). Several of its large cargo tanks were ruptured, immediately spilling large quantities of crude oil into the sea. Eventually, all of its cargo of crude was spilled into the entrance to the English Channel. During the following weeks, wind shifts dispersed the immense oil slick, moving it first to the beaches of Cornwall, then later across the English Channel to the shores of France (**Figure B15–2**).

FIGURE **B15-1**

The *Torrey Canyon.* This photograph shows the tanker *Torrey Canyon* impaled on the rocks of Seven Stones Reef, off the coast of Cornwall, England.

Once officials were notified of the wreck, a decision had to be made about how best to mitigate the ecological damage. It was obvious that two major economic enterprises of Cornwall—the fishing and tourist industries—would be severely damaged by the spilled oil. The choice for responding to the crisis seemed straightforward—the application of chemicals to disperse the oil slick. Unfortunately, treatment of oil spills with chemical dispersants is lethal to most marine biota, including fish. However, because revenues earned from tourism were an order of magnitude greater than those from the fisheries, officials decided to protect and clean up the Cornish beaches by the widespread, indiscriminate application of oil dispersants.

There were two surprising ecological outcomes to this decision. First, the plankton at the time of the disaster contained large numbers of pilchard eggs (pilchards are a small, commercially important fish related to the herring). There seemed no doubt to fishery biologists and to fishermen that the massive killing of these eggs by the chemical dispersants would have disastrous effects on future catches. Surprisingly, this fear proved unfounded; there was no noticeable reduction in the population of pilchards in subsequent years. So what was expected to be a major impact of the chemical treatment of the oil spill proved not to be the case.

Second, the extensive and long-term ecological damage to the habitats along the coast of Cornwall was caused not by oil itself as much as by the extreme toxicity of the chemical dispersants that were applied to the beaches to cleanse them of their gooey coating of crude oil. The dispersants caused the mass mortality of the limpet *Patella,* a dominant herbivore of the area. As a consequence, dense growths of seaweeds covered the area in subsequent years, preventing barnacles and limpets from settling and transforming the rocky shoreline into one dominated by seaweeds rather than by barnacles and limpets. In fact, it took four to five years for the limpets and barnacles to reestablish themselves as the dominant organisms of the strandline of Cornwall. Despite this apparent recovery, biologists maintain that the ecosystem still had far fewer species of invertebrates even ten years after the wreck of the *Torrey Canyon*. Because of this experience, chemical dispersants are now applied selectively and sparingly to habitats affected by oil spills, lest the "solution" to the problem worsen it.

Visit www.jbpub.com/oceanlink for more information.

FIGURE **B15-2**

Spread of the oil slick. After the *Torrey Canyon* was grounded, enormous oil slicks spread away from the wreck and drifted to the shores of Cornwall and of France, driven by winds that changed direction with time. [Adapted from R. B. Clark, *Marine Pollution* (Oxford: Claredon Press. 1989).]

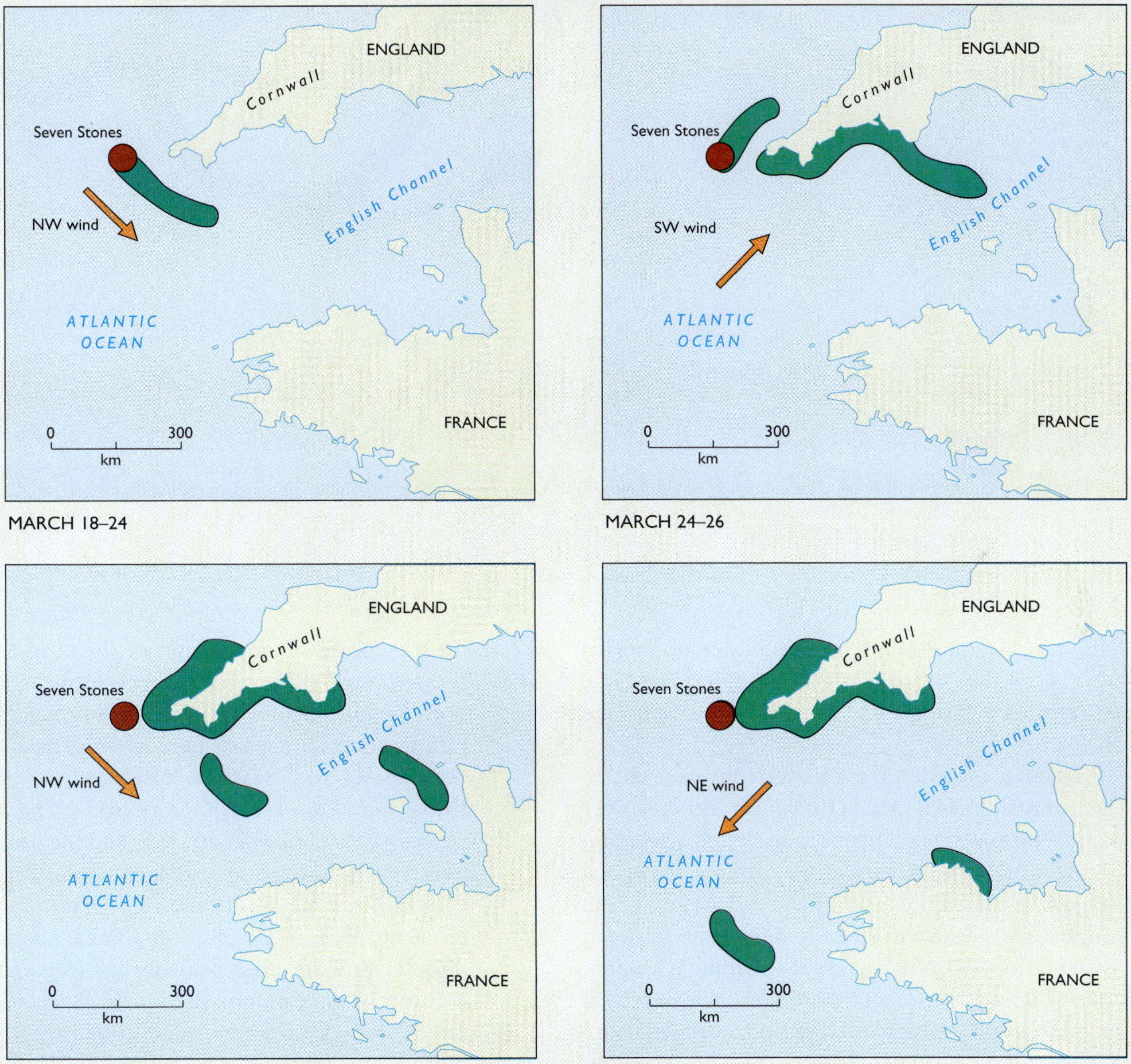

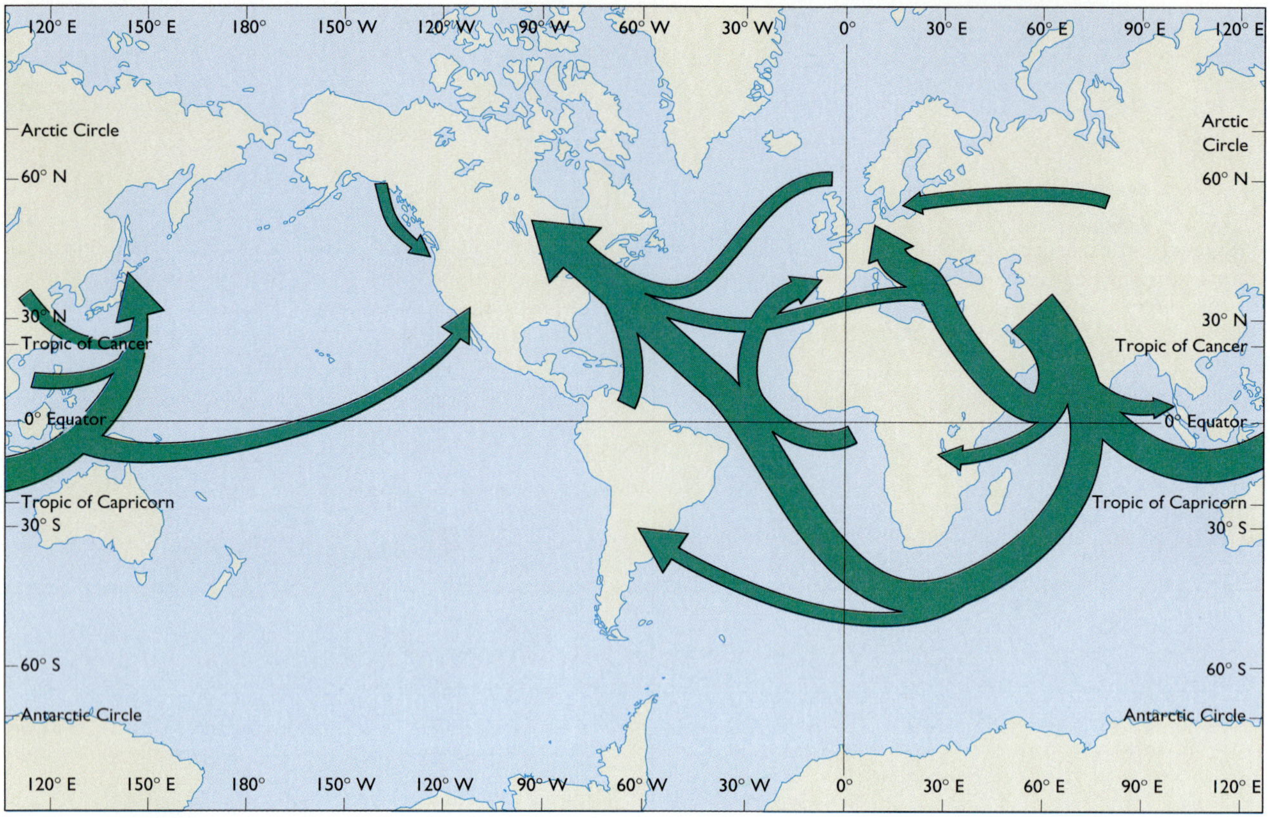

FIGURE **15-2**

Transport routes of crude oil. The thickness of the arrows indicates the relative volume of petroleum transported across the oceans. Note how dependent are the industrialized nations on Mid-East oil resources. [Adapted from R. B. Clark, *Marine Pollution* (Oxford: Claredon Press, 1989).]

the various markets. The remainder of the oil fluxes to the sea result from natural seepage, atmospheric fallout, and general ship traffic (see Table 15–2).

When oil is spilled at sea, various physical, chemical, and biological factors transform the slick over time. Natural spreading of the slick and *advection* (horizontal transport) caused by currents and winds (**Figure 15–3**) rapidly increase the surface area of the oil. In turn, various hydrocarbon fractions undergo evaporation, dissolution, **emulsification** (a suspension of liquid in another liquid, such as water in oil or oil in water), vertical mixing, and sedimentation. Concurrently, the degradation of the oil to carbon dioxide by microbes and its ingestion and subsequent metabolism by plankton and larger organisms biologically alter the composition of the oil and disperse the spill even more.

The actual rate and extent of dissipation and dispersal of the spill by natural processes depend on the specific composition of the crude oil, the weather, and the strength of the surface currents. For example, light oil, such as diesel fuel, spreads rapidly across the sea surface, whereas heavy Number 6 crude oil is viscous, barely staying afloat in the water. **Figure 15–4** summarizes the various processes that affect an oil slick, and the time scales over which they operate to degrade the oil spilled in the sea. In general, (1) the light fractions of the crude evaporate, (2) the water-soluble components dissolve and are mixed downward into the water column by mixing processes, and (3) the heavier, insoluble hydrocarbon residue emulsifies into globules. Many of these emulsified globules change with time into sticky "chocolate mousse," an apt description of their thick, puddinglike consistency, and then into massive tar balls. These tar lumps float along the sea and become denser with time, as they accumulate debris and lose lighter hydrocarbon molecules. Eventually they sink to the sea bot-

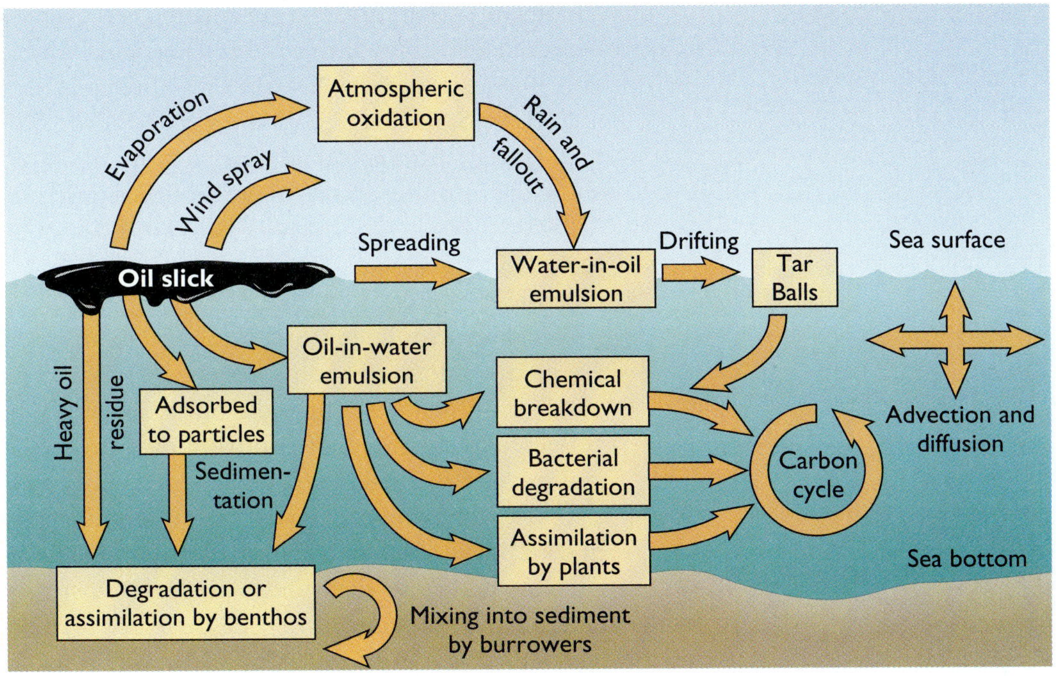

FIGURE **15-3**

The fate of an oil slick. An oil slick in the sea is affected by complex interactions among physical, chemical, and biological processes that cause weathering of the slick with time. [Adapted from R. Burwood and G. C. Spears, *Estuarine Coastal Marine Science* 2 (1974): 117–35.]

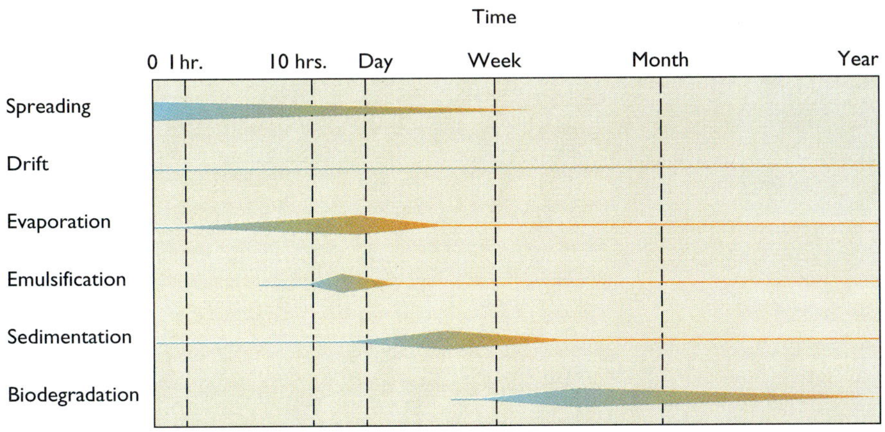

FIGURE **15-4**

The weathering of an oil slick. Different processes dominate at different times during the various stages in the natural weathering of an oil slick. [Adapted from R. B. Clark, *Marine Pollution* (Oxford: Clarendon Press, 1989).]

tom or wash up on the shore (**Figure 15–5**), where they slowly weather and decompose or get buried.

Oil compounds are toxic to marine organisms at all levels of the food web. The overall environmental impact of a specific oil spill depends on the habitat itself (**Table 15–3**). Certain marine ecosystems are much more susceptible to damage by oil than are others. Recovery following contamination also varies greatly with the specific nature of the habitat. For example, a large oil spill in a coastal ecosystem will typically kill the area's biota, both plants and animals, on a massive scale, either by poisoning or smothering them. Here, the biota is rich and the water is shallow, so that the benthos, as well as the nekton, plankton, and seabirds, are affected.

An oil slick washing up on a muddy intertidal flat has deadly effects on the invertebrate population, including snails, clams, oysters, crabs, and worms. During the fall of 1969, an oil barge spilled over 600 tons of diesel fuel into a corner of Buzzards Bay, Massachusetts, near Woods Hole Oceanographic Institute. The oil devastated the biota of the area. The average density of the benthic animals dropped from 200,000 organisms per square meter to 2 organisms per square meter, a reduction by five

FIGURE 15-5

Tar balls in the sea. The thick residue of crude oil in water will form into tar balls. Often these sink to the sea bed, and onshore currents transport them onto the shore.

orders of magnitude! The salt-marsh grass likewise was destroyed by the diesel fuel. Digging revealed that the oil had penetrated the substrate down to a depth of 1.5 meters (~5 feet).

Three years after the spill, recovery from the ecological catastrophe was well underway, even though clams were still tainted by hydrocarbon contamination. Noticeable behavioral abnormalities of marsh crabs that lived in the affected zone, such as a slow response time to threats and the construction of odd-shaped burrows, persisted for more than seven years after the spill. At present, however, the marsh has recovered and there appears to be no permanent damage to this local environment, although little is known about the chronic exposure of organisms to low concentrations of hydrocarbons.

Spills at sea seem to be less damaging than spills near the shore are because oil slicks in the open ocean spread out horizontally and vertically under the influence of currents, tides, wind, and waves. The slick disperses with time and becomes diluted and, hence, less toxic. Also, the sea bottom is deep and may not be affected by the surface slick. This means that, after an oil spill or blowout at an offshore drilling site, efforts to prevent the slick from drifting onto the shore must be started immediately after the accident, particularly if the wind is blowing onshore.

Various techniques are used to contain oil spills. Frequently, floating booms (**Figure 15–6a**) are set out to contain and channel the oil so that it can be collected by devices that pump out or skim the oil and transfer the residue to ships or barges. In practice, floating booms are only marginally effective in containing slick, particularly if the spill is large, the seas heavy, or the winds and currents strong. Chemical dispersants have been used, but many of them are more toxic than the oil itself and take a great deal of time to apply to large oil slicks. Furthermore, dispersants do not really remove the oil from the water. Rather, they cause the oil to sink below the surface out of sight. Burning the oil off has also been attempted (**Figure 15–6b**), but oil is difficult to ignite and, once lit, difficult to keep burning. Also, it is impractical if the slick breaks up and disperses. The best general strategy is to skim up the crude oil as fast as possible. If the oil slick washes up on the shore, the sticky mess needs to be cleaned up *without disturbing the sediment.* Some biologists are investigating the feasibility of **bioremediation**—stimulating the growth of microorganisms that actively feed on petroleum compounds (see the box, "Bioremediation).

Obviously, the best safeguard against local environmental degradation by oil pollution is the prevention of tanker collisions and accidents, a daunting task given the volume of oil that is transported annually across the oceans (see Figure 15–2), the impossibility of eliminating human error in ship handling, and the occurrence of severe weather at sea. This fact is driven home by the simple but powerful statistic that in 1981 more than 800 oil spills, most of them classified as minor, occurred in Chesapeake Bay. Try to imagine what the world's annual total of spilled oil is. The boxed feature, "The *Exxon Valdez* Oil Spill," describes a recent ecological disaster of major proportion created, in part, by the cleanup efforts of well-intentioned agencies.

TABLE **15-3** The environmental impact of oil spillage.

Ecosystem	Initial impact	Recovery
Wetlands, salt marshes, and mangroves	Heavy: Widespread mortality of plants and animals leading to decreases in population densities and changes in species abundances and diversity. Major impact on biological productivity.	Moderate to slow: Persistence of oil prolongs toxicity. Biological succession occurs at a moderate rate once oil is removed.
Estuaries, bays, and harbors	Moderate to heavy: Depends on the season (spawning, migration) and oil's persistence. Depresses populations and alters the species composition and abundance.	Fast to slow: Dependent on current flow, shoreline characteristics, and community stability.
Outer continental shelf	Light to moderate: Impact on plankton light and on fish larvae severe. Moderate impact on the benthos, if oil reaches the sea bottom.	Fast to moderate: Fast recovery for plankton because of rapid regeneration times. Moderate recovery rate of benthos, if affected.
Open ocean	Light: Many organisms avoid spill. Impact on plankton is local and depends on chance encounter with spill. Usually water is too deep for significant impact on benthos.	Fast: Rapid dispersion and degradation of oil.

Source: J. Hyland, *Bioscience* 26 (1976): 463–506.

FIGURE **15-6**

Containment of an oil slick. (a) Large, floating booms are used to contain oil slicks, the oil then being transferred to ships or barges. Booms are of very limited use when winds are strong or seas are rough. (b) Burning an oil slick is difficult and, in effect, transfers the pollution to the atmosphere. In this photo, the oil was intentionally ignited to minimize leakage into the water.

(a) BOOMS DEPLOYED TO CONTAIN AN OIL SLICK

(b) INTENTIONAL BURNING OF OIL REMAINING IN THE *NEW CARRISSA'S* HOLD AFTER IT RAN AGROUND ON THE OREGON COAST

THE OCEAN SCIENCES BIOLOGY

Bioremediation

Oil spills in the sea are likely to remain a problem for some time. Our energy-intensive industries require that petroleum be shipped from where it is produced to where it is needed to power machines (see Figure 15-2). Given that many tanker accidents are the result of human error, it is inevitable that spills, including very large ones, will occur in the future. It is imperative that agencies, both private and public, use all the available technology to contain the slick and to remove the oil from the contaminated area. One such innovative technology is bioremediation, the use of micro-organisms to degrade the contaminant naturally, whether it be oil or some other chemical toxin.

There are two basic approaches to bioremediation. First, the contaminated water can be fertilized with nutrients such as nitrate and/or phosphate. Nutrient enrichment then allows microorganisms already present in the environment to multiply, including those species that degrade the oil by natural means.

The second technique involves "seeding" the contaminated water with microorganisms that occur naturally with or without a supply of nutrients. This seeding technology may be revolutionized in the not-too-distant future by introducing bioengineered microorganisms into the tainted water. These organisms would not be natural, but genetically engineered to degrade specific types of crudes. However, for the moment there has been no attempt to release these engineered microorganisms into the environment, nor has there been serious consideration of this form of bioremediation. The long-term ecological risks of introducing alien, genetically engineered species into a natural habitat renders this technique too perilous for the short-term containment and cleanup of an oil spill. At present, most scientists favor nutrient enrichment of the natural population of microorganisms as the best and least problematical bioremediation technique.

Although bioremediation as a means of cleaning up an oil spill shows promise, much more research, both theoretical and applied, needs to be completed. Crude oil is a complex mixture of hydrocarbons with vastly different chemical properties. To expect a few species of microorganisms to degrade all these varieties of chemical compounds in a spill is unrealistic. The Alaska Bioremediation Experiment, an ambitious series of field experiments conducted by scientists from the Environmental Protection Agency, Exxon, and Alaska's Department of Environmental Conservation, has demonstrated the potential of nutrient enrichment as a bioremediation technique to clean certain habitats such as beaches that become covered with spilled oil. Several weeks after fertilizing beaches of Prince William Sound in Alaska that were contaminated with petroleum, there was a dramatic reduction in the amount of oil fouling the sand and rocks. Researchers are quick to point out, however, that the water of Prince William Sound contains large native populations of bacterial species capable of degrading hydrocarbons and that the porous beach sand and strong tidal currents aided tremendously in the biodegradation of the crude oil. It is not clear what the efficacy of bioremediation would be in areas where conditions are less favorable.

Visit www.jbpub.com/oceanlink for more information.

15-3

Municipal and Industrial Effluent

The variety of substances discarded by humans in the ocean is large and includes both natural (sewage, for example) and artificial (plastics, for example) materials. It is estimated that humans are disposing of 20 billion tons of waste each year and much of it ends up in the ocean. This is 20,000,000,000 tons of waste each year! Many of these effluents, such as heavy metals, are very toxic and even lethal at low concentrations. Others, such as raw sewage, promote the growth of **pathogens** (any disease-producing organism), including viruses, bacteria, and parasites that can infest coastal waters and cause diseases in humans that come in contact with them. A convenient way to discuss the general chemical and biological effects of effluents is to group them broadly into several types—sewage, metals, and artificial biocides.

Before considering the environmental effect of each of those effluents, it is worthwhile to review the general pathways of domestic and industrial effluents to the coastal ocean. Because of the high population densities along rivers and their convenience for disposing of wastes, a large proportion of contaminants found in harbors, estuaries, bays, wetlands, and the nearshore zone is introduced by river discharge. Most rivers are polluted with solid and dissolved material derived from the wastes of farmland, towns, cities, and industrial complexes and they in turn pollute the coastal habitats that receive their tainted water. Once in coastal waters, the river-supplied waste is diluted by mixing processes, unless the waters are still and stagnant. Dilution by a factor of a thousand or more in the first few hours after discharge into coastal waters is not unusual, provided that nearshore currents, tides, and waves are vigorous. Unfortunately, the recent increase in the volume of untreated municipal and industrial wastes created by high population densities up river is exceeding the "self-cleansing" capacity of these environments.

All of this is rather oversimplified, because solid wastes are quite heterogeneous in composition, consisting of particles that range widely in density and chemical properties. The result is uneven dilution, with lighter, smaller particles tending to float to the surface, and denser, larger particles tending to settle on the sea bottom.

A common means of disposing of effluents is to discharge them from underwater pipes (**Figure 15–7a**). Countless municipal and industrial pipelines along the shores of the United States unload effluents directly into estuaries where tidal flushing is expected to dilute and to dispose of the unwanted substances (**Figure 15–7b**). Depending on the temperature and composition of the effluent, the size and shape of the plume extending upward from the pipe opening varies considerably. Typically, the effluent will stream upward and widen as turbulent motion in the plume entrains seawater and dilutes the discharged solution. Particulate material will settle out as currents wane and become separated into fractions depending on the size, density, and shape of the solids. Part of the effluent may be dispersed horizontally along the pycnocline by

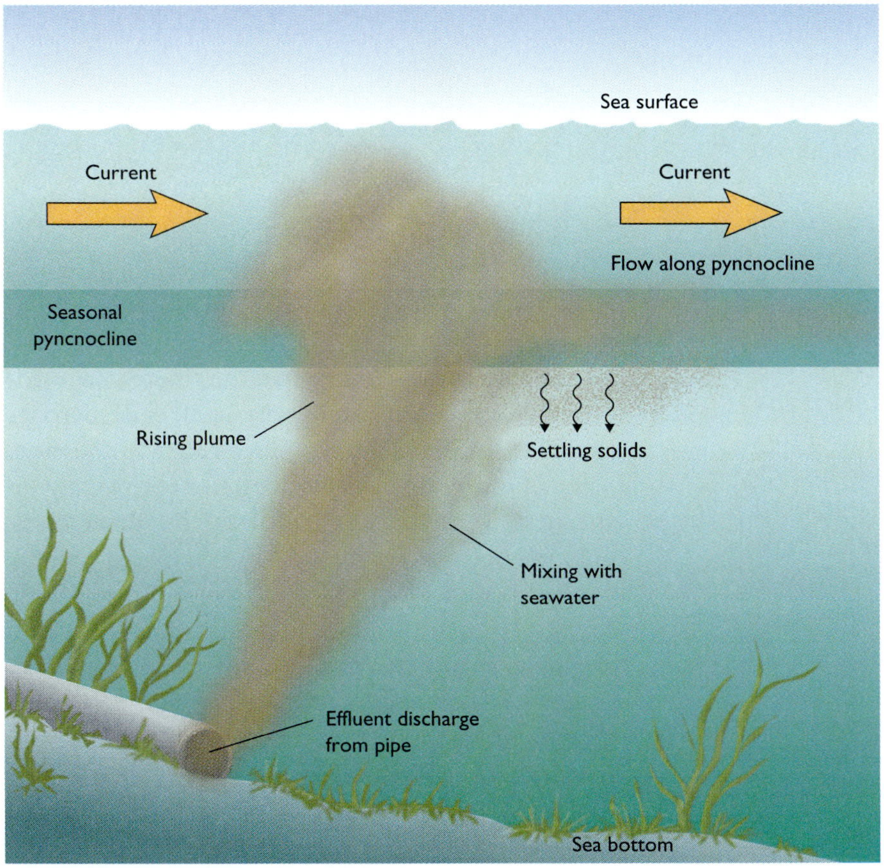

(a) AN EFFLUENT PLUME

FIGURE **15-7**

The fate of effluents. (a) Effluents discharged from a pipe tend to rise as a plume and spread out along pycnoclines. Solids tend to settle out in the downcurrent direction, which changes with the stage of the tide. [Adapted from U.S. Congress, Office of Technology Assessment, *Wastes in Marine Environments* OTA-0-334 (Washington, D.C.: U.S. Government Printing Office, 1987).] (b) Along shorelines there are many effluent pipes that disgorge wastes into the ocean. Many are exposed at low tide.

(b) AN EFFLUENT PIPE EXPOSED AT LOW TIDE

THE OCEAN SCIENCES BIOLOGY

The *Exxon Valdez* Oil Spill

A nightmare occurred on 24 March 1989. The supertanker *Exxon Valdez* went hard aground on submerged rocks in Alaska's pristine Prince William Sound. Over 35,000 tons (more than 10 million gallons) of Alaska crude poured out of the tanker's damaged hull and was dispersed by heavy seas and strong tides. Somewhere between 13 and 17 percent of the crude was contained by booms and recovered by the hard work of over ten thousand people. Between 20 to 40 percent of the oil evaporated, 8 to 10 percent was burned, and 5 or so percent was degraded by microbes. The remainder, estimated to be between 25 and 45 percent of the spill, drifted onto beaches or rocky shorelines and sank into sediment. About 400 miles of shoreline were fouled by the drifting crude oil.

The media coverage was intense. Something had to be done immediately to contain the oil and to clean up the shoreline in order to minimize damage to the fragile coastal ecosystems. So a plan was put in place. The guiding principle was "cleanup." The "rescue" teams relied primarily on washing the rock and sediment with high-pressure, hot-water hoses (**Figure B15–3**). The results were a disaster. Most of the small organisms at the base of the food web were literally cooked by the 65°C (105°F) water. The hot water washed the crude off the higher parts of the beaches to the lower parts, contaminating even more of the shoreline and killing plants and animals indiscriminately.

Treated beaches were soon "sparkling clean," but had virtually nothing living on them. They were sterile and antiseptic. Oil-stained beaches that were not treated began to recover naturally, and within two years of the spill the biodiversity of a few beaches was similar to that of the unpolluted stretches of the coastline. Ironically, then, the "cleanup" was more disastrous for the ecosystem than the spill itself, despite the expenditure of over $2 billion dollars! As was the case with the *Torrey Canyon* spill (see the boxed feature by that name), doing less would have been much better than doing even what was done.

Visit www.jbpub.com/oceanlink for more information.

currents (see Figure 15–7a). Some may reach the surface and become incorporated into the neuston microlayer. Also, the plume will swing upstream and downstream with the flood and ebb currents of the tides.

Over time, this outfall of effluent affects successive zones around the pipe. An interior zone near the outfall may consist of sludge, with few, if any, large infauna. This area grades outward into a zone with depleted benthic populations that include dwarfed members. In turn, this zone merges with an area where animal numbers increase dramatically, well above normal population densities. Beyond this is the unaffected sea bottom with normal-sized benthic communities. The overall effects of chemical contaminants on cells, organisms, populations, and communities are described in **Table 15–4**. With these preliminary ideas in mind, let's examine the nature and effect of various effluents in the coastal ocean.

SEWAGE

Municipal **sewage** consists of a messy sludge, a heterogeneous mixture of organic and inorganic chemicals. One of its principal ingredients is human waste, a mixture of organic matter, inorganic nutrients (such as nitrate and phosphate), and microbes (such as viruses and bacteria). A great deal of sewage, some of it untreated, is discharged into the water of the coastal zone. Because of their complicated circulation patterns, many estuaries trap and thereby concentrate discharge that they receive, including sewage effluent. Recall that nitrate and phosphate are critical nutrients for plant photosynthesis. Coastal water bodies, if suffused with massive dosages of nutrients, respond by supporting long-lasting, dense plankton blooms, a biological process that gorges the water with plant cells. The large quantities of algae eventually die and decompose.

FIGURE B15-3

The *Exxon Valdez* oil spill. The use of high-pressure hoses to wash the crude off sand and rocks with very hot water caused more damage to the ecosystem than did the oil itself.

HOT WATER FROM HIGH-PRESSURE HOSES

Decomposition is a process that takes up dissolved oxygen from the water column and from the pore water in the sediment. The *biological oxygen demand* (BOD) that results from excess nutrients can create **eutrophication:** the water body becomes hypoxic (having low levels of dissolved oxygen) and even anoxic (having no oxygen at all). Invertebrates and fish that graze plants cannot live in such a depleted-oxygen state, and the reduced grazing pressure allows even more algae to survive, creating an even greater BOD. Such excessive fertilization of water with nutrients, if persistent, eventually damages the food web, causes the mass killing of organisms, and undermines the chemical and biological balance of the ecosystem.

This is a question of overfertility and it causes deoxygenation in those parts of estuaries that are receiving regular and large supplies of sewage wastes (see the box, "Chesapeake Bay" in Chapter 12). Hypoxia as a result of dumping sewage and other human-created wastes has even affected the

TABLE 15-4

Responses to chemical contaminants.

Level	Responses
Cell	Toxication (poisoning) Metabolic impairment Cellular damage Detoxification (to rid of poison or effects of poison)
Organism	Physiological changes Behavioral changes Susceptibility to disease Reduced reproductive effort Decreased viability of larvae Readjustment of rate functions
Population	Changes in age, size, recruitment, mortality, and biomass Adjustment of reproductive output
Community	Changes in species abundance, distribution, and biomass Altered trophic interactions Ecosystem adaptation

Source: J. E. McDowell, *Oceanus* 36 (2) (1993): 56–61.

open waters of the continental shelf in places such as the New York Bight (**Figure 15–8a**), along the southern coast of Mississippi and Louisiana (**Figure 15–8b**), and even within parts of the Great Lakes themselves. The chemical poisoning of the water in the New York Bight area is so severe and long lasting that many of the bottom fishes in the area are diseased, suffering from "black gill" and "fin rot" (**Figure 15–9**).

The discharge of solid municipal wastes destroyed kelp beds that flourished off southern California (**Figure 15–10a**), eliminating the habitat of many fish and shellfish in the area. Apparently, the discharge of suspended solids and the highly toxic pesticide, DDT, a chlorinated hydrocarbon that will be discussed more fully below, caused the destruction of the kelp forests. There may be some hope, however. The combination of the reforestation of the kelp beds (**Figure 15–10b**) and stricter controls on the treatment of wastes and pollutants initiated in the late 1960s is permitting the kelp beds of southern California to rebound and to reestablish themselves on a modest scale (see Figure 15–10a).

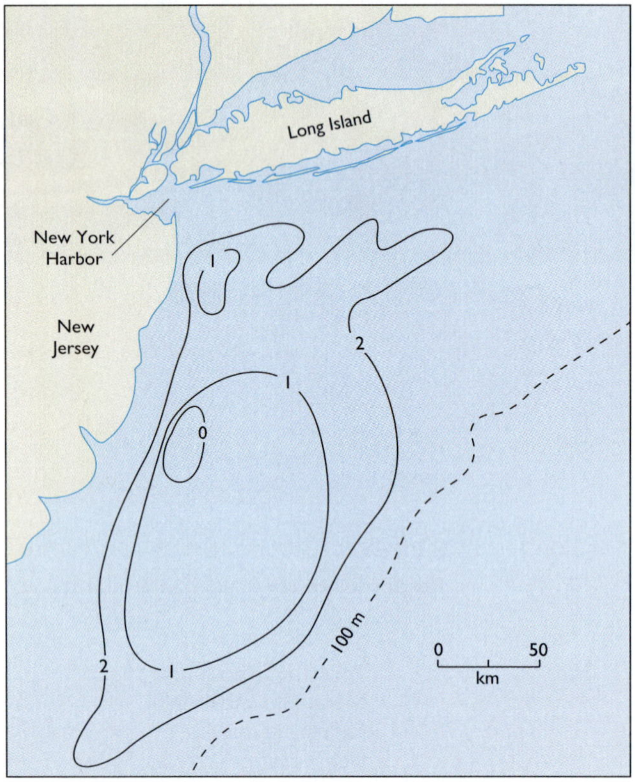

(a) OXYGEN (mg/l) IN BOTTOM WATER

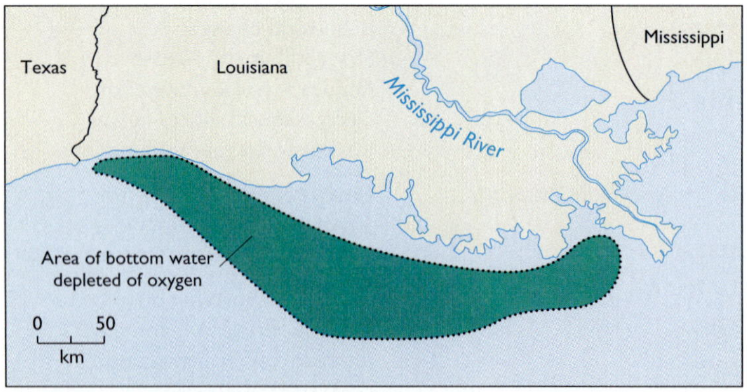

(b) HYPOXIC WATER OFF LOUISIANA

FIGURE **15-8**

Hypoxia on the continental shelf. (a) The New York Bight, located just to the south of New York Harbor, has been a long-standing dumping site, creating bottom water that has little dissolved oxygen. [Adapted from R. B. Clark, *Marine Pollution* (Oxford: Claredon Press, 1989).] (b) Bottom water along the continental shelf of Louisiana is hypoxic due to the high BOD of the organic discharge from the Mississippi River. River-supplied fertilizer contributes over 50% of the input of all nitrogen to the Gulf waters of Louisiana. [Adapted from N. Rabalais and D. F. Boesh, as shown in U.S. Congress, Office of Technology Assessment, *Wastes in Marine Environments* OTA-0-334 (Washington, D.C.: U.S. Government Printing Office, 1987).]

METALS

The term **heavy metal** is loosely applied to a collection of elements such as lead, mercury, cadmium, arsenic, and copper that typically occur in trace amounts in the environment. Most of the metals that occur naturally in seawater were supplied by rivers and represent the weathering products of rocks and sediment on land. Volcanoes also spew out large quantities of ash with their heavy-metal content, and winds can disperse these ash clouds to the most remote parts of the oceans. Over geologic time organisms in seawater containing trace amounts of these chemical compounds have evolved to such a point that heavy metals are vital to many biological processes, including photosynthesis and cellular metabolism. At high dosages, however, heavy metals are extremely toxic to life; they become lethal poisons. Not surprisingly, human activity often is responsible for introducing large quantities of metals in the environment. These metals are usually a waste by-product of some manufacturing or industrial process.

Discharge from industrialized or urbanized areas

FIGURE **15-9**

Fish disease. Fish that inhabit polluted water commonly show signs of disease. This fish shows "fin rot."

FIGURE **15-10**

Kelp deforestation. (a) Kelp beds growing near Palos Verdes, California, have been destroyed by the discharge of solid wastes. The kelp beds are rebounding, though, because of restoration efforts and treatment of wastes. [Adapted from J. C. Meistrell and D. E. Montagne, in *The Effects of Waste Disposal on Kelp Communities,* W. Bascom, ed. (LaJolla, Calif.: University of California Institute of Marine Resources, 1983).] (b) This underwater photograph shows the thick, luscious growth of kelp beds that grow along much of the California shoreline. These marine forests provide many habitats and niches for invertebrates and fish.

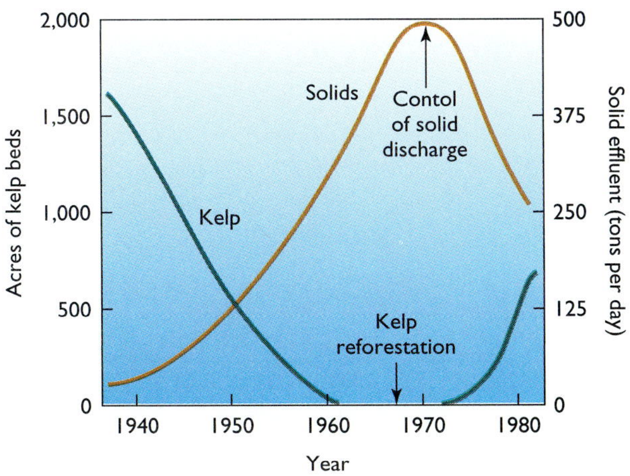

(a) KELP DEFORESTATION OFF SOUTHERN CALIFORNIA

(b) KELP BEDS

typically contains heavy metals that can augment by orders of magnitude the naturally low concentrations of these substances. For example, in Narragansett Bay, Rhode Island (**Figure 15–11a**), the dissolved levels of nickel build up from relatively low levels at the estuary's mouth to very high amounts at its head (**Figure 15–11b**). The same is true for many other metals in the bay's water as well. This trend of ever-increasing loads toward the head of the estuary reflects the highly industrialized nature of upper Narragansett Bay. The upper bay has served in the past as a convenient dumping site for huge quantities of pollutants from the city of Providence and neighboring municipalities. Narragansett Bay has also had a long history of receiving nutrients from the land and from the air. The input of phosphorus and nitrogen to the waters of the bay increased markedly beginning in the mid-nineteenth century as a result of human activity (**Figure 15–11c**). Phosphorus and nitrogen are used in farming (for both growing crops and raising livestock), in a variety of industrial enterprises, and in laundry detergents. All of these sources elevated the nutrient content of the bay water, causing massive plankton blooms and eutrophication of the water. The benthos, if present in the organic-rich muds, are variably contaminated with pollutants of all sorts that if consumed by humans cause various gastrointestinal ailments. Besides, fish and invertebrates tend to avoid excessively contaminated habitats. If they cannot, their behavior may become modified: they grow more slowly, become generally less active, or show signs of being diseased.

Mercury, particularly in the form of methyl

(a) NARRAGANSETT BAY, RHODE ISLAND

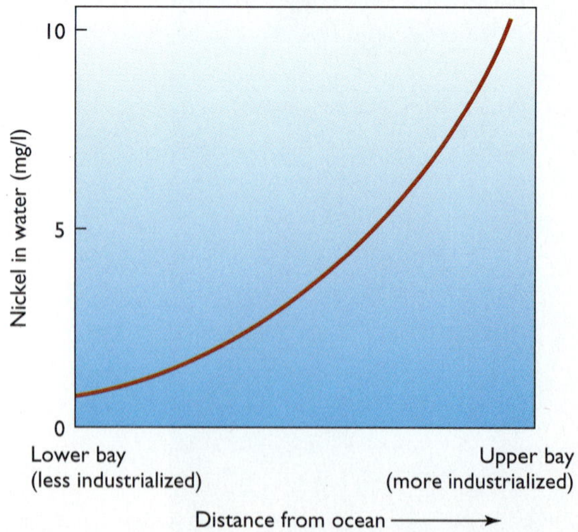

(b) NICKEL CONCENTRATIONS IN NARRAGANSETT BAY

FIGURE **15-11**

Nickel concentrations in Narragansett Bay, Rhode Island. (a) A satellite image of Narragansett Bay. (b) The dramatic increase in the nickel concentration of the waters of Narragansett Bay with increasing distance up the estuary indicates the source of the metal is at the head of the bay. [Adapted from *Down the Drain: Toxic Pollution and the Status of Pretreatment in Rhode Island* (Providence, R.I.: Save the Bay, Inc., 1986).] (c) The upper reaches of Narragansett Bay have received an ever-increasing load of nutrients from the surrounding land. This has led to eutrophication of the water and the creation of hypoxic and anoxic conditions. [Adapted from S. W. Nixon, *Oceanus* 36 (2) (1993): 38–47.]

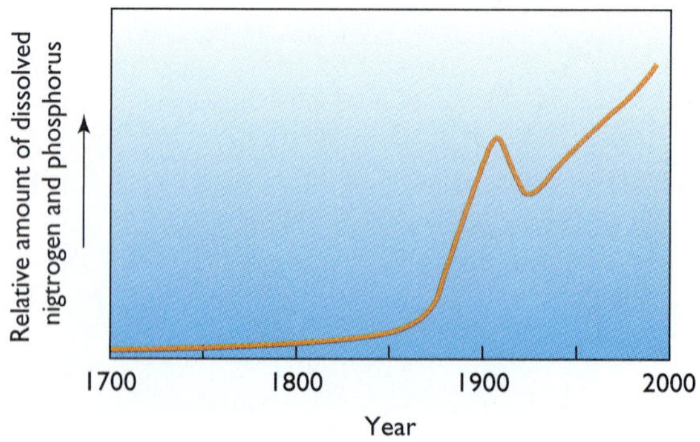

(c) NUTRIENTS IN NARRAGANSETT BAY

mercury (CH_3Hg^+), is deadly poisonous as it accumulates in cells and damages the central nervous system of higher organisms. It resists biodegration, in other words, microbes do not decompose it into simple compounds, and therefore it stays in the environment for a very long time. Mercury in nearshore marine environments tends to be produced by humans. It is a by-product of the chlorine-alkali and paper-pulp industries, a constituent of the antifouling paints that cover the hulls of ships, and a chemical ingredient in pesticides and some pharmaceutical products.

Once in the water, mercury tends to be adsorbed (gathers on the surface) to solid particles and settle out of the water into the sediment. Because of its propensity to complex with and form stable organic compounds, it is subjected to **bioaccumulation**, meaning that organisms can concentrate and build up methyl mercury in their tissues. Methyl mercury can undergo **biomagnification**, whereby the metal is concentrated in ever greater amounts at each step of the food web. Biomagnification signifies that herbivores have relatively low quantities of methyl mercury in their tissue, whereas first-order carnivores have much more, and second-order carnivores even more. The ingestion of mercury-tainted shellfish and finfish causes neurological dysfunctions and kidney malfunctions in humans, and, in cases of extreme contamination, even death. For example, mercury poisoning afflicted over two thousand people who had unknowingly consumed methyl-mercury-laced shellfish from Minamata Bay, Japan. Of those unfortunate people, forty-three died and seven hundred were permanently and seriously debilitated.

ARTIFICIAL BIOCIDES

It is estimated that more than seventy thousand chemical compounds are used by industries the world over, and about one thousand or so chemical substances are introduced every year. Their molecular structure, biodegradability, effects on organisms, including humans, and impact on and behavior in the environment vary enormously. A group of human-made chemicals that enters the sea is known as the *halogenated hydrocarbons* or *organochlorines*. Two members of this group are **DDT** (dichloro-diphenyl-trichloro-ethane) and **PCBs** (polychlorinated biphenyls). These hydrogenated hydrocarbons are complex molecules that contain chlorine as well as other more exotic elements, including fluorine, iodine, and bromine. They differ from natural petroleum hydrocarbons in one crucial respect—they are not readily degraded by bacteria or by chemical reaction. Thus, once released, they remain unaltered for long periods of time in the environment. Transported to the sea by rivers, they are chemicals that are adsorbed strongly to the surface and edges of silt and clay particles. Organisms that filter water or ingest mud inadvertently consume them and, because of their solubility in fatty tissue, have difficulty in excreting them. This necessarily leads to bioaccumulation and biomagnification in the food chain. Let's consider DDT and PCBs separately.

DDT is a highly toxic synthetic compound that was used extensively by farmers and forestry personnel as a pesticide from about 1939 until the late 1960s, when legislation was passed banning its use in the United States. In 1963 the book *Silent Spring* by Rachel Carson educated the public about the health and ecological dangers of DDT. Before then, its attributes—high toxicity to insects, persistence in soils, and cheapness to manufacture—led to its indiscriminate use all over the world to kill pests. Today, although it is outlawed in the West, there is little regulation of DDT usage in the rest of the world. In fact, the world's production of DDT has *increased substantially* since it was banned in the West!

DDT reaches the ocean by a variety of pathways. A common means of applying DDT to crops is aerial spraying. Unfortunately, only about 50 percent of the quantity of pesticide released ever reaches the ground. The remainder stays in the air as a fine aerosol and may be transported great distances by winds. Of the fraction that dusts the crops, much of it is washed off and adsorbed to fine sediment grains. During rainstorms, floodwaters erode silt and clay with their coating of attached pesticides from farm fields and disperse them to rivers where they eventually reach estuaries and the ocean as part of the suspended mud load. What has startled scientists is the rapid dispersal rate of DDT. It has been detected in muds of the deep sea and in the snow and ice of Antarctica, thousands of kilometers from where it was applied to cropland.

Like DDT, PCBs are synthetically produced. They have been used extensively since 1944 in the manufacture of electrical equipment, paints, plastics, adhesives, coating compounds, and other useful products. As a group, they comprise over two hundred separate chemical compounds, all highly toxic. What shocked scientists is the fact that PCBs are found everywhere in the ocean. This is perplexing, because PCBs are not pesticides and so they are not sprayed on farmland and then transported to the sea as an aerosol or as part of the suspended mud load of a river. Apparently, they are released into the environment by an alternate route, the unregulated incineration of discarded products, largely paints and varnishes, and electrical supplies. Then they stream across the globe on air currents and settle on distant land and water. Production of PCBs was banned in the United States in 1979.

Halogenated hydrocarbons are highly toxic and, once released into the environment, persist for long periods of time. They enter the food chain by first being absorbed by plankton and stored in their fat. It is these characteristics that make them a serious risk to biota and specifically to the health of humans. These poisonous substances build up in tissue as they are passed up the food chain. For example, mussels have concentrated and biomagnified DDT and PCBs by a factor as high as 690,000 over the measured quantities in the surrounding seawater! Known toxic effects include incomplete development of copepods and oysters, and the death of shrimp and a variety of fish. Fish-eating birds as well as sea lions (**Figure 15–12a**) have died because of a diet of organisms laced with high levels of DDT and PCB.

Figure 15–12b provides a conceptual model of the nature and effect of biomagnification of DDT in a food chain in Long Island Sound. The specific effects of halogenated hydrocarbons on humans are not known with certainty. However, shellfish beds in estuaries and bays contaminated with these artificial biocides—for example, New York Harbor and areas of Puget Sound—have been closed to harvesting in order to protect humans from the possible ill effects of eating tainted seafood. Also, fishing for bluefish, eels, and striped bass has been prohibited in the New York Bight area because of excessive PCB contamination of its water and sediment.

15-4

Ocean Dredging and Mining

In the preceding sections, the discussion focused on the ocean as a receptacle for the disposal of unwanted wastes, both solids and chemicals, that are the by-products of modernization (urbanization,

FIGURE **15-12**

Biomagnification of DDT. (a) Sea lions are top predators on a long food chain. Many of them have high levels of DDT and other contaminants in their fatty tissue, built up by biomagnification. (b) DDT, which is a fat-soluble compound, gets magnified in successively higher trophic levels of the food web. [Adapted from G. M. Woodwell et al., *Science* 156 (1967): 821–24.]

(a) SEA LIONS

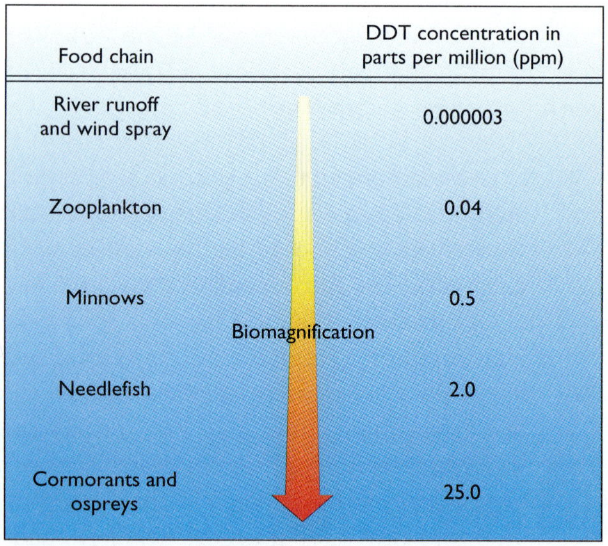

(b) BIOMAGNIFICATION IN LONG ISLAND SOUND

agribusiness, manufacturing, and industrialization). The ocean also is being exploited in a variety of ways by humans as well. In addition to commercial and recreational fishing, people are dredging harbors and navigable seaways, and are actively mining minerals from the sea floor. Although these activities benefit people, they harm the environment and its biota. The purpose here is not to review the reasons that people use to justify these acts (those reasons are obvious to most people), but to examine the drawbacks and environmental damage that result from such activity.

DREDGING

The U. S. Army Corps of Engineers has a congressional mandate to maintain over 40,000 kilometers of navigable channels in coastal marine waters and the Great Lakes of the United States. An enormous volume of sediment, some 230 million cubic meters, is removed annually by **dredging** in compliance with this federal decree at a cost of about $725 million each year (**Figure 15–13**). All these dredge spoils must be dumped somewhere, and this usually means somewhere in the sea where they are out of sight. So the ocean is affected by dredging, both at the site of removal and at the site of dumping. Dredged sediment represents some 80 to 90 percent by volume of the waste material that is routinely dumped into the ocean. It represents a problem that engineers and planners predict will only become more acute in the near future.

If the dredged material is uncontaminated with sewage, metals, and chemical toxins, its disposal is not deemed to be a serious problem for the ecosystem in the long run. Initially, the benthic communities at the dump site are exterminated by burial and smothering (**Figure 15–14**). Depending on the quantity of material and rate of dumping, some invertebrates will simply burrow up to the surface and suffer no apparent ill effects. Bottom-feeding fish abandon the site during the actual dumping activity, but tend to return soon thereafter, unless the area is severely altered. A few months to a few years after the dumping stops, bottom-dwelling organisms typically will have recolonized the disposal site.

Available studies of specific dumping sites indicate that the damage caused by waste disposal is apparently short lived. The species makeup of the benthos that recolonize the area may be different from the original occupants. For example, an important quahog (clam) fishery in the open water near Narragansett Bay, Rhode Island, was destroyed when large quantities of dredge spoils were dumped on Brenton Reef. With the termination of dumping, the rock reef has now become a major fishing ground for lobsters, but the quahog did not return.

FIGURE 15-13

Channel dredging. Keeping channels and waterways safe for boat and ship traffic, requires the continual dredging of sand.

In some cases, uncontaminated sand that is filling in a navigable coastal channel can be dredged and used artificially to nourish nearby beaches that are eroding (**Figure 15–15**). This is an ideal solution, because what is normally an expensive liability — the disposal of dredge spoils—becomes a valuable resource for beach sustenance. This process is known as *beach nourishment.*

Because dredging is done largely in coastal environments, some unknown fraction of the sediment removed is heavily contaminated with oil, metals, biocides, and a variety of human-manufactured wastes. When dumping of contaminated dredge spoils at a specific site is discontinued, some members of the bottom fauna that recolonize the area will concentrate some of the pollutants by bioaccumulation and biomagnification. Other possible environmental effects associated with ocean dumping include the ejection of pore water and their dissolved chemical load into the water column as the spoils compact and consolidate and the resuspension of sediment by waves during storms.

Because most (about three-fourths) of the dump-

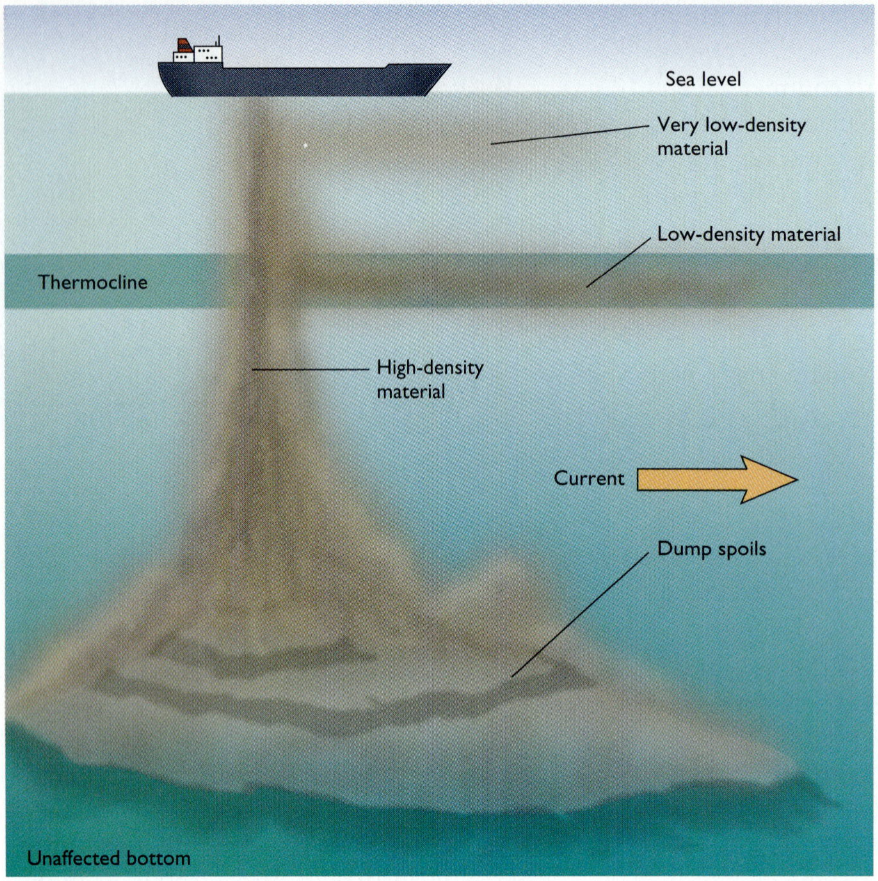

FIGURE 15-14

Ocean dumping. Solid wastes dumped into the bay spread out across the bottom and along pycnoclines. [Adapted from W. E. Pequegnat et al., *Procedural Guide for Designation Surveys of Ocean Dredged Material Disposal Sites,* EL 81-1, U.S. Army Corps of Engineers (Washington, D.C.: U.S. Government Printing Office, 1981).]

ing sites in the United States are in water shallower than 20 meters (~66 feet), coastal currents and wave activity that regularly stir up the bottom can disperse contaminated sediment elsewhere in the environment. Consequently, attempts are often made to isolate the spoils by constructing containment "walls" or by covering the site with uncontaminated sediment. Most organisms beyond the dumping locality do not seem to be affected by sediment suspensions or turbid plumes in any significant way. In the New York Bight area, which has been receiving dredge spoils for over a century, oceanographers estimate that about 85 percent of the discarded material has remained in the designated dumping grounds. The remainder apparently has drifted to the northwest, transported shoreward by bottom currents.

OCEAN MINING

The oceans contain a vast array of minerals and other resources that are potentially useful to humankind. Gravel and sand, salt, phosphorite, and tin have been exploited for centuries from beaches and the shallow water of the nearshore zone. Since the mid 1960s, oceanographers have looked to the deep sea for sources of minerals and chemicals. Techniques and equipment for finding, removing, and recovering economic mineral deposits from the sea floor that is thousands of meters below the keel of a ship are being developed, tested, and refined. The technological problems for doing this efficiently and economically are enormous, which makes the task even more compelling for oceanographic engineers.

Fields of manganese nodules with their content of manganese, copper, cobalt, and nickel blanket parts of the deep ocean floor (see Figure 4–15b in Chapter 4). Although their position is arguable, many resource planners believe that the mining of nodule deposits for their metals is crucial for the continued industrial health and national security of the United States. Let us disregard for the moment the legal quandary of who can legitimately claim resources that are located beyond the territorial waters of nations and consider the potential environmental effects of deep-sea mining.

FIGURE **15-15**

Sand nourishment of a beach. If dredge spoils offshore consist of clean sand, a means of disposing the material is to pump it to the updrift end of a beach. This process of so-called beach nourishment provides a supply of sand for currents to transport downdrift.

The problem has two aspects: (1) the techniques for removing and processing the ore, and (2) the natural oceanographic processes at the mining and disposal sites. Several engineering systems have been proposed for mining nodules. A common technique relies on a hydraulic pump system. Water, solids, and organisms are drawn (sucked) into the hydraulic lift system from the sea bottom, not unlike the way dirt from a carpet is sucked into a powerful vacuum cleaner. During the "vacuuming" process, various liquids and slurries are discharged near the bottom and at the sea surface. Let's examine some possible impacts of this type of mining on the open-ocean environment.

Hydraulic pump systems pull in all solids—both living and nonliving—indiscriminately, which obviously destroys the substrate and its benthic community. Such machines are capable of processing between 4,000 and 5,000 tons of sediment each day. Newly designed systems allegedly are quite efficient in separating the nodules from the unwanted sediment. The mass that moves up the pipe consists of about 3 percent solids, with a nodule-to-sediment ratio of around 4 to 1.

Field experiments are limited in number, and their results are contradictory. Some indicate that the sediment that is stirred up by the collector settles back to the bottom within a few tens of meters of the excavated furrow. Others suggest that enormous plumes of suspended sediment are created that affect the biota of the abyssal floor far removed from the mining operations. The slurry that reaches the water surface consists of water, sediment, organisms, and nodules. The nodules are retained, and the rest is discarded into the sea, forming a plume of water with a density some 1 percent or so greater than the surface water. Biologists worry that discharge of this sediment cloud will inhibit light penetration and reduce plant photosynthesis. Some scientists speculate that the heavy-metal content of the discharge water may enter the food chain and accumulate in fish species that feed in the mining area. Others believe that the plume of turbid water will remain near the sea surface for a few days at most, with virtually no effect on plant productivity and rendering the biomagnification of toxic metals by surface-dwelling organisms unlikely.

Several tests indicate that mining the metalliferous sediments of the Atlantis II Deep in the Red Sea is economically feasible. High-pressure water jets lowered from a mining vessel will convert the bottom sediment into a slurry, which would then be pumped up to the sea surface at the rate of about 200,000 tons each day. This enormous volume of material would have to be processed aboard the mining vessel while it is at sea. The material would be highly toxic because of its heavy-metal content. Once processed, the residue (tailings) would constitute a major waste-disposable problem. Engineers have developed a technique whereby only 1 percent of the metal concentrate would be transported to a smelter on land. The remaining 99 percent of the slurry would be diluted with seawater and treated with chemical reagents before being discharged into water deeper than 1,000 meters (~3,300 feet). Marine life is sparse at such depths in the Red Sea, a fact that, engineers reason, would minimize the impact of these metal toxins on the ecosystem.

15-5

Overfishing

Humans have been catching fish in the ocean for a very long time. This fact combined with the immensity of the oceans has created a strong conviction that fishing stocks are inexhaustible, just about impossible to deplete. Supporters of this claim point out that the global fish harvest has increased dramatically in the past fifty years, growing from a base of 20 million tons in 1950 to slightly more than 90 million tons in the mid 1990s (**Figure 15–16**). Unfortunately, this dramatic success is proving to be short lived, as heavily fished areas are yielding smaller and fewer fish despite increases in the fishing effort and improvements in the effectiveness of the gear to catch them (**Figure 15–17**). In fact, some traditional fishing grounds that have yielded a large fish bounty for centuries, such as Georges Bank off New England (see the box, "Collapse of the New England Fisheries"), have recently been closed in the hope that fish populations would rebound naturally once excessive predation by humans ceased. This tragic story has been repeated elsewhere: the Peruvian anchoveta fisheries (**Figure 15–18a**), the California sardine fisheries (**Figure 15–18b**), the Pacific herring and Japanese mackerel fisheries (**Figure 15–18c**), the Atlantic and Pacific salmon fisheries, and the Alaskan King crab fisheries, among others, have all collapsed. In fact, the United Nations Food and Agricultural Organization judges that worldwide about 69 percent of fisheries are overfished, entirely depleted, or are recovering from overfishing. What exactly has led to this sad state of affairs? Can something be done to prevent the collapse of important fisheries? We will explore both of these questions in the remainder of this section.

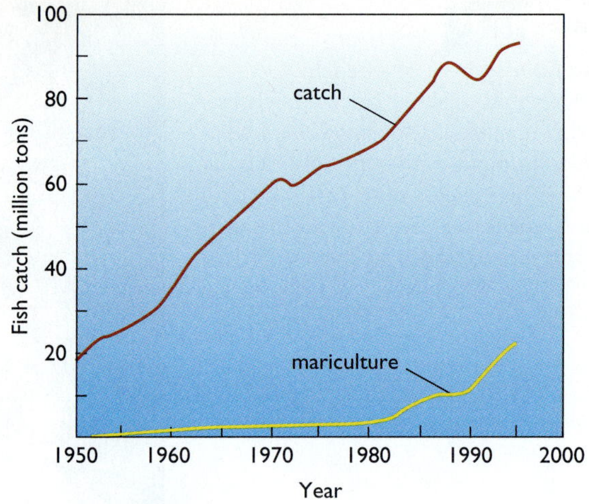

FIGURE **15-16**

Fish harvests. The global harvest of fish caught and raised by mariculture. [Adapted from A. P. McGinn, in *State of the World 1998* (New York: Norton, 1998).]

Fisheries collapse for various reasons, the most important being that species are harvested at rates that exceed their reproductive capacity. It's not merely a matter of too many fishermen. Fishing

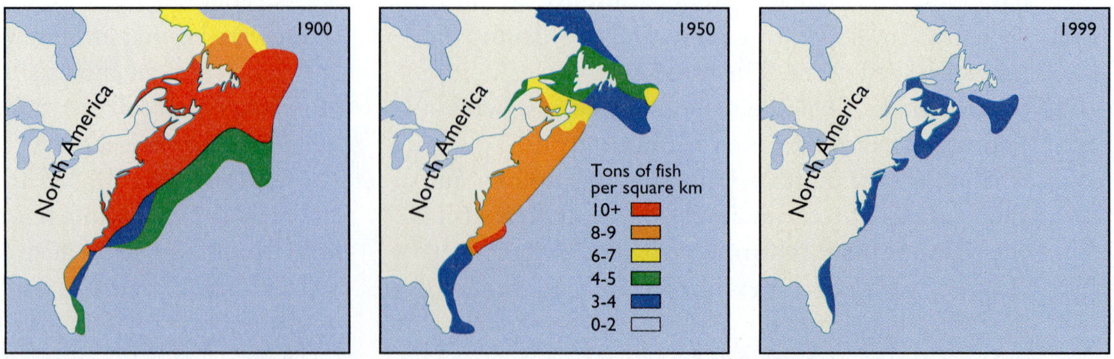

FIGURE **15-17**

Table fish biomass. There is a clear sharp decline in the biomass of commercial fish in the western North Atlantic ocean during the 20th Century. [Adapted from D. Malakoff and D. Pauly, *Science* 296 (2002): 458–61.]

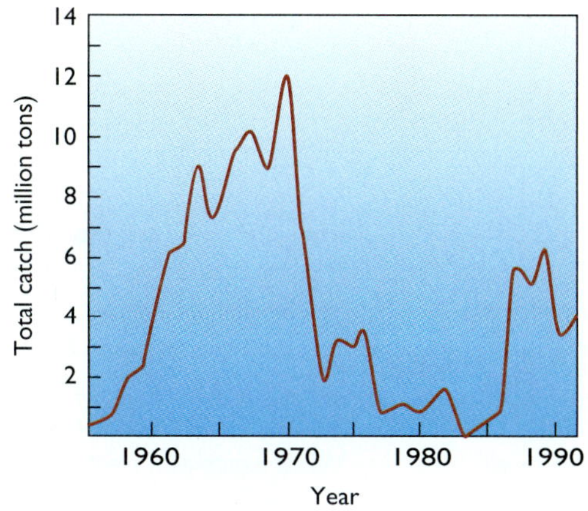

(a) PERUVIAN ANCHOVY

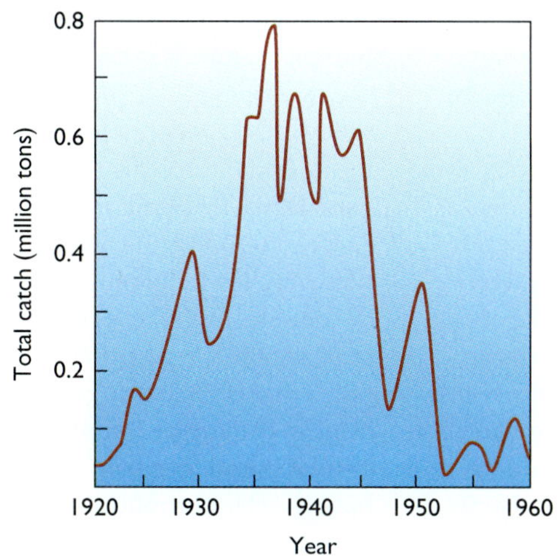

(b) SARDINE

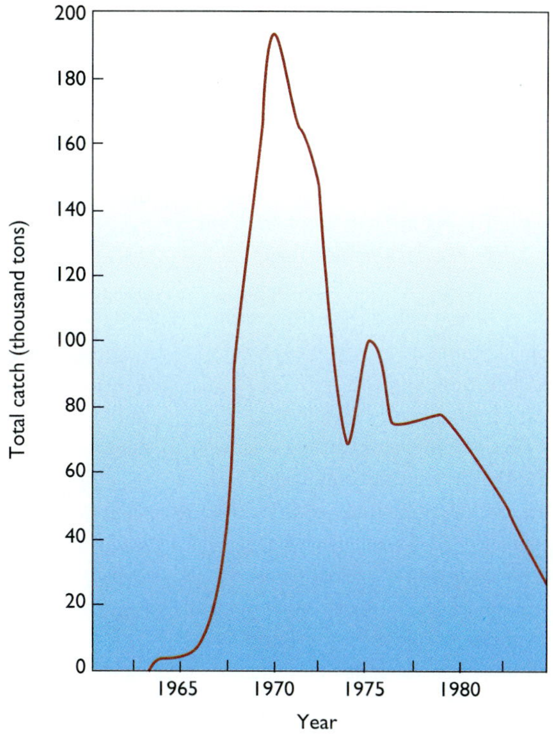

(c) PACIFIC HERRING AND JAPANESE MACKEREL

FIGURE **15-18**

Declining harvests. Decline in total fish catch for (a) Peruvian anchovy, (b) sardine, (c) Pacific herring and Japanese mackerel. [Adapted from M. Berrill, *The Plundered Seas: Can the World's Fish Be Saved?* (San Francisco, Calif. Sierra Club Books, 1997).]

technology has made it easier to locate fish schools and to capture them efficiently. The problem is often exacerbated by management policies that themselves are ineffective or inappropriate and that are resisted by fishermen who tend to be fiercely independent and resent government intervention in their livelihood. Boat owners incur a large debt in buying and maintaining a seaworthy vessel and all of the associated technology that is needed to hunt fish competitively. They resist the notion of a fish quota that will, they fear, reduce their ability to pay off their large debts.

During the last half of the twentieth century, fishery biologists and managers have relied heavily on the principle of **Maximum Sustainable Yield** (MSY), a theoretical maximum quantity of fish that can be harvested each year without significantly interfering with the regeneration of the fishing stocks. In essence, it is a laudable attempt to fish sustainably. This protects the fish stocks for future exploitation by setting yearly catch quotas that in theory will not cause the fish stocks to decline over time. Essentially, MSY is calculated from a complex mathematical formula based on biological factors such as population dynamics, food webs, spawning success, recruitment of new fish to the fishery, and fishing effort. By monitoring as closely as possible the population dynamics of fish stocks and the fishing effort expended to catch them, biologists can set catch quotas that will establish a theoretical equilibrium for each fishery, meaning that the biomass of fish harvested roughly balances the amount of new biomass added each year. Given the widespread use of MSY, why are fish stocks so commonly exploited?

Some of the more important reasons for the failure of MSY as a guiding principle for harvesting fish sustainably include:

Collapse of the New England Fisheries

There is a long tradition of deep-sea fishing in New England. One of the region's richest fishing grounds is Georges Bank (**Figure B15–4**), a broad submarine mountain some 300 kilometers (~93 miles) long and 150 kilometers (~186 miles) wide that is situated almost 200 kilometers (~124 miles) offshore of Cape Cod and straddles the border between the U.S.A. and Canada. Georges Bank is as shallow as 20 meters (~65.4 feet) in spots, despite its location far at sea. Here, a combination of tides, waves, currents, and winds churns up the water, and continually mixes nutrient-rich bottom water with surface water. Nutrients are retained over the bank by a large clockwise circulation gyre (see Figure B15–4) that isolates the region's water from the Gulf of Maine and the open North Atlantic Ocean. Phytoplankton respond to the continuous infusion of nutrients with huge blooms. These dense concentrations of plants support uncountable numbers of zooplankton that feed small fish that in turn nourish large fish. It is the latter that fisherman out of New Bedford catch (**Figure B15–5**) and bring to market. The main commercial species harvested from Georges Bank include groundfish such as cod, haddock, flounder, ocean perch, and pelagic fish such as herring, mackerel, and bluefin tuna.

Before the Magnuson Act was passed in 1976 and the United States claimed the exclusive right to manage fisheries within 200 miles of its national boundaries, the Georges Bank fisheries was world renowned as a prime fishing ground. Large factory fleets from the former U.S.S.R., Japan, Norway, and West Germany made regular forays onto Georges Bank to catch and process fish. Fishermen from New England were outraged, accusing foreigners of purposely and carelessly overfishing the grounds without any concern for sustaining the ecological health of this long-standing fishery. The Magnuson Act effectively prevented foreigners from fishing New England waters. The results were largely unexpected. The offshore New England fishing fleet modernized, expanded quickly, and began the large-scale commercial harvesting of Georges Bank's fish stocks, all without any significant restrictions. At first, the tonnage of cod, haddock, and flounder landed by New England fishermen doubled, but elation and prosperity were short lived. By 1984, the total harvest of groundfish had dropped off to the lowest tonnage for the past thirty years (**Figure B15–6**), in spite of a greater fishing intensity then in the past and the adoption of the latest technology for detecting and capturing fish.

Clearly, there was a problem. Put simply, Georges Bank had been overfished, not by foreigners but by American fishermen, and the grounds had to be closed if there were to be any hope for the recovery of the region's fish stocks.

What exactly caused the collapse of the fishery? Fishing pressure clearly was implicated, but was it the main cause? Because ecosystems are complex and are affected by both natural and human factors, it is hard to separate the two from each other. For example, yellowtail flounder reproduce well and are abundant in years that follow cold winters in New England. If there are a succession of warm winters, the numbers of yellowtail flounder plummet despite the fishing effort and catch quotas that are set either too high or too low. Also, naturally occurring variations in the size of fish populations are common and, because fish migrate and have patchy distributions, it is difficult to estimate the actual numbers of any

FIGURE **B15-4**

The general bathymetry and circulation on Georges Bank. Note the clockwise circulation pattern around the edge of Georges Bank.

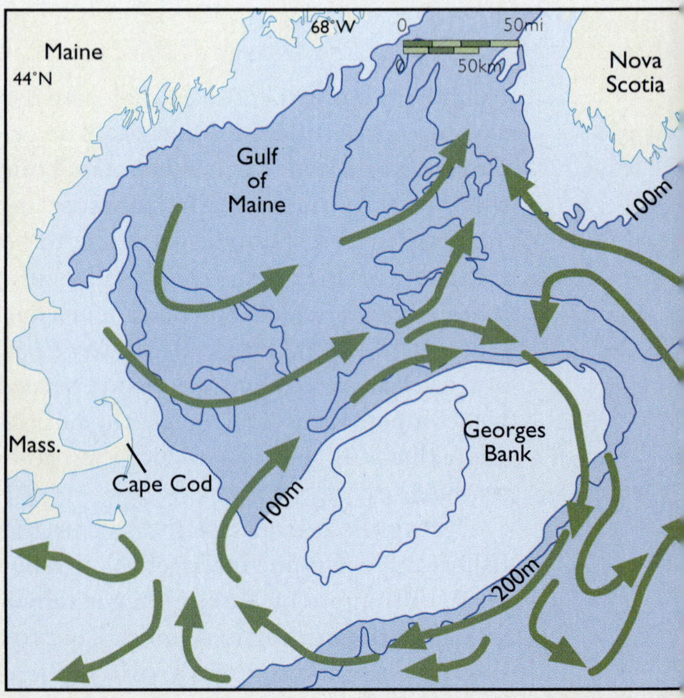

FIGURE **B15-5**

Fishing fleet. U.S. fishing trawler.

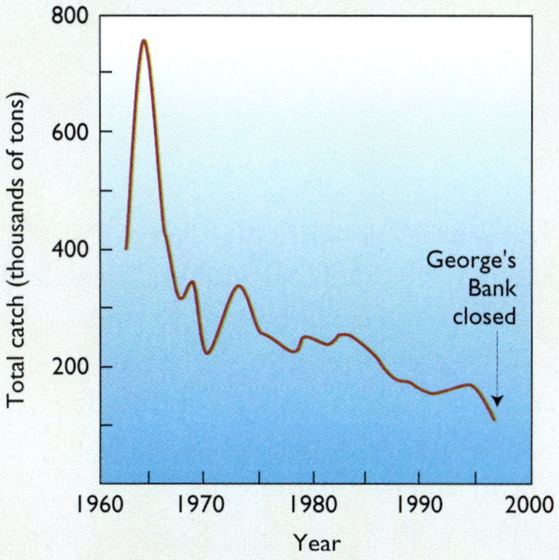

FIGURE **B15-6**

Collapse. The total catch of groundfish (cod, haddock, and flounder) offshore New England. [Adapted from B. Holmes, *Science* 264 (1994): 1252–53.]

fish species. Fishery biologists believe that population estimates of many fish stocks may only be within 30 percent of the actual population size. Obviously, it is difficult to establish fishing policy with such crude estimates of fishing stocks. For this reason, many believe that with the uncertainty in population estimates, fish quotas should err on the side of caution. Exactly the opposite had occurred in the past. Fishermen on Georges Bank have regularly harvested about 60 percent of the estimated population each year, an amount that is almost twice the size of the estimated maximum sustainable yield. In fact, scientists had been warning New England fishermen for over a decade that the collapse of the fishery was possible, if not imminent, and they urged that catches be severely curtailed in order to reduce predation pressure on this valuable resource. Although some were alarmed at the trends, this advice was generally ignored by the fishermen.

Whatever caused the collapse of the Georges Bank fishery, whether it was natural causes or chronic overexploitation or some combination of the two, the only recourse is to cease fishing for some undisclosed amount of time. Because the fish stocks of the region were clearly impoverished, few argued that this should not be done. Biologists believe that with adequate time fish populations on Georges Bank should rebound at least to some degree. Once they do, then management policies need to be conservative with fishing restrictions tight. Given the recent wholesale collapse, the Georges Bank fishery is amenable to more responsible management and regulation so that this important resource is sustained over the long term. Preliminary surveys taken since the fishing grounds closed indicate that some fish stocks are beginning to rebound on Georges Bank.

Data about fishing effort and catch tonnage are provided by the fishing industry itself. Although it is impossible to document accurately, fishery managers believe that unreported and misreported catches are the rule rather than the exception. This means that the information used to calculate MSY is inherently inaccurate.

Many fish are affected, sometimes dramatically, by population fluctuations of other species, particularly if the species is an important food resource for the commercial fish or if the species competes directly or indirectly with the fish being harvested.

Estimating fish stocks is very difficult for species that migrate extensively. Such commercial fish can move into or out of the fishing grounds depending on various ecological or behavioral conditions, making the size of the stock difficult to estimate accurately, which is essential for calculating MSY in order to regulate a fishery.

Discards, fish that are caught and thrown back into the ocean because of their small size or inferior quality or because they exceed catch quotas, may amount to 27 million tons worldwide. This figure, which is almost one-third of the reported global annual harvest of fish biomass, has major short- and long-term ecological consequences for marine ecosystems generally and for commercial fish stocks specifically.

Political rather than biological factors are often used by regulatory agencies to set fish quotas. In the past, it has been common practice to modify and even override MSY estimates made by fishery biologists. The European Union, for example, which sets the total fish catch for its member countries, instituted a 5 percent reduction in the 1995 hake harvest, despite the recommendation of the International Council for the Exploration of the Seas that the harvest be reduced by 40 percent.

It has become apparent to many scientists that reliance on the ability to "fine tune" the fishing industry according to MSY calculations about sustainability is seriously flawed, because it ignores other critical factors that affect fish stocks: the unpredictability of environmental stresses, complex and largely undiscovered relationships between predator and prey, unknown short- and long-term ecological interactions, and social and economic trends that are often impossible to anticipate despite sophisticated computer models and hardware. For example, on 2 July 1992, the Canadian Fisheries Minister closed the inshore northern cod fishery because, during 1991, unexpected "severe oceanographic conditions" reduced by half the total biomass and by almost three-quarters the spawning biomass of the northern cod. Obviously, it was prudent to close down the fishery; tragically this edict undermined the livelihood of twenty-five thousand fishermen and about ten thousand other workers in the industry.

To avoid the collapse of fisheries in the future, accompanied as such events are by severe economic and social hardship for fishing communities, some scientists reject MSY outright as an unworkable management strategy. They champion a new doctrine for fishery management, the *precautionary principle*. This guiding principle advocates restraints on those fishing practices that inadvertently reduce the fish stock for future generations of people, even when solid scientific proof of a decline in fish stocks is lacking. The scientists contend that, although it is not possible to decide on the right course of action given the complexity of marine ecosystems, it is feasible to identify practices that are known to damage fish stock; these wrong practices and policies should be avoided at all costs. Then, and only then, they contend, is responsible fishing that is sustainable over the long run possible. **Table 15–5** summarizes some precautionary measures that would go a long way in lessening the overexploitation of a fishery.

15-6 Climate Change

Although we tend to treat them separately, The Earth's oceans interact in numerous, important ways with the atmosphere, the lithosphere (the Earth's crust and upper mantle), the cryosphere (ice caps, glaciers, and sea ice) and the biosphere (living organisms in the sea and on land). These subsystems interchange matter and energy continually on a variety of time scales. Climate, for example, represents a dynamic state of the atmosphere and it influences and in turn is influenced by processes on land and in the ocean with response times that range from days to hundreds of thousands of years. The entire Earth system is very complicated, and marine scientists are striving to unravel the manner by which ocean processes are intermeshed with those of the climate. What are some of the interconnections between oceans and climate?

To begin, the ocean's water stores immense quantities of heat that is redistributed across the Earth's surface by geostrophic currents and thermohaline circulation. In effect, the ocean-atmosphere system is comparable to an enormous heat engine that exchanges the cold air and water of the high latitudes for the warm air and water of the low latitudes.

Let's begin by considering the surface flow of the immense circulation gyres. The Gulf Stream in the North Atlantic, for example, moves warm water from the tropics to cooler latitudes (see Figure 6–5a in Chapter 6) where it heats the atmosphere. This heat transfer creates a mild climate in Norway and Sweden despite their location at a latitude equivalent to that of the cold Arctic tundras of northern Canada. Heat is also moved from the water surface to the ocean depths by the sinking of dense water masses such as Antarctic and North Atlantic Bottom Waters (see Figure 6–15). These cold water masses form near the surface in the high latitudes, sink because of their high densities, and flow equatorward along the sea bottom where they remain for about a thousand years. The descending polar waters are replaced by upwelling deep water and by the surface inflow of warmer water from the temperate latitudes. Even coastal upwelling can dramatically influence local climate. Consider San Francisco, which is affected by the upwelling of cold water from depth. Here air in contact with the cold water at the sea surface is chilled, giving rise to air temperatures that are as low on average as are those along the coast of Ireland, despite San Francisco's being more than 1,500 kilometers (~930 miles) closer to the equator than Ireland is.

Today, it is apparent that human activity may be influencing climate, inadvertently inducing change on a global scale. Newspaper and magazine articles and radio and television reports warn of the impending warming of the Earth. Popularly known has the **greenhouse effect**, climactic warming may occur because of the rapid buildup of CO_2 in the atmosphere as a result of the widespread burning of fossil fuels (coal, oil, and natural gas) and slash-and-burn deforestation. Molecules of CO_2 permit sunlight to pass through the air and heat the Earth's surface, but they effectively absorb and trap heat that is radiated from the ground and the ocean's surface (somewhat comparable to a greenhouse and thus the phrase *greenhouse effect*). The higher the CO_2 levels of the atmosphere, the less heat escapes the Earth and dissipates into space. Consequently, global air and water temperatures gradually rise over time as the climate system establishes a new equilibrium.

Since 1958, atmospheric CO_2 levels have been measured at Mauna Loa, Hawaii, far out in the Pacific Ocean away from the principal industrial regions of the Western Hemisphere. This long record (**Figure 15–19**) shows the unmistakable impact of humans on the gas composition of the atmosphere, which has been climbing steadily upward with time. Annual ups and downs in the concentration of CO_2 result from natural biological cycling of carbon. Plant photosynthesis, which dominates the spring and summer, removes CO_2 from the air; animal respiration, which dominates the fall and winter, releases CO_2 to the air. Superimposed on this seasonal fluctuation is a clear, upward drift in the total concentration of atmos-

TABLE 15-5

Elements of the Precautionary Principle.

—Control access to new fisheries immediately.
—Establish a conservative cap on both fishing capacity and total fishing catch rate.
—Develop conservative catch limits and set an upper range that cannot be exceeded.
—If fish stocks decline, implement recovery plans without delay.
—Reduce government subsidies and promote development of fisheries that are economically self-sufficient.
—Establish data-collection and reporting systems that yield reliable information.
—Minimize bycatch through the use of more selective fishing gear and techniques.
—Establish protected areas as refuges for fish stocks and for restoring habitats.
—Develop management policies cooperatively with all stakeholders.

Source: Adapted from A. P. McGinn, *Rocking the Boat: Conserving Fisheries and Protecting Jobs,* paper 142 (Washington, D.C.: World Watch, 1998): 59.

pheric CO_2, which is the result of humans burning fossil fuels and forests. This rising trend has since been measured in Alaska and Antarctica, establishing the dependability of these data without any doubt. Today, the amount of CO_2 in the atmosphere is about 360 ppm (parts per million), which is very high compared to the average concentration of about 250 ppm in 1850. This change represents an increase of some 44 percent over merely a century and a half! Climatologists believe that the oceans currently absorb some 30 to 50 percent of the CO_2 emissions created by the burning of fossil fuels. Current atmospheric levels of this greenhouse gas would be much higher, probably in the order of between 500 and 600 ppm, if the ocean were not a large CO_2 sink. Energy experts expect that fossil-fuel emissions of CO_2 generated by human activity will continue to increase in the future, although at a lower average annual rate.

What concerns scientists is that changes in CO_2 levels in the geologic past have been associated with striking changes in climate. There seems little doubt that the current buildup of atmospheric CO_2 will perturb the climate, although it is not possible with any degree of assurance to know exactly how or when such changes will occur. The results of numerous experiments using elaborate climate models that are run on powerful computers suggest that the lower atmosphere will warm up by an average of 2°C within the next fifty to one hundred years, a rate of temperature change that has not occurred for at least the past nine thousand years. The increase, however, will not be uniform across the planet. Polar temperatures, for example, are expected to climb by as much as 6°C to 10°C, warming polar seas and melting ice sheets.

Some scientists suspect that warming has already begun. During the past one hundred years, the average global temperature has risen by about 0.5°C. The problem with attributing this temperature rise to the buildup of greenhouse gases is that this measured change is within the range of natural climactic variability. In other words, it is not possible to separate natural temperature variations from those that may have been induced by human activity. The rate of warming over the last one hundred years generally agrees, however, with the predictions made by climate models in which the CO_2 loading of the atmosphere by fossil-fuel emissions is examined. Those predictions raise the real possibility that global warming may be underway.

What are the possible environmental consequences of global warming, if it does occur in the twenty-first century as many climate models predict? If polar temperatures rise some 6 to 10°C, the Antarctic and Greenland ice caps will melt more

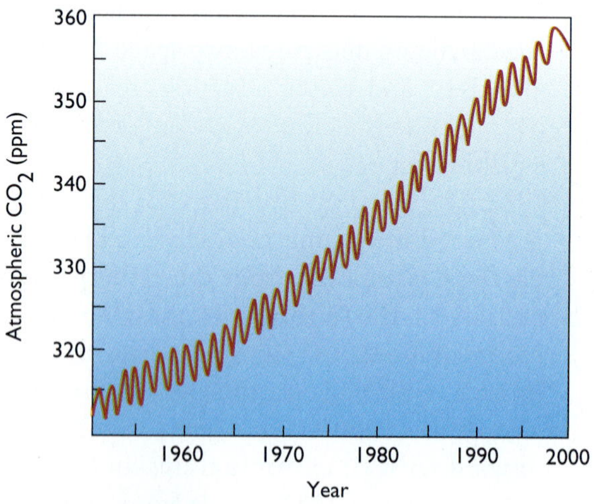

FIGURE **15-19**

Atmospheric CO^2. The 40-year record of CO^2 concentrations in the atmosphere measured at Mauna Loa Observatory in Hawaii. [Adapted from M. L. McKinney and R. M. Schoch, *Environmental Science: Systems and Solutions* (Sudbury, Mass., Jones and Bartlett, 1998).]

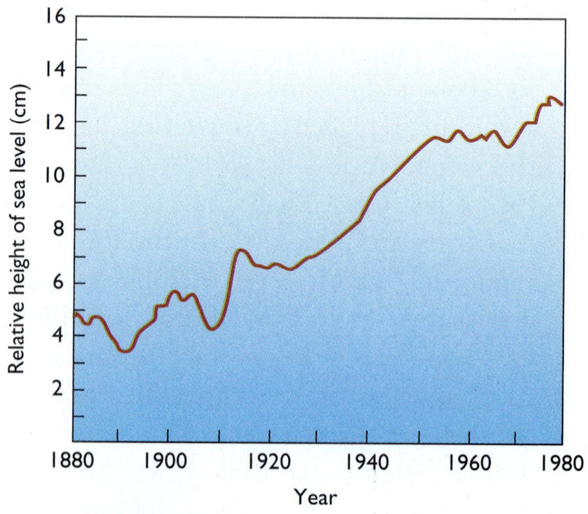

FIGURE **15-20**

Sea levels. The recent global rise of sea level. [Adapted from V. Gornits, S. Lebedeff, and J. Hausen, *Science* 215 (1982): 1611–14.]

rapidly and cause sea level worldwide to rise at an even faster rate than it does at present (**Figure 15–20**). The result—discussed in the box, "The Global Rise of Sea Level and Its Effects"—will be a widespread, gradual flooding of the world's coastal cities and low-lying islands, with all the attendant social, political, and economic repercussions. Because the heating will be greater in the high than in the low latitudes, the temperature contrast between the equator and the poles will diminish, causing global and regional wind and precipitation patterns to change accordingly (**Figure 15–20**). Some regions will benefit from longer growing seasons and more rainfall; others will suffer and become hotter and more arid. As climate modellers readily admit, it is difficult to ascertain exactly what will transpire because of the complex nature of the interactions among the lithosphere, cryosphere, atmosphere, hydrosphere, and biosphere, where feedback loops abound, each affecting another on a variety of time scales.

Are predictions of global warming valid or are they the exaggerated products of some overactive imaginations? Computer models of global systems are highly simplified versions of reality. Even a basic understanding of the exchange of energy and matter across ecosystems and among the Earth's spheres is quite limited and thus oversimplified in climate models. Nevertheless, regardless of these uncertainties, the collective wisdom of the international community of scientists is that an incremental warming of the Earth is highly probable in the very near future, although the exact time frame and possible ecological and social reverber-

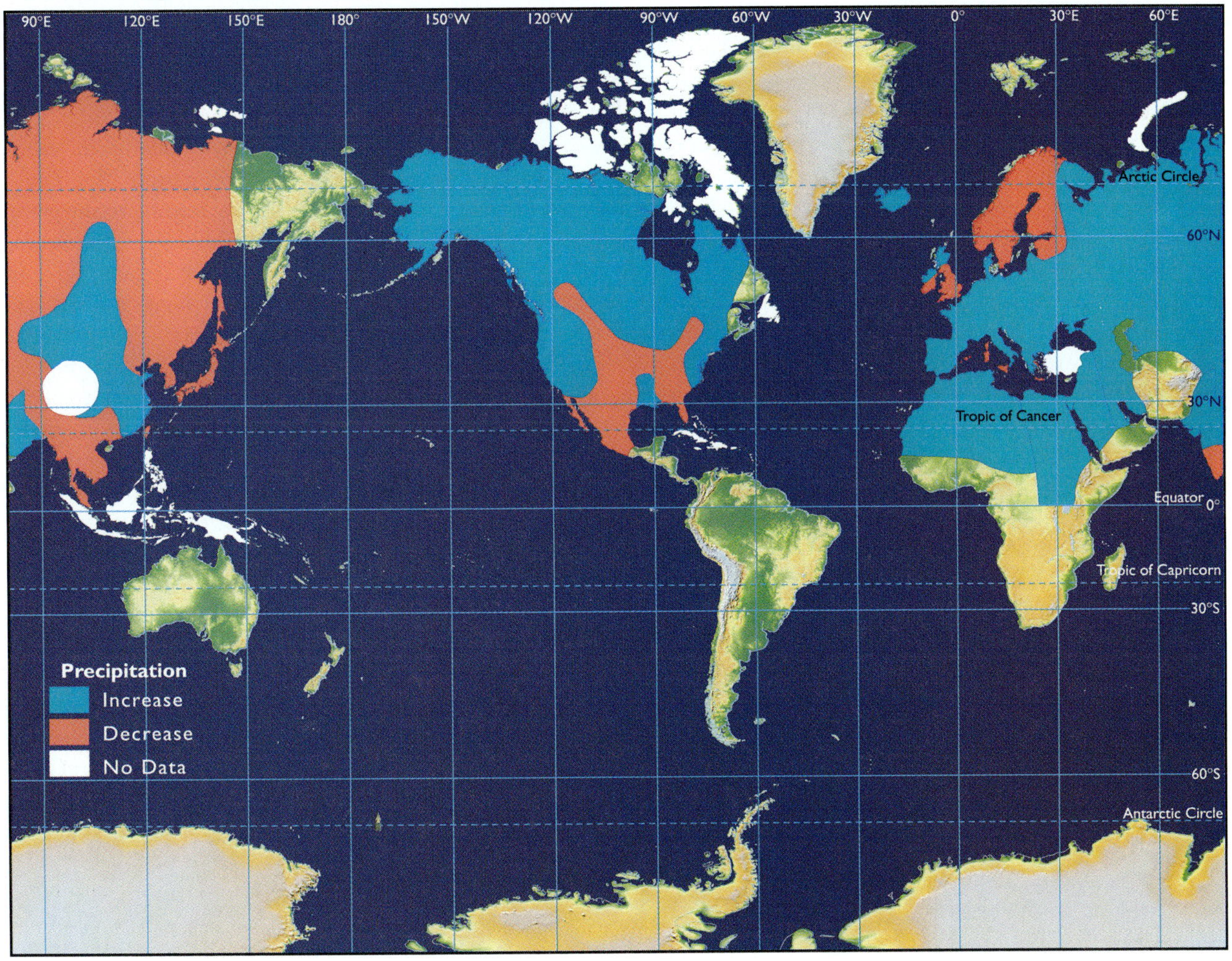

FIGURE **15-21**

Global warming. Possible changes in the pattern of precipitation that might be induced by global warming, based on a general circulation model. [Adapted from J. Gribbin, *Future Weather and the Greenhouse Effect* (New York: Delacorte Press/Eleanor Friede, 1982).]

THE PROCESS OF SCIENCE

The Global Rise of Sea Level and its Effects

There is consensus among climatologists that anthropogenic loading of the atmosphere with carbon dioxide (see Figure 15-18) and other heat-trapping gases will alter the Earth's future climate in some way. Although assumptions differ, models indicate generally that the Earth's air will be significantly warmed by some 2 to 6°C; the sharpest temperature rise is expected to occur in the polar latitudes. If so, what are some of the likely results of elevated air temperatures and, most importantly, can reasonable contingency plans that will mitigate these perturbations be developed? The answer to both questions is an emphatic yes. But scientists warn that we must act now and be proactive, rather than wait and respond after the fact.

Global warming will cause sea level to rise everywhere for two principal reasons: water expands as it is heated, and the Antarctic and Greenland ice caps will melt faster than they do presently. These effects, combined, will most likely raise sea level by about 15 centimeters (~5.9 inches) by the year 2050, and by about 35 centimeters (~13.8 inches) by the year 2100. Of course, more drastic changes are possible, but less probable. For example, there is a 10 percent chance that sea level during the twenty-first century will rise by about 65 centimeters (~2.1 feet) and a 1 percent chance that it will rise by one meter (~3.3 feet), but such changes are unlikely. This increase in the level of the sea will be superimposed on the current rise of sea level (see Figure 15-19), which averages about 0.1 centimeter per year (~0.4 inch per year). The point being made is that the sea will be at a much higher level by the end of the twenty-first century than it is at present, regardless of the exact magnitude of the change.

How should society respond? It is clear that a high-tech fix will not solve the problem. We cannot afford to hold back the seas everywhere by designing elaborate systems of costly dikes and seawalls. We could stabilize global emissions of heat-trapping greenhouse gases. Some analysts have estimated that, if we stabilize the input of carbon dioxide gases by the year 2050, we would substantially reduce the rate at which sea level rises and perhaps cut it back by as much as a third or even a half—a significant abatement. Some countries such as Norway, the Netherlands, Germany, and Sweden, among others have taken a bold initiative by setting stringent goals for carbon-dioxide emissions in order to help slow down the warming of the Earth. Although numerous other ecological, economic, and social disruptions will be brought about by air temperatures that shoot upward, the focus of this discussion is the likely repercussions of a global rise of sea level.

Sea level has fluctuated regularly during the Earth's long geologic past, most recently because of glaciation and deglaciation episodes during the Pleistocene epoch (the Ice Age). The most troubling aspect of the rise expected as a result of global warming is that shorelines and many low-lying islands are now densely populated. For example, almost 75 percent of Americans live within fifty miles of the shore, many in large cities such as New York, Miami, Los Angeles, and Seattle. As water floods the land, barrier islands, rivers, wetlands, and mudflats will be inundated by seawater. Rising seas will intensify coastal flooding and worsen the erosion of beaches, bluffs, and wetlands, as well as threaten jetties, piers, seawalls, harbor facilities, and waterfront property in general. Moreover, a rising sea level will result in the intrusion of salt water into estuaries, rivers, and streams, and seepage will contaminate coastal aquifers, raising the water table and rendering some groundwater supplies undrinkable. The seepage of saline groundwater into prime agricultural land will make the soil infertile.

If we fail to respond adequately to a higher ocean level, some researchers fear that by the end of the twenty-first century about one-third of the global cropland may be lost to cultivation and over one billion people will be displaced from their homelands. Ironically, the ten countries most vulnerable to a global rise in sea level rise (**Table B15–1**) are the least industrialized, poorer nations of the world and, as such, have contributed insignificant quantities of heat-trapping gases to the atmosphere.

Along barrier islands, the erosion of beachfront property by flooding water will be severe, leading to the greater likelihood of overwash during storm surges. Some houses, cottages, and hotels built on pilings that are driven deep into the sand will remain habitable for a while with water surging around and beneath the structures; eventually they will suffer storm damage and, if not destroyed outright, will have to be abandoned or relocated. Some coastal geologists favor engineering a retreat, whereby fill would be placed on the bay side of the barrier island to create new land artificially; this would offset the natural erosion occurring on the ocean side (**Figure B15–7**). People living in the front of the barrier could then relocate on the backshore region. The legal and political problems of an engineered retreat are formidable, but presumably resolvable given the seriousness of the situation.

Another remedial scheme involves raising the height of the land on the ocean side of the barrier island by pumping sand onto the beach and artificially elevating coastal roads as well as private and municipal properties (**Figure B15–8**). This is appealing to many beachfront occupants who would rather not have to abandon their property and move to less desirable land farther inshore. Even discounting the prohibitive cost of building land up, the catastrophic environmental impact of extensive dredging to get adequate fill to engineer the vertical buildup of the shorefront makes this option a difficult one to implement. There may, however, be no other recourse for people who insist on living on the front edge of barrier islands.

Although the human impact is severe, most ecologists believe that a rise sea level will cause a simple landward shift of coastal ecosystems along shores not stabilized and protected by humans. The redistribution of environments will be particularly marked along flat, low-lying shores such as those associated with river deltas, but be inconsequential over the short term (less than 100 years) along shores buttressed by steep, durable cliffs. The widespread extinction of coastal species is not expected even if sea level does rise.

Visit www.jbpub.com/oceanlink for more information.

TABLE **B15-1** Countries most vulnerable to rising sea levels.

Country	1990 Population (million)	Per-capita Income (in 1990 dollars)
Bangladesh	114.7	$ 160
Egypt	54.8	$ 710
The Gambia	0.8	$ 220
Indonesia	184.6	$ 450
Maldives	0.2	$ 300
Mozambique	15.2	$ 150
Pakistan	110.4	$ 350
Senegal	5.2	$ 510
Surinam	0.4	$ 2,360
Thailand	55.6	$ 840

Source: Adapted from J. L. Jacobson, *State of the World 1990* (London: Norton, 1990).

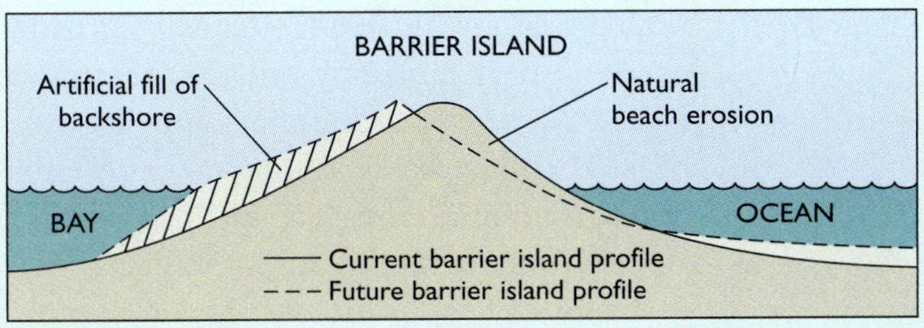

FIGURE **B15-7**

An engineered retreat. By backfilling the barrier island while normal erosion of the ocean side of the barrier island is allowed to occur as sea level rises, the island is preserved.

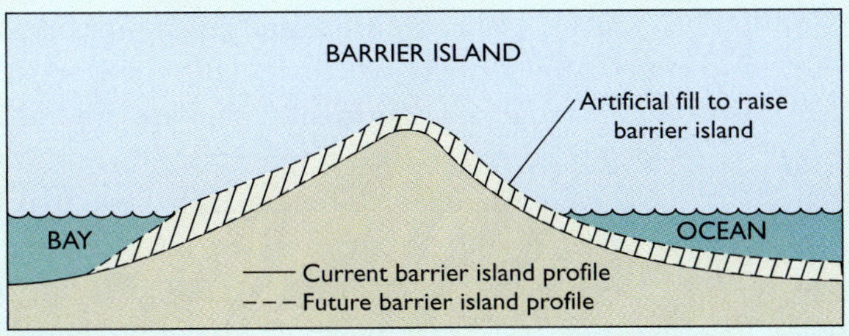

FIGURE **B15-8**

An engineered stand. The entire barrier island is raised with fill, keeping abreast of rising sea level.

(a) PLASTIC WASTES

(b) A SIGN OF THE TIMES

FIGURE **15-22**

Coastal pollution. (a) Many beaches are littered with garbage and wastes that are brought in by the tide. Many townships must regularly clean their beaches during the tourist season. (b) Numerous clamming grounds have been closed because of contamination with toxins. People who eat these shellfish will suffer a variety of gastrointestinal ailments, some of them very serious.

ations of this warming trend are debatable. According to Roger R. Revelle, the former director of Scripps Institution of Oceanography, we have embarked unknowingly on "man's greatest geophysical experiment."

15-7

The Ocean's Future

The signs abound that the land and the atmosphere of the Earth are besieged by the activity of humans, particularly the energy-intensive, industrial lifestyles of more developed nations. Unfortunately, the ocean is not the exception. A visit to a shore, particularly a metropolitan one, will reveal numerous examples of environmental abuse, ranging from the accumulation of discarded refuse and flotsam to the presence of toxic chemicals, from oil-stained rocks and bulkheads to closed shellfish grounds, from abandoned, leaking barges to crowded condominium developments and filled wetlands (**Figure 15–22**). These disturbing signs are, in large part, the result of a longstanding view that the ocean is a convenient receptacle, a garbage bin, for our industrial effluents. Also, some people believe that the ocean's capacity for accepting human refuse and for self-cleansing is limitless. It is not. The coastal ocean has a finite, natural capacity that in many cases has been exceeded.

The future prospects are as alarming, for it is apparent that environmental degradation by humans is not limited to the coastal fringe of the oceans but is spreading out to its remotest areas. **Figure 15–23** is a world map showing areas of the ocean that have been polluted to various extents by human contaminants. The most seriously polluted zones are the stretches of coastal and shelf waters that border heavily industrialized manufacturing sites and large urban centers. This is where refuse is produced in prodigious amounts in the manufacturing, transportation, and consumption of goods. Nearby waterways, particularly the waters of estuaries and the continental shelves, are convenient and cheap sites for dumping all of this solid and dissolved waste. Some biologists believe that the frequency and extent of some coastal phenomena such as red tides (see the box, "Red Tides") may be a direct consequence of unregulated contamination of nearshore water by humans.

But, alas, there is more to this problem than the despoilment of coastal water. Vast tracts of the open

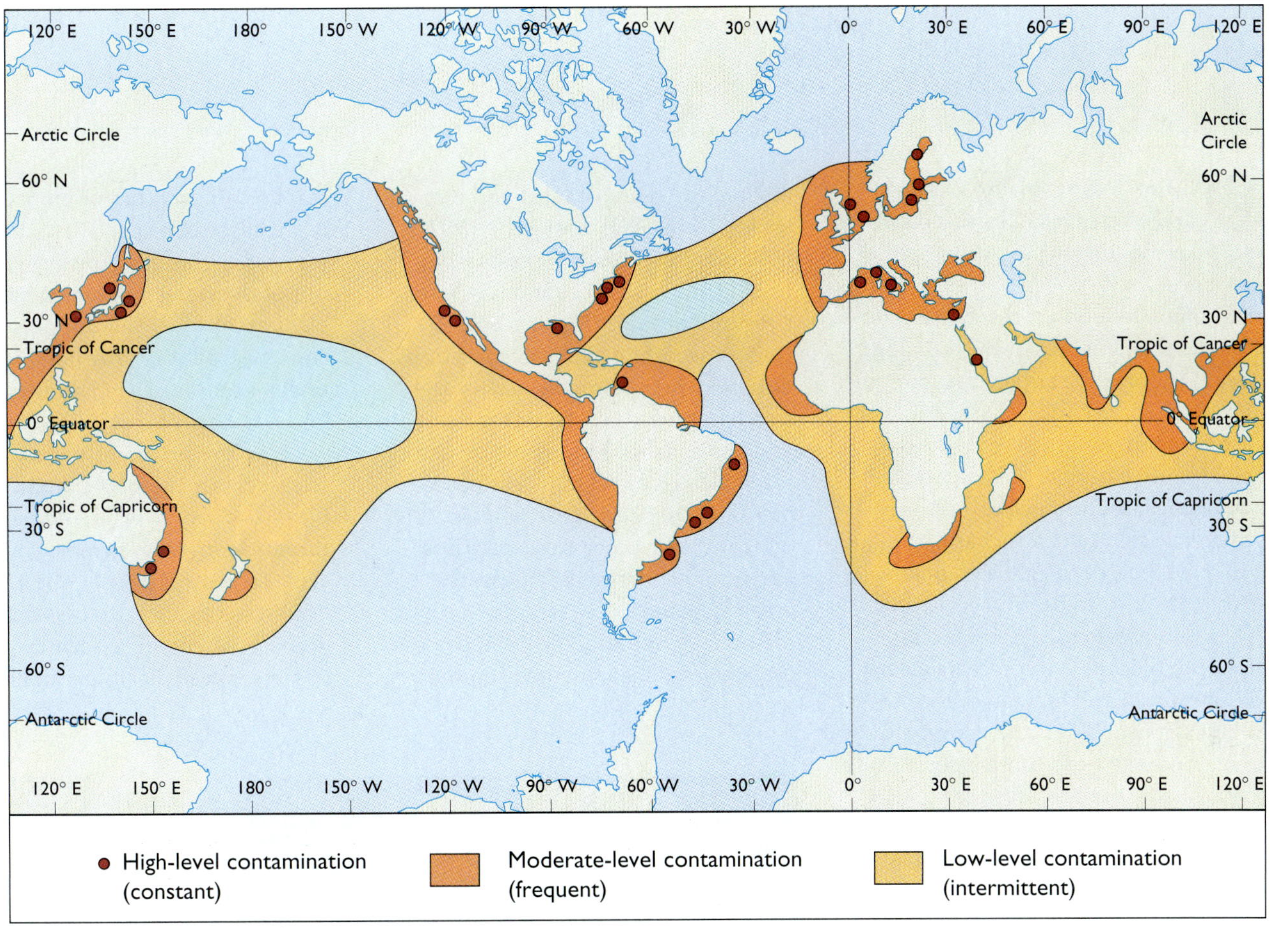

FIGURE **15-23**

The global spread of marine pollution. Coastal areas bordering industrial sites are heavily polluted. Even the waters of the open ocean show signs of human-caused contamination. [Adapted from A. Couper, ed., *The Times Atlas of the Oceans* (New York: Van Nostrand Rheinhold, 1983).]

ocean, thousands of kilometers from land, show unmistakable signs of human contamination. A comparison of Figures 15–2 and 15–23 explains in part how this has come about. Many areas of the open ocean that are polluted coincide with the major shipping lanes used by oil tankers to transport crude oil from where it is mined to where it is processed and consumed. We now understand that the water of all the oceans is exchanged regularly, so that contaminants introduced locally can be dispersed widely. Biomagnification leads to the build-up of toxic chemicals in the fatty tissue of animals that occupy successively higher levels of a food web, rendering many finfish and shellfish poisonous for predators, including humans. Many of these tainted animals, including marine birds, migrate long distances seasonally,transporting chemicals stored in their tissue to parts of the ocean far from industrial sites. There the chemicals are incorporated into the biogeochemical cycles of ostensibly remote ocean areas as the animals are eaten or die and decompose.

It is obvious that our collective activities are overpowering the resilience of some marine habitats and their biota, in some cases, irreversibly destroying them at an alarming rate. It is certain that the decision not to pollute or, more realistically, to mitigate despoilment of the ocean will be expensive. But there is no alternative in a world that is finite. It must be done; and many groups are banding together in good faith to see how this goal can be accomplished as quickly as possible so that ecosystems—both terrestrial and marine—maintain their natural complexity, resilience, and beauty. For example, we can take heart in the global morato-

Red Tides

Phytoplankton, mainly diatoms and dinoflagellates, are unicellular, floating algae. As primary producers, they photosynthesize and manufacture food for themselves and for all other animals, both small and large. They are the critical foundation on which marine food chains and webs are founded. Although most phytoplankton are benign, a few species of dinoflagellates produce toxins that when ingested can cause harm to many marine species and even to humans. When these toxin-producing dinoflagellates undergo plankton blooms, they discolor the water red (hence the term *red tide*) and may cause the mass deaths of fish, birds, and mammals. Such outbreaks of toxin-producing algae are referred to as Harmful Algal Blooms (HABs). Lately, HABs have become a common phenomenon in the shallow inshore waters of the Gulf of Mexico, especially along sections of coastal Texas and Florida.

Most red tides are small and of little consequence. For example, the Gulf Coast of Texas has experienced al least a dozen red tides during the past sixty years. A few HABs, notably those that occurred in 1935, 1986, and 1997, have, however, seriously affected marine life and humans. For example, fishery biologists estimate that more than 14 million fish, mainly mullet and menhaden, were killed during the 1997 HAB event of Texas. The culprit has been identified as the dinoflagellate *Gymnodinium breve* (**Figure B15–9**). Under the right ecological conditions, *G. breve,* which normally occurs in small numbers, proliferates rapidly, and concentrations in seawater can reach as high as 40,000 cells per milliliter (~40,000 cells per one-third of an ounce); these dense concentrations impart a distinctive red hue to the water (**Figure B15–10**) because of the red pigmentation in the algal cells. *G. breve* produces a neurotoxin that can be concentrated or biomagnified in shellfish such as oysters, clams, and mussels as these organisms filter food out of the water column. When eaten by humans, the resultant poisoning can cause nausea, dizziness, fever, and tingling sensations in hands and feet. When the cells burst in the turbulent water of the surf zone, the toxin is released as an aerosol that affects people in a variety of ways, inducing sore throats, coughing, sneezing, and itchy eyes. To protect people, the Texas Department of Health closes shellfish grounds whenever concentrations of *G. breve* exceed 5 cells per milliliter.

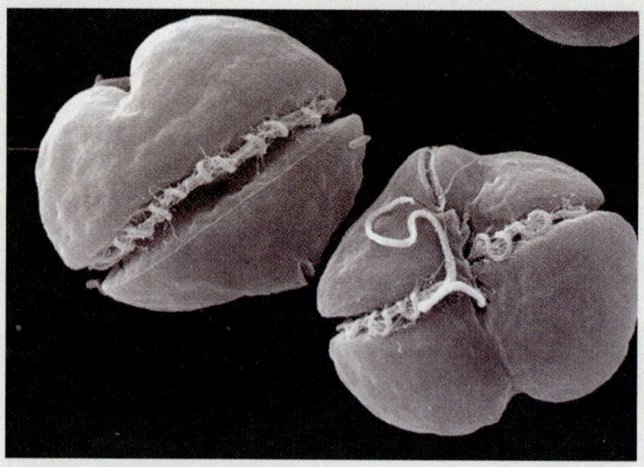

FIGURE **B15-9**

Plankton. Photomicrograph of the dinoflagellate *Gymnodinium breve,* a species that causes red tides in U.S. waters.

rium on the dumping of radioactive wastes into the ocean, a multilateral agreement that was reached in 1986. Also, in 1971 the Inter-Governmental Working Group on Marine Pollution, under the auspices of the United Nations, met to discuss the international regulation of ocean dumping of wastes. Over sixty-four nations have ratified an agreement known as the London Dumping Convention that addresses in a serious manner the entire problem of production, treatment, and disposal of wastes. Its recommendations are summarized in **Table 15–6**. This is a heartening beginning.

In 1989 the Group of Experts on the Scientific Aspects of Marine Pollution (GESAMP), an international committee sanctioned by the United Nations, met to evaluate the polluted state of the oceans. Some of the group's more important conclusions follow:

1. Most of the water of the open ocean is "clean" except in frequently used shipping lanes.
2. Coastal, nearshore, and shelf waters are contaminated just about everywhere. The human impact in these marine areas varies considerably and

The ban on shellfishing lasts longer than the duration of the actual HAB, because the neurotoxin is known to persists for up to two months in the tissue of bivalves that have ingested the algae.

This irksome, dangerous problem is not confined to Texas coastal waters. Between March and April of 1996, for example, HABs of *G. breve* are believed to have caused the death of about 150 manatees, which represents over 5 percent of the estimated population of these endangered mammals. Also, researchers have noted an increase in two major forms of cancer in Florida shellfish during the past twenty-five years, which they suspect is related to more frequent outbreaks of red tides in the region. HABs are now reported for the first time in the nearshore waters of Louisiana, Mississippi, and Alabama. Oyster fisheries were recently closed in Alabama because beach tourists suffered a flare up of respiratory ailments caused by the toxic aerosols associated with *G. breve*. Numerous reports of red tides during the past decade have affected the entire Atlantic seaboard of the United States, including Maine, Massachusetts, Rhode Island, and North Carolina. Economists have noticed a significant loss of tourist dollars locally in areas affected by red tides. Elsewhere the effect of red tides has been more tragic. HABs have been responsible for the poisoning of more than 125 people and the death of more than 30 in the Philippine Islands during the summer of 1996.

FIGURE **B15-10**

Plankton blooming. Photograph of red tide outbreak in Florida waters.

What causes HABs? Nobody knows for certain. This research topic is currently being funded generously in an effort to determine the specific ecological conditions that trigger blooms of toxic-producing dinoflagellates in coastal waters. By so doing, scientists hope to be able to predict the future occurrence of HABs and thereby mitigate their effects on marine life and people.

Visit www.jbpub.com/oceanlink for more information.

reflects the local and regional population density, degree of urbanization, agricultural practices, and shipping activity. Unfortunately, all of these activities will continue to increase in the future as human populations, resource exploitation, and developing and developed national economies expand at an ever-faster rate.

3. Coastal habitats—beaches, wetlands, bays, estuaries, coral reefs, mangrove forests—are being severely affected and even destroyed outright at an unprecedented rate. This is the direct result of human activity on land, particularly coastal development, tourism, urbanization, and industrialization.

4. Although there is an immediate concern about major pollutants—oil, plastics, sewage, heavy metals, radioactive contaminants, nutrient loading, and synthetic organic compounds—there may also be an unrecognized effect on the ocean's ecosystems that is associated with the long-term presence of contaminants that are present in sublethal amounts. What are the probable effects of low-level pollution over the long run? What is the collective effect of all of this material?

SCIENCE BY NUMBERS

Steady State

Let's assume that an estuary that is 10^4 meters long, 2×10^3 meters wide, and 10 meters deep is receiving an industrial pollutant that is highly soluble in water at a rate of 1.5 tons/day. Rivers pour into the estuary 2×10^5 m^3 of water each day. Biologists concerned about the well-being of the ecosystem want to calculate the steady-state concentration of the contaminant. In other words, what will be the contaminant's concentration in the water once the system is balanced and input equals output of the pollutant? This is not a difficult problem to solve as long as we do it systematically, understand the concepts of steady state and residence time, and make the appropriate assumptions.

When the system reaches steady state, the inflow and outflow of the pollutant are balanced, and we can calculate the residence time of the pollutant. If we assume that the contaminant is uniformly mixed in the water, then it follows that the residence time of the pollutant must equal the residence time of the water in the estuary. So we begin by calculating the residence time (*RT*) of the water. We first need to know the volume of water in the estuary. This is easily calculated from the estuary's dimensions as follows:

$$\begin{aligned}&\text{Volume of water}\\ &= (\text{length})\,(\text{width})\,(\text{depth})\\ &= (10^4\ \text{m})\,(2 \times 10^3\ \text{m})\,(10\ \text{m})\\ &= 2 \times 10^{(4+3+1)}\ \text{m}^3\\ &= 2 \times 10^8\ \text{m}^3.\end{aligned}$$

Now the residence time (*RT*) of water in the estuary is calculated by dividing its volume by the inflow rate of water.

$$RT = \text{volume of water in the estuary/inflow rate of water.}$$
$$RT = 2 \times 10^8\ \cancel{\text{m}^3}/2 \times 10^5\ \cancel{\text{m}^3}/\text{day} = 1 \times 10^{(8-5)}\ \text{days} = 10^3\ \text{days.}$$

TABLE 15-6 Guidelines for minimizing ocean pollution.

Recycle and reuse wastes.
Treat wastes that cannot be reused or recycled.
Use pesticides and fertilizers in such a fashion that they do not enter the marine environment.
Use sea disposal only for those materials that are compatible with the marine environment.
Use locations for sea disposal of wastes that will not interfere with other uses of the sea.
Use waste-disposal practices at sea that minimize local impacts at the point of disposal.
Monitor the health of the oceans on a continuing worldwide basis.
Manage the resources of the sea so as to prevent depletion of resources on a worldwide basis.

Source: I. W. Duedall, *Oceanus* 33 (2) (1990): 29–38.

GESAMP recommended in its report that research to study such questions be initiated.

5. Overall, according to GESAMP, it appears that "few" of the ocean's principal resources have been irreparably damaged. Most of the open ocean is "unpolluted," and the damage caused by pollution at some coastal localities has been countered and even reversed. GESAMP concluded strongly that too little is being done to curtail human activity on land, the origin of essentially all of the pollutants that are directly or indirectly despoiling marine ecosystems.

6. Nations must reduce waste, recycle materials, and conserve natural resources. Otherwise marine ecosystems worldwide will continue to be degraded and destroyed.

The United Nations declared 1998 the Interna-

We conclude that the residence time of water in the estuary is 1,000 days. If the pollutant is uniformly mixed, then it follows that its residence time likewise must be about 1,000 days.

Now let's answer the question about the concentration of the pollutant once it reaches steady-state conditions. We use the residence-time formula once again, but apply it to the contaminant as follows:

If RT = amount of pollutant in the estuary ÷ input rate of pollutant,
then the amount of pollutant in the estuary
$= (RT)$ (input rate of pollutant)
$= (10^3 \cancel{\text{days}})(1.5 \text{ tons}/\cancel{\text{day}}) = 1.5 \times 10^3 \text{ tons}.$

We've determined the steady-state amount of the pollutant in the estuary to be 1.5×10^3 tons. Let's convert this to another measure of concentration, such as parts per million (ppm). This is a straightforward determination, if we proceed systematically and think about what we have to do. We know the total amount of the pollutant in the estuary under steady-state conditions to be 1.5×10^3 tons. If we compare this amount to the total amount of water in the estuary, then we will have the pollutants' concentration.

The first step is to convert the volume of seawater in the estuary to a mass. By definition, a cubic meter of water weighs exactly 1 metric ton (m^3 of water = 1 ton). Therefore, 2×10^8 m^3, the volume of water in the estuary calculated above, equals 2×10^8 tons. Our steady-state concentration for the pollutant is

concentration of pollutant = amount of pollutant ÷ amount of water
$= 1.5 \times 10^3 \cancel{\text{tons}} \div 2 \times 10^8 \cancel{\text{tons}} = 1.5 \times 10^{(3-8)} \div 2$
$= 1.5 \times 10^{-5} \div 2 = 0.75 \times 10^{-5} = 7.5 \times 10^{-6}$

Hence, we conclude that the steady-state concentration of the contaminant is 7.5 ppm (7.5×10^{-6}).

tional Year of the Ocean (IYO) in order to educate people of the world about the importance, challenges, and opportunities provided by the Earth's many marine ecosystems. Programs of all sorts were developed to demonstrate that the resources of the sea are critical to the welfare of humans everywhere and that it is important to use them responsibly and sustainably. The aim of the programs was to show governments, businesses, organizations, and individuals how a scientific understanding of the workings of the oceans is essential for protecting and preserving these valuable environments. Other special programs were developed to foster cooperative action and procedures for mitigating the economic impact of future El Niños and for avoiding the unnecessary and unwanted collapse of commercial fisheries. The IYO fostered education programs, provided scientific assessments of ocean processes and marine ecosystems, gave awards to recognize special ocean-related work and achievements, sponsored a host of conferences on ocean–issues, designed and issued postage stamps to celebrate IYO, and produced many special publications on related issues of all sorts.

The main objective of IYO is to show how humans are tied to the sea and that the oceans are a commons that needs to be protected and nourished for all to enjoy in perpetuity. To appreciate these points, consider the following. One out of six jobs in the United States is marine related and more than half of the U. S. population lives within fifty miles of the shore. Also, about one-third of the gross domestic product of the United States is affiliated with the coast, involving recreation, fishing, transportation, and other shore-based industries. So, it is apparent to many people that we need to safeguard the resources of the sea, recognizing that unregulated human impact can alter ecosystems in ways that we want to avoid so that we do not bequeath a damaged and ill-functioning environment to future generations. This is the reason why programs such as the International Coral Reef Initiative (ICRI) are being organized to identify the extent and causes of coral reef degradation and to establish workable conservation efforts while there is still time to act.

15-8 Future Discoveries

Major collaborative efforts by ocean scientists, managers and policy decision makers are being developed to assure that renewable ocean resources are used sustainably and are extracted with the least possible impact on the environment. As marine fishery biologists obtain more information about the ecological needs of fish populations, guidelines and policy decisions for the wise management of these ecosystems are being prepared and enacted. Major research projects are underway in many regions of the ocean to study the biology of fish stocks and to see if any new species can be added to the fisheries. Many scientists are involved with climate studies to see how variations in regional and even global climates impact upwelling processes and therefore fish production. Other oceanographers are studying the effect of toxins and pollutants on the behavior and life habits of communities in shallow and deep water.

The projects are far ranging, many of them using novel approaches. For example, an icebreaker was frozen purposely into the Arctic sea ice. While drifting slowly with the currents, scientist living aboard the vessel gathered invaluable information about ice buildup and melting, heat exchange, and physical and chemical interactions among clouds, snow, ice, air and water. This invaluable information will provide a better understanding of the workings of the Arctic which in turn will be used to improve models for forecasting global climate.

STUDY GUIDE

KEY CONCEPTS

1. Pollution is the introduction by humans of material and energy from the environment such that concentrations of these substances are raised above natural levels to a degree that results in environmental degradation.

2. When spilled at sea, crude oil is affected by evaporation, dissolution, emulsification, spreading, vertical mixing, sedimentation, and biodegration. Because of the complex chemistry of hydrocarbons that is determined by its geologic origins, crudes will respond differently to these processes, making it difficult to predict the short- and long-term outcome of a particular spill.

3. Municipal and industrial effluents include sewage (human wastes), metals (lead, mercury), and artificial biocides (DDT, PCBs). Many of these effluents are toxic and as a persistent influx can be devastating to the biota of a local habitat. Some of these substances undergo bioaccumulation and biomagnification, whereby their concentrations are built up in the tissues of organisms that occupy the successively higher levels of the food chain.

4. Ocean dredging and mining remove large quantities of material from the ocean floor, seriously disrupting the benthic community. Furthermore, the dredge spoils and mining slurries must be disposed of. This affects another area of the sea floor,

particularly if the dumped material is contaminated with pollutants.

5. Environmental spoilage is occurring at an alarming rate as pollutants are spread across the globe, even to the most remote regions of the ocean. Ocean pollution is no longer merely a local or regional concern; it is a global one that will require multinational cooperation if it is to be resolved.

KEY WORDS

bioaccumulation
biomagnification
bioremediation
DDT
dredging
eutrophication
greenhouse effect
Harmful Algal Blooms
heavy metal
Maximum Sustainable Yield
pathogen
PCBs
pollution
oil
sewage

QUESTIONS

REVIEW OF BASIC CONCEPTS

1. What is the definition of pollution?

2. What exactly is crude oil?

3. To reduce significantly the oil pollution of the ocean, where should we concentrate our effort and funds? Explain your reasons.

4. What processes affect an oil spill in the open ocean? Which of these are most important in the first few days of the spill? Which are most significant several weeks after the spill? Explain.

5. What is eutrophication? How is it caused and how can it be prevented?

6. By what biological process can even low-level concentrations of metals and artificial biocides dissolved in water affect an ecosystem? Describe the process.

7. What specific environmental impacts are associated with mining the floor of the deep sea?

8. What is bioremediation, and what are its advantages and disadvantages for the clean-up of oil spills?

9. What combination of factors have caused fisheries to collapse?

10. What is the greenhouse effect and how does it connect to a rise in sea level?

CRITICAL-THINKING ESSAYS

1. The standing water in a pond of a remote and isolated salt marsh is naturally hypoxic most of the time? Is this pollution? Why or why not?

2. How can deepening of a tidal channel in an estuary by dredging be considered a form of pollution? Be specific.

3. What are the short- and long-term consequences of a large oil spill along a coastline? of discharging heavy metals in an estuary? of discharging raw sewage in a bay?

4. Why are dredge spoils usually not suitable for the sand nourishment of a badly eroding beach? Be specific.

5. Speculate about the long-term consequences of using genetically engineered microbes for bioremediation of a shoreline.

6. If you were in charge of sustaining the Georges Bank fishery, what would you do? What kinds of information would you need to decided on a management policy?

7. How are even the most remote parts of the ocean—the polar seas and the deep-sea floor—being affected by pollution?

8. Should municipalities and states start planning for the expected global rise in sea level? Or is it more prudent to wait and see whether a rapid rise of sea level will occur sometime in the twenty-first century?

DISCOVERING WITH NUMBERS

1. Assume that a metal dissolved in the water of a bay occurs in a concentration of 0.000005 ppm (parts per million). The concentration of this same metal in the tissue of ospreys that hunt fish in the area is 50 ppm. How many times has the metal been biomagnified?

2. Examine Figure 15–12b. How many orders of magnitude is DDT magnified in ospreys in comparison to DDT levels in the water?

3. If a machine that collects nodules processes on the average about 5,000 tons of sediment per day, how many tons of nodules are mined each day if the nodule-to-sediment ratio by weight is 4 to 1?

4. Examine Figure B15–2. Calculate the average rate of drift of the oil spill in centimeters per second and in knots (see Appendix II) between March 18 and 24.

5. Given the rate of drift calculated for the preceding question, estimate the average wind speed, assuming the spill was drifting at about 3 percent of wind speed.

SELECTED READINGS

Anderson, D. M. 1994. Red tides. *Scientific American* 271 (2): 62–68.

Atlas, R. 1993. Bacteria and bioremediation of marine oil spills. *Oceanus* 36 (2): 71.

Bascom, W. 1974. Disposal of waste in the ocean. *Scientific American* 231 (2): 16–25.

Butler, J. N. 1975. Pelagic tar. *Scientific American* 232 (6): 90–97.

Capuzzo, J. E. M. 1990. Effects of wastes in the ocean: The coastal examples. *Oceanus* 33 (2): 39–44.

Clark, W. C. 1989. Managing planet Earth. *Scientific American* 261 (3): 47–54.

Dauncy, G. and P. Mazza. 2001. *Stormy Weather: 101 Solutions to Global Climate Change.* British Columbia, Canada: New Society Publishers.

Grove, R. H. 1992. Origin of western environmental action. *Scientific American* 276 (1): 42–47.

Duffy, D., et al. 1987. Penguins and purse seiners: Competition or coexistence? *Research and Exploration* (National Geographic Society) 3 (4): 480–88.

Houghton, R. A., and G. M. Woodwell. 1989. Global climactic change. *Scientific American* 260 (4): 36–44.

Hull, E.W.S. 1978. Oil spills: The causes and the cures. *Sea Frontiers* 24 (6): 360–69.

MacDonald, I. R. 1998. Natural oil spills. *Scientific American* 279 (5): 56–61.

McGinn, A. P. 1998. Rocking the Boat: Conserving Fisheries and Protecting Jobs. Paper 142. Washington, D.C.: World Watch.

Nowadnick, J. 1977. There is but one ocean. *Sea Frontiers* 23 (3): 130–40.

Oceanus. 1982–83. Marine policy for the 1980s and beyond. Special issue 25 (4).

Oceanus. 1990. Ocean Waste Disposal. Special issue 33 (2).

Oceanus. 1996. Ocean and Climate. Special issue 39 (2).

Odell, R. 1972. *The Saving of San Francisco Bay.* Washington, D.C.: Conservation Foundation.

O'Hara, K. J., S. Iudicello, and R. Bierce. 1988. *A Citizen's Guide to Plastics in the Ocean: More Than a Litter Problem.* Washington, D.C.: Center for Environmental Education.

Parker, P. A. 1990. Clearing the oceans of plastic. *Sea*

Frontiers 36 (2): 18–27.
Pontecorvo, G. 1986. *The New Order of the Oceans.* New York: Columbia University Press.
Revelle, R. 1983. The oceans and the carbon dioxide problem. *Oceanus* 26 (2): 3–9.
Ruivo, M. (ed.) 1972. *Marine Pollution and Sea Life.* London: Fishing News Books.
Thorne-Miller, B. 1998. *The Living Ocean: Understanding and Protecting Marine Biodiversity.* Washington, D. C.: Island Press.
United Nations/GESAMP. 1990. *The State of the Marine Environment.* New York: United Nations Publications.
Wacker, R. 1991. The bay killers. *Sea Frontiers* 37 (6): 44–51.
White, R. M. 1990. The great climate debate. *Scientific American* 263 (1): 36–43.
Wilber, R. J. 1987. Plastics in the North Atlantic. *Oceanus* 30: 61–68.
Zorpette, G. 1999. To save a salmon. *Scientific American* 280 (1): 100–05.

SPECIAL BOOKS ON HUMANS, ECOLOGY, AND THE EARTH

Catton, W. R., Jr. 1982. *Overshoot: The Ecological Basis of Revolutionary Change.* Chicago: University of Illinois Press.
Ehrenfeld, D. 1993. *Beginning Again: People and Nature in the New Millennium.* New York: Oxford University Press.
Elgin, D. 1981. *Voluntary Simplicity: Toward a Way of Life That Is Outwardly Simple, Inwardly Rich.* New York: William Morrow.
Freyfogle, E. T. 1993. *Justice and the Earth: Images for Our Planetary Survival.* New York: The Free Press.
Gore, A. 1992. *Earth in Balance.* Boston, Mass.: Houghton Mifflin.
Hardin, G. 1993. *Living within Limits: Ecology, Economics, and Population Taboos.* New York: Oxford University Press.
Meadows, D. H., D.L. Meadows, and J. Randers. 1992. *Beyond the Limits.* Post Mills, Vt.: Chelsea Green Publishing.
Orr, D. 1994. *Earth in Mind: On Education, Environment, and the Human Prospect.* Washington, D.C.: Island Press.
Rasmussen, L. L. 1996. *Earth Community, Earth Ethics.* Maryknoll, N. Y.: Orbis Books.
Wilson, E. O. 1992. *The Diversity of Life.* Cambridge, Mass.: Belknap Press.
Worster, D. 1993. *The Wealth of Nature: Environmental History and the Ecological Imagination.* New York: Oxford University Press.

TOOLS FOR LEARNING

Tools for Learning is an on-line review area located at this book's web site OceanLink (**www.jbpub.com/oceanlink**). The review area provides a variety of activities designed to help you study for your class. You will find chapter outlines, review questions, hints for some of the book's math questions (identified by the math icon), web research tips for selected Critical Thinking Essay questions, key term reviews, and figure labeling exercises.

Appendix I

PROPERTIES OF THE EARTH

THE EARTH'S DIMENSIONS

Average radius	6,371 kilometers	3,956 miles
Circumference of equator	40,077 kilometers	24,902 miles
Amount of land	149 million square kilometers	58 million square miles
Amount of oceans and seas	361 million square kilometers	140 million square miles
Highest elevation (Mt. Everest)	8,848 meters	29,028 feet
Greatest depth (Mariana Trench)	11,035 meters	36,200 feet

THE EARTH'S OCEANS

Ocean (excluding seas)	Area	Ocean Area	Volume	Average Depth
Atlantic Ocean	82.441 x 10^6 km^2	29.4 %	323.6 x 10^6 km^3	3,926 m
	31.822 x 10^6 $miles^2$		77.7 x 10^6 $miles^3$	12,877 ft
Indian Ocean	73.443 x 10^6 km^2	20.6	291.0 x 10^6 km^3	3,963 m
	28.349 x 10^6 $miles^2$		69.8 x 10^6 $miles^3$	12,999 ft
Pacific Ocean	165.25 x 10^6 km^2	50.0	707.6 x 10^6 km^3	4,282 m
	63.79 x 10^6 $miles^2$		169.8 x 10^6 $miles^3$	14,045 ft

LANDSCAPES OF THE CONTINENTS AND OCEANS

Oceans	70.8 %
Continents	29.2
Midocean ridges	22.1
Ocean basin floor	29.8
Continental shelf and slope	11.4
Continental rise	3.8
Volcanoes, deep-sea trenches, abyssal hills	3.7
Continents excluding mountain ranges	18.9
Mountain ranges on continents	10.3

Appendix II

CONVERSION FACTORS

UNITS OF LENGTH

1 micrometer (μm) = 10^{-6} m = 10^{-3} mm = 0.0000394 in
1 millimeter (mm) = 10^{-3} m = 0.0394 in = 10^{3} μm
1 centimeter (cm) = 10^{-2} m = 0.394 in = 10^{4} μm
1 meter (m) = 10^{2} cm = 39.4 in = 1.09 yd = 0.55 fathom
1 kilometer (km) = 10^{3} m = 0.621 statute mile = 0.540 nautical mile

UNITS OF VOLUME

1 liter (l) = 10^{3} cm^{3} = 1.0567 qt (U.S.) = 0.264 gal (U.S.)
1 cubic meter (m^{3}) = 10^{6} cm^{3} = 10^{3} l = 10^{6} ml = 35.3 ft^{3} = 264 gal (U.S.)
1 cubic kilometer (km^{3}) = 10^{9} m^{3} = 10^{15} cm^{3} = 0.24 statute $mile^{3}$

UNITS OF AREA

1 square centimeter (cm^{2}) = 0.155 in^{2}
1 square meter (m^{2}) = 10.7 ft^{2} = 1.19 yd^{2}
1 square kilometer (km^{2}) = 0.306 statute $mile^{2}$ = 0.292 nautical $mile^{2}$
1 hectare = $10^{4}m^{2}$ = 2.47 acres

UNITS OF SPEED

1 centimeter per second (cm/sec) = 1.97 ft/min = 0.033 ft/sec
1 meter per second (m/sec) = 2.24 statute miles/hr = 1.94 kn (knot)
1 kilometer per hour (km/hr) = 27.8 cm/sec = 0.55 kn
1 knot (kn) = 1 nautical mile/hr

UNITS OF MASS

1 gram (g) = 10^{3} mg = 0.035 oz
1 kilogram (kg) = 10^{3}g = 2.205 lb = 35.28 oz
1 metric ton (MT) = 10^{6}g = 2,205 lb

UNITS OF TEMPERATURE

Conversion: °F = (1.8 x °C) + 32

°C	0	10	20	30	40	50	100
°F	32	50	68	86	104	122	212

USEFUL CONVERSION VALUES

To convert	Into	Multiply by
cm	in	0.3937
m	ft	3.28
km	statute miles	0.621
km	nautical miles	0.540
g	oz	0.035
kg	lb	2.205
m/sec	statute miles/hr	2.24
m/sec	kn	1.94
km/hr	kn	0.55

Appendix III

GEOLOGIC TIME

THE GEOLOGIC TIME SCALE

Era	Period	Epoch	Age (10^6 yr)	Life Forms	Geologic Events
Cenozoic	Quaternary	Recent	0.01	Age of humans	
		Pleistocene	1.6	Appearance of hominids	Glacial advances and retreats Gulf of California opens
	Tertiary	Pliocene	5		Rocky Mountains rise
		Miocene	24	Whales, grazing animals Appearance of apes	
		Oligocene	37		Rise of Alps and Himalayas
		Eocene	58	First primates	
		Paleocene	66	Dinosaur extinctions	Asteroid impact
Mesozoic	Cretaceous		144	Dinosaur zenith	South Atlantic begins to form
	Jurassic		208	First dinosaurs	
	Triassic		245	Primitive mammals Reptiles evolve	North Atlantic begins to form Appalachians formed Glaciation in Gondwanaland
Paleozoic	Permian		286	First reptiles	
	Pennsylvanian		320		
	Mississippian		360	First amphibians and forests	
	Devonian		408	First land plants	Mountain building in Europe and in eastern North America
	Silurian		438	First fishes and vertebrates	
	Ordovician		505		
	Cambrian		600	Age of marine invertebrates	
Proterozoic			2500	First life Oceans form	Extensive mountain building
Archeozoic			4500	Earth's crust forms	

Appendix IV

MAPS AND TOPOGRAPHIC PROFILES

Lines of latitude and longitude create a grid across the Earth's surface, such that any point on land and in the ocean has a unique set of coordinates: for example, 47°N latitude, 118°W longitude. How exactly has this grid network been set up? Let's find out.

If you pass an imaginary plane through the equator, the Earth is divided into two halves: the Northern Hemisphere and the Southern Hemisphere. The radius of this equatorial plane extends from the Earth's center to its surface. If you rotate the radius out of the equatorial plane upward by exactly 90 degrees, the radius will now connect the center of the Earth to the north pole. If you rotate the radius downward out of the equatorial plane by exactly 90 degrees, the radius will now connect the center of the Earth to the south pole. Any orientation of the Earth's radius between the equator and the two poles will have an angle that lies between zero degrees (the equator) and 90 degrees (the poles). **Latitude** is measured north or south of the equatorial plane from zero to 90 degrees. The latitude of two points, 45°N (the radius rotated up out of the equatorial plane by 45 degrees toward the north pole) and 30°S (the radius rotated down out of the equatorial plane by 30 degrees toward the south pole), are shown in Figure IV–1a. If you take the radius for both of these points and rotate it around the Earth, always maintaining the angles, you circumscribe lines called parallels of latitude, in this case, the 45°N parallel of latitude and the 30°S parallel of latitude. Note that all points on the Earth's surface that lie on any parallel of latitude are exactly the same distance from the equator.

With latitude, we can specify how far north or south any point on the earth's surface is from the equator. How do we distinguish the east or west position of a point? This is done with **longitude**, and we simply measure the angular distance to the east or west of a reference frame, which for historical reasons is Greenwich, England. The **meridians of longitude** are north-south lines that converge at the poles. Notice that, unlike latitude, they do not lie parallel to one another. If we pass a plane through the Earth that passes through Greenwich, England and the spin axis that connects the two poles, we separate the Earth into two halves, the Western and Eastern Hemispheres. The prime meridian is zero degrees, and it passes through Greenwich, England. Rotate the prime meridian 180 degrees and you are on the opposite side of the Earth. You can rotate 180 degrees to the west or east of the prime meridian. The prime meridian (0°) and the 90°W and 30°E longitudes are shown in Figure IV–1b.

There, we have it: a grid that divides the Earth's surface into specific and unique coordinates of parallels of latitude and meridians of longitude (Figure IV–1c). Each degree of latitude and longitude is divided into 60 minutes, and each minute into 60 seconds. Let's specify some coordinates: Boston: 42°12′N, 71°03′W (read as 42 degrees, 12 minutes north latitude; and 71 degrees, 3 minutes west longitude); Atlanta: 33°27′N, 82°14′W; Chicago: 41°30′N, 87°27′W; San Francisco: 37°47′N, 122°26′W. By definition, a **nautical mile** is equal to precisely 1 minute of latitude, which means that each degree of latitude consists of 60 nautical miles (each degree of latitude contains 60 minutes).

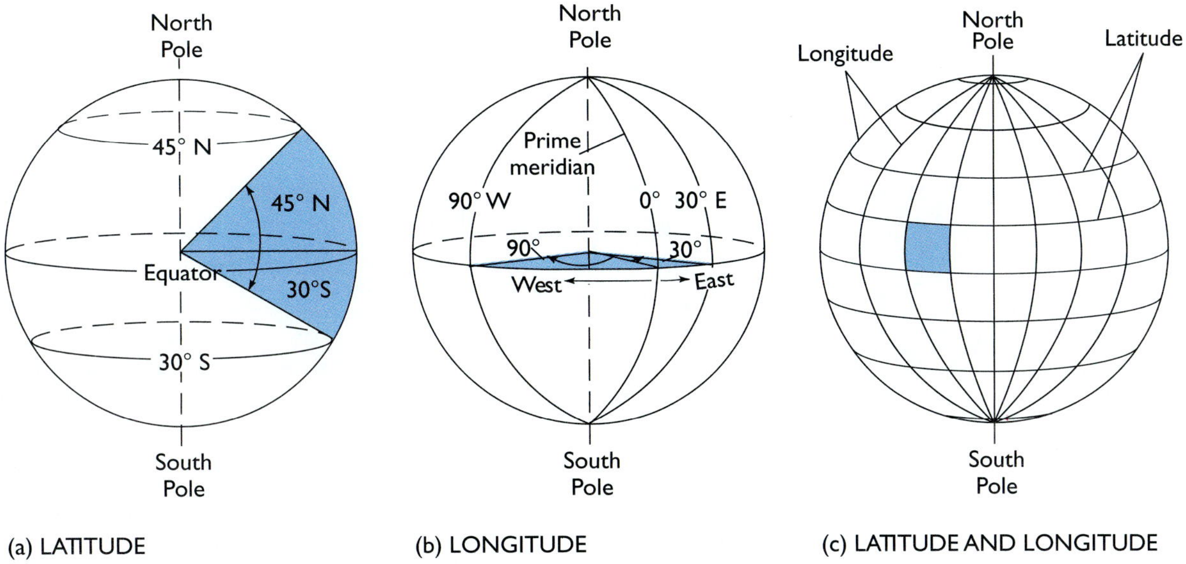

FIGURE **IV–1**

Latitude and longitude. (a) Latitude both to the north and to the south of the equator is a measure of the angle made by the Earth's radius when it is raised above or lowered below the equatorial plane. Rotating this angle 360 degrees around the Earth describes parallels of latitude. The 45°N and 30°S latitudes are shown here. (b) Longitude is the angular distance to the east or west of the prime meridian (0°) chosen to run through Greenwich, England. The meridians of longitude converge at the poles and assume a maximum value of 180 degrees. The 0°, 30°E, and 90°W longitudes are shown here. (c) The intersection of lines of latitude and longitude create a grid that uniquely defines coordinates for any area or spot on the Earth's surface.

CONTOUR MAPS

A road map displays essential features—roads, cities, towns, rivers, bridges—accurately located relative to one another. Topography and elevation, if they are denoted on road maps at all, are depicted crudely by shading or coloring, or by identifying hills or mountain peaks with symbols and numbers indicating their heights. Contour maps show much more than this. They display topography in its three-dimensional shape by the use of contour lines. A **contour line** connects all points having the same elevation. This means that if you follow the 100-meter contour line, you will be walking exactly at an elevation of 100 meters above sea level. If you move off the line to one side, you will rise in elevation; to the other side, you will drop.

The vertical distance separating adjacent contour lines is indicated by the **contour interval**. A contour interval of 10 meters, for example, indicates that 10 meters of elevation are gained or lost as you move to the next contour line. If you go downhill from the 100-meter contour line, you will have dropped to an elevation of 90 meters. If you climb uphill, you will gain 10 meters and be at an elevation of 110 meters. Contour lines delineate hills, valleys, slopes, and cliffs on a map. With practice, it is possible to visualize the details of the terrain without actually being there.

Examine Figure IV–2a which is a topographic map. Note that the contour interval is 50 meters, indicated in the inset at the lower left corner of the map. Imagine yourself moving from Point 1 to Point 2 on the map. Point 1 lies at an elevation of 400 meters. You begin hiking and, after hiking uphill for 1.5 kilometers, you've gained 50 meters and now are at an elevation of 450 meters. You continue for another 2.5 kilometers and gain 50 more meters of elevation, reaching 500 meters above sea level. Note that the walking was easier because the slope along your hiking path is gentler between 450 and 500 meters than it is between 400 and 450 meters. You can tell this easily from your map by noting how far apart the contour lines are. If you gain 50 meters in height while walking horizontally for 2.5 kilometers rather than for 1.5 kilometers, which of the two will be

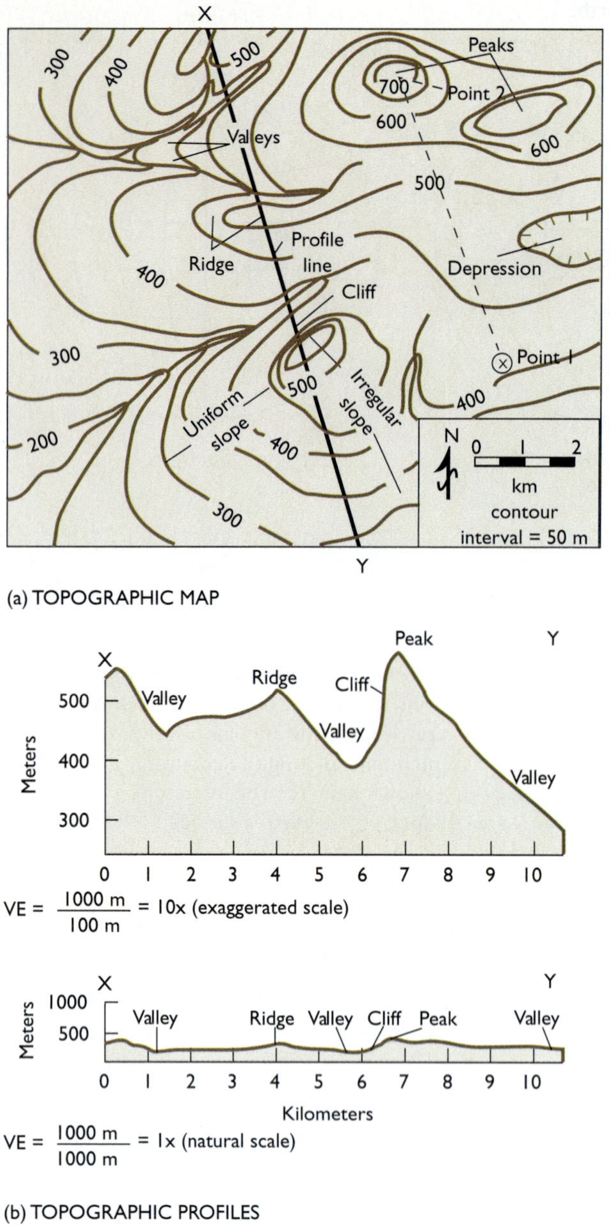

FIGURE **IV–2**

Topographic maps and profiles. (a) Contour lines connect points of equal elevation and are used to depict topographic features, such as valleys, hills, ridges, and slopes. Changes in the steepness of slopes are indicated clearly by the spacing of adjacent contour lines. (b) Two profiles delineate the topography along the X-Y line on the map in part a. Note that when the slope on the profile is steep, the contour lines on the map are closely spaced. The upper profile is drawn with a vertical exaggeration (VE) of 10x. The lower profile is drawn with no vertical exaggeration. Note how magnified the slopes are in the upper profile (an exaggerated scale) relative to the actual slopes in the lower profile (a natural scale).

TABLE **IV–1**

Important properties of topographic contour lines.

1. Every point along a contour line has the same elevation or depth.
2. Contour lines cannot cross one another, and they cannot merge or bifurcate (divide into branches).
3. The spacing of contour lines depends on the slope of the ground. If the slope is steeper, the contours are closer together; widely separated contour lines denote gentle slopes.
4. The regular spacing of a series of contour lines depicts a uniform slope; the irregular spacing of contour lines depicts an uneven slope.
5. Both ends of a contour line eventually must join together and form a continuous, unbroken line, either inside or outside the map boundaries.
6. A closed contour on a map indicates a hill or mountain, unless the contour line has hatch marks; then it represents a depression.
7. Contour lines bend up valleys and form V-shaped patterns, with the apex of the V pointing upvalley.

the steeper slope? Note how much steeper your hike is going to become as you climb to the top of the 700-meter peak, your destination. You can tell this quickly by the closeness of the contour lines as you get nearer to the peak.

Table IV–1 summarizes the important properties of contour lines that a map reader needs to understand in order to use the map properly. Marine charts depict the topography beneath the sea surface, which is called **bathymetry**. The contour lines on bathymetric maps, referred to as **isobaths**, depict depth below the ocean surface.

TOPOGRAPHIC PROFILES

A **topographic** or **bathymetric profile** is a topographic or bathymetric silhouette (Fig. IV–2b) along which variations of elevation are depicted accurately. It is like seeing dark mountain peaks against a light sky just as the sun is rising or setting. **Vertical exaggeration** is used to bring out relief along the profile. This exaggeration magnifies the vertical scale relative to the horizontal scale, so that vertical features are expanded (exaggerated). A vertical exaggeration of 10x indicates that the vertical dimension is expanded ten times relative to the horizontal dimension. This greatly distorts the topography, making even gentle slopes appear to be quite steep.

To appreciate this technique, compare the two topographic profiles in Figure IV–2b. The bottom profile has no vertical exaggeration (horizontal scale = vertical scale), whereas the top profile is exaggerated by a factor of 10 (horizontal scale = 10 x vertical scale). Note how deep the valley becomes and how steep-faced the peak appears.

OCEANOGRAPHIC PROFILES

An **oceanographic profile** is basically a cross section of some physical or chemical property of the water column. It is commonly used to display the vertical distribution of temperature, salinity, or the concentration of some chemical compound that is dissolved in the water. Contour lines are used to connect points in the water column that have equal values for some parameter, such as salinity and temperature. For example, an **isotherm** in a profile is a contour line that extends through water of equal temperature. In this case, the contour interval is in degrees of temperature, such as 5°C or 10°C. Figure IV–3 is an oceanographic profile that shows the distribution of water temperature. The spacing of the isotherms in this diagram reflects the vertical temperature gradient. The closer the contour lines are to one another, the more rapidly temperature changes with water depth. A sharp temperature gradient, a **thermocline,** appears as a series of closely spaced isotherms (see Figure IV–3). Contour lines for salinity are called **isohalines;** for density, **isopycnals.**

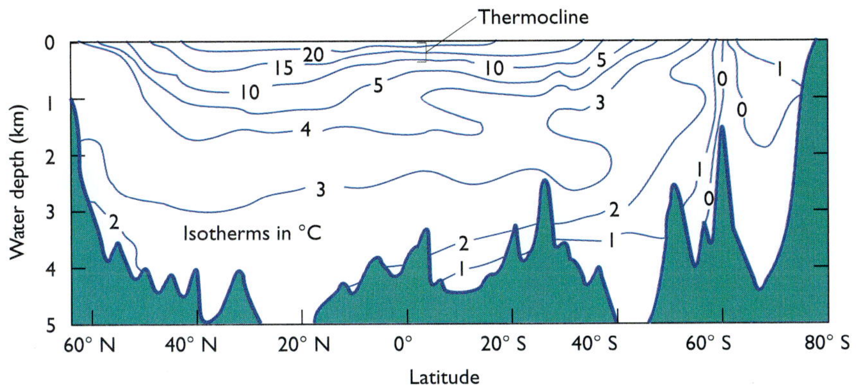

FIGURE IV–3

Oceanographic profile. Oceanographers employ contour lines to delineate variations in the properties of seawater, such as temperature. The isotherms shown in this temperature profile of the Atlantic Ocean are contour lines that connect points in the water column having equal temperature values. Note that a variable contour interval is used in the profile (a 1-degree interval for water colder than 5°C and a 5-degree interval for water warmer than 5°C). Also thermoclines (sharp vertical temperature gradients) are clearly evident in the upper levels of the profile, as indicated by the closely spaced isotherms. [Adapted from G. Deitrich, K. Kalle, W. Kraus, and G. Siedler, *General Oceanography: An Introduction* (London: John Wiley Interscience, 1980).]

Glossary

Abiotic Characterized by the absence of life.

Abyssal hill A relatively small hill, typically of volcanic origin, rising no more than 1,000 meters above the sea floor (Fig. 2-3).

Abyssalpelagic zone A subdivision of the pelagic province lying between the water depths of two and six kilometers (Fig. 9-1).

Abyssal plain A flat area on the deep-sea floor having a very gentle slope of less than one meter per kilometer, and consisting chiefly of graded terrigenous sediments known as turbidites (Figs. 2-3 and 2-6).

Abyssal zone A subdivision of the benthic province encompassing the sea floor between the depths of two and six kilometers (Fig. 9-1).

Accretionary prism The region of a subduction zone located between the deep-sea trench and volcanic arc, where thick sedimentary units and volcanic rocks are squeezed and wrinkled by compression to form thrust faults and tight folds (Fig. 4-7).

Acidic solution A liquid in which the concentration of H^+ exceeds the concentration of OH^-. A solution with a pH of less than 7 (Fig. 5-18).

Advection The horizontal transport of liquid and air in contrast to their vertical motion (convection).

Aerobic bacteria Bacteria that respire in the presence of free oxygen (O_2) (Fig. 9-2a).

Ahermatypic coral Coral that do not possess zooxanthellae (algae) in their tissue and, hence, can dwell in the dark waters of the aphotic zone. These do not form large reefs.

Algae Simple marine and freshwater plants, unicellular and multicellular, that lack roots, stems, and leaves. Examples include diatoms and seaweeds (Figs. 9-3 and 9-5).

Algal ridge An irregular, durable ridge composed of encrusting algae; it is located on the ocean side of many coral reefs where it is pounded regularly by big waves (Fig. 12-17).

Alkaline solution A liquid whose OH^- concentration exceeds its H^+ concentration, imparting a pH greater than 7 (Fig. 5-18).

Alluvial fan A cone-shaped deposit of sediment that forms at the mouth of steep mountain streams in desert regions (Fig. 4-11).

Altimetry A technique to measure the height of the sea surface from radar pulses emitted by a satellite.

Amphidromic point A nodal point in an amphidromic system from which cotidal lines radiate (Figs. 8-9 and 8-11).

Amphidromic system A depiction of the large-scale rotary motion of the tides in ocean basins and seas that results from the Coriolis deflection of the tides (Figs. 8-9 and 8-11).

Anaerobic bacteria Bacteria that respire in the absence of free oxygen (O_2) (Fig. B10-4).

Andesite A common volcanic rock found in the volcanic arcs of subduction zones; it is intermediate in composition between the granitic crust of the continents and the basaltic crust of the oceans (Table 2-2).

Angle of wave approach The angle subtended by a wave crest and the shoreline as the wave breaks (Fig. 11-1).

Anhydrite A type of evaporite mineral composed of calcium sulfate ($CaSO_4$).

Anion A negatively charged ion such as Cl^- (chloride) (Fig. 5-4).

Anoxic The absence of free oxygen (O_2) (Figs. 6-18 and 15-8).

Antinode The point on a standing wave where the maximum vertical displacement occurs (Fig. 8-17).

Aphotic zone The dark region of the ocean that lies below the sunlit surface waters (Fig. 9-1).

Ash The loose debris that is ejected from an erupting volcano (Fig. 3-9).

Aspect ratio As applied to the caudal fin of a fish, it is a ratio of the square of the height of the fin to its surface area. Darters and lungers have low aspect ratios, cruisers high aspect ratios (Fig. 9-22).

Asthenosphere The region of the Earth's upper mantle that, as a consequent of a small amount (<1%) of partial melting, deforms in a plasticlike manner; it extends downward for about 250 kilometers from the base of the lithosphere (Figs. 2-1 and 3-7b).

Atoll A ring-shaped coral reef that surrounds a lagoon (Fig. 12-20).

Atom The smallest component of an element, comprising neutrons, protons, and electrons (Fig. 5-1 and 5-3).

Authigenic deposits Material that formed *in situ*, typically by biochemical precipitation from seawater, as distinguished from material transported to the site of deposition such as sand particles (Fig. 4-15).

Autotroph Plants and bacteria that synthesize food from inorganic nutrients (Fig. 10-1).

Baleen A series of elastic, horny plates that grow in place of teeth in the upper jaw of a certain group of whales called Mysticeti.

Bar A linear ridge of sand in the nearshore zone that trends parallel or subparallel to the shore (Figs. 11-2 and 11-3).

Barrier island A long, often narrow accumulation of sand that is separated from the mainland by open water (lagoons, bays, and estuaries) or by salt marshes (Figs. 11-11, 11-12, B11-7, and 12-9a)

Barrier reef A coral reef growing around the periphery of an island, but separated from it by a lagoon (Fig. 12-17 and 12-20).

Basalt A dark, fine-grained igneous rock composed of minerals enriched in ferromagnesian silicates; it typifies the oceanic crust (Table 2-2).

Bathyal zone A subdivision of the benthic province encompassing the sea floor between the depths of 200 and 2,000 meters (Fig. 9-1).

Bathymetry The measurement of depth in the ocean in order to delineate the submarine topography (Fig. 2-2 and 2-4).

Bathypelagic zone A subdivision of the pelagic province lying between the water depths of one and two kilometers (Fig. 9-1).

Bathythermograph An instrument used to measure temperature as a function of water depth in the ocean.

Beach An aggregation of unconsolidated sediment, usually sand, that covers the shore (Figs. 11-1 and 11-11).

Beach face The section of the beach subjected to wave uprush (Fig. 11-3).

Beach nourishment Artificial replenishment of sand to a beach.

Bed load All sedimentary grains that move and roll along the bottom as they are transported by currents.

Benioff zone A tilted zone of earthquakes associated with collisional plate boundaries known as subduction zones (Fig. 3-6b).

Benthic province A primary subdivision of the oceans that includes all of the sea bottom (Fig. 9-1).

Benthos All organisms that live on or within the ocean bottom (Figs. 9-6 and 9-7).

Berm The flat accumulation of sand on a beach above the high-tide line (Figs. 11-2 and 11-3).

Bimodal distribution A frequency distribution possessing two distinct modes or maxima (Fig. 2-4).

Bioaccumulation The buildup of chemical substances in the cell or tissue of an organism.

Biomagnification The accumulation and amplification of chemical substances at each succeeding tropic level.

Bioremediation The use of organisms (bacteria) to "clean up" an oil spill.

Biogenic sediment Deposits consisting of > 30% by volume of particles produced by organisms, such as the shells of diatoms, radiolaria, and foraminifera (Fig. 4-4).

Bioluminescence The production of visible light by organisms (Fig. 13-7).

Biomass The quantity of living matter expressed as grams per unit area or unit volume (Figs. 10-11b and c and 15-17).

Biosphere An external shell or envelope of the earth that includes all organic matter, both living and nonliving.

Bloom *See* Plankton bloom.

Bore A steep wave that moves upriver during the flooding tide.

Bottomset beds Horizontal layers of fine sedimentary material deposited in front of a delta (Fig. 11-15).

Bottom water A general term applied to dense water masses that sink to the "bottom" of ocean basins (Fig. 6-15 and 6-16, and Table 6-2).

Brackish water Water of low salinity (typically between 5‰ and 25‰) produced by mixing of freshwater and saltwater.

Breaker A wave that is unstable and collapses or breaks at the shoreline (Fig. 7-9).

Breaker zone The region of the surf zone where shallow-water waves become oversteepened and break (Fig. 11-2).

Breakwater A massive structure erected offshore to protect a shoreline from the direct impact of incoming waves (Fig. 11-16c).

Brine Water having a much higher salinity than normal seawater.

Buffered solution A chemical solution that resists changes in pH despite the addition of small amounts of acids or bases.

Bulk emplacement Gravity-induced transport of sediment, such as slumping and turbidity currents (Figs. 4-9 and 4-10).

Buttress zone The seaward-sloping area of a coral reef (just beneath the algal ridge) that consists of alternating ridges and furrows (Fig. 12-17).

Calcareous Composed of calcium carbonate ($CaCO_3$).

Capillary wave A small wave with a wavelength <1.7 cm, which has surface tension as the dominant restoring force (Fig. B7-3 and Table 7-1.

Carbohydrates A family of organic compounds consisting of carbon, hydrogen, and oxygen, such as sugars and starches.

Carbon-14 A radioactive isotope with a half life of 5,560 years that is particularly useful for dating the remains of organisms.

Carbonate compensation depth (CCD) The depth in the ocean below which material composed of calcium carbonate is dissolved and does not accumulate on the sea floor (Fig. 4-17).

Carnivore An organism that feeds on animals (Figs. 10-1, 10-2, and 10-3).

Cation A positively charged ion, such as K^+ (potassium) and Na^+ (sodium) (Fig. 5-4).

Caudal fin The tail of a fish (Fig. 9-22).

Celerity The speed of a wave.

Celsius A temperature scale calibrated in such a way that (at standard atmospheric pressure) pure water freezes at 0°C and boils at 100°C.

Cement Minerals such as silica and carbonate that are chemically precipitated in the pores of sediments, binding the grains together and transforming them into rock.

Centric diatoms Diatoms with radial symmetry (Fig. 9-16b).

Centrifugal force An apparent force exerted outward from a rotating object. The faster the rate of rotation and the longer the radius of rotation, the stronger the apparent force (Fig. 8-5).

Chalk A white, soft limestone consisting dominantly of the shells of foraminifera.

Chemosynthesis The production of organic compounds from inorganic nutrients using energy from chemical oxidation rather than from sunlight (photosynthesis).

Chert A hard siliceous rock composed of opaline silica derived from the hard parts of microscopic plants (e.g., diatoms) and animals (e.g., radiolaria).

Chlorinity The chloride content, including all the halides, of seawater expressed as parts per thousand (‰).

Chlorophyll Green pigments found in plants that are essential for conducting photosynthesis (Fig. B10-6).

Cilia Short, hairlike features on many simple organisms that are used for locomotion, the generation of a current, or filter feeding.

Clastic sediments Deposits of fragments of preexisting rocks that have been transported from their point of origin.

Clay A term denoting a grain size between silt and colloids (Table 4-1).

Clay minerals Layered, platy minerals composed of hydrous aluminum silicates. Examples include kaolinite and chlorite (Fig. 4-13).

Coastal upwelling The upward flow of cold, nutrient-rich water that is usually induced by Ekman transport (Figs. 10-11a and 10-15b).

Coccolithophores Microscopic, single-celled, planktonic plants having an exoskeleton composed of tiny calcareous plates or discs called coccoliths (Fig. 4-14).

Cold-core ring Large eddies that have a core of cold water (Fig. 6-12).

Colloid Particles smaller than clay-size material (< 0.00024 mm) (Table 4-1).

Compensation depth, carbonate *See* Carbonate compensation depth.

Compensation depth, isostatic The depth in the earth or water column at which masses are balanced and pressures are equal.

Competitive exclusion The process whereby a species in the absence of predators outcompetes other organisms and becomes the dominant species of an ecosystem.

Compound A chemical substance consisting of two or more elements or atoms.

Compressional stress A force that pushes together parts of a rock or structure, decreasing its volume (Fig. 3-2).

Conduction The transfer of heat usually in solids whereby energy is passed from particle to particle by thermal agitation.

Conservative property A property of water that is unaffected by biological processes and, consequently, remains stable over time.

Constant proportion of seawater The abundances of the major ions in seawater occur in constant proportion relative to one another (Table 5-1).

Constructive wave interference The interaction of several waves that creates higher crests and/or deeper troughs as waves in phase combine with one another (Figs. 7-6b and B7-5).

Consumer A heterotroph that ingests an external supply of food (Fig. 10-2a).

Continental crust The light, buoyant granitic rock that underlies continental masses and averages about 35 kilometers in thickness (Figs. 2-4 and 2-5).

Continental drift The process whereby continents are and have been in motion relative to one another across the Earth's surface (Fig. 3-1).

Continental margin The drowned edges of continents consisting of the continental shelf, the continental slope, and the continental rise (Figs. 2-3 and 2-6).

Continental rise The enormous wedge of sediment deposited at the base of the continental slope (Figs. 2-3a and 4-10).

Continental shelf The shallow, near-horizontal sea floor extending from the coast to the upper continental slope (Figs. 2-3 and 2-7).

Continental slope The sloping sea bottom of the continental margin that begins at a depth of about 100 to 150 meters at the shelf edge and ends at the top of the continental rise or in a deep-sea trench (Fig. 2-3).

Contour line A line drawn on a map that connects all points of equal value (elevation, temperature, salinity, etc.) and that is used to portray the third dimension on a two-dimensional surface (Fig. IV-2 in Appendix IV).

Convection The vertical transport of a fluid or the transfer of heat in fluids.

Convergence The meeting of two opposing currents of water or air (Fig. 6-7).

Convergent plate boundary A compressional lithospheric plate boundary where one plate overrides another plate (Figs. 3-6b, 3-10, and 3-11).

Copepods Members that belong to an order of Crustacea that are shrimplike in appearance and that feed voraciously on phytoplankton (Figs. 9-19a and 10-2a).

Corange lines Contour lines on an amphidromic map connecting points having an equal tidal range (Figs. 8-9b and 8-13).

Core The innermost region of the earth that begins at the base of the mantle; it is composed of a very dense alloy of iron and nickel (Fig. 2-1).

Core of sediment A vertical, cylinder-shaped sample of sediment obtained by a sediment corer (Fig. B4-1).

Coriolis effect An apparent force that arises because of the Earth's spin about its axis. Freely moving objects are deflected to the right of their direction of motion in the Northern Hemisphere and to the left of their direction of motion in the Southern Hemisphere (Fig. 6-2b).

Cosmogenous sediment All sediment constituents of extraterrestrial origin.

Cotidal lines Contour lines on an amphidromic map that join all points where high tide occurs at the same time of the day (Figs. 8-9b and 8-13).

Covalent bond The linkage of atoms in a molecule by sharing electrons.

Crust The thin outermost sphere of the solid Earth consisting of either basalt (ocean crust) or granite (continental crust) (Figs. 2-1, 2-4, and 2-5).

Crystal lattice A regular arrangement of atoms and molecules in the structure of a crystal (Fig. 5-5a).

Curie temperature The temperature below which rocks become magnetized in the geomagnetic field.

Current A horizontal movement of water or air.

Debris flow A general term applied to the rapid downslope flow of unconsolidated debris; an example is the turbidity current (Fig. 4-10b).

Decomposers Microbes, primarily bacteria, that convert nonliving organic matter into inorganic compounds including nutrients and gases (Figs. 9-4 and 10-1).

DDT An abbreviation of dichlorodiphenyl-trichloroethane, a pesticide belonging to a chemical group called hydrocarbons.

Deep scattering layer A sound-reflecting layer filled with zooplankton and nekton that moves up at night and down each day (diurnal vertical migration) in response to changing levels of light.

Deep-sea trenches Long, narrow, and deep topographic depressions associated with volcanic arcs that together mark a collisional zone where one lithospheric plate is overriding another (Figs. 2-3, 3-6b, and 3-11).

Deep-water wave A wave travelling in water that is deeper than one half of its wavelength, such that the sea bottom does not interfere with its motion (Fig. 7-4a).

Delta A sediment deposit, typically triangular in shape, formed at the mouth of a river (Fig. 11-15).

Delta plain Plains formed by the accumulation of mud at river mouths (Fig. 11-15a).

Demersal Tendency of an organism to rest on the sea bottom, but swim and feed in the water column (Fig. 9-8).

Density The ratio of a mass to a unit volume specified as grams per cubic centimeter.

Density current A current powered by gravity such as a turbidity current (Fig. 4-10b).

Deposit feeder Organisms that extract food from sediment (Fig. 13-3).

Desalination The removal of salt from seawater, usually to make it drinkable (Figs. B5-3 and B5-4).

Destructive wave interference Several interacting waves that are out of phase so that the resultant is less than each wave constituent (Fig. 7-6c).

Detritivore Organisms that feed on nonliving organic matter.

Detritus Inorganic or organic debris.

Detritus food chain A trophic relation among a variety of organisms that is sustained at its base by organisms that gather bits (detritus) of nonliving organic material. This is to be distinguished from a grazing food chain.

Diatoms Microscopic, unicellular phytoplankton possessing silica valves; they are responsible for much of the ocean's primary production (Figs. 9-3c, 9-16, and 9-17).

Diffusion The dispersal of material by random molecular movement from regions of high concentration to those of lower concentration (Fig. 9-12a).

Dinoflagellates Microscopic, unicellular phytoplankton that engage in limited propulsion by the use of tiny whiplike flagella (Fig. 9-3b).

Dipole molecule A molecule such as H_2O that possesses a positively charged end and a negatively charged end (Fig. 5-4).

Discharge-river Rate of flow usually expressed as a volume of water per unit of time.

Discontinuity A marked or abrupt change in the property of a substance, such as water temperature, salinity, or density. Or a contact between different substances, such as the air-sea interface or the Moho, which separates the crust from the mantle.

Dispersion Outside of the fetch area, wind-generated waves sort themselves according to wave length and wave period, which control their speed in deep water. This sorting effect is dispersion (Fig. 7-7a).

Disphotic zone The part of the water column that is barely illuminated by sunlight from above. The "twilight zone" between the photic and aphotic zones (Fig. 9-1b).

Dissolution The act of dissolving a solid (Fig. 5-4).

Diurnal tide The occurrence of one high and one low tide during one lunar day (Figs. 8-3a and 8-6).

Diurnal vertical migration A habit of some zooplankton and nekton to descend out of the sunlit surface water during daytime and ascend to the surface during nighttime.

Divergence To move apart from a common source (Fig. 6-7).

Divergent plate boundary A lithospheric plate boundary at which plates move apart and new ocean crust (basalt) is created (Figs. 3-10 and B3-4).

Doldrums A belt of variable light winds near the equator.

Doppler acoustic current meter A bottom meter that determines the direction and velocity of currents by noting how the pitch of sound pulses changes when reflected from suspended particles moving with the water current (Fig. B6-1b).

Dorsal The upper or back surface of an animal.

Downwelling The sinking of a fluid (Fig. 6-7).

Dredge A metal collar and collecting bag that is dragged along the bottom to sample rock, sediment, or bottom organisms (Fig. B4-1).

Dynamic theory of the tides A predictive model of the tides that takes into account continents, ocean basins, and irregular bottom topography, and allows water to react dynamically rather than passively to the tide-raising forces (Fig. 8-8).

Earthquake A sudden trembling of the earth as a consequence of faulting or volcanic activity (Fig. B4-4).

Ebb current Water flow directed seaward as the tide drops (Fig. B8-1).

Echo sounding Determining water depth by measuring the time for a pulse of sound emitted near the surface to be reflected off the bottom and return to the surface (Fig. B2-1).

Ecology The study of the interaction of organisms with one another and with their physical and chemical environments.

Ecosystem A discrete ecological unit consisting of all of its constituent organisms and its total environment.

Eddy The circular or whirling flow of water generally found along the edge of the main current.

Ekman transport The net flow of water to the right of the wind in the Northern Hemisphere and to the left of the wind in the Southern Hemisphere which arises as a consequence of Coriolis deflection (Fig. 6-6).

Electron A negatively charged particle in orbit about the nucleus of an atom (Fig. 5-1).

Element A fundamental constituent of matter that cannot be broken down into simpler units by chemical reactions; an element consists of only one kind of atom.

Embayment An indentation in the shoreline that forms a bay.

El Niño An episodic (3–5 years), warm current that usually appears around Christmas off the coasts of Ecuador and Peru and that decimates the indigenous population of organisms (Figs. 10-15 and 10-16).

Encrustation A crustlike deposit or growth over rock, substrate, or other organisms (Fig. 12-18a).

Entrainment The act of resuspending or remobilizing a sediment grain resting on the sea bottom.

Eolian Describes sediment that was deposited by wind (Fig. 4-13).

Epicenter The point on the earth's surface directly above the spot in the earth (the focus) where the earthquake originated (Fig. 3-6a).

Epifauna Animals that live in contact with the sea bottom either moving freely over the substrate or attached to it (Figs. 9-7 and 9-24).

Epiflora Plants that live in contact with the sea bottom either attached or lying freely on the substrate (Figs. 9-7).

Epipelagic zone A subdivision of the pelagic province extending downward from the ocean surface to a water depth of approximately 200 meters (Fig. 9-1).

Epiphyte Organisms that grow attached to other organisms.

Epitheca The upper, larger silica valve of a diatom frustrule (Fig. 9-16a).

Equatorial upwelling The ascent of water along the equator that arises as a consequence of surface-current divergence induced by Coriolis deflection (Fig. 10-8a).

Equilibrium theory of the tide A simplified model of the tide whereby water is in equilibrium with the tide-raising forces and the earth is assumed to be covered by seawater of constant depth with no continents (Fig. 8-6).

Erosion The process of rock disintegration by either chemical or mechanical means (Figs. 4-1, B4-4, 4-9, and 5-6).

Estuary A partially enclosed body of water where freshwater is mixed with saltwater (Figs. 12-1 and 12-3).

Eukaryotes Cells with a true nucleus and other organelles.

Eustatic sea-level change A fluctuation of sea level that is worldwide rather than local or regional in extent (Fig. 4-2b).

Eutrophication A process whereby water becomes anoxic because it is choked with decomposing organic matter (Fig. B12-2).

Evaporation The physical process of transforming a liquid into a gas at temperatures below the boiling point of the liquid (Figs. B4-7, B5-5, 5-12, and 5-20, Table 5-11).

Evaporite A type of sediment precipitated from an aqueous solution, usually by the evaporation of water from a basin with restricted circulation.

Exclusive Economic Zone An agreement whereby coastal states regulate fishing, mineral resources, and scientific research in shelf waters extending out for 200 nautical miles from their shore (Figs. 14-1 and 14-2).

Exoskeleton A skeleton partially or completely covering the exterior of a plant or animal.

Exotic species Species of plants and animals that are not naturally indigenous to an ecosystem; they can outcompete the native species and displace them from their natural habitat.

Fan, deep-sea A large, cone-shaped sediment accumulation that is found at the lower end (mouth) of a submarine canyon. (Fig. 4-11)

Fat An organic compound composed of carbon, hydrogen, and oxygen that is insoluble in water.

Fault A fracture in rocks or sediments along which there has been some slippage (Figs. 2-11, B3-1, B3-2, and 3-6, Table B3-1).

Fauna The animals of a given region or period of time.

Fecal pellet A small aggregate of partially digested organic matter excreted by organisms, particularly invertebrates.

Ferromanganese nodules Chemically precipitated from seawater, these more-or-less spherical masses, a few millimeters to several centimeters in diameter, are composed of concentric layers of iron, manganese, and other metal oxides (Fig. 4-15).

Fetch The area and distance across which wind interacts with the water surface to generate waves (Fig. 7-7a).

Filter feeders Animals that feed by "sifting" small organisms or organic particles that are suspended in the water (Figs. 9-10, 10-3e, and 12-5a and b).

Fjord A narrow, deep estuary excavated by glaciers and that commonly has a sill at its mouth (Fig. 12-1b).

Flagellum A whiplike projection used by some organisms such as dinoflagellates for propulsion (Fig. 9-3b).

Flocculation A physicochemical process whereby clay particles in seawater aggregate into a clump or cluster and form a floccule.

Floe A detached piece of floating sea ice other than an iceberg.

Flood current The landward-directed flow of water associated with the rise of the tide (Fig. B8-1).

Flora The plants of a given region or period of time.

Food chain A simplified trophic relation whereby energy is passed in a stepwise fashion from primary producers to herbivores and to carnivores (Figs. 10-2a and 10-13).

Fold Wavy geologic structures formed by the compression and bending of sedimentary layers.

Food web A series of food chains that are interconnected, creating a mosaic of pathways for the transfer of energy through a biological community (Fig. 10-2b).

Foraminifera Planktonic and benthonic protozoans that have a test composed of calcium carbonate (Figs. 4-14 and 9-3a, Table 4-2).

Foreset beds A sequence of inclined beds that form on the frontal or seaward-facing slope of a delta (Fig. 11-15a).

Fossil The remains or traces of organisms preserved in rocks or ancient sediment.

Fracture zone A linear zone of highly irregular, faulted topography that is oriented perpendicular to ocean-spreading ridges (Fig. 2-3c).

Freezing point The temperature at which a liquid is transformed (freezes) into a solid (Fig. 5-8).

Frequency distribution The arrangement of data that shows the occurrence and range of the values of a variable (Fig. 2-4).

Frictional drag The retarding force associated with the surface of a body moving through a fluid or gas, or with a fluid or gas moving across a body surface. Sometimes called surface drag (Figs. 9-15 and 9-21).

Fringing reef A reef that is growing at the edge of a landmass without an intervening lagoon (Fig. 12-20).

Frustule The siliceous exoskeleton or valves of diatoms consisting of an epitheca and hypotheca (Fig. 9-16).

Fully developed sea A state of the sea whereby the energy in the waves has reached a maximum value for the given set of wind and fetch conditions.

Gas hydrates Individual molecules of methane (CH_4) that are trapped in cagelike structures associated with frozen water.

Geostrophic current A current controlled by a balance between a pressure-gradient force and the Coriolis deflection (Fig. 6-9).

Glacial-marine sediment Sediment transported by glaciers and deposited in a marine environment (Figs. 4-4 and 4-12).

Glacier Large masses of moving ice on land derived by the recrystallization of snow into ice under pressure.

Gondwanaland Refers to the large aggregation of continents—South America, Africa, India, Australia, Antarctica—that formed the southern half of the megacontinent Pangaea during the Mesozoic Era (Fig. 3-1).

Grab sampler A sampling device, often spring loaded, that bites a chunk of sediment out of the sea bed (Fig. B4-1).

Graded bedding Vertical grain-size gradation in a layer of sediment, typically with the coarsest particles at the bottom of the bed, the finest at the top. Turbidites display graded bedding (Fig. 4-10).

Granite A light-colored, coarse-grained, intrusive igneous rock composed mainly of quartz and feldspar and that typifies the continental crust (Table 2-2).

Gravity corer A metal pipe that is driven into soft sediment by a heavy weight as it freefalls a given distance to the sea bottom (Fig. B4-2).

Gravity wave A wave with gravity as the restoring force.

Grazers Animals that feed on plants.

Grazing food chain A food chain in which animals feed directly on plants. This contrasts with a detritus food chain (Fig. 10-2a).

Greenhouse effect The warming of the Earth's atmosphere caused by the absorption of infrared terrestrial radiation by gases such as CO_2 and H_2O.

Groin A small structure erected perpendicular to a beach and designed to trap the longshore drift of sand to promote beach accretion (Fig. 11-16a).

Gross primary production All of the organic material produced in an area by autotrophs.

Groundwater Water beneath the ground surface that seeped into the soil and rock from above (Fig. 5-21 and Table 5-10).

Guyot A seamount of volcanic origin with a flat top (Fig. 2-3).

Gypsum An evaporite deposit composed of hydrous calcium sulfate ($CaSO_4 \bullet 2H_2O$).

Gyre A large water-circulation system of geostrophic currents rotating clockwise (Northern Hemisphere) or counterclockwise (Southern Hemisphere) (Figs. 6-4, 6-5a, 6-7b, 6-9, and 6-10, and Table 6-1).

Habitat The place or site occupied by a specific plant or animal.

Hadal zone A subdivision of both the pelagic and benthic provinces (Fig. 9-1).

Halite A mineral composed of sodium chloride (NaCl) and commonly referred to as rock salt (Fig. 5-4).

Halocline A zone in the water column where the vertical change of salinity is relatively sharp (Fig. 5-13).

Headland A promontory of the coastline usually composed of durable rock that projects out into the ocean.

Heat capacity A measure of the quantity of heat needed to raise the temperature of one gram of a substance by 1°C.

Heavy metals Inorganic substances, such as lead, zinc, copper, that become highly toxic when concentrated in the environment.

Herbivore An animal that feeds principally on plants (Fig. 10-2).

Hermatypic coral Colonial, reef-building coral that have a mutualistic relationship with zooxanthellae (algae) and, hence, grow only in sunlit waters (Fig. 12-14).

Heterotroph Animals and bacteria that require prefabricated food for sustenance (Fig. 10-1).

High marsh The highest section of a salt marsh that is flooded by seawater only during spring tides or storms (Figs. 12-10c, B12-3 and B12-4).

Holdfast The "rootlike" structures that anchor seaweeds to the substrate (Fig. B9-4a).

Hot spot A zone on a lithospheric plate that overlies unusually hot asthenosphere (a mantle plume), and where large volumes of lava commonly are extruded, building a large volcanic pile on the sea floor (Fig. 3-9).

Hydration A chemical process that results in water being absorbed or added to a compound.

Hybrid water mass A water mass formed by the mixing of two other water masses.

Hurricane A large, powerful storm having winds that blow >120 kilometers per hour.

Hydrocarbon Organic compounds composed of hydrogen, oxygen, and carbon that are the main components of petroleum.

Hydrogen bonding A weak chemical bond that forms between dipolar molecules such as water molecules, and that greatly influences the physical and chemical properties of the substance (Fig. 5-3).

Hydrologic cycle The exchange of water among the ocean, atmosphere, and land by such processes as evaporation, precipitation, surface runoff, and groundwater percolation (Fig. 5-20).

Hydrolosis The chemical breakdown of a solid such as a mineral into other compounds by reactions with water.

Hydrosphere The envelope of gaseous, liquid, and solid water on the earth, including oceans, lakes, groundwater, ice and snow, and water vapor (Fig. 2-1).

Hydrostatic pressure The pressure at a specified water depth that is the result of the weight of the overlying column of water (Fig. 9-13a).

Hydrothermal deposits Minerals precipitated from or altered by very hot water circulating through rocks (Figs. B10-3a and b).

Hydrothermal vents An opening in the ground from which pour out hot, saline water solutions (Fig. B10-3).

Hypersaline water Water with a salinity much higher than normal ocean water.

Hypotheca The smaller, lower silica valve of a diatom frustrule (Fig. 9-16a).

Iceberg A large fragment of drifting ice that broke off the terminus of a glacier (Fig. 2-5).

Ice rafting The transport of sediment to the deep sea by floating icebergs which melt and drop their sediment load (Fig. 4-12).

Igneous rock A rock that crystallized from molten matter, either magma or lava.

Infauna Animals that live within or burrow through the substrate as distinguished from the epifauna, which live upon the substrate (Figs. 9-7 and 9-25a and b).

Infralittoral zone The lower zone of the intertidal zone that is exposed to air only during the very lowest spring tides (Fig. 9-24).

Infrared radiation The part of the electromagnetic spectrum that we perceive as heat.

In situ Latin; meaning in its original place.

Insolation Solar radiation reaching a body or area.

Interface The surface or boundary at which two different substances are in contact, for example, the air-sea interface.

Intermediate waves Waves moving in water where depths are between $\frac{1}{20}$ and $\frac{1}{2}$ of their wavelength; they have properties that are transitional between deep-water and shallow-water waves (Fig. 7-7b).

Internal wave A subsurface wave propagating along a density discontinuity (a pycnocline) (Fig. 7-12).

Interstitial water Water filling the pore spaces in a deposit of sediment.

Intertidal zone The area of the shore that lies between the highest normal high tide and the lowest normal low tide. Also called littoral zone (Fig. 9-1).

Inverse flow A pattern of current flow in lagoons whereby inflow occurs at the surface and outflow at the bottom, which is opposite (hence inverse) to water flow in many estuaries.

Ion An atom or molecule having an electrical charge because of a gain or loss of electrons.

Ionic bond A chemical bond created by the electrical attraction of anions and cations.

Irradiance A measure of the supply rate of radiant energy (Fig. 10-7a).

Island-arc system A linear to arcuate chain of volcanic islands associated with deep-sea trenches. They are formed by the partial melting of the lithosphere as a plate is subducted (Figs. 3-6 and 4-5).

Isobath A contour line connecting all points of equal depth below the water surface.

Isohaline A contour line of equal salinity (Fig. 5-13b).

Isopycnal A contour line of equal density (Fig. 5-14a).

Isostasy The regional mass balance of rocks in the earth's crust and uppermost mantle (Fig. 2-5).

Isotherm A contour line of equal temperature (Fig. 5-9).

Isotope Different forms of the same element related to variations in the number of neutrons in the nucleus (Fig. 5-1c).

Jetty A large structure extending seaward from the shore, which is erected to protect a harbor or inlet from shoaling by the longshore drift of sediment (Figs. 11-16b and B11-7).

Kaolinite A clay mineral composed of hydrous aluminum silicate that is a common weathering product in wet, hot regions (Fig. 4-13).

Kelp Large benthic species of brown algae (Fig. B9-4).

Keystone species A species that is critical in supporting a food web. Its elimination causes the food web to collapse and the ecosystem to change.

Kinetic energy Energy of motion.

Lagoon A shallow body of water that does not receive significant freshwater inflow and that is separated from the open ocean by a barrier island or coral reef (Fig. 12-20).

Langmuir circulation A series of parallel, counter-rotating circulation cells with long axes aligned parallel to the direction of the generating wind (Fig. 6-8).

Larva The immature form of an animal that differs significantly from the adult form (Figs. 9-9a and B10-7a).

Latitude The angular distance north and south of the equator with the equator being 0° and the poles 90° (Fig. IV-1 in Appendix IV).

Laurasia The aggregation of North America, Europe, and Asia into a large continental mass that comprised the northern half of the megacontinent Pangaea during the Mesozoic Era (Fig. 3-1).

Lava Molten rock that is extruded out of volcanoes (Figs. 3-4, 3-9, and 4-1).

Leptocephalus A leaf-shaped eel larva (Fig. B10-7a).

Levee The elevated part of a river bank or tidal channel that is produced by rapid deposition of sediment along the channel edges during flooding. Levees also occur along the side channels in the lower reaches of a submarine canyon.

Limestone A sedimentary rock composed dominantly of calcium carbonate, either precipitated from seawater (hydrogenous) or deposited as shell debris (biogenous) (Fig. 12-17a).

Lithification The process whereby loose sediment is compacted and cemented into rock.

Lithogenous sediment Sedimentary debris derived from the mechanical breakdown of rock.

Lithosphere The relatively cool, brittle, outer shell of the earth, some 100 kilometers thick, that includes the crust and uppermost mantle (Fig. 2-1).

Littoral zone A subdivision of the benthic province that lies between the high and low tide marks. Equivalent to the intertidal zone (Fig. 9-1).

Longitude The angular distance to the east or west of the prime meridian (0° longitude) that runs through Greenwich, England (Fig. IV-1 in Appendix IV).

Longshore bar A submarine sand ridge in the nearshore zone that is parallel or subparallel to the shoreline (Figs. 11-2, 11-3, and 11-12c).

Longshore current A shore-parallel current in the surf zone that is powered by breaking waves (Figs. 11-1 and 11-4).

Longshore drift The transport of sand in the surf zone parallel to the shoreline by longshore currents (Figs. 11-4, B11-1b, and B11-5c).

Low marsh The lower sector of salt marshes that is regularly immersed and emersed with the daily tides and that typically is colonized by the tall cordgrass *Spartina alterniflora* (Figs. 12-10e, B12-3, and B12-4). *See also* High marsh.

Macronutrients Nutrients, such as nitrate and phosphate, required by plants in relatively large quantities in order to photosynthesize and grow (Figs. 10-10 and 10-12b).

Macrotidal Tides having a range > 4m (Fig. 8-12).

Magma Molten rock within the Earth that crystallizes into igneous rock.

Magnetic anomaly A disturbance of the Earth's magnetic field created by magnetized rock in the Earth's crust (Figs. 3-3 and 3-5).

Magnetic anomaly stripe Long, linear bands of basalt crust that lie parallel to a spreading ocean ridge and that are alternately normally and reversely magnetized (Figs. 3-3 and 3-5a).

Mangrove forest A dense growth of mangrove trees in marshlike shoreline environments of the tropics and subtropics (Fig. 12-12).

Mantle The section of the earth's interior that extends from the base of the crust and to the top of the core and that is composed of ferromagnesian silicates (Fig. 2-1).

Mantle plume The upwelling of hot material into and through the lithosphere, with magma spilling out onto the earth's surface and building thick volcanic piles. If the lithosphere is moving relative to the plume or "hot spot," a linear trail of volcanoes is produced, a case in point being the Hawaiian Islands (Fig. 3-9).

Mariculture Marine agriculture, i.e., farming the seas to grow algae or to raise finfish and shellfish (Figs. 14-8 and B14-3).

Marsh A vegetated, intertidal flat (Figs. 12-10, B12-3, and B12-4).

Marsh zonation Bands of distinct plant and animal communities that reflect differences in ground elevation and the frequency and duration of salt water immersion by the tides (Fig. 12-9).

Mass transport The small net transfer of water in the direction of wave propagation (Fig. 7-3c).

Maximum sustainable yield A theoretical maximum tonnage of fish that can be harvested sustainably from year to year.

Meander Looplike bends or curves in the flow path of a current (Fig. 6-12).

Mesopelagic zone A subdivision of the pelagic province between the water depths of 200 and 1,000 meters (Fig. 9-1).

Mesosphere The deep, rigid portion of the mantle that lies between the plasticlike asthenosphere and the core (Fig. 2-1).

Mesotidal Tides with a vertical tidal range between two and four meters (Fig. 8-12).

Messenger A free-falling weight that is attached to a cable and used to trigger the closure of water-sampling bottles.

Metazoa Kingdom of many-celled animals.

Micronutrient Nutrients, such as iron, copper, and zinc, that are required in very small amounts by plants in order to photosynthesize, as distinguished from the macronutrients such as nitrate and phosphate, that are needed in large quantities.

Microtidal Tides having a vertical range < 2 meters (Fig. 8-12).

Midlittoral zone The portion of the intertidal zone that is covered and uncovered by water each day (Fig. 9-24).

Mid-ocean ridge The long, continuous mountain chain found in all oceans; ocean crust is created by the process of sea-floor spreading at its crest (Figs. 2-2, 2-3, and 3-2a).

Midwater fish Small, odd-looking fish that inhabit the dark waters located between 500 and 2,000 meters below the sea surface (Fig. 13-7).

Migration A term that refers to the habit of some animals whereby they travel long distances between feeding and spawning or breeding grounds (Figs. B10-7 and B10-8).

Mineral An inorganic substance that occurs naturally, usually in crystalline form, and has a distinctive chemical composition. An aggregation of different minerals constitutes a rock.

Mixed tide A tide with two unequal high waters and two unequal low waters each day (Figs. 8-3b and 8-4).

Mohorovičić discontinuity (Moho) A compositional and density discontinuity marking the interface between the rocks of the crust and the mantle (Fig. 2-5c).

Molecule Two or more atoms that constitute the smallest component of a compound while retaining its chemical characteristics (Fig. 5-3).

Molt The process of dispensing with an exoskeleton in order to secrete a bigger exoskeleton that will accommodate a larger body size.

Moraine Irregular, moundlike or ridgelike deposits of till laid down by glaciers (Fig. B11-1a).

Mud A mixture of silt and clay-sized particles (Table 4-1).

Mutualism A biological relationship whereby all organisms involved benefit from the association.

Neap tide The minimum range of tide in an area; it occurs when the moon is in its first and third quarters (Fig. 8-7).

Nearshore zone The inshore region of the coast between mean high tide and the breaker zone (Fig. 11-2).

Nekton Animals capable of swimming independently of current flow.

Neritic zone A subdivision of the pelagic environment; the water that overlies the continental shelf (Fig. 9-1).

Net primary production The quantity of energy remaining after plants have satisfied their respiratory needs.

Neutron A subatomic particle in the nucleus of an atom that has no electrical charge (Fig. 5-1).

Niche The specific ecological role of an organism in the life of the community and its position in the ecosystem.

Nisken bottle A common sampling container used to collect subsurface water for laboratory analysis.

Nitrate A macronutrient in the ocean that is essential for plant photosynthesis; its concentration may set a limit on the primary production of an area.

Nocturnal feeder Animals that feed at night.

Node The spot along a standing wave where there is no vertical displacement of the wave form (Fig. 7-11).

Nodule *See* Ferromanganese nodule.

Nonconservative property A property of water that changes over time because it is affected by biological activity. CO_2 and O_2, which are used and produced by organisms, are examples of nonconservative substances in seawater.

Normal fault High-angle faults with one block dropping down relative to another block; they denote tension and are found at the axial rift valleys of ocean-spreading ridges (Fig. 3-2)

Nucleus The cellular organelle of eukaryotic cells that contain chromosomes, or the central mass of an atom that consists of neutrons and protons (Fig. 5-1).

Nutrient Chemical compounds required by plants for normal growth; important nutrients are nitrate and phosphate (Fig. 10-10).

Oceanic crust The outermost rock shell of the earth, some 5 kilometers thick, that underlies ocean basins; it is composed of basalt and sedimentary layers (Figs. 2-4 and 2-5).

Oceanic ridge *See* Mid-ocean ridge.

Oceanic zone The deep water of the oceans beyond the shelf break (Fig. 9-1).

Oceanology A term denoting the science of the oceans.

Offshore zone The open water that lies seaward of the breaker zone (Fig. 11-2).

Oil The liquid form of petroleum consisting of a complex mixture of large hydrocarbon molecules.

Omnivore Animals that consume both plants and animals.

Ooze Pelagic sediment containing > 30% by volume shells of microorganisms (Figs. 4-14, 4-16, and Table 4-2).

Opal Amorphous (structureless) silica (quartz) secreted by organisms such as radiolaria and diatoms.

Opportunistic feeder Animals that consume most types of food that they encounter in their wanderings.

Orthoclase A potassium feldspar ($KAlSi_3O_8$) that is an important mineral constituent of granites.

Orthogonals Wave rays drawn perpendicular and spaced evenly along wave crests. Orthogonals are useful for examining the effect of wave refraction on the distribution of wave energy along a shoreline (Fig. 7-8).

Osmoregulation The process of controlling the amount of water in tissues and cells (Fig. 9-12).

Osmosis The diffusion of water through a semipermeable membrane from a solution with a lower to one with a higher solute concentration (Fig. 9-12).

Osmotic pressure The pressure that is needed to counteract the osmotic passage of water molecules across a semipermeable membrane into the more-concentrated solute.

Outcrop The exposure of bedrock at the Earth's surface.

Overfishing Harvesting fish at a rate that causes a fish stock to decrease alarmingly and if continued causes the fishery to collapse; a nonsustainable rate of fishing.

Oxygen minimum zone A layer of water between the depths of 500 and 1000 meters in which dissolved oxygen concentrations are lower than in the water above or below (Fig. 5-17a).

Pacific-type margin The edge of a continent that adjoins a tectonic plate boundary such as a subduction zone (Figs. 3-6 and 4-5).

Pack ice Numerous separate pieces of sea ice that have been packed together in dense concentrations.

Paleoclimatology Establishing the nature of ancient climates from fossils, rocks, or sedimentary deposits (Fig. 3-1b).

Paleogeography The reconstruction of the geography of the earth in the geologic past by interpreting rocks and their fossil contents (Fig. 4-5).

Paleomagnetic time scale A detailed chronology of the history of the polarity reversals of the earth's magnetic field (Fig. 3-4c).

Pangaea The megacontinent of the Mesozoic Era that consisted of all of the present-day continents joined together into a single unit (Fig 3-1).

Partially mixed estuary An estuary in which tidal mixing is a bit more influential than river inflow, such that vertical mixing creates a landward-directed bottom current and a seaward-flowing surface current (Figs. 12-2, 12-3, and B12-1).

Parts per thousand (‰) A unit of salinity; 35‰ indicates that 35 grams of salt are contained in 1,000 grams of seawater. In other words, salt comprises 3.5% by weight of a volume of seawater.

Passive continental margin A subsiding continental margin situated in a nontectonic setting away from a lithospheric plate boundary. An example of a passive continental margin is the Atlantic margin of North and South America (Figs. 2-3a and 4-5d).

Patch reef The small, localized growth of coral found in a variety of reef environments such as a lagoon (Fig. 12-17).

Patchiness A property whereby organisms are not uniformly distributed in a space but are clustered.

Pathogen A disease-causing agent, such as viruses and bacteria.

PCB Highly toxic and durable synthetic organic compounds that accumulate in the tissue of organisms.

Peat An organic deposit consisting predominantly of partly decayed plant matter.

Pelagic province All of the water of the oceans, including the neritic and oceanic realms (Fig. 9-1).

Pelagic sediment Deep-sea sediment that accumulates on the sea floor by the slow settlement of grains through the water column (Fig. 4-9 and Table 4-2).

Pennate diatom A bilaterally symmetrical diatom (Fig. 9-16c).

pH The negative log of the hydrogen ion activity; a pH value of 7 denotes a neutral solution, lower than 7, an acidic solution, and higher than 7, an alkaline or basic solution (Figs. 5-18 and 5-19).

Phosphate An essential plant macronutrient.

Phosphate deposit Accumulations of phosphoritic nodules in certain areas where coastal upwelling causes high biological production in the surface water.

Phosphorite A hydrogenous deposit consisting of nodules or crusts of P_2O_5.

Photophores Light-producing or luminous organs found in certain crustaceans and fishes (Fig. 13-7).

Photic zone The well-lit surface layer of the ocean where plants photosynthesize. Sometimes referred to as the euphotic zone (Figs. 91b and 13-8).

Photosynthesis Chemical reactions conducted in the presence of light by plants using chlorophyll whereby carbon dioxide and water are converted into carbohydrates and O_2.

Phytoplankton The plant members of the plankton community, which are the primary producers of the oceans (Figs. 9-3a, b, and c, 9-16, and 9-17).

Piston corer A long metal pipe that is driven into soft sediment with the help of a piston, yielding a long core of sediment (Fig. B4-2b).

Plankton Organisms that float or have weak swimming abilities, and are wafted by currents (Figs. 9-3, 9-9, 9-16, and 9-19).

Plankton bloom The sudden and rapid multiplication of plankton that results in dense concentrations of plant cells in the water (Fig. 10-10).

Plate tectonics The theory that the Earth's lithosphere is divided into broad, irregular plates that are either converging, diverging, or slipping by one another; these motions generate volcanism and earthquakes along the plate edges (Fig. B4-6).

Plunging breaker The classic "pipeline" breaker whereby the crest of the wave curls forward and collapses on itself (Fig. 7-9b).

Polychaete A class of annelid worms that includes most marine segmented worms (Figs. 9-6a, 9-7, and 9-25b).

Polymer As used in this book, a loose collection of water molecules that approximate the hexagonal crystalline structure of ice (Figs. 5-2b and 5-5b).

Polarity reversal, magnetic The periodic flipping of the Earth's magnetic field whereby the north magnetic pole changes position with the south magnetic pole (Fig. 3-4 and 3-5).

Population The collective total of all individuals of a particular species in an area.

Predator An animal that preys on other organisms.

Pressure-gradient force A force that arises as a consequence of a water slope; the steeper the water slope the stronger the pressure gradient (Fig. 6-9).

Primary consumer Plant grazers or herbivores (Fig. 10-2a).

Primary productivity The quantity of organic matter that is synthesized from inorganic materials by autotrophs (Figs. 10-10 and 10-11a).

Prime meridian The 0° line of longitude, which runs through Greenwich, England (Fig. IV-1 in Appendix IV).

Progressive wave A wave form that moves across space unlike a standing wave, which is fixed in space (Figs. 7-3 and 7-7).

Protein A family of complex organic compounds containing nitrogen and composed of various amino acids.

Protista A kingdom that encompasses all unicellular organisms that have cells with a true nucleus (Fig. 9-3).

Proton A positively charged particle in the nucleus of an atom (Fig. 5-1a).

Protozoa A phylum of unicellular animals such as foraminifera (Fig. 9-3a).

Pycnocline A zone having a marked change in water density as a function of water depth (Figs. 5-14b and 5-17a).

Radiolaria Microscopic, unicellular planktonic and benthic animals that possess siliceous (SiO_2) tests (Fig. 4-14e).

Red clay Descriptive term applied to pelagic or abyssal clay deposits of the deep sea; they range in color from red to brown and tend to accumulate slowly in the deepest and remotest parts of the oceans far from the influx of other types of sediments (Fig. 4-16 and Table 4-2).

Red tide Blooms of toxin-producing dinoflagellates that impart a red tint to the water (Figs. B4-4b and B15-10).

Reef A moundlike or ridgelike calcium-carbonate structure built by organisms such as coral (Figs. 12-16, 12-17, and 12-20).

Reflection The process by which wave energy is sent back offshore as the wave is partially deflected by a steep surface.

Refraction The process by which waves are bent and redirected as a consequence of wave interaction with bottom irregularities (Fig. 7-8).

Relict sediment Sediment on the shelf that was deposited during the recent geologic past and is not in equilibrium with the present-day environment (Fig. 4-3).

Relief The difference in elevation between the highest and lowest points in an area.

Residence time The average amount of time that an element remains dissolved in seawater assuming steady state conditions (Table 5-8).

Respiration A chemical process whereby organic matter is oxidized by organisms, releasing energy and CO_2.

Resting spore A dormant cell with the ability to spring to life when environmental conditions are favorable for its growth.

Reversing current A current that periodically changes its direction of flow, such as the ebb and flood currents of the tide (Figs. B8-1 and B8-2).

Rhizome A horizontal stem of some plants that is capable of producing new shoots as it grows.

Rift valley The fault-bounded valley found along the crest of many ocean ridges; it is created by tensional stresses that accompany the process of sea-floor spreading (Figs. 2-11 and 3-2a).

Rings Large whirl-like eddies created by meander cutoffs of strong geostrophic currents such as the Gulf Stream. They either have warm-water centers (warm-core rings) or cold-water centers (cold-core rings) (Fig. 6-12).

Rip current A narrow, swift, seaward-flowing current along the shore that drains water from the surf zone (Fig. 11-1d).

Riprap A pile of stones and boulders placed at the bottom of cliffs or on beaches to protect them from wave attack (Fig. B11-6).

Rogue wave An unusually large and dangerous wave, usually short-lived, that is created by constructive wave interference.

Rotary current The pattern described by tidal currents in the open ocean whereby the current flow shifts direction by 360° during one complete tidal period (Fig. 8-15).

Salinity A measure of the total concentration of dissolved solids in seawater usually expressed as parts per thousand (‰) (Fig. 5-12, Tables 5-1).

Salinometer An instrument that measures the salinity of seawater based on electrical conductivity.

Saltation The process by which wind cause sand grains to bounce and jump along the ground.

Salt marsh A vegetated intertidal mud or sand flat, often dominated by species of *Spartina* grasses (Figs. 12-9 and B12-3).

Salt-wedge estuary A type of estuary where river inflow dominates tidal mixing, producing a highly stratified water column with a sharp halocline (Figs. 12-2 and 12-3a).

Sand Clastic grains with diameters ranging between 62 and 2,000 µm (Table 4-1).

Sand budget A technique for documenting changes in the quantity of sand in a beach or nearshore system by comparing inputs and outputs of sand (Fig. 11-4).

Sand spit *See* Spit.

Sargasso Sea The central portion of the North Atlantic circulation gyre which contains large quantities of the floating brown alga *Sargassum* (Fig. 6-5 and 6-11).

Saturation A chemical state whereby the maximum amount of solute is dissolved under the given conditions.

Scavenger An animal that feeds on dead organic matter.

Schooling The tendency of many species of fishes and mammals to organize themselves into groups (Fig. 13-2).

Scientific method An investigative technique whereby data are collected objectively, interpreted as a hypothesis, and then tested by falsification.

Sea A term applied to the chaotic motion and irregular form of waves that are generated in the fetch area (Figs. 7-5a, 7-6a, and 7-7a).

Sea-floor spreading The process by which basaltic crust is created at the crest of ocean ridges where lithospheric plates are diverging (Figs. B3-3, B3-4, 3-5, and 4-17).

Sea ice Frozen seawater as opposed to glacial ice on land (Fig. B5-5).

Seamount A large, individual peak, volcanic in origin, with a crest that rises more than 1,000 meters above the surrounding sea floor (Figs. 2-3 and 2-10, Table 2-1).

Seawall A revetment or wall erected to prevent wave erosion and the encroachment of seas (Fig. 11-16d).

Sediment Grains or particles of either organic or inorganic origin deposited by air, water, or ice.

Sedimentary rocks Rocks that have formed by the compaction and cementation (lithification) of sediment.

Seiche The back-and-forth oscillation of a standing wave in a basin having a period ranging between a few minutes and several hours (Figs. 7-1b and 7-11).

Seismicity The distribution, frequency, and magnitude of earthquakes in an area (Figs. 3-6 and 3-7).

Seismometer An instrument for the detection and recording of earthquake waves.

Semidiurnal tide A tide characterized by two equal high waters and two equal low waters during a lunar day (Figs. 8-4 and 8-6).

Semipermeable membrane A membrane that permits the passage of water molecules but not solutes.

Sessile Benthic organisms that are attached to the substrate and, hence, fixed and unable to move across the sea bottom (Figs. 9-6b and e and 9-10).

Sewage Domestic, municipal, or industrial waste products disposed in the environment. When untreated, sewage can have serious impact on the quality of an environment and on the health of people.

Shallow-water wave A wave influenced by the sea bottom, usually in water that is shallower than

1/20 its wavelength (Figs. 7-4b and 7-7b).

Shelf break The steepening of the bottom that marks the seaward limit of the continental shelf and the beginning of the continental slope (Fig. 2-3).

Shoal A submerged bank or bar that is often a navigational hazard.

Shoreline The zone where the ocean is in contact with dry land.

Sigma-t (σ_t) Calculated by subtracting 1 from the specific gravity of sea water and multiplying by 1000; a seawater density of 1.028 g/cm^3 becomes a σ_t value of 28.

Significant wave height The average height of the highest one-third of the waves in an area of the ocean (Table 7-2).

Silica A compound with a composition of SiO_2, such as quartz in granite and opal in the shells of radiolaria.

Silicate Any compound that is complexed to the SiO_4^{4-} ion. These comprise the most abundant mineral group in the earth's crust.

Siliceous Material whose composition is silica (SiO_2).

Sill A shallow ridge that separates one basin from another, such as the Gibraltar sill at the mouth of the Mediterranean Sea or the sill at the mouth of a fjord (Figs. 6-17a and 12-1b).

Silt Sedimentary particles with diameters between 4 and 62 μm, making them larger than clay particles but finer than sand grains (Table 4-1).

Slumping The sliding of large, cohering blocks of sediment or rock downslope under the influence of gravity (Fig. 4-10a).

SOFAR channel A zone in the water column where sound waves attain a minimum speed such that refraction focuses and traps the them in a "channel" (Fig. B5-8b).

Solubility A measure of how readily one chemical substance can dissolve another.

Solute The chemical substances dissolved in a solution such as salts in seawater (Table 5-1).

Solvent The medium that disolves solutes such as the water in seawater.

SONAR An acronym for sound navigation and ranging; an instrument used to locate objects underwater by reflecting sound waves.

Sorting A measure of the range of grain sizes in a sedimentary deposit. Poorly sorted sediment has a large range of grain sizes, well-sorted sediment a limited range of grain sizes.

Sounding Determining the depth of water beneath a vessel (Fig. B2-1).

Specific heat The quantity of heat necessary to raise the temperature of one gram of a substance 1°C.

Spilling breaker A wave that breaks by having its crest spill down its face as it progresses through the surf zone (Fig. 7-9a).

Spit A narrow tongue of sand that extends from the shore and that is usually created by longshore drift (Figs. 11-12b and B12-4).

Spray zone The zone of the shore located above the highest high tide that is affected by the spray from breaking waves during storms.

Spring diatom bloom A distinct period of rapid biological productivity of diatoms that tends to occur during the spring season of temperate oceans (Fig. 10-10b).

Spring tide The maximum range of tide in an area; occurs twice a month when the moon is new or full (Fig. 8-7).

Standing crop The biomass of a population at any point in time.

Standing wave A wave that does not progress but oscillates up and down about a node with the crest changing into a trough and vice versa (Fig. 7-11).

Storm profile A beach profile characterized by a narrow berm, a gently sloping beach face, and submarine bars that is produced by storm erosion (Fig. 11-3).

Storm surge An unusually high stand of sea level produced by strong storm winds blowing water shoreward and by the ocean surface rising in response to low atmospheric pressure.

Stratigraphy The branch of geology that studies the relationship and significance of layered sedimentary rocks.

Stratum A layer or bed of sedimentary rock (pl. strata).

Strike-slip motion A high-angle fault, such as a transform fault, along which rocks move horizontally (laterally).

Stromatolite Banded rocks consisting of thin interbeds of mud and mats of benthic algae that thrive in hypersaline, shallow water.

Subduction zone A collisional plate boundary along which one lithospheric plate overrides another and produces a deep-sea trench, a volcanic arc, and seismicity (Figs. 4-7, B4-6, and 6-18).

Sublittoral zone A subdivision of the benthic province extending from the shoreline to a depth of 200 m (Fig. 9-1).

Submarine canyon Deeply incised, steep-walled valley, commonly V-shaped in profile, that is cut into the rocks and sediments of the outer continental shelf and the continental slope (Fig. 2-3 and Table 2-1).

Submersible A vessel that can submerge and operate underwater.

Subsidence The sinking of large portions of the Earth's crust.

Substrate A general term in reference to the surface on or within which organisms live.

Supersaturation A solution with a solute concentration that exceeds the amount dissolved at saturation for the given conditions.

Supralittoral zone The spray zone of the shore located above the highest high tide (Fig. 9-24).

Surf zone The section of the coastal zone between the shoreline and the breaker zone (Figs. 11-1 and 11-2).

Surface drag *See* Frictional drag.

Surging breaker Breakers that do not break entirely against the shore and are reflected back offshore (Fig. 7-9c).

Suspended load Sedimentary particles that are carried by currents above the sea bottom.

Suture zone A relatively narrow zone along collisional plate boundaries where two large continental masses are welded into one unit (Fig. 3-11).

Swash The rush of water up the beach face after a wave has broken (Fig. 11-2).

Swell Large, regular, long-period waves that result from wave dispersion outside of the fetch area (Figs. 7-1b, 7-5b, and 7-7a).

Swell profile A beach profile characterized by a broad berm and steep beach face that evolves during a stretch of fair weather when incoming waves move sand onto the beach (Fig. 11-3a).

Swim bladder A gas-filled pouch in many bony fishes that is used to attain neutral buoyancy by regulating the amount of gas in the bladder.

Tectonics Refers generally to the deformation and resultant structure of the Earth's crust and upper mantle; encompasses such geologic processes as folding, faulting, volcanism, and seismicity (Figs. B4-6 and 4-7).

Tensional stress A force that pulls apart rocks or parts of a structure (Fig. 3-2a).

Terrigenous sediment Sediments derived from the mechanical weathering of rocks on land; equivalent to lithogenous sediment (Figs. 4-4, 4-9, and 4-16a).

Tests The skeleton or shell of certain microorganisms.

Tethys Sea An immense seaway that separated Gondwanaland from Laurasia during the Mesozoic Era (Fig. 3-1b).

Thermocline A sharp, vertical temperature gradient that marks a contact zone between water masses having markedly different temperatures (Figs. 5-10 and 5-11).

Thermohaline circulation The vertical movement of water masses that results from differences in water density (Figs. 6-15 and 6-16).

Thrust fault A low-angle fault formed by compression, whereby the upper block is forced over the lower block causing the crust to shorten in length (Fig. 3-6).

Tidal bore *See* Bore.

Tidal inlet A channel or opening through a barrier island that admits the tidal flow of water (Figs. B8-1 and 11-13b).

Tidal period The elapsed time between successive high tides or low tides (Fig. 8-1).

Tidal range The vertical difference separating the water level between successive high and low tides; it is equivalent to the height of a wave (Figs. 8-3 and 8-12).

Tides The periodic rise and fall of the Earth's water surface as a consequence of the gravitational attraction of the Moon and the Sun (Figs. 8-1 and 8-4).

Till Poorly sorted sediment that is deposited by glaciers (Fig. B11-1).

Topset beds Sediment laid down in horizontal layers on the top of a delta (Fig. 11-15a).

Trace elements Elements that occur in seawater in tiny quantities, typically measured in concentrations of parts per billion (ppb).

Transform fault A steep boundary separating two lithospheric plates along which there is lateral slippage. The crest of ocean ridges commonly is offset along transform faults (Figs. 3-2 and 3-6).

Trench *See* Deep-sea trenches.

Trophic dynamics Refers to the complex biological processes whereby energy and matter are passed up to successive levels of food webs.

Trophic level A functional or process category for types of feeding by organisms (Figs. 10-2 and 10-6).

Tsunami Long-period water wave produced by tectonic effects such as earthquakes, volcanism, or slumping; sometimes called a seismic sea wave (Figs. 7-1 and 7-13).

Turbidite Sediment, typically with graded bedding, that is deposited by a turbidity current (Figs. 4-7, 4-10c, and 4-10d).

Turbidity A measure of the cloudiness of water, which is a function of the amount of suspended material, both organic and inorganic (Fig. 10-9).

Turbidity current A density-driven slurry of sediment-laden water which flows swiftly downhill displacing less dense water (Figs. B4-4a and 4-10).

Turbulent drag Irregular, chaotic fluid flow generated by a hydrodynamically inefficient body design that causes considerable drag on an object moving through water or air (Fig. 9-21).

Turbulent flow The irregular, chaotic flow of a fluid that results in random velocity fluctuations and in mixing.

Typhoon A tropical hurricane of the western Pacific region.

Undersaturation A solution that contains solutes in a concentration that is less than the maximum that can be dissolved under the given conditions.

Uplift The rising of one part of the Earth's crust relative to another part (Fig. 11-14d).

Upwelling The slow, upward transport of water to the surface from depth (Figs. 10-8, 10-11a, and 10-15).

Valve The shell or shells of certain organisms, such as snails and clams.

Vaporization The process by which a liquid is transformed into a gas.

Vascular system Plant tissue arranged into vessels that serve to conduct fluids from the roots to other parts of the plant (Fig. 9-14).

Ventral Of or pertaining to the underside of animals.

Vertical exaggeration The exaggeration of the vertical scale relative to the horizontal scale in a topographic profile or section (Fig. IV-2b in Appendix IV).

Vertical migration *See* Diurnal vertical migration.

Vertical zonation The arrangement and vertical succession of distinct bands of communities that are particularly common along rocky shorelines (Fig. 9-24).

Warm-core ring Large eddies that have a core of warm water (Fig. 6-12).

Washover fan The fan-shaped accumulation of sand on the landward side of a barrier island that is deposited by storm waves which overtop the island (Figs. 11-11b and 11-13a).

Water mass A large body of water with a distinctive set of properties, identifiable typically by its temperature and salinity (Fig. 6-15 and Table 6-2).

Wave (water) A disturbance that represents energy propagating or moving across the sea surface (surface wave) (Figs. 7-3 and 7-4) or along a density discontinuity within the water column (internal wave) (Fig. 7-12).

Wave base The level of the water surface equivalent to one-half the wavelength where wave-induced motion is absent (Fig. 7-3a).

Wave crest The highest part of a wave (Fig. 7-1a).

Wave-cut platform As a sea cliff is eroded back by waves, a platform is created that slopes gently seaward (Fig. 11-14c).

Wave height The vertical distance separating the wave crest from the wave trough (Figs. 7-1a and 7-7b).

Wavelength The horizontal distance between corresponding points on successive waves, such as from crest to crest or trough to trough (Figs. 7-1a and 7-7b).

Wave period The time required for two successive crests to pass a fixed point (Figs. 7-1b).

Wave ray *See* Orthogonals.

Wave set-up The mass transport of water into the surf zone by breaking waves, which can create localized surpluses of water along the shore that drive longshore currents.

Wave steepness A ratio of wave height to wavelength that measures the "peakedness" of a wave.

Wave trough The lowest part of the wave (Fig. 7-1a).

Weathering Of or pertaining to the breakdown of rocks and minerals by chemical and/or mechanical processes.

Well-mixed estuary An estuary in which mixing by the tides is so complete that vertical stratification of the water column is absent (Figs. 12-2c and 12-3c).

Western boundary intensification Describes the nature of the western currents of circulation gyres, which are swift, deep, and narrow in contrast to their eastern currents, which are weak, shallow, and broad (Fig. 6-10 and Table 6-1).

Wetlands *See* Salt marsh.

Windrows Linear accumulations of flotsam that are aligned with the wind and that typically result from the convergence of surface currents associated with Langmuir circulation (Fig. 6-8).

Wind-driven circulation Refers to the geostrophic currents of the large circulation gyres that are powered by the prevailing winds (Figs. 6-4, 6-5, and 6-10).

Zonation Recognizable bands or groupings of plant and animal communities in certain environments, such as within salt marshes (Fig. 12-9c) and along rocky shorelines (Fig. 9-24).

Zooplankton Animal plankton such as foraminifera, radiolaria, copepods, euphausiids, etc. (Figs. 9-3a, 9-13b, and 9-19).

Zoozanthellae Symbiotic algae (modified dinoflagellates) that live in the tissue of hermatypic corals (Figs. 12-14c and 12-15).

Index

■ Note: Page numbers in italics refer to illustrations; tables are indicated by the annotation *t* following the page number.

AABW. *See* Antarctic Bottom Water
AADW. *See* Antarctic Deep Water
AAIW. *See* Antarctic Intermediate Water
abyss, life in, 20
acceleration, centrifugal, 189
acid, nucleic, 328
acidity. *See* pH
Act
  Magnuson (1976), 498
  Marine Mammal Protection, 422
  Water Resources (1962), 393
adaptation, strategies for, 302–17, 441–42, 444
adsorption, 148
advection, oceanic, 162
Africa, plate tectonics and, 124, 216
air
  constituents of, 142*t*, 186
  currents. *See* wind
  density of, 186
  pressure, *See* pressure, atmospheric
aircraft, flight paths of, and Coriolis deflection, *188*, 189
AIW. *See* Arctic Intermediate Water
Alaska Bioremediation Project, 484
albatross, *438*, 469
algae, 291, *292*, 313
  blue-green (cyanobacteria), 288, 289*t*, 290, 312, 313, 333
  brown, 289*t*, 291, 292, 313, 446
  see also kelp, giant
  classification of, 288, 289*t*, 291
  on coral reefs, 423, 425
  in estuaries, 407
  green, 289*t*, 291, 292
  and the growth of stromatolites, 124
  life cycles of, 303–304
  mariculture of, 467
  nitrogen for, 339
  red, 289*t*, 291, 292, 313
  tides and, 277
  see also macroalgae; *Macrocystis*
Alps
  earthquakes in, 70
  origins of, 65, 108, 124
altimeter, *197*
altimetry, 46, *46*
aluminum, in seawater, *143*, 147, 149*t*
Alvin (submersible), *26*, 334, 449–50
Amazon rainforest, 102
Amazon River
  and ocean salinity, 154, 424
  as a source of marine sediment, *111*, 115
  tidal bore on, 274–75
Amchitka Island, Alaska, kelp forests off, 314–15, *315t*
ammonia
  oxidization of, 333
  in seawater, 339, 419
amphipods, 289*t*, 302, 441
  in the open sea, 441, 442, 443
  Parathemisto gaudi-chaudi, 296
  trophic dynamics and, 327, 443
analysis, harmonic, 236
Anarctica, deep-sea sedimentation around, *115*
anchor dredge, 297
Anchorage, Alaska, tides at, *261*
anchovy (anchoveta), *463*
  eggs of, *295*
  in a food chain, *326*
  harvests of, 354–55, *358*, 359, 418
  overfishing of, 496, *497*
  populations of, to estimate, *349*
  production, 349, *349*, 352, 356
Andes Mountains
  origins of, 108
  as a tectonic plate boundary, 72
andesite, *40*, 70, 72, 74
Andew, *See* Hurricane Andrew
anhydrite (calcium sulfate) sediments, 124
animals, marine
  body-fluid chemistry of, 299–301
  classification of, 289*t*, 291
anion, 139, 140
annelids, 289*t*, 291
anomaly, magnetic. *See* magnetism
anoxia, 160, 222, *332*, 333, 400, 406–407, *408*, 487
Antarctic Bottom Water (AABW), 209*t*, 212, 213, *214*, 501
Antarctic Deep Water (AADW), 209*t*, 212, *213*
Antarctic Intermediate Water (AAIW), 209*t*, 212, 213, *214*
Antarctic Peninsula, 468, 470
Antarctica
  krill population of, *468*
  ocean temperatures off, 151
  salinity off, 155
  sedimentation off, 119
  see also poles
antinode, of a standing wave, 247, 267, *267*
Appalachian Mountains, origins of, 65
arc (chain), volcanic
  and earthquakes, 70–72
  formation of, 75, *75*
  and sedimentation, *109*
  subduction zones, *105*
archeology, 23
Arctic Intermediate Water (AIW), 209*t*, 212
Arctic National Wildlife Refuge, Alaska, oil reserves in, 459
Arctic Ocean
  continental shelf of, 36
  evolution of, 70, 72, 76
  exploration of, 20
  primary production in, *344*
area to volume, ratios of, 304, *304*
argon, 142, 160, 186
arsenic, 489
Aru Island, villagers from, *22*
aspect ratio, 309, *309*, *310*
Assateague Island, Maryland, 390–91, *390*
asthenosphere, *33*, 34, *73*, 74
Atlantic Ocean
  bathymetry of, *35*
  benthic zonation of, 312, *312*
  continental drift and, 62
  currents in, 153
  earthquakes in, *102*
  evolution of, 70, 72, 76, *76*, 83, 104–106, *105*, *108*, 120
  physiography of the Western, 43, 47–57
  pollution in, 488, *488*, 494, 509
  salinity of, 157, *158*, *211*, 216
  sediment in, *116*, 121, *122*
  size of, 517
  temperatures of, 210, *211*
  water masses in, 212, 218–19
  see also North Atlantic Ocean; Sargasso Sea, South Atlantic Ocean
Atlantis II deep (Red Sea), 83, *83*, 496
atmosphere, *33*, 34, 500
  pressure and, 172, 186, 301, 356
  processes of, 186–91
  water in, 172*t*
atoll, 428, 429, *429*
  Pacific, 411, *412*
atoms, *described*, 134, *134*
Australia
  carbonate shelf off, 108
  stromatolites off, 124
autotrophs, 324, *324*, 335, 423
  and biomass, 346–48

back-island flat, 378, *378*, 379
Bacon, Francis, 388
bacteria, *290*
  aerobic, 331–33
  anaerobic, 160, 222, 333, 335–36, *336*
  as autotrophs, 335
  for bioremediation, 484
  chemical reactions of decomposition by, 333
  chemosynthetic, 333, 335–36
  classification of, 288, 289*t*

bacteria *(continued)*
as decomposers, 332–33, 339
extremophiles, 317
hydrostatic pressure and, 449–50, *450*
in photophores, 442
photosynthetic, 333
Bahama Islands, 108, 124
Baja California, lagoons of, 411, *411*
bar
longshore, 368, *369*, 379, *380*, 400
onshore, *370*
Barbados, 103
barium, in seawater, 149*t*
barnacles
classification of, *289t*, 291
in estuaries, 409, *410*
feeding, *329*, 330
habitat of, 312, 313
larvae of, 294
pollution and, 478
salinity and, 299
temperature and, 297, *298*
Barnstable, Massachusetts, evolution of salt marsh at, 415, *415*
barrier islands, 377–81, 390–91
geology of, *383*
landscape of, 378–79, *378*
origins of, 379, *380*
and a rise sea level, 504, 505, *505*
storms and, 379–81, *381*, 388
see also Padre Island, Texas
basalt
dating, 80–81, *81*
described, 39, *40*
isostasy and, 42–43
lava, 74, 76
magnetization of, 68–70
Red Sea, 82
basin
deep-ocean, 35, *37*, 38, 39–43
shallow, tides in, 272–75, *275*
bathymetry
dimensions in, *36t*
exploration of, 34–35, 44–46
frequency distribution of, 39, *40*
provinces of, 35–39, *35*
and sedimentation, 119
and water masses, 218
Bay of Bengal, storm surge in, 243
Bay of Fundy, Canada
calculating depth of, 278–79
tides in, 269, *269*, 272–73, *274*, *275*
bays
barrier islands and, 378, *378*
pollutants in, *483t*, 485
tidal inlets to, 377
tides in, 272
beach, 367–71
changes in, 367–68, 370, *370*, 388
composition of, 97–98
defined, 367
nourishment of, 391, 392–93, *393t*, 394, 493, *495*
pocket, 374, *374*
sea levels and, 98
beach face, 368
beach profiles, 367–70, *370*
Beagle, HMS, 18, *19*
bedding, graded, of sediments, 112, *113*
beds, sedimentary, 386
Benioff, Hugo, 72
benthic province, 286
salinity of, 299
zones of, 286, *287*
benthos
biomass of, *347*, 348
on the continental shelf, 436, 439–40
described, 291, 294, *294*
sampling, 296–97
sea floor and, 440, *441*
zones of, 312–13, 334
berm, 367, 368, *369*, 379
Bermuda Rise, 53
beryllium, 7
bicarbonate
in river water, 147*t*
in seawater, 140, *141t*, 147, *165*, 339
Bierce, Ambrose, 32
bioaccumulation, 491, 493
biocides, artificial, as pollutants, 419, 491–93
biogradation, 491
biology
at the sea-surface microlayer, 176–77
of the continental shelf, 436–40
of coral, 421, *423*
estuarine, 405–09
open ocean, 440–45
bioluminescence, 441–42
biomagnification, 419, 491, 492, *492*, 493, 507, 508
biomass
autotrophs and, 346–48
of estuaries, 409
food chains and, 333, 336, 337, 353
of the open sea, 445
bioremediation, 482, 484
biosphere, 34, *172*, 500
biota
classification of, 288–91, *288t*, 294
of the continental shelf, 436–40
effect of oil spills on, 481–82
El Niño and, *358*
of the open oceans, 441–45, 448–60
sampling, 296–97
birds, sea, *289t*, 291, 359, 437, 438, 492
Black Sea, water flows in, 216, 221–22, *221*
"black smoker," hydrothermal vent as, *334*, *335*
blowouts, of sand dunes, 375–76, *376*
bluffs, 373
BOD. *See* oxygen, biological demand for
bond
covalent, 137–38, *137*, *139*
ionic, 139
booms, floating, to contain oil spills, 482, *483*, 486
bore, tidal, 273–75, *275*
boric acid, *141t*
boron, 339
Boston, Massachusetts, tides at, 259, *261*
Brazilian Basin, 218
breakers, 237, *237*, 240, 242–43, *242*, 375
types of, 242–43, *242*, 366
breakwaters, *389*, 391
brine, *defined*, 148
Britannia, HMS, 449
brittle stars, *289t*, 448
bromide, 140, *141*
bryzoans, 409
bubbles, and the sea-surface microlayer, 159, 176–77
buffer, carbonate, 165–66, *166*, *167*
bulge, tidal, 262–63, *262*, 264, *264*, 265–66
declination and, 263
Buzzards Bay, Massachusetts, oil spill in, 481–82

cable, transatlantic telephone, earthquake damage to, 102–103
cadmium, 419, 489
calcium
in river water, *147t*
in seawater, 140, *141t*, *146t*, 148, 149, *149t*
calcium carbonate
coral, 421
in seawater, 299
sedimentary, 93, 100, 108–109, 116, 119
calcium sulfate. *See* anhydrite
calculations
of area-to-volume ratios, 304, *304*
conversions of measurements, 26–27, 518
of depth, 278–79
of doubling rates, 326–27
of global fish production, 349, *349t*, 352–54
of hydrostatic pressure, 301
of natural period of oscillation, 248
powers of 10, 41
of residence time, 430
of a sand budget, 392
of sea-floor spreading, *69*, 85
of sedimentation rates, 127
of steady state, 510–11
of tide celerity, 273
of volume transport, 210
of wave celerity, 234
Campobello Island, Canada, tides at, *259*
Cano, Sebastian del (Juan Sebastián de Elcano), 15–17
canyon, submarine, 36
dimensions of, *36*
origins of, 112
sedimentation and, 104, 112, *113*
see also Hudson Submarine Canyon
Cape Cod Bay Lobe, 372
Cape Cod, Massachusetts, evolution of, 372–73
carbon
in photosynthesis, 339, 345
species of, in seawater, *166*
transfer efficiency, and production, 349, *349*, 450, *450*
carbon dioxide
in air, 186, 501–502, *502*, 504
and global warming, 106–107, 501–503, *502*, 504
and pH, 163–67
in photosynthesis, *161t*, 165, 328, 501
in respiration, 165, 328, *450*
in seawater, 142, 142*t*, 160, *161t*, 162–67, 502
system, 166–67
see also photosynthesis; respiration
carbon-14 dating, 379
carbonate, in seawater, *166–67*
carbonate compensation depth (CCD), 116, 119, 120–21, 122, 166
carbonate shelf. *See* sediment, carbonate
carbonic acid, in seawater, 147
carcasses, as detritus, 445, *449*
Caribbean Sea, 102, *437*, *444*, 446
carnivores, 325, *327*, 330, 441
carrying capacity, 405
Carson, Rachel, *Silent Spring*, 491
Casco Bay, Maine, shellfish harvests in, 298–99, *299*
cation
described, 139
in seawater, 140, 148
CC. *See* current, Canary
CCAMLR. *See* Convention on the Conservation of Anarctic Marine Living Resources
CCD. *See* carbonate compensation depth
Ceará abyssal plain, 218
cell
circulation, 191, 201, *201*
organic, *487t*, 489
Challenger, HMS, expedition

of, 20, *21–22*, 145
channel, deep-sound (SOFAR [SOund Fixing And Ranging]), 170, *171*
Charleston, North Carolina, hurricane damage to, 243
chemicals
as pollutants, *476t*, *487t*, 491–93
to disperse oil slicks, 478–79, 482
chemistry
basic, 134–35
of the Black Sea, *221*
body-fluid, 299–301
chemosensors, 448
chemosynthesis, 333, 335–36
Chesapeake Bay
as an estuary, 406–408
oil spills in, 482
chimney, volcanic vent, 335, *335*
chloride
in river water, *147t*
in seawater, *139*, 140, *141t*, 146, *146t*, *147t*, 149t, *149*
chlorite
in sedimentary clay, 114, *116*
see also kaolinite/chlorite ratio
chlorophyll, 303, 324
chlorophyll a, 342, 343, *343*
chop, defined, 228, *228t*, 244
chordates, *289t*, 291
chromium, *149t*
ciliates, *289t*, 332
circulation
in estuaries, 400, 402, *403*, *404*, *407*
on Georges Bank, *498*
gyre. *See* gyre
in lagoons, 410–11
Langmuir, 201, *201*, 436
nearshore, 367
ocean. *See specific oceans*
in the surf zone, 366–67
thermohaline, 208, 210–15, *214*, 500
see also current, subsurface
cirri beats, 297, *298*
clams, *289t*, 291, 299, 330
coquina (*Donax variabilis*), 317
in estuaries, 409, 419
in hydrothermal vents, 334
mariculture of, 467
pollution and, 482, 493, *506*, 508
razor (*Ensis directa*), 313, 316, *316*
soft-shell, 298–99, *299*
classification, of organisms, 28–94, *288t*, *289t*
by lifestyle, 291, 294
clay, sedimentary, 93, *93*, 97, *116*, 438
red (brown; pelagic), 114, *117*, 119, 120, *120*, *121*
cliffs, erosion (recession) of, 374, 381–86, *384*
rates, *386t*
climate
change in, and the ocean, 500, 503
forecasting, 122, 222
and ocean currents, 192, 208, 501
and salinity of the ocean, 153–55, 156
variability in, 106–107
coast
defined, 399
production along, *349t*, 354
see also upwelling, coastal
water action and, 366–67
Coastal Zone Color Scanner (CZCS), 342–43, *342*
coastline
changes in, *99*
humans and, 388–93
cobalt
in ferromanganese nodules, 117, 461, 495
as a resource, 462–63
in seawater, *143t*, *149t*, 339
coccolithopores, *109*, 116, *117*, 289, *289t*, 441, *444*
cod, 298, *463*, 498, *499*, 500
colloids, 93, *93*
color
in lagoons, 411
in the oceans, 169–70, *169*, *170*, 216, 342, 345, 421
of sedimentary mud, 333
comb jellies, *289t*, 291, *293*
Common Water (CoW), *209t*, 212, 213
communities
benthic, 440, 441
chemical contaminants and, *487t*
composition, constant, 145
compound, organic, *described*, 142–43
computers, for oceanography, 25
conductivity, electrical, of seawater, 146
consumers, animals as, 325, 328–31, *424*
continental drift, 62–64, 104
Wegener's theory of, 62, 69
continents, distinguished from ocean basins, 39–43
contours, mapping, 521–22, *522*
convection, thermal at the sea-surface microlayer, 176
and plate tectonics, 74–75, *74*
Convention on the Conservation of Anarctic Marine Living Resources (CCAMLR), 468–69
Convention for the Prohibition of Long Driftnets, 466
convergence, 200
Antarctic, 210, 212, 213
conversions, of measurements, 26–27, 518
Cook, James, explorations of, 18
cooling, global, 106
copepods
classification of, *288t*, *289t*
on the continental shelf, 436
described, 306–307, *307*
fish production and, *349*
in the open sea, 441, 443
pollution and, 492
populations of, to estimate, *349*
temperature and, 408–09
trophic dyanmics and, 325, *326*, *327*, 328, 332
copper
in ferromanganese nodules, 117, 461, 494–95
in seawater, *143t*, 339, 489
coral, *289t*, 428
ahermatypic, 424
biology of, 421, *423*
brain, *423*, 425, *427*
branching, 425, *426*, *427*
encrusting, *427*
fan, *426*
hermatypic, 424, 425, 429
massive, *426*
sublittoral, 438
coral reef, 421, 423–29
algae on, 423
as an ecosystem, 295, 424, *424*
bleaching, 422, 427
Darwin's observations of, 18
decline, 103, 422, 427–28
described, 421
fish of, 428, *428*
geology of, 428–29, *429*
global distribution of, *425*
guyots and, 38
morphology of, *426*
primary production and, 344, *345t*
salinity and, 424
sedimentation and, 100, 104, 108, *108*, *109*
temperature and, 424
topography of, 424–25
trophic dynamics of, 421, *424*
see also reef
cordgrass (*Spartina alterniflora*), 412, *413*, 414, 416, *416*
core
of the Earth, 33, 34
sampling, 53, 94–95, 122, 382, 415, *415*
corers, for sampling, 94, *95*
Coriolis deflection, 186–89, *188*, 192
in estuaries, *404*, 405
and ocean currents, 197–99, *200*, *202*, *202*
tides and, 266, *267*, 270, *271*, 276
and upwelling, 340, 354
Coriolis, Gaspard-Gustave, 187
corralite, 421, *423*
costs
of cobalt, 463
of dredging, 493
of mariculture, 471
of sand nourishment, 391, 392, *393t*
CoW. *See* Common Water
crabs
blue, 407
classification of, *289t*, 291, *294*
feeding, *329*, 330
fiddler (*Uca pugnax*), 276–77, *276*, 412, *417*
green, and shellfish harvests, 299, *299*
horseshoe, *289t*
King, overfishing of, 496
in mangrove swamps, 417
Northern rock, *329*
pollution and, 482
Cretaceous period
North America in, *105*, *108*
sedimentation during, *109*
crust, of the Earth, 33, 39–40, 72, 76, *81*
boundary. *See* Moho, the
destruction of. *See* subduction zone
isostasy and, 42–43, *43*
probing, 44–46
crustaceans, 419, 441, 443
crysosphere, 500
CTD (conductivity, temperature, depth). *See* salinometer
current
Agulhas, waves and, 247
Antarctic Circumpolar, 212
Brazil, 208
Canary, *198*, 203, 204
convergent, 197, *200*, 201, *201*
and Coriolis deflection, 197–99, *200*, *202*, *202*, 206, *207*
deep-ocean. *See* current, subsurface
density and, 192, 208, 210, 214
development of, 199–201
direction of, 186
dispersal of pollutants by, 494
divergent, 197, *200*, 201, *201*
ebb, 268–69, 275
Ekman transport and, 197, 199–200
equatorial, 200
flood, 269, 275
flow patterns of, 276–77, *278*, 410
and gases in seawater, 161
geostrophic, 201–205, *202*, *206*, *207*, 210, 501
hydraulic, 268, *268*
inverse flow of, 411

current *(continued)*
Kuroshio, 208
longshore (shore-parallel), *338–39*, 366–67, 368, *369*, 382, 400
loops (circulation gyres) in, 192
measuring, 194–97
mixing rates and, 149
nearshore, 233, *369*
North Atlantic Central Water (NAC), *198*
North Atlantic Deep Water (NAC), *198*
North Equatorial (NEC), *198*, 200, 206–207
Peru, 354
prevailing, described, 192
rip (shore-normal), 367, 368, *369*, 374, 392
and sea-surface temperatures, *151–53*, 153–54
and sedimentation, 100, 104
shore-normal. *See* current, rip
shore-parallel. *See* current, longshore
South Equatorial (SEC), 200, 206–207
speeds of, 192, 199–200, 203, 206, 208
spiraling, 199–200, *199*
subsurface (deep-ocean), 192, 208–*17*, *215*
see also circulation, thermohaline
surface ocean, 192–208, *193*, 214, *215*
tidal inlet, 268–69
tides and, 101–102, 275–*78*, 437
turbidity, 102–103, *103*, 112, *113*, 445
vertical, 200–201
waves and, 246, *247*
winds and, 101, 192–208
cyanobacteria. *See* algae, blue-green
cycle, hydrologic, 173–74, *173*, *174*
Cyclothone, 444, 447
CZCS. *See* Coastal Zone Color Scanner

Darwin, Charles, 18, *19*, 428, 429
DDT, 177–78, 488, 491–92
Dead Sea, 148
declination, and tidal bulges, 263
decomposition
of artificial nutrients, 487
bacterial, 332–33, *332*, 417, 449–50
chemical reactions of, 333
and gases in seawater, 161, *161t*
Deep-Sea Drilling Project (DSDP), 24, 25
deflection. *See* Coriolis deflection
deglaciation. *See* glaciers, melting of
delta front, 386
delta plain, 386
deltas, 386–88, *387*, 400, 505
types of, 386–87, *387*
Denmark Strait, 218–19
density
of air, 186
of the Black Sea, 216
defined, 32
gradients of. *See* pycnocline
and hydrostatic pressure, 301
isostasy and, 42–43
of the Mediterranean Sea, 216
and ocean currents, 192, 208, 210, 214
of rock, 39–40, *40*
salinity and, 301
and temperature, 136, 157, 186, 301
of water, 139–40, 150–51, 157–59, 210, 301
deposition, 92, 96, *96*, 104, 120, 388
depth
compensation, 338, *338*
coral growth and, 425
and currents, 199, 208
Ekman spiral and, 200
and light penetration, 169–70, *169*, 286, 338–39, *338*
and ocean habitats, 286
and primary production, 338–39, *338*
of the Red Sea, 82
and salinity, *158*
sampling, 145
of the sea-surface microlayer, 176–77, *177*
and sedimentation, 97–99
and temperature, 152–153, *155*, 208–209, *209*
to calculate, 278–79
to measure, 44–46
and water (hydrostatic) pressure, 301–302, *302*
waves and, *233*, 235, 240, 248
zones by, 286
see also carbonate compensation depth
desalination, 148, 164–65
deserts
latitude and, 154
marine, 216, 344, 345, *348*, 352, 421, 441
dessication, in bethic organisms, 313
detritus
in the aphotic zone, 445, *447*, 450
in the dysphotic zone, 442
in estuaries, 408
sublittoral, 439, 440
see also food chain
detrivores, 412, 416, 417, 439, 440
diatoms
blooms of, 344, 345
centric, *305*
classification of, *288t*, 289, *289t*, 291
on the continental shelf, 436
described, *290*, 305, *305*, *306*
fish production and, *349*
fossilized, in sediment (biogenic ooze), 116, *117*, 149
Hantzschia virgata, 276
in the open sea, 441, 443
pennate, *305*
populations of, *349*, *444*
trophic dynamics and, 325, *326*, *327*, 328, 339, 508
Diaz, Bartholomew, 15
diffusion
molecular, at the sea-surface microlayer, 176
and osmoregulation, 300–301
turbulent eddy, 219
dinoflagellates
classification of, *289t*
on the continental shelf, 436, *444*
described, 289, 290, 306
in the open sea, 441, *444*
plankton blooms of, 328, 508–509
populations of, *444*
toxin-producing (*Gymnodinium breve*), 508–509, *508*
trophic dynamics and, 327, 328
distillation
of crude oil, 477
for desalination, 164, *165*
distribution
bimodal, 39, *40*
frequency, defined, 39
patchy, 436, 498
divergence, *200*
DNA, 317, 328
doldrums, 190
Doppler shift, 195
doubling, to calculate rates of, 326–27
downwelling, 200, 216
and primary production, 345, *348*, 441
drag, types of, and locomotion, 304, *304*, 308, *309*, 313, 317
drainage, and the stability of cliffs, 385
dredging
environmental degradation and, 493–94, 505
equipment for, 94, *94*
for sampling, *22*, *94*, 297
shorelines, 389
drift, longshore, *372*, 375, 381, *383*
see also continental drift
drift nets, use of, 465–66
drilling, deep-sea, 24, 25, 80, 94, *123*, 124
jurisdiction over, 456
for oil and gas, 458–59, *459*, 460–61, *461*
drought, El Niño and, 356
DSDP. *See* Deep-Sea Drilling Project
dunes, 371, 375–77
erosion of, 376–77, *377*
extent of, 371
field, 378, *378*, 379
formation of, 371, 376, *376*
migration of, 376, 379
profiles of, 371
ridges of, 371, *373*
dust storms, 102–03
Dutch Harbor, Alaska, tides at, *261*

Earth
composition of, 33
degassing of, 148
dimensions of, 517
evolution of, 32
magnetic (geomagnetic) field of, 66–68
rotation of
and Coriolis deflection, 187–89, *188*
tides and, 266
and western-boundary intensification, 204
structure of, 32–34, *33*
temperatures of, 106–107, *107*
earthquakes
along the Oregon coast, 374
and deep-sea sedimentation, 107
distribution of, 71, *71*
Grand Banks (1929), 102–103
Loma Prieta, California (1989), 77, 79
and midocean ridges, 38, 56–57, 65
and the San Andreas fault, 77, *77t*, 79
tectonism and, 70–72, 107
tsunamis and, 250, *251*, 252
types of, 70, 71
East Pacific Rise, 72, 80, 121, *162*, 334
East Wind Drift, 468
ECC. *See* Countercurrent, Equatorial
ecology
adaptive strategies for, 302–17
basic, 294–302
damage, from oil pollution, 478, 481–82
defined, 286
hydrostatic pressure and, 301–302
salinity and, 299–301
temperature and, 295–99

ecosystem
bottom feeding by whales as, 443
defined, 295, 324
education about, 394
energy in, 325–28
model of, 324–25, *324*
sea otter/sea urchin/kelp, *315t*
ECW. *See* Equatorial Central Water
eelgrass, *289t*, *292*
eels
American (*Anguilla rostrata*), 350, *350*
European (*A. anguilla*), 350
migrations of, 350–51, *350*
sand, trophic dynamics and, 327
EEZ. *See* zone, Exclusive Economic
effluent
dilution of, 485, *485*
effects of, 486–89
as a pollutant of the sea, 484–93, 506
Egypt, early mariners in, 5
Ekman transport, 197, 199–200, *199*, 201, 207
downwelling and, 345
upwelling and, 340, *340*, 354
waves and, 233
Ekman, V. Walfrid, 199
El Niño
cause of, 356–57
effects of, 356, 358–59, *358*, 375, *375*
investigation of, 25
SST anomaly in, *355*
upwelling during, 355, 358
water conditions and, *358*
El Niño Southern Oscillation (ENSO), 356, *357*
Elcano, Juan Sebastián de. *See* Cano, Sebastian del
electricity, tidal generation of, 279–80, *280*
electrodialysis, for desalination, 164
electron shells, 134, *134*
electrons, described, 134
Emperor Seamount Chain, *75*
energy
in an ecosystem, 325–28
bottom, 97, 437–39, *440*
kinetic, 135–36, 337
predation and, 436
pyramid, 325, 327–28
sediment deposition and, 93, 96–97, 98–99
transfer of, and food chains, 333, 336–38, 337, *337*
wave, 232, 240, 251, 437
England, sedimentation off, 101–102
ENSO. *See* El Niño Southern Oscillation
environment, degradation of, 506–12
and deep-sea mining, 495
from dredging, 493–94, 505
from oil spills, 478, 481–82, *483t*, 486
and mariculture, 470
and sand mining, 460–61
Environmental Protection Agency, 407, 484
enzymes
in bacterial decomposition, 332–33
synthesis of, and temperature, 296
epifauna, 294, 409, 437, 438, 439
epiflora, 294
epitheca, 305, *305*
equator
air pressure zones and, 189, *190*
currents and, 200, 206–207
sedimentation along, 100, 116, 119, 121–22
upwelling at, 200, 340, *340*, 343–44, 346
Equatorial Central Water (ECW), *213*
equilibrium. *See* steady state
erosion
of barrier islands, 391
of beaches, 367, 368, 370, *381*, 388, *388t*, 391–92
control, 377, *377*
of sand dunes, 376–77
sedimentation and, 92, 93, 96, *96*, 104, 105
estuary, 400–*10*
as an ecosystem, 295
bar-built, 400, *401*
biology of, 405, 408–09
circulation in, 400, 402, *403*, *404*, *407*
defined, 400
economic value of, 400
origins of, 400, *401*
partially mixed, 402, 403, *403*, *404*, 406, 418
pollutants in, 400, 407, 476, *483t*, 485, 487
production in, 339, 345, *345t*, 405
reclamation of, 418–20
salinity of, 403, *403*, 405, 406, *407*, 408
and salt marshes, 400, 416
salt-wedge, 402, *403*, *404*
sediment in, 402, 405
tectonic, 400, *401*, 418
tidal inlets to, 377
tides in, 268–69, 272, 403, 405
types of, 402–405, *403*
water diversion and, 419
well-mixed, 402, *403*, *404*, 405
see also Chesapeake Bay; San Francisco Bay
EUC. *See* Undercurrent, Equatorial
Euler, Leonhard, 194
Eulerian method for measuring currents, 194–95, *194*
euphausiids, trophic dynamics and, 327, 442, 443
European Union, fishing quotas and, 500
eutrophication, 419, 487, 490
evaporation, *175t*
defined, 136
and desalination, 164
and salinity, 148, 154, *156*, 210, 212, 410
evolution, Darwin's theory of organic, 18
exaggeration, vertical, 49, 522–23, *522*
exclusion, competitive, 448
exoskeleton, 421
exploration, of the oceans, 5–17, 23, 106, 456
chronology of, *8–13*
extinction, mass, theories of, 122
Exxon Valdez, oil spilled from, 486–87, *487*

fallout
and oil pollution, *477t*, 480
and the salinity of seawater, 147, *147*
fan
deep-sea, 51, 112, *114*
desert alluvial, *114*
washover (overwash), 377, 379, *381*
fault
geometry of, *79*
midocean ridges and, 64, 65
normal, 64, *64*, 65, 70, 76
San Adreas (California), 74, 77–79, *77t*, 418
slippage rate and, 77
transform, 38–39, 64, *64*, 65, 74, 77
faulting, 38
strike-slip, 65, 74, 77, *79*, 418
feces, pellets of, as sediment, 116, 336, 405, 445, *447*
feeding, methods of, 328–30
deposit, 317, 330, 439, 440, *440*
filter, 306, *307*, 330, 335, 405, 409, 439, *440*
grazing, 328–30, 486
predation, 330
scavenging, 330
feldspar, sedimentary, 100, 107, 147–48
Ferrel cells, *190*, 191
Ferrel, William, 191
ferromanganese nodules
composition of, 461–62
described, 116–18, *118*
mining, 456, 462, 495–96
as sediment, 93, 110, *462*
fertility, of estuaries, 405, 408–09
fertilizer, phosphate as, 463
fetch, wind-generated waves and, 229, 236, 245
field, geomagnetic. *See* Earth, magnetic field of
Fiji Basin, Benioff zone in, *71*
fins, aspect ratios of, 309–10, *310*
fish
adaptations by, and habitat, 441
classification of, *288t*, *289t*, 291, 310, *310*
commercial, 463–67
on the continental shelf, 436, *437*
coral reefs and, 428, *428*
deep-sea angler, 444, *447*
in estuaries, 409, 419
farming, 464–65
functional morphology of, 307–11, *310*
gas-bladders in, 302
global production of, to calculate, 349, *349t*, 352–54
ground, 463, *463*, 498, *499*
harvests of, 353, 354–55, *354*, 411, 466–67, 496–97, *496*, *497*, 498, *499*
lantern, 441, 443, *447*
midwater, 317, 443–44, *447*
osmoregulation in, 300–301, *300*
pelagic, 463, *463*, 498
in polluted waters, 488, *489*, 492
populations of, 448
as a resource, 463–67
sargassum (*Histrio histrio*), 446, *446*
schooling in, 436, *437*
speed of, 309–11
trophic dynamics and, 325, *331*
see also finfish; flatfish
fishery
collapse of, 353, *358*, 359, 496
global, *466*
improving, 353
precautionary principle of practices for, 500, *501t*
see also overfishing
fishing
as a business, 465, 466, 496
major fleets, *467t*
MSY and, 497, 499
Fitzroy, Robert, 18
fjord, 400, *401*
flagellates, 289, *289t*, 332
see also dinoflagellates; silicoflagellates
flatworm, 277, *289t*
flocculation, 405
flooding, effects of sea-level rise and, 502–503, 504–505, *505t*

Florida
carbonate shelf off, 108
hurricane damage to, 246
red tides off, 103, 508, *509*
Florida Keys, 103
flotation, mechanics of, 304
flounder, 294, *295*, 498, *499*
flow
model of geostrophic, 201–205, 209
turbulence in, *340*
fluoride, in seawater, 140, *141t*
food chain, 325, *326*, 333, *349*, 353
described, 325, *326*
detritus, 336–67, *336*, 408, 417
and energy transfer, 333, 336–38
grazing, 336, *336*
pollution and, 492, *492*
and productivity, 349, *349*, 353
food pyramid, 424, *424*
food, supply of and population size, 330–31, 449
food web, 325, *327*, 419, 441
foraminifera (forams)
classification of, 289, *289t*
described, 289, 290, 307, *307*
in the open sea, 441
in sediment (biogenic ooze), 116, *117t*, 120, 148
trophic dynamics and, 328
forams. *See* foraminifera
Forbes, Edward, 20
foredunes, 371
fossils, use of, for dating (fossil record), 80, 148
Fram, the, 20, 22
France, sedimentation off, 101–102
freezing, for desalination, 164
freezing point, and salinity, 151
frequency distribution, defined, 39
freshwater, converting seawater to, 164–65
front
oceanic, 218, *218t*
polar, 191
Frost, Robert, "Once by the Pacific," 381–82
frustule (shell), 305, *305*, *306*, 339
Fu-Ch'un River, China, tidal bore on, 275
fungi, *289t*, *291*
Fungi (Kingdom), *289t*, 291

Galapagos Ridge, 333, 334, *336*
Galveston, Texas
storm damage at, 243
tides at, *261*
Gama, Vasco da, 15
Ganges-Brahmaputra Delta, India, *387*
gas, natural (methane), as a resouce, 458–59, 460–61
gas hydrates, as a resource, 459
gases
in air, 186
in bottom-dwelling organisms, 301–302, *302*
measurements of, 158
regulation of, 159–61, *161t*
in seawater, 142, *143t*, 158–67, 221–22
sources of, *160*, 161–67
Geneva Convention on the Law of the Sea, 456
Georges Bank, North Atlantic
overfishing of, 496, 498–99
waves and, *241*
GESAMP. *See* Group of Experts
geyser, sea-floor, 85
Gibraltar waterfalls, *126*
glaciation
record of, and sedimentation, 103–104, *115*
and sea level, 104, 504
glaciers
deposits left by, 372
melting of (deglaciation), and changes in sea level, 98, 99, *99*, 393, 400
sediment and, 93
global warming, 500–06
Glomar Challenger, the, 24, *25*, 80, 94, 124
glucose, 328
gold, in seawater, 142, *143*
Gondwanaland, *63*
grab sampler, 94, *94*, 297
grain size, of sediments
as classification, 92–93, *93t*
and current velocity, 96–97, *96*, 438, *440*
and depth, 98–99
feeding habits and, 439, 440, *440*, *441*
and graded bedding, 112
granite
isostasy and, 42–43
properties of, 39, *40*
in sediment, 93, 100, 108
graphs, plotting, 17
grass
American beach (*Ammophila brevigulata*), 375
marram (*Ammophila arenaria*), 375
salt hay (*Spartina patens*), 412, *413*, *416*
salt-marsh, *289t*, 292, 379, 411–16, *416*, 482
saltwort (*Salicornia* sp.), 412, *413*, *416*
spike (*Distichlis spicata*), 412, *413*, *416*
see also cordgrass; eelgrass
gravel
erosion of, *386t*
mining, 460–61, *462*, 494
as a resource, 459–61, *462*
sedimentary, 93, *93*, 97, 102, 438–39, 459
gravity
and sedimentation, 110
tides and, 260–62, *262*, 266
grazing, and primary production, 338, 341
Great Salt Lake, 148
Greeks, as early explorers, 7, 14
greenhouse effect, 393, 501–503
groins, 385, 388, 389, 391, 393
groundwater
defined, 34
levels of, and the stability of cliffs, 385
quantities of, 172, *172t*
in the Red Sea, 83, *83*
and a rise in sea level, 504
Group of Experts on the Scientific Aspects of Marine Pollution (GESAMP), 476, 508–11
grunion (*Leuresthes tenuis*), tides and, 277, *277*
GS. *See* Gulf Stream
Gulf of California, 77, 334, *411*
Gulf Coast, sedimentation on, 101
Gulf of Mexico, *460*
cold seeps in, 334
oil in, 459, 460–61
red tides in, 508
Gulf of Saint Lawrence, Canada, tides in, 272, *274*
Gulf Stream (GS), *198*, 203, 204
flow pattern of, 205, *205*, 208, 218, 500
rings spawned by, 205, *205*, 208, 218
volume transport and, 210
gulfweed. *See Sargassum*
guyot, 38
gypsum, 148
gyre, circulation
Coriolis deflection and, 200, *200*
described, 192, *193*
as a geostrophic current, 202–205, *202*, *203*
and heat transfer, 501
North Atlantic, 197, *198*
primary production in, 344, 345, *348*
in the Sargasso Sea, *198*, 207, 446

habitat, oceans as, 286–87, *287t*
HABs. *See* Harmful Algae Blooms
haddock, *288t*, *354*, *463*, 498, *499*
Hadley cells, 190, *190*, 191
Hadley, George, 190
hake, 358, *358*, *463*, 500
halite (rock salt), *139*, 148
Halobates spp., *178*
haloclines,154–155, 299, *406*
halogen, in air, 186
Hampton Roads, Virginia, tides at, *261*
Harmful Algae Blooms (HABs), 508
Hatteras Island, North Carolina, *378*
Hawaiian Islands
cobalt deposits off, 463
formation of, 75, *75*
lava flow on, *75*
megaplumes and, 85
tsunamis off, 252
heat
capacity of water, 136, 138, 500
physics of, 135–36
helium, 142, 160, 186
Heraclitus, 365
herbivores, 325, *327*, 337
Herodotus, map of the known world by, 7, *7*, 14
herring, 327, 418, 496, *497*, 498
Hess, Henry, 68
heterotrophs, *324*, 325, 328, 333, 335
Hewitt, Roger, 470
hill, abyssal, *36*, 38
Himalayas
earthquakes in, 70
marine sedimentation and, 108, 110
origins of, 65, 74, 76, 80, *84*
Hjulström diagram, 96–97, *96*, 102
Hsü, Kenneth, 124, 126
Hudson Submarine Canyon, 36, *48*, 50–51, *52*, *53*
humans
classification of, *288t*
damage caused by, 377, 418, 448, 501
as omnivores, 325
pollution and, 476
as predators, 441
Hurrican Andrew, 206–207
hurricanes, 206–207, 243, 246, 356, 382, 388, 420
hydra, *289t*
hydration, described, 139, 151
hydrocarbons
halogenated (organochlorines), 491–93
petroleum, 457–58, 491
as pollutants, *476t*, 477, 480–83
hydrogen, 134, *134*, 135
bonding, 138
in seawater, 142, 147
hydrogen sulfide, in seawater, 160, 222, 332, 333, 336
hydrophone, 44, *44*
hydrosphere, *33*, 34, 172
hypotheca, 305, *305*
hypotheses, falsification of, and

the scientific process, 80–81
hypoxia, 406, 487–88, *488*

ice
pancake, 168, *168*
physics of, 135, *135*, 139, *140*
pressure ridges of, 168–69, *168*
water in, 168
see also sea ice
Ice Age, Little, 14, 106, *107*
Ice Age. *See* Pleistocene epoch
ice caps, melting of, 502, 504
ice floe, 169
ice lobe, 372
ice rafting, and sedimentation, 100, 112, *115*, 119
icebergs, 169
density of, and isostasy, 42, *43*
melting of, and sedimentation, 100, 112
IDOE. *See* International Decade of Ocean Exploration
IGY. *See* International Geophysical Year
incineration, as pollution, 492
Indian Ocean
bathymetry of, *35*
evolution of, 70, 72, 76
salinity of, 213
sedimentation of, 119
size of, 517
water masses in, 213, *213*
infauna, 294, 313, 330, 409, 439
inlets, tidal, 377, 379–80, *381*
currents through, 268–69
insolation, 151
intensification, western-boundary, 203–204, 205
Inter-Governmental Working Group on Marine Pollution, 508
International Council for the Exploration of the Seas, 500
International Decade of Ocean Exploration (IDOE), 24, 25
International Geophysical Year (IGY), 24
International Indian Ocean Expedition, 24
International Program of Ocean Drilling, 24
International Seabed Authority (ISA), 456
International Tsunami Warning System, 251–52
International Whaling Commission (IWC), 465
International Year of the Ocean (IYO), 511–12
iodine, in seawater, 142, *143t*
ions
conservative, 140
described, 135
hydration of, *139*, 150
nonconservative, 142
residence time, 149, *149t*
Ireland, sedimentation off, 101–102
Irish Sea, phytoplankton in, *437*
iron, in seawater, 103, 142, *143t*, 339
iron aluminum oxide, in river water, *147t*
iron oxide, 461
iron sulfide, *332*
ISA. *See* International Seabed Authority
island, barrier. *See* barrier islands
isohalines, *158*
isopods, *289t*
isostasy, 41–43, *43*
isotherms
described, 151, 523, *523*
in North Atlantic Ocean, 204–205
and species distribution, 296
isotopes, *described*, *134*, 135
IWC. *See* International Whaling Commission
IYO. *See* International Year of the Ocean

Janthina prolongata (snail), 177
Japan
extraction of manganese nodules by, 462
fishing by, 466, 468, 498
the Law of the Sea treaty and, 456
methane extraction by, 459
tsunamis in, 251, *251*, 252
whaling by, 465
jellyfish, *289t*, 441, 443
Jeremy Project, 23
jetties, 388–89, *389*, 390–91, *390*
Johansen, Frederick, 20
Joides Resolution, the, 24, *25*, 94
Juan de Fuca Ridge, 334, 374

kaolinite, in sedimentary clay, 114, 115, *116*, 147
kaolinite/chlorite ratio, 114–15, *116*
kelp, giant (*Macrocystis*), 314–15, *314*, 438, *489*
kelp forest
as an ecosystem, 295, 314–15, *314*, *315*
pollution of, 488–89, *489*
Ketchikan, Alaska, tides at, *261*
Key West, Florida, tides at, 259, *261*
Kilauea volcano, *75*
Komar, Paul, 374, 375
krill, Antarctic (*Euphasia superba*)
classification of, *289t*
described, 468, *468*
fishing of, 468–69, *470*
as food, 328, 468–70
populations of, *469*, 470
krypton, in air, 186
Kurile-Kamchatka Trench, *17*

La Rance River, France, tidal energy generation and, 280, *280*
lagoons, 410–12
barrier islands and, 378, *378*
circulation in, 410
coral reefs and, 425, *426*, 429
described, 400, 409–10, *411*
pollution of, 411
Lagrange, Joseph, 194
Lagrangian method for measuring currents, 194, 195, *195*
Laguna Madre (Padre Island), Texas, 382, 410
Lake Cape Cod, *372*, 373
lakes, water in, *172t*
Lamont Doherty Geological Observatory, 24
Lamont Earth Observatory, 24
lamp shells, *289t*, 291, *293*
landmass, effects of, on geostrophic flow, 202–203
Langmuir, Irving, 201
larvae, planktonic, 294, *295*
latitude, 520–*21*
Greek determination of, 7
horse, 190
ocean currents and, 200, 208
and primary production, 342
and salinity, 153–56
and sea-surface temperature, 152
and sedimentation, 100, *100*, 112, 114, 116, *116*
tides and, *264*
Vikings' use of, 14
and water density, 210
winds and, 190–91
Laurasia, *63*
lava, 74
magnetization of, 66–68
Law of the Sea, 456–57, 462
lead
in ferromanganese nodules, 117, 461
as a pollutant, 103, 419
in seawater, 103, 142, *143t*, 489
lead and line, to measure depth, 44
Leatherman, S. P., 391
lichens, *289t*
life, origin of, 85
lifestyle
changes during, 294
classification by, 291–94
light
and habitat zones, 286–87, *287*, 441, 451
and water depth, 157, 169–70, *169*, 286, 338–39, *338*
lime. *See* calcium carbonate
limestone
coral reefs and, 421, *426*
erosion of, 384, *386t*
in sediment, 93, 108, 120, *123*
limpets, *289t*, 478
line
corange, *267*, 270, *274*
cotidal, *267*, 270, *274*
Linnaeus, Carolus, 288
lithium, in seawater, *143t*, *149t*
lithosphere, *33*, 34, *72*, *73*, 74, 500
"fossil" slabs of, 85
lobsters, *289t*, 470, 493
London Dumping Convention, 408, 508
Long Island Sound, biomagnification in, 492, *492*
longitude, 7, 520–*21*
Louisiana
flooding in, 388
pollution off, 488, *488*, 509

MacIntyre, Ferren, 176
mackerel, 441, *463*, 496, *497*, *498*
Macoma, 440
macroalgae, *294*
macronutrients, 339, 344
macroplankton, 446
Macrosystis. *See* kelp, giant
Magellan, Ferdinand, 15, *16*
magma, 334
magnesium, 140, *141t*, *148*
magnetism, anomalous stripes of, 65–66, *65*, 67, 68, 69, 72, 80–81
magnetization, fossil (paleomagnetism), 68
magnetometer, 65
magnitude, orders of, 178–79
manatees, *289t*, 291, 417, 509
manganese
in nodules, 495
see also ferromanganese nodules
in seawater, 142, *143t*, 339
mangrove, *289t*, 291, 417, *420*
mangrove swamps, 416–17, 420, 470, *483t*
mantle, of the Earth, 33, 34, 45
mantle plume, 75, *75*, 81, 85
map
discussed, 520–22
air pressure and prevailing winds, *191*
bathymetric, *35*
biomass, *347*
of carbonate shelves, *109*
contour (topographic), 521–22, *522*
of deep-sea sedimentation, *120*
of the distribution of ferro-

manganese nodules, *118*
drift route of the *Fram,* *22*
of eel migrations, *350*
of Exclusive Economic Zones, *457*, *458*
Georges Bank, off New England, *498*
of global amphidromic tide systems, *271*
global distribution of coral reefs, *425*
of global pollution, *507*
of global production, *346*
of global seismicity, *71*
of global tides, *272*
of global warming, *503*
the known world (Herodotus), 7, *7*, 14
of lithospheric plates, *73*
of Magellan's voyage around the world, *16*
of the ocean crust, *81*
of ocean fisheries, *466*
ocean topography, *198*
of oil transport routes, *480*
of Pangea, *63*
of Polynesian settlement, *6*
of the route of the *Challenger* expedition, *21*
of the route of HMS *Beagle,* *19*
of salmon migrations, *351*, *352*
San Andreas fault, *78*
of sand and gravel deposits, *462*
of sea-surface temperatures, *152*
of sedimentation rates, *121*
of surface ocean currents, *193*
of thermohaline circulation, *214*
of transatlantic telephone cables, *102*
Viking, of the North Atlantic Ocean, *16*
of volcanic chains, *75*
of volcanic vent sites, *334*
margin, continental, 35, 36–38, *37*, 80
Atlantic-type, 104–106, *105*, *108*
dimensions of, *36*
Pacific-type, 104, *105*
Mariana Trench, 517
mariculture, 467, 470–71
marine archeology, 23
Mars (oil field), 459, 461, *461*
marsh. *See* salt marsh
mass balance, and isostasy, 41–43
matter
biochemical recycling of, 174–75, *175*
states of, *135*
Matthews, Drummond, 68, 80–81
Maury, Matthew Fontaine, 18
measurements
converting, 26–27, 518
of gases, 160
orders of magnitude as, 178–79
parts per thousand (ppt; ‰), 140, 141
parts per million (ppm), 142
parts per billion (ppb), 142
"Meddies," 219
"Medieval Little Optimum," 106
Mediterranean Intermediate Water (MIW), *209t*, 212, *213*, *214*, 216
Mediterranean Sea
density of, 216
drying up of, 47, 124–26
evolution of, 76, *221*
map of (Herodotus), 7
salinity of, 216, *217*
temperature and, *217*
water flow in, 215–*17*, 411
megaplumes, 85
mercury
as a pollutant, 103, 491
in seawater, 103, 142, *143t*, 489
mesosphere, *33*, 34
metals
from hydrothermal vents, 335
heavy, as pollutants, 419, *476t*, 484, 489–91, 495
in the Red Sea, 83
Metaphytae, *289t*, 291, *292*
Metazoa, *288t*, *289t*, 291, *293*
Meteor, the, explorations by, 20
meteorites, and mass extinctions, 122
meter
Doppler Acoustic Current, 194–95, *194*
Eulerian current, 194, *194*
methane. *See* gas, natural
methyl mercury, 491
Mexico
carbonate shelf off, 108
fishing by, 466
microlayer, sea-surface, 158, 176–78
temperatures (SST), 151–53
see also neuston; neuston layer
micronutrients, 339
microorganisms
and the bioremediation of oil spills, 484, 486
and salinity, 149
surface-water, and sedimentation, 110, 115–16
microwave, satellite, *197*
Mid-Atlantic Ridge, *49*, 64–65, 120, 334
described, 54–55, *56*
earthquakes and, 102
imaginary journey to, 47–57
magnetic properties of, 65–66, 68–69
normal faulting along, 64, *64*, 65, 70
Middle Ages, exploration during, 14
midlatitudes, primary production in, 344–45, *344*
migration
of barrier islands, 379, 391
diurnal vertical, 442–43, 468
of dunes, 376, 379
of longshore bars, 370, *370*
of marine animals, 291, 345, 350–52, 498, 500, 507
of shorelines, 388
of whales, 411
minerals
in the Earth, 33–34
evaporite, 149
mining, the sea, 83, 117, 456–57, 494–96
Miocene age
Mediterranean Sea during, 124–26
sediments of, 124
Mississippi, pollution off, 488, 509
Mississippi Delta, 386, *387*, 388
Mississippi River, 460, *488*
MIW. *See* Mediterranean Intermediate Water
mode, defined, 39
model
ecosystem, *315t*, 324–25, *324*
of global warming, 504
of tectonism, 72–75, *84*
of tides, 263–72, *266*
Moho, the, 40, 43, *43*
Mohorovičić, Andrija, 40
molecules
described, 134
dipole structure of, *137*, 138, 164
water, 136–43, *137*
molting, 306–307
molybdenum, in seawater, *143t*
Monera, 288, *289t*, *290*
Monitor, 23
monsoons, El Niño and, 356
Moon, tides and, 258–67, *261*, *265*, 266, 270–71, *270*
moraine, glacial, 400
recessional, 372–73
terminal, 372
motion, energy of. *See* energy, kinetic
mountain range
origins of, 62, *64*, 65, 108, 124
submarine, 64
Mozambique, Africa, waves off, 249, *250*
MSY. *See* Yield, Maximum Sustainable
mud
aerobic, *332*
anerobic, *332*, 333
barrier island, 379
in estuaries, 405
lime (calcium carbonate), 93
pelagic, 112, 114–15, 121, 122
plume, 218, 460
sedimentary, 93, *93*, 96, 98, 110, 112, 438, 440
silicaceous, 93
mudstone, 50
mussels
bacteria on, *450*
classification of, *289t*
culture of, 467
in estuaries, 409, 419
feeding methods of, 330
habitat of, 312, *313*
in hydrothermal vents, 334, *336*
pollution and, 492, 508
in salt marshes, 412
sublittoral, 438
mutualism, 421, 423
mysids, 443

NABW. *See* North Atlantic Bottom Water
NAC. *See* current, North Atlantic
NACW. *See* North Atlantic Central Water
NADW. *See* North Atlantic Deep Water
Nansen, Fridtjof, 20
Narragansett Bay, Rhode Island, pollution of, 490, *490*, 493
National Science Foundation (U.S.A), 24
Nauru, exploitation of phosphate rock by, 463
nautilus, *289t*, *448*
navigation, 7, 14–15
Neal, W. J., 393
NEC. *See* Current, North Equatorial
needlefish, redfin, 308, *309*
nekton, 291, 294
on the continental shelf, 436
functional morphology of, 307–11
sampling, 296, *297*
neon, in air, 186
neuston, 176
neuston layer, *177*, 476, 486
neutrons, *described,* 134, *134*, 135
New England
fisheries in, 353
sedimentation in, 101
New England Seamount Chain, *48*, *54*, *55*
New York Bight, pollution in, 488, *488*, 494
New York, New York, tides at, *261*
nickel
in ferromanganese nodules, 117, 494

as a pollutant, 490, *490*
Niskin bottles, 144, *144*
nitrate
bacterial decomposition and, 333, 339
in river water, *148*
in seawater, 339, 419
nitrite, 339
nitrogen
in air, 186
as limit in primary production, 339
in macronutrients, 339
in photosynthesis, 339
as a pollutant, 490, *490*
in seawater, 142, *142t*, 158, 333, 339
node, of a standing wave, 247, 267, *267*
nodules
manganese. *See* ferromanganese nodules
phosphoritic, 463
Norsemen. *See* Vikings
North America
continental margins of, 104–105
paleogeography of, *105*
North Atlantic Bottom Water (NABW), *209t*, *213*, 501
North Atlantic Central Water (NACW), *209t*, *213*
North Atlantic Deep Water (NADW), *209t*, 212, 213, *214*, 219
North Atlantic Ocean
circulation in, 197, *198*, 203, *203*, 204–205, 208, 220, 501
development of, *105*
primary production in, 343, *343*, 344, *344*
salinity of, 157, *220*
sedimentation of, 119, 120
temperatures of, 152
North Pacific Central Water (NPCW), *209t*, *213*
North Pacific Intermediate Water (NPIW), *209t*, 212, *213*
North Pacific Ocean, temperatures of, 152
North Pole, exploration of, 20
North Sea, fish harvests in, *354*
Northern Hemisphere
coastal upwelling in, *340*
Coriolis deflection in, 189, 197, 199, 200, 266, 270, 340
NPCW. *See* North Pacific Central Water
NPIW. *See* North Pacific Intermediate Water
nucleus, described, 134, *134*
numbers, to express large, 41
nutrients
concentration of, and primary production, 338, 339, 343, 498
in estuaries, 405, 407
for fertilizing oil spills, 484
from rivers, 440–41
as pollutants, 407, 419, *476t*, 486–87, 490–91, *490*
recycling of, 325, 332, *332*
in seawater, 142, *142t*, 216, 298, 304, 445, *447*
upwelling and, 340
see also trophic dynamics

ocean basins
described, 62
evolution of, 70, 75–76, *76*, 80, 82–83, *82*
faults in, 76
generation of. *See* sea-floor spreading
research on, 25
tides in, 270–71, *271*, *274*
see also basin, deep-ocean
Ocean City, Maryland, 390–91, *390*
ocean floor
consumption of, 72
habitat zones on, 286
physiography of, 34–39
pollutants on, 476
research on, 44–46
oceanography
careers in, 4
collaborative, 24
described, 4
early scientific investigation of, 5, 17–20
father of, 18
history of, 4–22, 24
modern, 5, 20, 24
satellite, 195–97t, 342–43, *343*
oceanology, 4
oceans
benthic province of. *See* benthic province
as a biogeochemical system, 174–75, *175*
biology of open, 440–45
carbon dioxide in, 502
climate changes and, 500, 503
color in, 169–70, *169*, *170*, 216, 342, 345, 421
deep layer of, 157, *159*, 161–62, 208–15, 301–302, *302*
dredging and, 492–94, *494*
drying-up of, 47, 83, 124–26
habitats in, 286–87, *287t*, 440–45
midwater region, 317
mining. *See* mining
oil spills in, 482, *483t*
ownership of, 456
pelagic province of. *See* pelagic province; water column
polar. *See* poles
pollutants in, *476t*, *483t*, 506–12
production in, 338–45, *345t*, *349t*, 352–53
properties of, 151–57, *159*, 168–71, 209, *209*
pycnocline layer of, 157, *159*, 161
sedimentation in, 97–122, 110–20, *111*
size of, 517
surface height of. *See* topography, sea-surface
surface layer of, 157, *159*, 161
surface layer of. *See* microlayer, sea-surface
volume of, *172t*
see also individual named oceans
octopus, *289t*, *448*
offshore, sand transport and, 367
oil
as a pollutant, 477–84, *477t*, 486–87
as a resource, 456, 457–59, 460–61
transport routes of, *480*
oil slicks
burning, 482, *483*, 486
containment of, 482–*83*
dispersal of, 478, 480, *481*, 486
drift of, 79, 80
washing with hot water, 486, *487*
omnivores, 325
ooze
biogenic sedimentary, 114, 115–16, *117*, 119, 120, *121*, 122, 149, 289
calcareous, 116, 119, *120*
carbonate, 124, 167
foraminiferal (foram). *See* ooze, biogenic
silicaceous, 116, *120*
orbit, of water particles, 232–33, *233*, 235, *235*
order of magnitude, 178–79
Oregon
coastline of, 374–75, *375*
sedimentation off, 101
organisms, marine
bottom-dwelling, 301–302, *302*, *447*
chemical contaminants and, *487t*, 491
classification of, 288–91, *288t*, *289t*, 325, *326*
in estuaries, 405
and gas concentrations in the sea, 159–61
growth rates of, 336
in the neuston, 177
thermohaline circulation and, 215
tides and, 276–77
trophic levels of, 325, *326*
organochlorines. *See* hydrocarbons, halogenated
orthoclase, weathering of, 147
orthogonals. *See* wave ray
oscillation, natural period of, to calculate, 248
osmoregulation, 300–301, *300*
osmosis, 300–301
reverse, for desalination, 164, *165*
ostracods, 441, 443
overfishing, 464–67, 496–500
overwash. *See* washover
oxidation, 160
oxygen, 134, *134*, 135
in air, 186
bacterial production of, 333, 336
biological demand for (BOD), 487
photosynthesis and, 328
in respiration, 328
in seawater, 142, 158–62, *161t*, *163*
solubility of, *158*
oysters
classification of, *289t*, 291
Crassostrea gigas, 409
cultivation of, 409, 467, *471*
in estuaries, 409, *410*
feeding, 330
larvae of, 294
pollution and, 407, 492, 508, 509
salinity and, 299
sublittoral, 438

Pacific coast, tides along, 259
Pacific Ocean
bathymetry of, *35*
benthic zonation of, 312, *312*
cobalt in, 462
continental shelf of, 36, 106–107
currents in, 153, 206–207, *206*
earthquakes and, 70–72
evolution of, 70, 72, 80
ferromanganese nodules in, *118*, 461
helium-3 in, *162*
salinity of, *211*
sedimentation in, 119, 121–22, *123*
size of, 517
stratigraphy of, 121–22, *123*
temperatures in, *154*, 210, *211*, *355*
water masses in, 212–13
Pacific Subarctic Water (PSW), *209t*, *213*, *214*
Padre Island, Texas, 382–*83*
Pakhoi, China, tides at, 263, *265*
paleogeography, of North America, *105*
paleomagnetism, 66–68
paleontology, and continental drift, 62
Palos Verdes, California, kelp beds off, *489*
Pangaea, 62, *63*, 69, 70, 104, *105*

parts per thousand, 140, 141
patchiness, of distribution, 436
PCBs, 419, 491, 492–93
peat, *378*, 379, 415, 417, 421
pelagic province, 286, *287*
penguins, 438–39, *439*, 469
Pensacola, Florida, tides at, 259, *260*, *261*
perch, 407, 498
periwinkles, 312, 313
Peru
El Niño and, 355–59
production off, 346, 354–59
pesticides. *See* biocides
petrel, 438, *438*, 469
petroleum hydrocarbons, 419, 457–58, 477
pH
defined, 162–63
scale, *164*
see also seawater, acidity of
Phoenicians, sea routes of, 7
phosphate
in the Mediterranean Sea, *217*
as a resource, 463
in seawater, 339, 419
phosphate nodules, 93
phosphorite, marine sedimentary, 118–19, 494
phosphorus
in macronutrients, 339, 344
in photosynthesis, 339
as a pollutant, 490, *490*
in seawater, 142, *142*, *146t*
photochemistry, at the sea-surface microlayer, 177
photophores, 441
bacterial, 442
photosynthesis, plant
bacteria and, 333
and carbon dioxide, 142, 159–60, 161, 165, 501
as a chemical reaction, *166*, 328, 345
and deep-sea mining, 495
described, 159–60, 324
gases and, 142, *161t*
heavy metals for, 489
light and, 157
marine, 303–304, *303*
oxygen and, 142, 160
phytoplankton and, 338
terrestrial, 303, *303*
physics, basic, 135–36
phytoplankton
on the continental shelf, 436, *444*
coral reefs and, 421, 423, 424, *424*
described, 291, 303, 304
in estuaries, 405, 408, 419
as food, 329, 330
in the open sea, 441
photosynthesis and, 338, 339
populations of, 343, *343*, *348*, *444*, 498
red tides and, 508–509
seasonal succession of, *343*, *437*
as sediment (biogenic ooze), 116
see also dinoflagellates
picoplankton, 359
Pilkey, O. H., 393
piloting, navigation as, 7
plain, abyssal, *36*, 38, *48*, *49*, 52–54, 119
plankton, 250, 291
on the continental shelf, 436
life cycles of, 303–307
in the open sea, 441, 443
pollution and, 476, 478, 480, 492
salinity and, 299
sampling, 296, *296*
plankton bloom
described, 305, *306*, 328
effect of, 339
and fish populations, 498
pollution and, 486, 490
seasonality of, 344, *344*, 408–409
see also diatoms, blooms of
plankton net, 296, *296*
plants
as autotrophs, 324, *324*, 325, *327*
as primary producers, 328
sand dunes and, 375
plate, tectonic
Atlantic, 121
boundaries of, 72–74, 77, *78*
composition of, 72
destruction of. *See* subduction zone
growth of. *See* sea-floor spreading
lithospheric, 72, *73*, 76, 80, 124, *124*, 374
mechanics of, 74–75, *74*
movements of, 70–81
North America, 77, *78*
Pacific, 77, *78*, 121–22
suturing of, 80, *84*
see also tectonism
platform
carbonate, 108–*109*
wave-cut, in cliffs, *384*, *385*, 386
Pleistocene epoch (Ice Age), 104, 106, 504
polarity
reversals of, *66*, *67*, 68, *69*
time scale of geomagnetic, *67*
poles
air temperatures over, 191, 502
primary production at, 339, *344*, 345, *346*
salinity of seawater at, 156
sedimentation at, 112, *115*, 119
water from, 209, 210, 212
politics, and fishing quotas, 500
pollution
coastal, *506*
defined, 476
in estuaries, 400, 406–408
guidelines for minimizing, *510t*
in the oceans, *476t*
and the sea-surface microlayer, 176–77, 476, 486
polychaete worm
classification of, *289t*, 291
on the continental shelf, 440, *441*
described, *293*, 317
in estuaries, 409, *410*
feeding, 316, 330
habitat of, *294*, *316*
polymers, in water, 140
Polynesia, migration from, as early exploration, 5, *6*
polyp, 421, *423*
population
chemical contaminants and, *487t*
estimates, *349*, 497, 499, 500
natural, and exotic species, 420
oil spills and, 481–82
size of, and food supply, 330–31
pore water, 450
porpoises, *289t*, 441, *445*
Port Adelaide, Australia, tides at, 263, *265*
Portugal, ocean exploration and, 15
potassium
in river water, *147t*
in seawater, 140, *141t*, *146t*, 148, *149t*
powers of 10, 41
precautionary principle, of fishing practices, 500, *501t*
precipitation, *175t*
climate change and, 503
and salinity, 154, *156*
and the stability of cliffs, 386
predators, 330–31, *331*, 441, 443, 444, *445*, 448
energy requirements of, 337–38
pressure
atmospheric, 90, 172, 186, *187*, 189, *190*, 301, 356
defined, 186
and the Earth's structure, 33–34
formula for, 186
and gases in seawater, *161t*
hydrostatic, and ecology, 301–302, *302*, 449–50, *450*
Southern Oscillation of, 356
and water density, 157
see also vapor pressure
pressure gradient
described, 186
tides and, 266, 268
water and, 192–93, 197, 268
and waves, 355–67
wind and, 187, 356
pressure sensors
to measure tides, 258
to measure waves, 230, *230*
Prince William Sound, oil pollution in, 484, 486
prism, accretionary, 107, *108*, *109*
prodelta, 386
producers, primary, 325, 328, *424*
production, primary
defined, 338
global patterns of, 345–49, *346*, 352–*54*
gross, *345t*
limits to, 338–41
measuring, 342, 349, 352–*54*
net, 338
in the oceans, 338–41, 345–48, *349t*, 352–53
and picoplankton, 359
rates of, 345–46
seasonal, *343*, 344, *344*
in upwelling waters, 354–55, 358–59
variations in, 341–45
profile
beach, 367–70, *370*
dune, 371
oceanographic, 523, *523*
salinity, *403*
seismic-reflection, *113*
storm, 368, *370*
swell, 368
temperature, *154*
topographic, 522–23, *522*
proportion, principle of constant (constant composition), 143, 145, 149
Protista, 288–89, *288t*, *289t*, *290*, 307
protons, described, 134
PSS78. *See* Scale, Practical Salinity Scale 1978
PSW. *See* Pacific Subarctic Water
pteropods, 116, *117*
Puerto Rico
desalination in, *165*
hurricane damage in, *243*
puffin, 438, *438*
purse seine, 296, *297*
pycnoclines, 157, *159*, 476, *485*
waves along, 249–50, *250*
Pytheas, 7

quartz, sedimentary, 100, 107

radar, satellite, *197*
radiation, solar, 210
and air currents, 186, *187*
ecosystems and, 324
and the hydrologic cycle, 172–74
and primary production, 338–39, 342, 345
radioactivity, decay of, and sea-

water gases, 160
radiolaria, 116, *117*, *289t*, 441
radiometer, *197*
radon, in seawater, 160
rainfall. *See* precipitation
Red Sea
evolution of, 76, *76*, 82–83
mining in, 83, 495
Red Sea Intermediate Water (RSIW), *209t*, 213
red tides, 103, 508–509
reef
barrier, *426*, 428–29, *429*
fringing, 428, *429*
patch, 425, *426*
see also coral reef
reef face, 425
reef terrace, 425
refineries, and oil pollution, 477, *477*, *477t*
reflection, seismic, 44–45, *45*, 110, 458
profile, *113*
refraction, seismic, 45, *46*
remotely operated vehicles, 23
reptiles, 291
residence time
of ions, 149, *149t*
to calculate, 430
of water masses, 216, 222
resonance, tidal
in the Bay of Fundy, 273
seiches and, 248–49, 273
resources, marine
fish, 463–67
jurisdiction over, 456
mineral, 457–63
respiration, 160, *161t*, 162, 328, *450*, 501
chemical reaction of, 165, *166*, 328
Revelle, Roger R., 503–504
Reykjanes Ridge, magnetic anomalies of, 65–66, *65*
ridge, algal, 424–25, *426*
ridge, midocean, 35, *37*, 38–39
as boundaries of tectonic plates, 72
dimensions of, *36*
sedimentation and, 119
spreading, 68–69, *70*
see also East Pacific Rise; Mid-Atlantic Ridge
rift valley
East Africa, 76, *76*
midocean ridges and, 38, *64*, *70*
ring
cold-core, 205, *205*, 208
warm-core, 205, *205*, 208
Rio Grand Rise, 218
Rio Grande, Texas, 382
riprap, 385
rise, continental
described, 36–38, 52
dimensions of, *36*
habitat zones on, 286
Riva, Joseph P., 458–59
rivers
constitutents of water in, *147t*
deltas of, 386–88
drowned valleys of, as estuaries, 400, *400*
estuaries and, 402
metals from, 489
minerals from, 382
nutrients from, 354, 440–41
and the salinity of seawater, 146, 146*t*, 154
as a source of ocean sediment, 92, 93, 100, 104, 107, 110–12, *111*, 115, 147, 341, 460
as sources of pollutants to the ocean, 477, *477t*, 485, 489, 491
water in, *172t*
see also estuary
RNA, 328
rock
bedded. *See* rock, stratified
cliff, and recession rates, 384, 386, *386t*
compression of, *64*, 65
density of, 39–40, *40*
igneous, 39
marine sedimentary, 148
phosphate, as a resource, 463
sedimentary, 374, 458
sources of sediment and, 93, 110, 489
stratified (bedded), 121
types of, in the Earth's crust, 39–40, *40*
volcanic, 39
weathering of chemicals out of, 147–48
rock salt. *See* halite
Rocky Mountains, origins of, 65, 108
Royal Society (London), 18, 20
RSIW. *See* Red Sea Intermediate Water
Ryther, John H., 349, 353

SACW. *See* South Atlantic Central Water
Sahara Desert, deep-sea sedimentation and, 102, 115
Saint John River, Canada, tides in, 269, *269*
salinity, 143–51
of the Atlantic Ocean, 155, *158*, *211*, 216
benthic, 299
of the Black Sea, 216
and body-fluid chemistry, 299–301
climate and, 152, 153–56
coral reefs and, 424
defined, 143
and density, 301
depth and, *158*
and ecology, 299–301
of estuaries, 403, *403*, 405, 406, *407*
and evaporation, 148, 154, *156*, 210, 212, 410
and gases in seawater, *161t*
gradients of (haloclines), 154–55, *158*, 208, 403
of the Indian Ocean, 213
of lagoons, 410
latitude and, 153–56, *156*
measures of, 140, 146, *146t*
of the Mediterranean Sea, 216, *217*
of the North Atlantic Ocean, 155, *220*
of the Pacific Ocean, *211*
profiles, *403*
and properties of water, 150–51, 153–56, 209, *209*
regulation of, 146–50
and solubility, 158
tolerance to changes in, 299
volcanoes and, *146t*
salinometer, 146
salmon
Atlantic (*Salmo*), 351, *352*, 464
mariculture of, 464–65, 467
migrations of, 351–*52*
North Pacific pink (*Oncorhynchus*), 351, *351*
overfishing of, 496
salt
absorption of, for desalination, 164
as a resource, 494
sources of, to the ocean, *146t*, 147–48
see also halite
salt marsh, 411–16
as an ecosystem, 295
barrier island, 378, *378*, 379
development of, *413*, 414–15, *414*
drainage in, 411–12, *413*
estuaries and, 400, 413
flooding in, 412
high, 412
low, 412, 413
plants of, 291
pollution of, 481–82, *483t*
production in, 412, 413
reclamation of, 416, 460
zones of, 412, *413*, *414*, *416*
saltation, 376
San Deigo County, California, cliff collapse in, 385, *385*
San Francisco Bay, environmental degradation of, 418–20
San Francisco, California
climate of, 501
reclamation of salt marshes and, 418–20
tides at, 259, *260*, *261*
San Joaquin River, California, 418
sand
beaches and, 267, 379
dunes. *See* dunes
grain size of, 368–69, *372*, 373
mining, 460–61, 494
nourishment, 391, 392–93, 460, 493, *495*
as a resource, 459–61, *462*
saltation of, 376
sedimentary, 47, 93, *93*, 96, 97–98, 110, 438–39, 459
sources of, 370
transport of, 96, *96*, 367, 370, 379, 388, 391
sand budget, 370–71, *371*, 392
to calculate, 392
sand dollars (*Mellita*), *289t*, *316*, 317
sand ridge, *380*
sand spits, 379, *380*
sandstone, 50, 119, *386t*
sardine (*Sardinops sagax*), 298, *358*, 496, *497*
Sargasso Sea
circulation in, *198*, 204–205, *204*, *205*, 207, 212, *348*, 446
phytoplankton in, *444*
Sargassum sp. (sargassum gulfweed), 204, *204*, 441, 446, *446*
satellites
for exploration, 25, *26*, 46, *82*, *84*, 195–97*t*, 218, *406*, *418*
oceangraphy and, 342–43
to measure waves, 231, *231*
saturation, 139, 148
saturation value, 158
Savannah River, Georgia, tides at entrance of, *261*
Scale, Practical Salinity 1978 (PSS78), 140
Scapa Flow Marine Archeology Project, 23
scavengers, 330
science, process of
defined, 14–15
and the falsification of hypotheses, 80–81
scour, tidal, and sedimentation, 101–102
Scripps Institution of Biological Research, 24
Scripps Institution of Oceanography, 24, 503
SCUBA diving, 233
sea anemones, *289t*, 438
sea cucumbers, *289t*, 291, *293*
sea ice
formation of, 168–69, *168*
seasonal, 195, *196*
Sea of Japan, methane from, 459
sea level
changes in, 98, *99*, 101, 104, 504
flooding and, 393, 502–503, *502*
and glaciation, 104
global rise of, 502–503, *502*, 504–505
Holocene rise of, *98*, 100, 379, *380*, 388, 400

low stand of, *99*, 104, 110, 112
measuring tides and, 258
sea lilies, *289t*, 291, *293*
sea lions, *492*
Sea of Marmara, 216
sea otters, *289t*, 314–15, *314*
sea, the
as a common heritage, 456, 457
defined, 216
fully developed. *See* waves, generation of
law of, 456
water flow in, 216–17, 221–22
sea urchins, *314*
classification of, *289t*, 291
feeding methods of, 328
habitat of, 313, 409, *441*
kelp forests and, 314–15, *315t*
populations of, *315*
sea wall
dunes as, 377
engineered, 385, *389*, 391–92
sea-floor spreading, 56–57, 64, 68–69, 70, 80–81, 120
and the evolution of ocean basins, 76, 80, 82–83
rate of, 69, *69*, 85
seals, *289t*, 291, 359, 469
seamount
cobalt on, 462
described, 38, *48*, 54
dimensions, *36*
sedimentation and, 119
seas. *See* wave interference, patterns of
SEASAT (satellite), 46
Seattle, Washington, tides at, *261*
seawater
acidity of, *161t*, 162–67, 333
analyzing, 145
carbon species in, *166*
density of, 157, *159*, 301
gases in. *See* gases
nutrients in, 142, *142t*
residence times of ions in, 149, *150t*
salinity of, 134, 139–40, 147–50, 154–57, *156*, 209
sampling, 144–46
solutes of, 140–*43*, 146–47, *146t*
temperatures of, *209*
trace elements in, 142, *143t*
seaway
passage through, 456
semienclosed, water flow in, 215–17, 221–22
seaweed
benthic, 313, 440
brown (*Laminaria*), 312, 313
ecological balance of, 478
holdfasts of, 304
Macrosystis, 314–15, *314*, 438, 489
Sargassum, 204, *204*
see also algae
SEC. *See* current, South Equatorial
sediment
anhydrite, 124
authigenic, 93, 110, *111*, 116–17
on beaches, 367
biogenic, 93, 100, *100*, 110, 115–16
bulk emplacement of, 110–12, *113*
carbonate (carbonate shelves), 107–10, *109*
see also calcium carbonate, sedimentary
classification of, 92–93, *93t*
continental margins and, 36–38, 104–107
continental shelf, 97–110, *98*, *100*, *120*, 459, 460
cosmogenous, 93
cycle of, 148–49
dating, 80–81
deep-sea, 97–122, 110–20, *111*
deltaic, 104, 112, 386, 388
distribution of, 99–100, 119–20, *120*
in estuaries, 400, 401, 402, 405
and the geologic time scale, 101–10
glacial, 11, *100*, 112, *115*, 119, *120*
graded bedding of, 112, *113*
infauna of, 313
and light transmission, 169–70, *169*
margin, *113*
Miocene, 124
pelagic, 110, *111*, 112, 114–16, *117t*, 119
prism, 107, *108*, *109*
processes of, 110–19
properties of, 93, 96–97
rates of, 97, 119–20, *121*, 127
in the Red Sea, 83
relict, 99, *99*, 100, 459
sampling, 94–95
sources of, 92, 93, 110, 119
stranded shelf, 47
stratigraphy of, 120–22
suspension of, by whales, 442–43
terrigenous, 93, 99, 100, *100*, 105, 107, 110, *111*, 112, 119, *120*, *121*, 460
thickness of, 70, 80, 120
turbulence and, 438
unconsolidated, *386t*
volcanogenic, 93, 107, *109*, 122
sedimentation
bathymetry and, 119
currents and, 100, 104
depth and, 97–99
earthquakes and, 107
equatorial, 100, 116, 119, 212–22
and erosion, 92, 93, 96, *96*, 104, 105
and latitude, 100, *100*, 112, 114, 116, *116*
rate of, 147
and tectonism, *105*, 106, *109*, 122
seepage, natural, and oil pollution, *477t*, 480
seiche, 228, *228t*, 247, 248–49, 273, 380
seismicity
global, 70–72
map of, *71*, *73*
measuring, 77
septa, 421
settling rates, of marine organisms, 304
sewage, pollution and, 407, 484, 486
shale
erosion of, *386t*
oil in, 458
sedimentary, 93, 119, 121
shark
great white, 330, *331*
as a predator, *308*, 330–31, 441, *445*
whale, 330, *331*
shelf break, 36, *48*, 49, 98
shelf, continental
biology of, 436–40
described, 36, 47–48, *48*, *49*
development of, 105–106
dimensions of, *36*, 97
habitat zones on, 286, 436
jurisdiction over, 456
polar, 100
pollution on, *483t*, 488, *488*
production on, 345–46, *345t*, 354
sedimentation of, 97–110, *98*, *120*, 459, 460
shellfish
anoxia and, 407
as detrivores, 412
mariculture of, 467
pollution and, 491, 508–509
salinity and, 299
yields of, and temperature, 298–99, *299*
shells
dissolution of, 166–67
as sediment, 93, 100
Shemya Island, Alaska, kelp forests off, 314–15, *315t*
shipping lanes, oil pollution along, 477, 480, *480*, 507
shoal, 402
shore, defined, 399
shoreline
engineered, 388–92, *389*
erosion of, *388t*
as habitat, 399–400
literature on the preservation of, 393, 396–97
migration of, 388
sedimentation and, 97
storms and, 243–46
zones of, 367
shrimp
classification of, *289t*, 291
El Niño and, *358*, 359
habitat of, 441
mariculture of, 467, 470
pollution and, 492
SICW. *See* South Indian Central Water
sidescanning. *See* sonar
silica
in river water, *147t*
in seawater, *146t*, 147, 148, 299
sedimentary, 93, 116
silicoflagellates, 339
silicon
in macronutrients, 339
in seawater, 142, *142t*, 147, 339
silt, sedimentary, 93, *93*, 97, *111*, 438
silver, in water, 419
sink
described, 148
of gases, *160*
seawater as, *146t*, 148
skeletons, as sediment, 93, 100, 115–16
slope, continental
described, 36, 49–50
dimensions of, *36*
habitat zones on, 286
slump, sedimentary, 50, 110, 112, *113*
and tsunamis, 250, *251*
slurries, sedimentation and, 112
snails
classification of, *289t*, 291
feeding methods of, 328, *329*, 330
habitat of, 312, 313, 409, 440
moon, *329*
in salt marshes, 417
Snider, Antonio, *63*
sodium
in river water, *147t*
in seawater, *139*, 140, *141t*, *146t*, 149, *149t*
SOFAR channel. *See* channel, deep-sound
Sohm Abyssal Plain, *48*, 52
solubility, 158
of oxygen, *160*
solutes, of seawater, 140–*43*, *141t*
solution
acidic, 163
basic, 163
neutral, 162

solvent, water as, 137, 138–39
sonar, sidescan, 44, *45*
sound
  transmission of, through water, 170–*71*
  velocity of, 170, *171*
sounding, technology of, 35, 44, *44*, 145
South Atlantic Central Water (SACW), *209t*, *213*
South Atlantic Ocean
  currents in, 208
  exploration of, 20
  ferromanganese nodules in, *118*
  salinity of, 156
South Indian Central Water (SICW), *209t*, *213*
South Pacific Central Water (SPCW), *209t*, *213*
South Pacific Ocean, water masses in, 212
Southern Hemisphere, Coriolis deflection in, 189, 192, 199, 200, 270, 340
spaghetti diagrams, Lagrangian, 195, *196*
SPCW. *See* South Pacific Central Water
species
  density, 441
  distribution of, 295–96, 493
  diversity of, 405, *409*, 418, 443, 445, 448
  exotic, and natural populations, 420
  keystone, 468, 469
speed
  of ocean currents, 192, 199–200, 203, 206, 208
  tide, 273
  of waves. *See* waves, celerity of
  wind, 192
spiral, Ekman, 199, *199*
sponges, *289t*, 290, 409
spraying, aerial, as pollution, 491–92
squid
  classification of, *289t*, 291
  described, 448–49
  giant (*Architeuthis*), 449, *449*
  habitat of, 441
  as a predator, *445*
SST. *See* temperature, sea-surface
starfish, *289t*, 291, *294*, 313, 409, *410*
steady state, 148, 370
  to calculate, 510–11
Stokes drift, 233, *233*
storm profile, 368, *370*
storm surge, 243, *243*, 248, 379
storms
  barrier islands and, 379–81, *381*
  beaches and, 368–69
  damage caused by, *381*
  and primary production, 345, 354
  and sedimentation rates, 101, 102
  shorelines and, 243–46
  and waves, 236, 237, 245
Strait of Gibraltar, 216, 219
stratification, water, 157
stratigraphy
  deep-sea, 120–22, *122*, *123*
  layer-cake, 121, *122*
strike-slip. *See* faulting, strike-slip
stromatolites, in the Mediterranean Sea, 124
strontium, in seawater, 140, *141t*, *149t*
structures, coastal-defense
  and a rise in sea levels, 504–505, *505*
  and the stability of cliffs, 385
  and the stability of shorelines, 388–92, *389*, 393–94
subduction zone, 70–72, 74, 80, 104, *109*, 122, *221*
  Mediterranean area, 124, *124*, *221*
  plate tectonics and, 70–72, 74, 80, 104
  and sedimentation, *105*, 106, *109*
  volcanoes and, *105*
submarines, 235
  effect of internal waves on, 250
submersibles, for exploration, 25, 26, 334, 449–50
subtropics, sedimentation along, 100
sulfate
  in river water, *147t*
  in seawater, 140, *141t*, 333
sulfides, 119, 333, *334*, 335
sulfur, seawater, 119
Sun
  heat from. *See* radiation, solar
  tides and, 254, 258–61, *265*
supersaturation, 148, 158
supertankers, 245–46, 486–87
  see also tankers
surface tension, 244, *244*
swell, ocean, 228, *228t*, 237, *237*, 244
  beaches and, 367–68
swell profile, 368

tankers, and oil pollution, 477–79, *477*, *477t*, 482, 507
tar balls, 480, *482*
taxonomy, 288
tectonism
  along the Oregon coast, 374–75, *384*
  cliffs and, *384*, 386
  and earthquakes, 70–72
  effects of, 92, 104–105, *105*, *374*, 375–75
  estuaries and, 400, *401*
  and the Mediterranean Sea, 124–26, *124*
  model of, 72–75, *84*
  San Francisco Bay and, 418
  and sedimentation, 122
  summarized, 80–81
temperature
  and air pressure, 186, *187*
  and biological activity, 295–96, 297–98, *298*, 409
  coral reefs and, 424
  and density, 136, 157, 186, 301
  depth and, 154, *154*, *155*, 157, 208–209, *209*
  of the Earth, 34, 106–107, *107*, 502
  and enzyme synthesis, 296
  and gases in seawater, *161t*
  and kinetic energy, 136
  and marine ecology, 295–99
  melting and boiling points of water and, 136, *137*, 138
  and primary production, 339
  profiles, *154*
  and salinity, 151, *151*, 157, 210
  sea-surface (SST), *152*, *153*, *299*, 354, *355*
  and solubility, 158
  and sound frequency, 145
  tolerance to changes in, 297–98
  see also freezing point
tension, midocean ridges and, 65
terrace
  wave-cut, 374, *374*
  see also reef terrace
Tethys Sea, 216
Texas, red tides off, 508
theca, 306
thermocline
  described, 152, *159*, 523, *523*
  permanent, 157
  in the Sargasso Sea, 204–205
  seasonal, 152–53, *155*, 344
  and sound transmission, 170
thermometer
  principles of, 136
  reversing, 145
Thomson, C. Wyville, 20
Thresher, USS, 250
tide table, 260
tides
  as amphidromic systems, 267, 270–71, *271*, 274
  and benthic communities, 312, *312*
  celerity (speed) of, to calculate, 273
  centrifugal force and, 260, 262–63, *262*
  characteristics of, 258–60
  classification of, 259
  coastal zone and, *369*
  components of, 270–71
  and Coriolis deflection, 266, 267, 270, 271, 276
  currents and, 101–102, 275–78, *340*
  deltas and, 386, *387*
  diurnal, 259, *260*, *264*, 267, 272
  electricity from, 279–80, *280*
  estuaries and, 402, 405
  gravity and, 260–62, *262*, 266
  as hydraulic currents, 268–69
  mixed, 259, *260*, *264*, 267, 272
  models of, 263–72
  the Moon and. *See* Moon
  neap, 259, 263–64, *265*
  origins of, 260–67, 270–71
  periods of, 258, 259, 264
  predicting, *270*
  Pythias's observations of, 7
  range of, 258, 259, *259*, 263, *265*, 269, 270, 272, 273
  records of, 259–60, *261*, 268, *268*
  salt marshes and, 411–12, *413*, 414
  semidiurnal, 259, *264*, 267, 272
  spring, 259, 263–64, *265*, 379
  the Sun and. *See* Sun, tides and
  to measure, 258, *258*, *260*
  as waves, *228t*, 258–59, 271
  see also red tides
till, glacial, 100, 104
time scale, geologic, *519*
  and sedimentation, 101–10
tin, as a resource, 494
Tonga Trench, plate tectonics and, *71*, 72
TOPEX/Poseidon satellite, 25, 26, 46
topography
  frequency distribution of, 39, *40*, 42
  mapping, 521–22, *522*
  sea-surface, 46, 192–93, 197, *198*, 202, *202*, 216
  submarine. *See* bathymetry
Torrey Canyon, oil spill from, 478–79
toxins
  algae producing. *See* red tides
  chemical, dumping of, 407
  in estuaries, 419
  halogenated hydrocarbons as, 492
  heavy metals as, 489–91, *490*
trace elements, in seawater, 142, *143t*, 419
transport
  mass, of waves (Stokes drift), 233, *233*

volume, to calculate, 210
see also sand, transport of
trawling gear, 296, *297*
treaty, about marine resources, 456, 468
trek, imaginary, across the sea floor, 47–57
trench, deep-sea, 38, *123*
dimensions of, *36*
earthquakes and, 70, 71
habitat zone in, 286, 301–302, *302*
sedimentation and, 107, *109*, 119, 122
trophic dynamics
bacteria and, 331–33
energy and, 325–28, 337–38, *337*
tropics, primary production in, 342–44, *344*
tsunami, 228, *228t*, 249, 250–52
tuna, *308*, 309, 310–11, *310*, *326*, 441, *445*, *463*, 498
turbidites, *109*, 112, *113*, 119
turbidity, of seawater
and density, 301
maximum, 405
and primary production, 338, *338*, 341, *341*
turbidity currents, 112, 113
turbulence, and primary production, 338, *340*, 341
typhoon, 206–07

undersaturation, 158
United Nations Organization
Food and Agricultural Organization, 496
international fishing and, 466
law of the sea and, 456
and oceanography, 24, 511
pollution and, 476, 508
United States
EEZ of, 457, *458*
the Law of the Sea treaty and, 456
upwelling, 200
and climate, 501
coastal, 340, *340*, 343, 346, 354, *355*
and Coriolis deflection, 340, 354
equatorial, 200, 340, *340*, 343, 346
and primary production, 338, 340–41, 343, 345, *345t*, 346, *349t*, 352–55
rates of, 354
and sedimentation, 119
U.S. Army Corps of Engineers, 391, 418, 493
U.S. Geological Survey, 418, 459
U.S. Navy, 24, 250
U.S.S.R. (former), fishing by, 468, 498

vapor pressure, 152–53
and salinity, 151
vaporization, 136
vegetation. *See* plants
vent, submarine volcanic (hydrothermal), 451
communities surrounding, 333, 334–36
metals in, 117, 119
sites of, *334*
Venus, 440
Vikings, exploration by, 14–15, *16*, 106
Vine, Fred, 68, 80–81
Vinland, 14
volcanoes
andesite, 70, 72
arcs of. *See* arc, volcanic
Cascade Range, 374
coral reefs and, 429
and gases in seawater, 161, *162*
heavy metals from, 489
Mediterranean region, 124
plate tectonics and, 74–75, *74*, 124
and the salinity of seawater, *146t*
sedimentation and, 93, 107, *109*, 122
submarine, 38, *54*, 57, 81, 85, 148
tsunamis and, 250
see also vents, submarine volcanic
volume transport, to calculate, 210
vorticity, and western-boundary intensification, 204

walruses, *289t*
warfare, modern, and scientific exploration, 24
warming, global, 106–107, 393, 470, 501–503, *503*
Washington State, sedimentation off, 101
washover (overwash), 379, 382, 504
waste
dumping of, 400, 411, 419, 484–86, 507–508
toxic, 83
water
anoxic, 160, 222, 400, 406–*408*
bonds in, 137–38, *137*, 139
circulation of. *See* circulation
cluster structure of, 138, *139*, 164
density of, 139–40, 151, 157, *159*
depths of. *See* depth
distribution of, on Earth, 31, 172–73, *172t*, 341
diversion of, and environmental damage, 419
flow patterns of, 185, *196*
fluxes of, 173–74, *175t*
gaseous (water vapor), 135, *135*, 136, 186
global cycle of, 173–74
liquid, 135, *135*
melting and boiling points of, *137*, 138
molecules, 136–43, *137*, 153
particles of, and wave motion, 232–33, *233*, 235
pH of, 162–67
photosynthesis and, 328
pressure. *See* pressure, hydrostatic
properties of, 136–40, 139, 150–51, 168–71
see also oceans, properties of
residence times of, 222, 501
in respiration, 328
solid. *See* ice
states of, *135*
subterranean. *See* groundwater
vapor. *See* water, gaseous
water budget, 209
water column
density and, 157, 301
height of, 193
mixing of, *340*, 341, 345
Pacific Ocean, 212
stratification of, 345, *348*, 402, 407
temperatures of, 210
thermocline and, 170, 343
turnover times, *221*
see also pelagic province
water mass, 208–209
Atlantic Ocean, 212, 218–19
bottom, *209t*, *213*
central, *209t*, *213*
deep, 212
distribution of, 212–15, 216
intermediate, *209t*, 212, *213*
listed, *209t*, 213
properties of, *209*, *209t*
residence times of, 216
structure of, *213*
as submarine "weather front," 218–20
waterfalls
reversing tidal, 269, *269*
submarine, 218–20, *219*
waters, jurisdiction over international, 456
wave base, 233
wave interference, 236, *237*, 249, 264, *265*
patterns of, 236, *237*, *238*
wave period, 228, 237, 248, *248*, 249
classification and, 228, *228t*, *229*
wave ray (orthogonals), 240, *241*, 366, 367
wave reflection, 392
wave refraction, 240, *241*, 271, *271*, 273, 366, 367
wave setup, 366, 367
wavelength, 228, *228t*, 230, 237, 250
waves
angle of approach of, 366, *368*
capillary, 228, *228t*, 244, *244*
causes of, *228t*
celerity (speed) of, 234, 235–36, 237, 240, 249, 250–51
deep-water, 235, *235*, 236
deltas and, 386, *387*
depth and, *233*, 235, 240, 248
described, 228
development of, 229, *231t*, 236
dispersion of, 237
energy and, 97, 101, 240, 388, 391
erosion by, 382
and gases in seawater, 159, *160*
giant, 245–46
gravity, 244
height of, 228, 230, 231, *231*, 232, 249
see also tides, range of
intermediate-water, 235, 236
internal, *228t*, 249–50, 402, *404*
life history of, 236–46
measuring, 230–31
waves *(continued)*
motions of, 232–36, *232*, *233*
oblique, *368*
ocean, 237
parallel, *369*
parameters of, *229*
progressive, 232–46, 249–52
properties of, 228–32
rogue, 236, 245–46, *246*
rotary, tides as, 267, *267*
shallow-water, 235, *235*, 236, 237–43, *245*, 250, 271
shoaling, 366
size of, 236, 249
solitary, 244, *245*
spectrum of, *229*
speed of. *See* wave, celerity of
standing, 246–49, 273
steepness of, 240, 242–43
tidal. *See* tsunami
transformations of, 239–40, *239*
of translation, 244–45, *245*
wind and, 192, 228–32
see also breakers
weather, underwater, 218–20
weathering
of oil slicks, *481*
of rocks, 93, 110, 147–48, 489
Wegener, Alfred, 62, 68
Wentworth grain-size scale, *93t*
western-boundary intensification, 203–04

wetland
loss of, 418, *419*, 470
pollution of, *483t*, 485
see also estuary; salt marsh
whales, 328
blue, 328
bottom feeding by, 442–43
classification of, *288t*, *289t*, 291
gray, 411, *411*, 442
killer (*Orcinus orca*), 311, *311*, 315
krill and, 468–70
sperm, *288t*, 449,
Whitman, Walt, 51
Wilson cycle, 75–76, *76*, 80
wind
air currents as, 186
circulation cells and, 187, *187*, 190–91, *190*
Coriolis deflection, 186
drag, 192
easterlies, *190*, 192
effects of, on ocean currents, 101, 192–208, 340
general patterns of circulation, 189–91, *190*, 503, *503*
Langmuir circulation of, 201, *201*
latitude and, 190–91
polar easterlies, *190*
prevailing, 190–91, *191*, 202
sediment and, 93
trade, 190, 192, 200, 201–202, *202*, 207, 356, *357*, 358
waves and, 228–32, *231t*
westerlies, 187, 192, 201–202, *202*
zonal flow of, 192
see also hurricanes
Woods Hole Oceanographic Institution, 24, *95*
worms
acorn, *289t*
arrow, *289t*, *327*, 441
in the open sea, 443
tube, 334, *336*
see also flatworm; polychaete worm

xenon, in air, 186

Yield, Maximum Sustainable, 497, 499

zinc
in ferromanganese nodules, 117, 461
in seawater, *143t*, *149t*, 339
zonation, vertical, 312, *312*
zone
abyssal, 286, *287*, 295, 449
abyssalpelagic, 286, *287*, 295
aphotic, 170, 287, *287*, 441–45, *447–50*
Arctic Convergence, 210, 212, 213
backshore, 367, *369*
bathyal, 286, *287*
bathypelagic, 286, *287*
beach, 378, *378*
Benioff, *71*, 72
breaker, 367, *369*
buttress, 425
coastal, *369*
dysphotic, 287, *287*, 441
epipelagic, 286, *287*, 295, 299
Exclusive Economic (EEZ), 456–57, *457*, 462, 463
fracture, 39, 64
hadal, 286, *287*, 294
hadalpelagic, 286, *287*
high-pressure, 186, *187*, 190–91, *190*, *191*
inclined, 71–71
intertidal, 286, 312–13, 440
littoral, 286, *287*
low-pressure, 186, *187*, 189–91, *190*, *191*
mesopelagic, 286, *287*, 299, 302
nearshore, 47, 367, *369*, 370, 378, *378*, 485
neritic, 286, 436–37
oceanic, 286
offshore, 367, *369*, 370
photic, *169*, 170, 287, *287*, 291, 304, 339
subduction. *See* subduction zone
sublittoral, 23, 286, *287*, 436, 437–40
surf, 366–67, *369*
swash, 367, *369*
zooplankton
biomass, 346–48, *347*
on the continential shelf, 436, 499
described, 291, 306–307, *307*, 328
distribution of, 348
in estuaries, 408–09
as food, *331*
food for, 332
in the open sea, 441, 442
in salt marshes, 417
zooxanthellae, *289t*, 421, 423, *423*, 424

PHOTO CREDITS [CONTINUED FROM PAGE iv]

Cover and chapter openers background: Digital Vision.

Image on title page: AbleStock.

Images on Contents: p. vii, Seventeenth Studios; p. ix, AbleStock; p. xi, Russell Illig/PhotoDisk, Inc.; p. xii, U.S. Fish & Wildlife Service; p. xii, AbleStock.

Images for box icons: The Ocean Sciences Physics, PhotoLink/PhotoDisc; The Ocean Sciences Geology, PhotoLink/PhotoDisc; The Ocean Sciences Biology, PhotoLink/PhotoDisc; The Ocean Sciences Chemistry, Phil A. Harrington/Peter Arnold, Inc.; The Process of Science, NASA; Science By Numbers, PhotoLink/PhotoDisc; Study Guide, PhotoLink/PhotoDisc

CO1: opener, NOAA; **1-2b**, Herb Kawainui Kane; **1-3**, The Granger Collection; **p. 8 top**, NOAA Central Library; **p. 8 middle**, Historical Picture Service; **p. 8 bottom**, Historical Picture Service; **p. 9 top**, H. Armstrong Roberts; **p. 9 middle**, Historical Picture Service; **p. 9 bottom**, Library of Congress; **p. 10 top left**, Historical Picture Service; **p. 10 top right**, NOAA Central Library; **p. 10 bottom left**, Historical Picture Service; **p. 10 bottom right**, Down House, Down, England/The Bridgeman Art Library; **p. 11 top**, Cobis/Bettman; **p. 11 middle**, Steve Nicklas, NOS, NGS/NOAA; **p. 11 bottom**, Steve Nicklas, NGS, RSD/NOAA; **p. 12 top**, San Diego Historical Society; **p. 12 bottom**, John Bortniak/NOAA; **p. 13 top**, Corbis/Bettman; **p. 13 middle**, Jet Propulsion Laboratory; **p. 13 bottom**, NASA; **1-6a**, The Granger Collection; **1-6b**, Down House, Down, England/The Bridgeman Art Library; **1-7a**, Steve Nicklas, NOS, NGS/NOAA; **1-8a**, The Natural Museum History, London; **1-8b**, Corbis/Bettman; **1-9a**, The Granger Collection; **B1-4**, NOAA; **B1-5**, Center for Coastal and Ocean Mapping/University of New Hampshire; **1-10a**, Scripps Institution of Oceanography University of California, San Diego; **1-10b**, Ocean Drilling Program, Texas A&M University; **1-11a**, NOAA; **1-11b**, NASA. **CO2: opener**, Tom Van Sant/Geosphere Project, Santa Monica/SPL/Photo Researchers, Inc.; **2.2**, NOAA; **B2-3**, Paul Pinet; **B2-5**, NOAA. **CO3: opener**, J. D. Griggs, Hawaiian Volcano Observatory/USGS; **B3-2a**, NOAA/NGDC; **B3-2b**, NGDC; **B3-3**, NOAA/NGDC; **B3-4a**, NASA; **3-11b**, NASA; **3-11c**, NASA. **CO4: opener**, Jacques Descloitres, MODIS Land Rapid Response Team, NASA/GSFC; **B4-1c**, Harley D. Nygren/NOAA; **B4-4ab**, Jacques Descloitres, MODIS Land Rapid Response Team/NASA/GSFC; **4-14b**, D.B. Scott/NOAA; **4-14c**, Kjell B. Sandved/Visuals Unlimited; **4-14d**, Woods Hole Oceanographic Institute; **4-14e**, Roger Klocek/Visuals Unlimited; **4-14f**, Neil Sullivan/NOAA; **4-15a**, UCSD Scripps Institution of Oceanography. **CO5: opener**, OAR/National Undersea Research Program/NOAA; **B5-1**, W.A. Banaszewski/Visuals Unlimited; **B5-2**, J. Swift/ UCSD Scripps Institution of Oceanography; **B5-3**, Dean DeChambeau, **B5-5ab**, Michael Van Woert/NOAA; **B5-7**, Jacques Descloitres, MODIS Land Rapid Response Team/NASA/GSFC. **CO6: opener**, NOAA Central Library; **B6-4 ab**, National Snow and Ice Data Center; **6-8b**, Courtesy Mobil Oil Corp.; **6-11a**, John D. Cunningham/Visuals Unlimited; **6-11b**, R.J. Goldstein/Visuals Unlimited; **6-12a**, NOAA; **B6-6**, Jacques Descloitres, MODIS Land Rapid Response Team, NASA/GSFC; **CO7: opener**, Mariners Weather Log/NOAA; **B7-2**, NASA; **7-5a**, Natlalie Fobes/Tony Stone Images; **7-5b**, Tony Arruza/Corbis; **7-5c**, Digital Vision; **7-9a**, Digital Vision; **7-9b**, Digital Vision; **7-9c**, Will and Deni McIntyre/Photo Researchers, Inc.; **7-10a**, NOAA; **7-10b**, NOAA; **7-13**, Harry Yeh/NOAA. **CO8: opener**, AbleStock; **8-2a and b**, Andrew J. Martinez/Photo Researchers, Inc.; **B8-3**, John D. Cunningham/Visuals Unlimited, **8-14**, Francois Gohier/Photo Researchers, Inc.; **B8-5**, Kerstick/Visuals Unlimited; **8-16a and b**, French Technology Office. **CO9: opener**, Florida Keys National Marine Sanctuary/NOAA; **9-2a**, Jeanne Poindexter/Visuals Unlimited; **9-2b**, T.E. Adams/Visuals Unlimited; **9-3a**, D.B. Scott/NOAA; **9-3bc**, David Phillips/Visuals Unlimited; **9-4**, William Jorgensen/Visuals Unlimited; **9-5a**, NOAA; **9-5bc**, OAR/NURP/NOAA; **9-5d**, Daniel W. Gotshall/Visuals Unlimited; **9-5e**, John Bruno/NOAA; **9-6a**, Casey Debenham/NOAA; **9-6b**, R. Griswold/OAR/NURP/National Marine Fisheries Service; **9-6c**, James R. McCullagh/Visuals Unlimited; **9-6d**, Kip Evans/NOAA; **9-6e**, OAR/NURP/NOAA; **9-8**, Dann Blackwood and Page Valentine/USGS; **9-9b**, R.J. Goldstein/Visuals Unlimited; **B9-1**, Philip Sze/Visuals Unlimited; **9-10**, Norbert Wu/Root Resources; **9-13b**, Kjell B. Sandved/Visuals Unlimited; **9-16b**, CBS/Visuals Unlimited; **9-16c**, John Cunningham/Visuals Unlimited, **9-18**, SeaWiFS Project/NASA/Goddard Space Flight Center and ORBIMAGE; **9-19b**, M.I. Walker England/Photo Researchers, Inc.; **9-20a**, James P. McVey/NOAA Sea Grant Program; **9-20b**, AbleStock; **9-21b**, Ken Lucas/Visuals Unlimited; **B9-3**, NOAA; **9-24a**, Gerald and Buff Corsi/Visuals Unlimited; **B9-4c**, Bill Curtsinger, **B9-4d**, NOAA, **B9-5**, National Ocean Service Photo Gallery; **9-25a**, Gary G. Gibson/Photo Researchers, Inc.; **9-25c**, Norbert Wu/Root Resources. **CO10: opener**, Paolo Curto/The Image Bank; **B10-1a**, Scot Anderson/NOAA; **B10-1b**, Marty Snyderman/Visuals Unlimited; **10-3a**, Jeff Rotman; **10-3b**, Neil McDaniel; **10-3c**, Florida Keys National Marine Sanctuary/NOAA;**10-3d**, George Harrison/U.S. Fish & Wildlife Service; **10-3e**, OAR/NURP/NOAA; **10-3f**, Daniel W. Gotshall/Visuals Unlimited; **B10-2b**, OAR/NURP/NOAA; **B10-4a**, Carl O. Wirsen/Visuals Unlimited; **B10-4b**, University of Delaware Graduate College of Marine Studies; **10-9**, R. DeGoursey/Visuals Unlimited; **B10-6**, NASA/Goddard Space Flight Center; **B10-7b**, David Schleser/Nature's Images/Photo Reseachers, Inc.; **B10-8b**, Paul McCormick/The Image Bank; **B10-9a**, William W. Hartley/U.S. Fish & Wildlife Service; **10-17**, Jacques Descloitres, MODIS Land Rapid Response Team/NASA/GSFC. **CO11: opener**, Russell Illig/PhotoDisk, Inc.; **11-1b**, Douglas Faulkner/Photo Researchers, Inc. **11-1d**, Sanford Berry/Visuals Unlimited; **11-3b**, David Weintraub/Photo Researchers, Inc.; **11-6**, Frank M. Hanna/Visuals Unlimited; **B11-2**, Jim Corwin/Stock Boston; **B11-3**, Pat and Tom Leeson/Photo Researchers, Inc.; **B11-4 ab**, Dennis Krohn/USGS; **11-7**, Jack Dermid/Photo Researchers, Inc.; **11-9**, David Weintraub/Photo Researchers, Inc.; **11-10**, Sapelo Island National Estuarine Research Reserve/NOAA; **11-11a**, Steve McCutcheon/Visuals Unlimited; **11-13ab**, Paul Pinet; **11-13c**, Ty Harrington/FEMA; **11-13d**, Albert Copley/Visuals Unlimited; **B11-5a**, Matt Bradley/Tom Stack and Associates; **11-14c**, Teresa Tamura/Seattle Times; **11-14d**, Michael Collier/Stock, Boston; **B11-6ab**, Gerald G. Kuhn; **11-15b**, NASA; **B11-7a**, KevinFleming/Corbis. **CO12: opener**, U.S. Fish & Wildlife Service; **B12-1a**, Jacques Descloitres, MODIS Land Rapid Response Team/NASA/GSFC; **12-5a**, Paul Pinet; **12-5b**, Mary Hollinger/NOAA; **12-5c**, NOAA; **12-5d**, NOAA; **12-7**, Gerald and Buff Corsi/Visuals Unlimited; **12-8**, G. Carleton Ray/Photo Researchers, Inc.; **12-9b**, NOAA; **12-10a**, NOAA; **12-10b**, NOAA; **12-10c**, John Bruno/NOAA; **12-10d**, Alison Robb/NOAA; **12-11a**, South Florida Water Management District; **12-11b**, Paul Pinet; **B12-5**, NASA; **12-12a**, NOAA; **12-12b**, OAR/NURP/NOAA; **12-13a**, NOAA; **12-13b**, Science Visuals Unlimited/Visuals Unlimited; **B12-7**, S. Miller/OAR/NURP/NOAA; **12-14a**, Brent Deuel/NOAA; **12-14c**, W. Ober/Visuals Unlimited; **12-17b**, Sean Linehan/NGS/NOAA; **12-17c**, Kjell B. Sandved/Visuals Unlimited; **12-18a**, Anthony R. Picciolo/NOAA; **12-18b**, Paige Gill/Florida Keys National Marine Sanctuary; **12-18c**, Frink/Waterhouse/H. Armstrong Roberts; **12-18de**, South Florida Water Management District; **12-19ad**, Mohammed Al Momany/NOAA; **12-19bc**, OAR/NURP/NOAA; **12-19e,** Florida Keys National Marine Sanctuary. **CO13: opener**, Francois Gohier/Photo Researchers, Inc.; **13-2a**, NOAA; **13-2b**, Florida Keys National Marine Sanctuary/NOAA; **B13-1a**, Barbara Gerlach/Visuals Unlimited; **B13-1b**, Prance/Visuals Unlimited; **B13-1c**, Fred Bavendam; **B13-2ab**, Michael Van Woert/ORA/NESDIS/NOAA; **B13-2c**, Sarah Masters/ PhotoNewZealand; **13-6a**, William Millhouser/NOS/NOAA; **13-6b**, OAR/NURP/NOAA; **13-6c**, Kip Evans/NOAA; **13-6d**, AbleStock; **B13-4a**, Bill Curtsinger; **B13-4b**, Larry Lipsky/Tom Stack & Associates; **B13-5a**, Fred Bavendam; **B13-5b**, Roger Klocek/Visuals Unlimited; **B13-5c**, Dave B. Fleetham/Visuals Unlimited;**13-9a**, Woods Hole Oceanographic Institution. **CO14: opener**, Danilo Cedrone/NOAA; **B14-1**, Goddard Space Flight Center/NASA, data from NOAA/GOES; **B14-2**, AP/Wide World Photos; **B14-3**, Natalie Forbes/Corbis; **14-5**, Frank Manheim/USGS; **B14-4**, Frank T. Awbrey/Visuals Unlimited; **14-8,** Doug Plummer/Photo Researchers, Inc. **CO15: opener**, Ablestock; **15-1a**, Keith Wood/Tony Stone Images; **15-1b**, AbleStock; **B15-1**, Corbis/Bettman; **15-5**, Sam Pierson, Jr./Photo Researchers, Inc.; **15-6a**, NOS/NOAA; **15-6b**, AP/Wide World Photos; **15-7b**, McCutcheon/Visuals Unlimited; **B15-3**, Exxon Valdez Oil Spill Trustee Council/NOAA; **15-9**, W.A. Banaszewski/Visuals Unlimited, Inc.; **15-10b**, National Ocean Service Photo Gallery; **15-11a**, NASA; **15-12a**, U.S. Fish & Wildlife Service; **15-13** and **15-15**, U.S. Army Corps of Engineers; **B15-5**, NOAA; **15-22a**, PhotoDisc, Inc.; **15-22b**, National Ocean Service Photo Gallery; **B15-9**, Fish and Wildlife Conservation Commission/Florida Marine Research Institute; **B15-10**, Sanford Berry/Visuals Unlimited.